ADVANCES IN STEEL STRUCTURES

Proceedings of the
Third International Conference on Advances in Steel Structures

9-11 December 2002, Hong Kong, China

Volume I

Elsevier Science Internet Homepage - http://www.elsevier.com

Consult the Elsevier homepage for full catalogue information on all books, journals and electronic products and services.

Elsevier Titles of Related Interest

CHAN & TENG
ICASS '99, Advances in Steel Structures. (2 Volume Set).
ISBN: 008-043015-5

FRANGOPOL, COROTIS & RACKWITZ
Reliability and Optimization of Structural Systems.
ISBN: 008-042826-6

FUKUMOTO
Structural Stability Design.
ISBN: 008-042263-2

HOLLAWAY & HEAD
Advanced Polymer Composites and Polymers in the Civil Infrastructure.
ISBN: 008-043661-7

KELLY & ZWEBEN
Comprehensive Composite Materials.
ISBN: 008-042993-9

KO & XU
Advances in Structural Dynamics. (2 Volume Set)
ISBN: 008-043792-3

LUNDQUIST, LETERRIER, SUNDERLAND & MÅNSON
Life Cycle Engineering of Plastics Technology, Economy and the Environment
ISBN: 008-043886-5

MAKELAINEN
ICSAS'99, Int. Conf. on Light-Weight Steel and Aluminium Structures.
ISBN: 008-043014-7

USAMI & ITOH
Stability and Ductility of Steel Structures.
ISBN: 008-043320-0

VASILIEV & MOROZOV
Mechanics and Analysis of Composite Materials.
ISBN: 008-042702-2

WANG, REDDY & LEE
Shear Deformable Beams and Plates.
ISBN: 008-043784-2

Related Journals

Free specimen copy gladly sent on request. Elsevier Science Ltd, The Boulevard, Langford Lane, Kidlington, Oxford, OX5 1GB, UK

Advances in Engineering Software
CAD
Composites Part A: Applied Science and Manufacturing
Composites Part B: Engineering
Composite Structures
Computer Methods in Applied Mechanics and Engineering
Computers and Structures
Computer Science and Technology
Construction and Building Materials
Engineering Failure Analysis

Engineering Fracture Mechanics
Engineering Structures
International Journal of Fatigue
International Journal of Mechanical Sciences
International Journal of Solids and Structures
Journal of Constructional Steel Research
Mechanics of Materials
Mechanics Research Communications
Structural Safety
Thin-Walled Structures

To Contact the Publisher

Elsevier Science welcomes enquiries concerning publishing proposals: books, journal special issues, conference proceedings, etc. All formats and media can be considered. Should you have a publishing proposal you wish to discuss, please contact, without obligation, the publisher responsible for Elsevier's civil and structural engineering publishing programme:

Keith Lambert
Senior Publishing Editor
Elsevier Science Ltd
The Boulevard, Langford Lane
Kidlington, Oxford
OX5 1GB, UK

Phone: +44 1865 843411
Fax: +44 1865 843931
E.mail: k.lambert@elsevier.com

General enquiries, including placing orders, should be directed to Elsevier's Regional Sales Offices – please access the Elsevier homepage for full contact details (homepage details at the top of this page).

ADVANCES IN STEEL STRUCTURES

Proceedings of the
Third International Conference on Advances in Steel Structures

9-11 December 2002, Hong Kong, China

Volume I

Edited by
S.L. Chan, J.G. Teng and K.F. Chung
The Hong Kong Polytechnic University

Organized by
Research Centre for Advanced Technology in Structural Engineering,
Department of Civil and Structural Engineering,
The Hong Kong Polytechnic University

Sponsored by
The Hong Kong Institution of Engineers,
The Hong Kong Institution of Steel Construction

2002

ELSEVIER

Amsterdam – Boston – London – New York – Oxford – Paris
San Diego – San Francisco – Singapore – Sydney – Tokyo

ELSEVIER SCIENCE Ltd
The Boulevard, Langford Lane
Kidlington, Oxford OX5 1GB, UK

© 2002 Elsevier Science Ltd. All rights reserved.

This work is protected under copyright by Elsevier Science, and the following terms and conditions apply to its use:

Photocopying
Single photocopies of single chapters may be made for personal use as allowed by national copyright laws. Permission of the Publisher and payment of a fee is required for all other photocopying, including multiple or systematic copying, copying for advertising or promotional purposes, resale, and all forms of document delivery. Special rates are available for educational institutions that wish to make photocopies for non-profit educational classroom use.

Permissions may be sought directly from Elsevier Science via their homepage (http://www.elsevier.com) by selecting 'Customer support' and then 'Permissions'. Alternatively you can send an e-mail to: permissions@elsevier.com, or fax to: (+44) 1865 853333.

In the USA, users may clear permissions and make payments through the Copyright Clearance Center, Inc., 222 Rosewood Drive, Danvers, MA 01923, USA; phone: (+1) (978) 7508400, fax: (+1) (978) 7504744, and in the UK through the Copyright Licensing Agency Rapid Clearance Service (CLARCS), 90 Tottenham Court Road, London W1P 0LP, UK; phone: (+44) 207 631 5555; fax: (+44) 207 631 5500. Other countries may have a local reprographic rights agency for payments.

Derivative Works
Tables of contents may be reproduced for internal circulation, but permission of Elsevier Science is required for external resale or distribution of such material.
Permission of the Publisher is required for all other derivative works, including compilations and translations.

Electronic Storage or Usage
Permission of the Publisher is required to store or use electronically any material contained in this work, including any chapter or part of a chapter.

Except as outlined above, no part of this work may be reproduced, stored in a retrieval system or transmitted in any form or by any means, electronic, mechanical, photocopying, recording or otherwise, without prior written permission of the Publisher.
Address permissions requests to: Elsevier Science Global Rights Department, at the fax and e-mail addresses noted above.

Notice
No responsibility is assumed by the Publisher for any injury and/or damage to persons or property as a matter of products liability, negligence or otherwise, or from any use or operation of any methods, products, instructions or ideas contained in the material herein. Because of rapid advances in the medical sciences, in particular, independent verification of diagnoses and drug dosages should be made.

First edition 2002

Library of Congress Cataloging in Publication Data
A catalog record from the Library of Congress has been applied for.

British Library Cataloguing in Publication Data
A catalogue record from the British Library has been applied for.

ISBN: 0 08 044017 7 (2 volume set)

♾ The paper used in this publication meets the requirements of ANSI/NISO Z39.48-1992 (Permanence of Paper).
Printed in The Netherlands.

The papers presented in these proceedings have been reproduced directly from the authors' 'camera-ready' manuscripts. As such, the presentation and reproduction quality may vary from paper to paper.

PREFACE

These two volumes of proceedings contain 9 invited keynote papers and 130 contributed papers presented at the Third International Conference on Advances in Steel Structures (ICASS '02) held on 9 - 11 December 2002 in Hong Kong. The conference was a sequel to the First and the Second International Conferences on Advances in Steel Structures held in Hong Kong in December 1996 and 1999 respectively.

The conference provided a forum for discussion and dissemination by researchers and designers of recent advances in the analysis, behaviour, design and construction of steel structures. The papers were contributed from over 18 countries around the world. They cover a wide spectrum of topics, reporting the current state-of-the-art and pointing to future directions of structural steel research.

The organization of a conference of this magnitude would not have been possible without the supports and contributions of many individuals and organizations. The strong support from Professor J.M. Ko, Associate Vice President and Dean of Faculty of Construction and Land Use, and Professor Y.S. Li, Head of the Department of Civil and Structural Engineering, have been pivotal in the organization of this conference.

We also wish to express our gratitude to the Hong Kong Institution of Engineers and the Hong Kong Institute of Steel Construction for sponsoring the conference, and also to the Conference Advisory Committee for mobilizing support from the local construction industry and various government departments.

Thanks are due to all the contributors for their careful preparation of the manuscripts and all the keynote speakers for their special support. Reviews of papers were carried out by members of the International Scientific Committee and the Conference Organizing Committee. To all the reviewers, we are most grateful.

We would also like to thank all those involved in the day-to-day running of the organization work, including members of the Conference Organizing Committee, and both the secretarial and the technical staff of the Department of Civil and Structural Engineering.

Finally, we gratefully acknowledge our pleasant cooperation with Keith Lambert, Noël Blatchford, Lorna Canderton and Vicki Wetherell at Elsevier Science Ltd in the UK.

S.L. Chan, J.G. Teng and K.F. Chung

INTERNATIONAL SCIENTIFIC COMMITTEE

H. Akiyama	University of Tokyo	Japan
F.G. Albermani	University of Queensland	Australia
D. Anderson	University of Warwick	UK
P. Ansourian	University of Sydney	Australia
R.G. Beale	Oxford Brookes University	UK
R. Bjorhovde	University of Pittsburgh	USA
M.A. Bradford	University of New South Wales	Australia
R.Q. Bridge	University of Western Sydney	Australia
C.S. Cai	Kansas State University	USA
C.R. Calladine	University of Cambridge	UK
W.F. Chen	University of Hawaii at Manoa	USA
Y.K. Cheung	University of Hong Kong	HKSAR, China
S.P. Chiew	Nanyang Technological University	Singapore
C.K. Choi	Korea Advanced Institute of Science & Technology	Korea
K.P. Chong	National Science Foundation	USA
M. Chryssanthopoulos	University of Surrey	UK
A. Combescure	Laboratoire de Mechanique et Technologie	France
J.G.A. Croll	University College London	UK
J.M. Davies	University of Manchester	UK
G.G. Deierlein	Stanford University	USA
S.L. Dong	Zhejiang University	China
P.J. Dowling	University of Surrey	UK
D. Dubina	University of Timisoara	Romania
M. Farshad	Swiss Federal Laboratories for Materials Testing & Research	Switzerland
F.C. Filippou	University of California at Berkeley	USA
Y. Fukumoto	Fukuyama University	Japan
H.B. Ge	Nagoya University	Japan
Y. Goto	Nagoya Institute of Technology	Japan
P.L. Gould	Washington University	USA
R. Greiner	Technical University of Graz	Austria
Q. Gu	Xian University of Architecture & Technology	China
J.F. Hajjar	University of Minnesota	USA
L.H. Han	Fuzhou University	China
G.J. Hancock	University of Sydney	Australia
J.E. Harding	University of Surrey	UK
J.F. Jullien	INSA Lyon	France
S. Kato	Toyohashi University of Technology	Japan
A.R. Kemp	University of Witwatersrand	South Africa
S. Kitipornchai	City University of Hong Kong	HKSAR, China
K.C.S. Kwok	Hong Kong University of Science & Technology	HKSAR, China
R.A. LaBoube	University of Missouri-Rolla	USA
T.T. Lan	Chinese Academy of Building Research	China
G.Q. Li	Tongji University	China
S.F. Li	Tsinghua University	China
R.J.Y. Liew	National University of Singapore	Singapore
J. Lindner	Technische Universitat Berlin	Germany
Xila Liu	Tsinghua University	China
Xiliang Liu	Tianjin University	China
L.W. Lu	Lehigh University	USA

INTERNATIONAL SCIENTIFIC COMMITTEE *(Continued)*

P. Makelainen	Helsinki University of Technology	Finland
P. Marek	Academy of Science of the Czech Republic	Czech Republic
J. Melcher	Technical University of Brno	Czech Republic
D.A. Nethercot	Imperial College of Science, Technology & Medicine	UK
D.J. Oehlers	University of Adelaide	Australia
G.W. Owens	The Steel Construction Institute	UK
J.M. Rotter	University of Edinburgh	UK
B. Samali	University of Technology, Sydney	Australia
H. Schmidt	University of Essen	Germany
G. Sedlacek	Institute of Steel Construction	Germany
S.Z. Shen	Harbin Institute of Technology	China
Z.Y. Shen	Tongji University	China
L.S. da Silva	Universidade de Coimbra	Portugal
T.T. Soong	State University of New York at Buffalo	USA
N.S. Trahair	University of Sydney	Australia
K.C. Tsai	National Taiwan University	Taiwan, China
C.M. Uang	University of California at San Diego	USA
T. Usami	Nagoya University	Japan
A.S. Usmani	University of Edinburgh	UK
A. Wada	Tokyo Institute of Technology	Japan
F. Wald	Czech Technical University	Czech Republic
E. Walicki	Technical University of Zielona Gora	Poland
C.M. Wang	National University of Singapore	Singapore
D. White	Georgia Institute of Technology	USA
F.W. Williams	City University of Hong Kong	HKSAR, China
Y. Xiao	University of Southern California	USA
Y.B. Yang	National Taiwan University	Taiwan, China
R. Zandonini	University of Trento	Italy
X.L. Zhao	Monash University	Australia
S.T. Zhong	Harbin Institute of Technology	China

CONFERENCE ADVISORY COMMITTEE

Chairman   J.M. Ko
                    The Hong Kong Polytechnic University

Members

Andrew. S. Beard	Mott Connell Limited
Francis S.Y. Bond	Maunsell Consultants Asia Limited
Andrew K.C. Chan	Ove Arup & Partners (Hong Kong) Limited
L.Y.K. Choi	Shui On (Contractors) Limited
K.P. Chong	Directorate of Engineering, National Science Foundation, USA
M. Hadaway	Gammon Construction Limited
J. Kong	BHP Steel Building Products Singapore Pte Limited
C.M. Leung	Buildings Department, HKSAR
A.Y.T. Leung	City University of Hong Kong
C.K. Lau	Civil Engineering Department, HKSAR
P.K.K. Lee	The University of Hong Kong
S.H. Ng	Icfox Hong Kong Limited
S.H. Pau	Architectural Services Department, HKSAR
S. Sin	Atkin China Limited
W. Tang	The Hong Kong University of Science & Technology
V.W.S. Tong	Housing Department, HKSAR
W.H. Wong	Meinhardt Engineering Limited
I. Kimura	Nippon Steel Corporation

CONFERENCE ORGANIZING COMMITTEE

Chairman **S.L. Chan**
The Hong Kong Polytechnic University

Co-Chairmen **J.G. Teng** and **K.F. Chung**
The Hong Kong Polytechnic University

Members

F.T.K. Au	The University of Hong Kong
C.M. Chan	The Hong Kong University of Science & Technology
T.H.T. Chan	The Hong Kong Polytechnic University
K.M. Cheung	Buildings Department, HKSAR
R.P.K. Chu	Meinhardt (C&S) Limited
G.W.M. Ho	Ove Arup and Partners (Hong Kong) Limited
M.K.Y. Kwok	Ove Arup and Partners (Hong Kong) Limited
E.S.S. Lam	The Hong Kong Polytechnic University
J.C.W. Lau	James Lau and Associates Limited
S.S. Law	The Hong Kong Polytechnic University
J.Q.S. Li	City University of Hong Kong
M.C. Luo	Ove Arup and Partners (Hong Kong) Limited
Y.W. Mak	Housing Department, HKSAR
Y.Q. Ni	The Hong Kong Polytechnic University
A.K. Soh	The University of Hong Kong
F.M.K. Tong	Architectural Services Department, HKSAR
K.Y. Wong	Highways Department, HKSAR
Y.L. Wong	The Hong Kong Polytechnic University
Y.L. Xu	The Hong Kong Polytechnic University
F.Y.F. Yau	Maunsell Structural Consultants Limited
B. Young	The Hong Kong University of Science & Technology

CONTENTS

VOLUME I

Preface	v
International Scientific Committee	vii
Conference Advisory Committee	ix
Conference Organizing Committee	x

Keynote Papers

Stability of High Strength G550 Steel Compression Members D. Yang and G. Hancock	3
The Application and Development of Pretensioned Long-Span Steel Space Structures in China S.L. Dong and Y. Zhao	15
Advanced Computer Calculations in the Design of Shell Structures J.M. Rotter	27
Exploiting the Special Features of Stainless Steel in Structural Design D.A. Nethercot and L. Gardner	43
Cassette Wall Construction: Current Research and Practice J.M. Davies	57
A New Issue in Plate and Box Girder Stability Design T. Usami and P. Chusilp	69
Monotonic and Hysteretic Behaviour of Bolted Endplate Beam-to-Column Joints R. Zandonini and O.S. Bursi	81
Design of Steel Arches Against In-Plane Instability M.A. Bradford and Y.-L. Pi	95
FEM Analysis of Steel Members Considering Damage Accumulation Effects Under Cyclic Loading Z.Y. Shen and Z.S. Song	105

Beams and Columns

A Review of Recent Developments on Design of Perforated Beams C.H. Ko and K.F. Chung	121
A New Derivation of the Buckling Theory of Thin-Walled Beams G.S. Tong and L. Zhang	129

Analysis of Strain Hardening in Steel Beams Using Mill Tests   139
    M.P. Byfield and M. Dhanalakshmi

In-Plane Ultimate Load-Carrying Capacity of Tapered I Columns   147
    Y.L. Guo and Y. Pan

Elastic Torsional-Flexural Buckling of Tapered I Beam-Columns   155
    Y.L. Guo, Y. Han, W.Q. Hao and T. Liu

Load-Carrying Capacity of Box Section Beam-Column   163
    T. Liu and Y.L. Guo

Multi-Directional Pseudo Dynamic Experiment of Steel Bridge Piers   171
    M. Obata and Y. Goto

Connections

Shear Lag in Double Angle Truss Connections   181
    D.B. Bauer and A. Benaddi

Structural Behaviour of Web Bolted Flange Welded Connection   189
    T. Emi, M. Tabuchi, T. Tanaka and H. Namba

Effects of Beam Flange Width-to-Thickness Ratio on Beam Flange Fracture Caused from Scallop Root   197
    T. Iguchi, M. Tabuchi, T. Tanaka and S. Kihara

Experimental Investigation of Slot Lengths in RHS Bracing Members   205
    T. Wilkinson, T. Petrovski, E. Bechara and M. Rubal

Experimental Study on Cyclic Behavior of Improved Beam-Column Connections   213
    Z.F. Li, Y.J. Shi, H. Chen and Y.Q. Wang

Repair/Upgrade of Steel Moment Frames in Low Rise Buildings   221
    J.C. Anderson, Y. Xiao and J.X.J. Duan

Ultimate Bearing Capacity of Welded Hollow Spherical Joints in Spatial Reticulated Structures   229
    Q.H. Han and X.L. Liu

Ultimate Strength of Welded Thin-Walled SHS-CHS T-Joints Under In-Plane Bending   237
    F.R. Mashiri, X.L. Zhao, L.W. Tong and P. Grundy

Tests and Design of Longitudinal Fillet Welds in Very High Strength (VHS) Steel Circular Tubes   245
    T.W. Ling, X.L. Zhao and R. Al-Mahaidi

Experimental Behaviour of End Plate I-Beam to Concrete-Filled Rectangular Hollow Section Column Joints   253
    L.C. Neves, L. Simões da Silva and P.C.G. da S. Vellasco

Composite Connections at Perimeter Locations in Unpropped Composite Floors   261
    M. Dhanalakshmi, M.P. Byfield and G.H. Couchman

Analysis of Steel and Composite Braced Frames with Semi-Rigid Joints A. Kozlowski	269
Numerical Evaluation of the Ductility of a Bolted T-Stub Connection A.M. Girão Coelho and L. Simões da Silva	277
Strength and Stress Analysis of Steel Beam-Column Connections Using Finite Element Method H. Chen, Y.J. Shi, Y.Q. Wang and Z.F. Li	285

Scaffolds and Slender Structures

Geometric Non-Linear Analysis of Flexible Supporting System Z. Wang, Y.Q. Wang and Y.J. Shi	295
Determination of the Factors of Safety of Standard Scaffold Structures B. Milojkovic, R.G. Beale and M.H.R. Godley	303
Sway Stability of Steel Scaffolding and Formwork Systems S. Vaux, C. Wong and G. Hancock	311
Second-Order Analysis and Design of Steel Scaffold Using Multiple Eigen-Imperfection Modes S.L. Chan, C. Dymiotis and Z.H. Zhou	321

Cold-Formed Steel

On the Distortional Post-Buckling Behaviour of Cold-Formed Lipped Channel Steel Beams L.C. Prola and D. Camotim	331
GBT-Based Distortional Buckling Formulae for Thin-Walled Rack-Section Columns and Beams N. Silvestre, K. Nagahama, D. Camotim and E. Batista	341
Testing and Numerical Analysis of Cold-Formed C-Sections Subject to Patch Load R.Y. Xiao, G.P.W. Chin and K.F. Chung	351
Torsional Buckling Experiments on Wide-Flange Thin-Walled Z-Section Columns R.A.D. Fish, M. Lee and K.J.R. Rasmussen	357
Structural Stability of Stainless Steel Compression Members Y. Liu and B. Young	365
Membrane Imperfections Measured in Cold Formed Tubes A. Wheeler and M. Pircher	375
Flexural Failure of Cold-Formed Single Channels Connected Back-to-Back M. Dundu and A.R. Kemp	383
Ultimate Strength Design of Bolted Moment-Connections Between Cold-Formed Steel Members J.B.P. Lim and D.A. Nethercot	391

Analysis of Cassette Sections in Compression 401
    P.A. Voutay and J.M. Davies

Performance of Wall-Stud Shear Walls Under Monotonic and Cyclic Loading 409
    L.A. Fulop and D. Dubina

Direct Strength Method for the Design of Purlins 421
    L. Quispe and G. Hancock

Cold-formed Purlin-Sheeting Systems 429
    F. Albermani and S. Kitipornchai

An Experimental Investigation into Lapped Moment Connections Between Z-Sections 437
    H.C. Ho and K.F. Chung

Practical Design of Cold-Formed Steel Z-Sections with Lapped Connections 445
    H.C. Ho and K.F. Chung

Destructive Mechanism of Large Span Cold-Formed Section Roof Truss 453
    Y.J. Guo, K. Li and X.X. Du

Sway Buckling of Down-Aisle Pallet Rack Structures Containing Splices 461
    R.G. Beale and M.H.R. Godley

Composite Construction

Composite Action in Non-Composite Beams 471
    R. Seracino and D.J. Oehlers

Effect of Concrete Infill on Non-Compact Tubes Subjected to Pure Bending 479
    A. Wheeler and R. Bridge

Simplified Elastic and Elastic-Plastic Analysis of Continuous Composite Beams 487
    P.A. Berry

Elastic Cross-Section Analysis of Continuous Composite Beams Affected by Web Slenderness 495
    P.A. Berry

Effects of Transverse Reinforcement on Composite Beams with Precast Hollow Core Slabs 503
    D. Lam and T.F. Nip

Shear Connection in Composite Beams Incorporating Profiled Steel Sheeting with Narrow Open or Closed Steel Ribs 511
    M. Patrick and R.Q. Bridge

Shear Connection in Composite Beams Incorporating Open-Trough Profile Decks 519
    M. Patrick and R.Q. Bridge

Research in Canada on Steel-Concrete Composite Floor Systems: An Update 527
    M.U. Hosain and A. Pashan

Early Age Shrinkage and Casting Sequence Effects in Composite Steel-Concrete Girders 535
    L. Dezi, G. Leoni and A. Vitali

Shear Strength of Prestressed Concrete Encased Steel Beams with Bonded Tendons 543
    S.C. Choy, Y.L. Wong and S.L. Chan

Instability Behavior of Prestressed Steel-Concrete Composite Continuous Beam 551
    Y. Han, Z.Z. Fang and Y.L. Guo

Evaluation of Simplified Superposition Design Method for Composite Columns 559
    J.H. Zhong and S.F. Chen

Tests on Concrete-Filled Double Skin (SHS Outer and CHS Inner) Composite Stub Columns 567
    X.L. Zhao, R.H. Grzebieta, A. Ukur and M. Elchalakani

Strength of Slender Concrete Filled Columns Fabricated with High Strength Structural Steel 575
    B. Uy, M. Mursi and H.B.A. Tan

Concrete-Filled Steel RHS Columns Subjected to Long-Term Loads 583
    L.H. Han, W. Liu and Y.F. Yang

Hysteretic Behaviors of Concrete-Filled Steel SHS Beam-Columns 591
    Z. Tao and L.H. Han

Experimental and Theoretical Studies on Steel-Concrete Hybrid Structures 599
    G.Q. Li, X.M. Zhou and X. Ding

Seismic Demand Evaluation Procedure for Concrete-Filled Steel Columns 607
    H.B. Ge, K.A.S. Susantha and T. Usami

VOLUME II

Preface v

International Scientific Committee vii

Conference Advisory Committee ix

Conference Organizing Committee x

Plates

Numerical Modelling of Stainless Steel Plates 617
    K.J.R. Rasmussen, T. Burns, P. Bezkorovainy and M.R. Bambach

Local Buckling of Biaxially Compressed Steel Plates in Double Skin Composite Panels 625
    Q.Q. Liang, B. Uy, H.D. Wright and M.A. Bradford

Ductility of High Performance Steel Rectangular Plates Under Uniaxial Compression 633
    K. Niwa, I. Mikami and Y. Miyazaki

Shear-Carrying Capacity of Steel Plate Shear Wall with Cross Stiffeners 641
    G.D. Chen and Y.L. Guo

Elastic Critical Moments of I Sections with Very Slender Webs 649
    A.J. Wang and K.F. Chung

Shells

An Efficient Strategy for the Evaluation of the Reliability of 3D Shells in Case of Non Linear Buckling 659
    A. Combescure and A. Legay

Case Study of a Medium-Length Silo Under Wind Loading 667
    M. Pircher, R.Q. Bridge and R. Greiner

Buckling of Thin Pressurized Cylindrical Shells Under Bending Load 675
    A. Limam and J.F. Jullien

Stability of Thin-Walled Cylindrical Shells Subjected to Lateral Patch Loads 683
    E. Feifel and H. Saal

Buckling of Circular Steel Silos Subject to Eccentric Discharge Pressures – Part I 693
    C.Y. Song and J.G. Teng

Buckling of Circular Steel Silos Subject to Eccentric Discharge Pressures – Part II 703
    C.Y. Song and J.G. Teng

Aspects of Corrugated Silos 713
    P. Ansourian and M. Gläsle

Buckling Experiments on Transition Rings in Elevated Steel Silos 721
    Y. Zhao and J.G. Teng

Buckling Strength of Cylinders with a Consistent Residual Stress State 729
    J.M.F.G. Holst and J.M. Rotter

Buckling Behaviour of Extensively-Welded Steel Cylinders Under Axial Compression 737
    X. Lin and J.G. Teng

Experiment on a Model Steel Base Shell of the Comshell Roof System 745
    H.T. Wong and J.G. Teng

Effect of Cracks on Vibration, Buckling and Parametric Instability of Cylindrical Shells 755
    A. Vafai, M. Javidruzi, J.F. Chen and J.C. Chilton

An Experimental Study for Seismic Reinforcement Method on Existing Cylindrical Steel Piers by Welded Rectangular Steel Plates 763
    K. Chu and T. Sakurai

Bridges

Metal Forms Replace Reinforcement in Bridge Deck Slabs 773
    B. Bakht, A.A. Mufti and G. Tadros

Analysis of the Camber at Prestressing of a New Kind of Composite Railway Bridge Deck 783
    S. Staquet, H. Detandt and B. Espion

Evaluation of Typhoon Induced Fatigue Damage Using Health Monitoring Data   791
    T.H.T. Chan, Z.X. Li and J.M. Ko

Fatigue Stress Analysis of Suspension Bridges Using FEM   799
    T.H.T. Chan, L. Guo and Z.X. Li

Curved Steel Box-Girder Bridges at Construction Phase   807
    G.C.M. Lee, K.M. Sennah and J.B. Kennedy

Numerical Study of Characteristic Behavior of Steel Plate Girder Bridges   815
    E. Yamaguchi, K. Harada, M. Nagai and Y. Kubo

Nonlinear Seismic Response Analysis of a Deck-Type Steel Arch Bridge   823
    T. Yamao, H. Harada and Y. Muramoto

The Unit Load Method - Some Recent Applications   831
    D. Janjic, M. Pircher and H. Pircher

Global Analysis of Steel and Composite Highway Bridges - Development of Improved Spatial Beam Models   839
    H. Unterweger

Dynamics

Field Comparative Tests of Cable Vibration Control Using Magnetorheological (MR) Dampers in Single- and Twin-Damper Setups   849
    Y.F. Duan, J.M. Ko, Y.Q. Ni and Z.Q. Chen

Evaluation of Ride Comfort of Road Vehicles Running on a Cable-Stayed Bridge Under Crosswind   857
    W.H. Guo and Y.L. Xu

Comparison of Buffeting Response of a Suspension Bridge Between Analysis and Aeroelastic Test   865
    Y.L. Xu, D.K. Sun and K.M. Shum

Dynamic Response of the Cable to Moving Mass   873
    Y.L. Guo, H. Wang and G.X. Ren

Traffic-Induced Microvibration Mitigation of High Tech Equipment Inside a Building Using Passive/Active Platform   881
    Z.C. Yang and Y.L. Xu

Dynamic Analysis of Coupled Train-Bridge Systems Under Fluctuating Wind   889
    Y.L. Xu, H. Xia and Q.S. Yan

Modal Parameter Identification of Tsing Ma Bridge During Typhoon Victor: EMD-HT Approach   897
    J. Chen, Y.L. Xu and R.C. Zhang

Dynamic Load from Pedestrian Footsteps   905
    S.S. Law

Frictional Joint in the Dynamic Analysis of a Portal Frame 913
S.S. Law, Z.M. Wu and S.L. Chan

Formulas for Vibration Period of Steel Buildings in Taiwan Derived from Ambient Vibration Data 921
L.J. Leu, C.Y. Liu, C.W. Huang and S.H. Yeh

Impact Mechanics

Some Recent Studies on Energy Absorption of Metallic Structural Components 931
G. Lu

Crash Analysis of Automobile Bumpers with Pedestrians 939
B. Wang and G. Lu

A Theoretical Model for Axial Splitting and Curling of Circular Metal Tubes 947
X. Huang, G. Lu and T.X. Yu

Experiment and Analysis of a Scaled-Down Guardrail System Under Static and Impact Loading 955
J.T.Y. Hui, T.X. Yu and X.Q. Huang

Crashworthiness of Motor Vehicle and Luminaire Support in Frontal Impact 963
M. Samaan and K. Sennah

Effects of Welding

Experimental and Numerical Uni-Axial Tests at High Temperature - Analysis of Models 973
Y. Vincent and J.F. Jullien

A Two Scale Model for the Simulation of Residual Stresses Due to Welding of a Metallic Multiphase Material 981
A. Combescure and M. Coret

Influence of Welding Details on the Performance of Beam to Column Connections of Steel MRFs in Seismic Areas 989
D. Dubina and A. Stratan

Fatigue and Fracture

Correlation of Fatigue Life of Fillet Welded Joints Based on Stress at 1mm in Depth 1001
Z.G. Xiao and K. Yamada

Fatigue Strength Prediction for Misaligned Welded Joints by Stress Field Intensity Method 1009
D.Q. Guan, W.J. Yi and L. Li

Failure of a Steel Plate Containing a Circular Rivet Hole with an Emanating Crack 1017
K.T. Chau and S.L. Chan

A Method to Estimate P-S-N Curve of Welded Joints Under General Stress Ratio 1025
D.Q. Guan, W.J. Yi and Q. Wang

Fatigue Crack Propagation of Tubular T-Joints Under Combined Loads 1033
    S.P. Chiew, S.T. Lie and Z.W. Huang

High-Cycle Fatigue Behaviour of Welded Thin SHS-CHS T-Joints Under In-Plane Bending 1043
    .R. Mashiri, X.L. Zhao, L.W. Tong and P. Grundy

On the Analysis of Fracture Phenomena Observed in Steel Structures During the Kobe Earthquake 1051
    H. Fujiwara, Y. Goto and M. Obata

Fire Performance

Assessment of Structures for Fire Safety - Insights on Current Methods and Trends 1061
    J.Y.R. Liew and H.X. Yu

World Trend for the Development of Performance-Based Fire Codes for Steel Structures 1071
    M.B. Wong

A New Method to Determine the Ultimate Load Capacity of Composite Floors in Fire 1079
    A.S. Usmani and N.J.K. Cameron

Graphical Method for Design of Steel Structures in Fire 1089
    M.B. Wong

High Temperature Transient Tensile Properties of Fire Resistant Steels 1095
    W. Sha and T.M. Chan

Mechanical Properties of Structural Steel at Elevated Temperatures 1103
    J. Outinen and P. Mäkeläinen

Structural Response of a Steel Beam Within a Frame During a Fire 1111
    Z.F. Huang, K.H. Tan and S.K. Ting

Effect of External Bending Moment on the Response of Boundary-Restrained Steel Column in Fire 1119
    K.H. Tan and Z.F. Huang

Concrete-Filled HSS Columns after Exposure to the ISO-834 Standard Fire 1127
    L.H. Han, J.S. Huo and Y.F. Yang

An Experimental Study and Calculation on the Fire Resistance of Concrete-Filled SHS and RHS Columns 1135
    L.H. Han, L. Xu and Y.F. Yang

Analysis and Design

A Unified Analysis Method to Predict Long-Term Mechanical Performance of Steel Structures Considering Corrosion, Repair and Earthquake 1145
    Y. Goto and N. Kawanishi

A Higher Order Formulation for Geometrically Nonlinear Space Beam Element 1153
    J.X. Gu, S.L. Chan and Z.H. Zhou

Unified Analytical Method of Gliding Cables in Structural Engineering – Frozen-Heated Method   1161
   Y.L. Guo and X.Q. Cui

Large Deflection Analysis of Tensioned Membrane Structures Allowing for Support Flexibility   1169
   J.J. Li and S.L. Chan

Torsional Analysis of Asymmetric Proportional Building Structures Using Substitute Plane Frames   1177
   W.P. Howson and B. Rafezy

On Some Problems of Analytical and Probability Approaches to Structural Design   1185
   J.J. Melcher

Design of Steel Frames Using Calibrated Design Curves for Buckling Strength of Hot-Rolled Members   1193
   S.L. Chan and S.H. Cho

Analysis of the Bending Strength of U-Section Steel Sheet Piles Crimped in Pairs   1201
   M.P. Byfield and R.J. Crawford

A Textbook for the New Canadian Standard - *Strength Design in Aluminum*   1209
   D. Beaulieu

Index of Contributors   I1

Keyword Index   I3

KEYNOTE PAPERS

STABILITY OF HIGH STRENGTH G550 STEEL COMPRESSION MEMBERS

D.Yang[1] and G.Hancock[1]

[1]Department of Civil Engineering, University of Sydney
NSW, Australia, 2006

ABSTRACT

High strength cold-reduced steel is typically of stress grade G550 (550 MPa nominal yield and tensile strength) and less than 1 mm thick. The steel has been used for many years for sheeting and decking but is now being used for structural members such as roof trusses and stud walls of steel framed houses. The steel has low strain hardening with a tensile to yield stress ratio of 1.0. The paper summarises a major research program on the stability of this steel which has been proceeding for several years at the University of Sydney. Short and long column compression members have been tested and compared with design standards and finite element analyses.

KEYWORDS

High strength steel, stability, structural members, compression, local buckling, overall buckling

INTRODUCTION

G550 sheet steels are manufactured by cold reducing mild sheet steels (f_y = 300 MPa) to a thickness which ranges from 0.42 mm to 1.0 mm. Cold reduction produces large deformations of the grain structure which cause an increase in yield stress and ultimate strength and a decrease in ductility. G550 sheet steels (0.48 mm $\leq$ t $\leq$ 0.90 mm) have recently been introduced for use as structural members in the Australian residential construction industry (Hancock and Murray, 1996), as well as for panel and deck sections in other types of building construction. The use of G550 sheet steels has been restricted to non-structural applications in most countries due to concerns regarding the ductility of members and connections, and possibly problems with stability due to low strain hardening. This paper focuses on recent research at the University of Sydney into the compression stability of G550 steel in members in the thickness range 0.42 mm to 0.60 mm.

Cold formed structural members are fabricated from sheet steels consisting of various material properties which must meet requirements prescribed in applicable design standards. The Australian/New Zealand Design Standard (AS/NZS, 1996) allows for the use of thin (t < 0.9 mm), high strength (f_y = 550 MPa) sheet steels in all structural sections. However, due to the lack of ductility exhibited by sheet steels which are cold reduced to thickness, the engineer must use a yield stress and ultimate strength reduced to 75% of the minimum specified values. The American Iron and Steel Institute (AISI) Design Specification (1996) further limits the use of thin, high strength steels to roofing, siding and floor decking panels.

Early research by McAdam et al. (1988) on steels with a low f_u/f_y ratio allowed full strength design to Clause A3.3.1 of the AISI Specification for purlins and girts which are principally flexural members.

The research described in the paper included several tests of compression members which showed results significantly below routine column design curves. For this reason, compression members were excluded from Clause A3.3.1 of the AISI Specification. Design of compression members was not permitted. Further research by Daudet and Klippstein (1994) at Dietrich Industries Indiana investigated the stub column strength of sections used for stud walls in residential construction. The material studied had a tensile to yield stress ratio close to 1.0 rather than 1.08 as required by Section A3.3.1 of the AISI Specification. The study showed that steels having an f_u/f_y as low as 1.01 and elongations as low as 3 percent can be conservatively designed in compression using a reduced yield stress of $0.75f_y$. Clearly further research is required into the reasons for the lower strength of compression members.

Research at the University of Missouri-Rolla by Wu, Yu and LaBoube (1996a, 1996b) on decking sections in flexure composed of ASTM A653 Structural Grade 80 steel has resulted in an exception clause being added to the latest update of the AISI Specification (AISI, 2000). This exception clause permits a reduced yield point ($R_b f_y$) for computing the nominal flexural strength of multiple web configurations. The reduced yield point depends on the plate slenderness but is generally greater than the 75% discussed previously.

Recent research by Rogers and Hancock at the University of Sydney has investigated ductility of thin G550 sheet steels in tension including perforations (Rogers and Hancock, 1997), bolted connections in G550 sheet steels (Rogers and Hancock, 1998), screwed connections in G550 sheet steels (Rogers and Hancock, 1999) and fracture in G550 sheet steel (Rogers and Hancock, 2001). A separate paper (Rogers, Yang and Hancock, 2001) summarises the more important findings and proposed design rules from those studies.

The stability of G550/Grade E steels with low strain hardening has only been investigated in the context of decking sections in bending. However, a new research program at the University of Sydney is investigating G550 steel in compression. This is being achieved by the manufacture of square box type sections with glued and screwed corner connections. The results of these tests are being compared in the first instance with AS/NZS 4600 using $0.75f_y$ and the $R_b f_y$ approach recently adopted in the AISI Specification for decking sections. The results of this project for stub columns and long columns in box shaped sections are described in this paper. New design rules are proposed.

MANUFACTURING PROCESS OF G550 SHEET STEEL

G550 Sheet Steels

The usual process for the production of these steels is cold reduction. This process can be used to increase the strength and hardness, as well as to form an accurate thickness for sheet steels and other steel products. Initially the sheet steels are rolled to size in a hot strip mill with finishing and coiling temperatures of approximately 940°C and 670°C, respectively. The hot worked coil of steel, typically 2.5 mm in thickness with a minimum specified 300 MPa yield stress, is uncoiled and cleaned in an acid solution to remove surface oxides and scale. The uncoiled strip is then trimmed to size and fed into a cold reduction mill, which may contain any number of stands. High compressive force in the stands and strip tension systematically reduce the thickness of the steel sheet until the desired dimension is reached, e.g. by approximately 85% for the 0.42 mm sheet steels.

The milling process causes the grain structure of cold reduced steels to elongate in the rolling direction, which produces a directional increase in the material strength and a decrease in the material ductility. The effects of cold working are cumulative, i.e. grain distortion increases with further cold working as a result of an increase in total dislocation density, however, it is possible to change the distorted grain structure and to control the steel properties through subsequent heat treatment. Various types of heat treatment exist and are used for different steel products. G550 sheet steels are stress relief annealed, i.e. the total dislocation density is reduced by annealing, although recrystallisation does not occur. Stress relief annealing involves heating the steel to below the recrystallisation temperature,

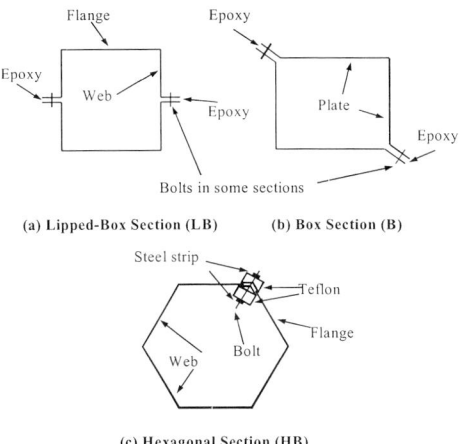

Fig.1 Test Sections

holding the steel until the temperature is constant throughout its thickness, then cooling slowly. Mild sheet steels of similar thicknesses are annealed to a greater extent in comparison with G550 sheet steels. Annealing is carried out in a hot dip coating line prior to application of either a zinc or aluminum/zinc coating. Upon final cooling the sheet steel is further processed through a tension levelling mill, e.g. 0.35% extension, to improve the finish quality and the flatness of the coil. The cold-reduced G550 sheet must be differentiated from other sheet steels whose high yield stress and ultimate strength values are obtained by means of an alloying process, i.e. high strength low alloy (HSLA) steels.

The material property requirements for G550 or Grade E sheet steels are specified in Australia by AS 1397 (1993) and in North America by the following ASTM Standards; A611, A653, A792 and A875. Material property specifications for HSLA sheet steels can also be found in ASTM Standard A653.

COMPRESSION STUDIES

Stub Column Test Specimens

The tests were performed on closed sections brake pressed from aluminum/zinc-coated Grade G550 structural steel sheet to AS1397. The sections tested are shown in Fig. 1. Epoxy was used to close the B & LB-sections and a Teflon/Steel Support was used to close the HB-sections. Bolts & clamps were also used on the B & LB-sections as discussed in Yang and Hancock (2002). The sections were fabricated from 0.42 mm and 0.60 mm steel sheets. The widths of the B-sections ranged from 20 mm to 100 mm for the 0.60 mm sheet steel and 14 mm to 70 mm for the 0.42 mm sheet steel. The widths of the LB sections ranged from 20 mm to 40 mm for the 0.60 mm sheet and 30 mm to 50 mm for the 0.42 mm sheet. The flats of the HB-sections ranged from 20 mm to 100 mm and the thickness was

0.60 mm. The holes/clamping configurations are shown in Fig. 1. The dimensions of the specimens are given Yang and Hancock (2002).

For the B-sections all sections had the same size lips of 7.5 mm. The lips of the LB-sections in 0.42 mm thickness were 6 mm. However, for the 0.60 mm thickness LB-sections, the lips varied from 6 mm to 10 mm. Two slotted pieces of Teflon and two long steel strips were used to join the two edges of the HB-sections together. The objective of the Teflon was to connect the longitudinal free edges without resisting axial compressive load. Further details are given in Yang and Hancock (2002).

Testing

The rig consisted of the Sintech/MTS-300kN testing machine with fix-ended bearings. The bottom bearing was adjustable so that it could orient specimens vertically. However, Pattenstone was needed at the top to ensure perfect contact. The load and shortening were recorded using the Sintech data acquisition system. The compressive deformation rate was 0.05 mm/min.

The ultimate loads (P_t) for all specimens are given in Figs. 3 & 4. The results have been non-dimensionalised with respect to the theoretical stub column strength (N_s) computed to AS/NZS 4600

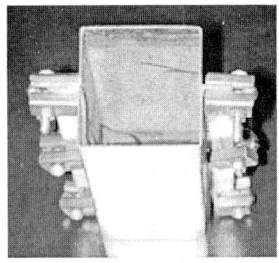
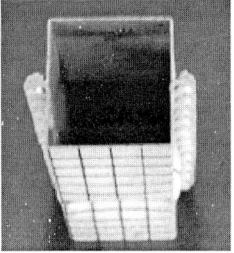

(a) Lipped-Box Section (LB) with Clamps   (b) Lipped-Box Section (LB) with Bolts

(c) Box Section (B) with Bolts   (d) Hexagonal Section (HB) with Teflon Support

Fig.2 Bolting & Clamping Configurations of Test Specimens

based on the measured data as discussed latter. A significant change occurs at b/t equal to about 50 (i.e. b=30 mm for t=0.60 mm or b=21 mm for t=0.42 mm), which is equivalent to a minimum value of the non-dimensional strength between 0.85 and 0.9. The full set of test results is given in Yang and Hancock (2002). The results of the HB sections are lower at high slenderness since the Teflon support was formed not to produce a perfect simple support but allowed some longitudinal edge distortion.

Comparisons with design standards & ABAQUS

The finite element non-linear analysis program "ABAQUS" was used to simulate the behavior of

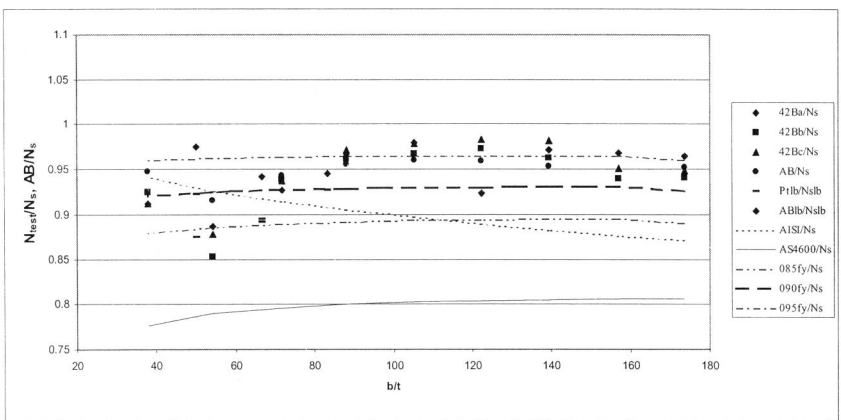

Fig.3 Comparison of B&LB-section Test Results with Design Standard (0.42 mm)

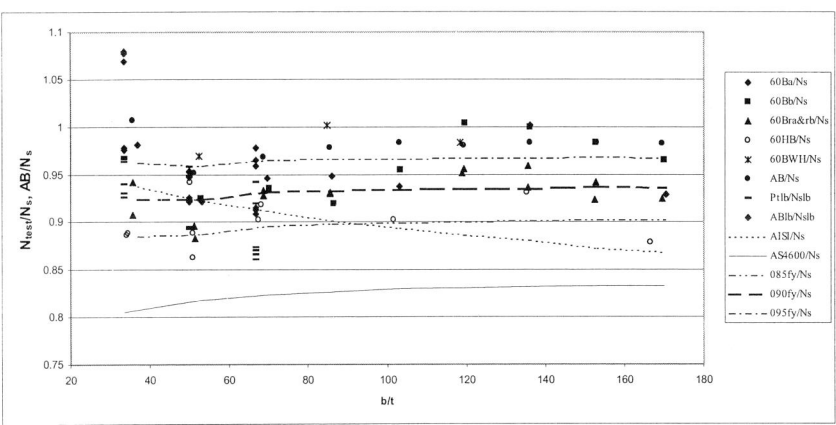

Fig.4 Comparison of B&LB&HB-section Test Results with Design Standard (0.60mm)

columns. The ratios of the ABAQUS simulations (AB) to the theoretical stub column strengths are also shown in Figs. 3 & 4. Further details are given in Yang and Hancock (2002), such as element type, material behavior, boundary conditions and geometrical imperfection.

The theoretical stub column strengths (N_s) were calculated according to AS/NZS 4600 (1996). The theoretical strengths were calculated using the average measured cross-section dimensions and the measured material yield stress. The test results (P_t) and ABAQUS (AB) results non-dimensionalised with respect to the theoretical stub column strength (N_s) are plotted against the plate slenderness ratios (b/t) in Figs. 3 & 4. Solid and dotted curves are also plotted in Figs. 3 & 4. The dotted curve is the ratio of N_{sRb}/N_s versus plate slenderness (b/t). N_{sRb} has been calculated based on $R_b f_y$ as included in Section A3.3.2 of AISI Specification Supplement No.1 (2000). The solid curve is the ratio of $N_{s0.75}/N_s$ versus the plate slenderness (b/t). $N_{s0.75}$ has been calculated based on $0.75 f_y$ as included in Clause 1.5.1.5(b) of AS/NZS 4600:1996.

In Figs. 3 & 4, it can be seen that the ratios of the test results and theoretical values became higher as the sections became more slender. The tendency of the test results was contrary to the results based on the AISI Specification Supplement No.1 (2000). Although the test results and the results based on

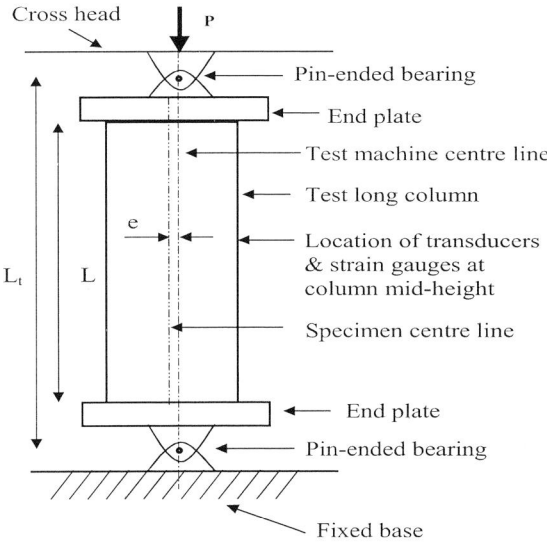

Fig. 5 Test Arrangement

(a) Local buckle at failure   (b) Local & Overall buckling (Clamps)

Fig.6 Buckling Mode

AS/NZS 4600 using a 75% reduction in the yield stress have the same tendency, those predicted results are too conservative to utilize fully the strength of material. From the test data, it appears that a modified reduction factor should be used. Three trial reduction factors (0.85, 0.90 and 0.95) were chosen to be used to calculate the section capacity. The results ($N_{s0.85}$, $N_{s0.90}$ and $N_{s0.95}$) based on those reduction factors were obtained respectively. The dash double-dot, dash and dash-dot curves are plotted in Figs. 3 & 4, which are the ratios of $N_{s0.85}/N_s$, $N_{s0.90}/N_s$ and $N_{s0.95}/N_s$ versus the plate slenderness (b/t).

As can be seen in Figs. 3 & 4, the dash curves based on the reduction factor of 0.90 fits the mean tests well. So the modified reduction factor 0.90 can be used to replace the reduction factor 0.75, which is specified for G550 steel with the thickness being less than 0.9 mm in AS/NZS 4600.

Long Column Test Specimens

In order to determine the capacity of long columns in compression, local and overall buckling and the interaction between them should be considered. For the high strength steel columns assembled from thin plates, the section configuration is an important factor affecting their behaviour. Thin plate elements will generally continue to carry load after local buckling into the post-buckling range so that local buckling does not mean failure of the whole column. However, a singly-symmetric section may have a neutral axis shift after local buckling occurs resulting in an additional moment. To eliminate this problem, in this test program, doubly-symmetric sections were chosen. These sections are the LB-sections previously described for stub columns and shown in Fig. 1.

The tests were performed on closed sections brake pressed from aluminium/zinc-coated Grade G550 structural steel sheet to AS1397 in 0.42 mm and 0.60 mm thickness. The sections tested are shown in Fig. 1. Epoxy was used to close the LB-sections. The lengths of the LB sections ranged from 450 mm to 1100 mm for the 0.60 mm sheet steel and 550 mm to 1700 mm for the 0.42 mm sheet steel. The ends of each specimen were milled flat and parallel to ensure full contact between specimens and end bearing. Bolts & clamps were also used on the LB-sections as shown in Fig. 1. Further details are given in Yang, Hancock and Rasmussen (2002).

Geometric imperfections were measured for all of the specimens. The full set of measurements and the maximum overall imperfections are given in Yang, Hancock and Rasmussen (2002). The values of the measured overall imperfections (out–of-straightness) varied from 0.2 mm to 1.5 mm. The material and its properties are fully described in Yang & Hancock (2002) and Yang, Hancock and Rasmussen (2002). The measured 0.2% proof stresses of the steel were 690 MPa and 711 MPa for 0.42 mm and 0.60 mm thickness respectively.

Testing

The arrangement is shown in Fig. 5. The rig consisted of the Sintech/MTS-300kN testing machine with pin-ended bearings. The load and shortening were recorded using the Sintech data acquisition system. The compressive deformation rate was 0.05 mm/min. Bolts were only located at the ends (four for each) and were only used for the 0.60 mm sections. Clamps were used on the lips of 0.42 mm sections.

The central deflections were measured using two transducers on two opposite sides of the LB-sections. Two strain gauges were attached at the centre of each side of each column. The transducers and strain gauges were connected to the SPECTRA data acquisition system. During the test, a trial axial load which was 1/15 or 1/10 of the estimated ultimate load was applied and readings of the strain gauges were obtained. Based on a calculation of the location of the initial central eccentricity (e) of the action line as shown in Fig. 5, the location of the column was adjusted at each end. This procedure was repeated until the value of the initial eccentricity (e) was approximately equal to the nominal value of L/1000 which is the maximum normally specified in structural design standards. The measured values and the method of calculation are given in Yang, Hancock and Rasmussen (2002).

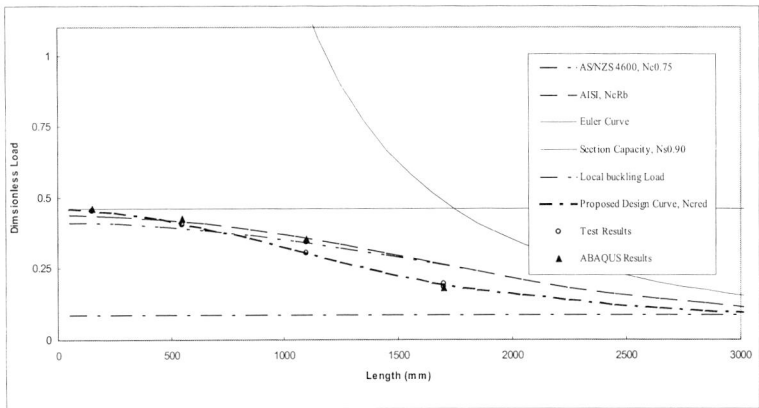

Fig.7 Comparison of Test & ABAQUS Results with Design Standard (0.42 mm)

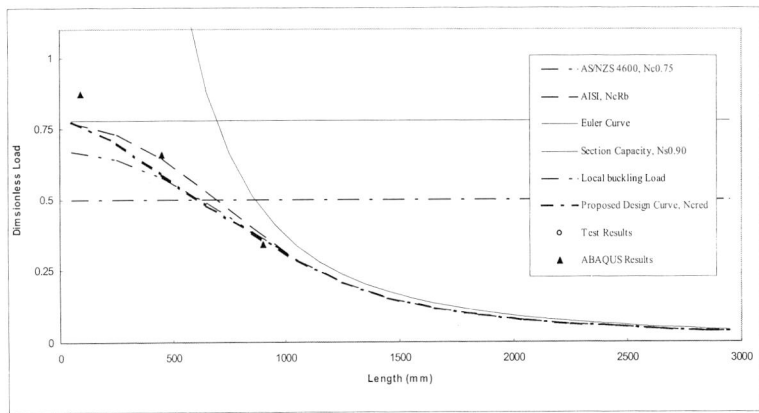

Fig.8 Comparison of Test & ABAQUS Results with Design Standard (0.60 mm)

Initially the columns remained elastic with the slope of the load-shortening diagram approximately constant after some initial take-up. The ultimate loads (P_t) are given in Yang, Hancock and Rasmussen (2002) for each specimen. For the 060LB20 and 060LB30 series, the ultimate load (P_t) was less than the theoretical local buckling load (N_{ol}) so that the slope kept approximately constant until the ultimate load was reached. The 450 mm long specimens in 0.60 mm material and the 550 mm long specimens in 0.42 mm material buckled inelastically soon after the ultimate load with a sudden drop in load. The 900 mm long specimens in 0.60 mm material and the 1100 mm long and 1700 mm long specimens in 0.42 mm material generally had a more gradual decrease in load until inelastic local buckling occurred at which point there was a sudden drop in load as for the 450 mm specimens. For the intermediate length columns in 0.42 mm thickness, the slope reduced continuously from the local buckling load until the ultimate load was reached. Typical buckling modes are shown in Fig. 6.

Comparisons with design standards & ABAQUS

Fig. 7 is a typical figure which shows the test strength (P_t) and the ABAQUS values (AB) compared with a range of design curves for 50 mm sections in 0.42 mm material. Fig. 8 is a typical figure which shows same for 30 mm sections in 0.60 mm material respectively. In Figs. 7 & 8, the test strengths (P_t) have been non-dimensionalised with respect to the squash load (P_y) as computed for the measured yield stress and dimensions. The full set of graphs are given in Yang, Hancock and Rasmussen (2002).

Dashed and dash-dotted curves are plotted in Figs. 7 & 8. The dashed curve is the ratio of N_{cRb}/P_y against the column length (l_x). N_{cRb} was calculated using the AISI Specification based on a yield stress $R_b f_y$ as included in Section A3.3.2 of the AISI Specification Supplement No.1(2000). The dash-dotted curve is the ratio of $N_{c0.75}/P_y$ against column length (l_x). $N_{c0.75}$ was calculated based on AS/NZS 4600 with a yield stress of $0.75 f_y$ as included in Clause 1.5.1.5(b) of AS/NZS 4600. The horizontal dashed line represents the ratio of the theoretical local buckling load to the squash load (N_{ol}/P_y). The horizontal solid line represents the ratio of the section capacity to the squash load ($N_{s0.90}/P_y$) against the column length (l_x). $N_{s0.90}$ was calculated based on AS/NZS 4600 with a yield stress of $0.90 f_y$ as described earlier in this paper and proposed in Yang and Hancock (2002). The heavy dash-dotted curve is the proposed design curve, which is the ratio of the nominal reduced member capacity to the squash load (N_{cred}/P_y) against the column length (l_x). The proposed design method is described following.

For all sections and lengths, when the elastic local buckling loads (N_{ol}) are lower than the test results (P_t), the test results are generally lower than the design curves at intermediate columns lengths where local and Euler buckling interact. For the sections of stockier cross-section, the AS/NZS 4600 design curve based on $0.75 f_y$ and the AISI design curve based on a stress $R_b f_y$ are generally slightly conservative.

The effect of local buckling on overall buckling behaviour has been studied in several research projects. On the basis of tests and analytical studies, DeWolf, Pekoz, Winter (1974) and Kalyanaraman, Pekoz, Winter (1977) conclude that a satisfactory approach is to calculate the overall buckling load using the effective radius of gyration and the effective area, both calculated at the overall buckling stress. The unified approach proposed by Pekoz (1986) was adopted by the AISI Specification (1986, 1991, 1996) and AS/NZS 4600 (1996) to determine the strength of columns where local and Euler buckling interact. This proposed approach does not take account of the reduction of the radius of gyration (flexural rigidity) resulting from local buckling which was proposed by the earlier researchers. However, for the high strength steel sections, the effect of the reduction in the radius of gyration on the interaction of local buckling and overall buckling is significant, especially buckling of these very slender sections at loads above the local buckling load.

To fit the test data, a new design approach is proposed. The proposed approach consists of two steps. Firstly, the reduced yield point $0.90 f_y$, which was determined by Yang and Hancock (2002) for stub columns described earlier in this paper, is used to replace the yield stress in Clause 3.4 of AS/NZS 4600 and Section C4 of the AISI specification. Secondly, a reduction factor γ is applied to the radius of gyration as defined in Eq.1. It is a function of the length varying from some limit γ_0 at length $l_x=0$ (here γ_0 taken as 0.65) to 1.0 at length $l_x=1.10*l_{x0}$. Here, l_{x0} is the length where the local buckling load equals the Euler buckling load as defined by Eq. 2. The reduction factor γ accounts for the loss of flexural rigidity due to local buckling. The value of reduced radius of gyration $\gamma*r_x$ is used in Clause 3.3.3.2(a) of AS/NZS 4600 to replace the normal radius of gyration r_x (Section C3.1.2, Eq. C3.1.2-8 of the AISI Specification). It can be seen in Figs 7 & 8 that the proposed design curves based on the reduced the radius of gyration γ fit the test data well.

$$\gamma = \gamma_0 + (1-\gamma_0) * l_x / (1.1 * l_{x0}) \quad (1)$$
$$l_{x0} = \pi * r_x * \sqrt{E/f_{ol}} \quad (2)$$

CONCLUSIONS

A range of stub columns in G550 sheet steel to AS1397 has been tested in compression to obtain the strength characteristics of this steel with low strain-hardening. The following detailed observations & conclusions can be made for the stub columns.

- The use of the finite element program ABAQUS for simulating the behaviour of the stub columns was successful since the ABAQUS results were generally in good agreement with experimental values. The results of ABAQUS were sensitive to the initial imperfections for the stockier sections but not for the slender sections. ABAQUS can be used for further work on such thin sheet steel sections.

- The results of the successful stub column tests have been compared with the design procedures in the Australian/New Zealand Standard for Cold-formed Steel Structures and recent Amendments to the American Iron and Steel Institute Specification. As expected, the greatest effect of the low strain hardening was for the stockier sections where material properties play an important role. For the more slender sections where elastic local buckling and post-local buckling are more important, the effect of low strain hardening does not appear to be as significant. This is contrary to recent design proposals in the USA (Wu, Yu and LaBoube) where it was believed that the more slender sections had been influenced.

- As shown in Fig. 3 & 4, the ratios of AB/N_s and P_t/N_s were mostly larger than 0.90 and higher than the results based on the AISI Specification Supplement No.1 (2000) and AS/NZS 4600 when the plate slenderness (b/t) was greater than 88. When the plate slenderness (b/t) was less than 50, the formulae used in AISI Specification Supplement No.1(2000) was slightly unconservative due to the imperfection sensitivity. The conclusion drawn from the test results is that for stub compression members, $0.90f_y$ may be used as the reduced yield point to determine the nominal strength of the stub columns.

Pin-ended column tests with box-sections and constructed from high strength G550 steel have been successfully performed. The plate slenderness (b/t) ranged from 33 to 119 and the column slenderness (L/r_x) ranged from 27 to 148. A load eccentricity which produced a column response equivalent to L/1000 was used for all tests. ABAQUS simulations of the test results with local imperfections and overall eccentricity were made.

- The ABAQUS results (AB) were generally in good agreement with the test results (P_t). The difference was on average less than 6% for all columns although the ABAQUS results were higher. More detailed investigation of imperfections to use in ABAQUS is required.

- The columns with stockier plate elements, which had high local buckling stresses (f_{ol}), failed by overall buckling. The test results and the ABAQUS results were close to the curves based on AS/NZS 4600 and the AISI Specification. For very long columns with slender plate elements which had lower local buckling stresses (f_{ol}), failure was still governed by overall buckling although local buckling occurred. However, for the intermediate length columns, the failure mode was governed by the interaction of local and overall buckling. The interaction of local and overall buckling reduced the column strength and made the test results lower than the design curves. The worst case had a difference between test results and the results based on AS/NZS 4600 and the AISI Specification of about 14%, which means that for the slender sections AS/NZS 4600 and the AISI method are unconservative.

- To account for the loss of flexural rigidity due to local buckling for the slender sections, a reduction of the radius of gyration is needed to take account of interaction buckling in the design curve. The proposed design curve based on a reduction factor (γ) fits the test data well, so that the reduced radius of gyration $\gamma \cdot r_x$ may be used in the design curve for the slender sections.

- Since ABAQUS results were in reasonably good agreement with test results, ABAQUS can be used to simulate long column tests which could not be performed in the available machine.

REFERENCES

American Iron and Steel Institute. (1997). "1996 Edition of the Specification for the Design of Cold-Formed Steel Structural Members", Washington, DC, USA.

American Iron and Steel Institute. (2000). "1996 Edition of the Specification for the Design of Cold-Formed Steel Structural Members, Supplement 1, July 1999", Washington, DC, USA.

American Society for Testing and Materials A611. (1997). "Standard Specification for Steel Sheet, Carbon, Cold-Rolled, Structural Quality", Philadelphia, PA, USA

Daudet,R., and Klippstein,K.H. (1994), "Stub Column Study using Welded, Cold-Reduced Steel", 12th International Specialty Conference on Cold-Formed Steel Structures, St Louis, Missouri, Oct 1994, pp 285 – 302.

Hancock, G.J. and Murray, T.M., "Residential Applications of Cold-Formed Structural Members in Australia, 13th International Specialty Conference on Cold-Formed Steel Structures, St Louis, Missouri, Oct 1996, pp 505 - 511.

Hibbitt, Karlsson & Sorensen, Inc., "ABAQUS/Standard User's Manual", Ver. 5.7, 1997

Kalyanaraman, V., Pekoz, T., and Winter, G., "Unstiffened compression elements", Journal of the Structural Division, ASCE, Sept., 1977.

Levy, S. Wooley, R.M., and Kroll, W.D., "Instability of simply supported square plate with reinforced circular hole in edge compression", Research Paper RP1849, Vol.39, Dec. 1947

McAdam, J.N, Brockenbrough, R.A., LaBoube, R.A., Pekoz, T, and Schneider, E.J., "Low Strain Hardening Ductile Steel Cold-Formed Members", 9th International Specialty Conference on Cold-Formed Steel Structures, St Louis, Missouri, Nov 1988.

Pekoz, T., "Development of a unified approach to the design of cold-formed steel members", Research Report CF87-1, American Iron and Steel Institute, 1987.

Rogers, C.A., Yang, D., Hancock, G.J., (2001), "Stability and ductility of thin high strength G550 steel members and connections", Proceedings of the 3rd International Conference on Thin-Walled Structures, Kracow, Poland, (published as Thin-Structures-Advances and Developments-Elsevier 2001, Eds Zaras, Kowal-Michalska and Rhodes)

Rogers, C.A., Hancock, G.J.. (1997), "Ductility of G550 Sheet Steels in Tension", Journal of Structural Engineering, ASCE, Vol. 123, No. 12, 1586-1594.

Rogers, C.A., Hancock, G.J.. (1998), "Bolted Connection Tests of Thin G550 and G300 Sheet Steels", Journal of Structural Engineering, ASCE, Vol. 124, No. 7, pp. 798-808.

Rogers, C.A., Hancock, G.J. (1999), "Screwed Connection Tests of Thin G550 and G300 Sheet Steels", Journal of Structural Engineering, ASCE, Vol. 125, No. 2, pp. 128-136.

Rogers, C.A. and Hancock, G.J. (2001), " Fracture Toughness of G550 Sheet Steels subjected to Tension", Journal of Constructional Steel Research, Vol 57, pp 71-89.

Standards Australia / Standards New Zealand. (1996), "Cold-formed steel structures - AS/NZS 4600", Sydney, NSW, Australia

Standards Australia. (1993), "Steel sheet and strip - Hot-dipped zinc-coated or aluminium/zinc coated - AS 1397", Sydney, NSW, Australia

Walker, A. C., "Design and analysis of cold-formed sections", International Textbook Company Limited, 1975

Wu, S., Yu, W.W, and LaBoube, R.A., (1996a), "Strength of flexural members using structural Grade 80 of A653 Steel (deck panel tests)", Second Progress Report, Department of Civil Engineering, University of Missouri-Rolla, November.

Wu, S., Yu, W.W, and LaBoube, R.A., (1996b), "Flexural members using structural Grade 80 of A653 steel (deck panel tests)", 13th International Specialty Conference on Cold-Formed Steel Structures, St Louis, Missouri, Oct 1996, pp. 255-274.

Yang, D., Hancock, G.J., "Compression tests of cold-reduced high strength steel stub column", Research Report No. R815, School of Civil and Mining Engineering, University of Sydney, 2002

Yang, D., Hancock, G.J., and Rasmussen, K.J.R. "Compression tests of cold-reduced high strength steel long column", Research Report No. R816, School of Civil and Mining Engineering, University of Sydney, 2002

ACKNOWLEDGEMENTS

The authors would like to thank the Australian Research Council and BHP Coated Steel Division for their financial support for these projects performed at the University of Sydney. The advice of Associate Professor Kim Rasmussen on the compression tests is gratefully acknowledged.

NOTATION

AB	ultimate load from ABAQUS
b	flange width
e	eccentricity
E	Young's modulus of elasticity
f_{ol}	elastic buckling stress
f_y	yield stress
f_u	ultimate stress
l_x, l_{x0}	length of column
L	length of column
L_t	length of pin-ended column
N_{cred}	reduced member capacity
N_{ol}	elastic buckling load
N_s	nominal section compression capacity
$N_{s0.75}$	nominal section compression capacity based on 0.75 f_y
$N_{s0.90}$	nominal section compression capacity based on 0.90 f_y
N_{sRb}	nominal section compression capacity based on $R_b f_y$
P_{cr}	elastic buckling load of test
P_t	ultimate load of test
P_y	squash load
r_x	radius of gyration of the cross-section
t_b	thickness of base metal
γ_0, γ	reduction factor of radius of gyration

THE APPLICATION AND DEVELOPMENT OF PRETENSIONED LONG-SPAN STEEL SPACE STRUCTURES IN CHINA

S. L. Dong and Y. Zhao

Space Structures Research Center, Department of Civil Engineering,
Zhejiang University, Hangzhou 310027, P. R. China

ABSTRACT

Pretensioned long-span steel space structures are a new type of hybrid structures resulting from combining modern prestressing technology with long-span structures such as space grid structures (including space trusses and lattice shells), cable-bar tensioned structures and spatial truss structures. These structures have been developed rapidly and have found many applications in China in the recent decade. This paper provides a review of applications and developments of pretensioned long-span steel structures in China. The following topics on new technologies and new structures are given emphasis: prestressed grid structures, cable-stayed grid structures, beam string structures and prestressed segmental steel trusses. Prospects for pretensioned steel space structures in the 21st century are presented at the end of the paper.

KEYWORDS

Pretensioned Structures, Long-Span Structures, Steel Space Structures, Application, Development, Prospects

INTRODUCTION

Pretensioned long-span steel space structures are a new type of hybrid structures resulting from combining modern prestressing technology with long-span structures such as space grid structures (including space trusses and lattice shells), cable-bar tensioned structures and spatial truss structures. These structures generally have the advantages of reasonable mechanical behavior, large structural rigidity, light self-weight and easy fabrication and erection. In China, these structures have been developed in the recent decade, and have found many applications in large public and industrial buildings, such as sports and exhibition halls, theaters, terminal buildings, aeroplane hangars and warehouses (Lu 1995, 1998; Dong et al. 1998). The prestressing technology in long-span steel space structures has the following characteristics and advantages.

(1) The pretensioning force can modify the load-bearing behavior of the structure. That is, it improves the structural rigidity and reduces the structural deformation and the peak value of internal forces.

(2) By using the prestressing technology, new structural systems or structural forms (such as cable domes) can be constructed.

(3) The prestressing technology can be adopted as a method for the assembly of prefabricated members, leading to new structures, such as prestressed segmental steel trusses described later.

(4) Hybrid space structures adopting prestressing technology can usually reduce the material consumption significantly, so that more economic structures can be achieved.

In general, there are two methods for creating prestress in steel space structures. The first is to tension the prestressing cables/bars directly. This method may either lead to the re-distribution of the internal force and consequently the improvement of the structural behavior, or result in a new structural form with a certain state of internal forces. The second method is to adjust the elevations of the supports for the constructed structure. This method may modify the reaction force at the supports, so as to re-distribute the internal forces and create prestresses in the structure. Either high-strength steel wire ropes or steel rods can be used as prestressing cables/bars.

PRESTRESSED GRID STRUCTURES

Combining prestressing technology with space grid structures (including space trusses and lattice shells) leads to prestressed grid structures. As mentioned before, two methods can be used to create prestress in space grid structures. One is to install prestressing cables on or below the space truss bottom chord plane (Fig. 1a), or to install cables at the perimeter of lattice shells (Fig. 1b). By this method, deflections and internal forces opposite in sign to that of the loading can be obtained by the tensioning of the cables. The other one is to force the space truss to be put on supports with different elevations (usually 'basin-pattern' supports are used in order to obtain nearly uniform reaction forces at the supports under service loading). The characteristics of prestressed grid structures can be summarized as follows.

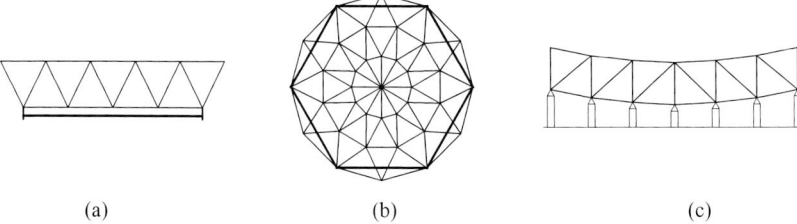

(a)   (b)   (c)

Figure 1: Methods for creating prestress in grid structures

(1) High-strength prestressing cables are used as main structural members in space grid structures to reduce material consumption.

(2) The principle of multi-stage tensioning and loading and the principle of multi-stage design can be adopted in order to get the best state of internal forces in the structure under service loading. A simple method for the analysis of prestressed grid structures was proposed by Dong and Deng (2001).

(3) The application of prestressing technology can improve the structural rigidity of space grids, leading to reduced deflection of the structure.

(4) For lattice shells, if suitable slippery supports are used, then the application of prestressing technology may lead to self-balanced structures without horizontal reactions.

(5) The method of changing the difference of elevation between supports and then re-distributing the internal force of the structure is the most economic method for creating prestresses, having no need of additional members or any other components.

More than 10 prestressed space grid structures have been built in China (e.g. Xiong et al. 1985; Ma et al. 1994; Yao et al. 1994; Yin et al. 1995; Chen et al. 1997; Yin et al. 2000). Some of the representative projects together with their technical parameters are given in Table 1. Fig. 2a shows the exterior view of Qingyuan Gymnasium at Guangdong province, while Fig. 2b shows a test model of the twist lattice shell for the roof structure of this gymnasium. This structure is a combination of six unisymmetrically double-layer twist lattice shells supported on six columns. Six prestressing cables are installed along the perimeter of the lattice shell, with a prestress value of 1600kN in each cable. It can be found from Table 1 that the application of prestressing technology may result in an obvious reduction in the steel consumption.

TABLE 1
REPRESENTATIVE PROJECTS OF PRESTRESSED GRID STRUCTURES

Project	Structural form of space grids	Plan dimensions and heights (m)	Pretressing technology	Steel consumption	Year of construction
Ninghe Gymnasium, Tianjin	orthogonal square pyramids	42.0×42.0×3.0	'basin-pattern' supports, a relative difference of elevation between supports of 9.0cm	28.5kg/m^2 12%	1984
Gymnasium of Chongqing First Middle School	diagonal square pyramids	37.8×37.8×2.34	'basin-pattern' supports, a relative difference of elevation between supports of 6.1cm	23.1kg/m^2 9%	1993
Gymnasium of Nankai Middle School	diagonal square pyramids	hexagonal planform 33.0×66.0×2.2	'basin-pattern' supports, a relative difference of elevation between supports of 7.3cm	19.8kg/m^2 11%	1993
Floors of Shanghai International Shopping Centre	composite space grids of orthogonal square pyramids	27.0×27.0	4 high-strength steel cables at a plane 20cm below the plane of bottom chords	48.0kg/m^2 32%	1993
Panzhihua Gymnasium	double-layer Geodesic braced dome	octagonal planform 74.8×74.8 8.89m in rise	supported on 8 columns, 8 prestressing cables at the perimeter, a prestress value of 700kN in each cable	49.0kg/m^2	1994
Qingyuan Gymnasium	combination of six double-layer twist lattice shells	regular hexagonal planform, 46.82m (side length), 8.0m in rise	supported on 6 columns, 6 prestressing cables at the perimeter, a prestress value of 1600kN in each cable	44.3kg/m^2	1995
Gaoyao Gymnasium	combination of four double-layer twist lattice shells	54.9×69.3	supported on 4 columns, 4 prestressing cables between columns, a prestress value of 1400kN in each cable	38.5kg/m^2	1995
Xinxing Gymnasium	combination of four double(single)-layer twist lattice shells	54.0×76.06	supported on 4 columns, 4 prestressing cables between columns	28.2kg/m^2 43%	1997
Suqian Culture and Sports Center	double-layer saddle lattice shell	elliptical planform 80.0×62.5×3.0	11 prestressing cables along the lower chords		1999

(a) Exterior view　　　　　　　　　　(b) The test model

Figure 2: Qingyuan Gymnasium

CABLE-STAYED GRID STRUCTURES

Cable-stayed space grid structures, which are generally composed of towers, cables and space grid structures, are a new type of hybrid space structures with an aesthetically pleasing form in long-span and medium-span buildings (Tang et al. 1992; Dong and Luo 1993; Zhou et al. 1997). Cable-stayed grid structures can enhance the internal space of the building, and provide a feature on the landscape with their novel appearance. The characteristics of cable-stayed grid structures can be summarized as follows.

(1) The advantage of the high strength of the steel cable can be fully utilized.

(2) More supports can be provided for space grid structures, leading to reduced deflection of the structure and smaller internal forces in the members.

(3) Pre-internal forces and inverted deflections can be established by tensioning of the cables, which can partially offset the structural internal forces and deflections under the superimposed loading.

(4) Cables should not be relaxed under arbitrary conditions of loading, therefore, prestressing force should be established in cables if necessary.

(5) For cable-stayed grid structures subject to both upward and downward wind loading, if the structural design is controlled by wind loading, then prestressing stability cables should be installed.

(6) Cables should be arranged in multi-directions. Planar cables or single-directional cables should be avoided.

(7) The angle of inclination of the cable should not be too small, and is usually greater than 25°. Otherwise, the elastic supporting effects provided by the cable will be reduced and difficulties in the construction of connections will arise.

Up to now, more than 10 cable-stayed grid structures have been constructed in China (e.g. Cui and Zhang 1991; Wu et al. 1994; Zhang et al. 1997; Jiao et al. 2000). Table 2 lists some of the representative projects. The roof for the main stadium of Dragon Sports Center in Zhejiang Province (Fig. 3) is one of the largest cable-stayed lattice shells in China (Jiao et al. 2000). The distance between two towers is 250m. In order to resist the upward wind loading, nine stability cables were arranged on the upper chord plane of the lattice shell (Fig. 3c). The stability cables were anchored at the inner and outer ring beam of the lattice shell respectively. The cable-stayed lattice shell for the Jiuguan tollhouse on the Taijiu express highway is the first cable-stayed grid structure with single tower and multi-directional cables (Zhang et al. 1997). Shenzhen Swimming and Diving Hall adopts a hybrid space structure composed of longitudinal and transverse spatial trusses, four masts and sixteen prestressing steel rods (Fig. 4, Space Structures Research Centre 2000).

(a) Photograph

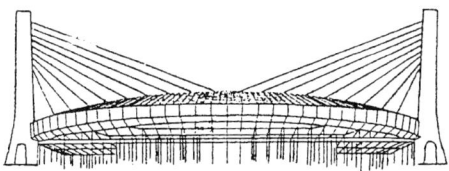 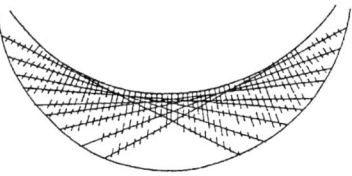

(b) Arrangement of stayed cables (c) Arrangement of stability cables

Figure 3: Cable-stayed roof for the main stadium of Dragon Sports Center

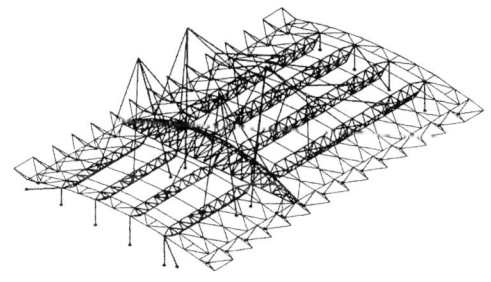 

(a) Structural layout (b) Photograph (under construction)

Figure 4: Shenzhen Swimming and Diving Hall

TABLE 2
REPRESENTATIVE PROJECTS OF CABLE-STAYED GRID STRUCTURES

Project	Structural form of space grids	Plan dimensions and heights (m)	Arrangement of towers and cables	Steel consumption	Year of construction
Sports Hall, National Olympic Sports Center	combination of two cylindrical lattice shells, diagonal square pyramids	70.0×83.2	2 towers at each end of the hall, 8 single-directional cables from each tower		1990
Stadium of Zhejiang University	orthogonal square pyramids	24.0×40.0×1.2	4 towers, 3 inclined cables and 1 horizontal cable from each tower, 14 cables in total		1993
Warehouse of PSA, Type A	orthogonal square pyramids	120.0×96.0 (4 pieces)	6 towers, 4 inclined cables from each tower, each cable composed of 4ϕ48 stainless steel rods	35.2kg/m^2 20-30%	1993
Warehouse of PSA, Type B	orthogonal square pyramids	96.0×70.0 (2 pieces)	4 towers, 4 inclined cables from each tower, each cable composed of 4ϕ48 stainless steel rods	20-30%	1993
Jiuguan Tollhouse on Taijiu Express Highway	two separate barrel vaults, orthogonal square pyramids	14.0×64.7×1.5	single tower, 28 multi-directional cables from the tower		1995
Stadium of Dragon Sports Center	cylindrical lattice shells, orthogonal square pyramids	crescent planform 244.0×50.0×3.0 (2 pieces)	2 towers, 9 cables from each tower, 9 additional stability cables	80.0kg/m^2	2000
Shenzhen Swimming and Diving Hall	spatial trusses	54.9×69.3	4 masts, 4 cables (steel rods) from each mast		2001

BEAM STRING STRUCTURES

Beam string structures are a new type of long-span steel structures developed in recent years. It can be adopted in either roof structures, floor structures, or wall structures (for example, supports for the curtain wall). The characteristics of beam string structures can be summarized as follows.

(1) Beam string structures are composed of upper chord beams, lower chord cables and vertical web members. Cables and web members are subjected to axial tensile and compressive forces respectively, while upper chord beams are subjected to combined axial compression and bending.

(2) Tensioning of the cables leads to upward forces in the vertical web members, consequently, the internal forces and deflections in the upper chord beams are opposite in sign to that of the loading. Therefore, the structural rigidity can be enhanced.

(3) The bracing system should be installed in the roofing system in order to ensure the out-of-plane stability of the beam string structure.

(4) The principle of multi-stage design should be adopted. The effect of geometrical nonlinearity should be taken into account in the analysis of beam string structures.

(5) Temporary or permanent constructional methods should be adopted at the supports, leading to self-balanced systems without horizontal reactions under the pretensioning and the superimposed loading.

(6) Upper chord beams can also be replaced by spatial trusses, leading to spatial truss string structures. Such structures are composed of bars and cables which only bear axial forces, both the analysis and the construction of the structure can thus be simplified.

(7) Spatial beam string structures (for example, two-way orthogonal beam string structures) can be developed from planar beam string structures.

The practical application of long-span beam string structures just started in China. One representative project is the roof for the terminal building of Shanghai Pudong International Airport built in 1999 (Wang et al. 1999). The whole roof covers four large spaces including entrance hall (R1), ticket hall (R2), shopping center (R3) and airport lounge (R4), with the horizontal projected spans of 49.3m, 82.6m, 44.4m and 54.3m respectively. The spacing between the string beams is 9m. The total longitudinal length of buildings R1, R2 and R3 is 402m, while that of R4 is 1374m. The upper chord beams are composed of three parallel steel tubes with rectangular/square sections, which are welded from cold-formed channels. The web members are steel tubes with circular sections, and the lower chord cables are high-strength cold-drawn galvanized wire ropes. Fig. 5a shows the interior view of the terminal building, while Fig. 5b shows the plan and sectional drawings of beam string structures for buildings R2 and R4. The string beams are supported by inclined steel I-columns with double web plates, with the spacing between columns being 18m. As the columns and the string beams are not in the same plane, a longitudinal spatial truss with a width of 1.7m and a height of 1.3m is installed between the columns to support the string beams (Fig. 5b). The bracing system for the roof is also shown in Fig. 5b. In order to resist the upward wind loading, the upper chord beams for building R2 are poured with cement mortar to increase the self-weight of the structure. The scheme of this structure was proposed by Andrew from France, however, great improvement was made by East China Institute of Architectural Design and Research.

Another representative project is the roof for Guangzhou International Convention and Exhibition Center (Fig. 6). As shown in Fig. 6a, a spatial truss with a triangular section is adopted in this project instead of the upper chord beam, leading to a spatial truss string structure with a long span of 126.5m. The spacing between the spatial trusses is 15m. The construction of this structure is expected to be finished by the end of 2002.

(a) Interior view

Figure 5: Beam string structures for the terminal building of Pudong International Airport

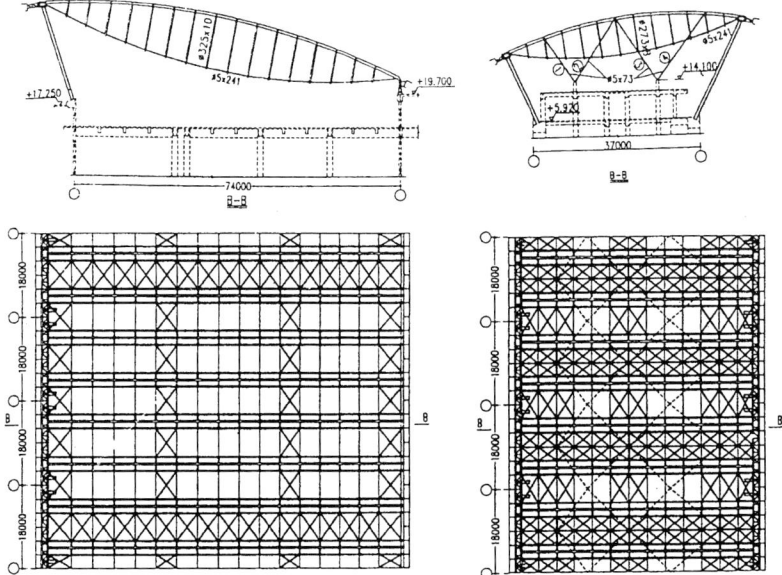

(b) Plan and sectional drawings of buildings R2 and R4

Figure 5: Beam string structures for the terminal building of Pudong International Airport (cont.)

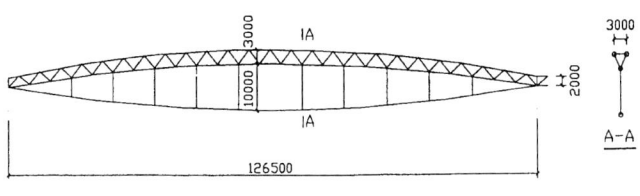

(a) Spatial truss string structure

(b) Photograph

Figure 6: Guangzhou International Convention and Exhibition Center

PRESTRESSED SEGMENTAL STEEL STRUCTURES

The prestressed segmental steel structure is a spatial lattice arch structure with a rectangular section. It was developed by Chinese engineers and researchers, and has been a patent of Beijing Zhiwei New Segmental Structures Company (Liu and Li 1995). The characteristics of prestressed segmental steel structures can be summarized as follows.

(1) Prestressed segmental steel structures are composed of prefabricated arch sheets, horizontal tie members, joint bodies as diaphragms and crossing tie rods (Fig. 7). A panel of lattice arch is composed of two pieces of arch sheets, horizontal tie members at upper and lower chord planes, and two pieces of joint bodies (Figs 7a & 7b).

(2) Arch sheets and joint bodies are usually pre-fabricated off-site and then assembled to segments on the ground on-site.

(3) Crossing tie rods are used as connecting bars between nearby panels during installation (Fig. 7c). On the other hand, prestressing is applied on the structure through these tie rods.

(4) By adding upper chord tie members (i.e. purlins) and longitudinal bracing members between prestressed segmental trusses, partial double-layer cylindrical lattice shells can be constructed.

(5) A whole process analysis should be carried out for the installation stage. At this stage, the structure can be analyzed as an arch structure with an increasing span simply-supported at both ends, while at the service stage, the structure is usually calculated as a hingeless arch fixed at both ends.

(6) Long-span structures can be assembled from small (light) components without the usage of large hoisting facilities. The construction of these structures is thus easy and fast.

(7) This type of structure can be adopted in either permanent buildings or temporary buildings, as it can be disassembled easily.

(8) Compared with braced barrel vaults with a same span, the steel consumption of prestressed segmental trusses is similar, but the horizontal reactions at supports are reduced significantly.

(9) All members adopt small square steel tubes and round steel bars, and no on-site welding work is needed, leading to low construction cost and good technical economy index.

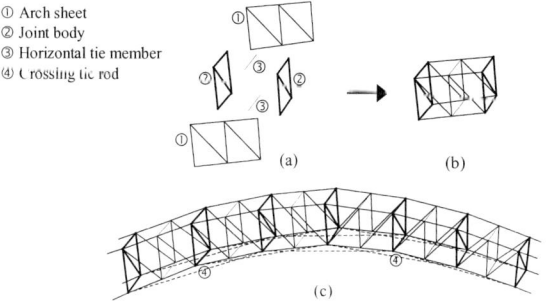

Figure 7: Schematic of prestressed segmental steel truss

About 20 prestressed segmental steel structures have been constructed in China (e.g. Wu et al. 1999; Zhang et al. 2000). Some of the representative projects are as follows: the roof for the indoor swimming pool at Urumqi Petrochemical Factory, with a span of 80m and a steel consumption of 43.5kg/m^2; the roof for the 8th hall of Beijing International Exhibition Center, with a span of 60m and a steel consumption of 31.7kg/m^2; and the roof for the indoor tennis court at Diao Yu Tai State Guesthouse, with a span of 40m and a steel consumption of 32.5kg/m^2. Fig. 8 shows the exterior and interior view of the indoor tennis court at Diao Yu Tai State Guesthouse. It has also been reported that a prestressed segmental steel structure with a long span of 165m was trial-assembled (Zhang et al. 2000).

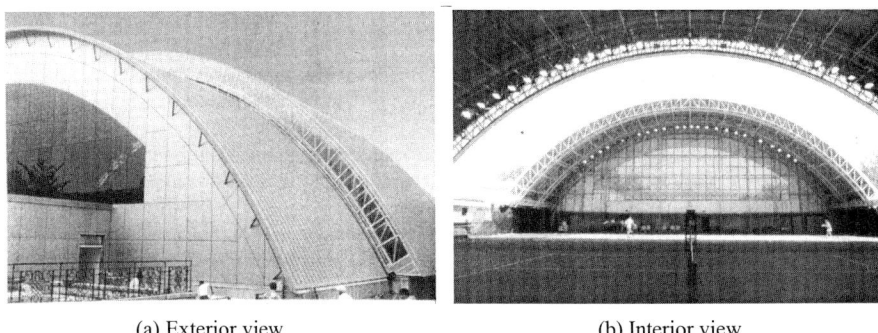

(a) Exterior view (b) Interior view

Figure 8: The indoor tennis court at Diao Yu Tai State Guesthouse

PROPECTS

By the end of the 20th century, about 80 pretensioned steel space structures have been constructed in China. Although the development and application of these structures in China has a history of only a little over ten years, a great vitality and a vast range of prospects have been manifested. It is believed that pretensioned steel space structures represent a new trend in the development of space structures. For pretensioned steel space structures in the 21st century, research on the following major subjects and leading topics should be carried out.

(1) Design and construction of pretensioned space structures with a span of 150m ~ 200m or even larger, because of the practical need;

(2) Practical forms of pretensioned space structures with super long spans of 200m ~ 500m;

(3) Study of cable-bar (or beam) hybrid structures and cable domes and their applications;

(4) Development of new materials (such as membrane fabric materials, large-diameter steel rods), new joints and new construction techniques;

(5) Calculating theory on wind-resistance and earthquake-resistance design and structural control techniques for long span and super long span pretensioned space structures, and practical methods and formulae for application in design;

(6) Development of new types of pretensioned space structures with aesthetically pleasing form and desirable mechanical behaviour under the close cooperation of structural engineers and architects.

REFERENCES

Chen L.C, Zhi Y.F., He P.B. and Xiong H.J. (1997). The Reasonable Application on Space Trusses Supported by Pattern Like a Basin. *Spatial Structures* **3:1**, 25-28.

Cui Z.Y. and Zhang G.Q. (1991). Roof Structure Design of the Gymnasium in National Olympic Sports Center. *Journal of Building Structures* **12:1**, 24-37.

Dong S.L. and Deng H. (2001). A Simple Computation Method of Prestressed Space Truss and Its Globle-Process Pretensioning Analysis. *Journal of Building Structures* **22:2**, 18-22.

Dong S.L. and Luo Y.Z. (1993). Simplified Calculation of Cable-Stayed Double-Layer Grid. *Building Structures* **8**, 28-30.

Dong S.L., Zhao Y. and Zhou D. (1998). New Technologies and New Structures in the Development of Steel Space Structures in China. *China Civil Engineering Journal* **31:6**, 1-11.

Jiao J., Song T., Zhao J.D., Qian J.H., Feng J.P. and Zhang W.Y. (2000). Structural Design of Cable-Stayed Lattice Shell for the Main Stadium of Dragon Sports Centre. *Proceedings of the Ninth National Conference on Space Structures*, Xiaoshan, China, 753-759.

Liu Z.W. and Li J.G. (1995). Research and Application of A Type of Creative Space Structure – Segmental Support Truss. *Proceedings of Application Technology of Steel Building Structures*.

Lu C.L. (1995). The Present Conditions and Developments of Prestressed Spatial Steel Structure. *Spatial Structures* **1:1**, 1-14.

Lu C.L. (1998). Development and New Achievements in Modern Steel Structures. *Advances in Modern Civil Engineering*, Southeast University Press, Nanjing, China, 157-166.

Ma K.J., Zhang X.G., An Z.S., Li X.H., Lin D.M. and Dai J.H. (1994). The Design, Structure and Mechanic Characteristics of a Large Span Combination Type Prestressed Twisted Net Shell Structure. *Spatial Structures*, First Issue, 55-63.

Space Structures Research Centre (2000). Reports on the analysis of steel roof in Shenzhen Swimming and Diving Hall. College of Civil Engineering and Architecture, Zhejiang University, China.

Tang C.M., Yan H. and Dong S.L. (1992). A Study on Static Characteristics of Cable-Stayed Space Truss. *Proceedings of the Sixth National Conference on Space Structures*, Earthquake Engineering Press, Beijing, China.

Wang D.S., Zhang F.L., Gao C.Y., Zhou J. and Chen H.Y. (1999). Research and Design of Steel Structure of Shanghai Pudong International Airport Terminal. *Journal of Building Structures* **20:2**, 2-8.

Wu J.Z., Zhang Y.G. and Shen S.Z. (1999). Study on Static Behaviour of Prefabricated Prestressed Square Steel Tube Structures. *Proceedings of 1999 Annual Meeting of Spatial Structures Association*, China Steel Structures Association, Xiamen, China, 198-205.

Wu Y.H., Zhang Y. and Chen Y.B. (1994). Design of Cable-Stayed Space Trusses for the Warehouse of Port of Singapore Authority (PSA). *Proceedings of the Seventh National Conference on Space Structures*, Wendeng, China, 467-471.

Xiong Y.C., Dong S.L., Yang Y.G. and Wu E.C. (1985). Design and Construction of Ninghe Gymnasium in Tianjin. *Building Structures* **6**, 19-23.

Yao N.L., Li L.Y., Jiang G.Y., Chan J., Liang B.R. and Pan D.Y. (1994). The Study, Design and Application of Prestressed Composite Space Truss with Threaded-Ring Connections. *Spatial Structures*, First Issue, 39-46.

Yin S.G., Fan D.R. and Zhou Y.G. (2000). Structural Design of A Hyperbolic Paraboloid Lattice Shell for Suqian Culture and Sports Center. *Proceedings of the Ninth National Conference on Space Structures*, Xiaoshan, China, 717-721.

Yin S.M., Hu Y.S., Gou K.C. and Dong S.Y. (1995). Static Characteristics of Prestressed Steel Lattice Shells. *Spatial Structures* **1:1**, 38-47.

Zhang P.J., Gao W.Y. and Liu Z.W. (2000). Design Study of Segmental Support Truss and Its Application in Beijing International Exhibition Center. *Proceedings of the Ninth National Conference on Space Structures*, Xiaoshan, China, 782-786.

Zhang Z.S., Wang K.W., Yan H., Luo Y.Z. and Dong S.L. (1997). The Structure Design of Cable-Stayed Reticulated Shell - Jiuguan Toll Station of Tai-Jiu Express Highway. *Spatial Structures* **3:2**, 40-45.

Zhou D., Dong S.L. and Deng H. (1997). Dynamic Characteristics of Cable-Stayed Lattice Shell and Nonlinear Seismic Response Analysis. *Proceedings of the Eighth National Conference on Space Structures*, Kaifeng, China, 208-214.

ADVANCED COMPUTER CALCULATIONS IN THE DESIGN OF SHELL STRUCTURES

J. Michael Rotter

School of Engineering and Electronics, University of Edinburgh, Edinburgh EH9 3JL, UK

ABSTRACT

The high efficiency of shell structures leads them to be widely used in a huge variety of applications: food and drink containers, cars and vehicles, aircraft and spacecraft, and large storage structures. Steel shell structures are most used in civil engineering as silos, tanks, pipelines, chimneys, towers and masts. Offshore structures present another example.

The new Eurocode 3 for the Strength and Stability of Shells is innovative in a number of ways. Amongst other features, it is the first standard of its kind to include codification of the use of computer analyses of different types, together with careful definitions of the manner in which the results must be interpreted into a limit state design framework. This paper describes some features of this regulation, and presents some of the difficulties that both the user and the code drafting panel must address.

KEYWORDS

Buckling, chimneys, design philosophy, European, limit state design, shells, silos, standards, tanks, towers.

INTRODUCTION

In many applications, a shell structure offers the most efficient usage of structural material, and produces very lightweight structures of considerable strength. In addition, shells are often seen as aesthetically pleasing, and they may be more widely used in architectural applications if difficulties of construction or fabrication can be overcome. However, shell structures display many complex responses: they are susceptible to sudden buckling failures, and are much affected by minor changes in geometry caused either by accident or as a result of loading conditions in service. The mathematics that governs the structural response of a shell is often far beyond the scope of hand calculation, and many practical structures and loading cases require a computer analysis, even to understand the stresses and deformations that will develop under normal service conditions. These considerations make the use of computer calculations very important in the design of shell structures, for all but the simplest of geometries and load cases.

The last forty years has seen a huge research effort in the formulation and commercialisation of numerical shell analyses. For many development analysts, the formulation itself and its computational verification were the primary goals. The difficulties that lie between the numerical output from the analysis and the determination of safe assessments of limit states for practical shells have not been widely discussed. The lack of good research in this area has caused difficulties in the adoption of computer analyses into design calculations, especially where the design must be regulated by a standard. Some of these difficulties are outlined in this paper.

Until recently, all standards for shell design presented only hand calculation assessments of the shell's strength. Sometimes, though rarely, a method of finding the stresses in the shell using hand calculation based on the linear bending theory of shells has also been provided (Baker et al., 1972), but the need for changed criteria of failure when a different calculation method is used has rarely been indicated. The first attempt to remedy these matters has been undertaken in the new Eurocode on Shell Stability and Strength (ENV 1993-1-6, 1999). However, since this standard has attempted to codify the use of many different types of shell analysis for application to a very wide range of different problems, it is likely that some omissions may have occurred and some unintended consequences may arise when the standard is applied to unusual designs.

Some researchers in applied mechanics often appear to think that a geometrically and materially nonlinear analysis (with imperfections included) is the only ones that is required, since it gives the "right answer" to the problem in question. It should therefore be clearly stated at the outset that these analyses do not solve problems in a general manner, because the significance of the result cannot be recognised or understood unless it is placed in the context of other reference loads. Thus the buckling stress of a column is only useful when its yield stress and slenderness are known. Transferring this concept to a general shell structure, the complete nonlinear analysis result is only useful if appropriate reference elastic buckling and plastic failure loads are known.

This paper provides an overview of the computer analysis features of the new standard. It is intended to guide users and researchers into understanding the thinking behind the standard. It is also intended to provoke discussion of how the provisions of the standard should be modified as further knowledge is gained in its application to a wider range of problems than those known to the drafting committee.

CONDITIONS THAT MAKE NUMERICAL ANALYSES NECESSARY FOR SHELL STRUCTURES

Throughout structural engineering, there is considerable potential to improve the accuracy and reliability of the design process by exploitation of computer calculations. However, shell structures present an especially fruitful field for the exploitation of numerical analyses in design. There are several reasons for this situation.

Complexity of the behaviour

First, the complexity of the behaviour of a shell structure means that hand calculations can only offer only a very approximate and often rather misleading image. The calculation procedures that might be considered begin with the membrane theory of shells (equilibrium alone), pass to the linear bending theory of shells (LA), but continue on to geometrically nonlinear analysis (GNA), materially nonlinear analysis (MNA) combined material and geometric nonlinearity (GMNA) and finally can include the effects of imperfections (GMNIA).

Hand design calculations of shells are normally based on the membrane theory of shells, which gives a reasonable picture of the membrane stresses in an axisymmetrically loaded and supported shell at points distant from boundaries and discontinuities. As soon as the loading becomes realistic (wind, earthquake, solids pressures in silos), significant unsymmetrical components of loading arise, and the membrane theory often gives a poor description. For a structure with a more complex geometry (e.g. torispherical heads, shell junctions), the application of membrane theory becomes more tedious and gives a poor representation of the stress state (Gould et al, 1985; Rotter, 1983, 2001). However, hand calculations for limit states are generally premised on the assumption that a membrane theory analysis has been performed, and this leads to difficulties when any other analysis is used.

The linear bending theory of shells appears to the structural engineer to provide the natural reference analysis, satisfying equilibrium and compatibility in an elastic structure, assuming small deformations (unchanged geometry). However, it is not a very useful tool for hand analysis, since the simplest cases of unsymmetrical loading (e.g. a cylinder) produce a set of simultaneous fourth order partial differential equations, whose solution is extremely onerous (Seide, 1975; Li and Rotter, 1996). Other geometries (e.g. cones) lead to even greater complexity, even when the loading is axisymmetric. Thus, hand analyses are generally not to be favoured, and even the linear response of a shell generally requires a numerical analysis.

The complexity of the behaviour is not only associated with analysis techniques to find the stresses. The failure conditions for shells are also rather more complicated. They can be related to those of frame structures, with which most structural engineers are far more familiar. Where the shell is very thin, buckling failures can occur at very low stress levels (e.g. 20MPa) and the yield stress of the construction material has no significance at all. However these buckling stresses are very sensitive to the form and amplitude of unavoidable minor geometric imperfections in the shell surface, so special procedures must be in place to ensure that the shell is constructed sufficiently accurately to achieve the intended strength. Buckling failures in shells can only occur in the presence of compressive membrane stresses. Where only tensile membrane stresses are present, failure by plastic collapse mechanism or rupture can occur. The plastic collapse mechanisms in shells are far more complex than those of frame or even plate structures, involving simultaneous bending and stretching of the shell, and usually achieving strengths that far exceed anything that can be estimated from a linear elastic bending theory analysis.

Both membrane theory and linear bending theory can only predict stresses: consequently criteria of failure must be described in terms of stresses or stress resultants. More advanced types of analysis appear to give a failure load, but in the following it will be seen that each of them needs further interpretation before the result can be used in design.

Shortage of experimental evidence

The second reason why numerical analyses are important in shell structures is that the total database of experimental investigations of shell structures is very limited. There have been many tests on very simple geometries (cylinders) under very simple load cases (axial compression, external pressure, torsion, global bending), but very few tests have explored many of the phenomena that the designer must incorporate into his evaluation.

Thus, geometries other than cylinders, cones and torispherical pressure vessel closures have not been not been widely studied, and the only non-symmetric loading case that has received significant experimental attention is the short cylinder (tank) under wind loading. With these exceptions, realistic design conditions, including load case combinations, lead to stress states in a shell that mostly cannot be related to the experimental database.

This poses two problems. First, many successful shell designs have grown up on a rather empirical basis (e.g. bracket supports under a storage vessel), and these cannot be justified using the very conservative simplistic criteria that relate to simple load cases. Second, geometric imperfections can dramatically affect the strength of shells when under relatively uniform loading, but the sensitivity to imperfections seems to be much less to stresses that are only locally high. In particular, it appears that the high stresses must extend over a zone of size comparable to the potential buckle before a limit state is approached (Cai et al, 2001). Much more testing is needed before the database can even be used adequately to verify numerical buckling predictions for realistic load combinations.

Design regulation for realistic conditions

Previous design rules cover only elementary cases of relatively uniform stress conditions in shell structures, being derived directly from empirical lower bounds on the results of simple tests. These simple safe rules relied on the use of simplified conservative approximations to represent the actions on the structure. However, the real forms of loads and their magnitudes have been much researched in recent times, and this improved realism can be exploited to achieve more economic designs with secure reliability. But realistic patterns of loading can only be used in a shell design assessment if the calculations are performed numerically. Thus, improved knowledge of loads on structures also demands the use of modern computational procedures in design.

Complex shell structures in 21st Century

There are many opportunities for architectural and technological innovation in the design of structures, and the aesthetic appeal of shell structures, which led to such structures as the Sydney Opera House and the London Millennium Dome, will continue. Such structures are generally far too complicated for their design to be based on knowledge gained from testing alone. Modern numerical analyses offer the opportunity to design these structures much more cost-effectively, but the relationship between the calculated stresses or strengths and those that should be used for a limit state assessments is less fully explored than is desirable. The new Eurocode offers the first attempt to ensure that numerical analyses of all classes can take their proper place in the design process.

ALTERNATIVE TYPES OF ANALYSES OF THE STRUCTURE

As indicated above, ENV 1993-1-6 (1999) codifies the use of a wide range of analyses. To this end, each is defined with care, using the abbreviations and meanings shown in Table 1.

Table 1. Structural analysis types

Type of analysis	Term	Shell theory	Material law	Shell geometry
Membrane theory of shells		membrane equilibrium	not applicable	perfect
Linear elastic shell analysis	LA	linear bending and stretching	linear	perfect
Geometrically non-linear elastic analysis	GNA	non-linear	linear	perfect
Materially non-linear analysis	MNA	linear	non-linear	perfect
Geometrically and materially non-linear analysis	GMNA	non-linear	non-linear	perfect
Geometrically non-linear elastic analysis with imperfections	GNIA	non-linear	linear	imperfect
Geometrically and materially non-linear analysis with imperfections	GMNIA	non-linear	non-linear	imperfect

Membrane theory uses only static equilibrium, predicts only membrane stresses and is a poor representation near discontinuities or for loading that is severely unsymmetrical, especially if this acts tangential to the plane of the shell surface. However, it is very widely used in design, and failure criteria for hand analysis of the shell are normally based on the assumption that a membrane theory analysis has been made.

Linear shell theory (LA) defines the normal operating condition of the shell under small displacement elastic conditions, and produces a complete membrane and bending distribution with six stress resultants at every point. Interpretation of the results into values relevant to limit states is not always easy, since very high local bending stresses can occur (e.g. near shell junctions) that are not found in a membrane theory analysis. Where the load case is to be repeated very many times, the question of fatigue naturally arises, but when the load case is an extreme event, the significance of high local bending stresses is not easily deduced. These are normally associated with satisfying boundary conditions and can be ignored in a ductile structure. But a change in the boundary conditions can easily produce similar bending stresses that are needed to satisfy equilibrium (e.g. Rotter, 1987). A linear analysis cannot distinguish between these two cases, so it is difficult to formulate general rules about how seriously bending stresses should be taken as being in the most general case.

Within the framework of linear theory is a Linear bifurcation or Eigenvalue Analysis (here termed LEA), performed on a perfect shell structure, yielding the elastic critical load for the system. It should be remembered that the critical load depends only on the membrane stresses and is unaffected by bending stresses.

The nonlinear analyses are identified according to the additional phenomena included. The materially nonlinear analysis (MNA) uses small displacement theory with plasticity and is only useful for determining plastic limit loads. The geometrically nonlinear elastic analysis (GNA) uses large displacement elastic theory and can trace nonlinear load paths. It is needed to detect elastic snap-though buckling events (elastic limit loads). It should also be able to follow the load path through a bifurcation onto the post-buckling path, but often this can only be achieved if an appropriate small perturbation is introduced which the numerical analysis can amplify into the new mode.

Geometrically and materially nonlinear analysis (GMNA) combines the above analyses, and provides good estimates of the strengths of perfect structures that are not imperfection-sensitive. However, when imperfection-sensitivity is present, this must be expanded to the GMNIA analysis, explicitly including geometric imperfections of appropriate amplitudes in appropriate modes. The latter task makes the use of GMNIA analysis far from straightforward.

REFERENCE CASES FOR THE EUROCODE DESIGN ASSESSMENT

Introduction

In beams, columns and frame structures, it is widely recognised that stocky construction produces structures that can attain a plastic collapse limit state (Fig. 1), dominated by material failure, relatively small displacements, and requiring ductility. Similarly, slender construction leads to buckling limit states (Fig. 1), with a variety of different forms of buckling (local, distortional, member, frame etc.) occurring according to the component that is slender. Economic frame design that exploits the material strength fully is generally found to produce designs where the buckling load and plastic failure load are similar. It may also be noted that geometric imperfections in these structures generally only cause a loss of strength by causing earlier yielding, not because the imperfection leads to altered destabilising stresses, as in shells.

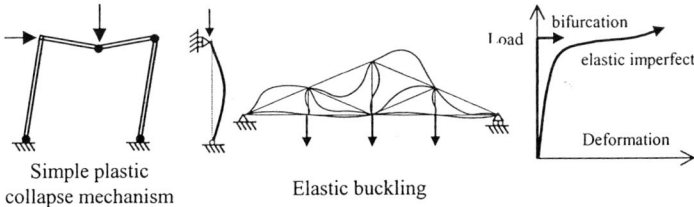

Fig. 1 The reference cases: plastic collapse and elastic stability

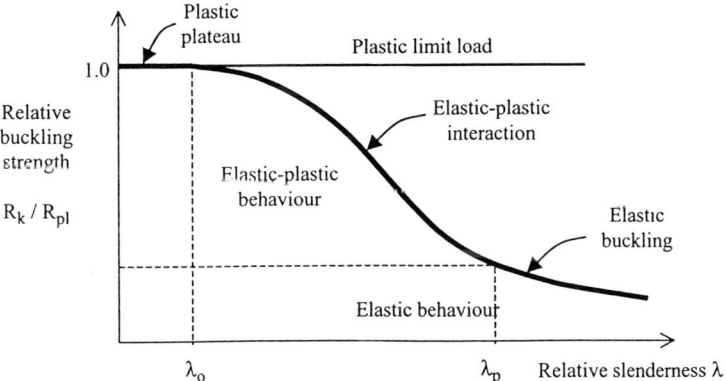

Fig. 2 Capacity curve for elastic-plastic buckling and collapse interaction

Where alternative types of analysis, many of which will exploit computers, are to be accommodated within the framework of the same design philosophy, it is necessary to address questions of plasticity, stability and imperfection sensitivity within the framework of a reliability assessment. The interaction between plasticity and stability is usually represented by the "capacity curve" (Fig. 2), which is sometimes referred to as a column curve. This defines the transition between a) fully plastic failure unaffected by geometric nonlinearity, and b) elastic stability failure unaffected by material nonlinearity.

The capacity curve is normally expressed in terms of the dimensionless or relative slenderness $\bar{\lambda}$

$$\bar{\lambda} = \sqrt{\frac{f_y}{\sigma_{cr}}} \tag{1}$$

in which f_y is the yield stress and σ_{cr} is the elastic buckling stress for a given geometry and loading condition. The definition of slenderness in terms of stresses arises from considerations of column buckling or local buckling, and it must be generalised further before it can properly be applied even to redundant beams and frames, since the evaluated stress depends on assumptions about the response of the structure to the loads.

A first generalisation might transform the stresses into stress resultants, but this is still insufficient, since attainment of a limiting stress resultant at one point in a structure does not normally correspond to plastic collapse. Thus the most general dimensionless slenderness is given instead in terms of load factors that achieve plastic collapse or elastic buckling in the complete structure. The plastic limit state (formerly the yield stress in Eq. 1) can be represented by the peak load factor on the design loads attained in a rigid-plastic or small displacement theory elastic-plastic analysis, here defined as the load factor R_{pl}. The elastic critical limit state (formerly the critical stress in Eq. 1) can be similarly represented as the bifurcation load factor on the design loads found from a linear eigenvalue analysis R_{cr}, though special provisions are needed where snap-through buckling is critical. The overall relative slenderness of the complete structure then becomes

$$\bar{\lambda} = \sqrt{\frac{R_{pl}}{R_{cr}}} \tag{2}$$

This revised definition permits all structures to be treated in a uniform manner, though it carries with it certain disadvantages, as noted below.

For shell structures, the same two limit loads can be usefully described, with a plastic limit state dominated by material failure and an elastic buckling limit state whenever compressive stresses are present. However, the optimal geometries are often governed by other considerations, so many shell structures are either very stocky or very slender according to this description.

The reference resistances R_{pl} and R_{cr}

The first step in defining the elastic-plastic interaction for shells is to define carefully the reference resistance loads R_{pl} and R_{cr}. The reason why a careful definition is needed is that geometric nonlinearity and geometric imperfections are both commonly thought to be important constituents of any good shell analysis, since they modify the buckling and plastic loads markedly in many cases. The Eurocode definitions (ENV 1993-1-6, 1999) are therefore stated here.

The primary definition of the elastic critical load R_{cr} is taken as the bifurcation buckling load of the perfect shell with ideal loading and boundary conditions found using a small displacement theory Linear elastic Eigenvalue Analysis (LEA). Geometric nonlinearity is excluded. There are, however, some classes of problem for which snap-through buckling occurs at a load far below the bifurcation load (e.g. shallow spherical caps, cylinders locally loaded normal to the surface), so that the bifurcation load is not a useful measure. For these, the snap-through buckling load is chosen as R_{cr}, so this requires a geometrically nonlinear analysis (GNA) of the perfect shell with ideal loading and boundary conditions found using a large displacement theory

The plastic reference load R_{pl} is similarly defined as the plastic limit load of the perfect shell with ideal loading and boundary conditions found using a small displacement theory materially nonlinear analysis (MNA). This corresponds to a classical plastic limit load (rigid-plastic behaviour), with the development of a kinematically admissible plastic strain velocity field. But lower bounds on this value can be obtained from other analyses.

Alternative choices for the reference resistances

For shell structures, the first question that is naturally asked concerns the relevance of the loads R_{pl} and R_{cr} to the real structural behaviour. Concerns can be expressed about both the adoption of each of these loads.

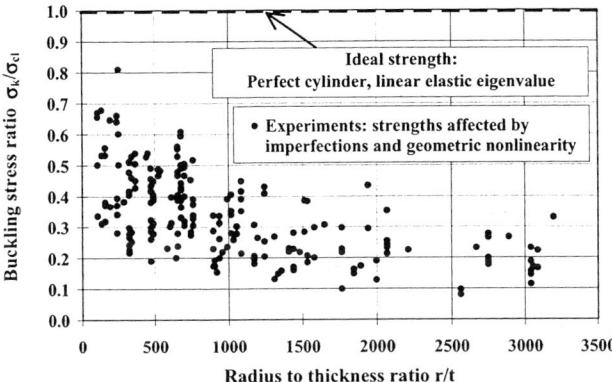

Fig. 3 Effect of geometric imperfections and geometric nonlinearity on cylinder axial buckling stress

Because shell buckling strengths are usually very sensitive to geometric imperfections (Fig. 3), and are also affected by prebuckling geometric nonlinearity, many would argue that the bifurcation load R_{cr} is of little significance. The ECCS standard (1988) used the imperfect elastic shell buckling strength in its evaluation of the relative slenderness for this reason. By choosing the imperfect buckling load, a capacity curve of the form of Fig. 2 approaches the elastic buckling condition at large slenderness, in line with the behaviour seen in columns, beams and frames. However, this choice carries a major disadvantage: the elastic buckling load, as reduced by geometric imperfections, is very dependent on the form and amplitude of the imperfections, and these may vary from one structure to another and with the quality of construction. If this choice is adopted, then it is not possible to plot a test result in a unique manner in Fig. 2, and the empirical elastic-plastic interaction becomes sensitive to the imperfection assumptions of the researcher (which often tend to be different). Thus the imperfect buckling strength is not a good choice.

Others might argue that the effects of geometric nonlinearity should be included (GNA) in determining the value of R_{cr}. However, this causes other difficulties. There are many geometries and loading conditions (e.g. some torispherical heads under internal pressure) where the change of geometry is beneficial, and the compressive stresses seen at small displacements are steadily reduced until they disappear, leaving no determinable load for the critical load R_{cr}. Since this is an elastic calculation, the real failure may have occurred at a much lower load where stability effects are still present, so it is not acceptable to have no critical load at all. Thus geometric nonlinearity is not acceptable in a general definition of an elastic critical reference load.

The plastic reference load involves similar difficulties. For the simplest cases (centrally loaded column, uniformly axially compressed cylinder, internally pressurised cylinder) a membrane state of stress is established, and the plastic reference stress is identical to the yield stress. However, for more realistic configurations (e.g. a local bracket support) (Holst et al, 2002), the plastic load demands the development of a full bending and stretching mechanism (Massonnet and Save, 1972). Since classical solutions exist for only the simplest geometries and loading conditions (mostly cylinders under axisymmetric loads), either an MNA analysis must be used, or a lower bound estimate found from an elastic analysis. The standard (ENV 1993-1-6, 1999) has adopted the Ilyushin yield criterion applied to the stresses at the most stressed point to try to permit some admission of plasticity in this assessment, but it is often very conservative indeed. There is no acknowledgement of the development of a mechanism, that this mechanism must be two dimensional, and that the kinematically admissible field generally involves both stretching and bending throughout it. This simple lower bound is satisfactory for simple uniform load cases, but it is difficult to find a less conservative interpretation of linear analyses that is also safe.

Again, some might argue that the effects of geometric nonlinearity should be included (GMNA) in determining the value of R_{pl} since the effects can be large in thicker shells. The current design rules for torispherical heads (ECCS, 1988) are equations that have been empirically fitted to calculations based on GMNA analyses, so this idea has been used. However, the same drawback occurs as for geometric nonlinearity in assessments of the elastic critical load: for all geometries in which the nonlinearity is stiffening, no limit load can be determined. This led Gerdeen (1979) to search for a more general definition. However, for some geometries, there is even the semblance of a knee in the curve (Teng and Rotter, 1989) so the inclusion of geometric nonlinearity takes away the possibility of a general definition, applicable to all problems, that can always be determined.

Thus, in conclusion, the small displacement theory perfect elastic shell linear bifurcation load and the small displacement theory plastic limit load represent simple loads that are always calculable and are relatively unaffected by choices made by the analyst. They represent the best choices for the reference loads. It should, of course, be noted that the bifurcation load is only calculable when compressive stresses are present.

Difficulties with the choice of reference resistances

The above may suggest that these reference resistances are universally useful. However, they still present problems in trying to maintain a good relationship between hand and computer calculations. For easy hand calculations, it is preferable to have reference loads that relate to simple load cases (uniform axial compression, uniform external or internal pressure, uniform torsion, global bending) and to adjust later parts of the calculation for the changes associated with local deviations (e.g. holes and penetrations, local supports etc.). However, the general definitions given above do not permit this, but instead deliver reference loads that can change significantly when quite small changes are made to the loading, geometry and boundary conditions. A simple example is the internally pressured thin imperfect cylinder under axial compression (Rotter, 1997). This difficulty will need to be considered carefully in future research studies.

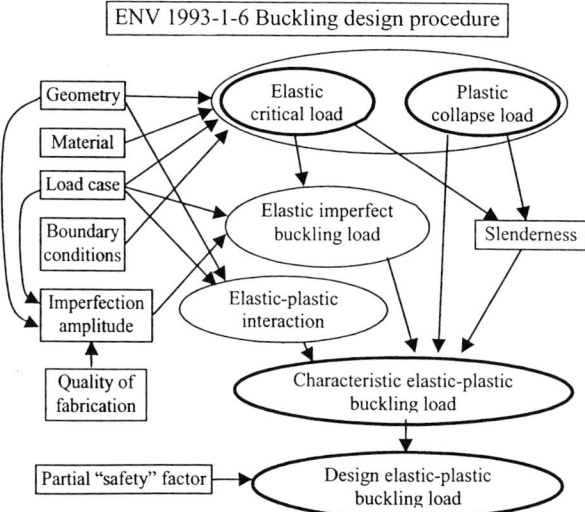

Fig, 4 Calculation concepts and the path to design strength

THE HAND CALCULATION PROCESS AND ITS PARTS

The hand calculation process defined in the Eurocode (ENV 1993-1-6, 1999) is relevant to computer calculations of shells because most of the computer calculations cannot be said to account for all the effects that are required to be considered in design. For this reason, each computer calculation is deemed to fulfil the requirements of part of the complete calculation, but hand adjustments must be made for the remainder. It is easiest to understand these hand adjustments in the context of the complete hand calculation.

The hand calculation of the imperfect elastic-plastic buckling resistance is separated into many different parts. The process is outlined in Fig. 4 and begins with the two reference loads that were defined above. These reference loads depend on the geometry, material properties, load case and boundary conditions. The relative slenderness is deduced directly from these loads.

The elastic imperfect buckling strength must next be found. This depends on the assumed imperfection amplitude, which is explicitly defined in ENV 1993-1-6 to give a strong relationship between design assumptions and the tolerances required in construction. The geometric imperfections that are most damaging depend very much on the load case (a useful estimated form being the linear eigenmode), so that a load case like external pressure of a cylinder requires very different tolerance control from that for axial compression. The elastic imperfection reduction for axial compression in cylinders is shown in Fig. 5.

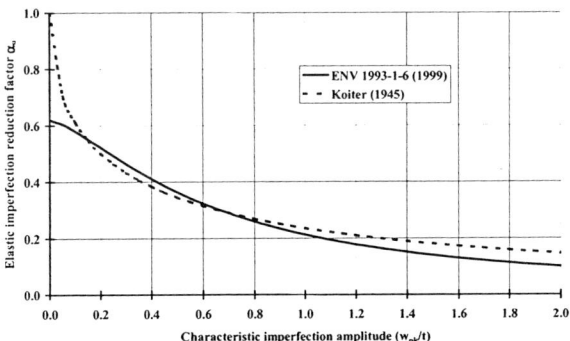

Fig, 5 Strength-imperfection relationship for elastic buckling

The next step requires the elastic-plastic interaction appropriate to this geometry and load case to be defined. The capacity curve is defined in a rather general manner (Rotter, 1999) in terms of the ratio of the characteristic resistance to the plastic reference resistance $\chi = R_k/R_{pl}$ as

$$\chi = 1 \qquad \text{when} \qquad \bar{\lambda} \leq \bar{\lambda}_0 \qquad (3)$$

$$\chi = 1 - \beta \left(\frac{\bar{\lambda} - \bar{\lambda}_0}{\bar{\lambda}_p - \bar{\lambda}_0} \right)^{\eta} \qquad \text{when} \qquad \bar{\lambda}_0 < \bar{\lambda} < \bar{\lambda}_p \qquad (4)$$

$$\chi = \frac{\alpha}{\bar{\lambda}^2} \qquad \text{when} \qquad \bar{\lambda}_p \leq \bar{\lambda} \qquad (5)$$

in which $\bar{\lambda}$ is the relative slenderness, α is the elastic imperfection reduction factor, β is the plastic range factor, η is the interaction exponent and $\bar{\lambda}_0$ is the squash limit relative slenderness, all of which are separately defined for each geometry and loading case.

The format of ENV 1993-1-6 has been arranged so that this interaction curve can be modified easily to take a different shape for different cases as more knowledge becomes available . Examples of the forms that that interaction curve can take are shown in Fig. 6.

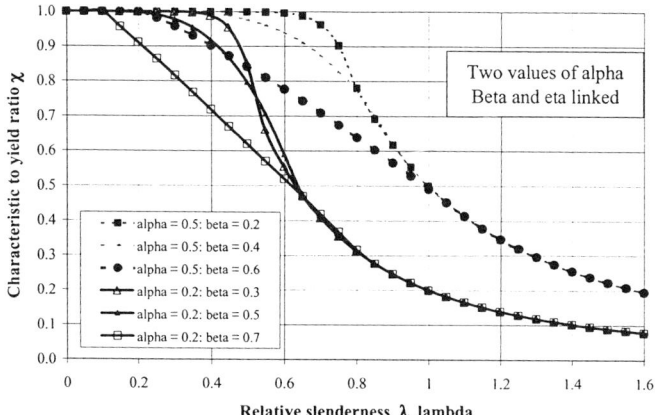

Fig, 6   Capacity curves for shells, using different values of the controlling parameters

The above calculation yields the characteristic elastic-plastic buckling resistance R_k, which must be reduced by the appropriate partial factor to achieve the design value R_d.

Different numerical analyses can contribute different parts of this complete assessment, but even the most sophisticated (GMNIA) needs careful calibration to ensure that its outcome is safe.

FEATURES THAT DIFFERENT ANALYSIS TYPES CAN ADDRESS

Introduction

An improved analysis can produce a more direct and accurate evaluation of individual parts of Fig. 4. Thus a MNA analysis may give a better plastic collapse load, and a GMNA analysis may give a better perfect shell strength, but the imperfection sensitivity must be separately evaluated.

Linear elastic shell analysis

When a linear elastic (LA) analysis is performed ("shell bending theory"), a much better description of the stresses in the structure is obtained than that of membrane theory, and unsymmetrical or local loads and discontinuities are all properly accommodated. The plastic limit load is often poorly estimated, as indicated above.

Where a linear eigenvalue analysis is undertaken, the elastic critical load can be determined precisely. The linear elastic eigenvalue analysis (LEA), based on LA stresses, is classed as part of the LA analysis, though it constitutes a considerable additional step. It accurately determines the elastic critical buckling load R_{cr} (Fig. 7). This is an important reference load for all analyses, since it is required in the formal definition of relative slenderness needed to interpret the results of other analyses (Fig. 4). Linear eigenvalue analysis cannot detect snap-through buckling, so cautionary notes are given concerning problems where snap-through phenomena are known to occur.

Materially nonlinear analysis (MNA)

This analysis uses small displacement theory with material yielding and its outcome R_{MNA} is an accurate determination of the classical plastic limit load of the system (Fig. 7) giving R_{pl} accurately. Although this type of analysis is not widely discussed any more in applied mechanics for a, for all problems involving complexities of geometry or loading, and where the structure may be sufficiently stocky to enter the elastic-plastic domain (Fig. 2), this is a very beneficial analysis to undertake, since it can greatly increase the estimate of R_{pl}. Strictly, an ideal elastic-plastic material model is needed. It is sometimes necessary to transform the data to obtain a precise estimate of the limit load (Holst et al, 2002). This analysis is also important in addressing the limit state known as cyclic plasticity (ENV 1993-1-6, 1999) which has not been described here.

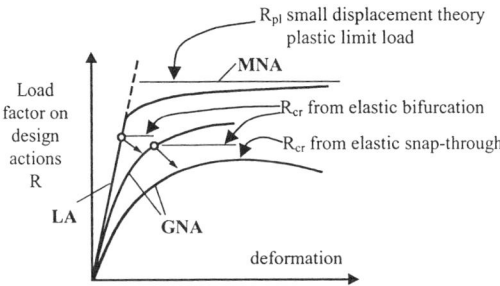

Fig. 7 Calculated load-displacement paths for LA, LEA, MNA and GNA analyses

Geometrically nonlinear analysis (GNA)

A geometrically nonlinear elastic analysis of the perfect structures (GNA) captures the large displacement elastic behaviour (Fig. 7). It is the minimum analysis required to predict snap-through buckling, and for structures susceptible to snap-through (arches, shallow spherical caps, shallow conical roofs, cylinders under local normal loads), the maximum load R_{GNA} is needed to define the reference critical load R_{cr} (Fig. 6).

This analysis may be used to explore bifurcation into non-symmetric modes, but the values so obtained are not very useful. The reason is that these values do not include imperfections, so they are only reference values for the perfect structure. Moreover, the reference buckling strength R_{cr} is based on the linear eigenvalue critical load, so the GNA analysis, which includes non-linear prebuckling stress redistributions, actually does not give the required result. Thus, on this occasion, the more "accurate" analysis is of less value. In geometrically hardening structures, the appropriate load cannot be found from this analysis.

Geometrically and materially nonlinear analysis (GMNA)

For structures that are not imperfection sensitive, a GMNA analysis provides a rather accurate evaluation of the true strength of the structure R_{GMNA}. It is therefore the reference precise analysis for frames and similar structures. It is an appealing analysis, because the difficulties of choosing appropriate imperfections are avoided, and there is only a single calculation in place of a whole series with different imperfection amplitudes needed for GMNIA.

However, since this analysis does not explicitly include geometric imperfections, and shell structures are generally imperfection sensitive, the standard requires that the maximum load determined from a GMNA analysis should be reduced according to the imperfection sensitivity factor $\alpha_{ov,GMNA}$ assessed using hand methods for similar structures, the similarity to take account of buckling modes, imperfection sensitivity, yielding sensitivity and postbuckling behaviour, together with the fabrication tolerance quality class.

Thus the current draft of the standard (1999) requires that the characteristic load R_k be taken as

$$R_k = \alpha_{ov,GMNA} \, R_{GMNA} \tag{6}$$

where R_{GMNA} is the calculated load. The reduction $\alpha_{ov,GMNA}$ is satisfactory for relatively slender structures, but it becomes very conservative for stocky shells. The reason is that it does not account for the reduced imperfection sensitivity of yielding structures (note the different imperfection sensitivity at large slenderness in Fig. 6 but the close proximity of the curves at small slenderness). For a very stocky shell, it returns the very conservative value of αR_{MNA}, and certainly needs improvement.

However, it is not easy to see how to make the required improvement. Unless the relative slenderness of the structure can be identified, the extent to which the imperfection sensitivity α should influence the calculation outcome cannot be determined. Thus, even a GMNA analysis is of very limited value for imperfection-sensitive structures unless it is accompanied by another means of determining the slenderness (either using MNA and LEA analyses, or an approximate evaluation from an LA analysis). Thus the two key reference loads R_{pl} and R_{cr} are invariably needed, even for this high quality analysis.

One last problem with the GMNA analysis is common with the GMNIA: in geometrically and strain hardening structures, it may be difficult to identify the maximum load. This is discussed below.

Geometrically and materially nonlinear analysis including geometric imperfections (GMNIA)

This very sophisticated analysis should, ideally, be able to identify the true characteristic strength of the structure R_k, and obtain calculated values that are rather close to the requirements of the design. However, several difficulties arise in seeking a description applicable to all shells. The different forms of load-deformation plot are shown in Fig. 8.

Three criteria are given for identifying the failure load:
a) the maximum load,
b) the lowest bifurcation load, and
c) the largest tolerable deformation.

The latter, needed for geometrically and/or materially hardening structures, is unfortunately difficult to define well, because no plateau load is achieved in a hardening structure, and the maximum load is rather arbitrarily defined (Fig. 8). Moreover tolerable deformations tend to be very problem-specific. It is important that the criterion is interpreted flexibly in terms of either displacements or rotations of the surface of the structure, and it should be noted that the latter are often more detectable by the naked eye. It may be that this restriction should be applied as a serviceability limitation at the operational load level.

The current draft standard is very unclear about the limitation that should be put on tolerable deformations. It is proposed here, following studies of bracket supported cylinders, that the limitation should be with respect to rotations of the surface, and that these should be limited to 0.1 radian. At this rotation, the deformation of the shell is very visible to the naked eye, and is often associated with significant deflections. By contrast, if a deflection limit is imposed, it must generally be defined as a relative deformation within the structure. Moreover, quite large relative deflections over the full length of a large structure are acceptable, but are completely unacceptable if occurring in a local dent. Thus the surface rotation limitation appears the most useful criterion.

The challenge of choosing an appropriate imperfection form and amplitude to capture an appropriately low strength in GMNIA analysis is not simple. The standard requires that the mode be chosen to give the most unfavourable effect on the buckling behaviour, and suggests both eigenmode imperfections (in the form of the critical mode from LEA analysis) and forms related to the fabrication method (e.g. weld depressions).

The amplitude of imperfection to be adopted is again related to the assumed tolerances through a quality class, and a warning is given that sometimes smaller amplitude imperfections can produce lower buckling loads (this is true for large amplitude imperfections on short shells).

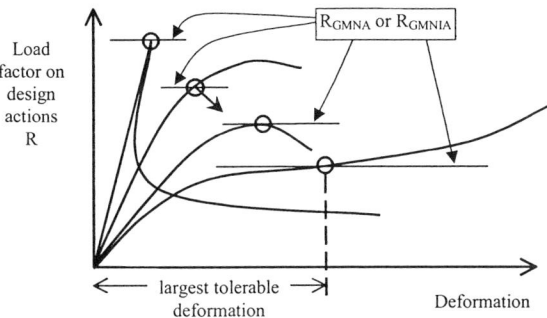

Fig. 8 Criteria of failure for GMNA and GMNIA analyses

Other forms of imperfection are always present: the load is not applied as perfectly as is commonly assumed, the boundary conditions are imperfect, and residual stresses always arise as a result of the fabrication process. Whilst the latter may not be very damaging (Holst et al, 2000), non-uniformities of load often affect the strength considerably. Thus, it must be recognised that the GMNIA analysis is a deterministic evaluation seeking to establish a characteristic value and it must be calibrated against similarly calculated known results for similar shells to ensure that an adequate margin is included for the scatter of strengths associated with unknown factors.

DIFFICULTIES IN THE STRUCTURE AND FORMAT OF THE BUCKLING CALCULATION

Despite the care and clarity of thought that has been put into devising the framework described above, there remain a number of difficulties that need resolution. Some of these have been brought out in the above discussion. However, three are outlined here as sources of particular difficulty.

Changing reference resistances with small loading changes

First, there is the sensitivity, in some situations, of the reference resistances R_{cr} and R_{pl} to small changes in the loading conditions. For simplicity and clarity, codified hand calculations need relatively fixed reference resistances (like the classical elastic critical stress under axial compression), and all existing hand calculation methods have followed this form when defining modified load case of internal pressurisation. But the computer calculations cannot retain these fixed reference resistances, but are modified by every small change. To bring the two back into line would require a considerable increase in the complexity of hand calculations, since both reference loads would need modification, whilst the current adjustments applied to the imperfection reduction factor α would still be needed. It would be most helpful if some way through this difficulty could be found.

Biaxial stress states

Second, the capacity curves of Figs 2 and 6 relate to stresses in a single direction. This arises from their origins in column, beam and uniaxially compressed plate buckling. A direct consequence of this single dominant stress is that the relative slenderness is a direct indicator of whether plasticity will affect the buckling stress or not: no situation can arise when very low compressive stresses can cause buckling when a different tensile stress is close to the yield value. However, most practical shell buckling conditions involve biaxial membrane stresses in the shell (e.g. axial and circumferential stresses in a cylinder). A factor that makes the consequent difficulty less obvious is that the tradition in shell structures has been to study reference load cases that involve uniaxial stresses (axial compression, external pressure, torsion, cylinder global bending). Yamaki (1984) termed these "fundamental loads". When the compressive stress acts in only one direction, the capacity curve can be used in the same manner as for columns: only stocky structures will experience plasticity and only slender structures will exhibit purely elastic behaviour.

However buckling is provoked only by compressive stresses, so when a state of biaxial stresses develops in the shell, it is quite possible that plasticity will develop when the shell appears to be slender. A simple example of this situation is the internally pressurised cylinder (Rotter, 1997), where "elephant's foot" plastic instabilities occur in very thin shells under very low compressive stresses.

A simple remedy is to use the von Mises stress to make the plastic limit load properly compatible with the computationally calculated value. However, this causes a different problem because the elastic-plastic interaction is required to accommodate two different phenomena: elastic-plastic buckling of geometrically imperfect thicker shells under uniaxial compression (distant from a boundary condition), and the elastic-plastic instability of very thin geometrically perfect cylinders (near a boundary condition) with high internal pressure. Because these phenomena are so different, two different elastic-plastic interaction curves will generally be needed. The current draft standard avoids this difficulty by placing the strength reduction for elastic-plastic "elephant's foot" buckling in the evaluation of the elastic imperfection reduction factor (Rotter, 1990), which leads to an accurate but rather strange evaluation process.

Snap-though buckling

The third problem relates to the uncertain definition of the elastic critical resistance R_{cr}. For most practical shell geometries and load cases, bifurcation buckling is the dominant consideration, so the simple choice of a linear eigenvalue analysis to determine R_{cr} is satisfactory. However, snap-through buckling not only occurs in the well known cases of spherical caps and arches, but appears to be quite common when very unsymmetrical or local loads are applied to curved shell surfaces. Thus, as more practical load cases are considered, snap-through may take on a more prominent role.

However, two difficulties arise: first, it is not easy to formally define in the standard when it is necessary to check for snap-through buckling. Further, since snap-though requires a GNA analysis in place of an LEA, this is not a requirement to be casually placed on the designer/analyst. It would be very useful if new criteria could be developed that would permit a better identification of conditions under which snap-though may become critical. Secondly, most of the literature on imperfection sensitivity relates to bifurcation buckling. The literature on imperfection sensitivity in snap-though buckling is much less extensive, and it is not yet clear whether the imperfection-sensitivity factors α developed from fundamental load cases for bifurcation buckling should be applied in these cases too. Much more information on the imperfection sensitivity of snap-though modes, the serious imperfection forms and appropriate methods of controlling them by tolerance measurements are needed.

CONCLUSIONS

The new European standard for the Strength and Stability of Shells Structures (ENV 1993-1-6, 1999) is the first to formally define the manner in which numerical analyses of different types should be used as part of the design process. It has produced a framework which allows different types of analysis with different fundamental assumptions to contribute to the overall design strength assessment, without each or any having to be comprehensive. The methodology of the hand analysis design process has been separated into its individual components to assist in this development of compatibility between hand and numerical analyses.

This paper has outlined the use of different types of computer analysis as part of the design process for metal shell structures. It has presented the basis for some of the critical choices that have been made, and has also outlined difficulties that arise at the next stage. Many challenging questions for potential future research that can improve the generality of the design process have been suggested.

The most exciting consequence of the connection that has been achieved between the hand and numerical analyses of shells is the potential for researchers everywhere to contribute new pieces to the complete jigsaw puzzle of the design calculation. The manner in which researchers should conduct their analyses has been set out in the standard and described above, but the opportunity now exists for new research on plastic limit loads, linear bifurcation loads, elastic imperfection sensitivity studies, snap-though buckling studies and local loading investigations to all produce vital pieces of the complete story and to be adopted into revisions of the standard in producing new values for the controlling parameters α, β, η and $\overline{\lambda}_0$ of the shell buckling capacity curve.

ACKNOWLEDGEMENTS

The author, Convenor of the Project Team CEN 250/SC3/PT4 who drafted the European standard ENV 1993-1-6, would like to express his sincere thanks to all members of the project team, together with the members of ECCS TWG8.4 Buckling of Shells, who also contributed greatly to the development. In particular, the special thanks is expressed for the major contributions of Professors Herbert Schmidt and Richard Greiner.

REFERENCES

Baker, E.H., Kovalevsky, L. and Rish, F.L. (1972) Structural Analysis of Shells, McGraw Hill.

Cai, M., Holst, J.M.F.G. and Rotter, J.M. (2002) "Buckling strength of thin cylindrical shells under localised axial compression", Proc. 15th ASCE Engineering Mechanics Conference, June 2-5, Columbia University, New York, NY.

ECCS (1988) European Recommendations for Steel Construction: Buckling of Shells, 4th edition, European Convention for Constructional Steelwork, Brussels.

ENV 1993-1-6 (1999) Eurocode 3: Design of steel structures, Part 1.6: General rules - Supplementary rules for the strength and stability of shell structures, Eurocode 3 Part 1.6, CEN, Brussels.

Gerdeen, J.C.A. (1979) "A Critical Evaluation of Plastic Behavior Data and a Unified Definition of Plastic Load for Pressure Vessel Components" WRC Bull. 254, Welding Research Council, N.Y. pp 1-64.

Gould, P.L., Lin, J.S. and Rotter, J.M. (1985) "Linear Stress Analysis of a Torispherical Head", Journal of the Engineering Mechanics Division, American Society of Civil Engineers, Vol 111, No EM10, Oct., pp 1295-1300.

Holst, J.M.F.G., Rotter, J.M. and Calladine, C.R. (2000) "Imperfections and buckling in cylindrical shells with consistent residual stresses", Journal of Constructional Steel Research, Vol. 54, pp 265-282.

Holst, J.M.F.G., Rotter, J.M., Gillie, M. and Münch, M. (2002) "Failure Criteria for Shells on Local Supports", Festschrift Chris Calladine, Celebration volume for the 60th birthday of Prof. C.R. Calladine, University of Cambridge, September.

Li, H.Y. and Rotter, J.M. (1996) "Algebraic analysis of elastic circular cylindrical shells under local loadings (Part 1 and Part 2)", Proc., International Conference on Structural Steelwork, Hong Kong, December 1996, pp. 801-807 and 808-814.

Massonnet, C.E. and Save, M.A. (1972) "Plastic Analysis and the Design of Plates Shells and Disks", North-Holland.

Rotter, J.M. (1983) "Effective Cross-Sections of Ringbeams and Stiffeners for Bins", Proc., International Conference on Bulk Materials Storage Handling and Transportation, Institution of Engineers, Australia, Newcastle, Aug., pp 329-334.

Rotter, J.M. (1987) Bending Theory of Shells for Bins and Silos, *Transactions of Mechanical Engineering*, Institution of Engineers, Australia, Vol. ME12 No.3 September, pp 147-159.

Rotter, J.M. (1990) "Local Inelastic Collapse of Pressurised Thin Cylindrical Steel Shells under Axial Compression", Journal of Structural Engineering, ASCE, Vol. 116, No. 7, July 1990, pp 1955-1970.

Rotter, J.M. (1997) "Pressurised axially compressed cylinders", Proc., Int. Conf. on Carrying Capacity of Steel Shell Structures, Brno, 1-3 October 1997, pp 354-360.

Rotter, J.M. (1999) "Proposal for generalisation of the elastic-plastic buckling interaction rule from Eurocode 3 Part 1.6", submission to CEN TC250/SC3/PT4 and ECCS TWG8.4 Buckling of Shells.

Rotter, J.M. (2001) "Guide for the Economic Design of Circular Metal Silos", Spon, London.

Rotter, J.M. (2002) "Shell Buckling and Collapse Analysis for Structural Design: The New Framework of the European Standard", Festschrift Chris Calladine, Celebration volume for the 60th birthday of Prof. C.R. Calladine, University of Cambridge, September 2002.

Seide, P. (1975) "Small Elastic Deformations of Thin Elastic Shells", Noordhoff, Leyden, Holland.

Teng, J.G. and Rotter, J.M. (1989) "Non-Symmetric Buckling of Plate-End Pressure Vessels", Journal of Pressure Vessel Technology, American Society of Mechanical Engineers, Vol. 111, No. 3, August, pp 304-311.

Yamaki, N. (1984) "Elastic Stability of Circular Cylindrical Shells", North Holland, Elsevier Applied Science Publishers, Amsterdam, 1984.

EXPLOITING THE SPECIAL FEATURES OF STAINLESS STEEL IN STRUCTURAL DESIGN

D. A. Nethercot and L. Gardner

Department of Civil and Environmental Engineering,
Imperial College, London, SW7 2BU, UK.

ABSTRACT

This paper outlines a new approach to structural stainless steel design that is based on exploiting the full deformation capacity of cross-sections, by adopting a continuous method of cross-section classification and member design, coupled with more accurate material modelling. Recently generated laboratory test results are presented, and it is shown how these have been used in combination with existing test data to validate the proposed method. A comparison of the prediction of test results is made for the proposed procedure and the Eurocode approach for square and rectangular hollow section stub columns, beams and pin-ended columns failing by flexural buckling. Average design advantages of around 20-25% are achieved.

KEYWORDS

Cross-section classification, ENV 1993-1-4, local buckling, stainless steel, structural engineering

INTRODUCTION

The use of stainless steel in construction is growing rapidly, yet its exploitation as a primary structural material remains rather limited, with the dominant applications being of a specialist or prestigious nature. Figures 1 and 2 show examples of landmark structures that have made use of stainless steel. Figure 1 shows the cladding on the recently completed Nagoya baseball stadium in Japan. Figure 2 shows The Gateway Arch (completed in 1965) in St. Louis, Missouri, which is the second largest structural application of stainless steel in the world. Extensive specialist use of stainless steel is made by the offshore and nuclear power industries to meet stringent safety and performance requirements whilst minimising the need for maintenance.

In addition to the clear aesthetic appeal of stainless steels there are other increasingly strong arguments for adopting the material in structures. The corrosion resistance of stainless steels makes them one of the most durable families of construction materials; a material with no need for protective coatings against corrosion has clear advantages in terms of economy, weight savings, reduction in environmental impact and lower maintenance costs. The fire resistance of stainless steel has been shown to be superior to that of carbon steel (Baddoo & Gardner, 2001), reducing or even eliminating the need for protective fire coatings to be applied to structural members. The combination of the residual value of the alloy content of stainless steel and the economic advantages gained from re-melting scrap in the dominant electric arc process has resulted in a high level of recycling of material.

Figure 1: Nagoya Dome, Japan.   Figure 2: The Gateway Arch, St. Louis.

There is a wide variety of grades of stainless steel that fall into three main groups; austenitic, ferritic and duplex. The most common grades for structural and architectural applications are the austenitic and duplex grades, with the duplex stainless steels offering higher strength and wear resistance than the austenitics, but at greater expense. Product forms include plate, sheet, tube, bar, cold-formed structural sections and hot-rolled structural sections, with the most commonly used products for structural applications being cold-formed (square, rectangular and circular) hollow sections. The current study is focused upon austenitic stainless steel, though the findings are applicable to any material that exhibits rounded stress-strain behaviour.

An important step towards the enhancement of understanding and use of stainless steel in structures has been the development of the design guidance given in the European document, ENV 1993-1-4 (1996) and provided by The Steel Construction Institute (Baddoo & Burgan, 2001). However, one of the principal drawbacks to these is that they were based on the rather limited amount of structural performance data available. Additionally, since they were 'first generation' design guides, an important factor was to ensure that a designer familiar with the carbon steel rules would be able to make a straightforward transition to stainless steel structural design. As a result, the authors were obliged to use a simplified elastic, perfectly plastic material model. This model is acceptable for carbon steel that exhibits a sharply defined yield point, followed by a plastic yield plateau. For stainless steel, though, where there is rounded stress-strain behaviour, no sharply defined yield point and substantial strain hardening is possible, this model leads to overly conservative designs.

LABORATORY TESTING PROGRAMME

General

A laboratory testing programme has been conducted at Imperial College, London to investigate the behaviour of stainless steel cross-sections and members, and to generate results that can be used to validate a proposed design procedure.

Tensile and compressive coupon tests were performed on flat material cut from the faces of RHS and SHS cross-sections to determine material stress-strain behaviour. Curved corner coupons were tested to investigate the effect of the forming process on material properties in these regions. A total of 37 stub column tests were conducted on square, rectangular and circular hollow sections, (SHS, RHS, and CHS respectively) to enable the development of a relationship between cross-section slenderness and deformation capacity, and to determine ultimate load carrying capacities. Member tests on 9 RHS and SHS beams and 22 columns were conducted to investigate interactions between local and global effects, and to determine ultimate load carrying capacities. Initial geometric plate imperfections were measured to aid the explanation of structural performance, and for use as data in numerical models.

Results

Full details of the laboratory testing programme have been reported by Gardner (2002). A selection of results is presented in this paper. Figure 3 shows deformed SHS 100×100×2 stub columns, which are of high cross-section slenderness, and the corresponding load-end shortening response, and Figure 4 shows similar details for RHS 100×50×6 stub columns of low cross-section slenderness.

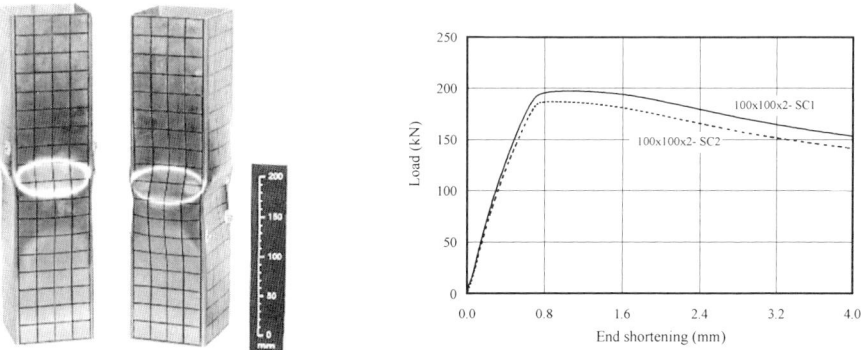

Figure 3: Deformed SHS 100×100×2 (β=2.18) stub columns with load-end shortening

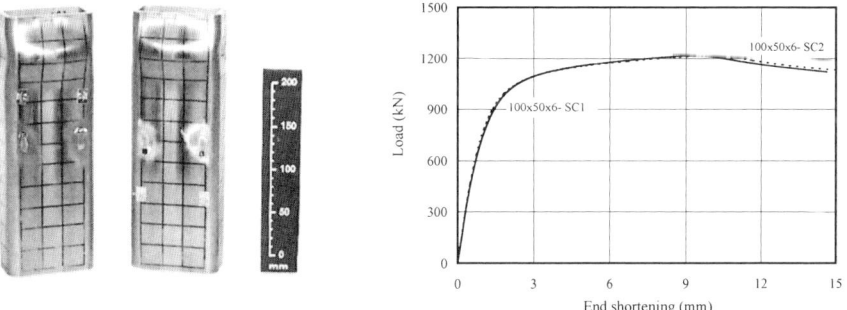

Figure 4: Deformed RHS 100×50×6 (β=0.77) stub columns with load-end shortening

STRUCTURAL PROPERTIES

Material Stress-Strain Behaviour

The rounded nature of the stainless steel stress-strain curve cannot be accurately reflected by the bi-linear elastic, perfectly-plastic model that is used for carbon steel and adopted in ENV 1993-1-4 (1996). For more efficient design a more accurate representation is required.

The expression most frequently used to describe non-linear material behaviour is the Ramberg-Osgood (1943) model as modified by Hill (1944), given in Eqn. 1.

$$\varepsilon = \frac{\sigma}{E_0} + 0.002 \left(\frac{\sigma}{\sigma_{0.2}} \right)^n \qquad (1)$$

where E_0 is the initial tangent modulus, $\sigma_{0.2}$ is the material 0.2% proof stress and n is a strain hardening exponent that defines the degree of roundedness of the stress-strain curve.

The basic Ramberg-Osgood formulation (Eqn. 1) gives excellent agreement with experimental stress-strain data up to $\sigma_{0.2}$. At higher strains however, the model generally overestimates the stress corresponding to a given strain (Gardner & Nethercot, 2001). Recent proposals have been made to give improved accuracy. Mirambell & Real (2000) proposed to use two adjoining Ramberg-Osgood curves. The basic Ramberg-Osgood expression from Eqn. 1 is used up to the 0.2% proof stress, and a modified Ramberg-Osgood expression, given in Eqn. 2, beyond the 0.2% proof stress. The modified expression re-defines the origin for the second curve as the point of 0.2% proof stress, and ensures continuity of the gradients.

$$\varepsilon = \frac{(\sigma - \sigma_{0.2})}{E_{0.2}} + \varepsilon_{pu} \left(\frac{\sigma - \sigma_{0.2}}{\sigma_u - \sigma_{0.2}} \right)^{n'} + \varepsilon_{t0.2} \qquad (\sigma \geq \sigma_{0.2}) \qquad (2)$$

where σ_u is the ultimate material strength, ε_{pu} is the plastic strain at ultimate strength, $\varepsilon_{t0.2}$ is the total strain at the 0.2% proof stress, n' is a strain hardening exponent that can be determined from the ultimate strength and another intermediate point, and $E_{0.2}$ is the stiffness at the point of 0.2% proof stress, and can be determined from Eqn. 3.

$$E_{0.2} = \frac{\sigma_{0.2} E_0}{\sigma_{0.2} + 0.002 \, n \, E_0} \qquad (3)$$

It is worth noting that the curve defined by Eqn. 2 produces a slight inconsistency in that it does not pass through the point of σ_u at ε_{tu}, (where ε_{tu} is the total strain at ultimate stress). However, due to the high ductility of stainless steels, the errors incurred are negligible. For consistency Eqn. 2 would be replaced by Eqn. 4.

$$\varepsilon = \frac{(\sigma - \sigma_{0.2})}{E_{0.2}} + \left(\varepsilon_{tu} - \frac{\sigma_u - \sigma_{0.2}}{E_{0.2}} - \varepsilon_{t0.2} \right) \left(\frac{\sigma - \sigma_{0.2}}{\sigma_u - \sigma_{0.2}} \right)^{n'} + \varepsilon_{t0.2} \qquad (\sigma \geq \sigma_{0.2}) \qquad (4)$$

Figure 5 demonstrates the improved accuracy at higher strains of the compound Ramberg-Osgood expression over the basic one for describing a typical experimental stainless steel stress-strain curve.

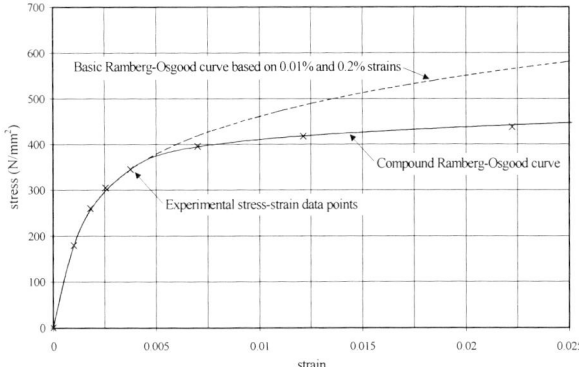

Figure 5: Comparison between compound and basic R-O models

For the description of stress-strain behaviour in compression, Eqns. 2 and 4 encounter difficulty because there is no ultimate stress in compression due to the absence of the necking phenomenon. It was initially proposed to simply adopt the ultimate tensile strengths and corresponding strains to represent the compressive behaviour (Gardner & Nethercot, 2001). In general this was an acceptable solution though increased errors were observed. It was therefore subsequently proposed (Gardner, 2002) to use the 1% proof stress instead of the ultimate stress to describe compressive stress-strain behaviour, and Eqn. 5 was derived. Eqn. 1 remained to be used for stresses up to $\sigma_{0.2}$.

$$\varepsilon = \frac{(\sigma - \sigma_{0.2})}{E_{0.2}} + \left(0.08 - \frac{\sigma_{1.0} - \sigma_{0.2}}{E_{0.2}}\right)\left(\frac{\sigma - \sigma_{0.2}}{\sigma_{1.0} - \sigma_{0.2}}\right)^{n'_{0.2,1.0}} + \varepsilon_{t0.2} \qquad (\sigma \geq \sigma_{0.2}) \qquad (5)$$

where $n'_{0.2, 1.0}$ is a strain hardening coefficient representing a curve that passes through $\sigma_{0.2}$ and $\sigma_{1.0}$.

Eqn. 5 was found to give excellent agreement with experimental stress-strain data, both in compression and tension, up to strains of approximately 10%.

Corner Properties

The properties of the corner regions in cold-formed stainless steel sections differ from the properties of the flat regions due to the material's response to deformation. Stainless steel exhibits pronounced strain hardening, resulting in corner regions of cold-formed SHS and RHS having 0.2% proof strengths commonly between 20% and 100% higher than the 0.2% proof strengths of the flat regions, accompanied by a corresponding loss in ductility.

Following analysis of available test data on material cut from the corner regions of cold-formed stainless steel cross-sections, it was seen that an accurate prediction of the corner 0.2% proof stress could be found using Eqn. 6, where $\sigma_{0.2,c}$ is the 0.2% proof stress of the corner material and σ_u is the ultimate strength of the flat material in the cross-section.

$$\sigma_{0.2,c} = 0.85\sigma_u \qquad (6)$$

The rationale behind the expression is that corner material is work hardened to strains between about 10% and 20%. This region of the stress-strain curve is relatively flat, so the stress is not sensitive to the exact level of applied strain. Between 10% and 20% strain, the stress is approximately 85% of the ultimate material strength. Despite the simplicity of the expression very good agreement has been observed for the available test data.

Residual Stresses

Residual stresses are introduced into cold-formed stainless steel members as a result of the deformations during the cold-forming fabrication process, and due to the thermal gradients that are induced during and after welding. Measurements of residual stresses in cold-formed stainless steel sections are scarce. Knowledge of the magnitude and distribution of residual stresses within a cold-formed stainless steel cross-section is therefore somewhat speculative. Some measurements were taken by Rasmussen & Hancock (1993) as part of an experimental programme on cold-formed stainless steel tubular members. Bredenkamp et al (1992) found that the magnitudes of residual stresses in built up stainless steel I-sections were of the same order as in an equivalent carbon steel section, whilst Lagerqvist & Olsson (2001) carried out a similar study and observed considerably higher residual stresses in the stainless steel sections. Gardner & Nethercot (2001) proposed an idealised means of accounting for residual stresses in finite element models of cold-formed stainless steel SHS and RHS.

Geometric Plate Imperfections

Initial geometric plate imperfections can have a significant effect on the load carrying capacity of structural members. Imperfection measurements were taken as part of the current study to use as data in finite element models. Analysis of the distribution of the imperfections highlighted that the ends of the cross-sections were flared outwards. This is believed to be due to the release of bending residual stresses introduced into the cross-sections during the cold-forming process. However, away from the end regions no clear initial imperfection distribution emerged. It was therefore decided to adopt the lowest buckling mode, determined from an elastic eigenmode and representing the simplest (and most onerous since it is coincident with the failure mode) imperfection distribution. To describe the magnitude of imperfections, re-calibration of a model proposed by Dawson & Walker (1972) proved to most accurately predict measurements. A graph of measured versus predicted imperfections is shown in Figure 6, and the re-calibrated model is given by Eqn. 7.

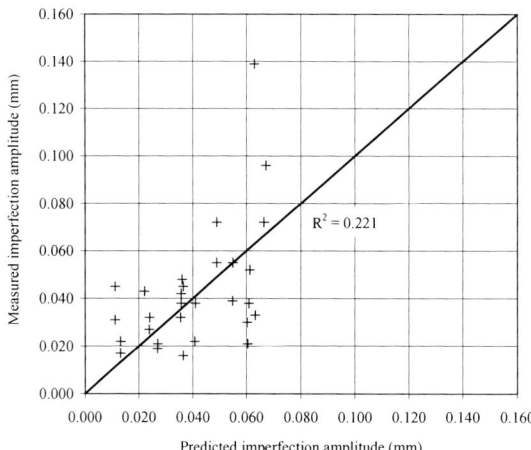

Figure 6: Measured versus predicted imperfection amplitudes for $\omega_0/t = 0.023(\sigma_{0.2}/\sigma_{cr})$

$$\omega_0/t = 0.023(\sigma_{0.2}/\sigma_{cr}) \tag{7}$$

where ω_0 is initial imperfection amplitude, t is plate thickness, $\sigma_{0.2}$ is material 0.2% proof stress, σ_{cr} is elastic critical buckling strength of the plate.

CROSS-SECTION BEHAVIOUR

General

ENV 1993-1-4 (1996) uses the concept of section classification (as adopted in corresponding carbon steel structural design guidance) to define the ductility of a cross-section. The classification is dependent upon the slenderness of the individual plate elements that make up the cross-section. From the elastic, perfectly-plastic material model four behavioural classes emerge. Class 1 and 2 cross-sections have high ductility and can develop the full plastic moment in bending. Class 3 cross-sections reach yield at the extreme fibres in bending, and the resistance of a Class 4 cross-section is limited by local buckling in the elastic range. The effect of the fundamental differences in material behaviour between stainless steel and carbon steel is to lessen the validity of the concept of a plastic and an elastic section.

In view of the continuous nature of the stainless steel stress-strain curve, it seems rational that a continuous, rather than a discretised section classification system should be adopted. Development of this idea is the focal point of this investigation. The four behavioural classes of cross-sections that are defined in the Eurocode approach have been replaced by a non-dimensional numerical value that is a measure of the deformation capacity of the cross-section. The deformation capacity is based upon the local buckling strain, ε_{LB} of a cross-section (taken as the strain at ultimate load), which is in turn dependent upon the slenderness of individual plate elements, and the interaction between elements within the cross-section.

Generation of Design Curves

To describe the local buckling behaviour of aluminium plate elements, Faella et al (1999) proposed an expression of the general form given in Eqn. 8, where E_0, $\sigma_{0.2}$ and ε_{LB} have been defined previously, ε_0 is the elastic strain at the material compressive 0.2% proof stress, taken as $\varepsilon_0 = \sigma_{0.2}/E_0$, and β is the slenderness of the most slender plate in the cross-section, $\beta = (b/t)\sqrt{\sigma_{0.2}/E_0}$. The plate width b will be measured between the centrelines of the adjoining plates, (i.e. (D-t) or (B-t), where t is the material thickness). The constants C_1, C_2, and C_3 were determined from a regression analysis of experimental points.

$$\frac{\varepsilon_{LB}}{\varepsilon_0} = \frac{C_1}{\beta^{C_2+C_3\beta}} \tag{8}$$

The right hand side of the expression was multiplied by χ^{C_4} by Faella et al (1999), where χ is the ratio of slenderness of the least slender element to the most slender element in the cross-section, (i.e. for a RHS of constant thickness and material properties, the aspect ratio of the cross-section), and C_4 was another constant that was determined experimentally to account for the greater edge restraint that the two longer faces of an RHS cross-section receive from the two shorter ones. With increasing cross-section aspect ratio there is clearly an increasing level of restraint.

A regression analysis of stub column test results yielded $C_1 = 7.07$, $C_2 = 2.13$ and $C_3 = 0.21$ for SHS. The experimental results indicated that the increased deformation capacity for the two longer faces of

RHS was less significant for sections with lower cross-section slenderness. To reflect this behaviour, it was therefore decided to multiply the right hand side of Eqn. 8 by $\chi^{C_4/\beta^{1/2}}$, where the constant $C_4 = -0.30$ was determined based on the experimental results for RHS with aspect ratios of 0.67 and 0.50. Substituting the derived constants into Eqn. 8 therefore yields Eqn. 9.

$$\frac{\varepsilon_{LB}}{\varepsilon_0} = \frac{7.07}{\beta^{2.13+0.21\beta}} \chi^{\frac{-0.30}{\beta^{1/2}}} \quad (9)$$

The resulting curves are plotted in Figure 7 for $\chi = 1.0$ (SHS), $\chi = 0.67$ and $\chi = 0.50$. It is worth noting that although the effect of the increased edge restraint for the RHS appears to be relatively small, it can lead to increases in cross-section compressive resistance of up to 10%.

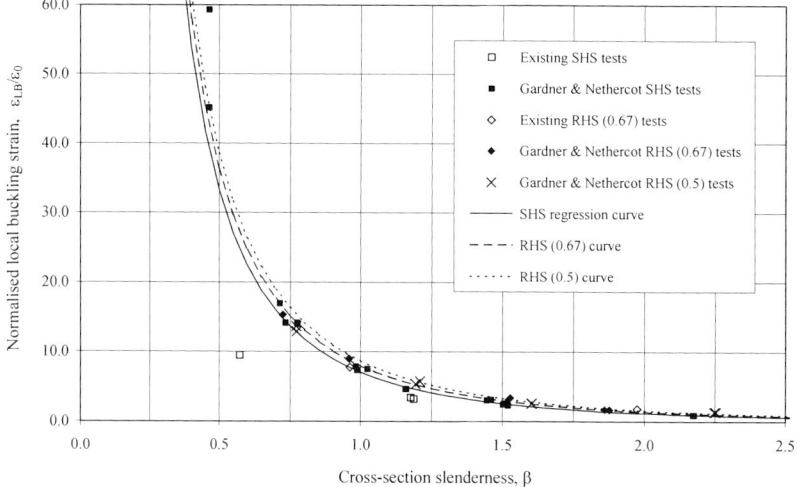

Figure 7: Cross-section deformation capacity versus cross-section slenderness

MEMBER DESIGN

A detailed account of the member design procedure for stainless steel hollow sections has been made by Gardner (2002). Member resistances can be determined using the cross-section deformation capacity (found from Figure 7) in conjunction with the material model given in Eqn. 5 and where necessary appropriate buckling curves. A schematic representation of the design procedure is given in Figure 8, where b and t are internal element width and thickness respectively, A is cross-sectional area, W_{el} is elastic section modulus, $\sigma_{0.2}$ is the material 0.2% proof strength and E_0 is material Young's modulus.

Compression

For a short column subjected to pure compression, there are 3 steps to determine member resistance:

- Calculate slenderness, β and restraint of most slender element in cross-section
- Determine compressive deformation capacity, $\varepsilon_{LB}/\varepsilon_0$ from chart (Figure 7)

- Member resistance is the product of the stress at this level of deformation and the gross cross-sectional area.

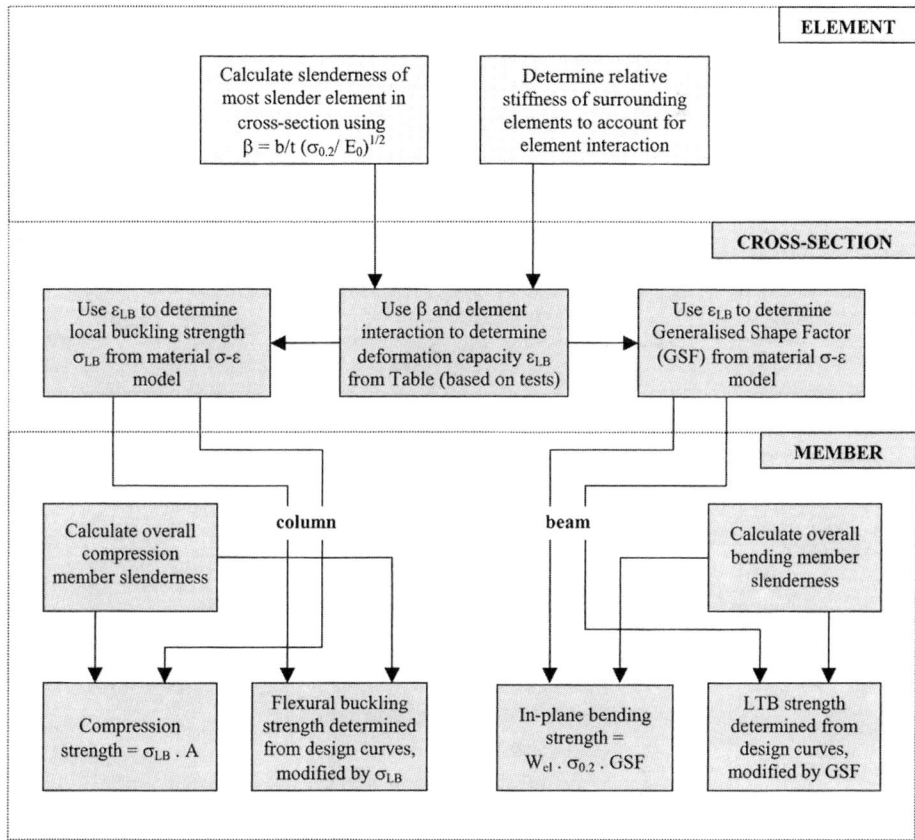

Figure 8: Schematic representation of design method

Flexural buckling

Flexural buckling involves an interaction between local and global effects. In Eurocode 3 the determination of flexural buckling resistance (Eqn. 10) is based upon the Perry-Robertson formula, whereby a compression member fails when the maximum stress at some point along the length reaches the material yield strength (or 0.2% proof strength) due to a combination of axial load plus bending moment.

$$N_{b.Rd} = \chi A \sigma_y \quad (10)$$

where χ is a reduction factor for the relevant buckling mode, A is gross area of the cross-section, σ_y is the material yield strength, and the reduction factor for Class 4 cross-sections and the material partial safety factor have been removed.

However, with no sharply defined yield point, the approach is less valid for stainless steel than for carbon steel. The initial proposal was to replace $\sigma_{0.2}$ in the buckling formula with σ_{LB}. However, a direct replacement would generally lead to over-predicted member resistances, since the Perry-Robertson formula is based on elastic material response, and (particularly for stocky cross-sections) σ_{LB} may only be reached following significant plastic straining. Following analysis of test results, it was found that modifying buckling strengths by a factor of $(\sigma_{LB}/\sigma_{0.2})^{0.32}$ provided good agreement between test and predicted values. Test results and design curves for varying values of β are shown in Figure 9.

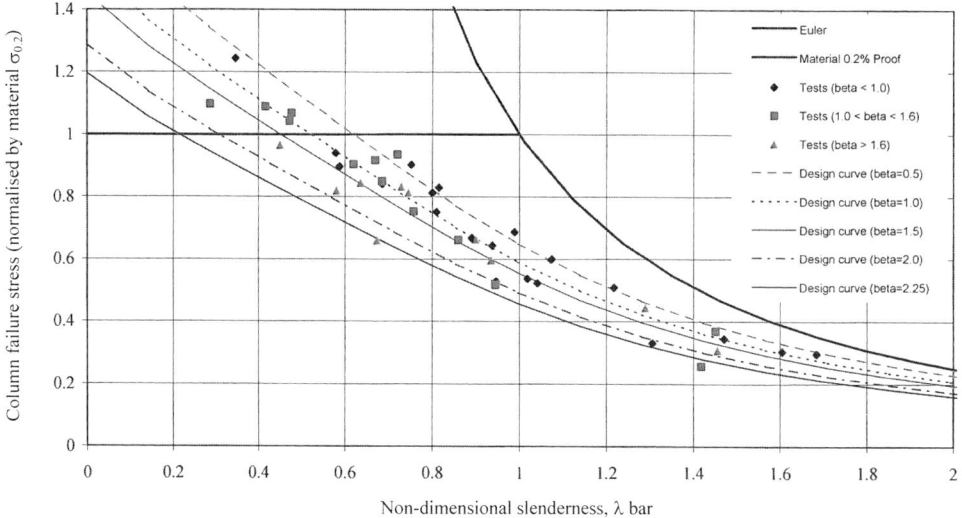

Figure 9: Flexural buckling test results and design curves

Bending

The failure of bending members is governed by an interaction between local buckling and global instability effects. However, for hollow sections global instability becomes less important (and of course does not occur in SHS and CHS members). The design procedure therefore currently only considers in-plane bending strength. Baddoo & Burgan (2001) have provided a formula to assess a limiting length below which RHS members are not susceptible to lateral-torsional buckling.

For a beam not susceptible to lateral-torsional buckling, there are 5 steps to determine member resistance:

- Calculate slenderness and restraint of the compression flange
- Determine deformation capacity of compression flange from chart or table
- Calculate geometric shape factor, equal to conventional plastic modulus divided by elastic modulus
- Determine generalised shape factor from corresponding chart or table
- Moment resistance is the product of design strength with the elastic modulus and generalised shape factor.

The concept of a generalised shape factor to determine the moment resistance of a beam formed from material with non-linear stress-strain characteristics was proposed by Mazzolani (1995). The generalised shape factor is essentially a means by which the material characteristics, as well as the geometric characteristics of a section can be incorporated into a single numerical value.

COMPARISON BETWEEN PROPOSED AND EUROCODE DESIGN METHODS

Table 1 summarises a comparison between the proposed procedure and the Eurocode design method for a series of tests on stub columns, beams and columns that failed by overall flexural buckling, where measured material properties have been adopted and all material and load factors have been set to unity. The results derive from a range of closed, longitudinally welded square and rectangular hollow sections. The values given in the third and fourth columns are the predicted failure loads divided by the test failure loads for the Eurocode and the proposed methods respectively. Test results are a compilation of those generated in the current study and as part of other investigations; Johnson & Winter (1966), Rasmussen & Hancock (1992 and 1993), Talja & Salmi (1995), Mirambell & Real (2000). The design advantage over the current Eurocode procedure is indicated in column 5.

TABLE 1
COMPARISON BETWEEN EUROCODE AND PROPOSED DESIGN METHOD

Structural Configuration	No. of tests	ENV 1993-1-4/ Test	Proposed/ Test	Proposed/ ENV 1993-1-4
Compression	24	0.75	0.97	1.29
Bending (No LTB)	18	0.70	0.90	1.28
Flexural buckling	44	0.93	1.00	1.08

CONCLUSIONS

It has been shown that by basing the structural design of stainless steel members on the actual material behaviour (rather than assuming an analogy with carbon steel) significant improvements in performance may be achieved. New test data on local buckling and plate element behaviour have been generated. Additional member tests on beams and pin-ended long columns have been conducted. Using a new proposed design approach, average improvements of around 20%, that still produce 'safe side' predictions of test data, are achieved. It is envisaged that the proposed design method will be incorporated into future revisions of Eurocode 3, bringing greater efficiency to structural stainless steel design and promoting more widespread use of the material.

ACKNOWLEDGEMENTS

The authors are grateful to EPSRC and the AvestaPolarit UK Research Foundation for the project funding, and would like to thank Nancy Baddoo and Bassam Burgan (The Steel Construction Institute) and David Dulieu (AvestaPolarit UK Research Foundation) for their technical support. Thanks should also be extended to Kim Rasmussen (University of Sydney), Asko Talja (VTT Building Technology Finland), and Esther Real (University of Catalonia), for their help in providing their test programme details.

REFERENCES

Baddoo, N. R. & Gardner, L. (2000). WP5.2: Member behaviour at elevated temperatures. ECSC project – Development of the use of stainless steel in construction. Contract No. 7210 SA/ 842. The Steel Construction Institute, UK.

Baddoo, N. R. & Burgan, B. A. (2001). Structural Design of Stainless Steel (P291). The Steel Construction Institute.

Bredenkamp, P. J., Van den Berg, G. J., & Van der Merwe, P. (1992). Residual stresses and the strength of stainless steel I-section columns. *Proceedings of the Structural Stability Research Council, Annual Technical Session*, Pittsburg, U.S.A.

Dawson, R. G. and Walker, A. C. (1972). Post-buckling of geometrically imperfect plates. *Journal of the Structural Division*, ASCE, **98: ST1**, 75-94.

ENV 1993-1-4. (1996). Eurocode 3: Design of steel structures - Part 1.4: General rules - Supplementary rules for stainless steel. CEN.

Faella, C., Mazzolani, F. M., Piluso, V. & Rizzano, G. (2000). Local buckling of aluminium members: testing and classification. *Journal of Structural Engineering,* **126:3,** 353-360.

Gardner, L. (2002). A new approach to structural stainless steel design. PhD Thesis. Structures Section, Department of Civil Engineering, Imperial College, London.

Gardner, L. & Nethercot, D. A. (2001). Numerical modelling of cold-formed stainless steel sections. *Proceedings of the Ninth Nordic Steel Construction Conference.* 781-789. Edited by Mäkeläinen et al. Helsinki, Finland.

Hill, H. N. (1944). Determination of stress-strain relations from the offset yield strength values. *Technical Note No. 927, National Advisory Committee for Aeronautics,* Washington, D.C.

Johnson, A. L. & Winter, G. (1966). Behaviour of stainless steel columns and beams. *Journal of the Structural Division, ASCE*, ST5, 97-118.

Lagerqvist, O. & Olsson, A. (2001). Residual stresses in welded I-girders made of stainless steel and structural steel. *Proceedings of the Ninth Nordic Steel Construction Conference.* 737-744. Edited by Mäkeläinen et al. Helsinki, Finland.

Mazzolani, F. M. (1995). *Aluminium Alloy Structures, 2nd Edition.* E & FN Spon, An imprint of Chapman and Hall.

Mirambell, E. & Real, E. (2000). On the calculation of deflections in structural stainless steel beams: an experimental and numerical investigation. *Journal of Constructional Steel Research,* **54**, 109-133.

Ramberg, W. & Osgood, W. R. (1943). Description of stress-strain curves by three parameters. *Technical Note No. 902, National Advisory Committee for Aeronautics*, Washington D.C.

Rasmussen, K. J. R. & Hancock, G. J. (1992). Stainless steel tubular beams – tests and design. *Proceedings of the Eleventh International Speciality Conference on Cold-formed Steel Structures,* 587-609. St. Louis, Missouri, U.S.A.

Rasmussen, K. J. R. & Hancock, G. J. (1993). Design of cold-formed stainless steel tubular members. I: Columns. *Journal of Structural Engineering, ASCE*, **119: 8**, 2349-2367.

Talja, A. & Salmi, P. (1995). Design of stainless steel RHS beams, columns and beam-columns. Research Note 1619, VTT Building Technology, Finland.

CASSETTE WALL CONSTRUCTION: CURRENT RESEARCH AND PRACTICE

J.M.DAVIES

Manchester School of Engineering, University of Manchester, Oxford Road, Manchester, M13 9PL

ABSTRACT

Light gauge steel cassette sections offer an alternative form of load-bearing wall assembly for use in low rise steel-framed construction. Consideration of the overall stability of the structure leads logically to the use of "stressed skin design" and "diaphragm" action. Cassette wall construction is particularly advantageous in this context so that the cassettes are subject to axial load, bending and in-plane shear. Outline design procedures for cassettes subject to these three load cases are given in Eurocode 3: Part 1.3 but these require some more detailed consideration and amplification. This paper reviews the practical design considerations in the light of recent research and development.

KEYWORDS

Buckling, cold formed steel, cassettes, compression, coupled instability, shear, walls.

INTRODUCTION

Light gauge steel cassettes (also known as structural liner trays) offer an alternative form of wall assembly for use in low-rise steel-framed construction and have been used successfully in a number of housing and commercial building projects (Davies 1998a, Davies 1998b, Davies 2000a). In the preferred arrangement, lipped C-shaped cassettes span vertically between top and bottom tracks to form panels which may be storey height or higher. The basic arrangement is then as shown in Fig. 1. When used in this way, the narrow flanges may point towards either the inside or the outside of the building. Usually, the narrow flanges to point inwards so that the wide outer flange immediately provides a weatherproof membrane and no second steel skin is necessary. The wall construction is then completed internally by insulation and a dry lining, an option often referred to as "cold-frame" construction. The alternative, with the insulation and a waterproof outer skin external to the cassette wall is "warm-frame" construction.

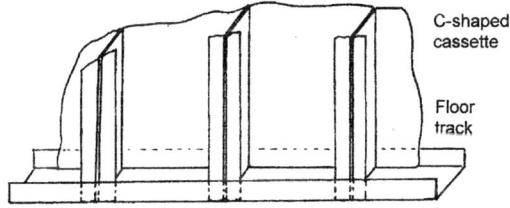

Figure 1: Cassette wall construction

The idea of cassette wall construction appears to have originated from Baehre in Stockholm in the 1970's. He envisaged the steel cassette as the primary structural element in modular construction as illustrated in Fig. 2. When used in this way, as shown in Fig. 3, cassette walls are subject to the three primary load systems of axial compression (from the storeys above), bending about the minor axis (from wind pressure and suction) and shear (from wind-induced diaphragm action). Accordingly, Baehre's research team investigated in some depth the performance of cassette sections when subject to each of these three actions individually (Thomasson 1978, Konig 1978, Nyberg 1976).

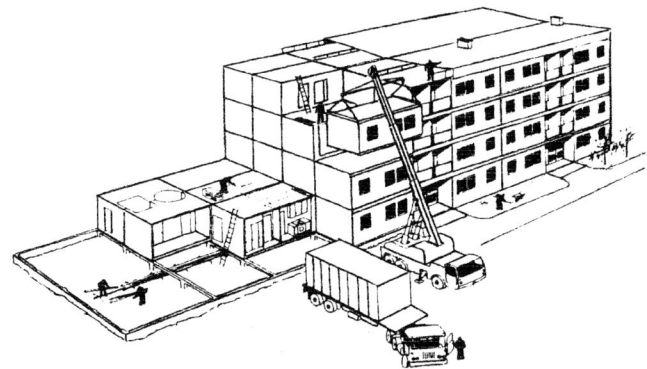

Figure 2: Modular construction with light gauge steel cassettes

Cassette wall construction may be viewed as being similar to conventional light gauge steel stud construction integrally combined with a metal lining sheet to provide a metal frame together with a flat steel wall. Its main advantages, in comparison with the alternative stud wall system, may be summarised as follows:

- Simple details and rapid construction
- The wall structure is immediately water-tight
- The stability problems of thin slender studs are avoided
- A rational provision for wind shear can be made without additional bracing members by utilising stressed skin action

The last bull point is particularly important and is discussed in some detail later.

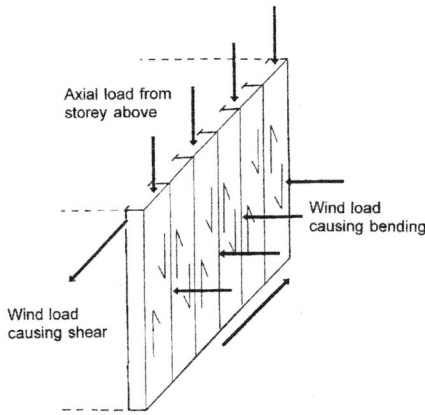

Figure 3: Load system in a cassette wall

As far as the author is aware, Baehre's ideas were not put into action until much later when the "Scanmodule" system (Davies et al 1995) used the construction method illustrated in Fig. 2. However, the more general use of the cassette wall system has recently been pioneered in France by the company 'Produits Acier Batiment' (PAB) under the name 'CIBBAP' and more than 20 projects have been completed using this concept.

In recent years, cassette sections have also been widely used in an alternative wall construction in which they span horizontally between structural frames. Here the cassettes interact with and are stabilised by a relatively light trapezoidally profiled metal outer skin which spans at right angles to the span of the cassette. For both thermal and acoustical reasons, the troughs are usually filled with insulating material. When used in this way to form a two-layer, built-up cladding system, cassettes are often termed "structural liner trays". Typical wall construction of this type is shown in Fig. 4 and a similar arrangement is also used in roof construction.

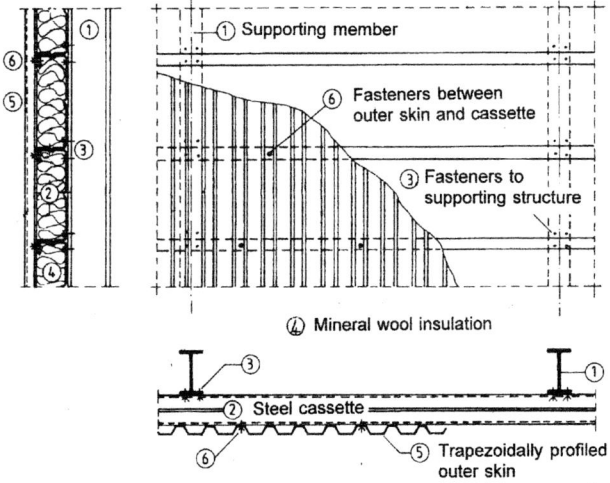

Figure 4: Two-layer built-up wall cladding system

Baehre, by this time Professor of Steel Construction at the University of Karlsruhe in Germany, then turned his attention to structural liner trays used in this way and carried out a significant number of tests (Baehre et al 1986, 1987, 1990). These formed the basis of the design clauses in Eurocode 3 : Part 1.3 (1996) which will be referred to as "EC3". As these design procedures are based on test results, EC3 places geometric restrictions on cassettes that appear to reflect the limits of Baehre's tests rather than any fundamental restrictions of the structural system.

Fig. 5 shows the geometry of a typical cassette together with the symbols used in EC3. These symbols will also be used in this paper. The elements of the section are described in this paper as the wide flange, the webs, the narrow flanges and the lips. The range of validity of the design procedures in EC3 is stated to be as follows:

$$0.75 \text{ mm} \leq t_{nom} \leq 1.5 \text{ mm}$$
$$30 \text{ mm} \leq b_f \leq 60 \text{ mm}$$
$$60 \text{ mm} \leq h \leq 200 \text{ mm}$$
$$300 \text{ mm} \leq b_u \leq 600 \text{ mm}$$
$$I_a/b_u \leq 10 \text{ mm}^4/\text{mm}$$

$$h_u \leq h/8$$
$$s_1 \leq 1000 \text{ mm}$$

Where I_a is the second moment of area of the wide flange about its own centroid as shown in the right hand insert on Fig. 5.

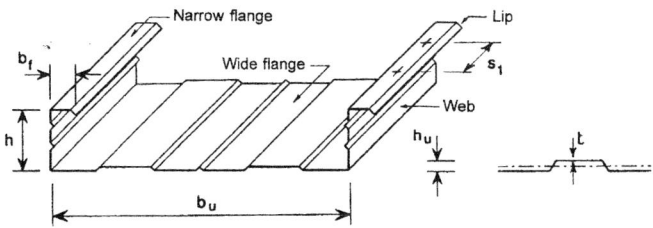

Figure 5: Typical cassette section

It may be noted here that, primarily because of "flange curling" which is discussed later, there is more scatter in the comparison of bending test results with the design procedures in EC3 than is customary with cold-formed sections in bending. EC3 therefore requires a material factor of $\gamma_{M2} = 1.25$ for bending strength in contrast to the more usual values of $\gamma_M = 1.0$ or 1.1. It is implicit, therefore, that significantly better bending strengths are likely to be obtained by test than by calculation and, for a mass-produced product, this is recommended.

As the clauses in EC3 explicitly require the stabilising effect of the second metal skin, it is necessary to reconsider their applicability to cassette wall construction where this second skin is unlikely to be present (or may be replaced by a much weaker material such as plasterboard).

DESIGN PROCEDURES FOR CASSETTES

Following Baehre's example, the author has set up a series of related research programmes which consider the performance of cassettes in wall construction subject to axial load, bending and shear. As well as utilising test results, these also take advantage of modern numerical analysis procedures. They consider the situation where the cassettes effectively act alone without any interaction with a second structural skin and also where they are filled with rigid thermal insulation. In the following sections, therefore, each of the basic load systems shown in Fig. 3 is considered in turn in the light of this research and the resulting clauses in EC3.

Axial compressive load

The axial compressive load case is not explicitly considered for cassettes in EC3. However, it is implicit that, in the absence of any longitudinal stiffeners in the wide flange, the cross-section model shown in Fig. 6 should be used in which the wide flange and web are treated by conventional effective width procedures and the narrow flange by the more complicated procedures for a flange stiffened by a lip. This requires an iterative procedure in order to determine the reduced effectiveness of the combination of the lip and the outer portion of the flange. This part of the design procedure for the axially loaded cassette is similar to that for bending with the narrow flanges in compression, which is given in some detail in EC3. The remainder of the design procedure then follows that for any thin-walled column bearing in mind that it is completely stable with respect to buckling in the plane of the wall. It is also necessary to consider distortional buckling (tripping) of the narrow flange. The considerations with regard to distortional bucking are similar to those considered in the next section.

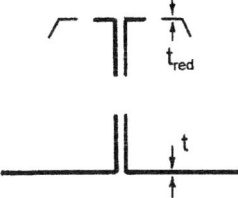

Figure 6: Design model for a cassette subject to axial compression

When the wide flange incorporates rolled-in stiffeners, the considerations become much more interesting because there is interaction between the plate buckling modes and stiffener buckling modes together with consideration of the post-buckling strength. This is the subject of another paper in this Conference (Voutay and Davies 2001) which concludes that the design procedures in EC3 are adequate for practical design but that the use of whole section buckling analysis together with "direct strength design" (Shafer 2001) is better.

Behaviour in bending with the narrow flange in compression

The behaviour of cassette sections in bending is characterised by all of the usual considerations of thin-walled, cold formed section construction with the addition that the wide flange tends to curl towards the neutral axis, as illustrated by Fig. 7, whether this flange is tension or compression. The available analytical procedures for dealing with "flange curling" are far from rigorous and the procedures in EC3 are largely empirical. Consequently there is a good deal of scatter of test results and a degree of in-built conservatism.

The design for bending with the narrow flange in compression is a particularly complicated design problem because there are three effects to consider:

➢ local buckling of the web and narrow flange.
➢ distortional buckling (tripping) of the narrow flange/lip combination
➢ "flange curling" of the wide flange (which is in tension)

The first and last of these are considered in detail in clauses 10.2 and 10.3 of EC3. Flange curling is treated by an effective width approach. In cassette wall construction, the absence of the second skin of cladding will not have any influence on these aspects of behaviour

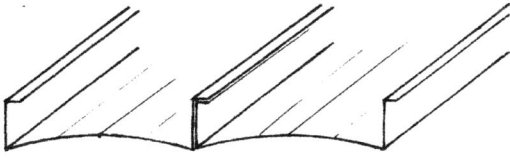

Figure 7: Flange curling

However, distortional buckling is more problematical because the minimum fastener spacings in EC3 clause 10.2.1 are clearly intended to prevent this phenomenon. We may note, however, that the requirements are very simplistic and presumably reflect the observation that distortional buckling was not critical in any of Baehre's test panels when this fastener specification was present. It does not automatically follow that these fasteners to an auxiliary bracing system are always necessary. What is required is a separate check for the distortional buckling stress of the unbraced cassette. It may also be born in mind that the web connections between adjacent cassettes, which are usually necessary in cassette wall construction, will also inhibit distortional buckling.

Accurate solutions for the distortional buckling stress of this type of section are available with the aid of, for example, the finite element method, the finite strip method or, particularly advantageously, second-order Generalised Beam Theory (Davies et al. 1994a, 1994b). The simplified design models proposed by Serrette and Pekoz (1995) are also available for this problem. In this simplified treatment, as shown in Fig. 8, the web and flange are, together, treated as a compression member which has both rotational and translational spring restraints which are continuous along the web-flange intersection. The rotational spring stiffness k_φ and the translational spring stiffness k_x represent the torsional and translational restraints respectively supplied, through the web, from the remainder of the section. In general, it has been found to be adequate to let the translational spring stiffness k_x be zero.

Models of this type have been found to be generally applicable to design for distortional buckling (Davies and Jiang 1998) although they are rather sensitive to the value assumed for the torsional stiffness k_φ which has to be chosen with care.

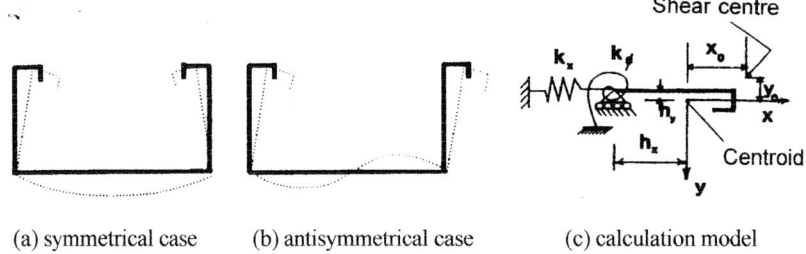

(a) symmetrical case   (b) antisymmetrical case   (c) calculation model

Figure 8: Calculation models for distortional buckling of the narrow flanges

When the behaviour of the complete section is considered, the two narrow flanges may buckle in either a symmetrical or an antisymmetrical mode, as shown in Fig. 8. It is necessary to consider both and to choose the most critical and Pekoz and Serrette give the equations for both cases. Some improvements and simplification of the Pekoz and Serrette method have recently been proposed by Davies and Jiang (1996) on the basis of studies using "Generalised Beam Theory" (GBT).

On the basis of such studies, it appears that, in many cases of cassette wall construction without a second skin, no reduction of the narrow flange compressive stress is required in order to cater for the possibility of distortional buckling.

Bending with the wide flange in compression

Here, the more complex narrow flange and lip assembly is in tension and does not buckle. The bending behaviour is, therefore, dominated by local buckling of the wide flange. However, flange curling as described above also occurs when the wide flange is in compression and interacts with local buckling. EC3 does not propose any rigorous treatment of this interaction and, indeed, this would appear to be exceedingly difficult. Instead, it suggests that the beneficial effect of intermediate stiffeners should be neglected and that the conventional effective width procedure should be used though with the material factor γ_m increased to 1.25 in order to deal with the additional uncertainty caused by flange curvature.

The absence of the second skin clearly has no influence when the narrow flange is in tension. However, significant benefits may be obtained when the cassette is filled with rigid thermal insulation (rigid plastic foam or mineral wool lamellas) as this will inhibit both local buckling and flange curling. Davies and Hakmi (1991) considered this interaction and gave a design equation for the effective width of a wide flange stiffened by such insulation. More recently, Zhao (2001) has given more detailed consideration, based on both tests and analysis, to the interaction between local plate buckling, flange curling and the stiffening influence of foam infill. However, his results have not yet been distilled into a simplified design approach.

Behaviour in shear

A cassette sub-assembly is also a ready-made shear panel or 'diaphragm' for stressed skin. Stressed skin design is explicitly allowed in EC3 and appropriate enabling clauses are included in section 9. EC3 also includes somewhat rudimentary provisions for cassettes acting as shear panels. These make it clear that the behaviour of a cassette wall panel in shear is not significantly different from that of a conventional shear panel comprising trapezoidal steel sheeting framed by appropriate edge members so that the procedures described in the definitive publications (e.g. Davies and Bryan 1981, ECCS 1995) may be used.

There are three main differences between cassette and liner tray systems and the trapezoidally profiled roof sheeting and decking for which the calculation procedures were originally devised:

> There is negligible flexibility due to shear distortion of the profile. This removes a design equation which tends to dominate the deflection calculation for trapezoidal profiles. Here it is possible to make a simplified estimate of deflections based on the assumption that the flexibility arises mainly in the fastenings.

> The strength calculation tends to be dominated by the tendency of the wide flange to buckle locally in shear before any of the more usual diaphragm failure modes (fastener failure, profile end failure or global shear buckling) are mobilised.

> There is often no separate edge member parallel to the cassettes. This means that there are no longitudinal edge fasteners to check and the web and narrow flange of the outermost cassette act as their own edge member which should be checked for the induced compressive force.

The first two of these considerations lead to the two equations given in EC3 for the ultimate and serviceability limit states respectively. These equations, and their deficiencies, have been discussed by Davies (1998a). Significantly, the wording of the clauses in EC3 may lead designers to overlook that fastener strength may also be critical. In addition to considering local buckling of the wide flange, it is essential also to consider the possibility of failure in each of the fastener failure modes considered in conventional stressed skin theory, namely:

> failure in the seam fasteners between adjacent cassettes
> failure in the fasteners connecting the ends of the cassettes to the foundation or the primary structure
> failure in the shear connector (longitudinal edge fasteners)

It appears clear to the author that the excessive simplicity in the approach in EC3 to the design of cassettes subject to diaphragm action is likely to lead to a lack of fundamental understanding and over-confidence. The inevitable result will be serious design errors.

The equation in EC3 for design with respect to local shear buckling of the wide flange is based on a simplified orthotropic plate theory with no allowance for boundary conditions, post-buckling etc (Baehre 1981, Davies 1998a). Davies and Fragos (2001) have described a comprehensive investigation into the local shear buckling of unstiffened wide flanges including the influence of rigid insulation infill. The basis of this investigation was a series of tests originally devised by Davies and Dewhurst (1997). The authors of the present paper have improved the test rig, which is shown in Fig. 9, and considerably expanded the original test series to include a wider range of geometries, infill materials etc.

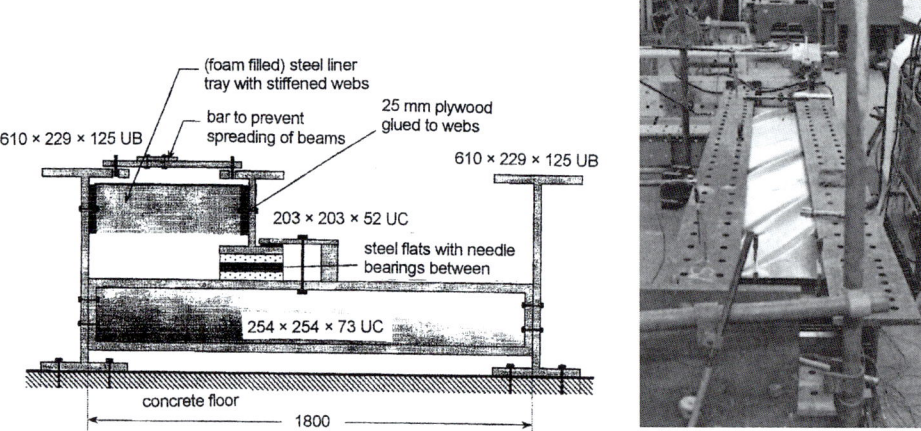

Figure 9: Cross-section through the test rig and view of specimen under test

The test series had idealised longitudinal boundary conditions that were intended to simulate full fixity together with some alternative boundary conditions along the short side. These test results were used to validate a comprehensive finite element model. This, in turn, allowed more realistic longitudinal edge boundary conditions to be considered in an extended parametric study. The outcome was the following equation which was found to give an accurate account of all of the buckling loads obtained:

$$\tau_{cr} = 10\frac{\pi^2 D}{b^2 t} + 2.9\frac{E_c}{t} - \frac{E_c}{at} \quad \text{where} \quad D = \frac{Et^3}{12(1-v^2)}$$

where: a = length of the cassette
b = width of the wide flange
t = thickness, net of coatings etc
E = Young's modulus of the cassette material
υ = Poisson's ratio of the cassette material

In the above equation, the stiffening effect of the foam is described in terms of the compressive modulus E_c of the insulant material. Various other core parameters were investigated and this was found to give the most consistent results. Evidently, the terms on the right hand side of this equation are not dimensionally consistent but the equation will be found to give correct results provided that the force units are in kN and the dimensions are in mm. Despite a detailed study, to date it has not been found possible to find a dimensionally consistent right hand side which takes into account variation of all three of the dimensional parameters, a, b and t.

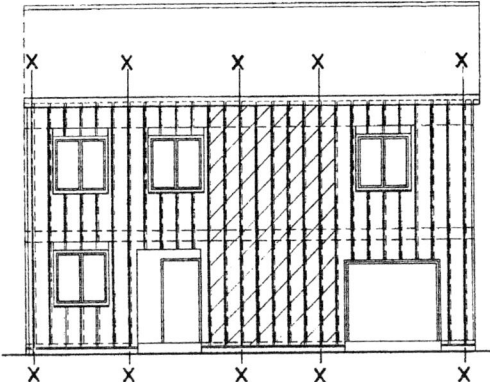

Fig. 10: Typical cassette wall in house construction

In cassette wall construction, it is expedient to space the holding down points as far apart as possible in order to reduce the shear forces in the wall and, at the same time, reduce the holding down forces into the foundations. This has implications for the architectural design so that early interaction between the Architect and the Engineer is required. The design of the holding down detail itself is another critical point in the structural design. Fig. 10 shows an elevation of a cassette wall as used in a typical house facade. The lines x-x show the division into prefabricated sub-panels for factory construction. The wind-shear diaphragm assumed in the design is shown by cross-hatching. Holding-down points to the foundations were provided at the extremities of the hatched area.

More recently, it has become clear that it is advantageous to design the cassette wall in its entirety when considering its resistance to the in-plane shear forces arising from diaphragm action. Fig. 11 illustrates the principles involved. The walls are considered to be held down to the foundations only at the corners of the buildings and the appropriate sub-structure is provided there. The cassettes provide the primary resistance to shear force but, where this shear flow has to bypass window and door openings, advantage is taken of the shear resistance of the roof and floor edge beams (usually channel sections of significant size) and the ground beam. This results in some additional connection forces which can, in the current state of the art, be determined by finite element analysis. A current research project is investigating whether a simplified design procedure may be possible.

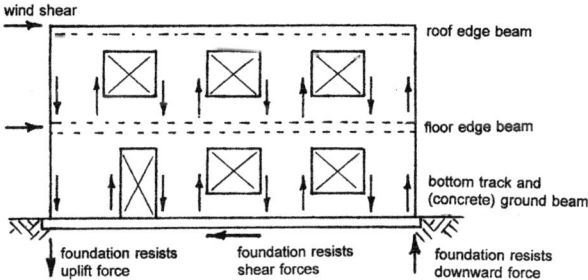

Figure 11: Alternative approach to the design of a cassette wall for in-plane shear

Combined effects

Although considered separately in this paper, axial load, bending and shear effects may evidently interact. Noting that there is no major axis bending or buckling, minor axis bending and axial load can be readily combined according to clause 6.5.1 of EC3.

The shear stresses in stressed skin action are very low and it is usual to neglect interaction between in plane shear and primary axial load and bending. However, the axial compressive forces arising from stressed skin action at the edge of a diaphragm (i.e. at the leeward holding-down point) must be combined with the axial force arising from load from the floors and roofs above. This will usually be of major design significance. Cassette wall panels can, therefore, be readily designed on the basis of EC3 together with the established procedures for stressed skin design and it is found that, for most low-rise construction, a standard panel and fastener specification is sufficient to carry the wind shear without any special provision other than for holding down forces at the leeward end of the diaphragm.

CONNECTIONS

Connections between individual cassettes and between cassettes and other structural elements may be made with any of the usual mechanical connections such as bolts, screws or blind rivets. Cassettes are substantial structural members and site connections are usually made by bolting. However, when forming prefabricated panels in the factory, a particular advantageous possibility is the use of press-joined connections (Davies et al. 1996) to form the seams between individual cassettes. Press-joining is quick, cheap and does not destroy the galvanising. The structural details tend to be simpler than those with stud construction so that erection is rapid. The CIBBAP cassette wall system uses large prefabricated units made in this way. The use of welding is not recommended as it destroys the galvanising leaving the wall more susceptible to corrosion damage.

CASE STUDY - HOUSES AT REIMS, FRANCE

Five houses were built at Reims as a demonstration project for the CIBBAP building system. Unlike most projects of this nature, the houses were built for the upper end of the market to a high quality on a prestige site. There was no difficulty in finding suitable customers. They were built using balloon construction so that full height wall panels, together with roof and floor panels, were prefabricated in the factory to give extremely rapid construction on site.

Figure 12: House at Reims: Artists impression and house under construction

Fig 12 shows some details of one of the houses. Fig. 10 is one of the less interesting elevations which, however, gives more detail of the method of construction. The cassette walls were exposed, though finished in an "architectural" coating.

CONCLUSIONS

Light weight steel cassette walls are an attractive alternative to steel wall studs for low rise steel framed construction. This paper has discussed how they may best be designed and given a recent example of their use. It is hoped that this may stimulate further interest in this form of construction.

REFERENCES

Baehre R. and Bucca J. (1986) "Die wirksame Briete des Zuggurtes von biegebeanspruchten Kassetten" (Effective width of the tension flange of cassettes in bending), *Stahlbau*, 9, pp 276-285.

Baehre R. (1987) "Zur Shubfeldwirkung und-bemessung von Kassettenkonstructionen" (On the behaviour and design of cassette assemblies in shear) *Stahlbau*, 7, pp 197-202.

Baehre R., Buca, J. and Egner, R. (1990) "Emfehlungen zur Bemessung von Kassettenprofilen" (Recommendations for the design of cassettes), *R. Schardt Festschrift*, University of Darmstadt.

Davies J. M. (1998a) "Light gauge steel cassette wall construction", *Nordic Steel Construction Conference 98*, Bergen, Sept. 14-16, pp 427-440.

Davies J. M. (1998b) "Light gauge steel framing for house construction", 2nd *Int. Conf. on Thin Walled Structures*, Singapore, Dec. 2-4, pp 17-28.

Davies J. M. (2000a) "Steel framed house construction", *The Structural Engineer*, Vol. 78, No.6, 21 March, pp 17-24.

Davies J. M. and Fragos A. S. (2001) "Shear strength of empty and infilled cassettes", 3rd *Int. Conf. on Thin Walled Structures*, Krakow, Poland, June 5-7, pp 3-18.

Davies J. M. and Bryan E. R. (1981) *"Manual of stressed skin diaphragm design"*, Granada.

Davies J. M. and Dewhurst D. W. (1997) "The shear behaviour of thin-walled cassette sections infilled by rigid insulation", Proc. Int. Conf. on Experimental Model Research and Testing of Thin-Walled Structures, Academy of Sciences of the Czech Republic, Prague, Sept., pp 209-216.

Davies J. M. and Hakmi M. R. (1991) "Postbuckling behaviour of foam-filled thin-walled beams", *J of Constructional Steel Research*, Vol. 20, pp 75-83.

Davies J. M. and Jiang C. (1996) "Design of thin-walled beams for distortional buckling", 14th Int. Speciality Conf. on Cold-formed Sections, University of Missouri-Rolla.

Davies J. M. and Jiang C (1998) "Design for distortional buckling", 2nd World Conf. on Steel in Construction, San Sebastian, 11-13 May.

Davies J. M. and Leach P. (1994a) "First-order Generalised Beam Theory", *J. of Constructional Steel Research*, 31, pp 187-220.

Davies J. M., Leach P. and Heinz D. (1994b) "Second-order Generalised Beam Theory", *J. of Constructional Steel Research*, 31 pp 221-241.

Davies J.M., Leach, P. and Kelo, E. (1995) "The use of light gauge steel in low and medium rise modular buildings", *Proc 3rd Int. Conf. on Steel and Aluminium Structures*, ICSAS '95, Istanbul, 24-26 May.

Davies R., Pedreschi R. and Sinha B. P. (1996) "The shear behaviour of press-joining in cold-formed steel structures", *Thin-Walled Structures*, Vol. 25, No. 3, pp 153-170.

Eurocode 3 (1996): *Design of Steel Structures - Part 1.3 : General rules Supplementary rules for cold formed thin gauge members and sheeting*, CEN ENV 1993-1-3, February 1996.

ECCS (1995) *"European recommendations for the application of metal sheeting acting as a diaphragm"*, European Convention for Constructional Steelwork, Publication No. 88.

Hancock, G. J., Rogers, C. A., Schuster, R. M. (1994) "Strength Design Curves for Thin-Walled Sections Undergoing Distortional Buckling". *J. of Constructional Steel Research*, 31(2-3), pp 169-186.

König J. (1978) "Transversally loaded thin-walled C-shaped panels with intermediate stiffeners", *Swedish Council for Building Research*, Document D7.

Nyberg G. (1976) "Diaphragm action of assembled C-shaped panels", *Swedish Council for Building Research*, Document D9.

Serrette R. L. and Pekoz T. (1995) "Distortional buckling of thin-walled beams/panels. I: Theory and II: Design methods", *J Struct Engrg.*, Vol. 121, No 4, April, pp 757-766 and 767-776.

Shafer B. W. (2001) "Thin-walled column design considering local, distortional and Euler buckling" *Structural Stability Research Council Annual Technical Session and Meeting.*

Thomasson P. (1978). "Thin-walled C-shaped panels in axial compression", *Swedish Council for Building Research.* D1: 1978, Stockholm, Sweden.

Voutay P.A. and Davies J. M. (2001) "Analysis of cassette sections in compression", *ICASS'02, 3rd Int. Conf. on Advances in Steel Structures*, Hong Kong, 9-11 Dec.

Zhao R. (2001) "Non-linear buckling analysis of C-shaped cassette sections in pure bending", PhD thesis, University of Manchester.

A NEW ISSUE IN PLATE AND BOX GIRDER STABILITY DESIGN

Tsutomu Usami[1] and Praween Chusilp[1]

[1] Department of Civil Engineering, Nagoya University
Chikusa-ku, Nagoya 464-8603, Japan

ABSTRACT

This paper addresses new considerations in design of plated structural members, such as box girders, plate girders, and shear-type hysteretic dampers, constructed in regions of high seismic risk. The topics discussed include the stiffener design methodology and stability and ductility design of plated members predominantly loaded in shear. The ultimate shear behavior is firstly investigated through experiments of three one-quarter scale, steel box girders. Influences of web slenderness and stiffener provision on the structural performance are discussed. Consequently, extensive numerical studies are carried on practical box and plate girders with key structural parameters varied over broad ranges. From the obtained results, reliable methods for estimating the shear and ductility capacities of plated members are presented. These methods can suitably be implemented in future seismic design specifications. Finally, based on the presented experimental and numerical investigations, suggestions for aseismic design of box girders, plate girders, and hysteretic dampers are made.

KEYWORDS

Plates, Box section, Girders, Ultimate behavior, Ductility, Shear, Buckling, Stability design

INTRODUCTION

Steel plated members subjected to predominant shear loading are commonly found in civil engineering structures and was recently proposed as a shear-type hysteretic damper to improve seismic performance of buildings (Nakashima 1995). Even in high seismic risk areas, these plated members are traditionally designed by considering only maximum strength while the cyclic and post-peak behaviors are disregarded. During the 1995 Hyogoken-Nanbu Earthquake, several box girders sustained significant shear damages (Figure 1), as a result of inadequate strength or ductility provisions. The stability and ductility problems of plated members seem to be closely correlated under an extreme loading condition. This observation reflects the need for a comprehensive stability design method that accounts for both strength and ductility, so that the structural members are provided with adequate margin of safety and the energy dissipation efficiency is enhanced.

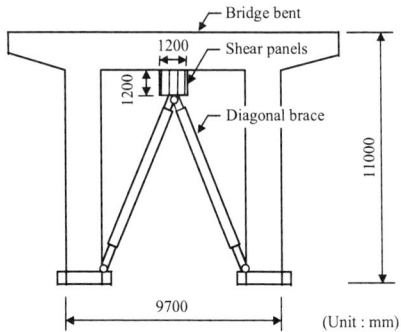

Figure 1: Shear buckling of box girder

Figure 2: Bridge bent with shear panels as dampers

This paper presents research developments concerning with stability design of plated structures, made of standard structural steel and predominantly loaded in shear, in regions of high seismic risk. The stiffener design method is firstly described. The performance of plated members under extreme cyclic shear is then investigated through experiments of box girders and numerical analyses of box and plate girders. From these investigations, practical methods for predicting the strength and ductility of plated members are described and suggestions for the design are made. The results presented herein would be useful for aseismic design of box and plate girders, as well as shear panels to be used as metallic-yielding hysteretic dampers for bridge structures as shown in Figure 2.

OPTIMUM DESIGN OF STIFFENERS

In design of stiffeners, it is generally accepted that stiffeners' cross-sectional dimensions be determined with regard to the optimum stiffener rigidity, γ_s^*, defined as the minimum stiffness that allows overall buckling of the stiffened plate and local buckling of the plate sub-panels to occur simultaneously. To ensure that the stiffeners remain straight up to the ultimate load (i.e., to achieve the highest load-carrying capacity of the stiffened plate), the optimum rigidity may be increased by some empirical factor as described by the European Convention for Constructional Steelwork (ECCS 1986). An appropriate value of this factor has been addressed by Chusilp & Usami (2002a).

Formulas for estimating the optimum stiffener rigidity have been proposed so far. For instance, the German stability design specifications, namely DIN 4114 (Deutsches Institut für Normung (DIN) 1953), suggested a set of design formulas for computing the optimum rigidity for various stiffener arrangements. It has been shown firstly by Klöppel & Scheer (1960) that these formulas may produce significant errors as a result of inaccurate approximation made in their theoretical derivation. A more accurate optimum rigidity can be interpolated from the design charts of the buckling coefficients proposed by Klöppel & Scheer (1960) and Klöppel & Möller (1968). However, such interpolation is available only at very few values of the plate aspect ratio. Due to difficulties in estimating the optimum rigidity, some design specifications, such as the British code BS5400 (British Standards Institution (BSI) 1982), abandoned this stiffener design concept and moved toward an ultimate design approach which treats a longitudinal stiffener as a strut. Provided that precise formulas for the optimum rigidity are available, the optimum stiffener design approach seems more attractive since the stiffness of the stiffener is clearly quantified with reference to the well-known index γ_s^*.

A study has been conducted by the authors to develop design formulas for the optimum stiffener rigidity of simply supported shear panels. Because web plates are practically reinforced with few stiffeners, plates with 1, 2, or 3 longitudinal and transverse stiffeners and those with 1, 2, or 3 longitudinal stiffeners only are considered. The buckling coefficients of these plates are determined numerically by means of the elastic buckling theory, using high degree of approximation regarding the deflection function of the plate. Consequently, the computed buckling coefficients are employed to determine the optimum stiffener rigidity. Approximate formulas for the optimum rigidity are proposed for plates with 1, 2, or 3 longitudinal and transverse stiffeners and for plates with 1, 2, or 3 longitudinal stiffeners as shown in Eqns. 1 and 2, respectively.

$$\gamma_s^* = \left(\frac{23.1}{n^{2.5}} - \frac{1.35}{n^{0.5}}\right) \frac{(1+\alpha^{3/n-0.3})^{2n-1}}{1+\alpha^{5.3-0.6n-3/n}} \quad (1)$$

$$\gamma_s^* = \frac{27.3n^{0.6}\alpha - 23.3\alpha}{0.20n^{0.7} - 0.60/\alpha + 0.52/\alpha^2} \quad (2)$$

where n = number of the parallel stiffeners; and α = plate aspect ratio (length/width). These equations are applicable to the cases of n = 1, 2, or 3 and $0.5 \leq \alpha \leq 2.0$. For plates with 1, 2, or 3 transverse stiffeners, the approximate formula can be derived by considering the symmetry [i.e., by replacing α in Eqn. 2 with $1/\alpha$ and multiplying the resulting expression by α]. Detailed developments of these formulas can be found elsewhere (Chusilp & Usami 2002b).

SHEAR STABILITY DESIGN OF BOX GIRDERS

In the light of aseismic design, the influence of the web slenderness and stiffener provision on the structural performance (i.e., strength, ductility, and energy dissipation capacity) is an important issue that needs to be considered. A preliminary experimental performance investigation of box girders under cyclic shear is presented in this section. Also described are relevant findings from an extensive parametric study conducted on over 100 box girders.

Experimental Verification of Cyclic Shear Behavior

The experiments were conducted on three one-quarter scale models of box girder at Aichi Institute of Technology under the joint research of Nagoya Highway Public Corporation, Nagoya University, and Aichi Institute of Technology. Each specimen was fabricated from JIS SM 490 (equivalent to ASTM A572) steel plates using welding connections. As illustrated in Figure 3, the specimen has the total length of 5700 mm, flange width b_f of 625 mm, and web width b_w of 500 mm. Four equidistant longitudinal stiffeners are provided at the top and bottom flanges so that severe flange buckling is inhibited. To ensure that the failure will occur with the middle portion of the girder where bending is minimum, other portions were fabricated from thicker plates and heavily stiffened by transverse diaphragms. The middle portion has the length of 550 mm and is enclosed by 6-mm thick diaphragms. The three specimens are designated as UW45, UW60, and SW60. The cross-section of the middle portion of each specimen is shown in Figure 3(b). Their structural parameters, including the slenderness parameter R and ratio of the relative flexural stiffener rigidity γ_s to its optimum value γ_s^*, are listed in Table 1. The subscripts w and f refer to web and flange, respectively. γ_{ws}^* and γ_{fs}^* are determined from the cases of a simply supported plate in pure shear (Eqn. 2) and that in uniform compression (Klöppel & Scheer 1960), respectively. R_w and R_f are computed by

$$R_w = \frac{b_w}{t_w}\sqrt{\frac{12(1-\nu^2)\tau_y}{k_s \cdot \pi^2 E}} \ ; \quad R_f = \frac{b_f}{t_f}\sqrt{\frac{12(1-\nu^2)\sigma_y}{k_c \cdot \pi^2 E}} \quad (3)$$

where t_w and t_f = web and flange thickness, respectively; k_s and k_c = elastic buckling coefficients of simply supported plates in shear and compression, respectively; τ_y = shear yield stress, corresponding to the shear yield strain γ_y; σ_y = tensile yield stress; ν = Poisson's ratio; and E = modulus of elasticity.

Each test girder was simply supported at a distant of 1813 mm from its ends and transversely loaded at each end through a hydraulic actuator (Figure 3(a)). At the middle web, a pair of displacement transducers was installed to measure the displacements in the web diagonal directions which can be converted to an average shear strain of the web, γ, (defined as the relative transverse displacement between the two ends of the middle web divided by the web length of 550 mm) by using a geometrical relation. To simulate the dead load of the superstructure, a gravity load of 130.3 kN was applied at each end of the girder through the actuators. The specimen was consequently loaded by quasi-static

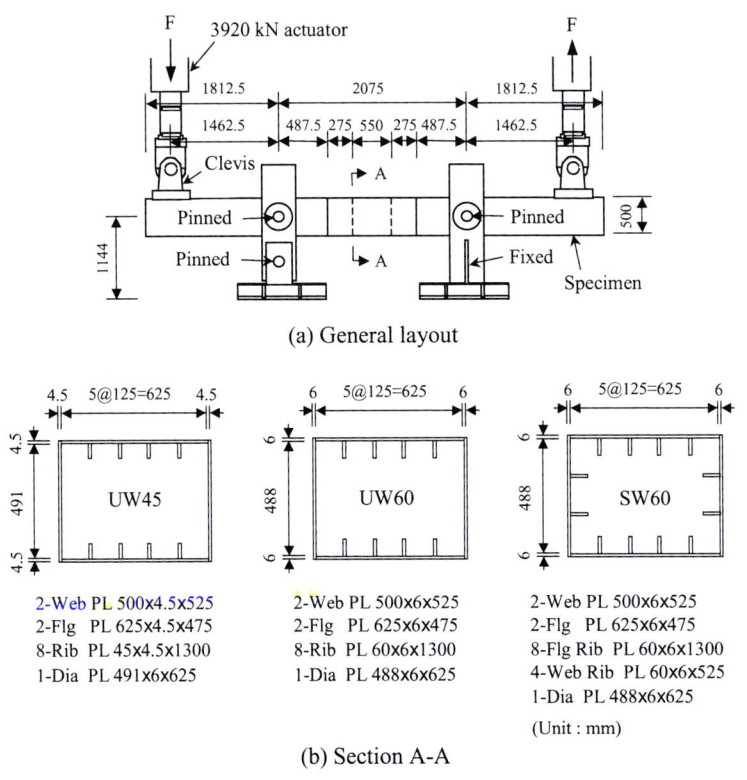

(a) General layout

(b) Section A-A

Figure 3: Setup of steel box girder specimen

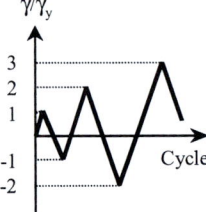

Figure 4: Cyclic loading history employed in tests

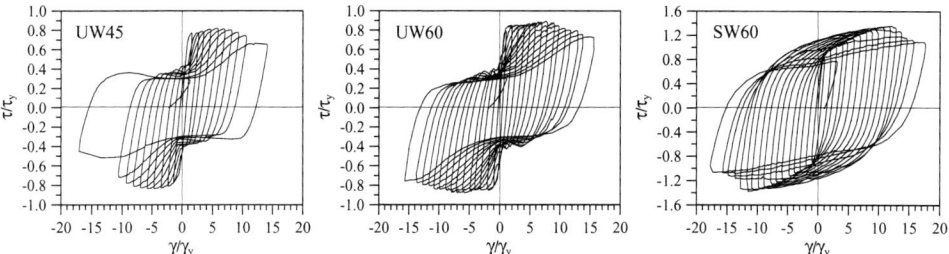

Figure 5: Shear stress versus shear strain relationship of test specimens

TABLE 1
STRUCTURAL PROPERTIES OF MIDDLE GIRDER PORTIONS

Specimens	R_w	R_f	$\gamma_{ws}/\gamma_{ws}^*$	$\gamma_{fs}/\gamma_{fs}^*$
UW45	1.31	0.64	-	1.18
UW60	0.99	0.49	-	1.58
SW60	0.41	0.49	0.46	1.58

TABLE 2
SUMMARY OF STRENGTH AND DUCTILITY

Specimens	τ_{cr}/τ_y	τ_m/τ_y	γ_u/γ_y
UW45	0.51	0.83	15.54
UW60	0.88	0.89	15.62
SW60	-	1.37	17.38

forces at the two ends, whose magnitudes are equal but in opposite direction. The loading history applied is shown in Figure 4.

The obtained normalized shear stress versus shear strain relationship of the middle portion of each specimen is presented in Figure 5, wherein τ = average shear stress of the webs determined by dividing the shear force acting on the middle portion by the transversal cross-sectional area of the two webs. Summary of the theoretical shear buckling strength, τ_{cr}, maximum shear strength, τ_m, and the maximum shear strain, γ_u, is given in Table 2.

Maximum shear strength

The maximum shear stresses (averaged from the positive and negative peaks) attained by the specimens UW45 and UW60 are only 60 and 65 per cent of that possessed by the specimen SW60. In the post-buckling range, the first two specimens resisted shear forces in part through a web membrane action, known as tension field action (see ECCS 1986; Galambos 1998), which could anchor against the flanges effectively. It should be noted that the flanges of these specimens are immediately stocky but of practical uses. If the flanges are more slender, the shear strength and collapse mechanism of the girder may depend strongly on the flexural rigidity of edge flange sub-panels and governed by local flange deformations imposed by the web membrane stresses (Rockey et al. 1973). Among the three specimens, the tension field action is the most prominent in the case of UW45, while the least pronounced case is SW60 in which the tension field developed after the web yielded in shear. The specimen SW60 would possess higher shear strength, as well as higher ductility and energy dissipation capacities, if the longitudinal web stiffeners are sufficiently stiff and remain straight until the girder collapses. The full shear resistance of a stiffened web can be exploited if the rigidity of the web stiffener γ_{ws} is not lower than its optimum value γ_{ws}^* (Chusilp & Usami 2002a).

Ductility capacity

The maximum shear strains averaged from the positive and negative peak strains observed in the specimens UW45 and UW60 are almost equal. The ductility capacity (γ_u/γ_y) of box girders with

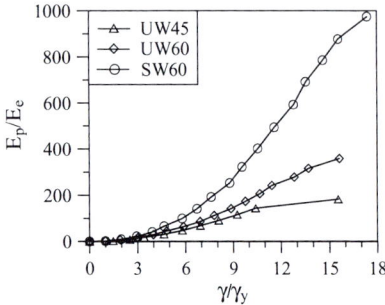

Figure 6: Cumulative energy dissipation

unstiffened web panels seems to be dominated by local damage (e.g., low-cycle fatigue cracks), rather than the web slenderness. Compared with the other two specimens, the specimen SW60 is capable of sustaining larger deformation. The ultimate shear strain of this specimen should be larger than $20\gamma_y$ if $\gamma_{ws} \geq \gamma_{ws}^*$ (Chusilp & Usami 2002a). On the other hand, less ductile behavior is expected if narrower and thinner flanges or weaker flange stiffeners are used. This is because the ductility capacity of box girders can be affected considerably by the flange slenderness as well as the rigidity of the flange stiffener (Chusilp & Usami 2002a).

Energy dissipation capacity

The energy dissipation is the other aspect that reflects the seismic performance of box girders. Figure 6 shows the plots of normalized cumulative energy dissipation versus average value of the positive and negative peak cyclic shear strains for the three specimens. In the figure, E_p denotes the cumulative plastic energy dissipated by the middle girder portion under cyclic shear and E_e denotes the elastic energy absorption capacity of the two middle webs in pure shear. The term E_p/E_e, therefore, equals to two times of the area inside the hysteresis loops shown in Figure 5 and is cumulated at the completeness of each cycle (i.e., at the end of negative unloading path).

As shown in Figure 6, the normalized cumulative energy at each cycle of the specimen UW45 is only one-half of that dissipated by the specimen UW60, although the ductility capacity of these girders are almost the same. The lesser amount of energy dissipation is attributed to the lower peak cyclic shear stresses and more severe pinching of the hysteresis loops. Until the last cycle, the normalized energies of the specimens UW45 and UW60 are only 19 and 37 per cent of the amount obtained in the case of SW60. This comparison sheds light on the effectiveness of the web stiffeners in enhancing the shear resistance and maintaining large hysteresis loops without severe pinching, leading to a large amount of energy dissipation.

Evaluation of Shear Strength and Ductility Capacity

Based on numerical investigations of monotonic and cyclic shear behavior of steel box girders with key parameters varied over practical ranges (Chusilp & Usami 2002a), the parameter R_w appears to be the most important parameter that governs the failure behavior of the girder. In general, the effects of shear reversals on the shear strength and ductility capacity may not be considered in design, except in the case of stocky webs ($R_w \leq 0.6$) where the stress reversal induces an early initiation of the strain hardening and results in a significant reduction of the ductility capacity (> 50% as compared to the monotonic case). It is found that when $\gamma_{ws} \geq \gamma_{ws}^*$, the influence of the web stiffener's rigidity on the shear strength and ductility capacity of box girders is very small. Also, the effect of the web aspect ratio on the ductility capacity is negligible, while it is already accounted for in most shear strength prediction methods. For the flange slenderness and flange stiffener's rigidity, their contributions to the shear resistance are not pronounced. These two parameters, however, considerably affect the ductility capacity of box girders. Realizing that the prime function of box girder flanges is to resist direct stresses and that excessive flange deformations may reduce its compression resistance, the contributions of the stiffened flanges to the shear resistance and ductility capacity should be ignored in design.

Shear capacity prediction

The application of available shear strength prediction methods, originally developed for plate girders under monotonic shear, to box girders is examined. The methods suggested by the European Convention for Constructional Steelwork (ECCS 1990) and the American Association of State Highway and Transportation Officials (AASHTO 1998) are chosen as candidates because they are simple, yet reflecting understandable shear-resisting mechanisms. Their key concept is that the maximum shear strength is determined from the theoretical shear buckling strength and web tension field action. According to ECCS, the shear strength model of Porter *et al.* (1975) is modified for box girders by neglecting the flange contributions and simplifying the inclination of the tension field stresses to one-half of the inclination of the web diagonal. The maximum shear strength is calculated from

$$\tau_m = \tau_{cr} + \sigma_{ty} \sin^2 \frac{\theta_d}{2} \cdot \left(\cot \frac{\theta_d}{2} - \cot \theta_d \right) \leq \tau_y \quad (4)$$

$$\sigma_{ty} = -\frac{3}{2}\tau_{cr} \sin 2\theta_d + \sqrt{\sigma_y^2 + \tau_{cr}^2 \left(\frac{9}{4}\sin^2 2\theta_d - 3 \right)} \quad (5)$$

where σ_{ty} = web membrane stress in the tension field that fulfills the yield condition in addition to the shear buckling stress; $\theta_d = \tan^{-1}(b_w/a)$ = web diagonal inclination; and a = web length. Based on the theories of Basler (1961) and Cooper (1967), AASHTO suggests the following equation for computing the shear capacity:

$$\tau_m = \tau_{cr} + \frac{\sqrt{3}(\tau_y - \tau_{cr})}{2\sqrt{1+\alpha_s^2}} \leq \tau_y \quad (6)$$

where α_s = aspect ratio of the web sub-panel. The shear strengths predicted by Eqns. 4 and 6 are limited to the shear yield strength so as to ignore the effects of the strain hardening. It is noted that these shear strength models are valid only if the longitudinal web stiffeners possess sufficient out-of-plane flexural stiffness so that they remain straight until the maximum load (Cooper 1967; Porter *et al.* 1975).

Figure 7 compares the normalized maximum shear strengths predicted by ECCS and AASHTO and the analysis results obtained under monotonic and cyclic loadings. Considering the cases of slender webs ($R_w \geq 1.1$), ECCS gives conservative results (12% for $R_w = 1.3$) as a consequence of neglecting all flange contributions to the shear strength. The method of AASHTO produces a greater underestimation (18% for $R_w = 1.3$), due primarily to an incorrect assumption made for the occurrence of the tension field. It is observed during the analyses that a single tension field apparently develops in the web independently of longitudinal stiffeners (Figure 8(a)), that is in agreement with the assumption of Porter *et al.* (1975) (Figure 8(b)), rather than in each individual web sub-panel like that assumed by Cooper (1967) (Figure 8(c)). For box girders with stockier webs having $\tau_m \leq \tau_y$, both ECCS and AASHTO produce some overestimations but, however, less than 6%. From this study, the method of

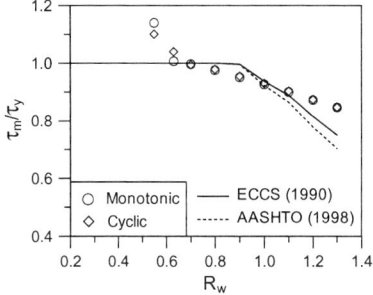 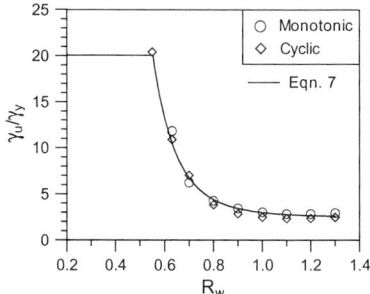

Figure 7: Shear capacity of box girders   Figure 9: Ductility capacity of box girders

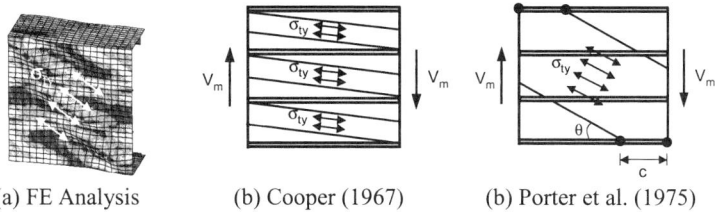

(a) FE Analysis   (b) Cooper (1967)   (b) Porter et al. (1975)

Figure 8: Development of tension field action in stiffened plate girders

ECCS seems appropriate for a practical design since it gives sufficiently accurate predictions while assumes the shear resisting mechanism which reasonably agrees with the actual behavior.

Ductility capacity prediction

The parametric study has evidenced that only the web slenderness should be considered in the ductility evaluation, provided that the web stiffener's rigidity is not lower than its optimum value. Based on the analysis results, a simple formula of the ductility capacity is proposed

$$\frac{\gamma_u}{\gamma_y} = 2.5 + \frac{0.5}{R_w^{6.0}} \leq 20.0 \qquad (7)$$

Eqn. 7 is applicable over broad parameter ranges of $\gamma_{ws} \geq \gamma_{ws}^*$ and $1.0 \leq \alpha \leq 2.0$, regardless of the flange slenderness and flange stiffener's rigidity. The proposed formula can also be applied to longitudinally stiffened box girders with any other practical flange geometries, considering large uncertainties in workmanship and material properties. The predicted ductility capacity is plotted against the analysis results in Figure 8. In any cases, the ductility capacity should be limited to 20.0 to inhibit excessive shear strain in local regions of the web and to avoid low-cycle fatigue problems. The proposed formula can be used to estimate the shear deformation at the maximum shear strength by multiplying the ductility capacity by a factor of 0.45.

SHEAR STABILITY DESIGN OF PLATE GIRDERS

To determine the shear capacity of plate girders, the method proposed by Porter *et al.* (1975) can readily be used because its theoretical background was developed particularly for plate girders and has been subjected to several experimental verifications (see Galambos 1998). However, to obtain a consistent design method, the simplifications of the shear strength theory made for box girders (ECCS 1990) may be applied to plate girders. Accuracy of this simplified method in predicting the shear strength of plate girders will be evaluated herein. Also, a reliable formula is needed in the assessment of the ductility capacity of plate girders. Eqn. 7 is likely applicable since it represents lower bound results for box girders of various flange slenderness and flange geometry. The application of Eqn. 7 to cases of plate girders will be verified.

Analytical Method

Figure 10 shows a finite element model of plate girders, in which the web width of 2000 mm and the flange width of 600 mm are assumed. The web, flange, and stiffener are discretized into 18×24, 18×12, and 18×7 meshes, respectively. All nodes on the plane $x = 0$ are simply supported, while on the plane $x = a$ transverse shear displacement δ will be specified. Inelastic large deformation analyses are performed by using program ABAQUS (*ABAQUS* 1998) incorporated with a constitutive model called two-surface model (Shen *et al.* 1995). A four-node doubly curved shell element (S4R) is used. Further details of boundary conditions, material properties (SS 400 steel), and geometric and material imperfections are similar to those adopted in the investigations of box girders (Chusilp & Usami 2002a). Only monotonic loading is considered, since preliminary analyses evidence that if the web is not too stocky ($R_w > 0.6$), the strength and ductility of plate girders are almost unaffected by the loading history.

Verifications of Shear Strength and Ductility Predictions

Analyses show that the effects of the web slenderness, stiffener rigidity, and web aspect ratio on the maximum shear strength and ductility capacity of plate girders are similar to those observed in box girders. Stiffening plate girders with $\gamma_{ws} = \gamma_{ws}^*$ is adequate to achieve the maximum shear and ductility capacities as shown in Figures 11 and 12. Only plate girders with the parameter R_w varied from 0.63 to 1.3 and the ratio t_f/t_w varied from 1.0 to 8.0 (practical value being just over 4.0) will be considered in verifications of the presented shear strength and ductility prediction methods. Other parameters are kept as $\gamma_{ws}/\gamma_{ws}^* = 1.0$ and $\alpha = 1.0$.

The normalized maximum shear strengths obtained from the analysis are compared with the predictions of ECCS's and AASHTO's methods (Eqns. 4 and 6) in Figure 13. The shear strengths of

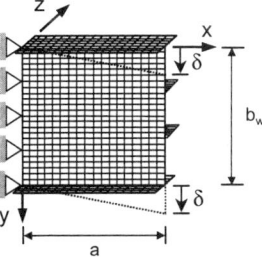

Figure 10: Finite element model of plate girders

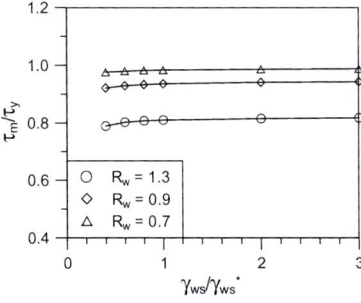
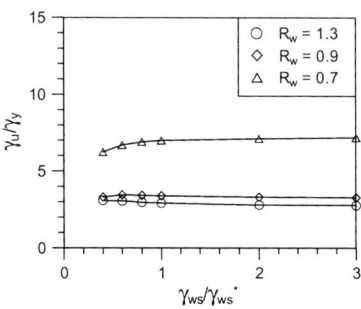

Figure 11: Shear capacity versus stiffener rigidity relationship of plate girders

Figure 12: Ductility capacity versus stiffener rigidity relationship of plate girders

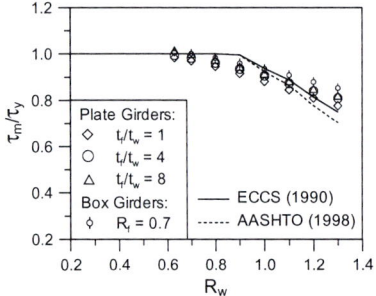
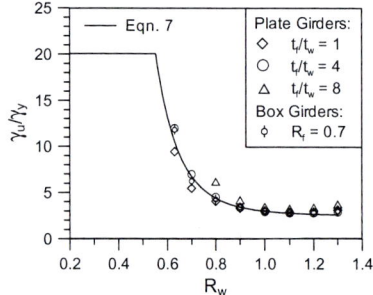

Figure 13: Shear capacity of plate girders

Figure 14: Ductility capacity of plate girders

box girders computed earlier are also given for comparison. It appears that the maximum shear strength of plate girders is weakly dependent on the flange stockiness, and ECCS's method can properly be applied to longitudinally stiffened plate girders.

Figure 14 presents the obtained ductility capacity and the prediction of Eqn. 7. For plate girders with stocky web ($R_w < 0.8$) and stocky flanges ($t_f/t_w = 8.0$), flanges play an important role in enhancing the deformation capacity of the girders. However, for practical plate girders ($t_f/t_w = 4.0$), the ductility capacity of plate girders lies close to the results of box girders over the whole considered range of R_w. Eqn. 7 is, therefore, valid for practical plate girders reinforced with two longitudinal stiffeners under shear. Similar to the case of box girders, the applicable parameter ranges of this formula are $\gamma_{ws} \geq \gamma_{ws}^*$ and $1.0 \leq \alpha \leq 2.0$, regardless of the flange stockiness.

DESIGN SUGGESTIONS AND CONCLUDING REMARKS

A new issue in stability design of box and plate girders, including considerations regarding the stiffener design and reliable estimation methods for the shear and ductility capacities of the girders, has been addressed in this paper. This is particularly important in aseismic design of the girders, as well as of shear-type hysteretic dampers, in which substantial inelastic excursions and a large amount of energy dissipation are expected and then the stability and ductility problems become closely correlated.

The presented numerical and experimental investigations suggested that the tension field action of webs should not be fully used. Such shear resisting mechanism usually associates with significant out-of-plane deformations which result in pinching of hysteresis loops and, thus, a degradation in the energy dissipation capacity. It is, therefore, recommended that at least one or two longitudinal stiffeners be provided in the webs once large energy dissipation is expected, in order to avoid severe pinching behavior. In design of the stiffener, caution should be taken to ensure that the web stiffeners are sufficiently stiff and remain straight up to the maximum load or, for better performance, up to the collapse stage. Accordingly, full shear resistance, ductility capacity, and energy dissipation capacity of the girder can be exploited. From numerical investigations presented herein, proportioning the stiffener with $\gamma_{ws} \geq \gamma_{ws}^*$ is adequate to achieve this goal.

Since the strength of plated members is likely unaffected by shear reversals, webs of plated members can be designed by using the shear strength prediction method suggested by ECCS (1990). However, when large inelastic deformations are encountered, ductility considerations must be made. In this regard, the web plate with the slenderness parameter $R_w \leq 0.55$ should be used so that the ductility capacity of, at least, 20 is obtained and the robustness of the girder (or hysteretic damper) is ascertained. The proposed empirical formula can also be used to evaluate the ductility capacity of existing plated members. Both strength and ductility prediction methods suggested herein can suitably be implemented in future seismic design specifications for steel structures.

In combined shear-bending loading cases, the strength prediction method developed for the pure shear case still forms the design basis. Further information on the strength prediction for plated members under combined shear and bending can be found elsewhere (e.g., ECCS 1986; Galambos 1998).

REFERENCES

ABAQUS/Standard User's Manual-Version 5.8. (1998). Habbitt, Karlson and Sorensen, Pawtucket, R.I.

American Association of State Highway and Transportation Officials (AASHTO). (1998). *AASHTO LRFD Bridge Design Specifications*, 2nd Ed., Washington, DC.

Basler, K. (1961). Strength of Plate Girders in Shear. *Journal of the Structural Division*, ASCE **87:ST7**, 151-180.

British Standards Institution (BSI). (1982). Steel, Concrete and Composite Bridges: Part 3, Code of Practice for Design of Steel Bridges, *BS 5400*, London.

Chusilp, P., and Usami, T. (2002a). Strength and Ductility of Steel Box Girders under Cyclic Shear. *Journal of Structural Engineering*, ASCE **128:9** (in press).

Chusilp, P., and Usami, T. (2002b). New Elastic Stability Formulas for Multiple-stiffened Shear Panels. *Journal of Structural Engineering*, ASCE **128:6** (in press).

Cooper, P. (1967). Strength of Longitudinally Stiffened Plate Girders. *Journal of the Structural Division*, ASCE **93:ST2**, 419-451.

Deutsches Institut für Normung (DIN). (1953). Stahlbau, Stabilitatsfalle (Knickung, Kippung, Beulung), Berechnungsgrundlagen. *DIN 4114, Blatt2*, Berlin (in German).

European Convention for Constructional Steelwork (ECCS). (1986). Behavior and Design of Steel Plated Structures. *Publication No. 44*, P. Dubas and E. Gehri, eds., ECCS Technical Working Group 8.3, Brussels.

European Convention for Constructional Steelwork (ECCS). (1990). European Recommendations for the Design of Longitudinally Stiffened Webs and of Stiffened Compression Flanges. *Publication No. 60*, ECCS Technical Working Group 8.3, Brussels.

Galambos, T. V., ed. (1998). *Guide to Stability Design Criteria for Metal Structures*, 5th Ed., Wiley, NY.

Klöppel, K., and Möller, K. H. (1968). *Beulwerte ausgesteifter Rechteckplatten II. Band*. Wilhelm Ernst, Berlin (in German).

Klöppel, K., and Scheer, J. (1960). *Beulwerte ausgesteifter Rechteckplatten*, Wilhelm Ernst, Berlin (in German).

Nakashima, M. (1995). Strain-hardening Behavior of Shear Panels Made of Low-yield Steel. I: Test. *Journal of Structural Engineering*, ASCE **121:12**, 1742-1749.

Porter, D. M., Rockey, K. C., and Evans, H. R. (1975). The Collapse Behavior of Plate Girders Loaded in Shear. *The Structural Engineers* **53:8**, 313-325.

Rockey K. C., Evans H. R., and Porter D. M. (1973). Ultimate Load Capacity of Stiffened Webs Subjected to Shear and Bending. *Proceedings of the International Conference on Steel Box Girder Bridges*, Institute of Civil Engineers, London, 45-61.

Shen, C., Mamaghani, I. H. P., Mizuno, E., and Usami, T. (1995). Cyclic Behavior of Structural Steels. II: Theory. *Journal of Engineering Mechanics*, ASCE **121:11**, 1165-1172.

MONOTONIC AND HYSTERETIC BEHAVIOUR OF BOLTED ENDPLATE BEAM-TO-COLUMN JOINTS

R. Zandonini and O.S. Bursi

Department of Mechanical and Structural Engineering, University of Trento, Trento, Italy

ABSTRACT

Semi-rigid frame design is one of the major developments in the structural design of steel and composite steel-concrete buildings. A vital requisite of this approach is the capability of approximating the response of beam-to-column joints. As to this aspect, the component method is increasingly accepted as a powerful and reliable tool. While its use for design under static loads is supported by many studies, its application to seismic design was not extensively investigated. This paper intends to discuss the main outcomes of a research work of the cyclic behaviour of end plate joints and their T-stub components. These results allow the component method to be appraised. Finally, possible future developments are outlined.

1. INTRODUCTION AND PURPOSE

If one item were to be selected to characterize the recent development in the design of steel framed structures, the philosophy which 'binds' frame and joint design is certainly a strong candidate. This methodology is frequently referred to as semi-rigid frame design. This definition is somehow inaccurate in the sense that it does not provide the full picture of the approach, which has a more general application and do not limit itself to frames with semi-rigid joints. The basic philosophy stresses that an efficient and effective frame design has to incorporate the joint response as an explicit parameter. In other terms it recognizes the important role of joints in assuring that a reliable, yet economical solution is achieved.

The importance of joint action was clearly perceived since the beginning of the last century. However, only in the 80's the tools available to the practitioners, and in particular the structural analysis programs, made the new methodology ready for practical use. Since then a number of research studies were carried on worldwide, aimed at building up the necessary knowledge to develop design tools ranging from general criteria to specific methods and rules for joints and frames (Bjorhovde et al. (1988), Narayanan (1988), Bjorhovde et al. (1992), Colson (1992), Lorenz et al. (1993), Wald (1994), Bjorhovde et al. (1996), Maquoi (1999) and Easterling and Leon (2002)).

All the key facets of the problem were investigated via experimental, numerical and theoretical analysis, including the behaviour of a wide range of joint types, the design models to approximate joint response and the influence of joint action on frames performance. Design criteria as well as specific recommendations were set up and included in Codes (Eurocode 3 (1993)) and design aids were developed in order to make the new philosophy to be accepted in practice. Studies also pointed out the potential benefits in terms of improvement of the cost-performance ratio (Jaspart (2002)).

The availability of prediction models, enabling, adequate approximation of the whole joint response in terms of stiffness, strength and rotation capacity is a vital pre-requisite to any design approach incorporating joint response as a key parameter. Many of the traditional methods were developed with the sole purpose of determining the connection resistance capacity. Furthermore, the complexity of the stress state in the nodal zone makes the range of application of most methods rather limited (Nethercot & Zandonini (1988)). An attempt to overcome this difficulties, and to provide a general and comprehensive tool is given by the so-called "component model", which identify the various elemental joint components, and build up the overall response of the joint on the individual response of these components. The advantages of this approach, schematically presented in figure 1, are multi-faceted: (i) the attention of design and research is focussed on the elemental components, the behaviour of which is easier to be determined (either experimentally or numerically) and modelled; (ii) the range of applicability is potentially 'unlimited', and actually bounded only by the range of geometrical and/or mechanical data, on which the component model is based; (iii) the response of the joint can be controlled in design via control of the critical component(s), i.e., of the component(s) governing the key aspect of the behaviour for the limit state considered.

A number of validation and calibration studies were carried out in the last decade, and design criteria and recommendations where developed and included in the Eurocodes 3(2001) and 4 (2002).

A similar development, as to the novel importance of joints in design, took place in earthquake engineering. The possible role of joints in the energy dissipation mechanism was in fact recognised and the potential of semi-continuous frames in seismic areas investigated. Design approaches based on conceptual design were also proposed for steel as well for composite frames. The limits of the traditional approaches for appraising the joint response become even more evident when cyclic loads are considered. A clear need for catching stiffness and strength deterioration, and possible pinching effects associated with buckling and fracturing of components as well as with the increase of lack-of-fit due to plastic deformations (in particular in bolted connections) makes the requirements to be met by prediction models to be more strict than in the case of static loading. The peculiar features listed above for the component method make it a fairly appealing solution for modelling joints also in seismic analysis. In particular, it should be stressed that it would allow to concentrate on the cyclic response of individual components, which can be investigated in a far more simple and economic way than the response of the whole joint.

This paper intends to discuss the main outcomes of a research work related to the experimental analysis of the cyclic behaviour of endplate joints and their T-stub components. The study aims to develop joint models enabling their seismic response to be captured in all its aspects, including the damage initiation and evolution up to failure. The results of the extensive experimental analysis were evaluated in terms of the main behavioural parameters of interest in seismic design. They were then used to appraise the component method, due to the particular interest in the extension of the model to the cyclic range. No definitive conclusion can yet be drawn. However, indications are presently available on which build up future studies.

2. ENDPLATE JOINTS: RESPONSE AND DESIGN MODELS

Endplate connections traditionally represent a viable alternative to welded beam-to-column joint solutions in moment resisting frames. Extended endplate as well as flush end-plate connections allow to realize joints covering a rather wide range of strength and stiffness. In many instances they also possess an adequate rotation capacity for plastic design. The complexity of the response, as influenced by the geometrical and mechanical parameters defining the key structural components (i.e., the endplate, the bolts, the column flange and web, the column panel zone), makes the experimental analysis to be a vital research tool. Numerous studies were also devoted to the simulation of the behaviour via numerical methods, in particular the FE analysis, and interesting results were obtained, which complemented the test outcomes permitting an extension of them to cover a larger range of cases of practical interest (Nethercot & Zandonini (1988), Bose et al. (1996), Bursi & Jaspart (1997),

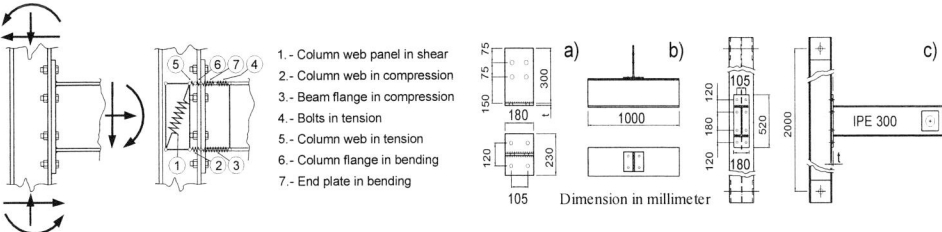

Figure 1. Component model of a bolted extended end plate joint according to Eurocode 3 (2001)

Figure 2. a) Isolated T-Stub; b) Coupled T-Stub; c) Complete Joint

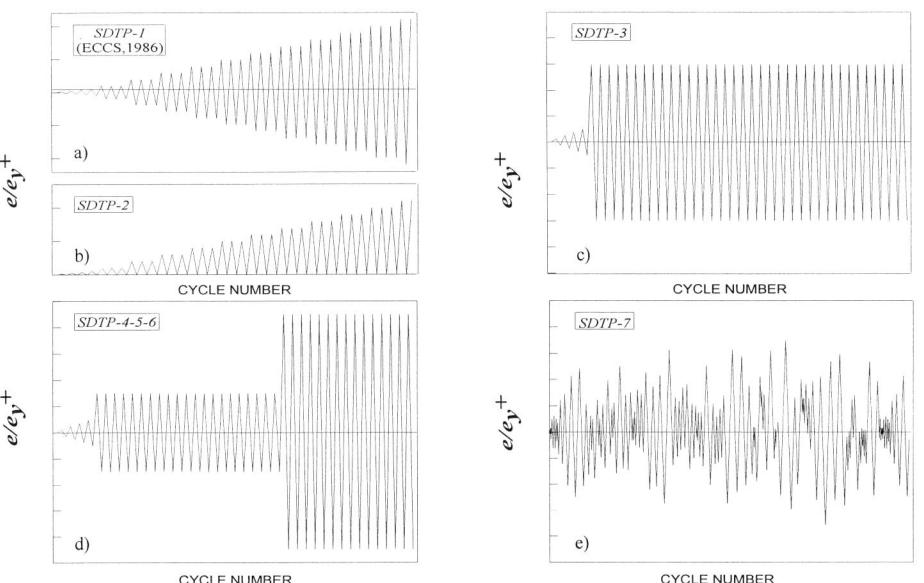

Figure 3. Controlled displacement test protocols: a) and b) Constant and variable amplitude cycles; c) Constant amplitude cycles; d) Large amplitudes cycles superimposed upon constant amplitude cycles; e) Random amplitude cycles

TABLE 1
PROPERTIES OF ISOLATED TEE STUBS

Specimen	t_p (mm)	Φ Bolt (mm)	Displacement test procedure
TM-1	12	16	Monotonic
TC-1	12	16	SDTP - 2
TM-2	12	20	Monotonic
TC-2	12	20	SDTP - 2
TM-3	18	20	Monotonic
TC-3	18	20	SDTP - 2
TM-4	18	24	Monotonic
TC-4	18	24	SDTP - 2
TC-5	25	20	SDTP - 2
TC-6	25	24	SDTP - 2

TABLE 2
PROPERTIES OF COUPLED TEE STUBS

Specimen	t_p (mm)	Φ Bolt (mm)	Column	t_{cf} (mm)	Displacement test procedure
C1B1-1	12	16	HEB180	14	SDTP - 2
C1B2-1	12	16	HEB280	18	SDTP - 2
C1A1-2	12	20	HEA180	9,5	SDTP - 2
C1A2-2	12	20	HEA280	13	SDTP - 2
C1B1-3	18	20	HEB180	14	SDTP - 2
C1B2-3	18	20	HEB280	18	SDTP - 2
C1B1-4	18	24	HEB180	14	SDTP - 2
C1A2-4	18	24	HEA280	13	SDTP - 2

Bursi & Jaspart (1998), Bahaari & Sherbourne (2000)). Despite important continuous advances in the FE analysis, which enable more and more refined approximations to be obtained, the reliability of the results is still remarkably affected by the importance of localised effects (e.g., evolution of plate contact area, bolt-plate interaction, low ductility of HAZ zones near welds, etc.). Besides, the associated burden makes this approach unfeasible in design practice, and its cost efficiency remains rather low even for research purposes. However, the knowledge of the behaviour of endplate connections under static loading was more than adequate to develop and validate design models, which meet the recent demand of semi-continuous frame design.

The behaviour of this type of connection under cyclic loading has been investigated, though at a lower extent than for the static case, and the main features clearly identified, including the stiffness and strength deterioration, the dependance on the loading history and the typical modes of failure (Ghobarah et al. (1990), Ghobarah et al. (1992), Bernuzzi et al. (1996), Kukreti & Biswas (1997), Adey et al. (1998)). The increasing degree of complexity of the phenomena involved with respect to the static case is apparent, as it is the higher difficulty in setting up a model capable to approximate the hysteretic behaviour under any type of loading history, with the accuracy level required in seismic design (Deng et al. (2000)). On the other hand, recent studies pointed out that moment resisting semi-continuous steel frames can possess a satisfactory seismic performance at competitive cost (Nader & Astaneh (1991)), which underlines once more the lack of simplified, yet reliable, design criteria and joint models. Such a limited knowledge hampers the practical use of semi-rigid frames in seismic areas, despite they are accepted by recent Codes (AISC (1997) and Eurocode 8 (2002)).

3. THE TRENTO'S STUDY PROGRAM

The Trento's study was devised to enable checking of the possible extension of the joint modelling technique by components to the approximation of the cyclic response, in view of its application to seismic design. The research work makes use predominantly of experimental analysis, due to the complexity of the problem, which cannot be approached with the necessary accuracy by the available numerical tools. In accordance with the Eurocode 3, the key component, on which the attention is focussed, is assumed to be the T-stub.

The overall program consists, up to now, of 36 tests on three sets of specimens of different complexity in terms of number of components involved: the first set comprises ten isolated tee stub components (Fig. 2a), the second 8 specimens coupling a T-stub with a column section (Fig. 2b), while the third includes 18 complete beam-to-column joints (Fig. 2c). The joints were designed to behave as semi-rigid partial strength joints and to collapse either at the beam-to-column connection or in the column panel zone.

The parameters investigated where:
- The joint geometry in terms of: end plate thickness (12 and 18mm), bolt diameter (16, 20 and 24 mm), column section (HEA180 and 280, HEB180 and 280), and
- The loading history: 7 histories were considered (see Fig. 3) in addition to the monotonic loading.

All bolts are 8.8 grade bolts preloaded to the 40% of the actual yield strength in order to roughly simulate the condition corresponding to the pretension induced by hand tightening up to the snug tight condition.

The main geometrical characteristics of the specimens are collected in Figure 2 and in the Tables 1 - 3, where also the loading history adopted for each specimen is specified.

The geometries of the components (T- stubs, bolts and column section) are coupled in such a way to cover a range of relative component stiffness (T-stub thickness to bolt diameter for the isolated T-stubs and T-stub thickness to column flange thickness for the coupled T-stubs and full joint specimens) of significance for practical interest.

TABLE 3
PROPERTIES OF COMPLETE JOINTS

Specimen	t_p (mm)	Φ Bolt (mm)	Column	t_{cf} (mm)	Displacement test procedure
JA1-2A	12	20	HEA180	9,5	SDTP - 1
JA1-2B	12	20	HEA180	9,5	SDTP - 1
JA1-2C	12	20	HEA180	9,5	SDTP - 4
JA1-2D	12	20	HEA180	9,5	SDTP - 5
JA1-2E	12	20	HEA180	9,5	SDTP - 6
JA1-2F	12	20	HEA180	9,5	SDTP - 3
JA1-2M	12	20	HEA180	9,5	Monotonic
JA1-2R	12	20	HEA180	9,5	SDTP - 7
JB1-3A	18	20	HEB180	14	SDTP - 1
JB1-3B	18	20	HEB180	14	SDTP - 1
JB1-3C	18	20	HEB180	14	SDTP - 4
JB1-3D	18	20	HEB180	14	SDTP - 5
JB1-3E	18	20	HEB180	14	SDTP - 6
JB1-3F	18	20	HEB180	14	SDTP - 3
JB1-3M	18	20	HEB180	14	Monotonic
JB1-3R	18	20	HEB180	14	SDTP - 7
JB1-4	18	24	HEB180	14	SDTP - 1
JA2-4	18	24	HEA280	13	SDTP - 1

TABLE 4
MECHANICAL PROPERTIES OF SPECIMENS

Component		t (mm)	ε_y (%)	ε_u (%)	$\varepsilon_u/\varepsilon_y$	f_y (MPa)	f_{max} (MPa)	f_{max}/f_y
IPE 300	Flange	10,2	0,16	43,58	272,4	307	471	1,53
	Web	7,1	0,17	40,88	240,5	328	477	1,46
HEA 180	Flange	9,4	0,18	38,99	216,6	317	471	1,49
	Web	6,0	0,18	37,14	206,3	373	494	1,32
HEB 180	Flange	13,8	0,12	43,61	363,4	292	478	1,64
	Web	8,3	0,17	43,01	253,0	316	493	1,56
HEA 280	Flange	12,4	0,17	34,61	203,6	413	528	1,28
	Web	8,6	0,23	31,64	137,6	428	545	1,27
HEB 280	Flange	16,7	0,16	34,61	216,3	266	440	1,65
	Web	10,6	0,10	45,39	453,9	270	462	1,71
End Plate		12,5	0,13	43,65	335,8	260	442	1,70
End Plate		18,0	0,16	22,14	138,4	318	441	1,39
End Plate		25,8	0,11	31,44	285,8	262	434	1,66
Weld Metal		5,4	0,19	26,76	140,8	355	489	1,38
Weld Metal*		5,4	0,18	3,9	21,7	441	528	1,20
Bolt		16	0,52	2,59	5,0	813	890	1,10
Bolt		20	0,43	4,50	10,5	888	948	1,07
Bolt		24	0,42	6,25	14,9	816	882	1,08

*Specimen with flaws

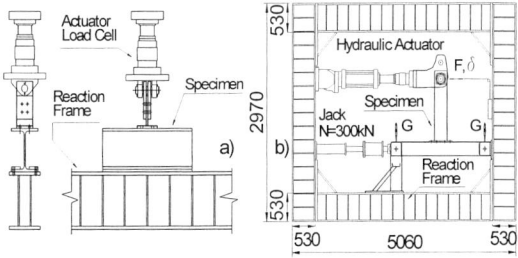

Figure 4. Test set-up and boundary conditions of:
a) Isolated Tee stubs and coupled Tee stubs specimens;
b) Complete Joint specimens

Figure 5. Details of the measuring apparatus for a Complete Joint

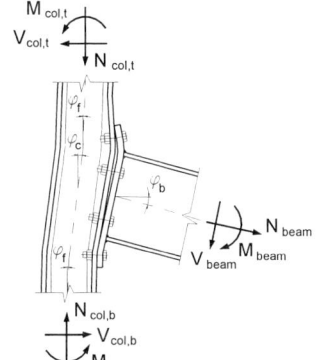

Figure 6. Definition of the rotations for a Complete Joint

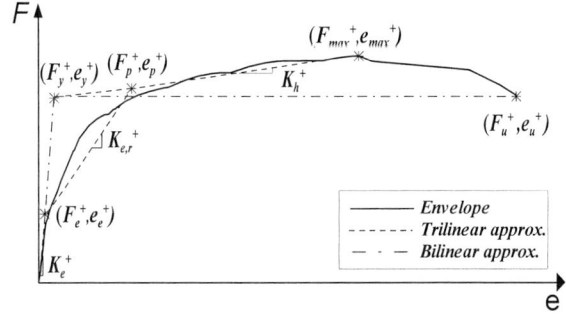

Figure 7. Bi- and trilinear fits of a force-displacement envelope

3.1 Material properties

Particular attention was paid to the mechanical and fracture-related properties of the structural components and of the fasteners (welds and bolts). The relevant results of tension coupon tests conducted on samples from end plates, bolts, profiles and fillets of weld are collected in Table 4. The strength properties of the structural elements are in satisfactory accordance with the nominal values for Fe 430 steel, whose yield strength f_y and ultimate tensile strength f_u are equal to 275 and 430 MPa respectively (Eurocode 3, 1993). The material performance generally meets the Codes' requirements both for strength and deformation capacity: the Eurocode 8 (2002) recommendation on the material over strength is fulfilled (i.e., f_y not higher than the nominal value of more than 37%); the strength f_{max}/f_y ratio exceeds the value of 1.25 specified by the seismic provisions of AISC (1997); the ultimate elongation e_u is greater than 20% and the ultimate material ductility (e_u/e_y) achieves fairly high values.

The performance of structural joints under recent strong earthquakes pointed out the significant importance of weld design and execution. Therefore, the beam stubs were connected to the end plates by means of fillet welds, executed with special care by licenced welders. The arc welding process was adopted with CO_2 shielding gas and no preheating. A filler metal was selected characterized by a nominal yield stress f_y equal to 420 MPa and a nominal ultimate strength f_u of about 520 MPa, respectively. This deliberate strength overmatch by the filler metal aimed at shifting the expected failure planes to the base metal adjacent to the weld. The test values of the yield and ultimate tensile strength for the weld metal samples extracted from virgin specimens comply well with the nominal ones (see Table 4).

A further important characteristic in seismic design is the material is toughness, in particular in the HAZ. In order meet this requirement a filler material was selected with nominal toughness, as obtained through a Charpy V-Notch impact energy test (ASTM, 1988), greater than 70 J at −20 °C and 50 J at −40 °C. Samples were then extracted from the weld metal and from the base metal in both the longitudinal and transverse direction. Three hardness Vichers (HV) tests and one Charpy V-Notch test were carried out on each sample. Tests were performed on samples obtained from both virgin and tested specimens. Results of the HV and CVN tests at ambient temperature of 20 °C meet the recommendations set up on the basis of specific studies carried out in the USA after the Northridge earthquake (FEMA, 1997). The weld metal showed a fracture surface with the internal part exhibiting the typical appearance of the transition from a fragile to a ductile behaviour, whilst the base metal exhibited the ductile fracture typical of low alloy structural steel.

3.2 Test equipment and measuring apparatus

All specimens were tested in the rigid counter frame illustrated in Fig. 4. The individual T-stub specimens were connected directly to the bottom beam of the frame, to which also the column stub of the coupled T-stubs specimens was attached as shown in Fig. 4a. The testing apparatus for the complete joint specimens is depicted in a greater detail in Fig. 4b. The column section is placed in horizontal position, hinged at both ends and subject to a constant moderate axial load of 300 kN. A servo-controlled hydraulic actuator enables application of any displacement sequence to the free end of the beam stub.

The measurement set-up was designed so that the overall response of the connection and the contribution of the main components are observed. The set up for the complete joint tests is illustrated in Fig. 5, which encompasses the set ups of the of the other series of test.

The displacement transducers (LVDTs) enable measurement of the following:
(i) Connection's vertical displacements at the beam flanges (LVDTs A); (ii) Connection's vertical displacements of the beam web at 20 mm above the end plate (LVDTs B); (iii) Horizontal displacements of one beam flange (LVDTs C); (iv) Column upper flange displacements at the end plate edge (LVDTs D); (v) Column web vertical displacements at the toe of the fillet radius (LVDTs E); (vi) Column bottom flange vertical displacements at 20 mm from the column web plane (LVDTs F); (vii) Vertical displacements of the column ends (see LVDTs G in Fig. 4b); (viii) Vertical

displacement of heads of bolts (LVDTs H). All the measurements are taken with respect to a separate reference frame.

Besides, the bolt shank elongation was measured in some few tests by means of LVDTs I as illustrated in Fig. 5b. As a matter of fact, the calibration of LVDTs I by means of companion bolts under a universal testing machine allows not only axial displacements but also bolt forces to be obtained.

The global response of a complete joint can be described by the moment-rotation relationship. The rotation of the joint can be determined as

$$\phi_j = \phi_b - \phi_f - \phi_r \qquad (1)$$

where ϕ_b represents the rotation of the beam at the end plate level (Fig. 6), ϕ_f denotes the elastic deformation of the column whilst ϕ_r denotes the rigid rotation of the column owing to the flexibility of the equipment supporting the column. Such a rigid rotation is detected by means of LVDTs G (Fig. 4b). Besides, the overall joint rotation ϕ_j defined in Eq.(1), the measuring apparatus allows the contributions of the connection (ϕ_{con}) and of the web panel zone (γ) to be estimated (Fig. 6):

$$\phi_{con} = \phi_b - \phi_c \qquad (2)$$

$$\gamma = \phi_c - \phi_f - \phi_r \qquad (3)$$

The rotation components due to the sole connection's elements (end plate and bolts), the column flange distortion and the column panel local deformation due to the pull-push action of the beam flange can also be determined (Deng et al. (1998)).

3.3 Testing Procedure

In accordance with the twofold purpose of the study, different loading histories were adopted in the cyclic tests, with histories consisting either of random amplitude sequences or of sets of constant amplitude cycles considered in addition to the ECCS approach (1986, fig. 3a). The latter enables assessment of the full range of the response, while the others provide the necessary data to calibrate damage models. Monotonic displacement histories were also applied to 6 specimens (four isolated T-stubs and 2 complete joints), identified by M in Tables 1 and 3. The monotonic tests aim to: (i) allow the performances of similar specimens under monotonic and cyclic loading to be compared (ii) define the conventional yielding displacement e_y on the monotonic load-displacement response F-e (Fig. 7) in accordance with the ECCS recommendations; (iii) complete the set of data needed for applying effectively the damage criteria.

The value of e_y can be defined on the basis of a bilinear approximation of the actual monotonic response curve or of a response envelope in the case of cyclic loading, as shown in Fig. 7. In the latter case, the bilinear approximation derives from a trilinear one suitably defined by imposing the condition of equal energy dissipation under the restraint of best fitting of the actual non-linear skeleton response. Thereby, stiffness, strength and displacement related parameters characterizing the specimen behaviour can be defined.

The ECCS sequential-phased displacement history considers alternate cycles of increasing displacement amplitude with at least three cycles at the same amplitude (SDTP-1 in fig. 3a). Such a sequence is similar to the one proposed by ATC 24 (1992). This law was adopted for all the cyclic tests on T-stubs (isolated and coupled) and for 6 complete joints (Table 3). The configuration of the isolated T-stubs imposed to modify the loading history by eliminating the reversals (SDTP-2 in Fig. 3b). The other loading histories considered (SDTP-2 to - 7 in fig. 3) are 'built up' to appraise the influence on the cyclic joint response, and in particular on the damage evolution, of different sequences of cycle amplitudes, simulating some features of the seismic input. Tests were carried out quasi-statically. Therefore, an increase in the ductility capacity compared to equivalent specimens loaded dynamically would be expected, based on the fact that fracture toughness of steel decreases with the strain rate growth. However, tests on welded beam-to-column connections performed by Suita et. al. (1998) pointed out the equivalence between quasi-static and dynamic tests with regard to ductility. Moreover, the strength as well as the absorbed energy is larger for dynamic loading than for quasi-static loading confirming that quasi-static test procedures lead to a conservative appraisal of these key parameters.

4. MAIN RESULTS AND FIRST SEISMIC ASSESSMENT

The results of the study are briefly presented in this section, focused on the experimental outcomes and their first assessment in the perspective of seismic design. The three sets of tests are here considered separately. The significance of these results in view of the validation of the component method is dealt with next.

4.1 Isolated T-Stubs

The applied load versus the upward displacement curves of the T-stub web represent an important overall indicator of the specimen's behaviour to be associated with the failure mode in order to understand the influence of the parameters investigated, i.e., the plate thickness, the bolt diameter and their ratio.

Figure 8 illustrates a typical T-stub response with reference to the case of 12 mm end plate thickness t and 20mm diameter bolts. Both the monotonic and cyclic responses are plotted, showing that the envelope of the latter one lies very close to the monotonic curve for a first significant portion of the loading process. Progressive strength deterioration is then occurring associated to the increase of end plate plastic deformations and damage at the weld toe.

As to the ultimate limit state, the TM-2 specimen was characterized by a collapse mechanism with four yield lines located at the bolt holes and at the weld toes in accordance with (i.e., the Mode 1 failure in the Eurocode 3 (2001)). The corresponding specimen subject to cyclic loading, TC-2, experienced premature plate fractures at a hot spot of plastic strain concentration located at weld toes. As a result, cyclic loading reduces significantly both the ultimate strength and the displacement ductility factors. Close observation of the plate failure reveals that brittle fracture evolved in three sequential phases: i) the initiation of a ductile crack at the steel surface owing to plastic strains; ii) a stable growth of a ductile crack in the plate thickness; iii) a sudden propagation of the crack in a brittle fracture mode. However, test indicates that the fillet welds performed well and were able to develop the required cyclic strength.

Similar considerations apply to the specimens with plate thickness equal to 18 mm. As expected, the specimen TM-3 was characterized by a Mode 2 failure according to Eurocode 3, with two yield lines located at the weld toes, whilst specimen TC-3 failed by plate fracture close to the weld toes following the same sequence described above for the TC-2 specimen.

A comparative assessment of the hysteretic performances of TC-2 and TC-3 specimens can be based (ECCS (1986)) on the relation between the mean energy ratio (i.e., the ratio between the absorbed energy, averaged on three cycles, and the energy dissipated per complete cycle by an equivalent elastic-plastic oscillator) and the maximum ductility e/e_y. Such a comparison pointed out that: (i) due to the Mode 1 failure pattern which involves four yield lines, TC-2 specimen is able to absorb greater amounts of energy also at smaller ductility ratios, (ii) the ultimate displacement ductility factor e_u^+/e_y^+ (see Fig. 7) are 73 and 53 for the TC-2 and TC-3 specimen, respectively. Thereby, the TC-2 specimen performs better both in terms of absorbed energy and ductility.

4.2 Coupled Tee Stubs

Different inelastic mechanisms were observed in coupled T-stub tests, associated with the different relative stiffness and strength of the relevant components: the T-stub, the column flange and the bolts. The total thickness of the connected parts (i.e., column flange and T-stub plate) also seems to affect the response.

Figure 9 compares the responses of specimens C1A1-2 and C1A2-2, different for the column section, which is a HEA180 in the former and a HEA280 in the latter test. It is apparent that specimen C1A1-2

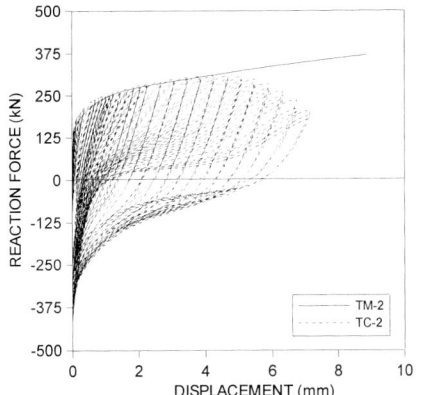

Figure 8. Experimental response of the Isolated T-Stubs TM-2 and TC-2

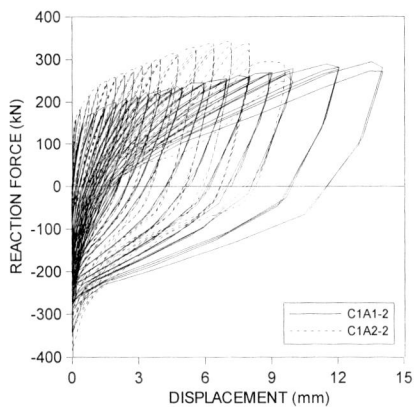

Figure 9. Experimental response of the Coupled T-Stubs C1A1-2 and C1A2-2

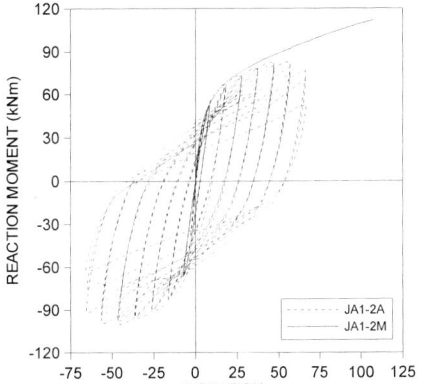

Figure 10. Experimental response of the Complete Joints JA1-2A and JA1-2M

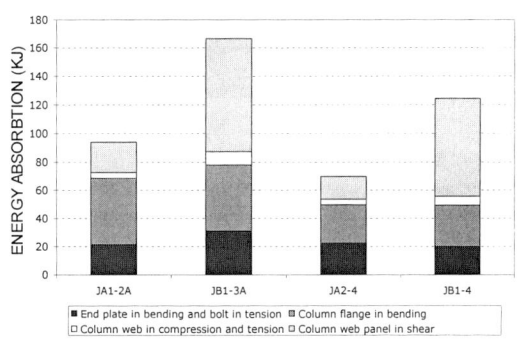

Figure 11. Shares of hysteretic energy dissipated among the components of some Complete joints

TABLE 5
ELASTIC STIFFNESS K_e AND PLASTIC FAILURE STENGHT F_p OF THE COMPONENTS

	EXPERIMENTAL K_e (kN/mm)					EUROCODE 3 $K_{e,code}$ (kN/mm)			$\frac{K_e}{K_{e,code}}$		
Component	TC-2	C1A1-2	JA1-2A	JA1-2B	JA1-2M	TC-2	C1A1-2	JA1-2	TC-2	C1A1-2	JA1-2
1	---	---	748	205	206	---	---	685	---	---	0,3
6	---	744	334	236	633	---	776	776	---	1,0	0,5
7	1393	795	1979	428	2539	378	378	378	3,7	2,1	4,4
Component	TC-3	C1B1-3	JB1-3A	JB1-3B	JB1-3M	TC-3	C1B1-3	JB1-3	TC-3	C1B1-3	JB1-3
1	---	---	266	286	256	---	---	796	---	---	0,3
6	---	1417	869	568	1093	---	3790	3790	---	0,4	0,2
7	3003	2363	2507	345	1931	1020	1020	1020	2,9	2,3	1,6
	EXPERIMENTAL F_p (kN)					EUROCODE 3 $F_{p,code}$ (kN)			$\frac{F_p}{F_{p,code}}$		
Component	TC-2	C1A1-2	JA1-2A	JA1-2B	JA1-2M	TC-2	C1A1-2	JA1-2	TC-2	C1A1-2	JA1-2
1	---	---	193	235	279	---	---	288	---	---	0,8
6	---	184	226	229	228	---	326	326	---	0,6	0,7
7	234	201	264	258	257	146	146	146	1,6	1,4	1,8
Component	TC-3	C1B1-3	JB1-3A	JB1-3B	JB1-3M	TC-3	C1B1-3	JB1-3	TC-3	C1B1-3	JB1-3
1	---	---	344	352	469	---	---	346	---	---	1,1
6	---	322	388	422	379	---	564	564	---	0,6	0,7
7	417	359	417	423	402	396	396	396	1,1	0,9	1,0

Components: 1. Column web panel in shear; 6. Column flange in bending; 7. end plate in bending is characterized by large energy absorption and displacement ductility, due to the extensive plastic deformation occurring in both the T-stub (t =12mm) and the column flange (t_f = 9.5mm). Failure was attained by brittle fracture of the T-stub weld toes after crack propagation. The C1A2-2 specimen has a thicker column flange (t_f =13mm), which caused inelastic phenomena to concentrate in the T- stub only. This results in lower displacement ductility and energy absorption, whilst pinching phenomena appear in last cycles.

As to the corresponding two 18 mm plate thickness specimens: (i) specimen C1B1-3 with a HEB180 column section (t_f =14mm) experienced inelastic phenomena in the sole column flange, which resulted in a limited energy absorption and maximum displacements capability, while (ii) in the specimen C1B2-3 with a HEB280 profile (t_f =18mm) plastic phenomena occurred both in the T-stub and in the column flange, enabling larger displacement ductilities to be achieved; pinching phenomena appeared in the last cycles, when brittle failure develop at the weld toes of the tee stub.

Pinching somehow depends on the relative thickness of the connection plate to the column flange, and increased with the increase of this ratio. However, all the coupled T-stub specimens showed less important pinching than the isolated corresponding isolated T-stubs.

The C1A1-2 specimen, which is associated with the coupling of two 'thin' elements with relatively heavy bolts, showed the best performance in terms of mean energy ratio. On the other hand, the coupling of a thick T-stub with a column flange of similar thickness (C1B2-3 specimen) ensues the lowest energy ratio. As to the positive ultimate displacement ductility ratios, the experimental values of 50 for the specimen C1A1-2 compares with the value of 28 for the specimen C1B2-3.

4.3 Complete Joints

Based on the results of the coupled T-stub tests, the attention are focused on joints JA1-2 and JB1-3, which appear to be more adequate in terms of seismic design requirements. In this section the monotonic tests will be discussed as well as the cyclic tests adopting the ECCS displacement procedure. The subsequent sets of 10 tests adopting the different procedures shown in Figure 3 aimed at providing data for the calibration of the damage models considered, and will not be discussed in this paper. The monotonic tests confirmed the high rotation capacity coupled with extensive yield of the main joint components. The cyclic moment- rotation response of joint JA1-2a is plotted in Fig. 10. Similar behavioural features were observed for the joint JB1-3A, and for the twin specimens JA1-2B and JB1-3B. These joints achieve values of plastic rotation greater than 35 mrad, implying a satisfactory ductile behavior for seismic applications (Astaneh-Asl (1995), Eurocode 8 (2002)). It should be noted that the contribution of the column web panel is in all cases significant (Fig. 11). Failure of all specimens occurred at weld toes in the extension of the end plate by fragile crack propagation. This indicates a satisfactory behaviour of the fillet welds. The mean energy ratios as well as the ultimate ductility ratios indicate a satisfactory cyclic performance. All joints can reach mean energy ratios greater than 0.5, while ultimate rotational ductilities $\phi_{j,u}/\phi_{j,y}$ range from 8 to 21. Joint JA2-4, which also reached plastic rotations greater than 30 mrad, was characterized by a mean energy ratio below 0.5.

As to the joint classification, the response envelopes lie within the semi-rigid range of Eurocode 3 (1993) for unbraced frames, when typical beam lengths ranging from 4 to 8 m are taken into account. Furthermore, all joints can be classified as partial strength joints.

5. VALIDATION OF THE COMPONENT METHOD

The complete set of measured parameters provides detailed information on the responses of the various components to the applied displacement history. These data enable a quantitative appraisal of the role played by each component to be obtained, pointing out the effect of the components' relative stiffness and strength associated with the specimen geometry and material. A thorough analysis of these results is out of the scope of the paper. The attention is hence focused on the use of these data for checking the general validity of the joint model by component with reference to the approximation of the cyclic response.

A first appraisal can be achieved by comparing the response of the same individual component as part of different specimens (isolated T-stub, coupled T-stub and complete joint). Such a comparison is made somehow difficult by the type of displacement history adopted, which determines the amplitude of the inelastic cycles as a multiple of the elastic limit displacement. This parameter depending on the specimen, a direct comparison of cyclic responses related to different specimens is not practicable. However, reference can be made to the maximum and minimum values of parameters such as forces and displacements, to ductility indexes and to the absorbed energy.

The endplate represents an important component for the joints considered. In Figure 12 the endplate responses are plotted of the isolated T-stub TC-2 and joint JA1-2A. As mentioned above, the testing procedure imposes different cycle amplitudes and affects the number of cycles at failure. However, the collapse mode is the same (i.e., by plate fracturing) and the evolvement of the hysteretic behaviour has similar features. A comparison of the maximum force and displacement show that reference to the T-stub response would imply an overestimate of the first parameter of 7% and an underestimate of the second of 13%. The influence of the interaction in the joint between the different components appears to be limited. The same comparison for joints JA1-2B, JB1-3, JB1-4 and JA2-4 with the corresponding isolated and coupled T-stubs showed similar results.

A further appraisal of the responses of the elemental components can be obtained if reference is made to key parameters, such as the initial elastic stiffness and the plastic failure strength, which characterize the envelope of the cyclic response. A conventional elastic stiffness K_e, a plastic failure strength F_p and an ultimate displacement ductility factor e_u/e_y were determined (see Fig. 7) by the bi- and tri-linear approximations of the envelope curve, traced on the basis of best-fitting and dissipated energy-equivalence criteria. The values of the elastic stiffness K_e for the components of joints JA1-2 and JB1-3 and of the corresponding isolated and coupled T-stubs are presented in Table 5. The elemental components have been considered, which give the most significant contribution. Joints JA1-2 and JB1-3 have been selected for their satisfactory performance and as representative of joints with thin and thick extended endplates respectively. Differences among the stiffness of the same elemental components as part of different specimen are noticeable. However, it has to be noted that the initial stiffness is very sensitive to boundary conditions and lack of fit. Table 5 gathers also the elastic stiffness $K_{e,code}$ computed in accordance to Eurocode 3 (2002). The stiffness ratios $K_e/K_{e,code}$ lie in a wide range, and the level of accuracy of the EC3 models does not seem satisfactory.

As to the experimental plastic failure strength F_p, as defined in accordance with the tri-linear approximation of the response envelope, the differences for the same component in different specimens are more limited than for the initial stiffness. Such differences are a consequence also of the interaction which affects the location and evolution of plastic zones. As to the Eurocode 3 prediction model, the plastic strength, $F_{p,code}$, was computed using measured material properties and no resistance factors. These strength values are also collected in Table 5. The Eurocode underestimates significantly the strength of thin end-plates, while tends to overestimate, even remarkably, most of the other components. This seems to depend on the coupling effects among different components, which are not considered in the code. Moreover, non seismic codes do not consider the development of low cycle fatigue phenomena, which lead to initiation and propagation of cracks, and affect the yield lines sequence.

A further important comparison referred to the absorbed energy, and was limited to the 6th elemental component "Column flange in bending" and the 7th elemental component "End flange in bending", which dissipate the most of the energy within the connection zone. The comparison is performed with reference to the relation between the mean energy ratio and the displacement ductility in the i-th cycle (ECCS, 1986). The elemental component "Column flange in bending" for component parts and joints embodying a thin (12mm) extended endplate showed similar mean energy ratios (approaching values of about 0.7) in the coupled T-stubs and in the complete joints, but ultimate positive partial ductilities e_u^+/e_y^+ vary between 13 and 23, with the higher value related to the coupled T-stubs. As to the 7th component, the ultimate partial ductility ranges from 9 (complete joint) to 13 (isolated T-stub), while the mean energy ratios vary between 0.6 (isolated T-stub) to 0.8 (coupled T-stub). A similar comparison for the 6th and 7th elemental components in joints embodying thick extended end plate (t=18 mm) shows greater differences of ultimate displacement ductility factors e_u^+/e_y^+ and maximum values of the mean energy ratios.

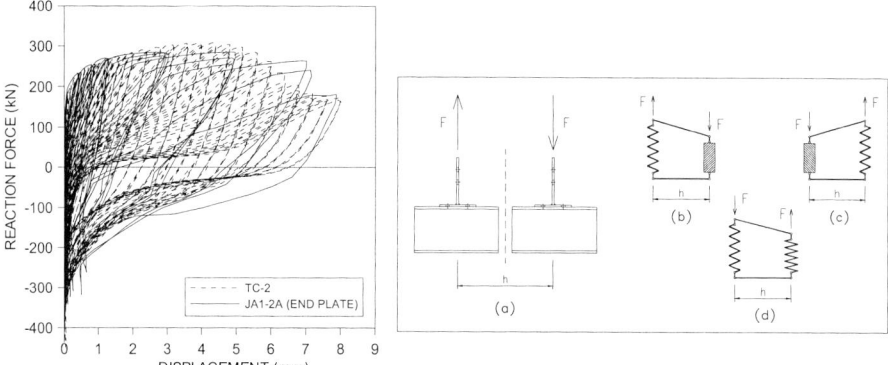

Figure 12. Experimental response of TC-2 and of the relevant component of JA1-2A

Figure 13. A Macrocomponent model of a Complete Joint without the "Column web panel in shear"

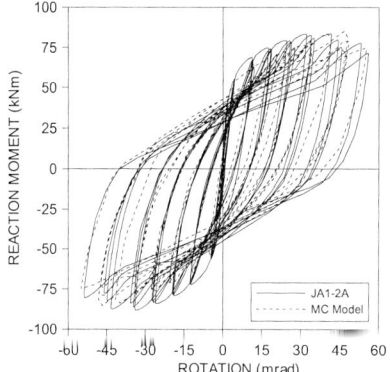
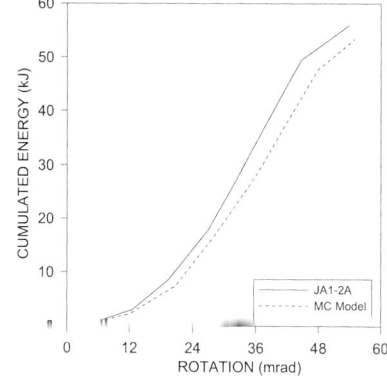

Figure 14. Response of the joint JA1-2A and of the relevant Macrocomponent model

Figure 15. Dissipated Energy of the joint JA1-2A and of the relevant Macrocomponent model

An evaluation of these results leads to consider the 'component method' not sufficiently accurate for approximating the full cyclic response, at least for the joint configuration considered in the study. The extension of this model for use in seismic analysis does not seem straightforward, when based on elemental components defined as in the Eurocodes. An alternative within the same philosophy was then explored, employing 'macrocomponents', as the coupled T-stubs, which take into account the influence of the key components in the compression and tension zone of the joint as well as of their interactions. The model of the connection zone is schematically represented in figure 13. The connection response in joints JA1-2A and JB1-3A were simulated via this model and compared with the experimental behaviour, modified for achieving consistency. Figure 14 shows such a comparison for the joint JA1-2A. The overall agreement all over the response is quite apparent, and is confirmed by the curves of the cumulated energy (Fig. 15). The macrocomponent model appears hence, from these first results, a viable tool for approximating the cyclic response of endplate joints.

6. SUMMARY AND CONCLUSIONS

The paper briefly presented the main outcomes of a research work of the cyclic behaviour of extended endplates beam-to-column joints. The study was mainly experimental and comprises of tests on full joints as well as on individual and coupled joint components. The attention was focussed on the responses and on the possible modelling approach. In particular, a first appraisal was attempted of the so-called 'component model', which was successfully developed and validated for static analysis.
Some few indications of general interest can be here summarized:

- When proper consideration is given to material selection and detailing, extended endplate joints show a cyclic performance adequate for seismic design. The joints considered exhibit a plastic rotation greater than 35mrad; therefore, they can be classified as 'ductile' in accordance to Eurocode 8. Besides, the contribution of the individual components to the hysteretic energy was determined, and dissipative components clearly identified.
- Cyclic tests on dissipative components enable identification of the failure mode and appraisal of the ultimate ductility. They allow hence the Code requirements on the rotational capacity of the joint to be checked.
- A component model, which approximates the full cyclic response of the joints on the basis of the responses of the elemental components, does not seem to possess sufficient accuracy for seismic analysis.
- The use of macrocomponents, incorporating some of the interaction effects among elemental components, appears to be more adequate.

These results are not definitive, and further analyses are in progress. The Authors are also exploring the numerical facet of the component method, which would advantageously reduce the computational effort required to perform parametric studies of joint response. Finally, the study is currently concentrating on the damage assessment aspect both for the components (individual and coupled) and the joint.

7. REFERENCES

Adey B.T., Grondin G.Y. and Cheng J.J.R. (1998), *Extended End Plate Moment Connections under Cyclic Loading*, J. of Construct. Steel Res., 46(1-3), Elsevier Applied Science, UK, pp. 435-436.
American Institute of Steel Construction (1992), *Seismic provisions for structural steel buildings*, Chicago, USA.
American Society of Testing and Materials (1988), *Standard test methods for notched bar impact testing of metallic materials*, E23-88, Philadelphia, USA.
Applied Technology Council (1992), *Guidelines for cyclic testing of components of steel structures*, Report 24, p. 57.
Astaneh-Asl A. (1995), *Seismic design of bolted steel moment-resisting frames*, in Steel Tips, Structural Steel Education Council, Moraga, USA, p. 82.
Bahaari M. R. and Sherbourne A.N. (2000), *Behavior of eight-bolt large capacity endplate connections*, Computers & Structures, vol 77, Pergamon Press, UK, pp. 315-325.
Bernuzzi, C., Zandonini, R. and Zanon, P. (1996), *Experimental analysis and modelling of semi-rigid steel joints under cyclic reversal loading*, J. of Construct. Steel Res., vol. 38(2), Elsevier Applied Science, UK, pp. 95-123.
Bjorhovde R., Brozzetti J. and Colson A.(1988), *Connections in Steel Structures: Behaviour, Strength & Design*, Elsevier Applied Science, London, UK, p. 395.
Bjorhovde R., Colson A., Haaijer G. and Stark J.W.B. (1992), *Connections in Steel Structures II: Behaviour, Strength & Design*, AISC, Chicago, USA, p. 464.
Bjorhovde R., Colson A. and Zandonini R. (1996), *Connections in Steel Structures III: Behaviour, Strength & Design*, Pergamon Press, London, UK, p.594.
Bose B., Sarkar S. and Bahrami M. (1996), *Extended endplate connections: comparison between*

three-dimensional nonlinear finite element analysis and full scale tests, Structural Engineering Review, vol. 8, No. 4, Pergamon Press, UK, pp 315-328.

Bursi O.S. and Jaspart J.-P. (1998), *Basic issues in the finite element simulation of extended end plate connections*, Computers & Structures, vol 69, Pergamon Press, UK, pp. 361-382.

Bursi, O.S., and Jaspart, J-P. (1997), *Calibration of a finite element model for isolated bolted end plate steel connections*, J. of Construct. Steel Res., 44(3),Elsevier Applied Science, UK, pp. 225-262.

Colson A. (1992), *First State of the Art Workshop,* EU Project COST C1: Semi-rigid Behaviour of Civil Engineering Structural Connections, European Commission, Belgium, p. 582.

Deng C.G., Bursi O.S. and Zandonini R. (1998), *Analysis of Moment Resisting Steel Connections under Reversed Cyclic Loading*, Proc. of the Fourth Int. Conf. on Computational Structures Technology, Edinburgh, UK.

Easterling S.W. and Leon R.T.,(2002), *Connections in Steel Structures IV: Behaviour, Strength & Design*, AISC, Chicago, USA, p. 473.

European Committee for Standardization (1993), *Eurocode 3: Design of Steel Structures. Part 1.1: General Rules and Rules for Buildings*, European Prestandard – ENV 1993-1-1, Belgium.

European Committee for Standardization (2002), *Eurocode 8: Design of structures for earthquake resistance Part 1: General rules, seismic actions and rules for buildings,*
Draft No. 5, Doc. CEN/TC250/SC8/N317, European Pr-EN 1998-1, Belgium, p. 213.

European Committee for Standardization (2001), *Eurocode 3 Design of Steel Structures. Part 1.8: Design of Joints*, European Prestandard – prEN 1993-1-8, Belgium, p.124.

European Committee for Standardization (2002), Eurocode 4: Design of composite steel and concrete structures, Part 1.1 General rules and rules for buildings, European Prestandard – prEN 1994-1-1, CEN/TC250/SC4 N259, Belgium, p.122.

FEMA. (1997), *Background reports: Metallurgy, fracture mechanics, welding, moment connections and frame system behavior*, Report 288, SAC (SEAOAC-ATC-CUREe) Joint Venture, USA, p. 315.

Ghobarah, A., Osman, A., and Korol, R.M. (1990),*Behavior of extended end plate connections under cyclic loading*, Engineering Structures, vol.12(1), pp.15-27.

Ghobarah, A., Korol, R.M., and Osman, A. (1992), *Cyclic behavior of extended end plate joints,* J. Struct. Engr., ASCE, USA, vol. 118(5), pp.1333-1353.

Jaspart J.P. (2002), *Design of structural joints in building frames*, Progress in Structural Engineering and Materials, vol. 4, N.o 1, Wiley, UK, pp. 18-34.

Kukreti A.R. and Biswas P. (1997), *Finite element analysis to predict the cyclic hysteretic behavior and failure of end-plate connections*, Computers & Structures, vol 65, Elsevier Science Ltd., UK, pp. 127-147.

Lorenz R.F., Kato B. and Chen W.F. (1993), *Semi-rigid Connections in Steel Frames,* Council on Tall Buildings and Urban Habitat, Mc Graw Hill, USA, p. 318.

Maquoi R. (1999), *Control of the semi-rigid behaviour of civil engineering structural connections*, Proc. of the Int. Conference, EU Project COST C1, Publication EUR 18854 EN, Belgium, p. 579.

Nader M.N. and Astaneh-Asl A. (1991), *Dynamic behavior of flexible, semi-rigid and rigid steel frames,* J. of Construct. Steel Res., vol. 18, Elsevier Science Ltd., UK, pp.179-192.

Nemati N.,Le Houedec D., Zandonini R. (2000), *Numerical modelling of the cyclic behaviour of the basic components of steel end plate connections*, Advances in Engineering Software, vol. 31, Elsevier Science Ltd., UK, pp. 837-849.

Narayanan R. (1988), *Structural Connections,* Stability and Strength series, Elsevier Applied Science, UK, p. 452.

Nethercot D. & Zandonini R. (1988), *Methods of prediction of joint behaviour* – Beam-to-column connections, Structural Connections – Stability and Strength , ed. R. Naranayan, Elsevier Applied Science, London, UK, pp. 23-62.

Suita, K., Nakashima, M., and Morisako, K. (1998). *Tests of welded beam-column subassemblies. II: Detailed behavior*, ASCE, J. Struct. Engr., vol. 124, No. 11, pp.1245-1252.

Wald F. (1994), *Second State of the Art Workshop,* EU Project COST C1: Semi-rigid Behaviour of Civil Engineering Structural Connections, European Commission, Belgium, p. 561.

DESIGN OF STEEL ARCHES AGAINST IN-PLANE INSTABILITY

M.A. Bradford and Y.-L. Pi

School of Civil and Environmental Engineering, The University of New South Wales,
UNSW, Sydney, NSW 2052, Australia

ABSTRACT

Steel arches may be subject to instability caused by in-plane buckling or out-of-plane flexural-torsional buckling. The in-plane buckling of flat steel arches is addressed in this paper, where the method of virtual work is used to determine the buckling loads of circular arches with either pinned or fixed ends and subjected to a uniform loading distributed radially around the arch, and to a point load at the crown. The formulation includes the nonlinear prebuckling configuration of the arch, in deference to the classical stability formulations that treat the prebuckling response as being linear. Design formulae for the elastic buckling loads of arches are derived, as are the limits of the arch modified slenderness that delineates between antisymmetric and symmetric (snap-through) buckling modes.

KEY WORDS

Antisymmetric, arches, buckling, elasticity, shallow arch, snap-through, symmetric, virtual work.

INTRODUCTION

Steel arches have been deployed in engineering structures over a long period of time, yet the analysis of their in-plane behaviour is quite complex and surprisingly little understood. The complexity of the analysis is most profound for shallow arches, and this is no doubt a consequence of the more widespread use of concrete arches that are usually not shallow, and for which the analysis is not as difficult. Steel arches may be subjected to instability by either in-plane buckling (Pi and Bradford 2002) or out-of-plane buckling (Bradford et al. 2002). This paper is concerned with the in-plane buckling of steel arches.

In routine buckling analysis, the buckling load that is sought is the load at the bifurcation from a prebuckling equilibrium path to an orthogonal buckling equilibrium path. Classical buckling theory (Trahair and Bradford 1998) considers the prebuckling equilibrium path to be linear, and so the stress resultants can be linearised, and this routine approach is suitable for the elastic buckling of columns, beams and frames. However, an arch in its prebuckling configuration is under axial compression and bending, and it may experience considerable transverse deformations prior to buckling. These deformations are quite nonlinear, and their effects need to be taken into account when considering the in-plane buckling of a steel arch.

For arbitrary loading, numerical techniques such as the finite element method need to be utilised to determine the buckling load, but many of these eigenvalue-type formulations treat the prebuckling behaviour as being linear. Discrepancies between classical solutions based on linear elastic analysis and test results have been identified (Gjelsvik and Bodner 1962, Dickie and Broughton 1971), as have discrepancies between finite element solutions and test results (Pi and Trahair 1998, Pi and Bradford 2002). These discrepancies arise owing to the linearisation of the prebuckling equilibrium path. It is not widely recognised that classical buckling theory cannot correctly predict the in-plane buckling load of shallow arches, and this paper presents elastic buckling loads that are sufficiently simple to be used for design, but which are based on nonlinear theory (Pi and Bradford 2002, Bradford et al. 2002). The analysis here is for circular arches that are either pinned or fixed, and that are restrained against out-of-plane (lateral) buckling. The analysis is also elastic, and considers an arch that is subjected to a central concentrated load at the crown, or to a load distributed uniformly around the arch in the radial direction, as shown in Fig. 1. It considers both antisymmetric (Fig. 2(a)) and symmetric (Fig. 2(b)) buckling modes.

EQUATIONS OF EQUILIBRIUM

In the virtual work procedure used by Pi and Bradford (2002) and Bradford et al. (2002), the membrane and bending strains at a point P on the arch are written as

$$\varepsilon_m = w' - \frac{v}{R} + \frac{1}{2}\left(v' + \frac{w}{R}\right)^2 ; \quad \varepsilon_b = -y\left(v'' + \frac{w'}{R}\right) \tag{1a,b}$$

where primes denote differentiation with respect to the arch coordinate s shown in Fig. 3. The nonlinear finite element results reported by Pi and Trahair (1998) indicate that the nonlinearity is largely attributable to large transverse displacements, and the axial displacements are quite small prior to buckling, and so their effects may be ignored. Because of this, the method considers the longitudinal normal strain ε at a point P to be the sum of the membrane and bending strains respectively in Eqn. 1, but simplified as

$$\varepsilon_m = w' - \frac{v}{R} + \frac{1}{2}(v')^2 ; \quad \varepsilon_b = -yv'' \tag{2a,b}$$

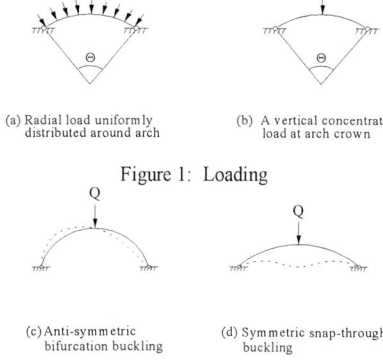

q
Θ
(a) Radial load uniformly distributed around arch
Q
Θ
(b) A vertical concentrated load at arch crown

Figure 1: Loading

Q
(c) Anti-symmetric bifurcation buckling
Q
(d) Symmetric snap-through buckling

Figure 2: Arch buckling modes

The principle of virtual work then requires that

$$\delta\Pi = \int_V \delta\varepsilon\sigma \, \mathrm{d}V - \delta W = 0 \tag{3}$$

for all sets of kinematically admissible virtual displacements δv and δw, where

$$\delta W = Q\delta v_0 \tag{4}$$

if the load is a point load Q at the crown ($s = 0$), or where

$$\delta W = q \int_{-S/2}^{S/2} \delta v \, \mathrm{d}s \tag{5}$$

S
o
f
y
Θ
s
R
L
(a) Geometry
Q
N
Θ
N

Figure 3: Arch geometry

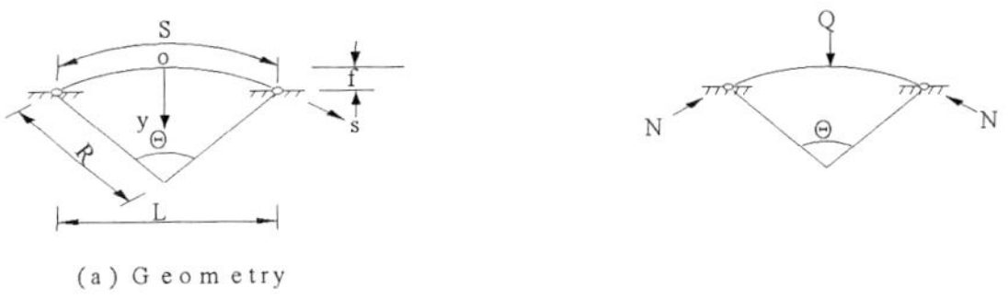

if the circular arch is loaded by the (uniform) radial load q. When $\sigma = E\varepsilon$, where E is Young's modulus of elasticity, is substituted into Eqn. 3 and the integration is performed by parts, it produces the differential equilibrium equation

$$-AE\varepsilon'_m = 0 \qquad (6)$$

in the axial direction, and

$$EIv^{iv} - AEv''\varepsilon_m - AE\frac{\varepsilon_m}{R} - AEv'\varepsilon'_m = 0 \qquad (7)$$

in the radial direction. Eqns. 6 and 7 may be solved by substituting the appropriate boundary conditions (Pi and Bradford 2002, Bradford et al. 2002). These produce

$$\frac{v^{iv}}{\mu^2} + v'' = \Omega \qquad (8)$$

in which

$$\mu^2 = \frac{\overline{N}}{EI_x} \qquad (9)$$

where $\overline{N}$ is the actual axial force in the arch (as distinct from its nominal value determined from linear elastic analysis), and where

$$\Omega = \frac{-1}{R}; \qquad \Omega = \frac{q}{\overline{N}} - \frac{1}{R} \qquad (10a,b)$$

for the point loading and radial loading cases, respectively.

The differential equilibrium equations may be solved when the boundary conditions appropriate to pinned and fixed arches are used, and these equations relate the external loading (q or Q) to the actual compression $\overline{N}$ within the arch. The equilibrium equations may be found in Pi and Bradford (2002) and Bradford et al. (2002).

BUCKLING EQUATIONS

The neutral equilibrium state defined is defined by

$$\delta^2 \Pi = 0 \qquad (11)$$

for the buckling displacements $v_b = \delta v$ and $w_b = \delta w$ that take place from the prebuckling equilibrium position v and w to the adjacent buckled position $v + v_b$ and $w + w_b$ under constant load. This requires vanishing of the functional

$$\delta^2\Pi = \int F(w'_b, v_b, v'_b, v''_b)\,ds = 0 \qquad (12)$$

where

$$F = \frac{1}{2}\left[AE\left(\varepsilon_{mb}^2 + \varepsilon_m v'^2_b\right) + EI v''^2_b\right] \qquad (13)$$

in which the membrane strain during buckling is

$$\varepsilon_{mb} = \delta\varepsilon_m = w'_b - \frac{v_b}{R} + v'v'_b \qquad (14)$$

Invoking the Euler-Lagrange equations of variational calculus for Eqns. 12 and 13 produces

$$\varepsilon'_{mb} = 0 \qquad (15)$$

for the axial direction (so that the membrane strain during buckling is constant), and

$$v_b^{iv} + \mu^2 v''_b = \frac{\varepsilon_{mb}}{r_x^2}\left(\frac{1}{R} + v''\right) \qquad (16)$$

in which $r_x^2 = I/A$. Eqn. 16 may be variously solved for fixed and pinned arches, using the equilibrium conditions derived in Eqn. 8 with appropriate boundary conditions.

Pin ended shallow arches

For antisymmetric buckling of pin-ended arches, it can be shown (Pi and Bradford 2002, Bradford et al. 2002) that at buckling

$$\sin(\mu S/2) = 0 \qquad (17)$$

which has the fundamental solution

$$\frac{\mu S}{2} = \pi \qquad (18)$$

so that the actual compressive force in the arch during antisymmetric buckling is

$$N_P - \overline{N} = \frac{\pi^2 EI}{(S/2)^2} \qquad (19)$$

which is the familiar second mode buckling load N_P of a pin-ended column under uniform compression. When this is substituted into the appropriate nonlinear equilibrium equation for distributed radial loading, then

$$(2\pi+15)\bar{q}_{ant}^2 + (4\pi^2+12)\bar{q}_{ant} + \frac{12\pi^4}{\lambda_s^2} = 0 \tag{20}$$

where

$$\bar{q} = \frac{qR - \bar{N}}{\bar{N}} \tag{21}$$

while substituting Eqn. 19 into the equilibrium equation for a point load produces

$$3\bar{Q}_{ant}^2 - 8\bar{Q}_{ant} + \pi^2 - \frac{2\pi^4}{3} + \frac{4\pi^6}{\lambda_s^2} = 0 \tag{22}$$

where

$$\bar{Q} = \frac{\pi^2 Q}{\Theta N_P} \tag{23}$$

in which Θ is the included angle (Fig. 3) and the arch modified slenderness parameter is

$$\lambda_s = \frac{\Theta(S/2)}{2} \frac{S^2}{r_x} = \frac{S^2}{4r_x R} \tag{24}$$

Solving Eqn. 20 for an arch under a radial load and its counterpart Eqn. 22 for an arch with a point load produces, for antisymmetric buckling

$$q_{ant} \approx \left(0.26 \pm 0.74\sqrt{1 - 0.63\frac{\pi^4}{\lambda_s^2}}\right)\frac{N_P}{R} \tag{25}$$

and

$$Q_{ant} \approx \left(1.33 \pm 4.5\sqrt{1 - 0.65\frac{\pi^4}{\lambda_s^2}}\right)\frac{\Theta N_P}{\pi^2} \tag{26}$$

Real antisymmetric buckling solutions of Eqns. 25 and 26 exist when $\lambda_s \geq 7.83$ for a radial load and $\lambda_s \geq 7.96$ for a point load, respectively.

Quadratic equations of the type in Eqns. 20 and 22 may be established for symmetric snap-through buckling of shallow pin-ended arches (Pi and Bradford 2002, Bradford et al. 2002). When this is done, the symmetric mode counterparts to Eqns. 25 and 26 are

$$q_{sym} \approx \left(0.15 + 0.006\lambda_s^2\right)\frac{N_P}{R} \tag{27}$$

and

$$Q_{sym} \approx \left(1 + 0.03\lambda_s^2\right)\frac{\Theta N_P}{\pi^2} \tag{28}$$

When the symmetric and antisymmetric buckling solutions are solved simultaneously, it can be shown that the switch between symmetric and antisymmetric buckling takes place at $\lambda_s \approx 9.38$ for a radial load and $\lambda_s \approx 9.80$ for a point load, and so both antisymmetric and symmetric buckling are possible in the range $7.83 \leq \lambda_s \leq 9.38$ for a radial loading and $7.96 \leq \lambda_s \leq 9.80$ for a point load. It was shown by Pi and Bradford (2002) that symmetric buckling occurs first when $7.83 \leq \lambda_s \leq 9.38$ for a radial load, and then antisymmetric buckling occurs on the descending branch of the load-deflection curve. In a similar fashion, Bradford et al. (2002) showed that the same situation takes place when $7.96 \leq \lambda_s \leq 9.80$ for a point load.

The radial displacement at the crown of a pinned arch with a radial loading can be obtained from the equilibrium formulation in Eqn. 8. Its value at buckling can be determined when $\mu S/2 \rightarrow \pi/2$ as

$$\lim_{\mu S/2 \rightarrow \pi/2} v_c = \lim_{\mu S/2 \rightarrow \pi/2} \frac{\bar{q}}{\mu^2 R}\left[\sec(\mu S/2) - 1 - \frac{(\mu S/2)^2}{2}\right] = \frac{4S^2}{\pi^3 R}\left(1 \pm \sqrt{1 - \frac{\pi^6}{64\lambda_s^2}}\right) \tag{29}$$

which is complex when $\lambda_s < \pi^3/8$. Because of this, a radially loaded arch does not buckle when $\lambda_s < 3.88$, and it can be shown similarly for a point load that buckling does not occur when $\lambda_s < 3.91$.

Fixed-ended shallow arches

Fixed arches may be treated in a similar way to pinned arches. For antisymmetric buckling, the characteristic equation becomes

$$\tan(\mu S/2) - \mu S/2 \tag{30}$$

which has the lowest solution

$$\frac{\mu S}{2} \approx 1.4303\pi \tag{31}$$

and so the actual axial force at buckling is

$$N_F = \overline{N} \approx \frac{(1.4303\pi)^2 EI}{(S/2)^2} \tag{32}$$

where N_F is the second mode buckling load of a fixed column under uniform compression. The buckling equation (that corresponds to Eqn. 20) for a uniform load distributed radially is

$$5\overline{q}_{ant}^2 + 4\overline{q}_{ant} + \frac{12(1.4303\pi)^2}{\lambda_s^2} = 0 \tag{33}$$

which can be solved to give

$$q_{ani} \approx \left(0.6 \pm 0.4\sqrt{1 - 30.686\frac{\pi^2}{\lambda_s^2}}\right)\frac{N_F}{R} \tag{34}$$

and which is has a real solution for antisymmetric buckling when $\lambda_s \geq 17.40$. The buckling solution for an arch under a point load (that corresponds to Eqn. 22) is

$$6.22\overline{Q}_{ant}^2 - 13.98 \cdot (1.4303\pi)\overline{Q}_{ant} + \frac{(1.4303\pi)^4}{3} + \frac{4 \cdot (1.4303\pi)^6}{\lambda_s^2} = 0 \tag{35}$$

which can be solved to give

$$Q_{ant} \approx 1.4303\pi \left(1.12 \pm 0.18\sqrt{1 - 15\frac{\pi^4}{\lambda_s^2}}\right)\frac{\Theta N_F}{\pi^2} \tag{36}$$

and which has a real solution when $\lambda_s \geq 38.15$. Eqn. 36 can be reduced to that of Schreyer and Masur (1966) for a fixed shallow arch with a rectangular solid cross-section subjected to a central concentrated load.

Quadratic equations may again be established for symmetric buckling of the two types of loaded arch. Under radial loading, symmetric buckling may occur when $\lambda_s \leq 18.60$ at the load

$$q_{sym} = \frac{(0.36 + 0.0011\lambda_s^2)N_F}{R} \tag{37}$$

and for which it can be shown that buckling is not possible when $\lambda_s \leq 9.87$.

For a fixed arch subjected to a concentrated load at the crown, on the other hand, when the expressions for the symmetric and antisymmetric buckling loads are equated, it is not possible to produce a real-value solution for λ_s. This indicates that symmetric buckling always governs. The buckling load for symmetric buckling is very complicated, and a simplified value suitable for design is

$$Q_{sym} \approx \begin{cases} (3.30 + 0.17\lambda_s - 0.002\lambda_s^2)(\Theta N_F / \pi^2) & 11 < \lambda_s \leq 38 \\ (5.88 + 0.03\lambda_s - 0.0001\lambda_s^2)(\Theta N_F / \pi^2) & \lambda_s > 38 \end{cases} \qquad (38)$$

Furthermore, it can be shown that the buckling displacement at the crown is complex when $\lambda_s < 11.07$, so that a fixed arch subjected to a concentrated load at the crown will not buckle when its modified slenderness is less than 11.07.

VALIDATION OF THEORETICAL SOLUTION

The value of the theoretical treatment is that it has presented and solved a generic model for the buckling of shallow arches, which is lost when computer studies are performed. However, in order to validate the theoretical solutions and design approximations, numerical calibration has been undertaken with the finite element package ABAQUS (1998) as well as the numerical model of Pi and Trahair (1998). This calibration was undertaken by Pi and Bradford (2002) for arches subjected to radial loading, and by Bradford et al. (2002) for arches with a concentrated load at the crown. These studies have shown very good correlation between the design equations developed in this paper and the numerical solutions.

CONCLUSIONS

The analytical treatment of buckling considered in this paper has produced design equations for the elastic in-plane buckling of shallow arches. For pinned arches under a radial loading, symmetric (snap-through) buckling takes place first with the modified arch slenderness in the range $3.88 \leq \lambda_s \leq 9.38$ and antisymmetric buckling first when $\lambda_s > 9.38$, while for pinned arches with a point load at the crown, symmetric buckling takes place first when $3.91 \leq \lambda_s \leq 9.80$ and antisymmetric buckling first when $\lambda_s > 9.80$. For fixed arches under a radial loading, symmetric buckling takes place first in the range $9.87 \leq \lambda_s \leq 17.4$ and antisymmetric buckling first when $\lambda_s > 17.4$, while for fixed arches with a point load at the crown only symmetric buckling takes place first when $\lambda_s > 11.07$. Arches with modified slenderness values outside of these ranges do not buckle elastically in-plane.

ACKNOWLEDGEMENT

The work reported in this paper was supported by the Australian Research Council under its Large Grants Scheme.

REFERENCES

ABAQUS Standard Users Manual Version 5.8 (1998). Hibbit, Karlsson and Sorensen Inc., Abaqus, Pawtucket, Rhode Island, USA.

Bradford M.A., Uy, B and Pi, Y.-L. (2002). In-plane elastic stability of arches under a central concentrated load. *Journal of Engineering Mechanics*, ASCE, **128**:7, 710-719.

Dickie, J.F. and Broughton, P. (1971). Stability criteria for shallow arches. *Journal of the Engineering Mechanics Division*, ASCE, **97**:EM3, 951-965.

Gjelsvik, A. and Bodner, S.R. (1962). Energy criterion and snap-through buckling of arches. *Journal of the Engineering Mechanics Division*, ASCE, **88**:EM5, 87-134.

Pi, Y.-L. and Bradford, M.A. (2002). In-plane stability of arches. *International Journal of Solids and Structures*, **39**:1, 105-125.

Pi, Y.-L. and Trahair, N.S. (1998). Non-linear buckling and postbuckling of elastic arches. *Engineering Structures*, **20**:7, 571-579.

Schreyer, H.L., and Masur, E.F. (1966). Buckling of shallow arches. *Journal of the Engineering Mechanics Division*, ASCE, *92*:EM4, 1-17.

Trahair, N.S. and Bradford, M.A. (1998). *The Behaviour and Design of Steel Structures to AS4100*, 3rd Australian edn, E&FN Spon, London.

FEM ANALYSIS OF STEEL MEMBERS CONSIDERING DAMAGE ACCUMULATION EFFECTS UNDER CYCLIC LOADING

Z. Y. Shen and Z. S. Song

State Key Laboratory for Disaster Reduction in Civil Engineering, Tongji University, Shanghai, 200092, China

ABSTRACT

Due to the steady growth of micro level deterioration, the properties of structural steel, such as yield stress, Young's modulus and hardening coefficient will deteriorate under cyclic loading, and fatigue failure maybe take place at a low number of cycles. Based on continuum damage theory and experimental results, a cumulative damage mechanics model, as measured by effective plastic strain, is suggested. A general FEM analysis procedure on damage cumulation is presented and a damage crack criterion is proposed. Cantilever members subjected to cyclic loading are numerically simulated with a FEM program which takes damage cumulation into consideration. The analytical results agree very well with the experiments, which demonstrates that the cumulative damage mechanics model presented in the paper is accurate enough for simulating the behavior of steel structures under seismic loading.

KEYWORDS

low cycles fatigue, damage cumulation, finite element method

INTRODUCTION

The properties of structural steel, such as yield stress, Young's modulus and hardening coefficient will deteriorate under cyclic loading. For understanding the performance of steel structures under cyclic loading, a damage mechanics method has been used to study the damage and its cumulation in steel members. Due to the complication of factors related to damage, different definitions of the damage index D have been proposed. Kachanov (1986) was the first one who pointed out that the damage can be described as the ratio of effective area to real area. Rabotnov (1969) defined the damage index by the ratio of damaged area to real area. Park and Ang (1985), Park, et al (1985) used a linear combination of deformation and energy to express the damage index. Based on the proposal of Park and Ang (1985), Kumar and Usami (1994) suggested an improved expression. As for the rule for the cumulation of damage, linear and nonlinear cumulation equations have been proposed by Miner, Chaboche and Coffin-Manson (Yu and Feng, 1997).

But all of these results are very difficult to use in practical structural analysis. Recently, based on experimental results, Shen and Dong (1997) suggested a uniaxial cumulative damage mechanics model for steel subjected to cyclic loading, which can be applied in steel structural analysis. In this paper, the model is broadened to fit triaxial stress conditions and a FEM analysis procedure taking the damage cumulation into consideration is presented.

CUMULATIVE DAMAGE MECHANICS MODEL OF STRUCTURAL STEEL UNDER CYCLIC LOADING

The model established by Shen and Dong (1997) can be expressed as follows.

The damage index D is computed according to the plastic strain and hysteretic energy dissipation of steel

$$D = (1-\beta)\frac{\varepsilon_m^p}{\varepsilon_u^p} + \sum_{i=1}^{N} \beta \frac{\varepsilon_i^p}{\varepsilon_u^p} \tag{1}$$

where N is the total number of half cycles which cause plastic strain, β is the weighted value of the ith half cycle, ε_i^p is the plastic strain during the ith half cycle, ε_m^p is the largest plastic strain during all half-cycles and ε_u^p is the ultimate plastic strain of the material.

The effects of damage on Young's modulus, yield strength and strain hardening coefficient can be expressed as

$$E^D = (1-\xi_1 D)E_0 \tag{2}$$

$$\sigma_s^D = (1-\xi_2 D)\sigma_{s0} \tag{3}$$

$$k = k_0 + \xi_3 \sum_{i=1}^{N} \frac{\varepsilon_i^p}{\varepsilon_u^p} \tag{4}$$

where E_0 and E^D are the Young's modulus in respect of $D=0$ and D respectively, σ_{s0} is the initial yield stress when $D=0$, σ_s^D is the yield stress in respect of D, k_0 and k are the hardening coefficients in respect of $D=0$ and D and ξ_1, ξ_2, ξ_3 are material parameters.

Figure 1 is the hysteretic model of steel allowing for damage cumulation given by Shen and Dong

(1997)

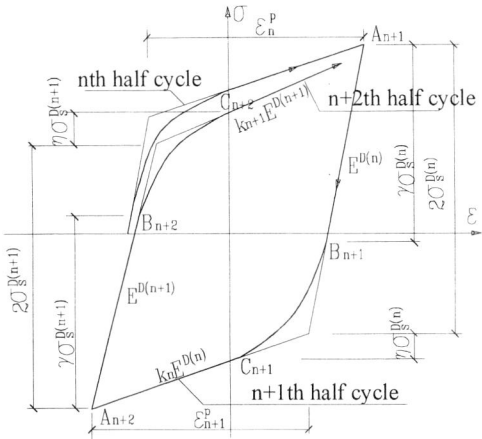

Figure 1: Hysteretic model of steel considering damage cumulation

The hysteretic models of steel shown in Figure 1 and Eqn. 1 to Eqn. 4 are derived from cyclic loading tests in one-dimension. For the condition of triaxial stresses and proportional loading, the above results can also be valid if ε_i^p and ε_u^p are substituted with equivalent plastic strain p_i and p_u, and σ is substituted with equivalent stress σ_{eq}.

$$D = (1-\beta)\frac{p_m}{p_u} + \beta \sum_{i=1}^{N} \frac{p_i}{p_u} \tag{5}$$

$$p_i = \int_{ith-halfcycle} dp \tag{6}$$

$$dp = \left(\frac{2}{3} d\varepsilon_{ij}^p d\varepsilon_{ij}^p\right)^{1/2} \tag{7}$$

$$\sigma_{eq} = \sqrt{\frac{1}{2}\left[(\sigma_1-\sigma_2)^2 + (\sigma_2-\sigma_3)^2 + (\sigma_3-\sigma_1)^2\right]} \tag{8}$$

According to the energy damage theory, the relationship between p_u and ε_u^p is (Yu and Feng, 1997)

$$p_u = \varepsilon_u^p / R_v \tag{9}$$

R_v is called the triaxial stress index, which reflects the effect of triaxial stress ratio.

$$R_v = \frac{2}{3}(1+v) + 3(1-2v)(\frac{\sigma_m}{\sigma_{eq}})^2 \tag{10}$$

where, v is the Poisson ratio of the material, σ_m is the hydrostatic stress and $\sigma_m = \sigma_{kk}/3$. If $\sigma_1 = \sigma_2 = \sigma_3$, from Eqn. 8, then $\sigma_{eq} = 0$ and $R_v = \infty$. In this case, the damage index should not be calculated using Eqn. 5.

Based on the plastic damage theory, the release rate of damage strain energy can be expressed as (Yu and Feng, 1997)

$$-Y = \frac{1}{2E(1-D)^2}\left[\frac{2}{3}(1+v)\sigma_{eq}^2 + 3(1-2v)\sigma_m^2\right] \qquad (11)$$

If $\sigma_m = \sigma_1 = \sigma_2 = \sigma_3$, $\sigma_{eq} = 0$, then Eqn. 11 becomes

$$-Y = \frac{3(1-2v)\sigma_m^2}{2E(1-D)^2} \qquad (12)$$

and for uniaxial tension, since $\sigma_{eq} = \sigma$ and $\sigma_m = \sigma/3$, Eqn. 11 becomes

$$-Y = \frac{\sigma^2}{2E(1-D)^2} \qquad (13)$$

Assuming that the value of $-Y$ is the same when damage fracture occurs in uniaxial and triaxial conditions, then the hydrostatic stress σ_m at fracture is

$$\sigma_{mu} = \frac{\sigma_u}{\sqrt{3(1-2v)}} \qquad (14)$$

where, σ_u is the ultimate strength of the material.

Let damage index be 1 at fracture and assuming its value is proportional to the hydrostatic stress, then the increment of damage index under triaxial stress conditions can be calculated from

$$\Delta D = \frac{\Delta \sigma_m}{\sigma_{mu}} = \frac{\Delta \sigma_m \sqrt{3(1-2v)}}{\sigma_u} \qquad (15)$$

In practical analysis, Eqn. 15 will be employed if R_v is larger than a specified number.

CONSTITUTIVE RELATION OF STRUCTURAL STEEL WITH DAMAGE CUMULATION

If the material is of the Von Mises-type, the subsequent yield function can be expressed as

$$F = f - h \qquad (16)$$

$$f = \frac{1}{2} s_{ij} s_{ij} \qquad (17)$$

$$h = \frac{1}{3} \sigma_s^2(p) \qquad (18)$$

where, s_{ij} is the deviatoric stress tensor and $\sigma_s(p)$, i.e. σ_s^D is the current yield stress.

It is assumed that the total strain tensor can be separated into elastic and plastic components, ε_{ij}^e and ε_{ij}^p, respectively, and the differential of the plastic strain tensor is given by the plastic flow rule

$$d\varepsilon_{ij}^p = d\lambda \frac{\partial f}{\partial \sigma_{ij}} \qquad (19)$$

By differentiating Eqn. 16, the following expression is obtained

$$\frac{\partial f}{\partial \sigma_{ij}} d\sigma_{ij} - \frac{2}{3} \sigma_s^D \frac{d\sigma_s^D}{dp} dp = 0 \qquad (20)$$

where

$$\frac{\partial f}{\partial \sigma_{ij}} = s_{ij} \qquad (21)$$

$$\frac{d\sigma_s}{dp} = E^p \qquad (22)$$

E^p is the plastic modulus of the material, which may be expressed as

$$E^p = \frac{E^D E^{tD}}{E^D - E^{tD}} = \frac{E^D \cdot kE^D}{E^D - kE^D} = \frac{k}{1-k} E^D \qquad (23)$$

where E^{tD} is the tangent modulus in respect of the damage index D.

Substituting Eqn. 19 into Eqn. 7, then

$$dp = (\frac{2}{3}d\varepsilon_{ij}^p d\varepsilon_{ij}^p)^{1/2} = d\lambda \, (\frac{2}{3}\frac{\partial f}{\partial \sigma_{ij}}\frac{\partial f}{\partial \sigma_{ij}})^{1/2} \qquad (24)$$

and Hook's law of material with cumulative damage may be written as

$$d\sigma_{ij} = \widetilde{C}_{ijkl} d\varepsilon_{kl}^e = \widetilde{C}_{ijkl}(d\varepsilon_{kl} - d\varepsilon_{kl}^p)$$

$$= \widetilde{C}_{ijkl} d\varepsilon_{kl} - \widetilde{C}_{ijkl} d\lambda \frac{\partial f}{\partial \sigma_{kl}} \qquad (25)$$

where

$$\widetilde{C}_{ijkl} = \frac{E^D}{1+v}\delta_{ik}\delta_{jl} + \frac{v \cdot E^D}{(1+v)(1-2v)}\delta_{ij}\delta_{kl} \qquad (26)$$

δ_{ij} is the Kronecker delta.

Putting Eqn. 3, 22, 23, 24, 25 into Eqn. 19, the scalar function $d\lambda$ can be solved

$$d\lambda = \frac{\frac{\partial f}{\partial \sigma_{ij}}\widetilde{C}_{ijkl} d\varepsilon_{kl}}{\frac{\partial f}{\partial \sigma_{ij}}\widetilde{C}_{ijkl}\frac{\partial f}{\partial \sigma_{kl}} + \frac{2k}{3(1-k)}(1-\xi_1 D)(1-\xi_2 D)\sigma_{s0} E_0 \left(\frac{2}{3}\frac{\partial f}{\partial \sigma_{ij}}\frac{\partial f}{\partial \sigma_{ij}}\right)^{1/2}} \qquad (27)$$

substituting Eqn. 27 into Eqn. 25, then

$$d\sigma_{ij} = \widetilde{C}_{ijkl}^{ep} d\varepsilon_{kl} \qquad (28)$$

where

$$\widetilde{C}_{ijkl}^{ep} = \widetilde{C}_{ijkl} - \widetilde{C}_{ijkl}^{p} \qquad (29)$$

$$\widetilde{C}_{ijkl}^{p} = \frac{\widetilde{C}_{ijmn}\frac{\partial f}{\partial \sigma_{mn}}\frac{\partial f}{\partial \sigma_{rs}}\widetilde{C}_{rskl}}{\frac{\partial f}{\partial \sigma_{ij}}\widetilde{C}_{ijkl}\frac{\partial f}{\partial \sigma_{kl}} + \frac{2k}{3(1-k)}(1-\xi_1 D)(1-\xi_2 D)\sigma_{s0} E_0 \left(\frac{2}{3}\frac{\partial f}{\partial \sigma_{ij}}\frac{\partial f}{\partial \sigma_{ij}}\right)^{1/2}} \qquad (30)$$

$\widetilde{C}_{ijkl}^{p}$ is named as plastic tensor.

NONLINEAR FINITE ELEMENT FORMULATIONS BASED ON CUMULATIVE DAMAGE MECHANICS MODEL

Taking damage cumulation, elastoplasticity and large displacements into account, the updated Lagrangian finite element formulation can be written

$$\int_{t_V} \tilde{C}^{ep}_{ijkl}\, _te_{kl}\delta\, _te_{ij}\,^t dV + \int_{t_V}\, ^t\overline{\sigma}_{ij}\delta\, _t\eta_{ij}\,^t dV = ^{t+\Delta t}R - \int_{t_V}\, ^t\overline{\sigma}_{ij}\delta\, _te_{ij}\,^t dV \tag{31}$$

where

$$^t\overline{\sigma}_{ij} = \frac{^t\sigma_{ij}(1-{}^tD-\Delta D)}{1-{}^tD} \tag{32}$$

$$_te_{ij} = \frac{1}{2}(_tu_{i,j} + _tu_{j,i}) \tag{33}$$

$$_t\eta_{ij} = \frac{1}{2}\, _tu_{k,i}\, _tu_{k,j} \tag{34}$$

and u_i is the displacement at node i

FEM ANALYSIS OF STEEL MEMBERS SUBJECTED CYCLIC LOADING

To verify the cumulative damage mechanics model and constitutive relations suggested, a nonlinear program was developed in which a 20-nodes solid element was employed and used to provide numerical results which were compared with the results obtained from experiments conducted by Chen (2000).

Description of Test Specimens

To understand the effects of damage cumulation on steel members, two series of cyclic tests, S1 and S2 respectively, were conducted in the State Key Laboratory for Disaster Reduction in Civil Engineering, Tongji University. Outline of a specimen is shown in Figure 2 and design dimensions are shown in TABLE 1.

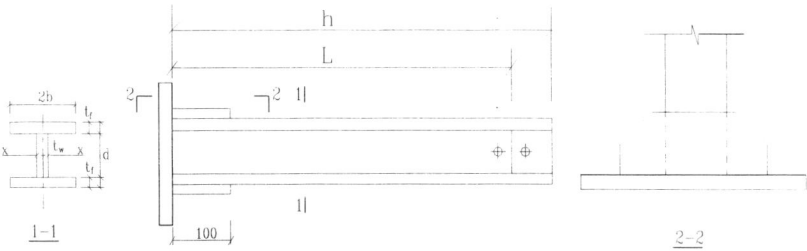

Figure 2: Outline of specimen

TABLE 1
DESIGN DIMENSIONS OF TEST SPECIMENS

Specimen	H (mm)	L (mm)	b (mm)	d (mm)	t_f (mm)	t_w (mm)
S1	1300	1100	54	140	6	6
S2	1300	1100	42	120	6	6

Cyclic loading was applied to the top of the specimen along the weak direction and the testing procedure was controlled by the displacement measured at that point, which is shown as Figure 3.

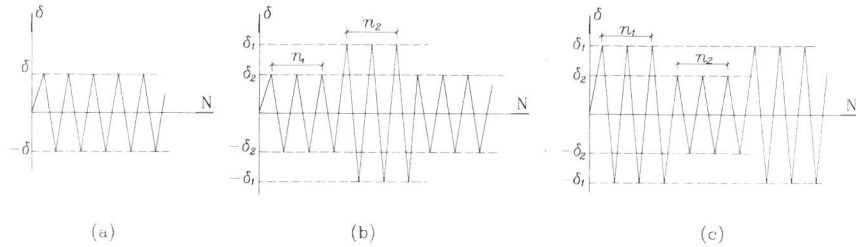

Figure 3: Testing Procedure

S1-1: constant displacement amplitude, δ=50 mm;
S1-2: constant displacement amplitude, δ=60 mm;
S1-3: constant displacement amplitude, δ=40 mm;
S1-4: inconstant displacement amplitude, δ=50 mm(5 cycles)～δ=60 mm(5 cycles)～δ=50 mm(to failure);
S1-5: inconstant displacement amplitude, δ=60 mm(5 cycles)～δ=50 mm(5 cycles)～δ=60 mm(to failure);
S2-1: inconstant displacement amplitude, δ=40 mm(5 cycles)～δ=60 mm～δ=40 mm(to failure);
S2-2: constant displacement amplitude, δ=50 mm;
S2-3: constant displacement amplitude, δ=60 mm;
S2-4: constant displacement amplitude, δ=65 mm;
S2-5: inconstant displacement amplitude, δ=70 mm(5 cycles)～δ=65 mm(5 cycles)～δ=70 mm(to failure);

Finite Element Subdivision

To simplify the computation, only that part above the stiffened plate at the bottom of the column was considered in the FEM analysis. The boundary conditions are assumed as fixed. Due to symmetry, the finite element model was generated for only half of the specimen and necessary constraints are imposed on the plane of symmetry. The material properties of steel and the parameters in the cumulative damage model used in the FEM analysis are listed in TABLE 2 and TABLE 3, respectively.

TABLE 2
MATERIAL PROPERTIES OF STEEL

σ_{s0}(Mpa)	σ_u(Mpa)	E_0(Mpa)	ε_u^p (%)
310	445	196784	24.185

TABLE 3

PARAMETERS IN THE CUMULATIVE DAMAGE MODEL

β	ξ_1	ξ_2	ξ_3	k_0	γ	η
0.0081	0.227	0.119	0.000073	Eqn. 35	1.44	0.041

$$k_0 = 0.014 - 0.165|\varepsilon_m| + 1.12|\varepsilon_m^2| - 2.88|\varepsilon_m^3| \tag{35}$$

Criterion of Damage Crack and Treatments

Damage in steel was found to along with the increase of plastic strain and the number of cycles. This increase is irreversible. According to the cumulative damage mechanics model presented in this paper, a damage crack occurs at a gauss point if the value of the damage index at this point is greater than 1.

Comparisons between FEM and Experimental Results

The number of half cycles during which a damage crack was initiated is shown in TABLE 4, and hysteretic curves of FEM analysis and experiment before the crack occurs are shown in Figure 4 to Figure 13

TABLE 4

NUMBER OF HALF CYCLES DURING WHICH DAMAGE CRACK INITIATED

Specimen	Calculated values	Experimental results
S1-1	25	24
S1-2	20	14
S1-3	40	32
S1-4	22	12
S1-5	21	16
S2-1	51	48
S2-2	44	16
S2-3	30	24
S2-4	22	16
S2-5	16	8

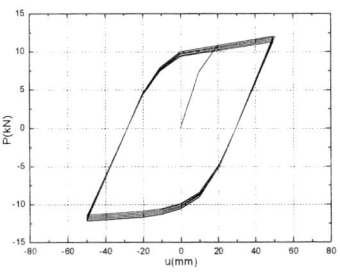
(a) experimental curve

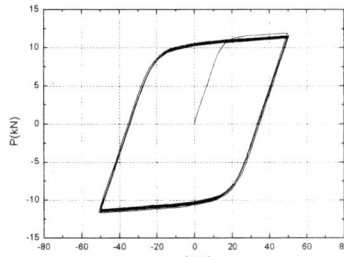
(b) FEM curve

Figure 4: Hysteretic curves of specimen S1－1

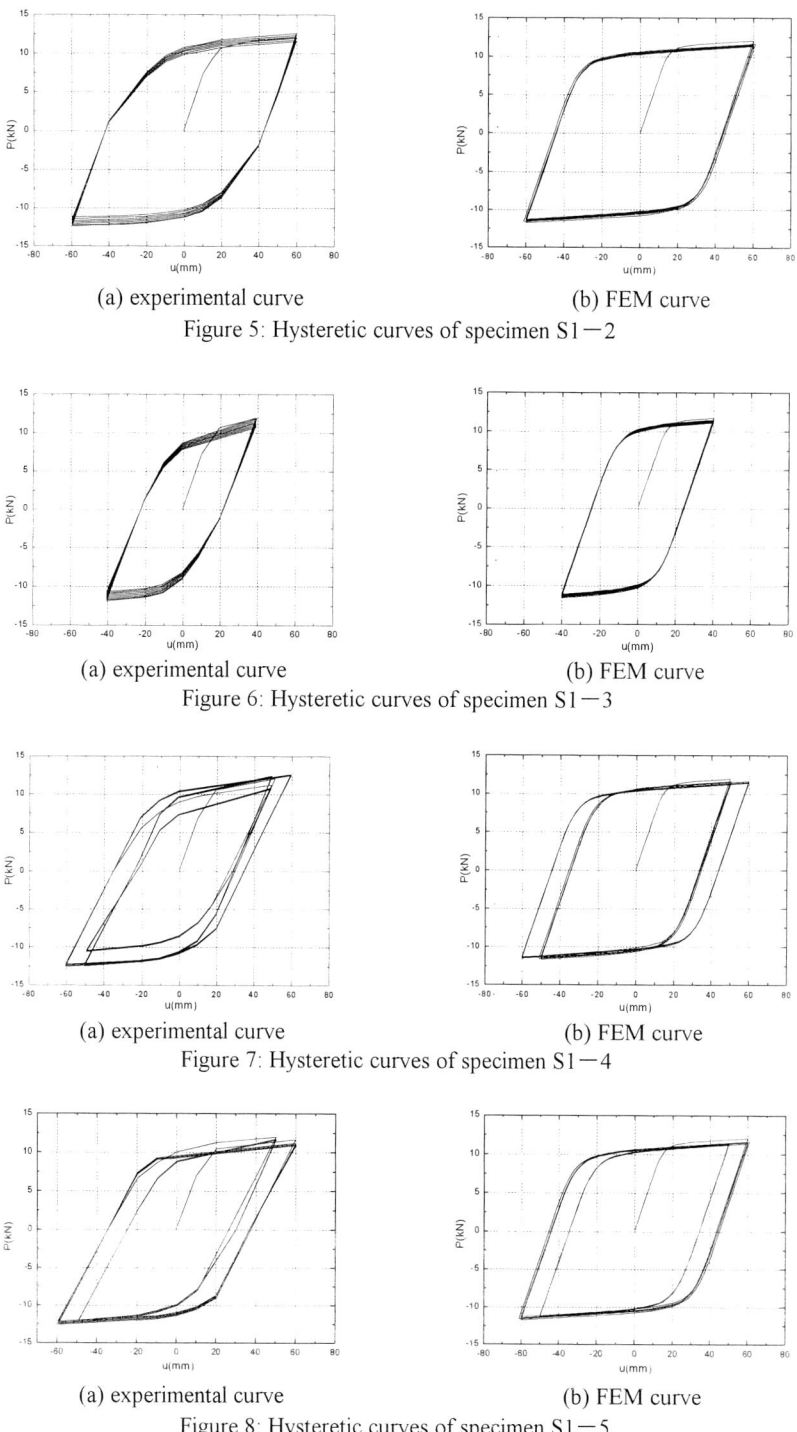

(a) experimental curve   (b) FEM curve
Figure 5: Hysteretic curves of specimen S1－2

(a) experimental curve   (b) FEM curve
Figure 6: Hysteretic curves of specimen S1－3

(a) experimental curve   (b) FEM curve
Figure 7: Hysteretic curves of specimen S1－4

(a) experimental curve   (b) FEM curve
Figure 8: Hysteretic curves of specimen S1－5

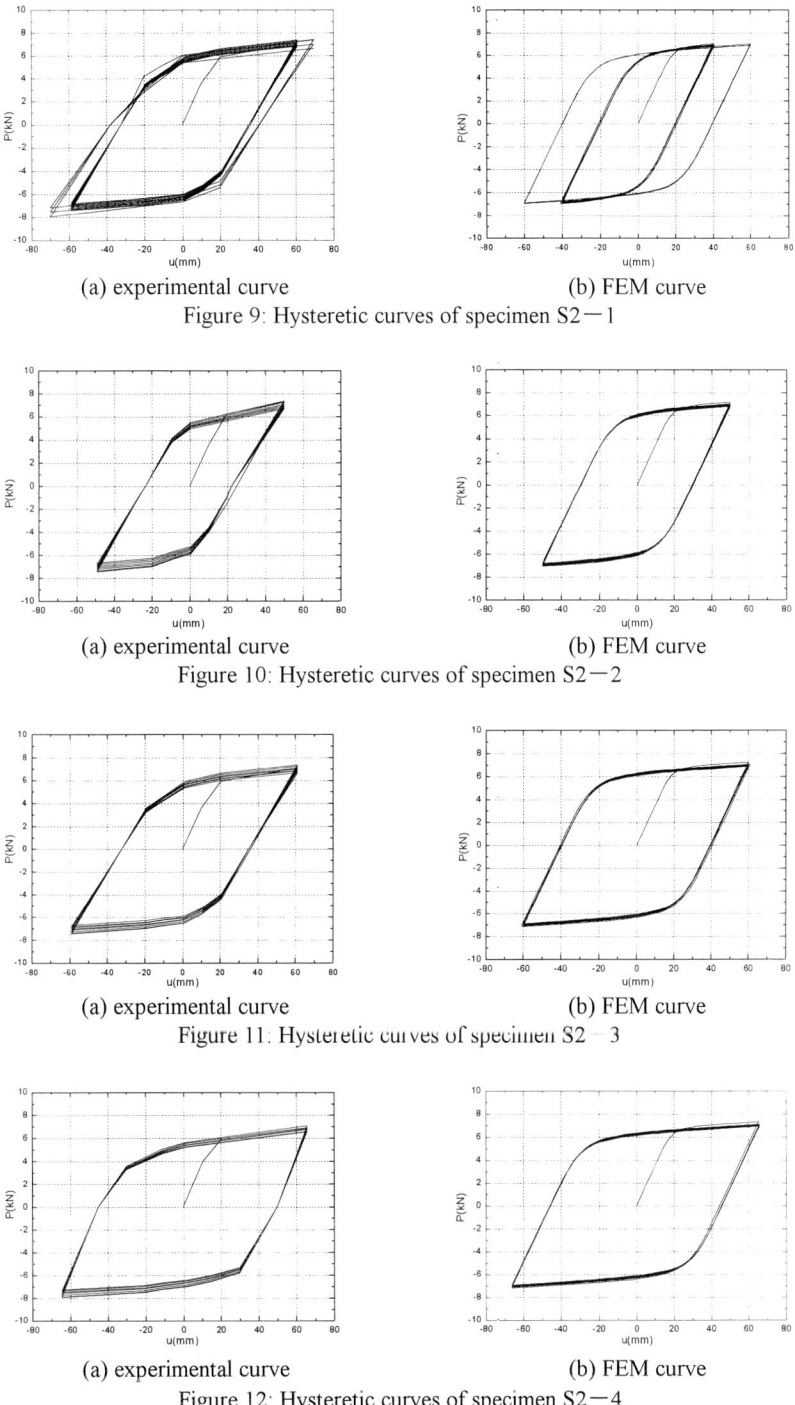

(a) experimental curve (b) FEM curve
Figure 9: Hysteretic curves of specimen S2−1

(a) experimental curve (b) FEM curve
Figure 10: Hysteretic curves of specimen S2−2

(a) experimental curve (b) FEM curve
Figure 11: Hysteretic curves of specimen S2−3

(a) experimental curve (b) FEM curve
Figure 12: Hysteretic curves of specimen S2−4

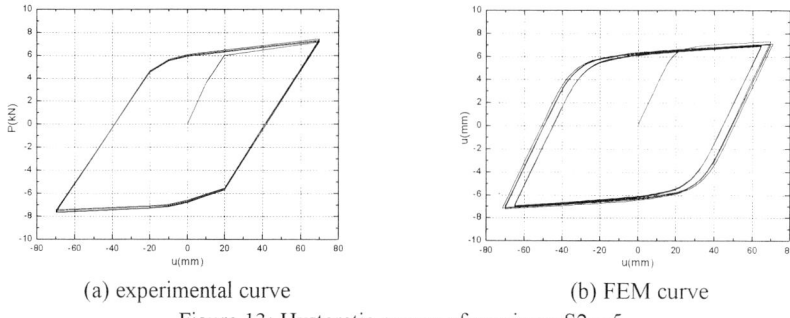

(a) experimental curve (b) FEM curve

Figure 13: Hysteretic curves of specimen S2—5

From TABLE 4 and Figure 4 to Figure 13, it can be seen that the FEM analytical results are in good agreement with the experimental ones. The cumulative damage mechanics model suggested in this paper is accurate enough to evaluate the damage cumulation in the structural steel and to predict initiation of damage cracks under cyclic loading. The seismic behavior of steel structures, therefore, can be simulated with an acceptable error for an earthquake, and repeated earthquakes by using of damage model presented.

CONCLUSION

Based on the experiments and energy damage theory, a cumulative damage model which caters for triaxial stress conditions is proposed in this paper. A general procedure for computing damage cumulation in steel members with finite element method is suggested. Cantilever members subjected cyclic loading at their ends were simulated numerically with a FEM program. The analytical results showed good agreement with the experimental results.

ACKNOWLEDGEMENT

The project is financially supported by National Science Foundation of China as a key project (59895410).

References

Chen, R. Y. (2000). *Damage Cumulation Analysis of Tall Steel Mega-Structures under Seismic Actions*. Doctoral Dissertation, Tongji University, (in Chinese)

Kachanov, L. M. (1986). *Introduction to Continuum Damage Mechanics*, Martinus Nijhoff Publishers, Dordrecht

Kumar, S. and Usami, T. (1994). A Note on Evaluation of Damage in Steel Structures under Cyclic Loading, *JSCE Journal of Structural Engineering*, 40A, 177-188

Lemaitre, J. and Chaboche, J. L. (1985). *Mécanique des Matériaux Solides*, Dunod, Paris (Chinese Translation, 1997)

Park, Y. J. and Ang, A. H. S. (1985). Mechanistic Seismic Damage Model for Reinforced Concrete,

ASCE Journal of Structural Engineering, 111: 4, 722-739.

Park, Y. J., Ang, A. H. S. and Wen, Y. K. (1985). Seismic Damage Analysis of Reinforced Concrete Buildings, *ASCE Journal of Structural Engineering*, 111: 4. 740-757.

Rabotnov, Y. N.(1969) Creep Rupture. *Proceeding of the Twelfth Internal Congress of Applied Mechanics*, IUTAM, 342-349.

Shen, Z. Y. and Dong, B. (1997). An Experiment-Based Cumulative Damage Mechanics Model of Steel under Cyclic Loading. *Advances in Structural Engineering*. 1:1, 39~46

Yu, S. W. and Feng, X. Q. (1997). *Damage Mechanics*. Tsinghua University, Beijing, China(in Chinese)

BEAMS AND COLUMNS

A REVIEW OF RECENT DEVELOPMENTS ON DESIGN OF PERFORATED BEAMS

C. H. Ko and K. F. Chung

Department of Civil and Structural Engineering,
The Hong Kong Polytechnic University, Hung Hom, Hong Kong

E-mail: cekchung@polyu.edu.hk

ABSTRACT

In modern commercial or residential buildings with high specifications in building services, a common method of incorporating services within the floor-ceiling zone of buildings is to create large openings in the webs of beam members. The openings are most likely to be rectangular or circular, and may be in the form of discrete openings, or a series of openings, along the member length of the beam. Over the past few decades, the structural behaviour of beams with web openings was a popular research topic, and numerous research projects were executed and reported including experimental investigations, finite element studies, and also development of design rules. This paper presents a review on recent developments on design of perforated beams, and recommendations for future development are also provided.

KEYWORDS

Perforated beams, design development, tee section approach, perforated section approach, shear-moment interaction curves, and web openings of various shapes and sizes.

1. INTRODUCTION

Large individual openings are often formed in the webs of steel beams to provide passage of building services at specific locations. The presence of web openings may have a severe penalty on the load carrying capacities of floor beams, depending on the shapes, the sizes, and the locations of the web openings. Due to the presence of the web openings, three different modes of failures at perforated section are possible: i) shear failure, ii) flexural failure, and iii) '*Vierendeel*' mechanism.

1.1 Beams with single web openings

A large number of tests have been carried out on steel and composite beams with discrete rectangular openings, notably those in UK [1], in USA [2-4], in Canada [5-9], and in Australia. An interesting series of tests on 21 m span beams was carried out at the University of Kaiserslautern, Germany [10], as a prototype to the construction of the Kommerzbank in Frankfurt.

There are also a number of design recommendations [11-14] available in the literature for both steel beams and composite beams with rectangular web openings. Most of the work is only intended for small openings with opening dimensions (along section depth and member length) less than half of the section depth of the structural members. An overall study on the design recommendations shows that in general, there are two design approaches in assessing the structural behaviour of beams with rectangular web openings:

- *Tee section approach*
  In this approach, the structural adequacy of a beam with a web opening depends on the section resistances of the tee sections above and below the web openings under co-existing axial forces, shear forces and local moments; all of these local forces and moments are due to global bending action. The accuracy of the design methods depends on the accuracy of a number of design rules against respective failure modes. Moreover, it should be noted that there are a number of different ways in allowing for the effect of co-existing axial forces in assessing the moment resistances of tee sections. The calculation procedures are usually complicated and they differ significantly among each other, depending on the design methodology adopted, and also the accuracy and the calculation efforts involved. It should be noted that the design methods are often very general, and applicable in principle for beams with web openings of various sizes and shapes.

- *Perforated section approach*
  In this approach, the perforated cross-section is the critical section to be considered in design. The structural adequacy of the beams depends on the section resistances of the perforated sections under co-existing global shear force and bending moment. In general, the design procedures for both the shear and the moment resistances of perforated sections are relatively simple and similar among different design methods. However, the '*Vierendeel*' moment resistances of the perforated sections are evaluated implicitly based on various assumptions on the effects of co-existing shear forces and moments. Simplifications are usually made to those design rules derived from the *Tee section approach*, and thus, empirical global shear-moment $(V-M)_o$ interaction curves are often provided to engineers for practical design.

For beams with circular web openings, most of the current design methods for beams with rectangular web openings are applicable through the use of an equivalent rectangular opening of modified dimensions as suggested by [15]. However, due to the simplistic approach, the load carrying capacities of those beams are always under-estimated significantly.

1.2 Beams with multiple web openings

Castellated beams were developed in 1940's to enhance the flexural performance of universal beams with profile cutting and welding of webs. They are effectively used in highly serviced buildings and they are also popular in exposed roof and floor systems for aesthetic reasons. With careful optimisation on the opening configuration (shapes, sizes and intervals), castellated beams may be designed with the simple rules for common universal beams using appropriate section properties. The use of castellated beams with expanded web-posts was developed in Europe based on elastic design method in 1970's. No reinforcements are normally provided. In additional to those failure modes associated with steel beams with single web openings, buckling of web posts may be critical in castellated beams when the openings are closely spaced. Moreover, additional deflection due to the presence of web openings should also be considered but the calculation is usually very laborious and not practical.

In 1990's, there was a castellated steel section with multiple circular web openings, or 'cellular beam', developed in UK for composite construction. These beams have limited shear resistance and are best used as long span secondary beams or where loads are relatively low. No optimisation on opening configuration was reported and the design rule for the 'Vierendeel' mechanism is 'under-developed' and based on elastic design method. In general, there are a number of design guides and recommendations developed for beams with multiple web openings based on analytical investigations and experimental tests on specimens of specific sizes and opening configurations. Some of those design rules are simple, empirical and 'product-specific'; their applications are very restrictive and are invalid for large web openings with different opening configurations.

In order to assess the load carrying capacities of beams with multiple circular web openings in an explicit manner, a design method [16] was developed in the UK in 1990. The method was later incorporated into Amendment A2 of Eurocode 3: Part 1.1: Annex N in 1998 [17] after minor modification. However, for steel beams with individual circular web openings, the use of a different set of approximate design rules was recommended in Annex N.

2. CURRENT SITUATIONS

The current situations on the design of perforated beams may be summarized as follows:

- For beams with single web openings, many design rules are developed for single small openings, i.e. the opening dimensions along section depth and member length are less than half of the section depth.
- For beams with multiple web openings, many design rules developed in 1970's are 'product-specific' with simple empirical rules based on elastic design principles.
- Design assumptions and methods are very different from one another. In some cases, different sets of design rules are proposed by the same authors for different ranges of opening dimensions.
- Design methods are only applicable for beams with web openings of limited shapes.
- Little work on composite sections with large web openings and asymmetric fabricated I-sections was found in the literature.
- Deflection calculation is rather labor-intensive using either frame analysis or even finite element technique, which are very tedious and time-consuming for practical engineers.
- There is a lack of unified approach in both strength and stiffness analysis.

3. RECENT DEVELOPMENTS

3.1 Design methods for various structural forms of beams

In 1987, a SCI design guide on steel and composite beams with web openings [11] was published and the design method was calibrated against full-scale tests in 1992 [1]. With the release of Eurocode 4, the design method was re-presented in the format of application rules to Eurocode 4 for detailed design of composite beams with large web openings [18]. Furthermore, in order to provide advice to engineers at the scheme design stage, general information on sizing of openings was also presented as a function of the utilization of the shear and the moment resistances of composite beams. The effect of these openings on deflection was estimated by a simple factor, which was dependent on the size and the location of the openings. Typical design tables for composite beams with large rectangular openings were also available. Design rules for other forms of construction such as circular openings and notched beams were also provided together with general detailing rules in the literature. It is interesting to note that for steel beams with web openings, the design principles and application rules was covered in Eurocode 3 Part 1.1 Annex N [17].

In 1995, a research project [19,20] on the structural performance of cold-formed steel members with large single and multiple web openings was reported. The basic structural design principles were similar to those presented in the SCI design guide [11] on steel and composite beams with large web openings except that effective section properties were used in cold-formed steel sections. The proposed design rules were calibrated with a total of 137 tests on cold-formed steel sections with large rectangular and circular web openings. The design rules were demonstrated to be safe and structurally economic in predicting the load carrying capacities of cold-formed steel sections with circular and rectangular web openings up to 75% of the section depth.

Consequently, a number of design rules for hot-rolled steel beams, steel-concrete composite beams and cold-formed steel beams are available. It is interesting to note that since all these design rules are based on structural design principles with similar analytical expressions on combined actions against axial forces, shear forces and local bending moments, it is highly desirable to harmonise all three sets of design rules for increased user-friendliness to designers.

3.2   Beams with circular web openings

In the current British design method for steel beams with circular web openings [16], the load carrying capacities of the beams is assumed to be limited by the formation of plastic hinges in the top tee sections at the low moment side (LMS) of the web opening. Moreover, a linear interaction formula is used to assess the moment resistance of the tee sections above and below the web openings under co-existing axial and shear forces. The method is regarded to be conservative as the formation of plastic hinges in the top tee sections at the LMS of the web openings does not cause failure to the perforated section. The beams are capable to carry additional load until other plastic hinges at critical locations of the perforated sections are formed to initiate the '*Vierendeel*' mechanism. Moreover, the reduction in the moment resistances of the tee sections under co-existing axial and shear forces is less severe than that anticipated in the linear interaction formula.

An investigation [21] on the Vierendeel mechanism in steel beams with circular web openings based on analytical and numerical studies is reported in the literature. The current British design method is examined in details with plastic hinges formed at the LMS and the HMS of the web openings separately. A finite element model is also established with both material and geometrical non-linearity so that load re-distribution across the web openings may be incorporated. Moreover, the moment resistances of the tee sections above and below the web openings may be properly evaluated in the presence of co-existing axial and shear forces in the finite element model, and thus, the load carrying capacities of typical universal steel beams with circular web openings are presented. An empirical global $(M-V)_0$ interaction curve at the perforated sections is also suggested for practical design of steel beams with circular web openings.

It is shown in Figure 1 that shear yielding in steel beams with circular web openings is very important as there is extensive shear yielding at the tee sections with minimum web depth which reduces the effectiveness of load re-distribution across the web openings. Such effect is not significant in steel beams with rectangular web openings where local Vierendeel moment is often dominant. After calibration against the results of the finite element study, it is shown that the circular web opening may be readily replaced by an equivalent octagonal opening so that the current design rules may give load carrying capacities of perforated sections close to the finite element results. It should be noted [22] that the dimensions of the equivalent octagonal openings relate not only to the diameter of the circular web openings, but also to the specific design procedure adopted.

3.3   Design methods for steel beams with web openings of various shapes and sizes

A comprehensive and systematic parametric study on steel beams with web openings of various shapes and sizes using finite element technique is conducted [23], and the primary structural characteristics of those steel beams are examined in details. A number of common opening shapes are covered in the study including circular, hexagonal, octagonal, square, rectangular and elongated circular openings; the opening depth ranges from 50%, 67% to 75% of the section height of the steel beams. It is found that all these steel beams behave similarly to each other in terms of deformed shapes under a wide range of applied moments and shear forces, as shown in Figure 2. The three typical failure modes in perforated sections are common to all beams. Moreover, the yield patterns of those perforated sections at failure are also similar to each other. In general, plastic hinges are always formed at both ends of the tee sections above and below the web openings. Comparison on the global $(M-V)_o$ interaction curves of the beams shows that they are similar to each other in shape, as shown in Figure 3. Thus, it is possible to derive empirical global $(M-V)_o$ interaction curves to assess the load carrying capacities of all these steel beams using harmonized design rules. Furthermore, it is shown that for all web openings of various shapes and sizes considered in the study, the most important parameter in assessing the structural behaviour of the perforated sections is the critical opening length, c, which controls the magnitude of local Vierendeel moments acting on the tee sections. Based on the finite element results, a simple empirical design rule is developed and fully presented in [24].

4. FUTURE DESIGN DEVELOPMENT

In view of the current situations on the design of perforated beams, it is recommended to develop a unified design approach for beams with large web openings based on plastic design method and formulated in accordance with analytical structural design principles. The following areas of interests will be covered:

- Steel sections and composite sections with single and multiple web openings of different shapes and dimensions under different loading and support conditions.
- Hot-rolled or fabricated steel sections with different cross-section geometries, such as symmetric or asymmetric I-sections, prismatic or tapered beams, fabricated I-sections with expanded webs and extended web-posts.
- Optimization of opening configuration (shape, dimensions and interval) and cross-section geometry for both steel sections and composite sections under practical design requirements.
- Different levels of design procedures appropriate to manual or computer calculations depending on time constraint and accuracy required.
- Simple design rules for deflection assessment.

5. CONCLUSIONS

It is shown that since all design rules on perforated beams are based on structural design principles with similar analytical expressions on combined actions against axial forces, shear forces and local bending moments, it is highly desirable to harmonise these design rules for increased user-friendliness to designers. Moreover, these design rules will be further developed to design beams with web openings of various sizes and shapes. They will enable engineers to design modern floor and roof systems fully integrated with building services for composite, hot-rolled or fabricated steel beams and cold-formed steel structures.

ACKNOWLEDGEMENTS

The research project leading to the publication of this paper is supported by the Research Grants Council of the Government of the Hong Kong Special Administrative Region (Project No. PolyU5085/97E).

REFERENCES

1. Lawson R. M., Chung K. F., and Price A. M. (1992). Tests on Composite Beams with Large Web Openings to Justify Existing Design Methods. *The Structural Engineer* **70:1**, 1-7.
2. Bower J. E. (1968). Ultimate Strength of Beams with Rectangular Holes. *Journal of the Structural Division, Proceedings of the ASCE* **94:ST6**, 1315-1337.
3. Clawson W. C. and Darwin D. (1980). Composite Beams with Web Openings. *SM Report No. 4*. University of Kansas Center for Research, Lawrence, Kansas, USA.
4. Donahey R. C. and Darwin D. (1986). Performance and Design of Composite Beams with Web Openings. *SM Report No. 18.* University of Kansas Center for Research, Lawrence, Kansas, USA.
5. Cho S. H. and Redwood R. G. (1992). Slab Behavior in Composite Beams at Openings II: Tests and Verifications. *Journal of Structural Engineering, ASCE* **118: ST9**, 2304-2322.
6. Redwood R. G. and McCutcheon J. O. (1968). Beam Tests with Unreinforced Web Openings. *Journal of the Structural Division, Proceedings of the ASCE*. **94:ST1**, 1-17.
7. Redwood R. G. and Wong P. K. (1982). Web Holes in Composite Beams with Steel Deck, *Proceedings of the Eighth Canadian Structural Engineering Conference*. Canadian Steel Construction Council, Willowdale, Ontario, Canada, February 1982.
8. Redwood R. G. and Poumbouras G. (1983). Tests of Composite Beams with Web Openings. *Canadian Journal of Civil Engineering* **10:4**, 713-721.
9. Redwood R. G., Baranda H., and Daly M. J. (1978). Tests of Thin-Webbed Beams with Unreinforced Holes. *Journal of the Structural Division, Proceedings of the ASCE* **104:ST3**, 577-595.
10. Bode H., Stengel J., and Zhou D. (1996). Composite Beam Test for A New High-rise Building in Frankfurt. *AISC Conference on Composite Construction in Steel and Concrete*, Irsee/Germany, 1996, 14-19.
11. Lawson R. M. (1987). Design for Openings in the Webs of Composite Beams. *CIRIA Special Publication and SCI Publication 068*, CIRIA/The Steel Construction Institute, UK.
12. Darwin D. (1990). Steel and Composite Beams with Web Openings. *Steel Design Guide Series No. 2*, American Institute of Steel Construction, Chicago, IL, USA.
13. Redwood R. G. and Cho S. H. (1993). Design of Steel and Composite Beams with Web Openings. *Journal of Constructional Steel Research* **25**, 231-248.
14. Oehlers D. J. and Bradford, M. A. (1995). *Composite Steel and Concrete Structural Members: Fundamental Behaviour*, Pergamon.
15. Redwood R. G. (1969). The Strength of Steel Beams with Unreinforced Web Holes. *Civil Engineering and Public Works Review* **64:755**, 559-562.
16. Ward J. K. (1990). Design of Composite and Non-Composite Cellular Beams. *SCI Publication 100*, The Steel Construction Institute, UK.
17. British Standards Institution. ENV 1993-1-1: 1992/A2: 1998, *Amendment A2 of Eurocode 3: Annex N – Openings in Webs, 1993.*
18. Chung K. F. and Lawson R. M. (2001). Simplified Design of Composite Beams with Large Web Openings to Eurocode 4. *Journal of Constructional Research* **57:2**, 135-163.
19. Chung K. F. (1995). Structural Performance of Cold Formed Sections with Single and Multiple Web Openings. Part I: Experimental Investigation. *The Structural Engineer* **73:9**, 141-149.
20. Chung K. F. (1995). Structural Performance of Cold Formed Sections with Single and Multiple Web Openings. Part II: Design Rules. *The Structural Engineer* **73:14**, 223-228.
21. Chung K. F., Liu T. C. H., and Ko A. C. H. (2001). Investigation on Vierendeel Mechanism in Steel Beams with Circular Web Openings. *Journal of Constructional Steel Research* **57**, 467-490.
22. Ko C. H. and Chung K. F. (2000). A Comparative Study on Existing Design Rules of Steel Beams with Circular Web Openings. In Yang, Y. B., Leu, L. L., and Hsieh, S. H., (eds.),

Proceedings of the First International Conference on Structural Stability and Dynamics, Taipei, Taiwan, December 2000, 733-738.
23. Liu T. C. H. and Chung K. F. (in press). Steel Beams with Large Web Openings of Various Shapes and Sizes: Finite Element Investigation. *Journal of Construction Steel Research.*
24. Chung K. F., Liu T. C. H. and Ko A. C. H. (in press). Steel Beams with Large Web Openings of Various Shapes and Sizes: An Empirical Design Method Using A Generalized Moment-Shear Interaction Curve. *Journal of Construction Steel Research.*

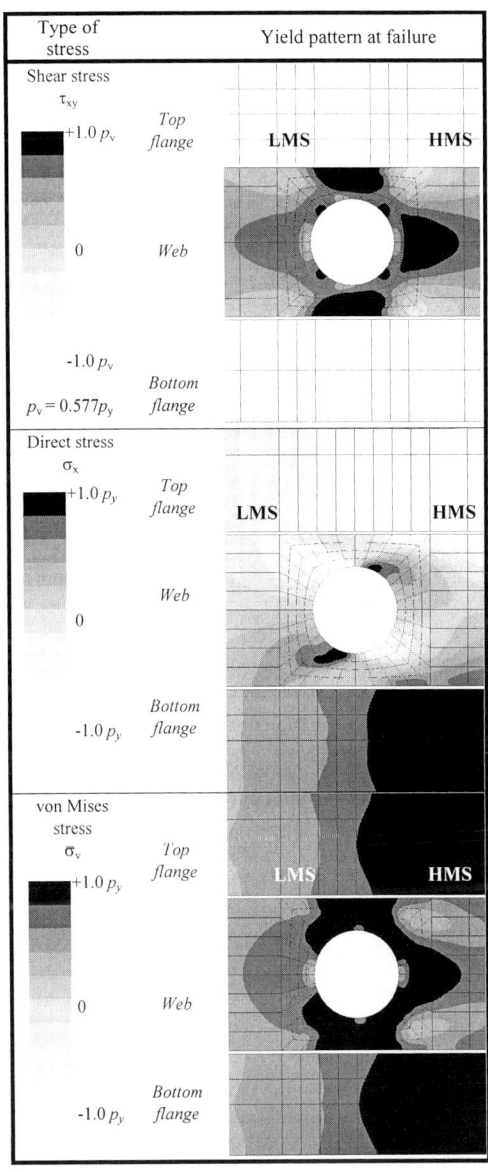

Figure 1: Stress distribution at typical perforated section with circular web opening
(Stresses are shown at the mid-thickness of the steel elements)

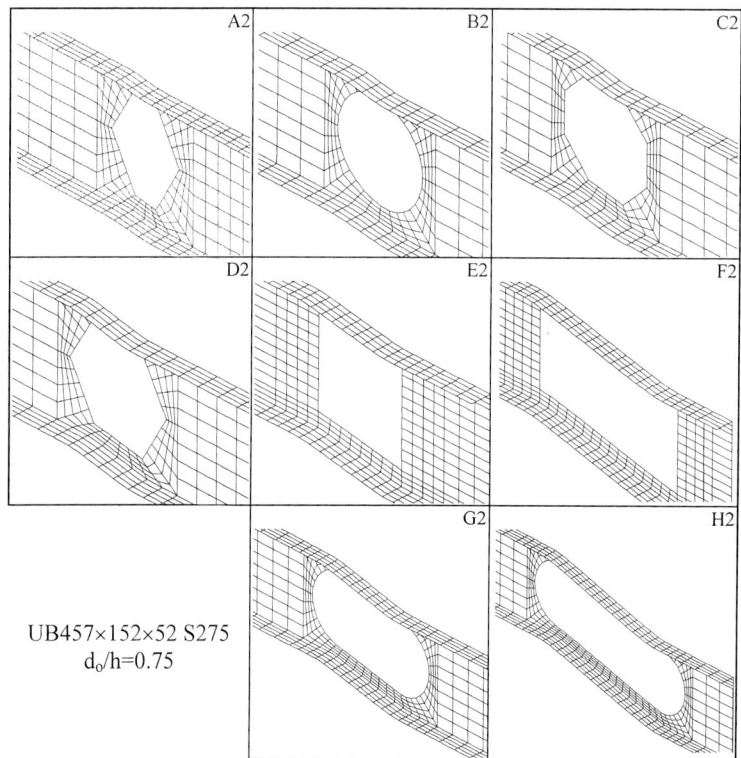

Figure 2: Deformation at failure for openings under large global shear and moment

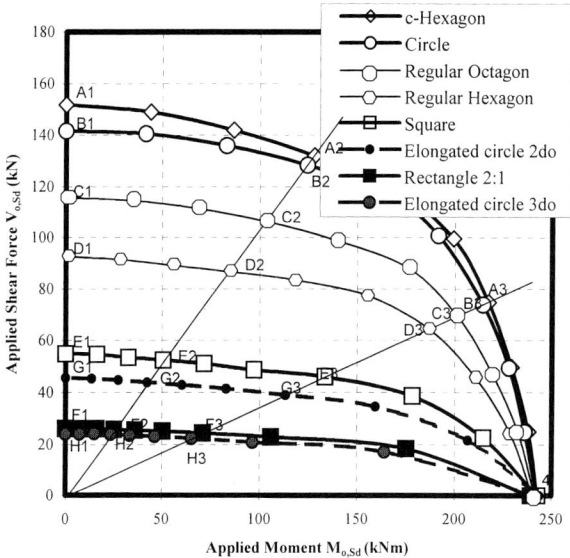

Figure 3: Shear-moment interaction curves for perforated sections
(UB457x152x52 S275 with $d_o/h = 0.75$)

A NEW DERIVATION OF THE BUCKLING THEORY OF THIN-WALLED BEAMS

Gengshu Tong and Lei Zhang

(Zhejiang University, Hangzhou, 310027)

ABSTRACT

There existed two formulae at present for calculation of flexural-torsional buckling moments of thin-walled beams with monosymmetric I-sections. This paper compared the stability theories from which these formulae were derived. This paper presented a buckling analysis of I-sectional beams based on the stability theory of shells, which is certainly superior to the beam theories because fewer assumptions are adopted to establish the shell buckling theory. Totally 36 I-sectional beams were analyzed using the FEM program ANSYS, among which 12 have doubly symmetrical sections, 12 tensile flange strengthened and 12 compressed flange enlarged. The element SHELL63 was used in the analysis. It was found that the shell buckling analysis verified the traditional formula, not the newer, seemly more rational theory. This paper presented a new derivation of the flexural-torsional buckling theory of thin-walled beams. This new derivation, which is in agreement with the results of shell buckling analysis, verified the traditional formula.

KEYWORDS

Stability, Thin-walled section, Beam, Critical moment

1. INTRODUCTION

The buckling moments of beams with monosymmetrical I-sections under transverse loads are calculated by (Chen, 2001)

$$M_{cr} = C_1 \frac{\pi^2 EI_y}{l^2}[-C_2 a + C_3 \beta_y + \sqrt{(-C_2 a + C_3 \beta_y)^2 + \frac{I_\omega}{I_y}(1 + \frac{GJl^2}{\pi^2 EI_\omega})}\,] \tag{1}$$

in which $I_y, I_\omega, J, l, a, G, E$ are inertia moment of the cross-section about the weak axis, warping inertia moment, torsional action constant, unbraced length of the beam, distance from the loading point to the shear center (a is positive when the loading point is above the shear center), and Young's modulus, shear modulus of the steel, and $\beta_y = \frac{1}{2I_x}\int y(x^2+y^2)dA - y_s$, β_y is positive when the compressive flange is strengthened. C_1, C_2 and C_3 are coefficients: for simply supported beams under uniformly distributed loads

$$C_1 = 1.13, C_2 = 0.45, C_3 = 0.53 \tag{2a}$$

for simply supported beams under central point loads

$$C_1 = 1.35, C_2 = 0.55, C_3 = 0.41 \tag{2b}$$

In 1980s~90s, a modification to the coefficient C_3 appeared in the literature (Galambos,1999; Guo, 1990; ISO/TG167,1996), for the above two load cases, C_3 is changed to $C_3 = 1$.

For beams with monosymmetrical cross sections, this change brought great differences in the buckling moments. For example, for the sections showed in Fig1a, 1b, the differences in buckling moments using $C_3 = 1$ and eqns. (2a),(2b) are shown in Table 1 (the slenderness ratio of beams about the weak axis is 80 ($L = 966.2t$)).

Table 1 Comparison of critical moments

Section	Loading	Load Position	C_3	M_{cr}	$\frac{M_{(C3=0.53,0.41)} - M_{(C3=1)}}{M_{(C3=0.53,0.41)}}$
Fig.1a	Uniformly distributed load	Top Flange	0.53	1.875Et^3	21.7%
			1.0	1.468Et^3	
		Bottom Flange	0.53	5.024Et^3	40.2%
			1.0	3.005Et^3	
Fig.1b	Concentrated load at mid-span	Top flange	0.41	22.473Et^3	-98.80%
			1.0	44.675Et^3	
		Bottom Flange	0.41	76.026Et^3	-33.20%
			1.0	101.276Et^3	

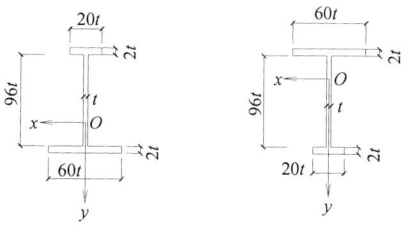

1a   1b
Fig.1 Monosymmetrical I-beams

2. THE CURRENT THEORETICAL BASIS OF THE BUCKLING MOMENT

2.1. THE THEORETICAL BASIS OF $C_3 = 0.53$ AND $C_3 = 0.41$

Eqs. (2a),(2b) were obtained from the following total potential energy:

$$\Pi_1 = \frac{1}{2}\int [EI_y u''^2 + EI_\omega \theta''^2 + GJ\theta'^2 + 2\beta_y M_x \theta'^2 + 2M_x u''\theta - qa\theta^2]dz \quad (3)$$

where u is the lateral displacement of the shear center, θ is the twisting angle of the section during buckling, M_x the bending moment, q the uniformly distributed load acted at point $(0,a)$. The fifth term in eqn(3) may be seen as the work done by the bending moment component $M_x \theta$ on the buckling curvature u'', the sixth is the potential energy of the load.

The total potential energy may also be derived using the nonlinear stress-strain relation (Bleich,1952). Denoting the shear center coordinate is (x_s, y_s), the nonlinear longitudinal strain is:

$$\varepsilon^N = \frac{1}{2}\{[u' - (y - y_s)\theta']^2 + [(x - x_s)\theta']^2\} = \frac{1}{2}[u'^2 - 2(y - y_s)u'\theta' + \rho_s^2 \theta'^2] \quad (4)$$

in which $\rho_s^2 = (x - x_s)^2 + (y - y_s)^2$. The strain energy related to the strain ε^N is $\int \sigma \varepsilon^N dA dz$
Summation of this to the linear part leads to

$$\Pi_2 = \frac{1}{2}\int [EI_y u''^2 + EI_\omega \theta''^2 + GJ\theta'^2 + 2\beta_y M_x \theta'^2 - 2M_x u'\theta' - qa\theta^2]dz \quad (5)$$

compared with eqn. (3), the inter-influenced term is $-2M_x u'\theta'$, not $2M_x u''\theta$. The difference is the 2nd term in the following equation:

$$-\int 2M_x u'\theta' dz = -2M_x u'\theta \Big|_0^l + \int 2(M_x u')'\theta dz = \int 2(M_x u'' + Q_y u')\theta dz \quad (6)$$

If the load is acted at the shear center, the critical load obtained from eqn(5) is nearly doubled as compared to that from eqn(3). This is obviously incorrect. Therefore taking only consideration of the nonlinear longitudinal strain is not enough. This was perhaps why Bleich (1952) used eqn(3) for flexural torsional buckling of beams, while in formulating a total potential for flexural-torsional buckling of eccentrically compressed columns, he used the second techniques.

2.2. THEORETICAL BASIS OF $C_3=1.0$

In Lu.*et al.* (1983) we found the term which must be added to eqn (5): the strain energy corresponding the nonlinear shear strain on the midsurface of the thin-walled cross section.:

$$U_3 = \frac{1}{2}\int (2Q_y \beta_y \theta\theta' - 2Q_y u'\theta)dz \quad (7)$$

The second form in eqn(7) will cancel the second term in eqn(6) when added together. The total potential energy becomes:

$$\Pi_3 = \frac{1}{2}\int [EI_y u''^2 + EI_\omega \theta''^2 + GJ\theta'^2 + 2\beta_y M_x \theta'^2 + 2M_x u''\theta + 2Q_y \beta_y \theta\theta' - qa\theta^2]dz \tag{8}$$

The difference between eqn(3) and eqn(8) may be seen more clearly when the equilibrium equations are derived. Equations derived from eqn(3) are

$$EI_y u'''' + (M_x \theta)'' = 0 \tag{9a}$$
$$EI_\omega \theta'''' - (GJ + 2M_x \beta_y)\theta'' - 2Q_y \beta_y \theta' + M_x u'' - qa\theta = 0 \tag{9b}$$

and eqn(9b) is changed to eqn(10) if eqn(8) is used

$$EI_\omega \theta'''' - (GJ + 2M_x \beta_y)\theta'' - 2Q_y \beta_y \theta' + M_x u'' - q(a - \beta_y)\theta = 0 \tag{10}$$

comparing these two sets of equations tells the only difference in the load q terms.. When eqn(10) is used, one obtains $C_3(\text{eqn}(10)) = C_2 + C_3(\text{eqns}(9a,b)) = 1.0$.

The total potential eqn(3) of the traditional theory was not theoretically complete, because it is based on a summation of different terms, it is very likely that other terms might have been omitted. On the other hand, the total potential eqn(8) seems more rational, because it is based on the nonlinear strain energy, but the physical meaning of $q(a - \beta_y)\theta$ in eqn(10) is not clear.

In the newer theory, the nonlinear shear strain must be included in eqn(8) to obtain a rational critical moment.. In the theory of thin-walled members, the linear strain has already been neglected, why should the nonlinear part been considered?

The transverse linear normal strain is also assumed to be zero because of the assumption of rigid profile of the cross-section during lateral torsional deformation, should we also consider the nonlinear transverse normal strain in the total potential energy? This nonlinear strain may be calculated as:

$$\varepsilon_s = \frac{\partial v_s}{\partial s} + \frac{1}{2}(\frac{\partial v_n}{\partial s})^2 = \frac{1}{2}(\frac{\partial v_n}{\partial s})^2 = \frac{1}{2}\theta^2 \tag{11}$$

These questions are difficult to answer within the framework of the thin-walled members theory. Therefore, in this paper the shell buckling theory is used, which is obviously closer to reality because of their fewer assumptions used in its establishment.

3. RESULTS OF SHELL BUCKLING ANALYSIS

The program ANSYS is used here to analyse the buckling of thin-walled beams. as a shell problem. The element shell63[6] is used, the geometric stiffness is based on the following nonlinear expressions:

$$\varepsilon_x = \frac{\partial u}{\partial x} + \frac{1}{2}\left[(\frac{\partial u}{\partial x})^2 + (\frac{\partial v}{\partial x})^2 + (\frac{\partial w}{\partial x})^2\right], \quad \gamma_{xy} = \frac{\partial u}{\partial y} + \frac{\partial v}{\partial x} + \frac{\partial u}{\partial x}\frac{\partial u}{\partial y} + \frac{\partial v}{\partial x}\frac{\partial v}{\partial y} + \frac{\partial w}{\partial x}\frac{\partial w}{\partial y}, \text{etc.}$$

When comparison is made between the results from FEM shell buckling analysis and those of beam theories, should the results of shell buckling analysis be modified to eliminate poisson's ratio effect? Because in beams, the stress is mainly in longitudinal direction, the transverse stress is usually in one order less than the longitudinal stress. The longitudinal strain is therefore

$$\varepsilon_z = \frac{\sigma_z - \mu\sigma_s}{E} \approx \frac{\sigma_z}{E}$$

that is the longitudinal stiffness is nearly unaffected by the transverse stress, therefore the poisson's ratio effect need not be considered, the results of shell buckling analysis may be directly compared with those of beam theory.

Table 2 gives the results of FEM analysis and the critical moments from eqn(1) using different values of C_3. 6 cross-sections are selected, each has 3 types of loading acted at top and bottom flanges respectively. Totally 36 beams are analysed. The cross-sections、span、loading and the loading point in the analysis are given in the tables. In order to keep as close as possible the rigid cross-section hypothesis during the deformation of thin-walled member, 7 transverse stiffeners are added along the beams which are modeled as plane stress elements in the analysis.

The first and second beams have doubly symmetrical I-sections. For these beams the above two theories gave the same critical moments. These beams are analyzed in order to verify that the ANSYS can predict the overall lateral-torsional buckling moments of beams accurately, i.e., it can predict the correct buckling mode, reflect the effect of loading position and the load distribution (shape of the moment diagrams),…etc. From table 2 it is seen that the results of ANSYS have very good agreement with those from the beam theory. ANSYS can be relied upon in the beam buckling analysis. Because both linear and nonlinear shear deformations are included in the element SHELL 63 of ANSYS, but have not been considered in the beam theory, the agreement between the beam theory and the ANSYS shows the negligible effect of the shear deformation on the buckling of these two beams.

Table2 Comparison between results from ANSYS and the beam theories

	Loading	Loading position	Beam theory M_{mcr}	Ansys M_{pcr}	$\|M_{mcr} - M_{pcr}\| / M_{mcr}$	Beam size
1	Uniformly distributed load	TF	560374.413	560597.767	0.04%	$b_T = b_B = 40$
		BF	1124225.989	1099395.937	2.2%	$t_T = t_B = 2$
	Concentrated load at midspan	TF	627505.474	630590.800	0.49%	$h_w = 96, t_w = 1$
		BF	1432930.243	1442703.500	0.68%	$L = 2000$
	Concentrated load at L/4,3L/4	TF	531047.975	528940.950	0.40%	$\lambda_y = 219$
		BF	1004865.537	977115.150	2.76%	
2	Uniformly distributed load	TF	481982.558	489871.967	1.64%	$b_T = b_B = 30$
		BF	812715.240	802678.31	1.23%	$t_T = t_B = 2$
	Concentrated load at midspan	TF	547931.602	558077.28	1.85%	$h_w = 46, t_w = 1.5$
		BF	1020361.414	1023636.18	0.3%	$L = 1200$
	Concentrated load at L/4,3L/4	TF	453586.38	459839.31	1.38%	$\lambda_y = 173$
		BF	731508.727	720386.67	1.52%	

*Note: elastic model of material =206000, Poisson's ratio of material =0.3
TF(BF)=loaded at top (bottom) flange; b_T (b_B)=width of top (bottom) flange; t_w=thickness of web; h_w=net high of web
λ_y=the slenderness ratio of beams about the weak axis; t_T (t_B)=thickness of top (bottom) flange; L=length of beams

The third and fourth beams have the tensile flange strengthened cross-sections. For these beams the two beam theories give quite different critical moments as shown in Table 3. The newer theory, which considered the nonlinear shear deformation on the middle surface gave lower critical moments. The ANSYS results are very close to those of the traditional theory. Very great difference exists between ANSYS and the newer beam theory.

Table 3 Comparison between results from ANSYS and the beam theories

	Loading	Loading position	C_3	Beam theory M_{mcr}	Ansys M_{pcr}	$\frac{|M_{mcr}-M_{pcr}|}{M_{mcr}}$	Beam size
3	Uniformly distributed load	TF	0.53	199724.02	201056.84	0.67%	$b_T=20$ $b_B=40$ $h_T=h_B=2$ $h_w=96, t_w=1$ L=1600, λy=214.6
			1	163866.17		22.70%	
		BF	0.53	418443.70	407843.65	2.53%	
			1	309002.43		31.99%	
	Concentrated load at mid-span	TF	0.41	227807.88	224326.17	1.35%	
			1	180014.88		24.62%	
		BF	0.41	554343.50	543420.08	1.97%	
			1	375999.74		44.53%	
	Concentrated loads at L/4, 3L/4	TF	0.57	188906.72	189981.37	0.57%	
			1	156854.92		21.12%	
		BF	0.57	371436.98	356623.22	3.99%	
			1	282096.81		26.42%	
4	Uniformly distributed load	TF	0.53	169100.95	169756.19	0.39%	$b_T=20$ $b_B=30$ $t_T=t_B=2$ $h_w=46$ $t_w=1$ L=1800, λy=284.7
			1	155131.14		9.4%	
		BF	0.53	245411.45	240519.92	1.99%	
			1	223360.42		7.68%	
	Concentrated load at mid-span	TF	0.41	195996.94	193045.66	1.51%	
			1	176215.20		9.55%	
		BF	0.41	305767.76	303735.00	0.66%	
			1	271645.38		11.81%	
	Concentrated loads at L/4, 3L/4	TF	0.57	158154.24	159106.60	0.60%	
			1	146067.46		8.93%	
		BF	0.57	222282.93	217721.47	2.05%	
			1	203950.80		6.75%	
5	Uniformly distributed load	TF	0.53	60889328.9	61496509.8	1.00%	$b_T=150$ $b_B=110$ $t_T=t_B=10$ $h_w=250$ $t_w=6$ L=5500 λy=177.7
			1	68378182.9		10.06%	
		BF	0.53	106684500	105915772.	0.72%	
			1	117831037.		10.11%	
	Concentrated load at midspan	TF	0.41	68317942.8	69839506.	2.23%	
			1	79019952.4		11.62%	
		BF	0.41	133281004.	132122867.	0.87%	
			1	150433852.		12.17%	
	Concentrated loads at L/4, 3L/4	TF	0.57	57439460.3	57724536.4	0.50%	
			1	63867170.2		9.62%	
		BF	0.57	95945253.6	94086352.	1.94%	
			1	105207996.		10.57%	
6	Uniformly distributed load	TF	0.53	272427.75	276276.45	1.41%	$b_T=30$ $b_B=10$ $t_T=t_B=2$ $h_w=38$ $t_w=1.5$ L=1600 λy=273.8
			1	310681.69		11.07%	
		BF	0.53	376619.49	375102.0	0.40%	
			1	423804.15		11.49%	
	Concentrated load at mid span	TF	0.41	313372.36	322971.58	3.06%	
			1	369776.46		12.66%	
		BF	0.41	461541.20	459968.16	0.34%	
			1	533735.94		13.82%	
	Concentrate loads at L/4,3L/4	TF	0.57	254013.60	256911.72	1.14%	
			1	286414.67		10.30%	
		BF	0.57	341510.43	337323.95	1.23%	
			1	380831.39		11.42%	

*Note: Refer to the Note of Table 2

The fifth and sixth beams have the compressive flange strengthened cross- sections. For these beams the two beam theories give also quite different critical moments as shown in Table 3.once again the ANSYS results are very close to those of the traditional beam theory. The newer theory predicted much higher critical moment in these cases than ANSYS. Therefore it is concluded that ANSYS, which should be more accurate because it uses fewer assumptions than any beam theory, verified the traditional beam theory, not the newer theory.

4. A NEW FORMULATION OF FLEXURAL-TORSIONAL BUCKING THEORY FOR THIN-WALLED BEAM

After comparing the plate buckling theory and the newer beam theory, and excluding the possible effect of linear shear deformations, it is found that the only difference between them is that the shell buckling theory includes the nonlinear transverse strain energy. The nonlinear transverse strain is given by eqn(11) ,and the corresponding strain energy is

$$U_4 = \frac{1}{2}\int\int\sigma_s\varepsilon_s dAdz = \frac{1}{2}\int\int(\sigma_s^L + \sigma_s^N)(\varepsilon_s^L + \varepsilon_s^N)dAdz$$

because the rigid contour assumption used, $\varepsilon^L = 0$, neglecting the higher order term, and using

$\sigma_s^N \varepsilon_s^L = \sigma_s^L \varepsilon_s^N$, one obtains $U_4 = \frac{1}{2}\int\int(\sigma_s^L \varepsilon_s^L + \sigma_s^L \varepsilon_s^N + \sigma_s^N \varepsilon_s^L)dAdz = \int\int \sigma_s^L \varepsilon_s^N dAdz$

substituting eqn(11) into the above equation:

$$U_4 = \frac{1}{2}\int[\int\sigma_s^L dA]\theta'^2 dz \tag{12}$$

The traditional linear theory of the thin-walled members pays no attention to the transverse normal stresses, It considered only the equilibrium of the longitudinal equilibrium of the thin-walled elements from which the shear stresses on the cross-section is obtained. But eqn (12) requires the transverse normal stresses to be calculated. For the sake of brevity, the detailed procedure will not be given here, the final result of the integration is as follows:

$$\int_A \sigma_s^L dA = -\beta_y q - qa$$

The nonlinear strain energy of the transverse normal stress becomes:

$$U_4 = -\frac{1}{2}\int_L (q\beta_y + qa)\theta'^2 dz \tag{13}$$

The total potential is:

$$U = \frac{1}{2}\int_L \left[EI_y u''^2 + EI_w \theta''^2 + (GJ + 2\beta_y M_x)\theta'^2 - 2M_x u'\theta' + 2Q_y \beta_y \theta\theta' - 2Q_y u'\theta - (\beta_y + a)q\theta^2\right]dz \tag{14}$$

For simply supported beam, the boundary conditions are: $z = 0, L : \theta = 0$, and $Q_y' = -q$, therefore

$\int_0^L 2Q_y\beta_y\theta\theta' dz = Q_y\beta_y\theta^2\Big|_0^L - \int_0^L (Q_y\beta_y)'\theta^2 dz = \int_0^L q\beta_y\theta^2 dz$. The total potential for simply supported beam becomes:

$$U = \frac{1}{2}\int_L \left[EI_y u''^2 + EI_w \theta''^2 + (GJ + 2\beta_y M_x)\theta'^2 - 2M_x u'\theta' - 2Q_y u'\theta - aq\theta^2\right]dz \tag{15}$$

Making use of eqn(6), one obtains the potential of the traditional theory eqn(3). For beams loaded by concentrated forces P_i at $z = L_i$, the integration of eqn(13) becomes

$$U_4 = -\frac{1}{2}\sum P_i(a+\beta_y)\theta^2\Big|_{z=L_i} \tag{16}$$

one obtains also the potential of the traditional theory. Therefore we have confirmed the validity of eqn(3) through our new derivation.

The load potential was added to the total potential in all the current buckling theories of thin-walled beams, why eqn.(15) does not include this potential? For this problem, one can remember the flexural buckling of the column and the buckling of the plates. For axially compressed column ,if the nonlinear strain energy is calculated , then the nonlinear load potential should not appear in the total potential(Trahair,1993), while in buckling of the plate, the work done during buckling by the boundary loads is never mentioned if the nonlinear strain energy is calculated。 It will lead to erroneous results if both the nonlinear strain energy and the nonlinear load potential are summed together. From Washizu (1982),one finds the following general variational principle for the stability of solid structures:

$$\iiint_V [\sigma_{ij}\delta\varepsilon_{ij} + k\sigma_{ij}^0 u_{k,j}\delta u_{k,i}]dV = 0 \tag{17}$$

in which σ_{ij}^0 ($i,j=1,2,3$) are the initial stresses which are in a equilibrium state with the external loads. σ_{ij} ($i,j=1,2,3$) are the incremental stresses due to the buckling displacements u_i, $\varepsilon_{ij} = \frac{1}{2}(u_{i,j}+u_{j,i})$, k is a load parameter. It is seen that the variational principle for buckling has no relation with the external loads, so the total potential corresponding to eqn.(17) should not include a nonlinear load potential.

5. CONCLUSION

This paper made a comparative study of two different theories for the buckling of thin-walled beams. One is the traditional; the other is the newly proposed. There existed great differences in buckling moments of simply supported beams obtained from these two theories, when the beams have monosymmetric cross-sections and are loaded by transverse loads. The cause leading to this unusual inconsistency is analyzed. This paper made also an analysis of the buckling of the beams using FEM based on the shell buckling theory. Totally 36 simply supported beams were analyzed, among which 12 had doubly symmetrical sections, 12 with their top flange strengthened, and 12 the bottom flange enlarged. The beams were acted by uniformly distributed load, midpoint concentrated load or concentrated loads at quarter points respectively. Comparison with the two member buckling theories showed that the results of shell buckling theory were in agreement with those from the traditional theory, although the newer theory seemed to be theoretically closer to the shell buckling theory in its derivation, because it included, as in the shell theory, the nonlinear shear strain on the mid surface of the cross-section. In order to explain this disagreement between the shell buckling theory and the newer beam theory, the present paper made a new derivation of the buckling theory of the thin-walled beams based on the general variational principle for stability of solid structures proposed by Washizu(1982). This new derivation is different from either of the above two theories in that the new derivation did not include a load potential $-\frac{1}{2}\int_0^l qa\theta^2 dz$, and that the new derivation presented considered the nonlinear transverse normal strain in the thin walled cross sections. Although the derivation presented in this paper was new, but the result is the same as the traditional theory.

REFERENCES
Attard,M.M., Bradford M.A.(1990). Bifurcation experiments on mono-symmetric cantilevers.*12th Aust. Conf. On Mechanics of Structures and Material*, 207~213.
Bleich, F(1952)., *Buckling Strength of Metal Structures*, McGraw-Hill Book Company.
Chen Ji (2001), *Stability of Steel structures: Theory and Application*, 2nd. Ed, Publishing House of Science,Beijing.(in Chinese)
Guo.Y.J. (1997), *Theory and Application for Cantilevers*, Publishing House of Hua Zhong University of Technology, Wuhan. (in Chinese)
Guo.Y.J, Fang.S.F (1990). Analysis of Lateral-Torsional Buckling of Steel Beams. *Journal of Building Structures*, **11:3,** 38-44.(in Chinese)
ISO/TG167, *Steel Structures, Material and Design*, 1996.
Lu L. W. Etc (1983), *Stability of Steel Structural Members*, Publishing House of Chinese Construction Industry, Beijing. (in Chinese)
Anderson, M. and Trahair, N.S. (1972). Stability of Monosymmetric Beams and Cantilevers. *J Struct Div, ASCE*, **98:1,** 269~285.
S.P.Timoshenko J.G.Gere (1961), *Theory of Elastic Stability*, 2nd Edition, McGraw Hill.
Tong Gengshu, Zhang Lei(2002), A Controversy and its Settlement in the calculation of Buckling moments of thin-walled beams with monosymmetrical I-sections under distributed loads, *Journal of Building Structures,* Vol.23,No.3.(in Chinese)
Trahair,N.S.,(1993), *Flexural-Torsional Buckling of Structures*, E & FN SPON,1993,London.
T.V Galambos(1998), *Guide to Stability Design Criteria for Metal Structures*, 5th Ed, John Wiley & Sons,Inc.
Washizu,K(1982), *Variational Methods in Ealsticity and Plasticity*, Pergamon Press Oxford, 3rd, Edition.
Xia.Z.B. Etc. (1988), *Theory of Structural Stability*, Advanced Education Press. (in Chinese)

ANALYSIS OF STRAIN HARDENING IN STEEL BEAMS USING MILL TESTS

M.P. Byfield and M. Dhanalakshmi

Cranfield University, RMCS Shrivenham, Swindon, SN6 8LA, UK

ABSTRACT

Experimental bending tests carried out to assess the reliability of the plastic moment of resistance formula revealed experimental bending strengths that typically exceeded their theoretical plastic moment capacities by an average of 18%. These findings are based on tests carried out on lightweight sections, whose stress strain characteristics showed strain hardening immediately after yielding, with no significant zone of plasticity. It is not understood whether the results are transferable to heavier sections and this work attempts to answer the scale effects question. A method for predicting the full (elastic and post-plastic) moment-rotation curves for laterally restrained steel beams, based on stress-strain data from mill tests is explained. A comparison between experimental and theoretical moment vs. end rotations shows that the method is capable of predicting strength and ductility with a high degree of accuracy. Furthermore, a survey of mill test results from a variety of section sizes is presented, the results of which are used to characterise strain-hardening effects for sections with differing span to depth ratios. Thus, the effect that section size and span to depth ratio has on strength and ductility is assessed.

KEYWORDS

Materials testing, mathematical modelling, numerical integration, mill tests, steel structures, steel beams, strain hardening, stress analysis.

INTRODUCTION

Conventional plastic design theory assumes that the stress strain curve for steel approximates to that relationship sketched in Figure 1a, whereby an extended region of plasticity is followed by strain hardening. Early work demonstrated that the plastic moment capacity, M_p, leads to a marginal underestimation of strength because it ignores the effect of strain hardening, Baker (1963). This conservatism was shown to be off-set by the reduction in the collapse load of frames due to the slight changes in geometry caused by deflections, also known as second-order effects, that occur during the formation of plastic hinges in continuous portal frames. Calculations based on the elastic-plastic stress-strain relationship in Figure 1b were justified because the conservatism of the M_p formula is cancelled by the reduction in load capacity due to second order effects.

Work carried out to assess the reliability of the M_p formula by Byfield and Nethercot (1998) revealed that the sections tested demonstrated no significant zone of plasticity, with strain hardening beginning shortly after initial yielding. This early onset of strain hardening was reflected in experimental bending strengths significantly higher than M_p. A total of 12 tests were carried out on beams fully restrained against lateral movement and demonstrated that the experimental strength exceeded the theoretical strength by an average of 18%. A further 20 sections were tested. Lateral restraints were positioned such that the non-dimensional slenderness ($\bar{\lambda}_{LT}$) was set to just less than 0.4 (this corresponds to the limit of applicability of the M_p resistance function in accordance with Eurocode 3: Part 1.1, and thus corresponds to the lower bound of resistance). These specimens exceeded their theoretical bending strength by an average of 14%. In short, modern hot-rolled sections were shown to exhibit an earlier onset of strain hardening than observed by the pioneers of plastic design theory, resulting in a significant underestimation of strength for the sections tested.

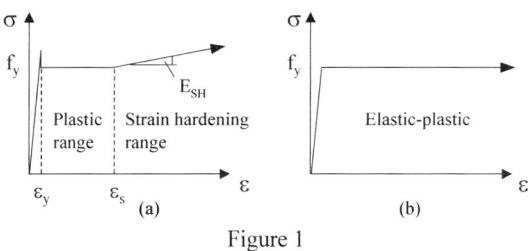

Figure 1

The 4-point bending tests upon which these findings are based were carried out on light-weight sections (203x102x23UB and 152x152x30UC) and it is not understood whether the results are transferable to the heavier sections more commonly found in practical frames. Scale effects are particularly important since the conservatism inherent in M_p formula is due to the stress-strain curve observed in the small sample of sections analysed for the project, and it is well known that the stress-strain relationship is sensitive to material thickness. This design conservatism is compounded by the mill test of steel often being significantly higher than the nominal yield stress assumed during design. A survey based on over 7000 material and geometric properties of structural steel, Byfield and Nethercot (1997), showed that the average mill stress of steel was some 16% higher than the nominal yield stress, where flange thickness was greater than 10mm. When flange thickness was less than 10mm the average mill stress exceeded the nominal yield stress by 37%. The survey also showed that geometric properties of steel are close to the nominal values and have little impact on strength.

A novel numerical modelling technique for predicting the moment vs. deformation for steel sections is presented. The technique uses neither stiffness matrices nor finite element analysis but numerical integration based on the full mill test stress strain curves. Using the stress-strain values as a polynomial function, the moment capacity for a given value of curvature is defined by integrating the product of stress, area and lever arm throughout a given cross section. Thus, the moment curvature relationship of a cross section is established. The curvature distribution along a member is integrated to define the slope distribution, followed by further integration to define the deflected shape. The process is repeated incrementally to define the full moment vs. end-rotation and moment vs. mid-span deflection graphs. The model is constructed from first principles and achieves a high degree of accuracy.

The method does not account for the weakening effects of lateral torsional and local buckling on the moment curvature relationship. This decision is justified on the evidence from bending tests on class 1 cross-sections presented by Byfield and Nethercot (1998), which showed that lateral torsional and local buckling have no significant effect on the moment vs. curvature relationship where the end rotation is limited to 6 degrees and where the non-dimensional slenderness is below the limit allowed for plastic design ($\bar{\lambda}_{LT} = 0.4$). Techniques are available for calculating the available rotation capacities of class 1 and 2 cross-sections for use in portal frame structures *inter alia*, Kemp and Dekker (1991), and can be used to limit the analysis to calculated maximum rotations, beyond which buckling will limit bending strength and thus cause the model to overestimate strength. No attempt is made to account for the weakening effects of local buckling, the method is therefore only justified for use with class 1 (plastic) cross-sections, as defined in Eurocode 3 Part 1.

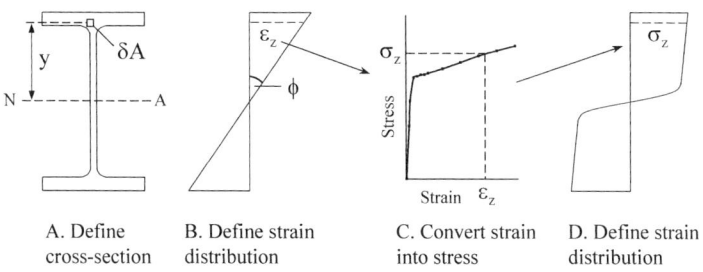

A. Define cross-section   B. Define strain distribution   C. Convert strain into stress   D. Define strain distribution

Figure 2: Description of steps for calculating M-ϕ relationship

MODELLING OF MOMENT VS. CURVATURE USING MILL TESTS (M-ϕ)

Figure 2A shows a cross-section subjected to a pure sagging moment applied about the major axis. A fibre of the beam (δA) is located a distance y from the neutral axis, which is located at half the depth of the cross-section assuming that: (a) the section is symmetrical, (b) zero axial load, and (c) that the stress strain relationship for steel is equal for compression and tension. Given that plane sections can be assumed to remain plane, then the strain (ε_z) in the fibre (δA) is equal to:

$$\varepsilon_z = \phi y \qquad (1)$$

Where ϕ is the curvature, see Figure 2B. The mill test stress strain relationship is converted into a polynomial function, and used to define the stress σ_z for the fibre (δA), Figure 2C. By this means the distribution of stress for the given value of curvature can be derived, Figure 2D. The moment due to the normal force in a fibre (δA) is equal to the product of force and distance from the neutral axis. The integral of all such moments across the entire cross-section provides the total moment, i.e.:

$$M = \int_A \sigma_z y \, dA \qquad (2)$$

The curvature is increased gradually and the corresponding moment for each curvature is calculated. Using this technique the moment curvature relationship is characterised for a given section.

MODELLING OF MOMENT VS. END ROTATION (M-θ)

Once the M-ϕ relationship for the cross-section is determined it is relatively simple to establish the distribution of curvature. For every loading increment, the distribution of bending moments is calculated, Figure 3A, which is converted into the distribution of curvature using the M-ϕ relationship,

Figure 3B. The curvature distribution is then integrated to define the distribution of slope, Figure 3C, which is integrated again to define the distribution of deflection, Figure 3D, i.e.

$$\theta = \int \phi \, dz \qquad (3)$$

and

$$v = \iint \phi \, dz \qquad (4)$$

Where θ is the end rotation and v is the deflection. The process is repeated incrementally to define the full moment vs. mid-span deflection or end-rotation for a member. The bending moment diagram shown in Figure 3A is symmetrical. This is a convenient situation to analyse since the end rotation is equal to the integral of curvature between the support and the centre of the span. Where loading is non-uniform it is easier to double integrate curvature using the support displacements as boundary conditions, in order to define the displacement distribution. Slope distribution can then be defined by differentiating the deflection.

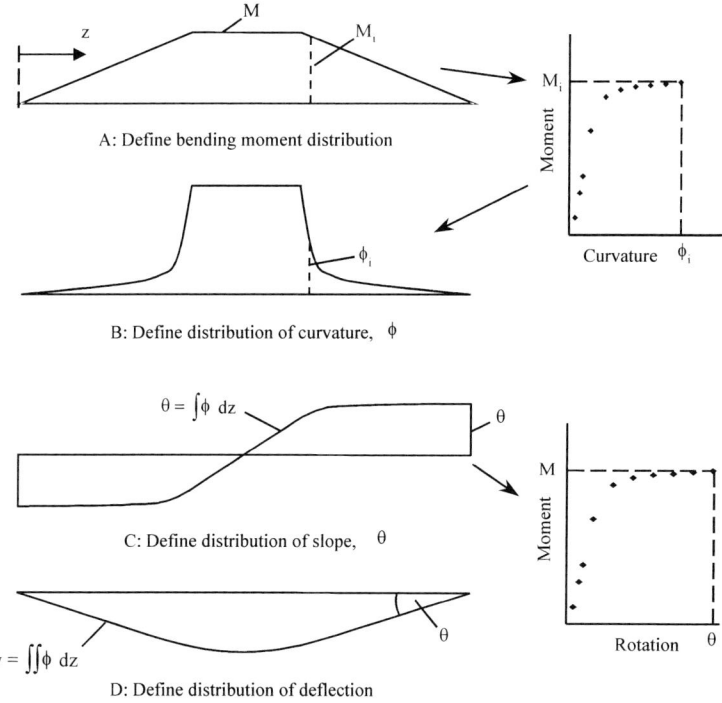

Figure 3: The calculation of slope and deflection

SURVEY OF MILL TESTS

A survey of 50 mill tests taken from hot rolled I and H sections with varying thickness and steel grade has been carried out as part of this research, see Table 1. The tests were carried out at the Corus (British Steel) Lackenby Mill and used steel produced using the Concast process. Table 1 lists average values for 0.5RT taken for the various steel grades and flange thicknesses surveyed. The 0.5RT is the material strength that corresponds to the stress at 0.5% total strain. This is the stress value that is usually quoted by manufacturers as the yield stress, since manufactures tend not to specify either the upper nor lower yield points. Mill tests are carried out at a relatively high rate over the first 1% of strain; thereafter the strain rate is not particularly high in comparison with low strain rate tests. The tests were in accordance with the British Specification BS EN 10025, which specifies that test coupons must be cut from the flanges. This is in contrast with US practice, where coupons are normally taken from web material.

TABLE 1
RESULTS FROM THE SURVEY OF MILL TESTS

Steel grade	Sample size	Mean flange thickness (mm)	Mean 0.5RT (N/mm^2)	E_{SH} (N/mm^2)
S275	10	10.4	326	2502
S275	8	20.0	302	2435
S275	7	34.9	293	2554
S355	8	11.1	399	2963
S355	9	20.0	395	2235
S355	8	37.8	374	2527

Figure 4 shows the stress strain curves from the 50 mill tests surveyed. Interestingly the tests showed all the samples exhibited roughly the same rate of strain hardening beyond strains of 1 to 1.5%. Table 1 lists the average strain hardening modulus E_{SH} derived from the tests, where E_{SH} is the slope taken between 1.5% and 4% strain. The results reveal material thickness and rate of strain hardening has no distinct influence on each other. Furthermore, the rate of strain hardening seems to differ little between the S275 and S355 steel grades. A clear link exists between the yield point and material thickness, although this link is already well documented by Byfield and Nethercot (1997), amongst others.

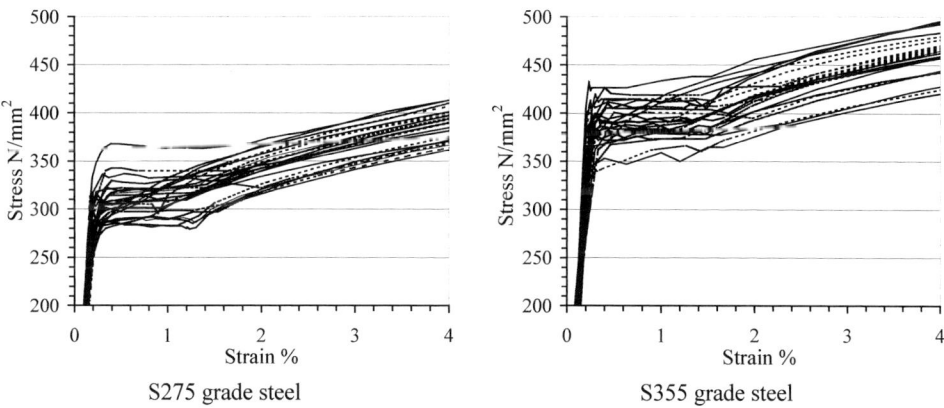

Figure 4: Combined results from 50 mill tests

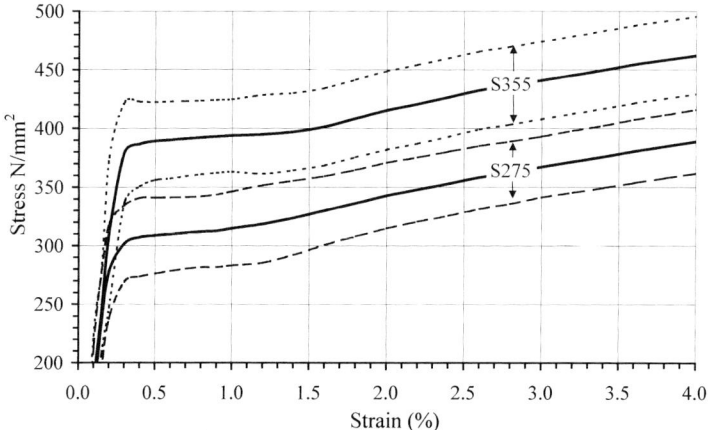

Figure 5: Stress strain curves - Mean(solid lines), upper and lower 95% confidence limits

Figure 5 shows the mean values of stress strain curves (together with the 95% confidence limits) based on the data presented in Figure 4. It is generally accepted that nominal values of yield stress are characteristic values, i.e. the lower 95% confidence limit. Figure 5 confirms this assumption, with the characteristic stress at 0.5% strain corresponding almost exactly to the nominal values of 275N/mm^2 and 355N/mm^2. Using these data the characteristic value of strain hardening modulus, E_{SH}, was found to be 2700N/mm^2 and strain hardening was found to commence at a strain equal to 6 times the yield strain, i.e., $\varepsilon_s = 6\varepsilon_y$, see Figure 1a. These values are compared with those found by other researchers in Table 2, which shows good agreement with Lay and Smiths work, although significantly less strain hardening than predicted by Horne.

TABLE 2
COMPARISON OF STRAIN HARDENING PROPERTIES

Reference	E_{SH}	ε_y
Horne (1981)	$\cong 0.05E$ (10000 N/mm^2)	6-10 ε_y
Lay and Smith (1965)	2550N/mm^2	11ε_y
Byfield and Dhanalakshmi	2700N/mm^2	6ε_y

VERIFICATION OF NUMERICAL MODEL

The capability of the model to produce accurate predictions of large deformations is shown in Figure 6, where the model predictions are compared with the experimental moment vs. end rotation for restrained beams, Byfield and Nethercot (1998). The experimental data are based on the average results from 6 identical bending tests. Theoretical predictions used the mill test stress strain curves taken from the 6 test specimens. Thus, statistical variations are reduced which allows the method to fully demonstrate the degree of accuracy with which it is able to predict large deformations. Also shown on Figure 6 is the predicted M-θ relationship for the tests based on the characteristic stress strain relationship shown in Figure 5 for S275 steel. Excellent conformity between experimental and theoretical M-θ values is demonstrated again, although the divergence between experimental and theoretical values below 1 degree of rotation is due to the characteristic value of elastic modulus being significantly below 205000N/mm^2. This is because mill tests are a poor method of measuring elastic modulus, which creates a non-linear prediction of M-θ within the elastic range.

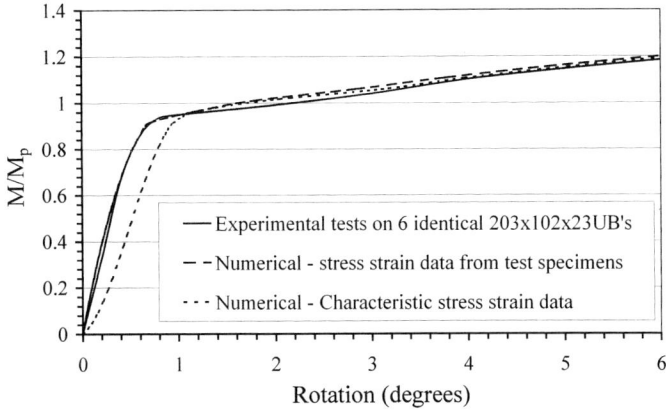

Figure 6: Comparison between experimental and predicted M-θ

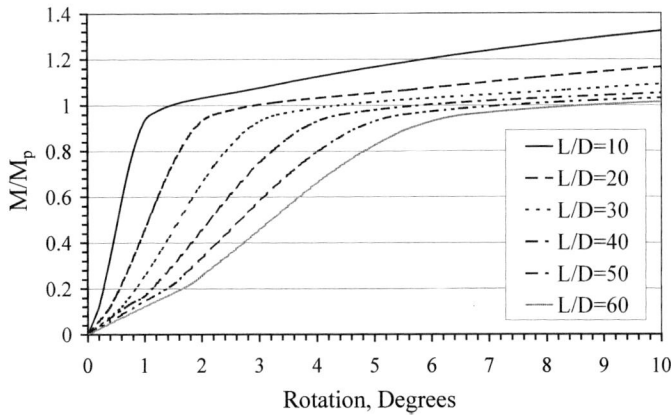

Figure 7: Normalised M-θ curves for different L/D ratios

INFLUENCE OF SPAN TO DEPTH (L/D) RATIO ON STRAIN HARDENING

Analysis of the predicted M-θ curves for sections of different weight but identical L/D ratio shows that section size does not significantly affect the capacity for strain hardening. However, L/D ratio does influence the capacity for strain hardening, as is shown in Figure 7, in which predictions were based on uniformly distributed loading on a simply supported beam, using the characteristic stress strain curve for S275 steel shown in Figure 5. This comparison of normalised moment vs. rotation for different L/D ratios shows that the higher L/D ratio, the higher the rotation required to achieve 1.0M_p. For this particular case of loading and boundary conditions, the curves imply that for each increment of 10 in the L/D ratio, there is an increment of approximately 1.5 degree in the rotation required to develop M_p. This factor may be different for different load cases and boundary conditions. The moment vs. rotation behaviour can be seen to be non-linear below the elastic moment. This is due to the slightly non-linear stress strain relationship recorded below the yield stress during the mill tests. As stated before, mill tests are an inaccurate method of measuring elastic modulus.

The bending tests carried out by Byfield and Nethercot (1998) were on a simply supported beam with a L/D ratio of 10. Figure 7 explains why these tests demonstrated such a pronounced capacity for strain hardening. If the same sections had been tested with a L/D ratio of 60, then 10 degrees of end rotation would have been required to develop $1.0M_p$. Span to depth ratios of this order are typically found in pitched roof portal frames, where it is generally assumed that second-order destabilising effects are to some extent offset by conservative assumptions regarding strain hardening, Baker (1963). Results from this investigation suggest that considerable end rotation is required in portal frames before strain hardening will begin to influence strength.

CONCLUSIONS

A survey of 50 different mill tests shows that the onset of strain hardening and the strain hardening modulus are independent of section size and steel grade. In addition, a model for predicting moment vs. deflection and moment vs. end rotation using mill test data is presented. Comparison between model predictions and experimental test data shows that the technique is capable of predicting strength and ductility for steel beams with good accuracy. The model was used to define the moment vs. end rotation characteristics of beams with varying span to depth ratios, using the characteristic values of stress vs. strain data obtained from the survey of mill tests. Results show that the ability to strain-harden is closely related to the span to depth ratio. Low L/D ratio beams can be expected to significantly exceed their plastic moment capacities. In comparison, beams with high L/D ratios, such as those found in pitched roof portal frames, will require of the order of 10 degrees of end rotation in order to develop their plastic moment capacity alone. Such beams cannot be expected to have enhanced strengths due to strain hardening, unless considerable end rotation can be accommodated before failure via either local or lateral torsional buckling type mechanisms.

ACKNOWLEDGEMENTS

This paper reports certain aspects of a project that was financed by the Engineering and Physical Sciences Research Council.

REFERENCES

Byfield, M.P. and Nethercot, D.A. (1998). An analysis of the true bending strength of steel beams. *Structures and Buildings - Proc. Instn of Civ. Engrs*, **128**, 188-197.
Byfield, M.P. and Nethercot, D.A. (1997). Material and geometric properties of structural steel for use in design. *The Structural Engineer* Vol. 75:21 363-373.
Baker J.F. (1963). *Plastic Design in Steel to BS 968*. Publication No. 21, British Constructional Steelwork Association, London.
Kemp A.R. and Dekker, N.W. (1991). Available rotation capacity in steel and composite beams. *The Structural Engineer*, Vol. 69:5, No. 5, 5 March.
Lay G. and Smith P.D. (1965). Role of strain hardening in plastic design.*Journal of Structural Divison, ASCE*. ST3, 25-43.
Horne M.R. (1981). *Plastic design of low-rise frames*. Granada Publishing Ltd, UK.

IN-PLANE ULTIMATE LOAD-CARRYING CAPACITY OF TAPERED I COLUMNS

Yan-Lin Guo   Yong Pan

Department of Civil Engineering, Tsinghua University, Beijing 100084, P.R.China

ABSTRACT

It is well known that the buckling of thin-walled steel columns is generally divided into three modes: local buckling, overall buckling and local-overall interactive buckling. Many previous research works have been carried out to investigate the local and overall interactive buckling behavior and ultimate load-carrying capacity of thin-walled columns. But all their works are focused on the steel prismatic columns with constant sections longitudinally.

This paper is intended to present a study on the local-overall interactive buckling behavior and the ultimate load-carrying capacity of tapered I-section steel columns. Local buckling and post local buckling of plate components are considered by using large-deflection elasto-plastic shell elements. The behavior of local-overall interactive buckling is investigated in large deformation and elasto-plastic range by using shell element provided by ANSYS. Based on the nonlinear finite element structural analytic method, the effects of parameters on the ultimate load-carrying capacity, including width-thickness ratio of web and flange plates, tapering ratio and load eccentricity, are considered in the analysis. From the results obtained, it can be concluded that these parameters significantly affect both the buckling failure modes and the ultimate load-carrying capacity of tapered I-section columns.

By comparing the results obtained with those of current Chinese code, some valuable conclusions are drawn and some advice is proposed for the design of tapered I-section steel columns.

KEYWORDS

Interactive Buckling, Local Plate Buckling, Ratio of component plates, Tapering Ratio, Tapered I-Columns, Ultimate Load Capacity

1 INTRODUCTION

Lightweight steel portal-frame structure has been widely used in recent years. To reduce the consumption of the steel, cross-sections of members are always tapered longitudinally according to the variation of the bending moment. And the current Chinese code[1] relaxes the restriction of the web plate width-thickness ratio, and thus the post-buckling strength of web plate can be fully used.

It is well known that the buckling of thin-walled structures is generally divided into the following three buckling modes: (1) Local plate buckling (2) Overall member buckling (3) Interaction of local and overall buckling. The study of interactive buckling in thin-walled columns began in the 1950s(Bijlaard and Fisher[2] 1953). Since then a large number of analytical studies and experiments have been carried out on the interactive behavior of local and overall buckling. Hancock[3] studied the interactive behavior of I beams using finite strip method. Little[4] proposed a method to compute load deflection curves of locally buckled columns by integrating, along the member axis, moment-thrust-curvature relations which had been obtained from the results of large deflection elastic-plastic analysis of isolated plate elements. Usami and Fukumoto[5] investigated the interaction of local and overall buckling of welded box columns. Yanlin Guo[6] investigated the interactive behavior of short struts in elasto-plastic range by employing nonlinear finite strip method. And M-P-Φ curve obtained from short struts are applied to the analysis of long columns and beams. But most researches focused on the members with uniform cross-sections longitudinally, and researches on the tapered members are few. Therefore it is necessary to study the interactive behavior of tapered columns. Chinese code employs effective width method to consider the interaction of local and overall buckling. Its formula checking in-plane stability is:

$$\frac{N_0}{\varphi_{xy} A_{e0}} + \frac{\beta_{mx} M_1}{(1 - \frac{N_0}{N_E} \varphi_{xy}) W_{e1}} \leq f$$

where: N_0——Design value of axial compressive force on the smaller cross-section; M_1——Design value of bending moment on the larger cross-section; A_{e0}——Effective area of the smaller cross-section; W_{e1}——Section modulus of effective area of the larger cross-section; φ_{xy}——Coefficient of stability of columns subjected to axial compressive force (by slenderness ratio obtained according to the smaller cross-section); β_{mx}——equivalent coefficient of bending moment; N_E——Euler load obtained according to the smaller cross-section.

2 COMPUTATIONAL MODEL

2.1 Boundary Conditions

In general, if there is no crane load, side columns of lightweight steel portal frames are always simply supported at the bottom end. And the cross-sections of side columns are tapered according to the variation of bending moment. The bending moment diagram is shown in Fig.1 when portal frame is subjected to uniformly distributed load on roof.

The computational model of tapered columns is also shown in Fig.1. Both ends of tapered columns are simply supported. The larger cross-section of tapered columns are subjected to both axial compressive force and bending moment, while the smaller cross-section is only subjected to axial compressive force, as shown in Fig.1. All columns have end plates at both ends to simulate actual columns. Shell element provided by ANSYS is employed in analysis to consider the local buckling of component plates. Since this paper only study in-plane ultimate load capacity of tapered I columns, enough lateral bracings are applied to all the examples to prevent them from lateral instability. For convenient manufacture and construction, only the width of web plate varies linearly, while the thickness of component plates and the width of flange plate are kept unchanged. This paper will just study this kind of members.

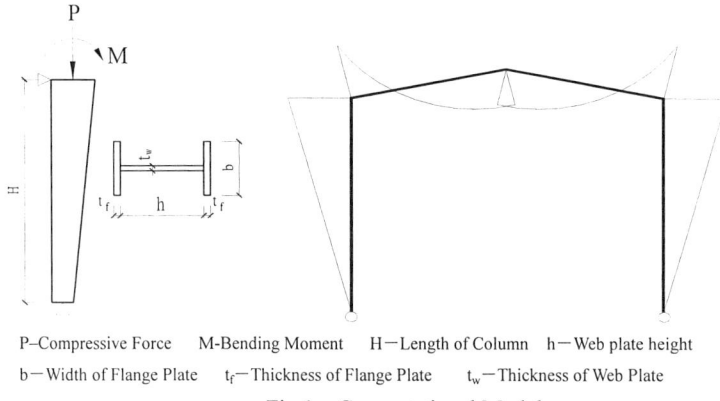

P−Compressive Force　　M−Bending Moment　　H−Length of Column　　h−Web plate height
b−Width of Flange Plate　　t_f−Thickness of Flange Plate　　t_w−Thickness of Web Plate

Fig 1　Computational Model

2.2 Material and Geometric Non-linearity

Generally speaking, the yielding of material will significantly reduce the stiffness of the tapered I columns and enlarge the buckling displacement. To consider the material non-linearity, the structural steel is assumed to be elastic-perfectly plastic isotropic hardening material in this paper, where $\sigma_y = 235 MPa$. Geometric nonlinear analysis is required for the consideration of local buckling and post buckling strength of plate components. Arc length method is employed to solve the nonlinear equilibrium equation and trace a full load-displacement curve.

2.3 Initial Geometric Imperfection

Initial geometric imperfection exists in all kinds of I columns. In this paper it is assumed that the columns to be studied have both plate initial geometric imperfection and axis initial geometric imperfection in its in-plane direction. The axis initial geometric imperfection is assumed as half sine-wave shape and its maximum value of imperfection is 0.001H, where H is the column length. Plate initial geometric imperfection is assumed as multiwave shape and its maximum value is 0.01 times web plate width-thickness ratio at the larger cross-section.

3 NUMERICAL STUDIES

The interactive buckling behavior and in-plane ultimate load capacity of tapered I columns are affected by many factors, such as width-thickness ratios of component plate, tapering ratio, load eccentricity, column slenderness ratio, residual stress and initial geometric imperfection. This paper will focus on web plate width-thickness ratio, flange plate width-thickness ratio, tapering ratio and load eccentricity.

In this paper, load eccentricity is defined as $e = M/P$, where M is the bending moment at the larger cross-section and P is the axial compressive force.

3.1 Web Plate Width-thickness Ratio

Due to the linear variation of the web plate, the width-thickness ratio of web plate is not a constant longitudinally. In this paper, the width-thickness ratio refers to that of web plate at the larger cross-section. Dimensions of examples for web plate width-thickness ratio are shown in table 1.

Table 1 Dimension of examples for web plate width-thickness ratio

Length (m)	Flange plate (mm)	Web plate (mm)	Tapering ratio (d_1/d_0-1)
8	200×12	300~600×3.0	0.926
8	200×12	300~600×4.0	0.926
8	200×12	300~600×6.0	0.926
8	200×12	300~600×8.0	0.926
8	200×12	300~600×10.0	0.926

Where d_0, d_1——height of the smaller cross-section and the larger cross-section respectively.

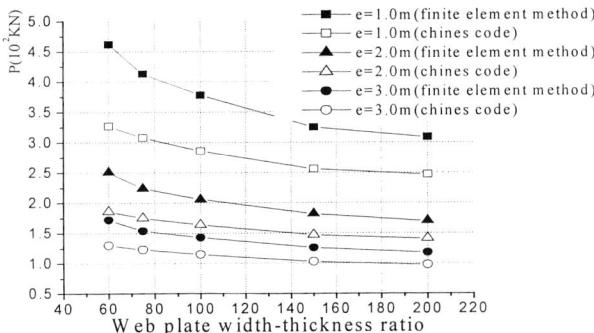

Fig.2   Effects of Web Plate Width-thickness Ratio

The effects of web plate width-thickness ratio on the ultimate load capacity are shown in Fig.2. It can be observed that the larger the web plate width-thickness ratio is, the lower the ultimate load capacity of the columns is. And with the increase of load eccentricity, the effects of web plate width-thickness ratio become less significant. The reason for this is that when load eccentricity increases, the flange plate may buckle firstly, and thus cause the failure of the columns. By comparing the results obtained by using finite element method with those of Chinese code, it can be concluded that Chinese code is conservative.

3.2 Flange Plate Width-thickness Ratio

In this paper, flange plate width-thickness ratio is defined as $(b-t_w)/2/t_f$. Dimension of examples for flange plate width-thickness ratio are shown in table 2.

Table 2 Dimension of examples for flange plate width-thickness ratio

Length (m)	Flange plate (mm)	Web plate (mm)	Tapering ratio (d_1/d_0-1)
8	200×12.0	300~600×3.0	0.926
8	200×10.0	300~600×3.0	0.926
8	200×8.0	300~600×3.0	0.926
8	200×6.0	300~600×3.0	0.926
8	200×4.0	300~600×3.0	0.926

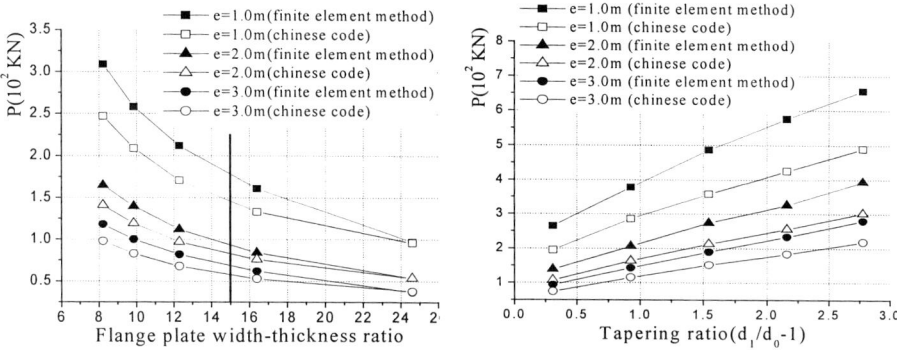

Fig.3 Effects of Flange Plate Width-thickness Ratio　　　　Fig.4 Effects of Tapering Ratio

3.3 Tapering Ratio

Tapering ratio is a significant factor determining the ultimate load capacity of tapered I columns. In Chinese code it is defined as: $\gamma = (d_1/d_0) - 1$, where d_1 is the height of the larger cross-section and d_0 is the height of the smaller cross-section. Only the larger cross-section of tapered columns is subjected to both compressive force and bending moment. So increasing the height of the larger cross-section will greatly enhance its inertia to resist bending moment. To investigate the effects of tapering ratio on the ultimate load capacity the height of the larger cross-section is variable, and the height of the smaller cross-section remains constant. Dimensions of columns for tapering ratio are shown in table 3.

Table 3 Dimension of columns for tapering ratio

Length (m)	Flange plate (mm)	Web plate (mm)	Tapering ratio (d_1/d_0-1)
8	200×12	300~400×6.0	0.309
8	200×12	300~600×6.0	0.926
8	200×12	300~800×6.0	1.543
8	200×12	300~1000×6.0	2.160
8	200×12	300~1200×6.0	2.778

The effects of tapering ratio on ultimate load capacity are shown in Fig.4. It can be found that the ultimate load capacity of tapered I columns can be greatly enhanced by increasing the height of the larger cross-section. But with the increase of the load eccentricity, the magnitude of increase becomes small.

3.4 Compressive Force and Bending Moment Interactive Curve

Generally, the design formulae are expressed as compressive force and bending moment interactive curve, and the designers can use them conveniently. The interactive curve is an effective demonstration to study the behavior of columns subjected to both compressive force and bending moment. Dimension of examples is shown in table 4.

Table 4 Dimension of columns for compressive force and bending moment interactive curve

Length (m)	Flange plate (mm)	Web plate (mm)	Tapering ratio (d_1/d_0-1)
5	200×12	200~400×6.0 (8.0)	0.893
5	200×12	200~600×6.0 (8.0)	1.786
5	200×12	200~800×6.0 (8.0)	2.678
7	200×12	300~600×6.0 (8.0)	0.926
7	200×12	300~900×6.0 (8.0)	1.852
7	200×12	300~1200×6.0 (8.0)	2.778

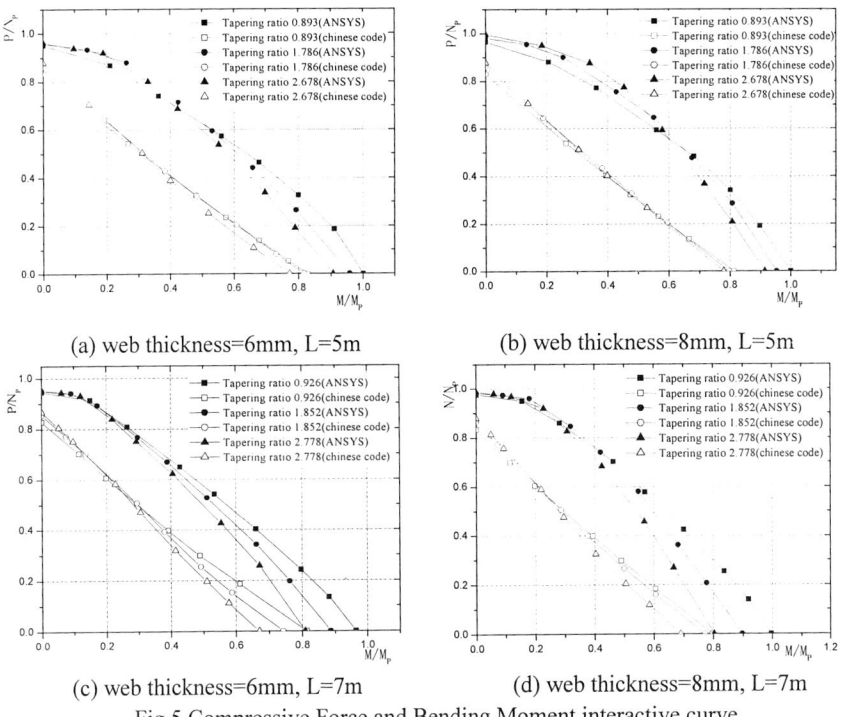

(a) web thickness=6mm, L=5m (b) web thickness=8mm, L=5m
(c) web thickness=6mm, L=7m (d) web thickness=8mm, L=7m
Fig.5 Compressive Force and Bending Moment interactive curve

Compressive force and bending moment interactive curve are shown in Fig.5, where N_P is the ultimate load when the full smaller cross-section subjected to pure compressive force yields and M_P is the ultimate bending moment when the full larger cross-section subjected to pure bending moment yields. It can be observed that when load eccentricity is very small, the ultimate load capacity of columns is almost the same because the yielding of the smallest cross-section is the main cause of the failure of the tapered columns. But with the increase of the bending moment, the relative ultimate load capacity of columns with larger tapering ratio and larger web width-thickness ratio is lower. The reason lies in the fact that the greater the tapering ratio is, the more easily the web plate of the larger cross-section buckles. So local buckling is much easier to take place for larger tapering ratio columns. And its relative ultimate load capacity is smaller for larger load eccentricity. It can also be seen that interactive formulae employed in Chinese code to check in-plane stability are conservative and almost linear.

3.5 Failure mechanism

The buckling deformation is shown in Fig.6, which shows the effects of web plate width-thickness ratio, load eccentricity and tapering ratio on ultimate load capacity of tapered columns. All the deformation is magnified 15 times. Dimensions and load eccentricities are shown in table 5.

Table 5 Dimension of tapered columns

Serial number	Length (m)	Flange plate (mm)	Web plate (mm)	e(m)	Tapering ratio
A	8	200×12	300~1200×6.0	2.0	2.778
B	8	200×12	300~600×3.0	1.0	0.926
C	8	200×12	300~600×6.0	2.0	0.926
D	8	200×12	300~600×6.0	1.0	0.926

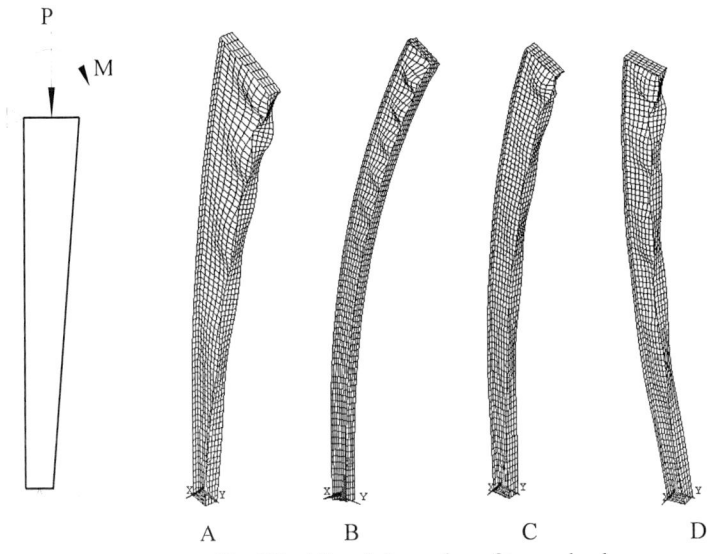

Fig.6 Buckling deformation of tapered columns

From the buckling deformation diagram, plate local buckling can be obviously observed at larger compressive component plates near the larger cross-section when the columns reach their ultimate loads. And the overall buckling of the columns is also obvious. So local buckling and overall buckling are interactive. When load eccentricity is small, and flange plate is thick and web plate is thin, the local buckling of web plate is more obvious, as shown in Fig.6 (B) and Fig.6 (D). On the contrary, the local buckling of flange plate is more obvious, as shown in Fig.6 (A) and Fig.6 (C). But local buckling of the smaller cross-section can hardly take place. Therefore, the local buckling of web plate and flange plate near the larger cross-section, as well as the interaction of local buckling and overall buckling, finally cause the failure of tapered I columns.

Because the web plate width-thickness ratio of the larger cross-section is generally larger, local buckling is easier to take place at the larger cross-section. At the same time, due to the effects of the bending moment, the compressive stress of component plate near the larger cross-section is much larger. Therefore the component plates near the larger cross-section buckle more easily, which speeds the failure of overall stability of tapered I columns in the end.

4 CONCLUSIONS

This paper investigated the effects of width-thickness ratio of web plate and flange plate, tapering ratio and load eccentricity on ultimate load capacity of tapered columns. The following conclusions can be reached:

(1) Width-thickness ratios of web plate and flange plate both have significant influence on in-plane ultimate load capacity of tapered columns, but the latter effects are greater;

(2) By increasing the height of the larger cross-section, the in-plane ultimate load capacity of tapered I columns can be greatly enhanced;

(3) with the increase of load eccentricity, local buckling of the flange plate and web plate at the larger cross-section is more obvious. In this case the failure of tapered I columns is mainly caused by the local buckling of compressive component plates under large load eccentricity.

(4) In-plane design formulae employed in Chinese code to check the in-plane stability of tapered columns are safe and reliable.

By comparing the results obtained by using finite element method and those of Chinese code, it is suggested that it may reach a good economic efficiency for the tapered I column with larger web plate and smaller flange plate width-thickness ratio. Thicker flange plate can both enhance the overall stability of tapered columns and prevent local buckling of component plates. As a result, it can prevent the failure of tapered columns from the buckling of the flange plate.

REFERENCES

[1] Chinese code of lightweight steel portal frame, Beijing, 1999
[2] Bijlaard, P.P. & Fisher, G.P., Column strength of H-sections and square tubes in post-buckling range of component plates. Technical Note 2994, NACA, USA, 1963
[3] Hancock G J. "Local Distortional , and Lateral Buckling of I Beams," J. Struct. , Div ,ASCE 1978, 104(11), 1787~1798
[4] Little, G. H., "The Strength of Square Steel Box Columns—Design Curves and their Theoretical Basis," The Structural Engineer, Vol. 57, Feb., 1979, pp. 49-61
[5] Usami, T., and Fukumoto, Y., "Local and overall buckling of welded box columns," J. Struct. Div., ASCE, 1982, 108(3), 525~542
[6] Yanlin Guo., "Local and Overall Interactive Instability of Thin-Walled Box-Section Colums," J. Construct. Steel Research., 1992, 1~19

ELASTIC TORSIONAL-FLEXURAL BUCKLING OF TAPERED I BEAM-COLUMNS

Y. L. Guo, Y. Han, W.Q. Hao, T. Liu

Department of Civil Engineering, Tsinghua University,
Beijing 100084, CHINA

ABSTRACT

This paper presents the study of elastic torsional-flexural buckling of longitudinally tapered I section column and the steel portal frame formed by tapered members, respectively. The tapered columns are subjected to both axial compression and moment generally.
The main research work has been focused on the investigation of parameter influence of tapered ratio on the torsional-flexural buckling of tapered column and the portal frame respectively. The adopted finite element model is based on thin-walled beam theory and is efficient in the analysis of bifurcation buckling of tapered column where the end section warp of the column is considered as fully free and fully constrained. Then a finite element program developed is employed to compute the elastic torsional-flexural buckling loads of tapered beam-columns and steel portal frame. Numbers of numerical studies have been carried out. From the obtained results presented in graphic form, a set of simplified interaction curves that can predict more accurately the elastic torsional-flexural buckling loads of tapered beam-columns under both axial forces and moment is conducted. Finally the influence of the out-of-plane bracings on the elastic torsional-flexural buckling loads of the tapered columns is analyzed theoretically, and the out-of-plane elastic buckling behavior of global portal frame is also involved in the analysis as well.

KEYWORDS: torsional-flexural, buckle, tapered, beam-column

INTRODUCTION

The steel portal frame has found its wide application in industrial building around the world, and the tapered members have been widely adopted in the frame because of their high economic efficiency.

The tapered members in portal frame are generally treated as the members where both axial compression and moment are acted on them, and the torsional-flexural buckling frequently controls their design generally.

Unfortunately, the previous research work mainly concentrated on the behavior of torsional- flexural buckling of prismatic columns, but few researchers have focused on the torsional-flexural buckling of longitudinally tapered section columns. Of interest are the works by Kitipornchai and Trahair(1972,1975) on the stability of tapered mono- and bisymmetric I-beams, in which the method of finite integrals is used to get clear physical meanings for the bending and torsion. Starting from the energy considerations, critical loads for the torsional-flexural buckling of tapered I-beams was determined by Brown(1981) using the finite difference scheme. Further work done by Yang(1987) derived the elastic and geomdetric stiffness matrices of tapered bisymmetric I-beams.

In this paper, an attempt is made to study the behavior of elastic torsional-flexural buckling of tapered beam-columns subjected to the combined action of axial compression at two ends and bending moment at one end. The objective to the analysis of tapered columns is to produce an interaction formula based on a lot of numerical investigation, which may directly be employed in the design. And the elastic lateral buckling behavior of global portal frame is also involved in the paper as well.

THEORETICAL ANALYSIS

In the following statement, the different geometric parameters of the two element at the ends are reflected as the geometric properties to discuss the varying cross sectional.

Assumption

For the present purposes, a tapered thin-walled member of an I-section will be considered (Fig.1(a-b)). In analysis, the following assumptions are made: (1) The material is elastic and homogeneous; (2) the length L of the member is large compared with the maximum cross-sectional dimension B, i.e. $B/L \leq 0.1 \sim 0.2$; (3) the section is thin-walled, the thick - width ratio satisfy $t/b \leq 0.1$; (4) every cross section is rigid in its own plane under the small deformational condition and permits warp in the direction of out of plane; (5) shearing deformation of the middle surface of the member is negligible.

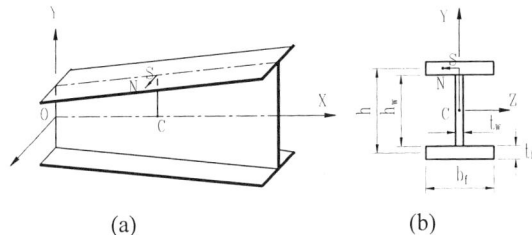

Figure 1: Tapered I-beam (a) Coordinates; (b) Section

3-Dimensional varying cross sectional thin-walled beam element

The elastic stiffness matric of this kind of element can be derived on the basis of the uniform thin-walled beam, considering the differences of geometric relationship of the deformation and strain between them. A linear displacement field is chosen for the axial displacement, u(x), and a cubic field for other displacements, i.e.

$$u(x) = [n_1]\{\bar{u}\}, \quad v(x) = [n_3]\{\bar{v}\}, \quad w(x) = [n_3]\{\bar{w}\}, \quad \theta(x) = [n_3]\{\bar{\theta}\} \tag{1}$$

where, using $i = x/L$

$$[n_1] = [(1-i)i], \quad [n_3] = [(1-3i^2+2i^3) \quad (i-2i^2+i^3) \quad (3i^2-2i^3) \quad (i^3-i^2)] \tag{2}$$

Then the elastic stiffness expression and the details of the elastic stiffness matix element can be found in Ref. 1.

Geometric stiffness matrix

If the nonlinear deformation-strain relationship is considered, the incremental virtual work equation of equilibrium may be writtten as

$$\int_V \left[Ee_{xx}\delta e_{xx} + 4Ge_{yx}\delta e_{yx} + 4Ge_{zx}\delta e_{zx} \right] dV + \int_V \left[\sigma_{xx}\delta\eta_{xx} + 2\sigma_{yx}\delta\eta_{yx} + 2\sigma_{zx}\delta\eta_{zx} \right] dV = R \tag{3}$$

in which the nonlinear components of the Green-Lagrange strain tensor may be written as

$$\eta_{mn} = \tfrac{1}{2}(u_{,m} \cdot u_{,n}); m,n = x,y,z. \tag{4}$$

If the stress resultants on a cross section of the tapered I-beam are defined as

$$F_x = \int_A \sigma_x dA; \quad F_y = \int_A \tau_{yx} dA; \quad F_z = \int_A \tau_{zx} dA; \quad B = -\int_A \sigma_x \omega_T dA;$$

$$M_x = \int_A (\tau_{zx} y - \tau_{yx} z) dA; \quad M_y = \int_A \sigma_x z dA; \quad M_z = -\int_A \sigma_x y dA; \tag{5}$$

substitution of Eqs. (4) and (5) into (3) yields the geometric stiffness matrix of the element according to the mentioned displacement model.

The modification of the geometric stiffness matrix

In order to prove the continuity and balance of the rotation of the joint, the geometric stiffness matrix $[k_g]$ should be added up a correctional stiffness matrix, which details in the Ref.1.

The effect of the location of the transverse load

The amended geometric stiffness matrix can reflect the effect of the shear center. The additional work caused by the transverse load owing to the moving of the shear center is:

$$V_a = \frac{1}{2}\sum_i P_{yi} a_{yi} \phi_i^2 + \frac{1}{2}\sum_i P_{zi} a_{zi} \phi_i^2 \tag{6}$$

where the directions of P_{yi}, P_{zi} are positive with the same direction of the Y, Z axis, and a_{yi}, a_{zi} are the coordinates of the force action point in local coordinate system.
The amended geometric stiffness matrix can be rewritten when the equation (6) was added in the corresponding joint rotational item of the geometric stiffness.

Solution

After assemble the global stiffness, one can then obtain the equations of equilibrium for a tapered I-beam in a matrix form as

$$([K_e]+[K_g])\{u\} = \{F\} \tag{7}$$

In a bifurcation buckling analysis, an attempt is made to detect the instability load without calculating the absolute values of displacements. Usually, the linear stiffness matrix is assumed to be constant, and the geometric stiffness matrix is considered simply as a multiple of its initial value obtained for a reference load. At the buckling load, the equations of equilibrium reduce to

$$\left|[K_e]+\lambda[K_g^*]\right| = 0 \tag{8}$$

where $[K_g^*]$ represents the initial geometric stiffness matrix compared for the reference load distribution; λ= load factor. Eq.(8) represents one form of the eigenvalue problem, which may be solved, by an inverse vector iteration in the paper.

NUMERICAL STUDIES

An interactive computer program has been performed to study the bifurcation buckling. In the following, it is assumed that the columns are supported simply at both ends. But the end sections are considered as two cases for discussion as follows respectively:
   a. Warp fully free of the support section:
      when x=0 or x=L, $u_y = u_z = \theta = 0$, $u_y'' = u_z'' = \theta'' = 0$
   b. Warp fully constrains of the support section:
      when x=0 or x=L, $u_y = u_z = \theta = 0$, $u_y'' = u_z'' = \theta' = 0$, where θ is the torsion angle.

Defining eccentricity distance e=M/P, the effects of the elastic torsion-flexural buckling of varying section column shown in Fig.2 are then discussed.

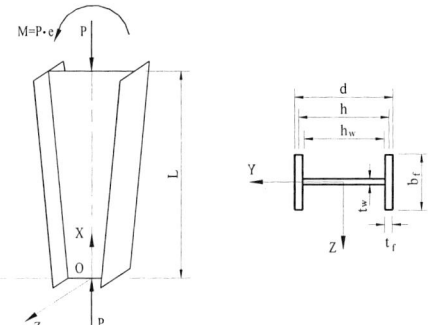

Figure 2: Torsion-flexural of the varying section column

Effect of tapering ratio γ

The following dimensions with successive varying smaller web value in table 1 were adopted to analyze the torsion-flexural instability and the height of the columns is all 8m. The P_{crl} represents critical loads at the larger section and only one of the results is shown in Fig.4 because of the identical tendency of others.

TABLE 1
THE SECTION DIMENSION (unit: mm)

	Flange	Larger web	Smaller web	I_y	Tapering ratio γ
C1	200x10	800x6	100~800x6	1.335×10^7	0 ~ 5.83
C2	250x10	800x6	100~800x6	2.606×10^7	0 ~ 5.83
C3	300x10	800x6	100~800x6	4.501×10^7	0 ~ 5.83

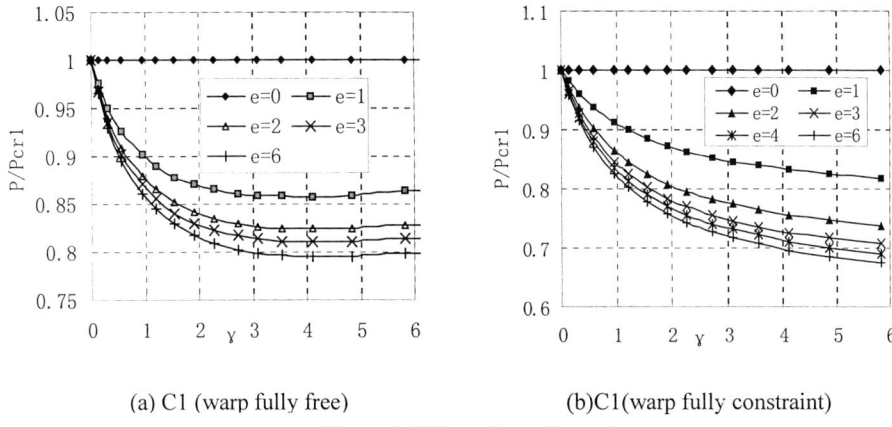

(a) C1 (warp fully free) (b) C1 (warp fully constraint)

Figure 3: Effect of tapering ratio

It can be seen from the Fig.3 that: (1) The critical load of minor axis remains the minimum value when $\gamma<6$ and the change of γ the will not influence the flexural stiffness of minor axis EI_y; (2) When the value of eccentricity (e) is large, the critical load reduce quickly with the growth of γ ; (3) The varying section will cause the decrease of the critical load when the warp of the support section is constrained; (4) The decrease of the critical load is more distinct when γ changes from 0 to 1 under the condition of $e \neq 0$.

Effect of the out-of-plane rigid bracing

Numerical solutions have also been obtained for the critical loads of the C2 (warp fully constraint) in table 1. The column height is 8m. The tapering ratio is $\gamma= 0.952, 2.727$, corresponding to the small web depth 800mm, 400mm, 200mm respectively. The location of the lateral brace is at x=4m, 5m, 3m and 6m from the column bottom end (see Fig. 2). The results obtained from two cases are drawn in Fig. 4, in which d_1 = 820mm is the height of the larger web section, and P_{cr1} is the critical load without bracing under the same condition.

As can be seen from the Fig.4 that: (1) when the eccentricities are small, the axial compression P is the main factor affecting column torsion-flexural buckling. However, if the moment (M = P· e) is predominate, the conclusion is opposite; (2) effects of the bracing at x=4m and x=5m are constansant if the tapering ratio is large. However, the effect of bracing at x=5m is stronger than at x=4m when load eccentricity is enough large.

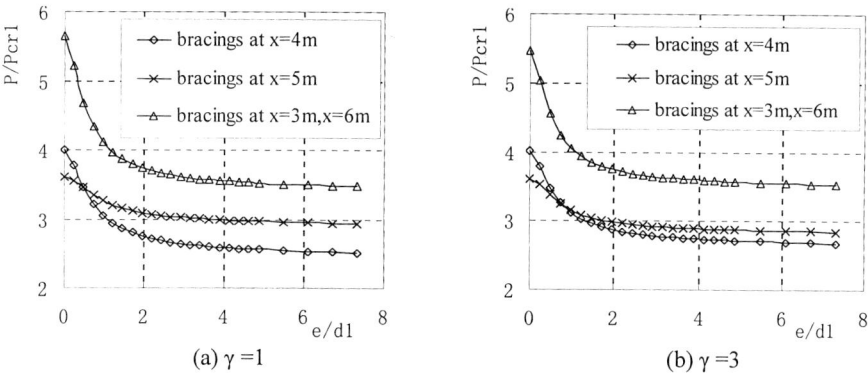

Figure 4: Effect of the brace of the varing section

Interative design curve

Again analyse the member shown in Fig.2 under two different boundry condition mentioned above while changing the ratio of M/P from 0 to 6 in table 1.

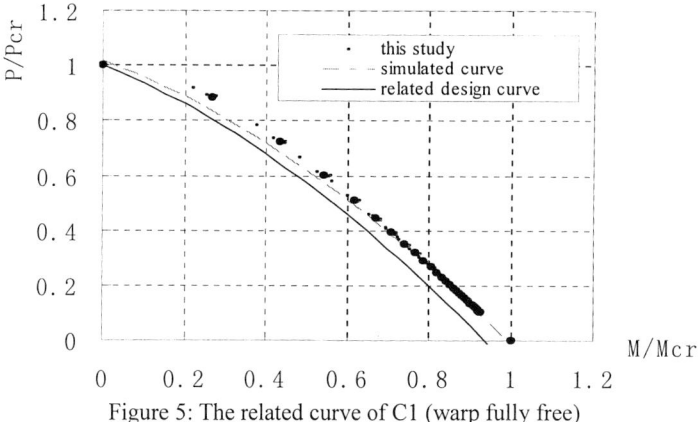

Figure 5: The related curve of C1 (warp fully free)

From the calculated results, the fitting curves can be drawn and the interaction formulae for predicting the flexural-torsional buckling load of the column with tapered section are suggested as follows:

$$\frac{P}{P_{cr}} + 0.5(\frac{M}{M_{cr}})^2 + 0.6\frac{M}{M_{cr}} \leq 1 \quad \text{(warp fully free)} \quad (9)$$

$$\frac{P}{P_{cr}} + 0.9(\frac{M}{M_{cr}})^2 + 0.15(\frac{M}{M_{cr}}) \leq 1 \quad \text{(warp fully constraint)} \quad (10)$$

where P_{cr} is the critical flexual load about the minor axis of varying section column with the axial compression; $P_{cr} = \frac{\pi^2 EI_y}{L^2}$, M_{cr} is torsional-flexural critical load of varying section column when moment is only acted on the large section and is expressed as[7]:

$$M_{cr} = \frac{1}{\beta_t}\left(\frac{\pi^2}{(\mu_s L)^2} EI_{y0} GI_{t0} + \frac{\pi^4}{(\mu_\omega L)^4} EI_{y0} EI_{\omega 0}\right)^{\frac{1}{2}}$$

$$\mu_s = 1 + 0.023\gamma \sqrt{Ld_0/A_{f0}}$$

$$\mu_\omega = 1 + 0.00385\gamma \sqrt{L/i_{y0}}$$

$$\beta_t = 0.6(1 + 0.25\sqrt{\gamma})$$

$$\gamma = (d_1 - d_0)/d_0$$

Subscript 0 represents the section property of the smaller section of varying section and subscript 1 represents the property of the larger section. The results of column (C1) shown in Fig.5 indicate that the suggested formulation is reasonable. The same conclusion can be indicated from the results of other columns that don't mension here.

Out-of-plane stability of portal frame

A typical portal frame (Fig.6) is adopted to analyse the out-of-plane instability. In the figure, L=36m, i=1:12. Referent load P=1kN, q=0.1kN/m. All the members belong to I section, which properties are: b_f=300mm, t_f=12mm, t_w=8mm. The section depth is: d_1=624mm, d_2=824mm and d_0 is variable with respect to the tapering ratio $\gamma = d_0/d_1 - 1$ (special for the tapering ratio of the column). The locations of the lateral bracings are also shown in Fig.6: (1) 3 point lateral bracings, one at the ridge of roof and the other two at top of columns; (2) 5 point lateral bracings, one at the ridge of roof, two at top of columns and two at quarter of the beam; (3) 7 point lateral bracings, one at the ridge of roof, two at the top of columns and four at the beam. The bracing of the column is at the half height of the column. In order to consider the effect of the lateral bracing on the out-of-plane stability, two steel portal frames of column lever height H=6m(GJ1) and H=12m(GJ2) is calculated.

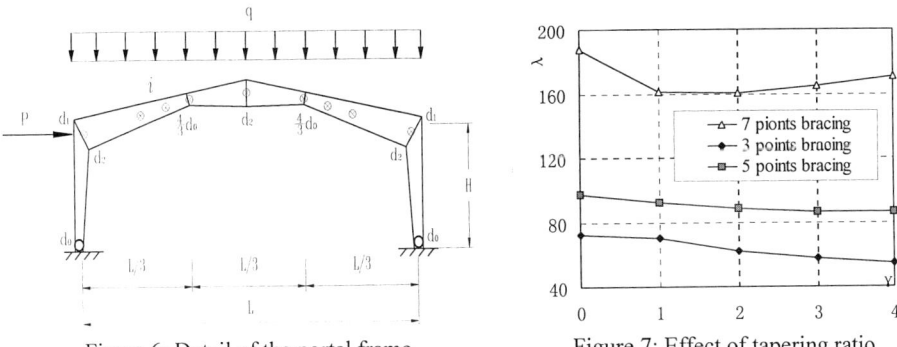

Figure 6: Detail of the portal frame   Figure 7: Effect of tapering ratio

The results obtained from the above cases were plotted in Fig.7 and Fig.8, in which the longitudinal coordinate $\lambda=P_{cr}/P$ or $\lambda=q_{cr}/q$, where P_{cr}, q_{cr} is the corresponding critical load. As expected, the effect of tapering ratio γ is not obvious to the out-of-plane stability. Furthermore, the bracing of the column at half height of the column is more efficient to increase the out-of-plane load carrying capability of the portal frame when the height of the column is large.

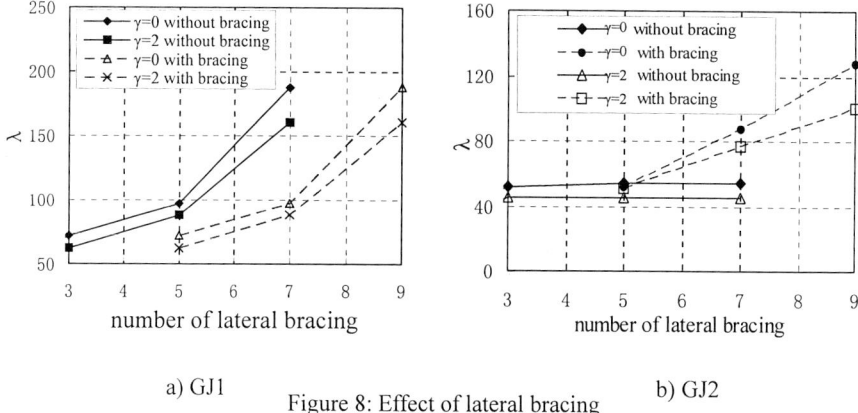

a) GJ1 Figure 8: Effect of lateral bracing b) GJ2

CONCLUSIONS

A 3-dimension thin-walled beam element analysis program is set up to accurately analyze the light gauge steel structure with tapered members. Numbers of examples are presented to study theoretically the elastic torsional-flexural buckling loads of tapered beam-columns and steel portal frames respectively. The interaction formulae that are used to predict the flexural-torsional buckling critical load of the column with tapered section are proposed, which may be helpful in the design of the portal frame with tapered I column.

It is also concluded from the out-of-plane stability study of tapered column that the reasonable location of the lateral bracing is at about the one-third height of the column from the top end when the end moment of the column is predominate.

REFERENCE

[1] W.Q.Hao(1999). Instability Behavior of Light Gauge Steel Frame Structures Graduate Thesis for Master Degree of Tsinghua University
[2] Yang Y.B.,Yau, J.D.(1987). Stability of Beams with Tapered I-Sections. Journal of Engineering Mechanics, ASCE, Vol.113(9),1337-1357.
[3] John C. Ermopoulos(1986).Buckling of Tapered Bars Under Stepped Axial Loads. Journal of Structural Engineering, ASCE, Vol112(6), 1346-1354.
[4] Kitipornchai, S., and Trahair, N.S.(1972). Elastic stability of tapered I-beams. Journal of structure Div., ASCE, Vol 98(3), 713-728.
[5] Kitipornchai, S., and Trahair, N.S.(1975). Elastic behavior of tapered monosymmetric I-beams. Journal of structure Div., ASCE, Vol 101(8), 1661-1678.
[6] Brown, T.G..(1981).Lateral-torsional buckling of tapered I-beams.Journal of Structure Div., ASCE, Vol 107(4), 689-697.
[7] Lee, G.C., M.L.Morrel, Ketter, R.L.(1972). Design of Tapered Member. Weld. Res. Counc. Bull, No.173, June, 1-32.

LOAD-CARRYING CAPACITY OF
BOX SECTION BEAM-COLUMN

T. Liu and Y.L. Guo

Department of Civil Engineering, Tsinghua University, Beijing, 100084, China

ABSTRACT

A finite element analysis with the shell element is employed to study more accurately the load-carrying capacity of box-section column where the component plates forming the column are suffered plate local buckling. With the application of shell element, the geometric shape and the boundary conditions of the box-section column can be exactly simulated. The initial plate imperfection of the component plates and the initial overall imperfection of column can be conveniently incorporated into the analysis.

Numbers of parameter analysis have been carried out to investigate the ultimate strength of the box-section columns that are subjected both the axial compressive force and the bending moment. The results obtained indicate that the plate width-thickness ratio and the column slenderness ratio are main factors that can affect its ultimate strength. These parameter influences have been emphasized in the paper. The failure modes of the columns with different width-thickness ratio and slenderness ratio are also studied. The results obtained show that the failure modes are quite different for the columns with various width-thickness ratios of component plates and the slenderness ratio of the member.

KEYWORDS: Finite shell element, box-section, beam-column, width-thickness ratio, slenderness ratio

1. INTRODUCTION

Steel frame structure has been widely used in the dormitory, supermarket, emporium and storehouse. The box-section columns are employed in the steel frame structure because of their good behavior in carrying biaxial axial eccentric loading. So the box-section column has found its wide application in engineering structures.
Traditional design method of the frame-column doesn't allow the local buckling of component plates before the member's overall failure. So the width–thickness ratio of the component plate is strictly restricted. Actually, the buckling of the component plates usually doesn't mean the member's failure, and the post-buckling strength of component plates is usually utilized in the design of thin-walled structure.
From the 1980s, many scholars have done a lot of research work on the ultimate strength of the

box–section column and employed many methods to consider the influence of plate local buckling. Generally speaking, these methods can be categorized into the following types. They are effective-width method (Dewolf. 1974), numerical integration method (Siu Lai Chan, Sritawat Kitipornchai and Faris G. A. Ai-Bermani, 1991; N. E. Shanmugam, J. Y. Richard Liew and S. L. Lee, 1989), finite strip method (Hancock 1981, Sridharan and Ali 1988, Guo and Chen 1989) and the curved shell finite element method (Shen and Zhang 1989). Among the above methods, the curved shell finite element method is the most effective one and can be used to predict accurately the behavior of locally buckled column because using shell element can exactly simulate the geometry shapes, boundary conditions, deformation and component plates buckling of the box-section column. Shen and Zhang (1989) have done some research work by utilizing this method, but their work needs to be expanded in many fields.

With the help of ANSYS software package where four node shell element is employed in this study, the load-carrying capacity of box-section column and its failure mechanism are investigated theoretically by varying its geometric parameters.

2. ANALYTICAL MODEL

Boundary Conditions

The column is supported simply and is subjected to a combination of axial force and bending moment at the ends simultaneously, as shown in the Fig. 1. It is assumed that the end sections of the column are fully warp constrained.

Initial Geometry Imperfection

The initial local imperfection of the component plate is assumed to be the same as the first order local buckling modal of the column when only the local buckling of the component plates occurs. The maximal amplitude of the local imperfection is taken as 0.01b, where b represents the width of the column section. An initial overall axis imperfection with the shape of half sine wave is adopted in this paper. Statistic imperfection amplitude employed in current Chinese code is L/1000 where L represents the length of the column.

Member Geometric Dimension

Both the width of the column section are 400 mm and the length of the column is variable to achieve different slenderness ratio, as shown in Fig. 1.

Finite Element Model

The computational mode of the finite element analysis of the box-section column is shown in Fig. 2. Two rigid plates are added at both ends of the column and the axial force is applied at the centroid of the rigid plate; the bending moment is applied by a pair of pressure force distributed along the two opposite direction on the end plate edges of the column.

Four nodes shell element is employed in the analysis. There are six degrees of freedom at each node, three translations in x, y and z directions and three rotations about the x, y and z-axis. This element is well suited for linear, large strain and large rotation nonlinear analysis. The material is assumed to be elasto-perfectly plastic, and the yield stress is 235N/mm².

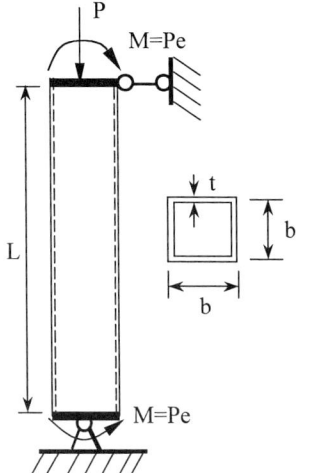
Fig. 1. Boundary condition of the box column

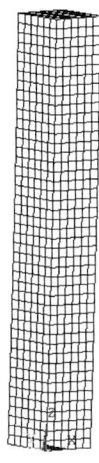
Fig. 2. FEM model of the column

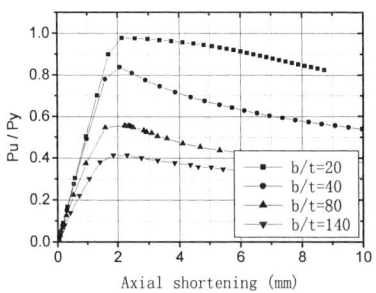
Fig. 3. Load-axial shortening interaction curves of column (λ =10)

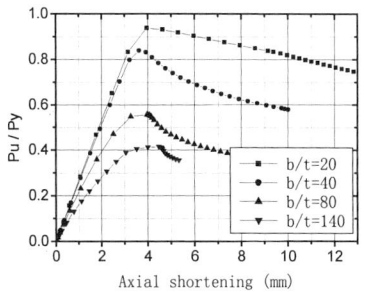
Fig. 4. Load-axial shortening interaction curves of column (λ =20)

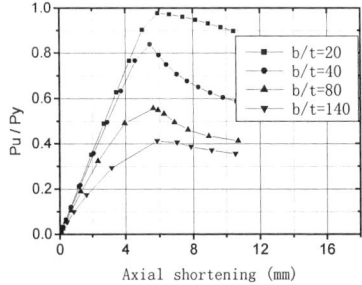
Fig. 5. Load-axial shortening interaction curves of column (λ =30)

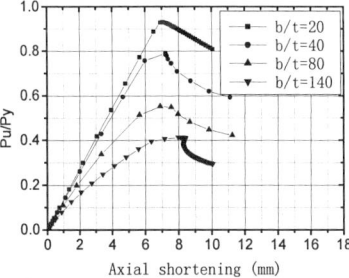
Fig. 6. Load-axial shortening interaction curves of column (λ =40)

3. NUMERICAL RESULTS

Numbers of parameter analysis has been carried out to reveal the load-carrying capacity of the box–section column. Generally speaking, the main parameters that may influence the ultimate strength of the column are width-thickness ratio of the component plates, the slenderness ratio of the column, residual stress and the initial imperfection. Among these parameters, the effects of width-thickness ratio of the component plates and the slenderness ratio have been focused on in this paper.

Ultimate Strength of Box-Section Column Loaded Axially

The load-axial shortening interaction curves of the columns with different slenderness ratio (λ =10, 20, 30 and 40) and width-thickness ratio (b/t =20, 40, 80 and 140) of component plates are shown in Fig. 3- Fig. 6.

Ultimate Strength of the Column Subjected to Eccentric Axial Compressing Force

A set of columns with the b/t=20,40,80,140 and different slenderness ratio (λ =10,20,30,40) are studied under the axial force and the bending moment. Two relative eccentricity is represented by e* where e*=e/b=2.5 and 5, and b is the width of the box-section and kept a constant in the study. The load-lateral deflection curves are presented in the Fig. 7 - Fig. 10

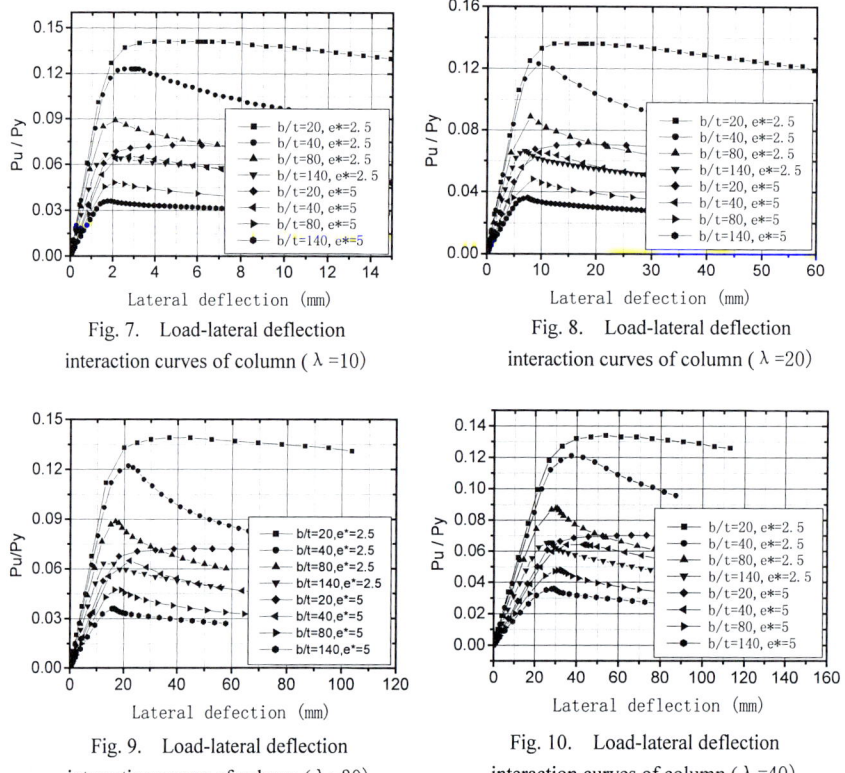

Fig. 7. Load-lateral deflection interaction curves of column (λ =10)

Fig. 8. Load-lateral deflection interaction curves of column (λ =20)

Fig. 9. Load-lateral deflection interaction curves of column (λ =30)

Fig. 10. Load-lateral deflection interaction curves of column (λ =40)

Failure Modes of Columns

The above interaction curves show that the ultimate strength of the column decreases as the increasing of the width-thickness ratio of component plates under axial and eccentric compressive loads. But the interaction curves are quite different from each other, especially at the range of post limit point of the column. The main reason for this is that the failure modes of the columns are different from each other, which is result from different width-thickness ratios of component plates.

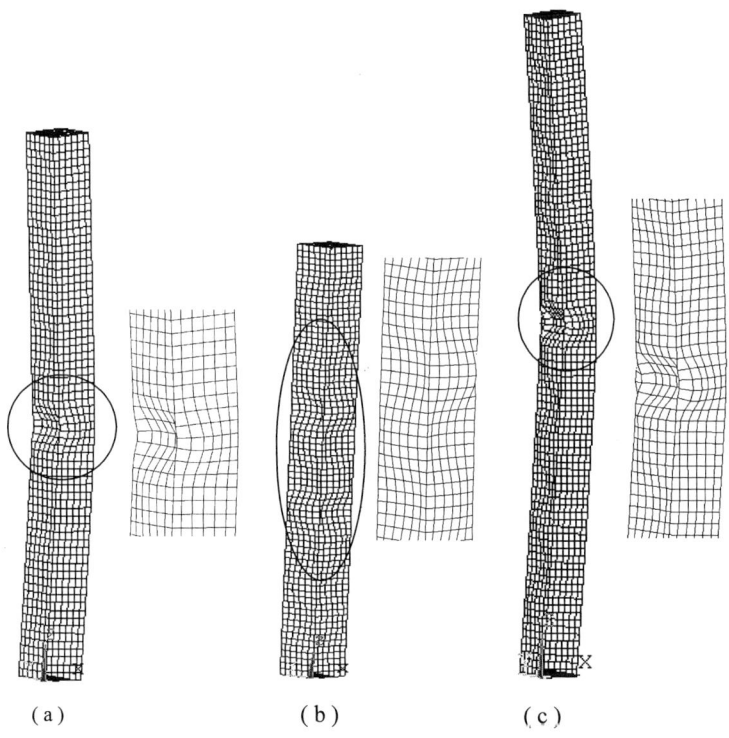

(a)　　　　　　　　　(b)　　　　　　　　　(c)

Fig. 11.　Failure modes of the columns

(a) λ=30, b/t =20.　(b) λ=20,b/t=140　(c) λ=40,b/t=140

For the columns with different slenderness ratio, those with larger ones tend to fail in the overall buckling, those with smaller ones tend to fail in the yielding of the whole section or the local buckling of the component plates, and the others with middle ones will fail in the interactive buckling.

The failure mode is also different for the column with different width-thickness ratio, those with smaller width-thickness ratio and smaller slenderness ratio usually fail mainly in the yielding of the whole section, but those with larger slenderness ratio tend to fail mainly in a single wave overall buckling along the axis of the column where the local buckling is predominate at the middle-length section of column (shown as the Fig. 11(a).). The columns with larger width-thickness ratio and smaller slenderness ratio tend to fail accompanied by a multi-wave local buckling of the component plates along the axis of the column (shown as the Fig. 11(b)), while those with larger width-thickness and larger slenderness ratio will fail accompanied by a main single-wave overall buckling and a set of subsidiary multi-wave local buckling (shown as the Fig. 11(c)), namely interactive buckling.

It is well known that the local buckling of the component plates will reduce the effective stiffness of the member and then decrease the load-carrying capacity. It can be seen that the larger the plate width-thick ratios are, the more sharply the ultimate loads drop, as shown in fig.12, fig.13 and fig.14. In fact, the columns shown in the figs may be considered as stubs because their slenderness ratio ranges only from 10-40.

4. CONCLUSION

A finite element method with shell element has been employed to investigate the parameter influence on the ultimate load-carrying capacity of the box-section column subjected to the axial force and the combined axial force and bending moment. The following conclusions could be reached:

(1) The width-thickness ratio of the component plates has great effect on the ultimate strength of the box-section column (shown as the Fig. 12 - Fig. 14.).

(2) The failure mode of the column has been significantly influenced by the width-thickness ratio of the column as shown in Fig. 11. For the columns with different width-thickness ratios, the load-carrying capacities vary a lot from each other. The designers can obtain some helpful advice from the analytic result of this paper.

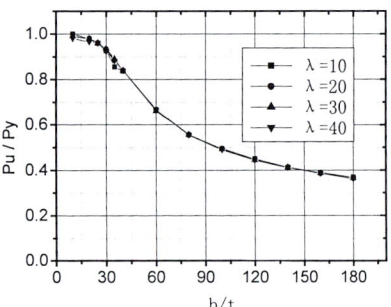

Fig. 12. Pu/Py-b/t interaction curves (e*=0)

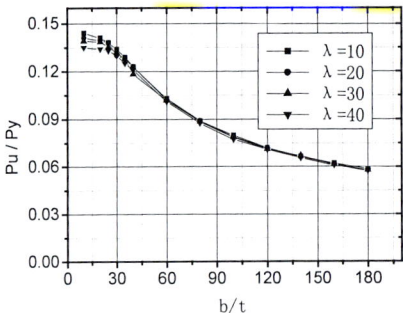

Fig. 13. Pu/Py-b/t interaction curves (e* =2.5)

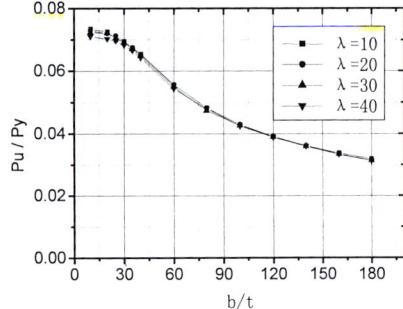

Fig. 14. Pu/Py-b/t interaction curves (e* =5)

REFERENCES

[1]. Shen, Zuyan and Zhang, Qilin. (1989). Interaction of Local and Overall Instability of Compress Box Columns. Journal of Structural Engineering, 117:11, 3337-3755.

[2]. Siu Lai chan, Sritawat Kitipornchai, and Faris G. A. Ai-Bermani. (1991). Elasto-Plastic Analysis of Box-Beam-Columns Including Local buckling Effects. Journal of Structural Engineering. 117:7, 1946-1960.

[3]. J.L. Meek and W.J.Lin. (1990). Geometric and Material Nonlinear Analysis of Thin-Walled Beam-Columns. Journal of Structural Engineering. 116:6, 116:6.

[4]. N.E.Shanmugam, J.Y.Richard Liew, and S.L.Lee. (1989). Welded Steel Box-Columns under Biaxial Loading.. Journal of Constructional Steel Research. 12, 119-139.

[5]. Dewolf, J. T., Pekoz, T.. (1974). Local and Overall Buckling of Cold-formed Members. J. Struct. Engrg. Div., ASCE. 100:10, 2017-2036.

[6]. Guo.Y. and Chen, S. (1992). Local and Overall Interactive Instability of Thin-Walled Box-Section Columns. Journal of Constructional Steel Research. 22:1, 1-19.

[7]. Hancock, G. J. (1981). Interaction Buckling in I-section Columns. J. Struct. Engrg. Div., ASCE. 107:1, 165-179

[8] Sridharan, S. and Ahraf Ali, M. (1988). Behavior and Design of Thin-walled Columns. Journal of Structural Engineering., ASCE. 114:1

MULTI-DIRECTIONAL PSEUDO DYNAMIC EXPERIMENT OF STEEL BRIDGE PIERS

M. Obata and Y. Goto

Department of Civil Engineering, Nagoya Institute of Technology,
Nagoya 466-8555, Japan

ABSTRACT

The control of the local damage to prevent overall failure of steel or composite structure has drawn interests in the seismic design of steel highway bridge piers. In such a case, it is critical to know the ultimate behavior of the structures during severe earthquake. Also important is to consider the effect of multi-directional, or the three-dimensional ground motion. Despite this inherent nature of earthquake waves, the vast majority of the experiments and analytical efforts are devoted to the unidirectional behaviors because the current design codes ignore the interaction of the structural response in the different directions. Although some researchers have been studying the multi-directional seismic response of structures to remedy the oversimplification in the design code, there still needs in-depth understanding of these behaviors to establish the more enhanced design code. Regarding these problems, recently, the authors have developed the testing system that is capable of the multi-directional pseudo-dynamic test as well as fully three-dimensional loading with the help of the unique three-dimensional hinge. In this work, we carry out pseudo dynamic tests of circular steel bridge piers using about 1/8 scaled model specimens and discuss the effects of multi-directional seismic loading on the damages of them.

KEYWORDS

seismic design of bridge, pseudo-dynamic test

INTRODUCTION

Steel columns are often used as bridge piers of elevated highways in the urban area of Japan. Steel columns have the widely recognized advantage of lightweight, short construction period, and ductility to the reinforced concrete columns. However, the extensive damage of steel columns in the 1995 Hyogoken Nanbu Earthquake (JSCE, 2000) prompted the re-consideration of the seismic design codes. The great progress has been made in the understandings of the ultimate behaviors of steel bridges during strong earthquake both in numerical and experimental viewpoints. For all these comprehensive research efforts, only small attention has been paid to the multi-directionality of ground motion despite the essential nature of earthquake waves. This ignorance is partly due to the fact that the modern seismic design codes have failed to consider this multi-directionality explicitly.

Currently, the effects of multi-directionality are investigated predominantly by numerical analysis such as FEM or a simplified model. Although FEM analysis with proper material constitutive relation is a versatile tool, its applicability and effectiveness still needs thorough verification by experiments. In this point of view, the applicability of the numerical analysis is limited by the lack of the results of multi-directional loading experiment.

The intrinsic nature of seismic loading is summarized as alternating direction, dynamic property, and multi-directionality. The majority of research works have made great efforts on the first feature. As for the second one, it has been considered by pseudo-dynamic (hybrid) test, which mimics a dynamic response of structure by applying quasi-static load, since its size often prohibits the use of a shaking table. Despite the great advantage of pseudo-dynamic test as alternatives to a shaking table test, thus far, the absence of the results of multi-directional loading cases sets the bounds to its applicability. The objective of this work is to carry out the multi-directional pseudo dynamic test and investigate the effect of multi-directional loading on the ultimate behaviors of steel bridge piers.

EXPERIMENT

3 dimensional structural testing system

The newly developed 3 dimensional structural testing system is used in the present experiments. The control and observation system is build around the core of MTS TestStarII®. It features 3 hydraulic actuators and a unique 3 dimensional hinge (ball joint) to apply arbitrary 3 dimensional load to a specimen (Fig. 1). The critical aspect in the 3 dimensional structural testing system is a joint that transmits the load from actuators to a specimen and measurement of 3 dimensional displacement. A three dimensional object has 6 degrees of freedom (3 translation and 3 rotation). Introduction of the 3 dimensional hinge leads to the elimination of the three rotational degrees of freedom. The appearance of only 3 translational degrees of freedom simplifies the apparatus and the control and

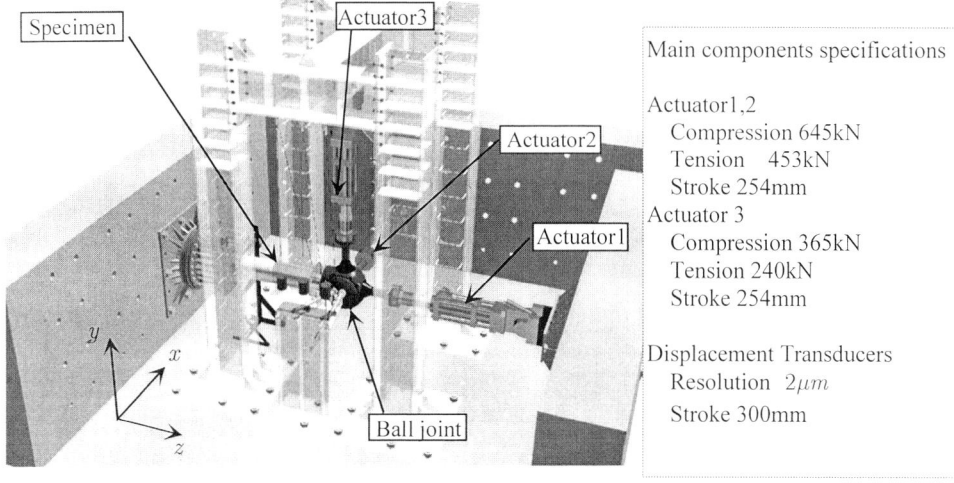

Figure 1 Three dimensional structural testing system

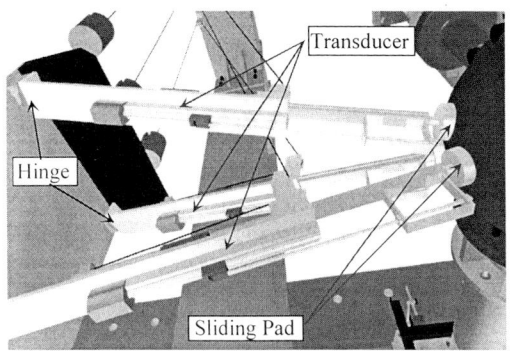
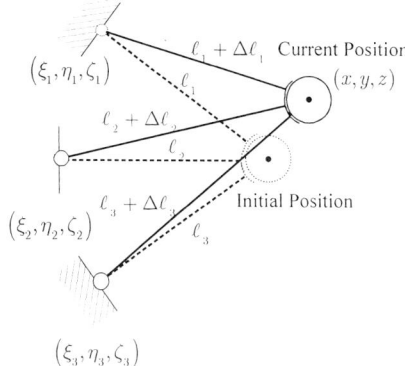

Figure 2 Spatial Transducer   Figure 3 Measurement of spatial displacement

observation system. Otherwise, despite its fidelity, full 6 degrees of freedom require at least 6 hydraulic actuators that would complicate the control system. As for the latter aspect, we developed a spatial truss type transducer (Fig. 2). It comprises 3 conventional linear differential variable transducers that make a spatial truss. With the help of the nature of the ball joint, the location of the center of the joint (x, y, z) is given as the solution of

$$\left(x - \xi_i\right)^2 + \left(y - \eta_i\right)^2 + \left(z - \zeta_i\right)^2 = \left(\ell_i + \Delta\ell_i\right)^2, \quad (i = 1, 2, 3) \tag{1}$$

where $\Delta\ell_i$ is the expansion of each transducer (Fig. 3). See Obata *et al.* (2001) for the details of the testing system.

Multi-directional pseudo dynamic test

A pseudo dynamic test, also known as a hybrid test or a computer on-line test, is a quasi-static displacement control test based on the real time numerical analysis (Hakuno *et al.*(1969) and Takanashi *et al.* (1975)). A specimen is regarded as a concentrated mass spring model in a computer (Fig. 4). Time integration of the following equations yield the response of the model to the specified earthquake waves.

$$M\ddot{x} + C_x\dot{x} + R_x = -M\ddot{x}_g, \quad M\ddot{y} + C_y\dot{y} + R_y = -M\ddot{y}_g \tag{2a,b}$$

where $\ddot{x}_g$ and $\ddot{y}_g$ are the acceleration waves of the earthquake in the x and y directions respectively. Then, they are transferred to the specimen by controlling hydraulic actuators. Since the characteristics of the spring is determined by real time observation of the reaction, R_x and R_y, the plastic behaviors of the specimen is included in a reasonable way. As far as the concentrated mass spring model is a reasonable representation of the modeled structure, thus controlled response of the specimen would reflect the one by a shaking table. A bridge pier that supports superstructures usually satisfies the condition of a concentrated mass. It should be mentioned that the present structural system in which the only 3 translational degrees of freedom appears has great advantage for the multi-directional pseudo-dynamic test. Since the model of a concentrated mass and spring has no rotational degree of freedom, the physical and the numerical model are theoretically identical. In contrast, full 6 degrees of freedom system makes the control of the pseudo dynamic test very complicated.

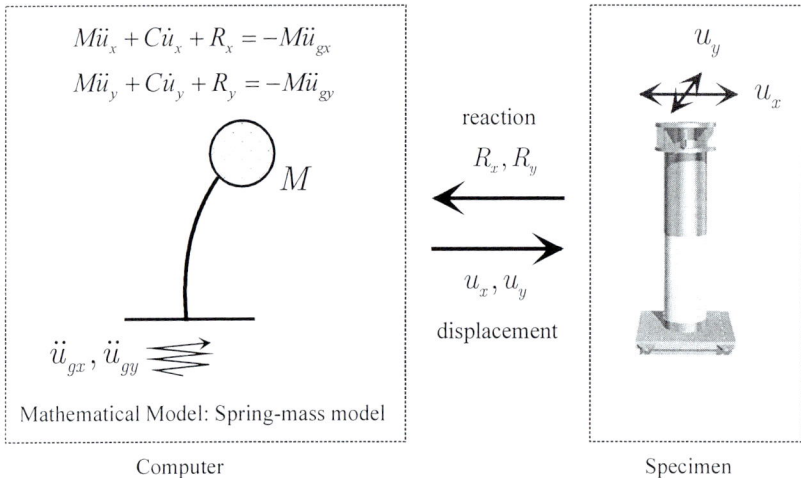

Figure 4 Pseudo dynamic test

Specimens and test set up

The specimens are modeling the hollow circular bridge piers of cantilever type (Fig. 5) They are made of a structural carbon steel pipe(STK400) of 6.6mm thickness. Geometric and material parameters are summarized in Table 1. Among the various parameters, the non-dimensional parameters R_t, and μ mainly govern the ultimate behaviors of circular steel piers with uniform cross sections. They are called radius-to-thickness-ratio parameter and axial force ratio respectively. In Japan, R_t of bridge piers are usually between 0.07 and 0.17. In the 1995 Earthquake, the damages of local buckling were found especially in the circular piers with $R_t \geq 0.085$. In order to set R_t in the above-mentioned range, the lower half of the pipe is scraped. The thickness in the table corresponds to that of the scraped part. The control parameters of the present experiment is R_t, and multi-directionality of the earthquake waves. In each type of specimen, we conducted the pseudo-dynamics test in unidirectional and in multi-directional waves.

The experiment is performed with scale factor S being 8, so that the real scale pier has about 2m diameter and 10m height. The mass is determined from axial force ratio, μ. Dumping ratio is ignored and only the hysteretic dumping is considered. When large scale yielding occurs as in these cases, hysteretic damping predominates over the viscous damping (Takanashi and Nakashima (1987)). In the scaled experiment, the computer assumes a real scale prototype and solves the differential equation for it, while the specimen is treated as a scaled model. To keep the equivalence of the resultant stresses of the prototype and those of the specimen, the reaction R_x and R_y of the specimen are interpreted to $S^2 R_x$ and $S^2 R_y$ of the prototype. The displacement x and y of the numerical model, on the other hand, is transferred to the specimen after dividing by the scale factor S. The axial force of the specimen is given using $\mu = 0.15$, not on the basis of the scaled mass. The related discussion of the scaled pseudo dynamic test is found in Kumar *et al.* (1997).

In the present testing system, the 3 actuators and a ball joint form a spatial truss structure. Since it is a statically determined structure, strictly speaking, displacement control can be achieved only when all the 3 actuators are in displacement control mode. Thus, it pauses a difficult problem when we

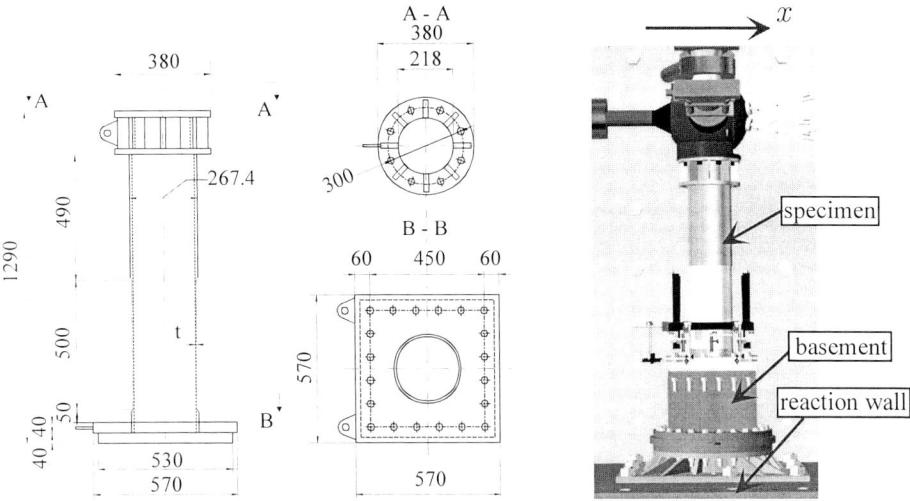

Figure 4a Specimen   Figure 5 Set up of a specimen

Table 1 Specimen specification

Specimen Type	Height (mm)	Radius (mm)	Thickness (mm)	R_t	H_{yield} (kN)	d_{yield} (mm)	σ_y (MPa)	E (GPa)	ν
T4.5	1040	131.6	4.5	0.07	41	5.8	290	206	0.3
T3.5	1040	130.6	3.5	0.09	31	5.5			

$R_t = \dfrac{R}{t}\dfrac{\sigma_y}{E}\sqrt{3(1-\nu^2)}$, H_{yield} :Initial Horizonal Yield Load, d_{yield} :Initial Horizontal Yield displacement

rigorously control the displacement in the two directions(x- and y-direction) and the force in the other direction(z-direction) as in this case. We employ a primitive approach to this problem. Since the actuator 1 and the specimen is nearly parallel and the specimen keeps a certain level of stiffness, the actuator 1 is set in a load control mode throughout the experiment. When the displacements in the x- and the y- direction is given, the zero incremental displacement in the z direction is assumed to compute the extension of the actuators. In each loading step, the errors in axial force and the location are monitored and corrected if necessary. As far as incremental displacement is small, the present procedure should be acceptable. As shown in Fig. 5, the specimen is attached to the reaction wall via the basement. Since the basement has only finite rigidity, its deformation must be counted to secure the accuracy of the experiments (Fig. 6). To this end, the displacement is continuously monitored at the top surface of the displacement using 8 transducers, 4 for translation and 4 for rigid body rotation, to make a proper coordinate transformation in each loading step.

The input earthquake wave is the one recorded at Japan Meteorological Agency in the 1995 Kobe earthquake (Fig. 7). Its maximum acceleration is 818gal in the North-South direction and 617 gal in the East-West direction. The record up to 25sec is used. Central time difference method is adopted to the step-by-step integration of Eqs. (2) with incremental time $\Delta t = 0.05 \sec$.

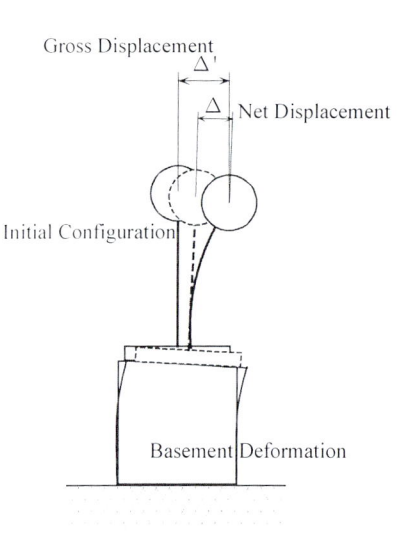

Figure 6   Correction of displacement due to basement deformation

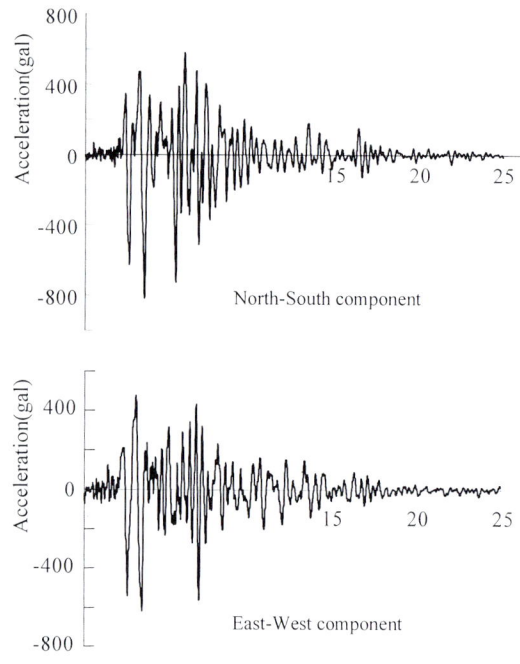

Figure 7   Acceleration waves recorded at JMA in the 1995 Hyogoken Nanbu Earthquake

RESULTS AND COMMENTS

Unidirectional test

The JMA NS wave is input to the x-direction. The time response and the load-displacement curves of T4.5 are shown in Figs. 8. The results of T3.5 are similar. These curves shows the qualitatively agrees with the analytical result with similar condition (Nakamura, H. (1996), see also Jiang et al.(2002)). The displacement and the load are normalized by the quantity at the initial yielding of the specimen. As shown in Fig. 8, the large displacement response appears in the first 5 seconds. The local buckling occurs at about 3 second and that virtually determines the residual displacement. Due to the local buckling of elephant foot type, the elastic stiffness decreased thereafter. After around 8 seconds, no more residual deformation accumulated. In the unidirectional test, the alternating load often refrains from the rapid evolution of the elephant foot bulge, because the reversal load stretches the bulge. Consequently, unidirectional test is likely to underestimate the damage by the earthquake.

Multi-directional test

In the multi-directional test, in addition to the JMA NS wave, the EW wave is input to the y-direction simultaneously. The time response curves and the buckled part are shown in Figs. 9. As shown in trajectory curve, a specimen deforms mainly in the Northwest direction. After the local buckling, the behavior becomes complex and a trajectory curve becomes like circular.

Roughly speaking, the input of waves in two different direction results in doubling the input energy. Thus, larger residual deformation appeared in this case. In contrast to the unidirectional loading

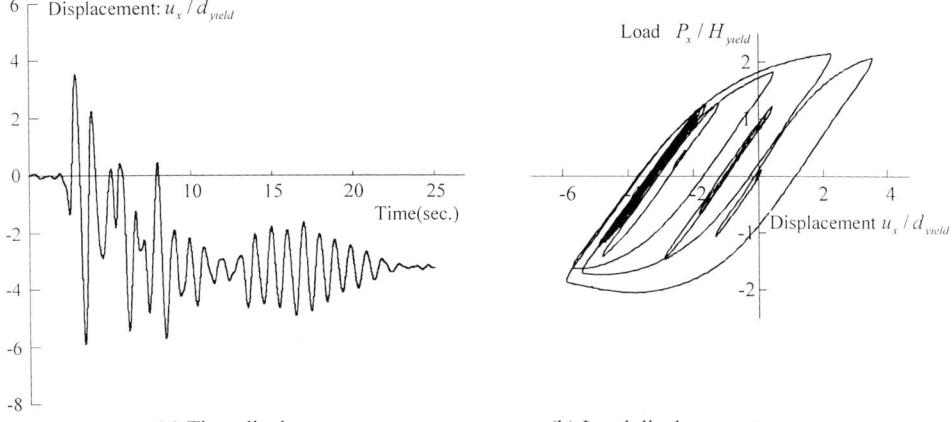

(a) Time displacement curve (b) Load displacement curve

Figure 8 Unidirectional test results (T4.5, H_{yield}: Initial Horizontal Yield Load, d_{yield}: Initial Horizontal Yield displacement)

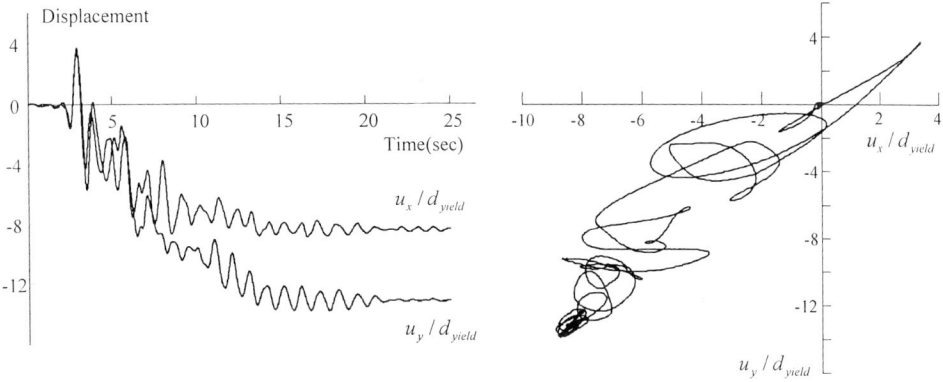

(a) Time displacement curves (b) Displacement trajectory

Figure 9 Multidirectional test results (T4.5, d_{yield}: Initial Horizontal Yield Displacement)

case, once local buckling occurs, bulged part is likely to develop monotonically without healing by the reverse loading. The restriction of the unidirectional deformation is non negligible and adverse to the seismic response of structures. These effects of multi-directional loading are also found in the static experiments (Watanabe et al. (2000)). It should be added that the effect of multi-directional input may be exaggerated since the weight of the ball joint is not perfectly countered. In this type of experiments, small perturbation often yields big difference.

The effect of multi-directional loading test shows that the estimation on the result of unidirectional tests fails to estimate the accurate seismic performance of structure. The results of the present pseudo-dynamic test reveal these adverse effects specifically. For the more refined seismic design, the establishment of the numerical analysis considering the effects is necessary. The present method should serve as a suitable basis for the rational seismic design when combined with the advanced numerical analysis and modeling. Finally, regarding the 3rd component of earthquake wave, up-down wave, the study is under the course. Because of the high stiffness in the axial direction, the

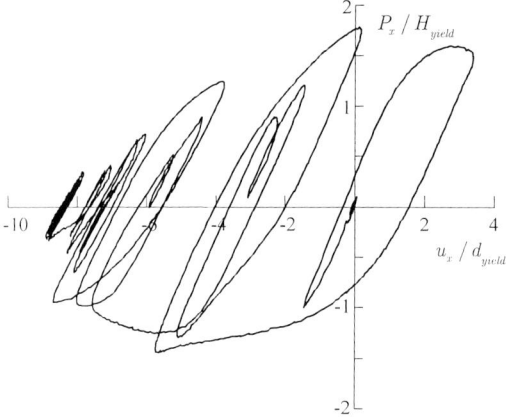 

(a) Load displacement curve (b) local buckling (at the end of the test)

Figure 10 Multidirectional test results (T4.5, H_{yield} :Initial Horizontal Yield Load , d_{yield} :Initial Horizontal Yield displacement)

usual experimental scheme is not applicable. Some revision or compromise must be made to perform the full three dimensional pseudo-dynamic test.

Acknowledgment

This study was partly supported by Tatematsu Zaidan. One of the authors, MO, appreciate the support during 2001 fiscal year.

References

Hakuno, M., Shidawara, M., Hara, T.(1969), Dynamic destructive test of a cantilever beam controlled by an analog computer, Trans. Japan Soc. Civil. Engrs, No. 171, pp.1-9(in Japanese)

JSCE subcommittee on investigation of seismic damage of steel structure(2000), Investigation of causes of damage to steel structure on Hanshin-Awaji Earthquake Disaster, J. Struct. Mech. and Earthquake Engrg, No.647/I-51, pp.17-30(in Japanese).

Jiang, L., Goto, Y., Obata, M.(2002), Hysteretic modeling of thin-walled circular steel columns under biaxial bending, J. Struct.Engrg, Vol. 128, pp. 319-327

Kumar, S., Itoh, Y., Saizuka, K., Usami, T.(1997), Pseudodynamic testing of scaled models, J. Struct. Engrg., Vol. 123, pp. 524-526.

Nakamura, H. (1996), Elastic-plastic dynamic buckling analysis of a steel circular pier, J. Struct.Mech and Earthquake Engrg, No.549/I-37, pp. 205-219(in Japanese)

Obata, M., Matsuo, N., Goto, Y. (2001), Experimental study on strain localization and ductile fracture of steel pipe due to cyclic horizontal loading, Proc. ISCSC01, Vol. 2, pp. 1533-1540.

Takanashi, K. (1975), Non-linear earthquake response analysis of structures by a computer actuator on-line system(part 1 details of the system), No. 229, Trans. Architectural Institute of Japan, 77-83

Takanashi, K., Nakashima, M.(1987), Japanese activities on on-line testing, J. Struct. Engrg, Vol. 113, pp. 1014-1032.

Watanabe, E, Sugiura, K., W. Oyawa(2000), Effects of multi-directional displacement paths on the cyclic behaviour of rectangular hollow steel columns, J. Strut. Mech. Earthquake Engrg, Vol. 17, pp. 69s-85s.

CONNECTIONS

SHEAR LAG IN DOUBLE ANGLE TRUSS CONNECTIONS

D.B. Bauer[1] and A. Benaddi[2]
[1,2] Department of Construction Engineering, École de technologie supérieure,
Montréal, Canada H3C 1K3

ABSTRACT

Shear lag can be described as a phenomenon that creates a loss in resistance in a tension member connected through only part of its cross-section. It is a complex problem which has been under study for many years by researchers. Parameters that influence the shear lag phenomenon are many and difficult to assess: type and size of cross-section, type of connection, length of welds, length of member, joint eccentricities, etc. Connection between double-angle web members and chords in trusses or open web steel joists are considered herein. Resistances and modes of failures are described as determined experimentally. Yield and ultimate loads are compared with the values calculated using the design guidelines of the Canadian Standard, which are reviewed in the article. Finally, tentative conclusions are drawn regarding the influence of shear lag in double angle truss connections.

KEYWORDS

Shear lag, steel structure, truss, welded connection, double-angle, tension web member.

INTRODUCTION

Steel trusses fabricated with double-angle web members are considered herein, in which joints are made economically by welding one leg of the web angles directly to the chord member. See Figure 1. Tension web members can be designed using Clause 12.3.3. of Canadian Standard S16.1-94 Limit States Design of Steel Structures (CSA 1994) in which shear lag is taken into account. Shear lag is a phenomenon that affects tension members connected at one or both ends through only part of the cross-section. Tensile stresses are transferred from the member into the connected parts and the stress distribution along the connection is non linear, resulting in a loss of strength. Yet, according to certain fabricators, the strength of such connections could be higher than suggested by current Standards.

LITTERATURE REVIEW

The amount of shear lag present in a connection is calculated using the $1-\bar{x}/L$ function, where $\bar{x}$ and L are an eccentricity and a characteristic length of the connection, respectively. See Figure 2. This function was presented by Munse and Chesson (1963) in an article dealing mostly with riveted and bolted connections. In his book on steel structures, McGuire (1968) computes bending stresses due to joint eccentricities for angle web members connected through one leg. He explains that the connection is strong enough for the gross section to reach yield and that, because of stress redistribution, the ultimate strength of the connection is only slightly affected. According to McGuire, this justifies that no reduction in the net area was required at the time in the AISC Standard. He mentions the work of Munse and Chesson for built-up sections, but questions the relevance of using the $1-\bar{x}/L$ function for connections that are not solicited in fatigue. In a recent study, Kirkham and Miller (2000) comment the AISC Standard (1993) regarding the treatment of shear lag. The authors point out that the definition of the eccentricity, $\bar{x}$, is not clear and should be revised. They mention that parts of the design recommendations are in fact based on extrapolations of previous research findings or empirical rules that have not been verified experimentally. They observe that most tests have been done on samples that were too short to develop a uniform stress distribution away from the connection according to St-Venant's principle; the short length of samples affected the stress distributions and was a source of errors. Furthermore, tested samples did not cover the full range of cross-section sizes. The authors recommend that tests be done on full size specimens in order to verify whether tests on small samples are valid, as well as tests and parametric studies on specimens covering a wide range of cross-section sizes.

THE CURRENT CANADIAN CODE

The resistance of a member subjected to uni-axial tension is given in Canadian Standard S16.1-94 Limit States Design of Steel Structures (CSA 1994). For double-angle members with welded connections, the factored tensile resistance, T_r, is to be taken as the least of (Clause 12.2 a)):

$$T_r = \phi A_g F_y \tag{1}$$

$$T_r = 0.85 \phi A'_{ne} F_u \tag{2}$$

where ϕ is a performance factor, A_g is the gross cross sectional area, A'_{ne} is the effective net area reduced for shear lag, F_y is the yield strength and F_u is the ultimate strength. Eqn. 1 represents full plasticity along the entire length of the member occurring when the gross cross-section reaches yield, which is a desirable mode of failure, and no reduction due to shear lag is included. Eqn. 2 represents rupture of the effective net area reduced in order to take into account shear lag occurring at the joint. The 0.85 factor takes into account the fact that there is no reserve in strength beyond fracture. For welded connections, the reduced effective net area is computed as:

$$A'_{ne} = A_{ne1} + A_{ne2} + A_{ne3} \tag{3}$$

where A_{ne1}, A_{ne2} and A_{ne3} are the effective net area of the connected parts computed in the following manner:

a) for a part connected by a transverse weld

$$A_{ne1} = wt \tag{4}$$

where w is the width and t is the thickness of the part.

b) for a part connected by longitudinal welds along two parallel edges

(i) when $L \geq 2w$, $A_{ne2} = 1.00wt$
(ii) when $2w > L \geq 1.5w$, $A_{ne2} = 0.87wt$ (5)
(iii) when $1.5w > L \geq w$, $A_{ne2} = 0.75wt$

where L is the average length of welds on the two edges.

c) for a part connected by a single line of welds

$$A_{ne3} = \left(1 - \frac{\bar{x}}{L}\right)wt \tag{6}$$

where $\bar{x}$ is the eccentricity of the weld with respect to the centroid of the part and L is the length of the weld in the direction of the loading.

NUMERICAL EXAMPLE USING THE CANADIAN CODE

Check the tensile load resistance of a 2-L76x51x4.8 double-angle member with short legs back to back, connected with longitudinal welds. CSA-G40.21 380W steel with F_y = 380 MPa and F_u = 480 MPa. In order to calculate the required length of weld, L, it is assumed that the member must carry a factored load, T_f, equal to the factored yield resistance, T_r, of the member:

$$T_r = \phi A_g F_y = 0.9 \times 2 \times 582 \text{ mm}^2 \times 380 \text{ MPa} = 398 \text{ kN} \tag{7}$$

As is often the case in trusses and due to fabrication constraints, the welds will be longitudinal and of equal length on either side of the connected leg, and no transverse weld will be used. With a weld size of 5 mm and E480XX electrodes, the required length of weld, L, is equal to:

$$L = \frac{398 \text{ kN}}{0.762 \text{ kN/mm} \times 2 \text{ angles} \times 2 \text{ sides}} = 131 \text{ mm} \tag{8}$$

Since the length of the welds, L = 131 mm, is more than twice the width of the connected leg, there is no reduction due to shear lag (see Eqn. 5 above). Therefore:

$$A_{ne2} = 1.00 \times (51 - 4.76) \text{ mm} \times 4.76 \text{ mm} = 220 \text{ mm}^2 \tag{9}$$

For the outstanding leg, the eccentricity, $\bar{x}$, is equal to $76/2 = 38$ mm and the effective net area of this leg is equal to (from Eqn. 6 above):

$$A_{ne3} = \left(1 - \frac{38}{131}\right) 76 \text{ mm} \times 4.76 \text{ mm} = 257 \text{ mm}^2 \tag{10}$$

Hence, the reduced effective net area of a complete angle is equal to:

$$A'_{ne} = 220 \text{ mm}^2 + 257 \text{ mm}^2 = 477 \text{ mm}^2 \tag{11}$$

The ultimate tensile resistance of the double-angle member is equal to:

$$T_r = 0.85\phi A'_{ne} F_u = 0.85 \times 0.90 \times 2 \times 477 \text{ mm}^2 \times 480 \text{ MPa} = 350 \text{ kN} \tag{12}$$

The latter value is lower than 398 kN found using Eqn. 7, hence it governs. The reduction in the ultimate resistance of the cross-section due to shear lag is equal to:

$$\% \text{ reduction} = 1 - \frac{A'_{ne}}{A_{ne}} = 1 - \frac{477 \text{ mm}^2}{582 \text{ mm}^2} = 18\% \tag{13}$$

The length of weld required to resist a load of 350 kN is somewhat smaller than 131 mm calculated with Eqn. 8 above. A shorter length of weld would lower the A_{ne3} value as well as the ultimate tensile resistance, T_r, calculated using Eqn. 12, and this again would require a smaller length of weld. After a few cycles of such calculations, the ultimate resistance is found to be equal to 336 kN, the required weld length is equal to 110 mm and the percent reduction due to shear lag is equal to 21 %. The effect of shear lag reduces the design member resistance only if Eqn. 2 governs over Eqn. 1. For 300W steel with F_y = 300 MPa and F_u = 450 MPa, this will be the case when the reduction in area due to shear lag is greater than 22 %. With higher strength steels, for example 380W with F_y = 380 MPa and F_u = 480 MPa, Eqn. 2 will govern if the reduction in area due to shear lag is greater than 7 %.

LABORATORY TESTS

Test samples

Laboratory tests on simplified specimens were carried out recently by the authors. Six specimens were tested with double-angle members ranging in size from 2-L38x38x4.8 up to 2-L76x76x4.8. The main test parameters are shown in Table 1. All specimens had an overall length of 2500 mm, the longest possible that would fit conveniently in the testing machine, which left a clear length for the double-angle member of over one meter between connections. Specimens were built symmetrical, except for the welds that were sized such that one end only would fail during the tests. Welds at the joints under study were longitudinal and of equal length, and no transverse weld was used, as it is the preferred practice for truss joints. The centroid of the double angles was aligned with the centerline of the end gusset plates in order to eliminate eccentricity measured parallel to the gusset plates. The other eccentricity, measured perpendicular to the gusset plates, is unavoidable for that type of connection. Two coupons per specimen were prepared and tested according to Canadian Standard CAN/CSA-G40.20-98 General Requirements for Rolled or Welded Structural Quality Steel (CSA 1998). Average measured yield and ultimate strengths are given in Table 1. Measured values match those of CSA G40.21 Grade 300W steel with a minimum specified yield value F_y = 300 MPa and an ultimate strength F_u = 450 to 620 MPa, rather than Grade 380W which was initially assumed in the design of the test specimens. Cross sectional areas calculated using measured widths and thicknesses were 1.5 % on average higher than the nominal areas, equal to $(b_1 + b_2 - t) \times t$ where b_1 and b_2 are the width of the legs and t is the thickness. Because the actual cross sectional area is difficult to measure yet close to the nominal value, the nominal cross sectional areas were used in calculating the expected yield and ultimate loads of the specimens. Table 2 shows calculations of the effective net area for both legs of the angles. For all specimens, the width of the connected leg is smaller than twice the weld length and hence there is no reduction in area due to shear lag for that leg. For the outstanding leg, all specimens have an area reduced to a value between 72 % and 78 %. The predicted reduction in area for the complete angles varies between 12 % and 17 %. The weld length was calculated in order to resist the double-angle member yield load assuming F_y = 380 MPa and this lead to somewhat long welds. Further tests with smaller assumed loads and shorter welds could be needed.

Test set-up and Instrumentation

Tests were done in a MTS universal testing machine with a capacity of 11 000 kN, located in the Structures Laboratory of the Department of Civil Engineering and Applied Mechanics at McGill University. See Figure 3. The specimens were loaded in tension under quasi-static conditions. Loading was displacement controlled at a fixed rate of 0,01 mm/sec up to the beginning of strain hardening and then was increased to 0,1 mm/sec up to failure. The applied load was measured with the load cell integral with the testing machine. The displacement of the loading end was measured with the LVDT integral with the machine. Eight strain gauges were placed on each specimen at the connection under study, two on the connected leg and six on the outstanding leg. Strains were measured up to 1.5 %, that is, about ten times the yield strain. The measured strains will be compared elsewhere with strain distributions obtained from finite element models of the joints. Two LVDT's were placed perpendicular to the double-angle member at mid-height in order to measure transverse deformations. On each specimen, one angle was covered with whitewash that revealed zones of high strains by spalling during the tests. A Vishay 6000 data acquisition system was used to record all the loads, displacements and strains.

Test Results and Discussion

The load versus overall deformation curves for the six specimens are shown in Figure 4. All specimens underwent yielding over their entire length, strain hardened and failed with large plastic deformations. Specimen No. 2 reached 18 % overall strain and the other specimens reached about 10 % strain. Specimen No. 2 failed in a tensile mode with the final break located towards the center of the double-angle member. The other specimens failed also in tension, either very close or right at the connection under study. The failure mode of specimen No. 4 is shown in Figure 5. Predicted as well as experimental yield and ultimate loads are given in Table 3. Loads were predicted using measured values of F_y and F_u and the reduced effective net areas from Eqns. 5 and 6. Comparisons between these loads are given in the same Table. Regarding the yield loads, only specimen No. 2 tested 10 % lower than the predicted value. All other specimens yielded at about the predicted load with specimen No. 3 being 10 % stronger than predicted. Concerning the behaviour at ultimate load, all specimens were between 3 % and 20 % stronger than the $A'_{ne}F_u$ values, with an average of 12 %. Test results were on average only 3 % lower than $A_g F_u$ values, specimen No. 2 being 9 % lower and specimen No. 3 being 2 % higher. These latter values, i.e. $T_u/A_g F_u$ ratios given in Table 3, are experimental reduction factors which can be compared directly with the predicted reduction factors given in Table 2 for the complete angles.

CONCLUSIONS

This section summarizes the conclusions of this paper. These are briefly given below:

1. Six double-angle members, with sizes ranging from 2 L38x38x4.8 up to 2-L76x76x4.8 and connected with equal length longitudinal welds, were tested in tension. All specimens yielded over their entire length, strain hardened and broke with a final overall strain of about 10 % (18 % for specimen No. 2).

2. For all specimens, there was no reduction in the yield load from that predicted using measured values of the yield strength, except specimen No. 2 for which there was a 10 % reduction. This agrees with the recommendation found in current Standards that no reduction due to shear lag need be considered for calculating the yield resistance of tension members.

3. The ratio of experimental to predicted ultimate load, $T_u/A'_{ne}F_u$, was 1.19 and 1.20 for specimens Nos. 3 and 5 with double angles connected with the short legs back to back. For equal leg double

angles, the ratio was between 1.09 and 1.11 for specimens Nos. 1, 4 and 6, and 1.03 for specimen No. 2.

4. The experimental ultimate load was 3 % on average less than the predicted load, $A_g F_u$, based on the gross cross sectional area of the member. In other words, a 3 % reduction in area due to shear lag was found experimentally, compared to a value of 13 % predicted using current design recommendations.

It can be concluded that, with the limited test data presented here, design recommendations found in the Canadian Standard are adequate regarding the yield resistance and somewhat conservative regarding the effect of shear lag and the ultimate resistance. In order to get a better understanding of the joint behaviour, further analyses of the test results will include comparison between strain gauge readings and stresses calculated using finite element models of the joints, as well as calculations of the bending stresses due to joint eccentricities. Also, a second series of tests on complete truss panels is being prepared in order to determine the influence of more realistic loading conditions on the behaviour of double-angle tension web member joints.

ACKNOWLEDGMENTS

The present research was made possible by grants from the Steel Structures Education Foundation of the Canadian Institute of Steel Construction for the years 2001 and 2002, from the National Sciences and Engineering Research Council of Canada (N° 238967-01), and from the École de technologie supérieure (PSIRE-Research 2001 and 2002). The authors would like to thank Canam Steel for giving the test samples, and the Department of Civil Engineering and Applied Mechanics of McGill University for the complimentary use of their universal testing machine.

REFERENCES

AISC (1993). *Load and Resistance Factor Design Specification for Structural Steel Building, 2nd Ed.* American Institute of Steel Construction, Chicago, IL.

CSA (1994). *CAN/CSA-S16.1-94 Limit States Design of Steel Structures*. Canadian Standard Association. Toronto, Canada.

CSA (1998). *CAN/CSA-G40.20-98 General Requirements for Rolled or Welded Structural Quality Steel*. Canadian Standard Association. Toronto, Canada.

Kirkham W.J. and Miller T.H. (2000). Examination of AISC LRFD Shear Lag Design. *Engineering Journal* **37:3**, 83-98.

McGuire W. (1968). *Steel Structures*. Prentice-Hall, Englewood Cliffs, New-Jersey, U.S.A.

Munse W.H. and Chesson E. (1963). Riveted and Bolted Joints: Net Section Design. *Journal of the Structural Division*. Proceedings of the American Society of Civil Engineers **89:ST1**, 107-126.

Figure 1: Typical truss with double-angle web members (Canam Steel)

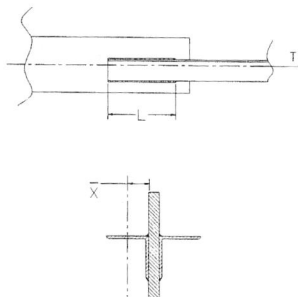

Figure 2: Connection with equal length longitudinal welds

Figure 3: Test set-up

Figure 5: Failure of specimen No. 4

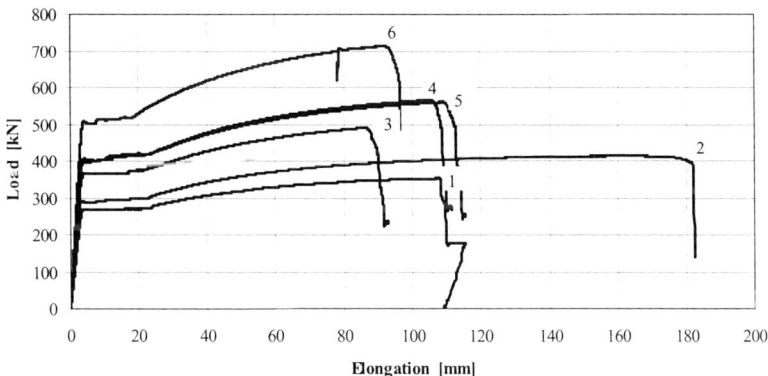

Figure 4: Load versus overall displacement for specimens Nos. 1 to 6

TABLE 1
TEST PARAMETERS AND AVERAGE MEASURED COUPON STRENGTH

No.	Size [mm x mm x mm]	Arrangement	Gross Area of One Angle [mm^2]	Weld Size [mm]	Measured Weld Length [mm]	Yield Strength F_y [MPa]	Ultimate Strength F_u [MPa]
1	2L-38 x 38 x 4.8	Equal Legs	340	5	87	393	531
2	2L-51 x 51 x 4.8	Equal Legs	461	5	112	349	492
3	2L-64 x 51 x 4.8	Short Legs Back to Back	521	5	122	318	461
4	2L-64 x 64 x 4.8	Equal Legs	582	5	136	345	499
5	2L-76 x 51 x 4.8	Short Legs Back to Back	582	5	138	339	487
6	2L-76 x 76 x 4.8	Equal Legs	703	5	169	348	527

TABLE 2
CALCULATION OF THE EFFECTIVE NET AREA REDUCED DUE TO SHEAR LAG

No.	Connected Leg			Outstanding Leg			Complete Angle	
	w_2 [mm]	Reduction Factor	A_{ne2} [mm^2]	w_3 [mm]	Reduction Factor	A_{ne3} [mm^2]	A'_{ne} [mm^2]	Reduction Factor [%]
1	38	1.00	158	38	0.78	141	300	0.88
2	51	1.00	220	51	0.77	187	408	0.88
3	51	1.00	220	64	0.74	225	445	0.85
4	64	1.00	282	64	0.76	233	515	0.88
5	51	1.00	220	76	0.72	262	482	0.83
6	76	1.00	339	76	0.78	280	620	0.88
							Average	0.87

TABLE 3
PREDICTED AND EXPERIMENTAL STRENGTHS OF SPECIMENS

No.	Predicted Strengths				Experimental Strengths		Ratios of Experimental / Predicted Strengths		
	$A_g F_y$ [kN]	$0.85 A'_{ne} F_u$ [kN]	$A'_{ne} F_u$ [kN]	$A_g F_u$ [kN]	T_y [kN]	T_u [kN]	$\dfrac{T_y}{A_g F_y}$	$\dfrac{T_u}{A'_{ne} F_u}$	$\dfrac{T_u}{A_g F_u}$
1	267	271	318	361	268	353	1.00	1.11	0.98
2	322	341	401	453	288	414	0.90	1.03	0.91
3	331	348	410	480	365	490	1.10	1.20	1.02
4	402	437	514	581	398	566	0.99	1.10	0.97
5	394	399	470	567	401	561	1.02	1.19	0.99
6	489	555	653	741	502	713	1.03	1.09	0.96
						Average	1.01	1.12	0.97

STRUCTURAL BEHAVIOUR OF
WEB BOLTED FLANGE WELDED CONNECTION

T. EMI, M. TABUCHI, T. TANAKA and H. NAMBA

Graduate School of Science and Technology, Kobe University,
Kobe, 657-8501, Japan

ABSTRACT

This study is planned to examine the effects of the slip load of the bolt connection on the elasto-plastic behaviour of the beams. Three specimens which have different slip load were tested. The slip load was varied by adjustment of the cramping forces of the bolts. The clamping forces (T) were adjusted to give $T/T_0 = 0.8$, 1.0 and 1.2, here, T_0 meant the standard cramping force.

Further finite element analyses (FEA) were carried out to investigate for detailed study on the effect of the slip load. FEA results show approximately similar to test results.

The results are briefly shown below: when the width to thickness ratio (D/t) of the column is large, it has little effect on the behaviour of the beam to vary the slip load of the bolt connection on the beam web.

KEYWORDS

WBFW Connection, High-strength Bolt, Cramping Force, Slip Load, FEA

1. INTRODUCTION

Recently, the steel structures, whose beam webs are bolted and flanges are welded by full penetration welding (WBFW) at beam to column connections, have been widely used in Japan. In WBFW connections, the beam webs are bolted to the shear plates by friction-type connection with high-strength bolts and the flanges are welded at the site. The high-strength bolts on the beam web are tightened before the beam flanges are welded to the columns. For this type connection, several experimental studies have been carried out, for example, Kohzu (1997) and Nakano et al. (2001). However, the effect of the slip behaviour of the friction joint in the beam web has not been wholly cleared.

In this paper, the effect of the slip load on the elasto-plastic behaviour of the beams is examined. Three specimens were prepared for this study. The test variable was the slip load of the beam webs.

The slip load was varied by adjustment of the cramping forces of the bolts. The clamping forces (T) were adjusted to give T/T_0 = 0.8, 1.0 and 1.2, here, T_0 meant the standard cramping force. To make clear the friction joint behaviour of the beam web, finite element analyses were also conducted.

2. LOADING TEST

Figure 1 shows the test specimen and the test set-up. The test specimen is composed of a rolled H-beam (H-500 x 200 x 10 x 16, SN490B) and a cold-formed square hollow section column (□-350x350x12, BCR295). The beam flange is welded to the through diaphragm (PL-19, SN490B) by full penetration welding, and the beam web is bolted to the shear plate (PL-16, SN490B) welded to the column. In this test, F10T 6-M20 high strength bolts are used for the friction joint of the beam web. Three specimens which have different slip load were tested as shown in Table 1. The high-strength bolts of the specimen B10G29 were cramped with the standard cramping force (T_0). The bolts of the

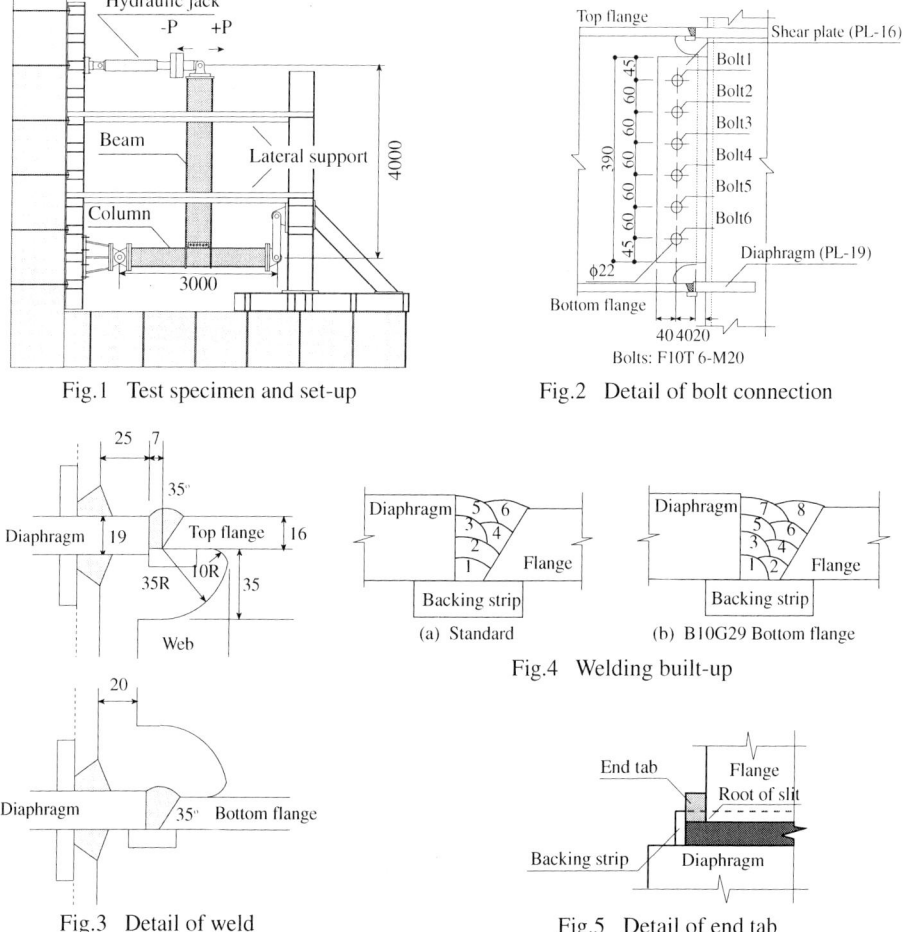

Fig.1 Test specimen and set-up

Fig.2 Detail of bolt connection

Fig.3 Detail of weld

Fig.4 Welding built-up

(a) Standard

(b) B10G29 Bottom flange

Fig.5 Detail of end tab

specimen B08G29 and B12G29 were cramped with $0.8T_0$ and $1.2T_0$, respectively. The cramping force of the bolt was measured by strain gauge mounted on the bolt. Mechanical properties of the materials are shown in Table 2

Details of the bolt connection and the weld connection are shown in Fig.2 and Fig.3, respectively. The welding built-up as shown in Fig.4 (a) is adopted for the specimens except the bottom flange of the specimen B10G29 which has a large root gap (=11 -12mm) because of the fabrication error. For this reason, the welding built-up shown in Fig.4 (b) is adopted for the bottom flange of the specimen B10G29.

Figure 5 shows the detail of the steel end tab, which isn't cut off in common practice. The existence of the end tab makes a slit composed between the edge of the beam flange and the end tab. The root of this slit becomes the source of the crack initiation (see Fig.5).

In this test, the friction surfaces between the beam webs and the shear plates were blast-cleaned surfaces with no coatings. The friction surfaces were prevented from rust to keep the slip coefficient being constant.

The slip tests were carried out to determine the slip coefficient. The slip test is figured in Fig.6 and the results of the slip tests are shown in Table3. The specimen Slip1 was loaded on the day when the loading test specimens were fabricated. The specimens Slip2 and Slip3 were tested just before the first loading test. The μ_n means the slip coefficient, which varies 0.61 to 0.67 except one result (0.47).

Table 1 Specimens of WBFW connection

Specimen	Cramping force	Column	Beam
B10G29	T_0	□ -350 x 350 x 12	H-500 x 200 x 10 x 16
B08G29	$0.8 T_0$		
B12G29	$1.2 T_0$		

Table 2 Mechanical properties

Location	Steel grade	σ_y	σ_u	ε_u	El.	YR	$_vE_0$	B_0	T_r
		N/mm²			(%)		J	%	°C
Beam flange	SN490B	365	525	18	45	69	208	12	-40
Beam web		413	551	16	37	75			
Shear plate		385	554	17	41	69			
Diaphragm		363	542	19	43	67	-	-	-
Column	BCR295	385	480	14	39	80			
High-strength bolt	F10T	1034	1077	5	12	96			

σ_y: Yield stress, σ_u: Tensile strength, YR; Yield ratio, ε_u: Uniform elongation
$_vE_0$: Charpy absorbed energy at 0 °C, B_0: Percent Brittle Fracture at 0 °C,
T_r:Transition Temperature

Table 3 Slip test

Specimen	μ_1	μ_2	Date of test
Slip1	0.61	0.47	Fabricated day
Slip2	0.67	0.63	Just before loading test
Slip3	0.63	0.66	(13 day after fabricated)

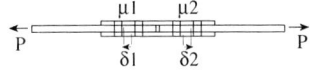

Fig.6 Slip test

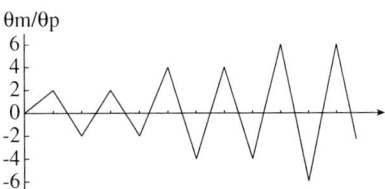

Fig.7 Loading program

Loading program is shown in Fig.7. Here, θ_p means calculated bending and shearing deflection of the beam at the full plastic moment (M_p). In this study, the positive loading cycle is defined as bottom flange is tensile side.

3 TEST RESULTS

Figure 8 shows the applied moment of the beam end versus beam rotational angle θ_m curves. Summary of the test is as follows:

(1) B10G29: In the process of $2\theta_p$ (-1), the initial ductile crack was observed at the root of the slit on the top flange. In the process of $4\theta_p$ (+1), the ductile crack progressed about 50 (mm) from the root of the slit on the bottom flange, and finally the brittle fracture propagated to the weld metal. Figure 9 (a) shows the brittle fracture on the bottom flange.

(2) B08G29: In the process of $2\theta_p$ (-2), the initial ductile crack was observed at the root of the slit on the top flange. In the process of $6\theta_p$ (+1) the local buckling was observed on the top flange. In the process of $6\theta_p$ (-1), the crack at the root of the slit progressed about 15 (mm) and the crack on the root of the scallop occured on the top flange. Finally the crack of the root of the slit and the root of the scallop were connected with ductile manner. Figure 9 (b) shows the ductile fracture on the top flange.

(3) B12G29: In the process of $2\theta_p$ (-2), the initial ductile crack was observed at the root of the slit on the top flange. In the process of $6\theta_p$ (-1), the crack at the root of the slit progressed about 20 (mm) and the crack at the root of the scallop occurred on the top flange and the local buckling was also observed clearly. Finally the specimen failed due to the local buckling in the process of $6\theta_p$ (+2). Figure 9 (c) shows the local buckling of the top flange.

Table 4 shows the summary of the test results. The specimen B10G29 has the poorest plastic rotation capacity, which defined in Fig.10. The premature failure of B10G29 may the result from the difference of the weld condition for the bottom flange (see Fig.4). Namely, the excessive root gap causes higher interpass temperature, and reduces the strength of the deposit metal.

Figure 11 shows the moment versus the slip displacement (δs) curve for the specimen B10G29. The slip displacement (δs) was defined as the relative displacement between the beam web and the shear plate, which measured by displacement transducers as shown in Fig.12. As shown in Fig.11, hysteresis loop presented steady behaviour. The skeleton curves on the positive loading cycle are shown in Fig.13. As shown in Fig.13, it has little effect on the slip behaviour to vary the cramping force.

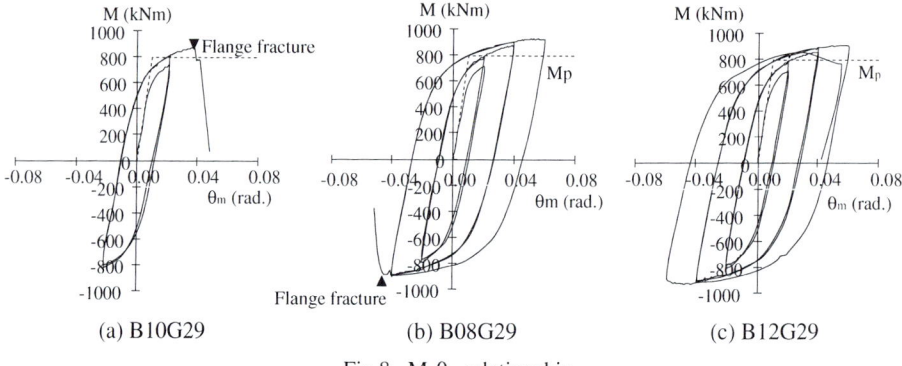

Fig.8 M-θ_m relationship

(a) Brittle fracture (B10G29)　　(b) Ductile fracture (B08G29)　　(c) Local buckling (B12G29)

Fig. 9　Failure mode

Table 4　Test result

Specimen	Failure mode	Crack initiation	Failure cycle	M_{max}/M_p (+)	M_{max}/M_p (-)	$\Sigma\theta_{pi}$	η
B10G29	Brittle fracture on bottom flange	$2\theta_p(-1)$*	$4\theta_p(+1)$	1.09	-1.04	0.11	11.2
B08G29	Ductile fracture on top flange	$2\theta_p(-2)$	$6\theta_p(-1)$	1.16	-1.13	0.42	41.1
B12G29	Local buckling of top flange	$2\theta_p(-2)$	$6\theta_p(+2)$	1.14	-1.17	0.47	50.5

*$2\theta_p(-1)$: the first cycle of the negative side at twice of θ_p

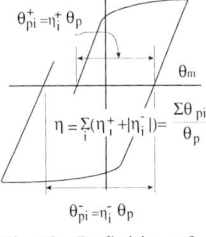

Fig.10　Definition of plastic rotatation capacity

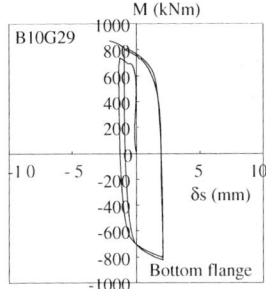 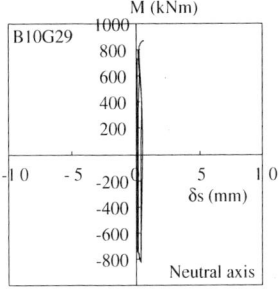 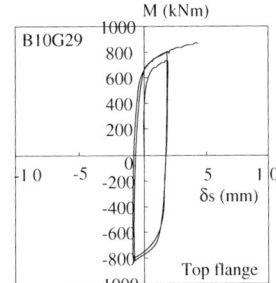

Fig.11　M-ds relationship

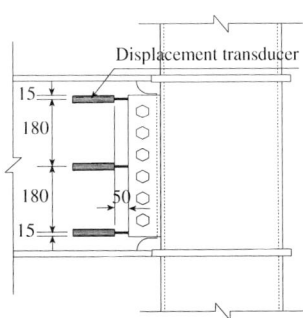

Fig.12　Measurement of slip displacement (δs)

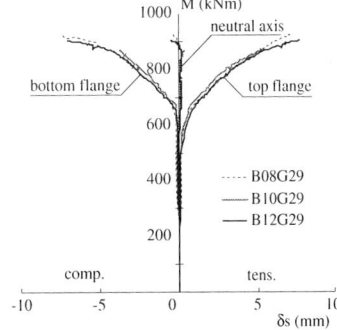

Fig.13　Slip displacement

4. FINITE ELEMENT ANALYSIS

Numerical analysis using ABAQUS ver. 6.2 is conducted to make clear the behaviour of the friction joint. Finite element meshing is shown in Fig. 14 (a). To make clear the effect of the slip behaviour, the welded type connection (WWFW) is also modeled as shown in Fig.14 (b).

The models are composed of eight-node and six-node solid elements near by the beam-to-column connection, and bar elements in the other parts. In these models, the contact behaviour is modeled by using the surface-based contact. Each surface between the beam web and the shear plate, the bolt and the beam web, the bolt and the shear plate is defined as the contact surface. Stress-strain curves of the materials are simulated by multilinear approximation to the coupon test results. The slip coefficient between the beam web and the shear plate is assumed as $\mu = 0.65$ from the slip test result.

Before applying the load at the beam end, the cramping force of the bolts, which corresponds to the forces measured at the loading tests, are introduced.

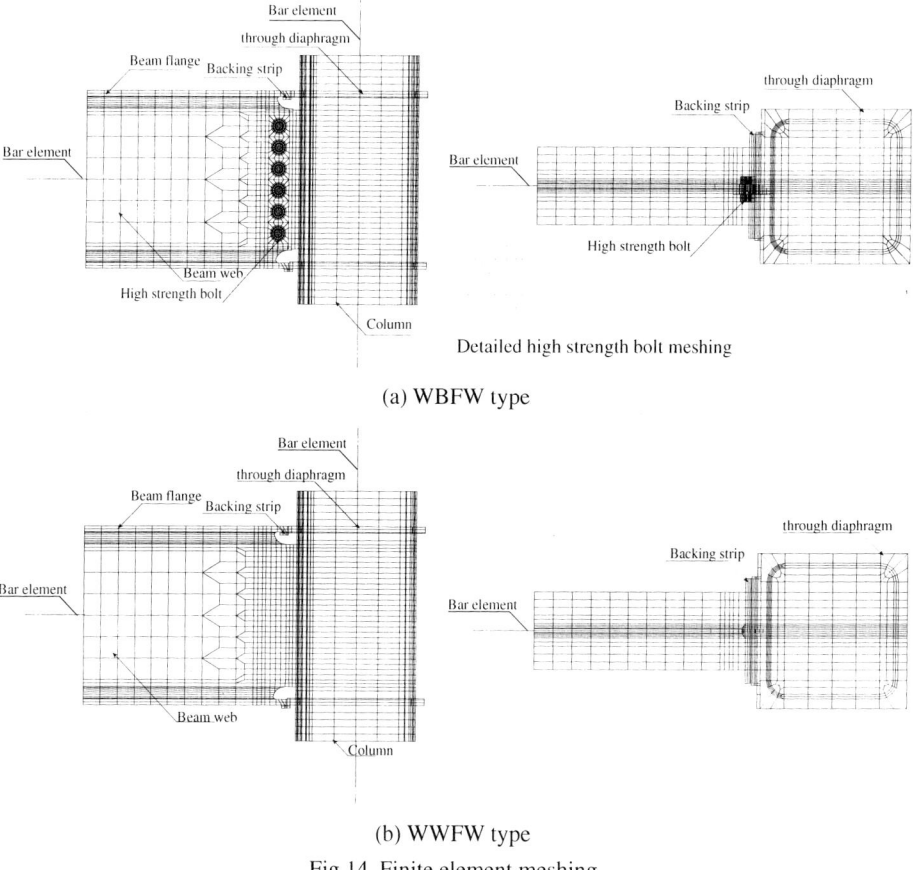

(a) WBFW type

(b) WWFW type

Fig.14 Finite element meshing

5 DISCUSSION

Figure 15 shows comparison with the test results and the FEA results. Each FEA result shows good correspondence with the test's elasto-plastic behaviour.

Figure 16 shows the moment versus the slip displacement δs curves. The FEA results well simulate the tendency of the test results. The slip load in tensile side occurs earlier than that in compressive side. As the cramping force is smaller, the occurrence of the slip becomes earlier.

Figure 17 shows M_w/M_p versus θ_m relationship. Here, M_w denotes the bending moment born by the shear plate (WBFW) or the beam web (WWFW), which is calculated by the nominal stress distribution of the shear plate or the beam web as shown in Fig.18. Figure 18 shows the nominal stress distribution at $\theta_m=2\theta_p$. The center part of the shear plate or the web has little stress and only the extreme outer parts bear the bending moment. This phenomenon is caused by the out-of-plane deformation of the square hollow section column which has large width-to-thickness ratio as this test (D/t=29). As the cramping force is smaller, the slip load becomes smaller in Fig.17. However, the moment versus beam rotation relationships in Fig.19, have no effect of the cramping force. Because, the difference of the Mw affected by the slip load was 2% of the Mp at the most.

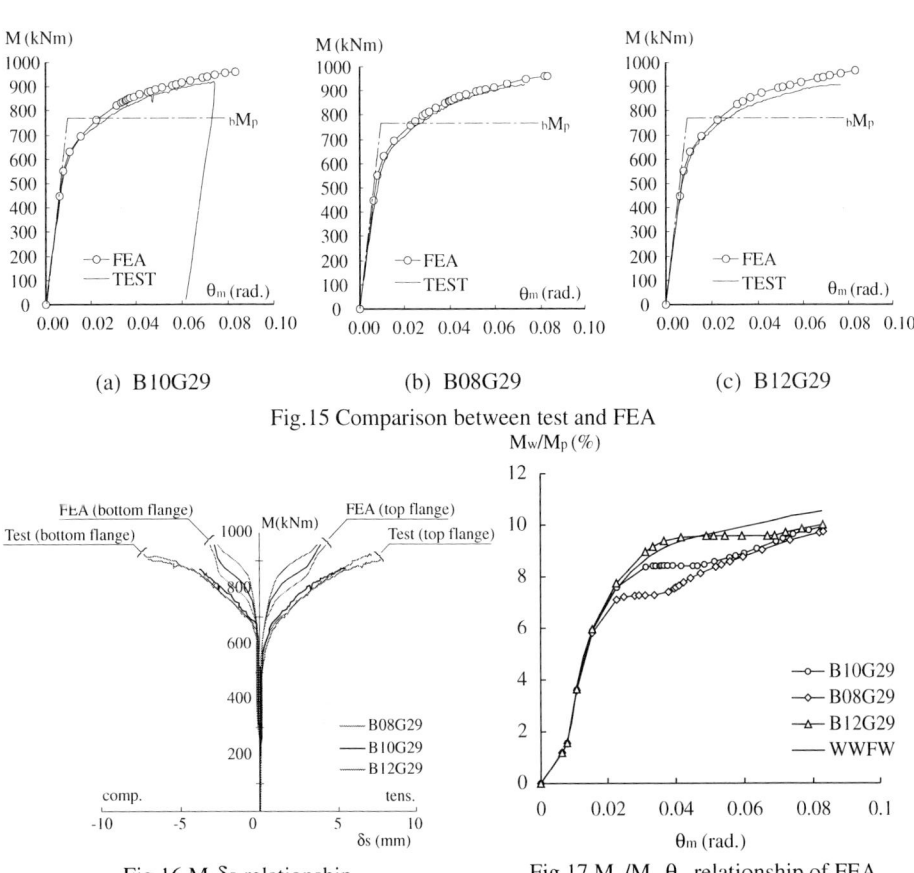

Fig.15 Comparison between test and FEA

Fig.16 M-δs relationship

Fig.17 M_w/M_p-θ_m relationship of FEA

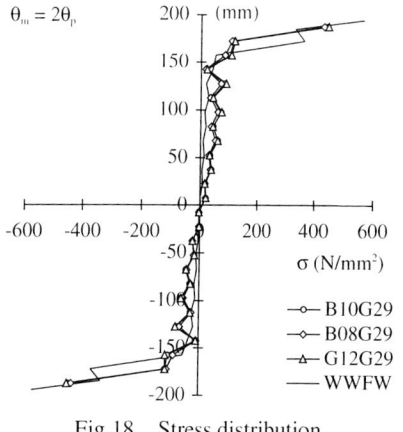
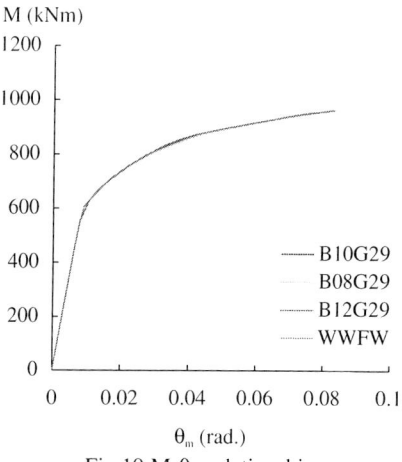

Fig.18 Stress distribution in shear plate or beam web

Fig.19 M-θ_m relationship

6. CONCLUSIONS

The effect of the slip load on the elasto-plastic behaviour of the WBFW Connections have been investigated. It has little effect on the beam behaviour to vary the slip load in this test results and this FEA results. Because, when the width to thickness ratio of the column is large, the bending moment born by the beam web is little. This phenomenon is caused by the out-of-plane deformation of the column. So, the difference of the slip load has little effect on the beam behaviour.

Reference

1. Kohzu I. (1997). Seismic performance of Welded Flange-Bolted Web Connections In Steel Moment Resisting Frames. *Behaviour of Steel Structures in Seismic Areas*, 614-621
2. Nakano T., Hiroshi M. and Tanaka A. (2001). Study on the Static Characteristics of WBFW Type Beam-to Column Connections. *Proceedings of Sixth Pacific Structural Steel Conference* :**2**, 713-718

EFFECTS OF BEAM FLANGE WIDTH-TO-THICKNESS RATIO ON BEAM FLANGE FRACTURE CAUSED FROM SCALLOP ROOT

T. IGUCHI, M. TABUCHI, T. TANAKA and S. KIHARA

Graduate School of Science and Technology, Kobe University,
Kobe, 657-8501, Japan

ABSTRACT

In steel building structures, the plastic rotation capacity of H-shaped beam welded to the column is affected by failure mode. When width-to-thickness ratio of beam flange (b/t) is large, local buckling of the beam flange governs the rotation capacity of the beam. On the other hand, when stocky beam section is used, fracture of the beam flange or weld at the stress concentration point of the beam end may occur. In this case, premature fracture of the beam flange or weld may occur before the local buckling of the beam flange, when the ductility and the toughness of the beam flange material is poor.

In this study, four specimens with b/t = 5.0, 6.7, 8.3 and 10.0 are tested to investigate the boundary between the local buckling and the fracture of the beam flange. To clarify the failure mode, elasto-plastic FEA is also conducted.

The results of this study are summarized as follows;
1) The failure mode of the specimen with b/t = 10.0 was the local buckling of beam flange.
2) The specimens with less b/t ratio were fractured caused by the ductile crack at the scallop root.
3) A specimen with b/t = 8.3 showed both failure mode and the maximum rotation capacity of the beam.

KEYWORDS

Ductile Crack, Local Buckling, Width-to-Thickness Ratio, Weld Connection, FEA

1. INTRODUCTION

In steel building structures, the plastic rotation capacity of H-shaped beam welded to the column is affected by failure mode. One of characteristic damage in Hyogoken-Nanbu earthquake in 1995 is the brittle fracture of the beam flange caused by the ductile crack at the scallop root (Tanaka and Tabuchi, 1997). The brittle fracture shall decrease in the plastic rotation capacity of the beam. When the width-to-thickness ratio (b/t) becomes large, the local buckling of the beam flange governs the rotation capacity of the beam. On the other hand, when the b/t ratio becomes small, the fracture of the beam flange may occur before the local buckling. So, it is important to make clear the boundary between the fracture and the local buckling. The weld details have been improved after the earthquake in 1995, so the fracture of the beam flange is hard to occur. The purpose of this study is to know the behaviour of the beam welded to the column with conventional weld detail before the earthquake. So, the weld detail and the end tab detail, which were usually used in middle and low rise steel buildings constructed before the earthquake, were adopted for specimens. Four specimens with the different width-to-thickness ratio were tested. To clarify the failure mode, elasto-plastic FEA is also conducted.

2. TEST

2.1 *Specimens*

A list of specimens is shown in Table 1. The only difference between these four specimens is the width of the beam flanges. Figure 1 shows geometrical configuration of the specimens. The columns of all specimens are □-350x350x12 (BCR295). The beam for specimen BT10.0 is H-434x299x10x15 (SS400). The beams for specimens BT8.3, BT6.7 and BT5.0 are cut off the both side of the beam flange to the own width, 250mm, 200mm and 150mm, respectively. Figure 2 and Figure 3 show the weld detail and the end tab detail respectively, which usually used in middle and low rise steel buildings constructed before Hyoguken-Nanbu earthquake in Japan.

Table 1 List of Test Specimens

Specimen	Beam		Diaphragm	Column	b/t
BT10.0		H-434x299x10x15			10.0
BT8.3	SS400	H-434x250x10x15	19 mm	□-350x350x12	8.3
BT6.7		H-434x200x10x15	(SS400)	(BCR295)	6.7
BT5.0		H-434x150x10x15			5.0

b : Half of beam flange width   t : Beam flange thickness

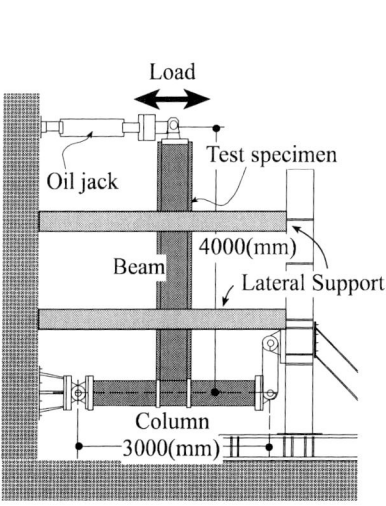

Fig.1 Test Specimen

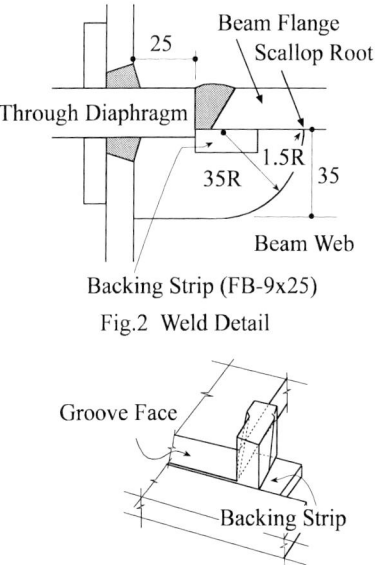

Fig.2 Weld Detail

Fig.3 Ceramic End-Tab

The full penetration welding between the beam flange and the diaphragm was performed by CO_2 gas-shielded semi automatic arc welding. The welding was carried out by four-pass per four-layer.

2.2 Mechanical Properties

Table 2 shows the mechanical properties of the materials used for beams, diaphragms and welds. The position of tensile and Charpy impact test pieces are shown in Fig.4. Charpy absorbed energy (at 0 ℃) of the beam flange is 31 (J).

2.3 Loading Program

Figure 5 shows the loading program, (1) two cycles of loading at twice of the rotation angle of the beam, θ_p, calculated on bending and shearing deflection of the beam at the full plastic moment (Mp), (2) two cycles of loading at four times of θ_p, and (3) two cycles of loading at six times of θ_p.

3. TEST RESULTS

3.1 Moment-Rotation Curves

Figure 6 shows non dimensional moment-rotation hysteresis curves. Specimens BT8.3, BT6.7 and BT5.0 fractured in the beam flanges caused by the ductile crack at the scallop root. A specimen BT10.0 failed due to the local buckling of the beam flange. Table 3 summarizes the test results. Figure 7 shows the definition of the cumulative plastic deformation capacity (η). The plastic rotation capacity of a specimen BT8.3 was superior to any other specimens.

Table 2 Mechanical Properties

		σ_y (N/mm²)	σ_u (N/mm²)	YR (%)	ε_u (%)	vE_0 (J)	B_0 (%)
Beam	Flange	300	472	64	19.0	31	93
Beam	Web	370	492	79	21.0	—	—
Beam	Fillet	—	—	—	—	25	90
Diaphragm		285	447	64	20.5	—	—
Welds		421	529	80	18.5	—	—

σ_y : Yeild strength σ_u : Tensile strength
YR : Yeild ratio ε_u : Uniform elongation
vE_0 : Charpy absorbed energy at 0℃
B_0 : Percent brittle fracture at 0℃

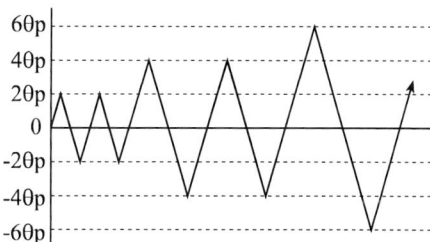

Fig.5 Loading Program

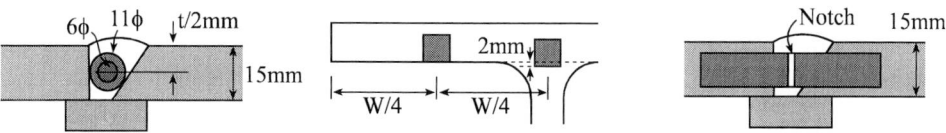

(a) Welds (Tensile Test)   (b) Flange and Fillet (Charpy Impact Test)   (c) Welds (Charpy Impact Test)

Fig.4 Test Pieces

3.2 Failure Modes

(1)BT10.0 : The initial ductile crack was observed at the root of the scallop on inner surface of the tensile beam flange in the process of $4\theta p(+1)$, the local buckling was observed in the compressive beam flange at the peak of $4\theta p(+1)$. The specimen failed due to the local buckling in the process of $6\theta p(+1)$. Figure 8 shows the failure mode of specimen BT10.0.

(2)BT8.3 : The initial ductile crack was observed at the root of the scallop in the process of $4\theta p(+1)$, the local buckling was also observed at the peak of $4\theta p(-1)$. The ductile fracture occurred in the tensile beam flange in the process of $6\theta p(-1)$. Figure 9 shows the failure mode of specimen BT8.3.

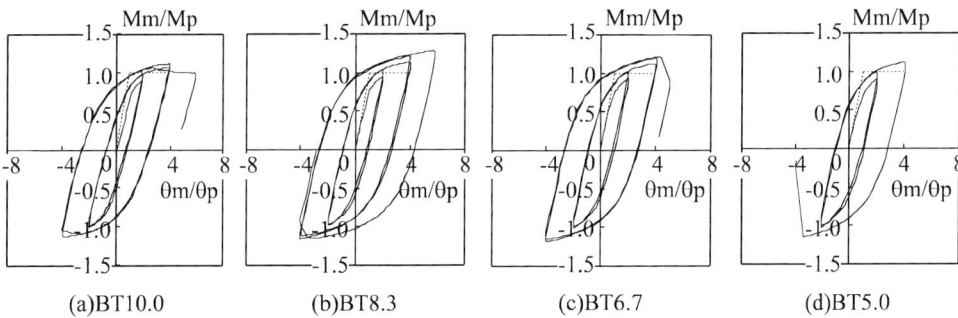

(a)BT10.0   (b)BT8.3   (c)BT6.7   (d)BT5.0

Fig.6 Moment-Rotation Curves

Table 3 Test Results

Specimen	Failure*	Failure mode	Ductile crack initiation	η
BT10.0	$6\theta p(+1)$	Local buckling	$4\theta p(+1)$	30
BT8.3	$6\theta p(-1)$	Ductile fracture	$4\theta p(+1)$	40
BT6.7	$6\theta p(+1)$	Ductile fracture	$4\theta p(+1)$	32
BT5.0	$4\theta p(-1)$	Brittle fracture	$2\theta p(+2)$	16

* "$6\theta p(+1)$" means the first cycle of the positive side at six times of θp

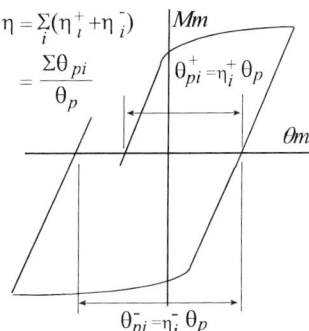

Fig.7 Cumulative Plastic Deformation Capacity (η)

Fig.8 Local Buckling (BT10.0)

Fig.9 Ductile Fracture (BT8.3)

(3)BT6.7 : The initial ductile crack was observed at the root of the scallop in the process of $4\theta p(+1)$. The ductile fracture occurred in the tensile beam flange in the process of $6\theta p(+1)$. Figure 10 shows the failure mode of specimen BT6.7.

(4)BT5.0 : The initial ductile crack was observed at the root of the scallop in the process of $2\theta p(+2)$. The brittle fracture occurred in the tensile beam flange in the process of $4\theta p(-1)$. Figure 11 shows the failure mode of specimen BT5.0.

4. FEA

4.1 *Analysis Model*

The numerical analyses were carried out using the general purpose finite element program ABAQUS ver. 6.2. Two types of FEA models shown in Fig.12 were adopted. The solid model is used to investigate the ductile crack initiations and the shell model is used to investigate the local buckling occurrences. The solid model, which is shown in Fig.12(a) is composed of eight-node or six-node solid elements near by beam-to-column connection and bar elements in the other parts. The solid elements near by the root of scallop are regular hexahedrons, whose size is 1x1x1 mm, which are minimum elements size in this model. The shell model, which is shown in Fig.12(b), is composed of four-node shell elements near by beam-to-column connection and bar elements in the other parts. Stress-strain curves of the materials are simulated by considering the coupon test results.

Fig.10 Ductile Fracture (BT6.7)

Fig.11 Brittle Fracture (BT5.0)

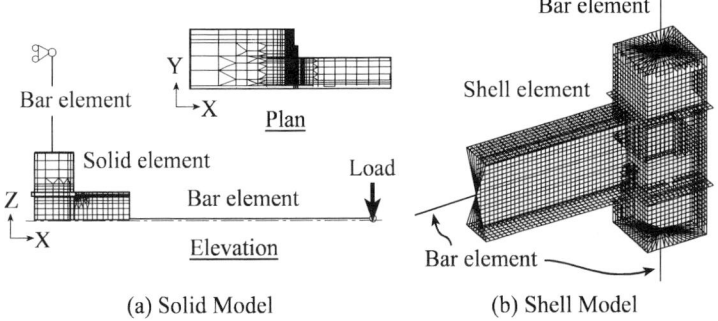

(a) Solid Model (b) Shell Model

Fig.12 FEA Model

4.2 Comparison between Test Result and FEA Result

Figure 13 shows the moment-rotation curves obtained by the solid models, the shell models and test results converted to the skeleton curves. Figure 14 shows the way of the conversion of hysteresis curves (test results) to the skeleton curves. The analytical results agree well with the test results. The shell model analyses, which simulate BT10.0 and BT8.3, show the local buckling and degrading of the load carrying capacity before reaching the maximum strength in the tests.

4.3 Crack Initiation

Figure 15 shows relationships between equivalent plastic strain(ε_{eq}) at the root of the scallop, where the strain concentrate extremely, and deformation ratio($\theta m/\theta p$). As the width-to-thickness ratio of the beam flange is larger,

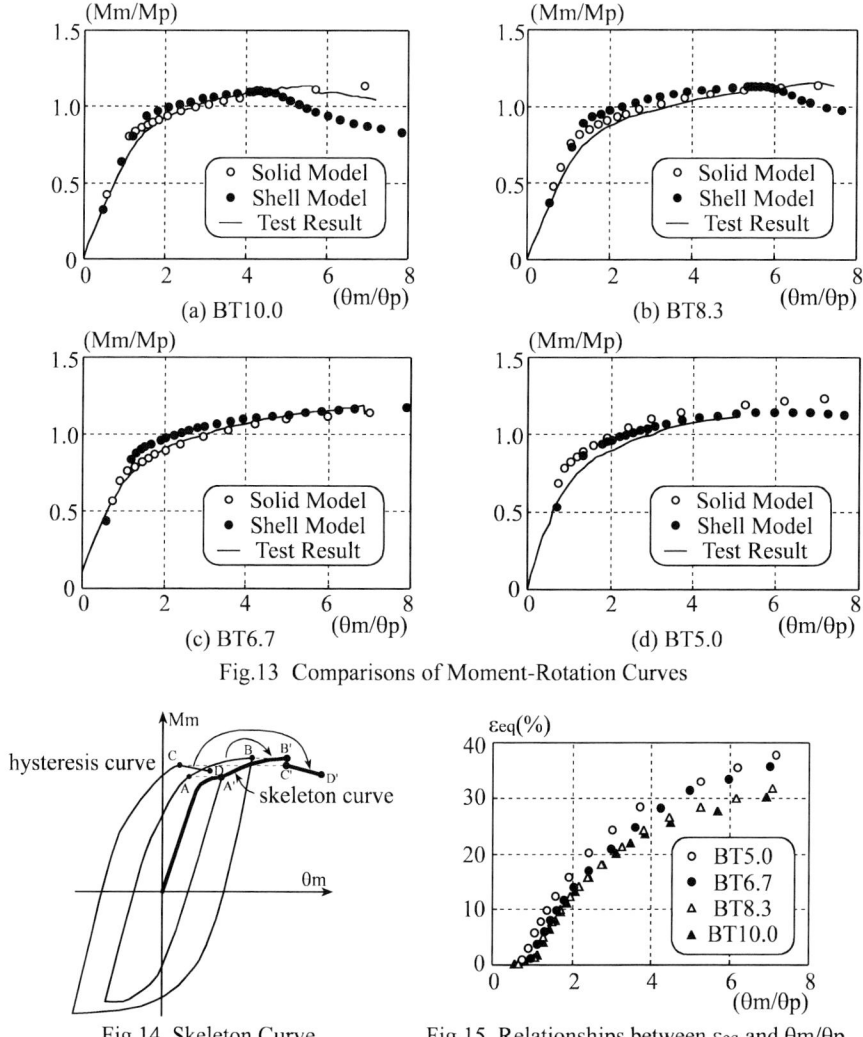

Fig.13 Comparisons of Moment-Rotation Curves

Fig.14 Skeleton Curve

Fig.15 Relationships between ε_{eq} and $\theta m/\theta p$

the equivalent plastic strain(ε_{eq}) at the root of the scallop is smaller.

It is assummed that the ductile crack at the root of the scallop initiates when ε_{eq} reaches the following condition ε_{ep} (Kuwamura and Yamamoto, 1995).

$$\varepsilon_{ep} = \varepsilon_u / \tau^2$$
$$\tau = \sigma_h / \sigma_{eq}$$
......(1)

where ε_{ep} = crack initiation strain
  ε_u = uniform elongation of the beam flange
  τ = stress triaxiality ratio
  σ_h = hydrostatic stress
  σ_{eq} = Von Mises equivalent stress

Figure 16 shows conditions of ductile crack initiations, which have two ε_{eq}-τ curves. The line with some circles shows the condition at the root of the scallop and the number in the circle means the value of the deformation ratio of the beam ($\theta m/\theta p$). Another line shows the criterion curve for crack initiation (eq.1). The ductile crack initiates when the two lines cross each other as shown in Figs.16(a)~(d). As the width-to-thickness ratio of the beam flange is larger, the crack initiation at the root of the scallop is later.

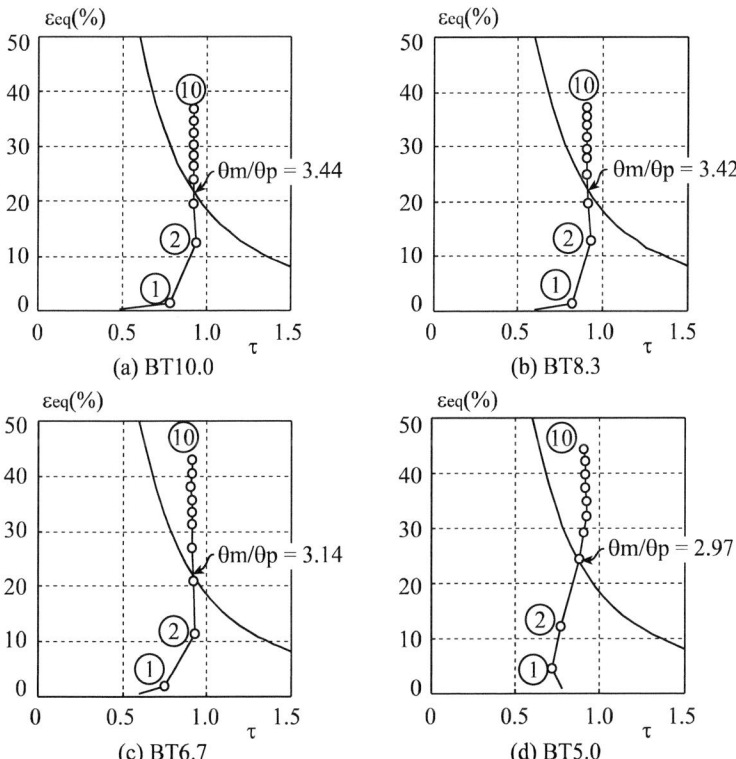

Fig.16 Ductile Crack Intiation

4.4 Local Buckling Occurrence

Figure 17 shows the example for BT10.0 of the relationship between the strain at the peak of the local buckling mode on the compressive beam flange and the deformation ratio($\theta m/\theta p$). It is assumed that the local buckling occurs when the slope of the ε-$\theta m/\theta p$ relationship turns from negative to positive. In the case of BT10.0, the local buckling occurs at $\theta m/\theta p=2.94$.

Figure 18 shows relations of width-to-thickness ratios to the local buckling occurrences and the ductile crack initiations. As width-to-thickness ratio of the beam flange is larger, the local buckling occurrence is earlier. In the case of BT10.0, the local buckling occurrence is earlier than the ductile crack initiation. In the other case, the local buckling occurrence is later than the ductile crack initiation. The specimen BT8.3 have the possibility to take both failure mode. This tendency agreed with the experimental failure modes in the tests.

5. CONCLUSIONS

In this paper, the specimens with the different width-to-thickness ratio were tested and two types of FEA were conducted. The following results were obtained.
1) The failure mode of the specimen with b/t = 10.0 was the local buckling of the beam flange and the specimens with less b/t ratio were fractured caused by ductile crack at the root of the scallop.
2) A specimen with b/t = 8.3 shows both failure mode and the maximum rotation capacity of the beam.
3) The failure mode which depend on the width-to-thickness ratio of the beam flange clearly explained by FEM analysis.
4) The boundary of the width-to-thickness ratio of both failure modes was b/t ≒ 9.

REFERENCE

1) H. Kuwamura and K. Yamamoto (1995). Criterion for Ductile Crack Initiation in Structure Steels under Triaxial Stress State. J. Struct. Constr., AIJ, No.477, Nov., 129-135. (in Japanese)
2) T. Tanaka and M.Tabuchi (1997). Fracture on SHS Column to H-beam Connections by Hyogoken-Nanbu earthquake, Behaviour of Steel Structures in Seismic Areas. Stessa '97, Aug., 866-873

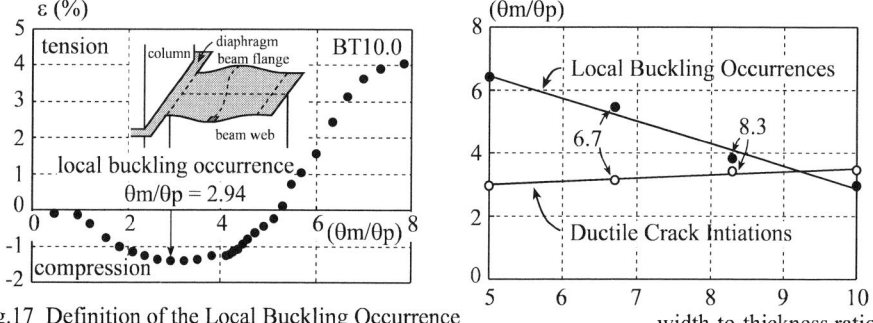

Fig.17 Definition of the Local Buckling Occurrence

Fig.18 Comparions with Local Buckling Occurrences and Ductile Crack Intiations

EXPERIMENTAL INVESTIGATION OF SLOT LENGTHS IN RHS BRACING MEMBERS

T. Wilkinson, T. Petrovski, E. Bechara and M. Rubal

Department of Civil Engineering, The University of Sydney, NSW, 2006, Australia

ABSTRACT

For slotted end plate connections in RHS, it is usual that the slot should extend into the RHS for a length at least equal to the depth of the RHS section. This paper describes tension tests on a series of slotted end plate connections in which the slot lengths were less than the depth of the section. Two thicknesses of Grade C450 RHS were used, and two different fillet weld leg lengths were considered. It was found that the full strength of the weld, according to Australian Standard AS 4100 predictions, was met in each case. Most importantly, on several occassions, the failure load was higher than the predicted shear lag failure modes. This suggests that the correction factors to account for shear lag may be unecessarily conservative. However, since the failure mode was weld shear and not shear lag tension, it not possible to quantify the extent of this conservatism.

KEYWORDS

Cold-formed steel, structural hollow sections, connections, welding, tension, fracture.

INTRODUCTION

Rectangular, square and circular hollow sections (RHS, SHS and CHS) are often used as bracing elements in steel frames. SHS and CHS are particularly practical as their compression capacities are the same for buckling about any set of axes. However, there is always difficulty in joining hollow sections as bolting can only ever be done from one side. A common end connection is the slotted end plate. Generic design guidance for such connections is given by Packer and Henderson (1997) and Syam and Chapman (1996). Previous studies on similar connections have been performed by Zhao and Hancock (1995), Korol (1996), Cheng, Kulak and Khoo (1998) and Zhao, Al-Mahaidi and Kiew (1999).

Figure 1 shows a typical slotted end plate connection in an RHS and defines the various symbols applicable to such a connection. An RHS has typical cross sectional dimension $d \times b \times t$. The slotted end plate has dimensions $d_p \times b_p \times t_p$. A slot of length L_s is cut into the RHS, and the plate is fillet welded into the slot. Since the plate is (usually) welded on both sides the total length of weld metal is (usually) given by $L_w = 4L_s$ plus any additional end return weld. A typical fillet weld cross section is also shown, usually denoted by its leg length (t_w – thickness of the weld). The weld throat thickness (t_t) is a significant strength parameter. For RHS the slotted plate is most commonly oriented along the weak of the section (parallel to the longer side).

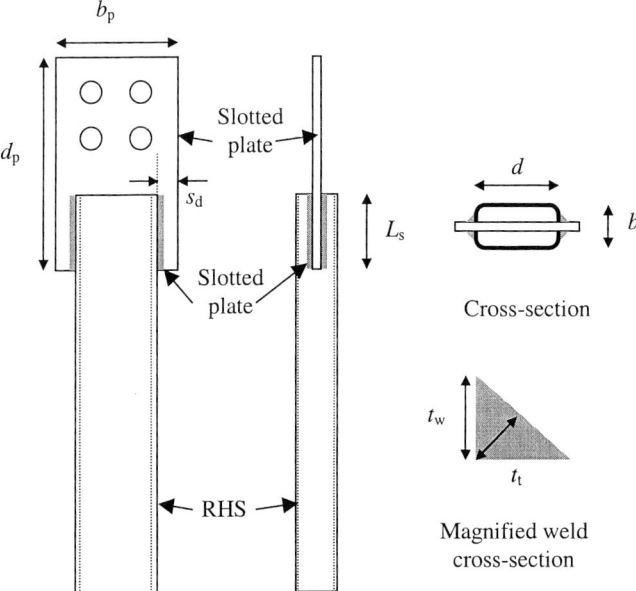

Figure 1: Typical layout and notation of a slotted end plate connection

Packer and Henderson (1997) give design guidance that the slot length should, as a minimum, be the same length as the distance between the welds (on the cross-section). In most cases, this corresponds to d.

The aim of this test series is to test slotted end plate connections with short slot lengths ($L_s < d$) to determine if the minimum slot length requirement is appropriate.

TEST PROGRAM

Specimens

Two RHS were used, 100 × 50 × 4 and 100 × 40 × 6, Grade C350/C450 to AS 1163 (Standards Australia 1991). These tubes are called "DualGrade" and are manufactured by Palmer Tube Mills in Queensland Australia. The plate for the slotted end plates was Grade 350 to AS 3678, and the weld metal used was W503H (nominal strength $f_{uw, nom}$ = 480 MPa) to AS 2717.1 (Standards Australia 1984).

Material Properties

Coupons were taken from the flats and corners of each tube. The coupons were prepared and tested in accordance with AS 1391 (Standards Australia (1991)) in a 250 kN capacity INSTRON Universal Testing Machine. Since the steel was cold-formed, the yielding was gradual. Accordingly the yield stress (f_y) used is the 0.2% proof stress. The 100 × 50 × 4 exhibited an average yield stress of f_y = 456 MPa and ultimate strength f_u = 563 MPa, while the corresponding values for the 100 × 50 × 4 section were 477 MPa and 556 MPa. A full weld metal tensile test to AS 2205 Part 2.2 was also performed to determine the tensile properties of the weld metal. An ultimate strength f_{uw} = 482 MPa was obtained. Full details on coupon tests, including stress strain curves are given in Bechara (2001), Rubal (2001) and Petrovski (2001).

Weld Details

Different weld sizes, lengths and orientations were chosen as given in Table 1. Welding procedure sheets were maintained as part of qualification requirements to AS/NZS 1554.1 and welding procedures are given in Bechara (2001), Rubal (2001) and Petrovski (2001).

Test Procedure

The connection tests were performed in a 2000 kN capacity DARTEC testing machine, using a servo-controlled hydraulic ram as shown schematically in Figure 2. Tensile loading was applied with a constant cross head speed. LVDTs measured the extension across the weld. Full testing procedures are given in Bechara (2001), Rubal (2001) and Petrovski (2001).

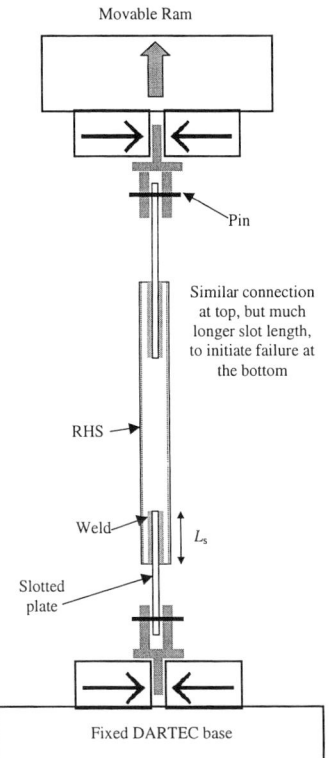

Figure 2: Test Setup

RESULTS

Table 1 summarises the results of each test. The table gives the relevant dimensions for each test, and the maximum load. Load deflection curves are given in Figure 3, and typical failure shown in Figure 4.

TABLE 1
SUMMARY OF DIMENSIONS AND MAXIMUM LOADS

Notation	RHS size	Orientation	Nom. weld length $L_{s,nom}$ (mm)	Meas. weld length $L_{s,meas}$ (mm)	Nom. weld thick. $t_{w,nom}$ (mm)	Nom. weld throat thick. $t_{t,nom}$ (mm)	Meas. weld throat thick. $t_{t,meas}$ (mm)	Capacity (kN)
Slot 4_40	100×50×4		40	45.67	4	2.8	4.03	245
Slot 4_60	100×50×4		60	68.49	4	2.8	3.54	350
Slot 4_60OD	100×50×4		60	64.66	4	2.8	4.28	337
Slot 4_80	100×50×4		80	89.55	4	2.8	4.10	478
Slot 6_40	100×50×6		40	47.01	6	4.2	4.24	391
Slot 6_60	100×50×6		60	68.59	6	4.2	4.20	547
Slot 6_60OD	100×50×6		60	67.46	6	4.2	4.40	503
Slot 6_80	100×50×6		80	87.15	6	4.2	4.33	677

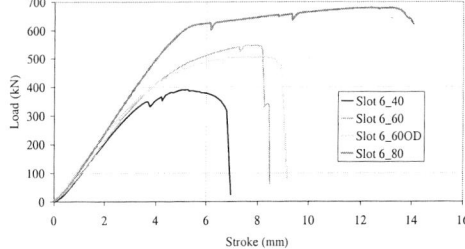 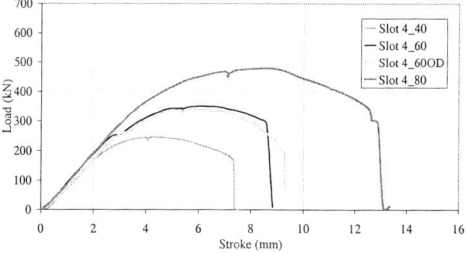

Figure 3: Load extension plots

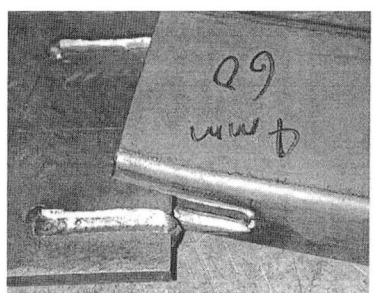 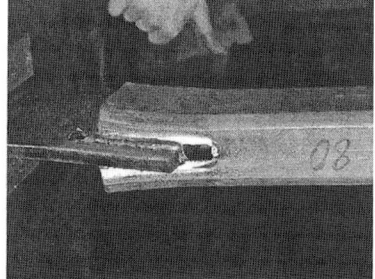

(a) 100×50×4 RHS – 60 mm weld length   (b) 100×50×6 RHS – 80 mm weld length

Figure 4: Typical Failure Modes

COMPARISON WITH CURRENT DESIGN MODELS

The (Australian) AISC (Syam and Chapman 1996) has collated various design equations for the various failure modes applicable to slotted end plate connections in tension. These are presented below in Table 2. The nomenclature is consistent with Australian Standard AS 4100 (Standards Australia 1998), but can equally be applied to other design standards. The most significant parameters are the factors k_t, k_{ti} and k_{to} which account for shear lag effects.

TABLE 2
SUMMARY OF DESIGN FORMULAE

Failure Mode	Design Equation (Nominal Capacity)
Hollow section shear lag	$N_{ts} = 0.85 A_n k_t f_u$
Plate yield or fracture	$N_{tp} = A_{gp} f_{yp}$, or $0.85 A_{np} k_{tp} f_{up}$
Plate shear lag	$N_{tps} = 0.85(A_{ni} k_{ti} + 2 A_{no} k_{to}) f_{up}$
Weld in shear	$V_w = 0.6 f_{uw} t_t k_r$
Bolt bearing/tearout/block shear	not applicable – bolts not considered
$k_t = 1.0$ (for $L_s/w > 2.0$) $k_t = 0.87$ (for $2.0 > L_s/w > 1.5$) $k_t = 0.75$ (for $1.5 > L_s/w > 1.0$) $k_t = 0.5$ (for $1.0 > L_s/w > 0.5$) w = half perimeter of RHS – slot width $k_r = 1.0$ A_{ni} = slotted plate area between welds A_{no} = slotted plate area outside welds $\varphi = 0.9$, $\varphi_{weld} = 0.8$ (capacity reduction factors apply for design purposes)	$k_{ti} = 1.0$ (for $L_s/d > 2.0$)* $k_{ti} = 0.87$ (for $2.0 > L_s/d > 1.5$)* $k_{ti} = 0.75$ (for $1.5 > L_s/d > 1.0$)* $k_{ti} = 0.5$ (for $1.0 > L_s/d > 0.5$)* * use b rather than d for slots oriented oppositely $k_{to} = 1.0$ (for $L_s/s_d > 4.0$) $k_{to} = 0.87$ (for $2.0 > L_s/s_d > 3.0$) $k_{to} = 0.75$ (for $1.5 > L_s/s_d > 2.0$) $k_{to} = 0.5$ (for $1.0 > L_s/s_d > 1.0$)

Table 3 compares the experimental results to those predicted by the various equations in Table 2. For brevity, the experimental results are compared only to the most relevant design equations: hollow section shear lag (N_{ts}), plate shear lag (N_{tps}), and weld in shear (V_w).

TABLE 3
COMPARISON OF EXPERIMENTAL RESULTS TO PREDICTIONS

Notation	RHS size	Experiment N_{exp} (kN)	RHS lag N_{ts} (kN)	$\dfrac{N_{exp}}{N_{ts}}$	Plate lag N_{tps} (kN)	$\dfrac{N_{exp}}{N_{tps}}$	Weld shear V_w (kN)	$\dfrac{N_{exp}}{V_w}$
Slot 4_40	100×50×4	245	291	0.840	359	0.682	**186**	**1.313**
Slot 4_60	100×50×4	350	291	1.200	419	0.835	**246**	**1.423**
Slot 4_60OD	100×50×4	337	**291**	**1.156**	419	0.805	297	1.134
Slot 4_80	100×50×4	478	**291**	**1.641**	448	1.068	379	1.260
Slot 6_40	100×50×6	391	345	1.134	354	1.103	**196**	**1.993**
Slot 6_60	100×50×6	547	345	1.585	414	1.322	**292**	**1.875**
Slot 6_60OD	100×50×6	503	345	1.458	413	1.216	**305**	**1.646**
Slot 6_80	100×50×6	677	**345**	**1.965**	442	1.533	401	1.691
Mean				**1.372**		**1.070**		**1.542**

Note: Critical failure mode shown in bold.

DISCUSSION

Firstly, all slot lengths were deliberately chosen to be low. In every case $L_s/d < 1$ and for several cases $L_s/d < 0.5$. The correction factor to account for shear lag caused by short slot lengths is given by $k_{ti} = 0.5$ (for $1.0 > L_s/d > 0.5$), and no value is specified for the very short slots ($L_s/d < 0.5$). For the calculation of the capacity for these cases with very short slots, it has been assumed that $k_{ti} = 0.5$.

It is evident that the specimens exhibited capacities that exceeded the critical failure load in all cases. In each case, failure was associated with either shear failure of the weld, or shear (pullout) failure of the heat affected zone adjacent the weld. A tensile fracture associated with shear lag was not observed in any case.

For each case, the experimental load exceeded the predicted load for weld shear failure. This is most likely caused by the nature of the fillet welds. All fillet welds were convex. The capacity was these welds was determined to AS 4100 which defines a throat thickness as given in Figure 5 for convex fillet welds. However, it is clear that the "true" throat thickness is notably larger than this definition allows, resulting in higher than predicted capacities. AS 4100 recommends against including the "return weld" (around the bottom of the slot plate) when determining the length of weld that resists the load. This will also increase the capacity beyond predictions.

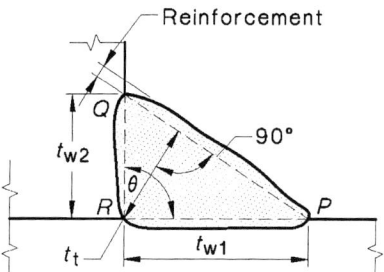

Figure 5: Definition of throat thickness for convex fillet weld (Fig 9.7.3.1(b) of AS 4100)

In some cases, the failure load was less than the predicted shear lag failure modes. However, since the failure mode was weld shear and not shear lag tension, it not possible to say that the predicted shear lag capacities were incorrect in those cases.

Most importantly, on several occassions, the failure load was higher than the predicted shear lag failure loads. This suggests that the correction factors to account for shear lag may be unecessarily conservative. However, since the failure mode was weld shear and not shear lag tension, it not possible to quantify the extent of this conservatism.

It is suggested that further investigation be required to establish the potential degree of conservatism of the shear lag correction factor.

There was no significant difference between the behaviour of the two sets of specimens with different orientations of the slotted end plate (either through the strong or the weak axis of the cross-sections). In these cases there was no sigificant difference between the maximum load reached or the deformation capacity of the sections.

SUMMARY

This paper has described tests on welded slotted end plate connections in cold-formed rectangular hollow sections. The slot lengths were deliberately small, with lengths less than those recommended in current design guides. The aim was to examine the capacity of these sections with respect to the weld strength, and the reduction factor to account for shear lag effects with short slot lengths.

All specimens experienced failure associated with weld shear or shear failure in the heat affected zone near the weld. The convex shape of the welds, which is not accounted for in the determination of the weld throat thickness, is the main reason behind this under prediction.

In some instances, the failure load was higher than the predicted shear lag failure loads, suggesting that the shear lag correction factors may be conservative. Since the failure mode was weld shear and not shear lag tension, it not possible to quantify the extent of this conservatism. Further investigation be required to establish the potential degree of conservatism of the shear lag correction factor.

ACKNOWLEDGEMENTS

The results presented in this paper were obtained by authors 2 – 4, as part of their undergraduate thesis in The Department of Civil Engineering at The University of Sydney. The first author was the thesis supervisor. Tube specimens were provided by Palmer Tube Mills Ltd. The experiments were carried out in the J. W. Roderick Laboratory for Materials and Structures, Department of Civil Engineering, The University of Sydney. Thanks to Mr Grant Holgate for the preparation of the specimens.

REFERENCES

Bechara, E., (2001), "Tests of Slotted End Plate Connections in RHS", *Thesis*, Department of Civil Engineering, The University of Sydney, Australia.

Cheng, R. J. J., Kulak, G. L. and Khoo, H. A., (1998), "Strength of Slotted Tubular Tension Members", *Canadian Journal of Civil Engineering*, Vol 25, 1998, pp 982 – 991.

Korol, R. M., (1996), "Shear Lag in Slotted HSS Tension Members", *Canadian Journal of Civil Engineering*, Vol 23, 1996, pp 1350 – 1354.

Packer, J. A. and Henderson, J. E., (1997), *Hollow Structural Section Connections and Trusses – A Design Guide*, Canadian Institute of Steel Construction, 2nd edition, Willowdale, Ontario, Canada.

Petrovoski, A., (2001), "Tests of Slotted End Plate Connections in RHS", *Thesis*, Department of Civil Engineering, The University of Sydney, Australia.

Rubal, M., (2001), "Tests of Slotted End Plate Connections in RHS", *Thesis*, Department of Civil Engineering, The University of Sydney, Australia.

Standards Australia / Standards New Zealand, (1995), Australian / New Zealand Standard *AS 1554.1 Structural Steel Welding, Part 1: Welding of Steel Structures*, Standards Australia, Sydney, Australia.

Standards Australia, (1984), Australian Standard *AS 2717.1 Welding - Electrodes - Gas metal-arc, Part 1: Ferritic Steel Electrodes*, Standards Australia, Sydney, Australia.

Standards Australia, (1990b), Australian Standard *AS 3678 Structural Steel: Hot-rolled Structural Plates, Floorplates and Slabs*, Standards Australia, Sydney, Australia.

Standards Australia, (1991), Australian Standard *AS 1163 Structural Steel Hollow Sections*, Standards Australia, Sydney, Australia.

Standards Australia, (1991), Australian Standard *AS 1391 Methods for Tensile Testing of Metals*, Standards Australia, Sydney, Australia.

Standards Australia, (1998), Australian Standard *AS 4100 Steel Structures*, Standards Australia, Sydney, Australia.

Syam, A. A. and Chapman, B. G., (1996), *Design of Structural Steel Hollow Section Connections – Volume 1: Design Models*, Australian Institute of Steel Construction, Sydney, Australia.

Zhao, X. L. and Hancock, G. J., (1995), "Longitudinal Fillet Welds in Thin Cold-Formed RHS", *Journal of Structural Engineering*, American Society of Civil Engineers, Vol 121, No 11, November 1995, pp 1683-1690.

Zhao, X. L., Al-Mahaidi, R. and Kiew, K-P., (1999), "Longitudinal Fillet Welds in Thin-Walled C450 RHS Members", *Journal of Structural Engineering*, American Society of Civil Engineers, Vol 125, No 8, August 1999, pp 821-828.

EXPERIMENTAL STUDY ON CYCLIC BEHAVIOR OF IMPROVED BEAM-COLUMN CONNECTIONS[*]

Z.F. Li, Y.J. Shi, H. Chen, Y.Q. Wang

Department of Civil Engineering, Tsinghua University, Beijing 100084, China

ABSTRACT

After the Northridge earthquake in 1994, some researches in USA have identified three critical issues that significantly affect the strength and ductility of beam-column connections. They are: fracture toughness of weld metal, geometry and size of weld access hole, and control of panel zone deformation. This paper reports the experimental results from the second phase of a beam-column connection research project supported by Natural Science Foundation of China. The experiments focus on the geometry and size of weld access hole. Three new types of connection details are proposed and eight full size specimens are tested under cyclic loading. The stress distribution, stress concentration, and hysteretic behavior of the connections are investigated. The influence of connection details on ultimate strength and premature fracture of the connections are also analyzed and advices on improvement of connection details are given. The experimental results provide valuable information for the updating of seismic design code for high-rise steel buildings.

KEYWORDS

Beam-column connection, experimental research, fracture failure, steel structure

INTRODUCTION

Web-bolted and flange-welded (WBFW) beam-column connection is the most widely used connection type in steel frames and was once believed to have good ductility in earthquakes. In Northridge earthquake, however, brittle failures were found in the connections of many steel structures (Miller D.K., 1998; Popov E.P. *et al.*, 1998). Some researches in USA after the earthquake have identified three critical issues that significantly affect the strength and ductility of beam-column connections. They are: fracture toughness of weld metal, geometry and size of weld access hole, and control of panel zone deformation (Lu L.W. *et al.*, 2000). Further researches have introduced some methods to improve connection ductility. One method is reinforcing the connection, such as adding cover plates or haunches. Another is weakening the beam to take the plastic hinge away from the face of the column, such as Reduced Beam-section Connection (RBC) and web-slotted connection. Improved Weld Access

---

[*] This project is supported by Natural Science Foundation of China through Grant No. 59878026

Holes (WAHs) are also introduced in order to release local stress concentration.

Some improved WAH details can take plastic hinge away from the column face and introduce better connection ductility, with little loss on connection strength. A new type of WAH geometry is recommended in FEMA-350 (FEMA, 2000). Seismic Design Code for Buildings of China (Ministry of Construction P. R. China, 2001) also recommends a new type of connection geometry, which is similar to the TS3 specimen in this article. This paper reports the experimental results from the second phase of a research project in Tsinghua University (Zhao D.W, *et al.*, 2000). The experiments focus on the geometry and size of weld access hole. Three new types of connection details, which are called Enlarged Weld Access Hole (EWAH) connections, are designed and compared with the 'Standard' WBFW connection, which is widely used in China.

THE EXPERIMENTAL PROGRAM

The Specimens

Eight specimens of four types of connection details are tested in the experiments. They are standard connection (TS1, TS1a), EWAH I connection (TS2, TS2a), EWAH II connection (TS3, TS3a), and EWAH III connection (TS4, TS4a). Two specimens, one of which has supplemental weld between the shear tab and beam web, were tested for each connection detail. The specimens are made of Q235 structural steel with a yielding strength of 260MPa and a Young's modulus of 2.0×10^5 MPa.

All specimens have same beam section and same column section. The beam is 400mm high and 150mm wide. The beam flange is 12mm thick and the web is 8mm. The beam flanges are welded to the column using full penetration welds and stiffeners are provided on the column web to improve stiffness of the panel zone. Beam web is connected to the shear tab with four M20 high strength bolts (Grade 10.9). The connection details are shown in figure 1.

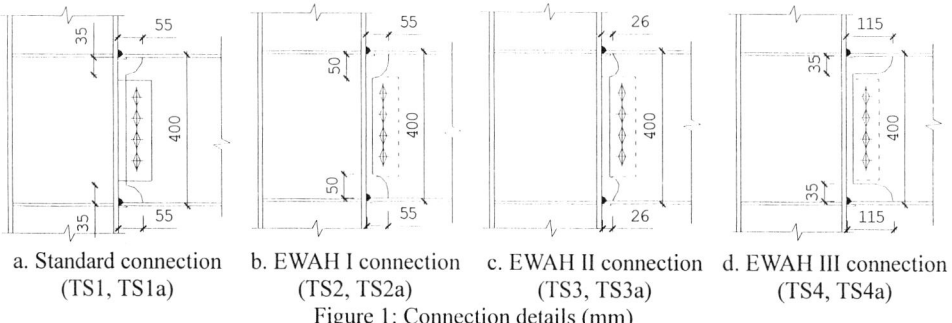

a. Standard connection (TS1, TS1a) b. EWAH I connection (TS2, TS2a) c. EWAH II connection (TS3, TS3a) d. EWAH III connection (TS4, TS4a)

Figure 1: Connection details (mm)

Experimental Set-up

The typical experimental set-up is shown in figure 2. The column is horizontally placed on the ground and a horizontal cyclic load is applied at the top end of the beam by an actuator. As for experimental conditions, no axial load is applied to the column or the beam. The following parameters are measured during the loading process: the applied force, the displacement of the loading point, the relative rotation between the beam and the column, the deformation of the panel zone, the relative displacement between the shear tab and the beam web, the strain at several points on the surface of the beam flange and beam web.

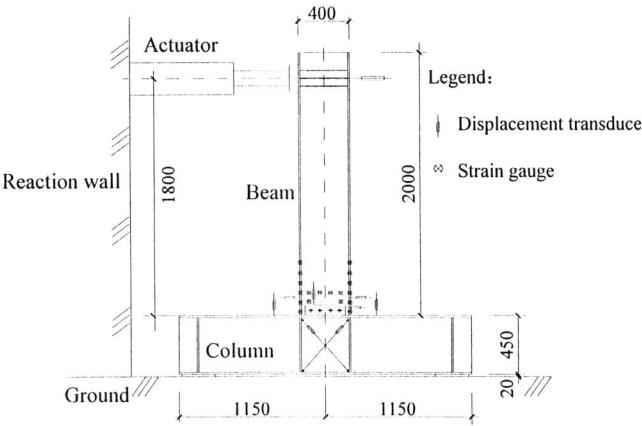

Figure 2: Experimental set-up (mm)

Loading Schemes

Four loading schemes, each including elastic and plastic stage, are used in the experiments, as shown in table 1. The applied force is taken as the controlling parameter during the elastic stage and the displacement of the loading point is controlled for the plastic stage. Scheme I is applied to specimen TS3 and TS4a, scheme II is applied to TS1, scheme III is applied to TS2a and scheme IV is applied to TS1a, TS2, TS3a and TS4. In table 1 the number of loading loops is given in the brackets and the specimen name shows that it fails during that loading stage.

TABLE 1 LOADING SCHEMES

Scheme	Elastic stage – force (kN)				Plastic stage – displacement (mm)					
	1	2	3	4	5	6	7	8	9	10
I	-------	-------	-------	-------	±42(n) TS3,TS4a	-------	------	-------	-------	-------
II	±40(1)	±80(1)	±100(1)	±120(1)	±22(2)	±32(2)	±42(2)	±32(2)	±22(2) TS1	-------
III	±120(2)	±100(2)	------	------	±32(2) TS2a	±12(2)	±32(2)	±42(2)	±52(2)	±62(2)
IV	±40(1)	±80(1)	±100(1)	±120(1)	±22(2)	±32(2) TS3a	±42(2) TS1a, TS2,TS4	±52(2)	±62(2)	±72(2)

EXPERIMENTAL RESULTS AND FAILURE MODES

Testing Results

The failure modes of the specimens are shown in figure 3. Obvious buckling of the flanges can be observed, especially in TS4 and TS4a.

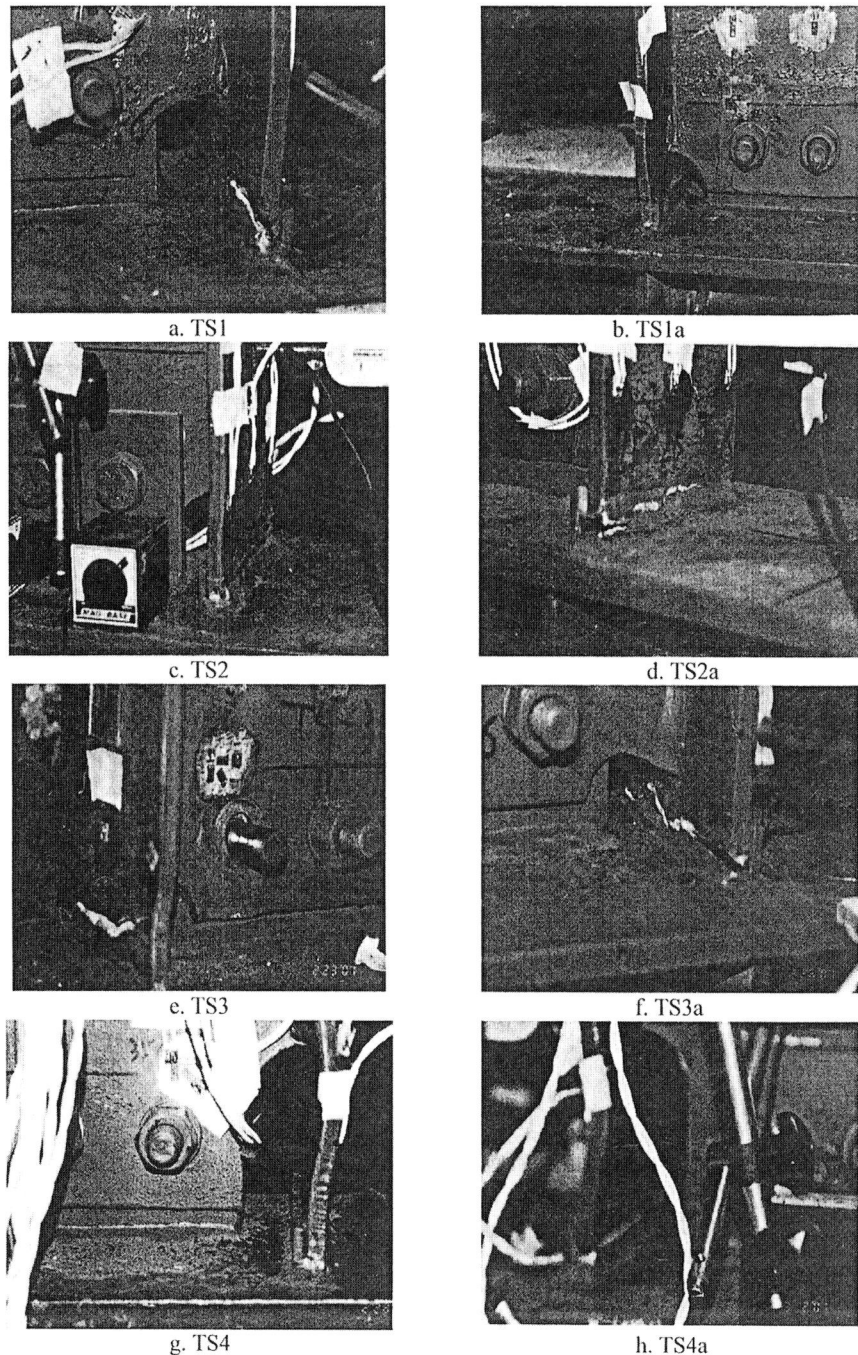

Figure 3: Failure modes of the specimens

In the plastic loading stage tearing is observed between the beam flange and web and cracks occur at the welds of the beam-column interface. Sliding is observed between the shear tab and beam web. Finally the weld of the bottom flange almost cracks in its whole length and the connection fails. Local buckling of the beam web of TS1a is observed in the plastic loading stage. The supplemental weld cracks and the loading point has an out-plane displacement of 7mm. The weld at the beam-column interface cracks suddenly and the connection fails, with tension fracture in bottom flange and severe buckling in top flange.

Sliding of the shear tab occurs early in TS2. The weld of the bottom flange cracks and the crack propagates from the flange edge to the center. Finally the connection fails because of this crack. Sliding of the bolts can be observed in TS2a during the plastic loading stage. The connection fails abruptly because of the fracture of the tension flange. Supplemental weld cracks and local buckling is observed on the flanges.

Displacement at the beam tip is taken as the controlling parameter for TS3. When the displacement is 44mm, sliding is observed between the web and shear tab and local plastic development occurs in the beam web and flange. Finally the beam web tears the top flange open at the toe of WAH and the supplemental weld cracks. TS3a fails because of the failure of bottom flange weld.

Connection stiffness deduction is observed in TS4 during the plastic loading stage. Buckling occurs in both flange and shear tab sliding is obvious. The connection finally fails because of the failure of weld of the bottom flange.

Failure Modes of the Specimens

Experimental results show that some issues have significant effect on the failure mode of the connection, such as geometry of the WAH, welding quality and loading scheme.

WAH geometry has significant influence on the stress distribution near the beam-column interface, which affects the failure mode of the connection. This is obvious in TS4 and TS4a, which have long access holes. The beam flange nearby the access hole behaves as a thin plate under alternate tension and compression. Local buckling in the flanges is obvious (Fig. 3g, Fig. 3h). The hysteretic loops are very smooth and no fracture failure occurred.

In all specimens stress concentration occurs at the toe of WAH where the web intersects with the flange. This has great effect on the failure mode of the connection. Most specimens have cracks at this position and some of them fail quickly because of these cracks. The standard connection has a quadri-circle curve at the toe of the hole and the web cuts the flange in right angle, which intensifies the stress concentration. A new type of WAH geometry is recommended in FEMA-350. The angle of the access hole cut to the flange surface is limited to 25 degree maximum, in order to release the stress concentration and improves connection ductility. Seismic Design Code for Buildings of China also introduces a new type of connection detail, which is similar to specimen TS3 and TS3a in this paper. This type of connection detail does not give limitation on the angle of the access hole cut to the flange surface. As can be seen from the experimental results, TS4 and TS4a provide an alternative way to improve connection ductility. By changing the WAH size rather than the geometry, TS4 and TS4a change the failure mode and improved connection ductility significantly.

Poor welding quality creates obvious welding defeats and is very harmful for a ductile failure to occur. Obvious welding defeats are often the places where cracks begin to develop. Specimen TS2a has a welding defeat at the end of a weld at the beam-column interface and a crack occurs here and propagates quickly in the cyclic loading procedure. The specimen finally failed because of this crack.

Loading schemes also influence failure modes of the specimens. Cyclic loading makes the cracks open and close repeatedly and propagate quickly under low stress. This makes the connection fails early under a low loading. Loading scheme III is applied to specimen TS2a, which finally fails at the displacement of 32mm, smaller than the maximum displacement of TS2, which is 42mm. TS3a also fails early than TS3 because more loading loops is applied to TS3a, although in the elastic loading stage.

ANALYSIS ON THE EXPERIMENTAL RESULTS

Hysteretic Properties of the Connections

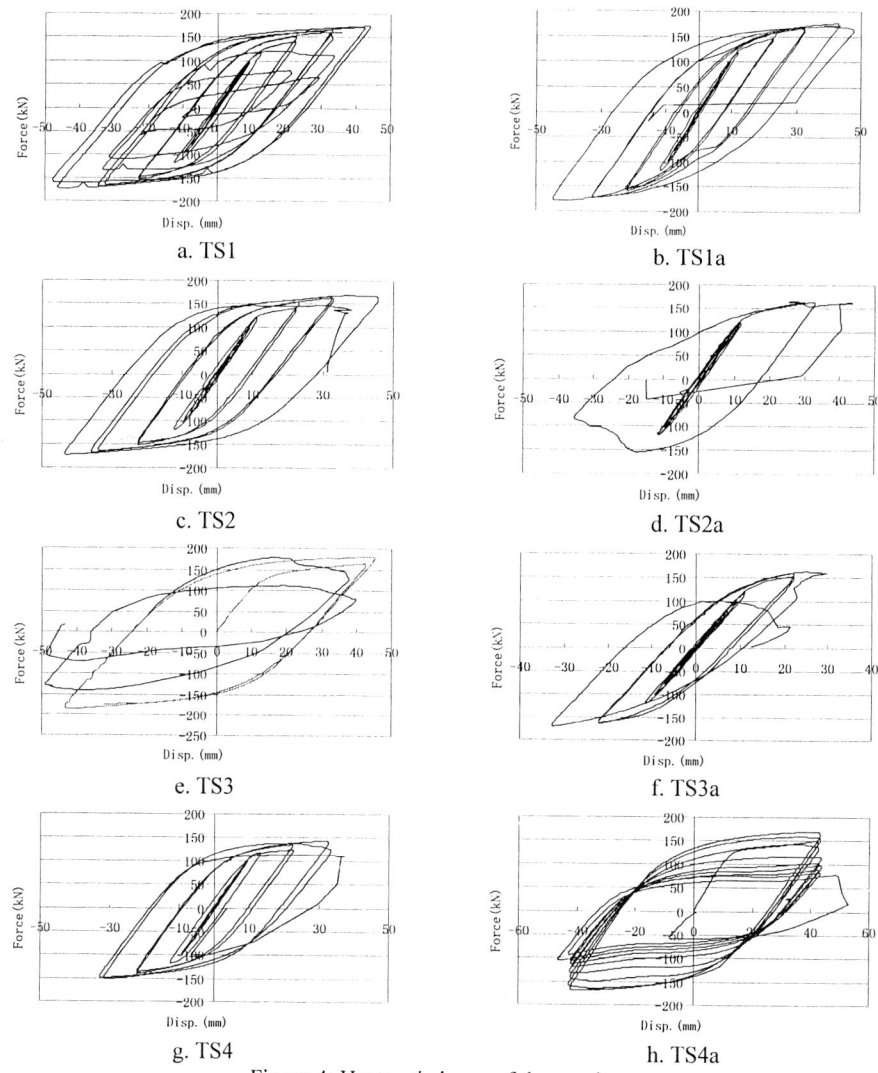

Figure 4: Hysteretic loops of the specimens

The specimens have similar hysteretic circle areas except TS3a and TS4, which have less hysteretic circles because they failed when the beam-tip displacement is 32mm. The standard connections (TS1 and TS1a) have slightly larger hysteretic circle areas but the difference is rather small. It is reasonable to make the conclusion that these specimens have similar ability on energy absorbing.

For the specimens with supplemental welds, the panel zone has larger deformation because the supplemental welds help to transmit the shearing force to the panel zone. More plasticity development at the panel zone can be observed and the elastic strength of the connection is higher. But according to the hysteretic loops in figure 4, significant improvement on the ability of energy absorbing is not observed.

The stress concentration at the toe of WAH causes some specimens, such as TS1 and TS3, to crack. The beam web compresses the flange and tears it open at the intersecting point. This type of failure may increase the ability of energy absorbing but it is not favorable. Failure at a place of severe stress concentration is often unpredictable. In specimen TS1 and TS3, the fatigue cracks almost cut through the whole section of the beam flange. Under earthquake loading, the toe of WAH is a very potential place to initiate brittle failure.

TABLE 2 EXPERIMENTAL RESULTS

Connection	Specimen	Strength /kN	Max. Disp. /mm	Drift angle /rad	Elastic strength /kN	Elastic drift angle /rad
Standard	TS1	175.469	47.657	0.0235	108.653	0.511
	TS1a	179.507	47.802	0.0236	115.542	0.548
EWAH I	TS2	173.195	45.179	0.0223	121.598	0.551
	TS2a	166.355	33.814	0.0167	118.041	0.545
EWAH II	TS3	184.407	44.899	0.0222	111.024	0.546
	TS3a	171.000	32.737	0.0162	126.067	0.572
EWAH III	TS4	151.586	33.078	0.0163	111.485	0.552
	TS4a	169.364	42.564	0.0210	120.501	0.618

Strength of the Connection

The length of WAH has significant effect on the strength of the connection. EWAH II (TS3, TS3a) connection has a short access hole and its strength is the highest among the connections. EWAH III (TS4, TS4a) connection has a long access hole and severe buckling of the beam flange occurs in the experiment. This makes TS4 and TS4a have the lowest strength. EWAH I (TS2, TS2a) connection has a wider access hole and the length of the hole is the same with the standard connection. The strength of TS2 and TS2a is close to that of standard connection.

TABLE 3 ULTIMATE STRENGTH BY FEA AND EXPERIMENTAL RESULTS

Specimen	FEA strength P_u /kN	Experimental strength P_t /kN	P_u/P_t
TS1	156.5	177.5	0.88
TS2	156.3	169.8	0.92
TS3	156.2	177.7	0.88
TS4	153.6	160.5	0.96

The specimens are analyzed with a general-purpose finite element analysis (FEA) program ANSYS. Table 4 lists the specimen strength given by ANSYS and experimental results. In both FEA and experimental results, TS4 has the smallest ultimate strength among the connections. The other three specimens have similar strengths in the FEA analysis. This shows that in case of ductile failure, WAH geometry has little effect on reducing the plastic ultimate strength of the connection although they have significant influence on the stress concentration near the beam-column interface. In connection design the geometry of access holes can be modified to achieve better ductility while not affecting the strength of the connection.

CONCLUSIONS

Experimental results show that the most important issue in preventing fracture failure is to assure good welding quality. Obvious welding defeats cause cracks to develop early and propagate quickly, resulting in poor ductility and energy absorbing ability.

WAH geometry and size also have great effect on connection ductility. In the experiments of this paper, the access hole have quadric circle curve at the end and the beam web cuts the flange in a right angle. The stress concentration here caused the flange torn open at this place. This should be avoided in connection design. Improved WAH geometry can be found in FEMA-350 and Code for Seismic Design of Building.

Supplemental welds between beam web and shear tab can improve connection stiffness and elastic strength, while having little effect on plastic strength of the connection. In steel structures that work in normal load cases, the beam-column connections work in elastic stage and supplemental welds are favorable for the connections.

REFERENCE

FEMA (2000), *Recommended Seismic Design Criteria for New Steel Moment-Frame Buildings*, Federal Emergency Management Agency, USA.

Gates W.E., Morden M. (1996). Professional Structural Engineering Experience Related to Welded Steel Moment Frames Following the Northridge Earthquake. *The Structural Design of Tall Buildings* **5:1**, 29-44.

Lu L.W. *et al.* (2000). Critical issues in achieving ductile behavior of welded moment connections. *Journal of Constructional Steel Research* **55:1-3**, 325–341.

Miller D.K. (1998). Lessons Learned from the Northridge Earthquake. *Engineering Structures* **20:4-6**, 249-260.

Ministry of Construction P. R. China (2001). *Code for Seismic Design of Building GB50011-2001.*

Popov E.P., Yang T.S. and Chang S.P. (1998). Design of steel MRF Connections before and after 1994 Northridge Earthquake. *Engineering Structures* **20:12**, 1030-1038.

Zhao D.W., Shi Y.J., Chen H. (2000). Experimental research on beam-column connections under low-cycle cyclic loading. *Building Structures* **30:9**, 3-6.

REPAIR/UPGRADE OF STEEL MOMENT FRAMES IN LOW RISE BUILDINGS

J. C. Anderson [1], Y. Xiao [1,2] and Jean X.J. Duan [1]

[1] Department of Civil Engineering, University of Southern California, Los Angeles, USA
[2] Cheung Kong Scholar, Hunan University, China

ABSTRACT

Experimental studies were conducted on four repair/upgrade details which have been suggested for welded beam to column moment connections. The simplest of these removes the old (cracked) weld material and replaces it with ductile weld material. Another detail considers the addition of horizontal reinforcing plates (cover plates) to the beam flanges, a third adds vertical triangular (fin) plates to both beam flanges and the fourth uses a weld overlay to reinforce the existing welds. These details are shown to provide varying degrees of improvement in connection behavior with plastic rotation capacities between two percent and four percent being achieved.

KEYWORDS

Damage, welded connection, repair, retrofit, seismic behavior, weld overlay.

INTRODUCTION

The most important design issue to be recognized following the Northridge earthquake was the seismic vulnerability of steel frames with welded moment connections. Damage to these connections was not recognized immediately after the earthquake because the fireproofing and partitions tended to hide the steel frame from visual inspection. Initial indications of cracking in the connections were discovered in buildings that were either under construction of permanently deformed [Bertero, et al., 1994]. As more thorough inspections of the buildings were completed, more cracked connections were uncovered with the current count being more than two hundred. In an effort to meet the immediate to develop a criteria for the repair/upgrade of these buildings, an experimental program was conducted to test the inelastic, cyclic behavior of several repair/upgrade procedures.

The most common type of cracking was found to occur in the welds that connect the bottom flange of the beam to the column flange. Cracks in this area tended to start at the beam web and propagate outward. This behavior is influenced by several factors which include the following: (1) there is a high stress concentration at this location causing high stress and strain gradients, (2) the weld is discontinuous giving rise to porosity and slag inclusions, (3) the web cope in the beam is flame cut and often not ground smooth resulting in a rough surface which is ideal for crack formation, (4) the so called "k-line" area of the beam web has increased hardness due to the rolling process and (5) the weld may be subjected to biaxial tension forces which restrict plastic deformation. A successful repair/upgrade procedure must neutralize the negative effects of these factors by improving the welds and lowering the stresses and stress concentrations in this critical region. The following is a summary of some of the results reported in detail elsewhere [Anderson and Duan, 1998].

EXPERIMENTAL PROGRAM

In order to evaluate the cyclic behavior of these repair/upgrade details, cyclic load tests were conducted on full scale test specimens representative of the smaller steel sections used in low rise buildings on the west coast. The test specimen shown in Figure 1 consisted of a W21x68 beam welded to a W12x106 column using a standard weld detail representative of welded moment connections prior to the Northridge earthquake. A constant axial compression load of 20 kips was applied to the 9 ½ foot column and a cyclic load under displacement control was applied at the end of a 6 foot beam. All columns had continuity plates across the column web and four of the columns had an additional plate (doubler) on one side of the column web extending eight inches above the top continuity plate and eight inches below the bottom continuity plate. The steel was specified as A36 although the measured yield strength was 47.5 ksi. The test specimens were fabricated using Flux Core Arc Welds (FCAW) and are representative of pre-Northridge moment connections. All repairs/upgrades were made in the laboratory using Shielded Metal Arc Welds (SMAW) with E7018 electrodes. The loading sequence in general followed the ATC test protocol [ATC, 1992], however, modifications were made to accommodate the test configuration used in the study and to assure sufficient data points would be obtained. The test specimen positioned in the load frame is shown in Figure 2.

Perhaps the easiest way to repair a welded connection is to remove the backup bar at the bottom of the flange welds, gouge out any inclusions or cracks in the weld material and weld again using a notch tough (ductile) weld material. This technique was used for many repairs following the Northridge earthquake. Another common method of improving the behavior of welded moment connections is the addition of horizontal reinforcing plates to the beam flanges. In this application, both the existing beam flange and the additional plate were welded to the column flange. In this configuration, the horizontal plates on the flanges are called "cover plates." The purpose of the cover plates is threefold. First, the centroid of the reinforced section is moved further from the neutral axis of the beam section so that the moment capacity is increased. Second, with the increased moment capacity, inelastic behavior (plastic hinging) is moved away form the connection region and into the beam, away from the crack sensitive region of the weld access hole. Third, the increased beam flange thickness at the column flange reduces the stress concentration in this area

In this study, the cover plate is the same width and thickness as the beam flange. The sides of the plate are beveled to permit a partial penetration groove weld along the sides instead of the more common fillet weld. The connection to the column is made with a full penetration grove weld. The existing weld was ground flush with the top (bottom) flange of the beam to receive the beveled plate. A full penetration weld using SMAW connected the cover plate to the column flange. Hence the existing FCAW weld was overlayed with a SMAW weld. The connection at the end of the cover plate was made with a 5/8 inch fillet weld. At the bottom flange, the cover plate is welded in the overhead position, allowing the entire weld to be made without interruption. The standard cover plates for these tests are 8 inches wide by 14 inches in length by ¾ inches thick. For one specimen a half size (mini-plate) was used which was 4 inches wide by 7 inches in length by ¾ inches thick.

A third type of repair/retrofit detail is the addition of a vertical triangular plate (fin) above the top beam flange and below the bottom beam flange in line with beam web. This diverts some of the force in the beam flange around the connection to the column flange. In addition, a circular hole was drilled in the center of the fins to reduce the stiffness and allow the hole to act as a type of shock absorber. Stresses in the welds are reduced by the additional weld material and by the resulting increase in the reactive moment arm of the modified connection.

The principle of the weld overlay is to deposit a more ductile weld material that is resistant to crack initiation and propagation on top of an existing weld that is less resistant and may contain small cracks and other defects. Some of these defects may be non-detectable even with the most sensitive testing methods. The overlay must be able to immobilize existing defects in the weld material, heat affected zone (HAZ) and parent material and exert a positive influence on the weld access hole and k-line regions. The quality of the overlay is primarily dependent upon the purity of the weld metal and control of trace elements before and during deposition of the weld material. In general, the Gas Tungsten Arc Weld (GTAW), the Gas Metal Arc Weld (GMAW) and the Shielded Metal Arc Weld (SMAW) can produce very high, high and medium quality welds respectively. Although GTAW and GMAW are the preferred processes, SMAW is more readily used for steel fabrication and was therefore the process used in the development of the repair/modification overlays. In a later study [Anderson, et al., 2002], the Flux Core Arc Weld (FCAW) procedure was used with a ductile, E71T-8, electrode and no difference in performance of the overlay welds was observed.

EXPERIMENTAL RESULTS

Gouge Out and Replace. One of the specimens with an additional plate on one side of the column web was tested to failure in the "as received" condition, repaired in the laboratory and retested to failure. The specimen sustained 11 displacement cycles prior to failure. On the third cycle at a displacement of 1½ inch, the bottom flange of the beam suffered a pullout type of failure during an upward displacement cycle. The total rotation was just over 2 percent as shown in Figure 3. However, the plastic rotation was only ½ % and behavior of the panel zone was elastic with peak strains of 0.13 percent. The failure mode was typical for this type of connection with a pull out occurring at the bottom flange (Figure 4).

The specimen was repaired by a combination of grinding out the divot in the column flange, removing the backup bar, and placing new weld material using SMAW. The shear tab was welded on the long side with a fillet weld and end returns. Since there was no visible damage to the weld at the top flange, nothing was done to this weld. All welds were ultrasonically tested and found to be sound. Note that the weld at the top flange was made with FCAW and the one at the bottom flange with SMAW.

Following repair, the connection was able to sustain 17 displacement cycles and develop a total rotation of 4.1% in both directions prior to the initiation of a crack in the top flange of the beam. Plastic rotation was 2 percent. The load versus plastic rotation hysteresis curve for the repaired specimen is shown in Figure 5. Whitewash on the beam gave a clear indication of the formation of a plastic hinge as shown Figure 6. During the last full cycle, at a displacement of 2 ¾ inches, both top and bottom beam flanges buckled.

Flange Cover Plates. The specimen having a column doubler plate and flange plates on the top and bottom beam flanges was subjected to fifteen displacement cycles increasing from ½ inch to three inches. The connection reached a total rotation of 4 ½ percent (3 percent plastic) before the test was halted due to severe deformation of the beam. The plot of moment versus plastic rotation is shown in Figure 7. Both beam flanges developed buckles as shown in Figure 8. The beam web also buckled out of plane, however, there was no visible sign of any cracking of the connecting welds. A plastic hinge formed at a distance of 16 inches from the face of the column resulting in little or no plastic deformation in the connection region and none in the area of the weld access holes. The moment versus rotation plot indicates some unloading of the specimen due to the local buckling during the last three cycles at a displacement of three inches, resulting in a reduction in moment capacity of 32 percent.

One specimen having a column doubler plate, was tested with half-size cover plates (mini-plates) on the top and bottom beam flanges. The 4 x 7 x ¾ mini-cover plates were welded to the beam and column flanges in the same manner as the full size plates. The specimen sustained 19 displacement cycles and developed a plastic rotation of 4.1 percent as shown in Figure 9. The test was stopped due to severe buckling of the beam flanges and the web (Figure 10) that resulted in a 32 percent reduction in moment capacity at maximum displacement. There was no visible cracking in the test specimen; however, it appeared that the plastic hinge region had just reached the face of the column. Therefore, it does not seem advisable to reduce the length of the flange plates any further.

Vertical Triangular Plates (Fins). The test of the specimen with a doubler plate in the column panel zone had triangular fins containing a 1 ½ in diameter hole to increase the flexibility of the fin and absorb some of the deformation. The specimen sustained 19 displacement cycles and reached a plastic rotation of 2.6 percent as can be seen in Figure 11. The test was stopped at this point due to severe deformation in the beam. The top and bottom beam flanges developed buckles of 2 inches and 1 ¾ inches respectively causing a small crack to form at the toe of the bottom fin. The plastic hinge developed 17 inches from the face of the column (Figure 12) thereby eliminating problems with the weld access holes. It should also be noted that the addition of the fin increased the moment capacity of the connection by approximately 20 percent.

Weld Overlay. This specimen was not tested prior to applying the weld overlays and no defects were known to be present in the existing FCAW welds. In this case the specimen represents a field upgrade of a connection fabricated with FCAW and E70T-4 electrodes. Overlays were applied to the welds on both sides of the top and bottom flanges following the detail shown in Figure 13. The completed overlay on the bottom flange of the beam is shown in Figure 14. The results of tests on specimens that were upgraded using overlays placed with FCAW and E71T-8 electrodes are discussed elsewhere [Anderson, et al., 2002].

CONCLUSIONS

The results of this experimental program of investigation into the cyclic behavior of welded moment connections leads to the following conclusions:

The original specimen with welded connections representative of pre-Northridge practice had a very low plastic deformation capacity of approximately 0.5%.

All of the repair/upgrade procedures discussed in this paper improved the original connection performance, however, the amount of improvement was variable.

The use of the weld replacement procedure returned the specimen to its original strength but with only a small increase in plastic rotation capacity.

The use of triangular fins increases both the moment capacity and the plastic deformation capacity that reaches a maximum of 2.6%.

The connection with full size cover plates developed a plastic rotation capacity of 3%, which meets current design requirements.

The connection with the mini (half-size) cover plate developed a plastic rotation capacity of 4.1% thereby exceeding current design requirements by 37%.

The specimen upgraded with the overlay welds was able to develop plastic rotation capacity of 3.2%, which also exceeds current design requirements.

REFERENCES

Anderson, J. C. and Duan, X. (1998). "Repair/Upgrade Procedures for Welded Beam to Column Connections," Report No. PEER 98/03, Pacific Earthquake Engineering Research Center, Richmond, Calif.

Anderson, J. C., Duan, J., Xiao, Y., and Maranian, P. J. (2002). "Cyclic Testing of Moment Connections Upgraded withy Weld Overlays," Journal of Structural Engineering, ASCE, Vol. 128, NO. 4, 509-516.

Bertero, V. V., Anderson, J. C. and Krawinkler, H. (1994). "Performance of Steel Building Structures During the Northridge Earthquake," Report No. UCB/EERC 94/09, University of California, Berkeley.

ACKNOWLEDGMENTS

Funding for this study was provided by the National Science Foundation. Material for the test specimens was provided by Nucor Yamata Steel Company and fabrication fo the test specimens was provided by the American Institute of Steel Construction, the Structural Steel Education Council and the Structural Shape Producers Council. This help is gratefully appreciated.

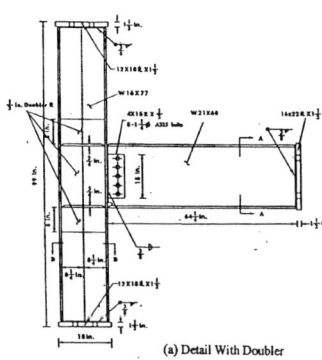

Fig. 1. Small Test Specimen   Fig. 2. Test Setup, Small Specimen

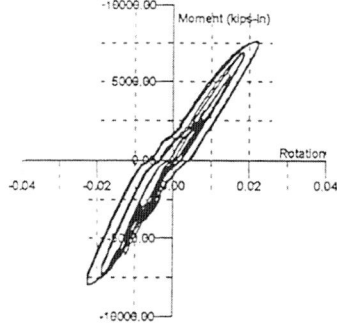

Fig. 3. Moment vs. Total Rotation, As Received

Fig. 4. Bottom Flange Pullout

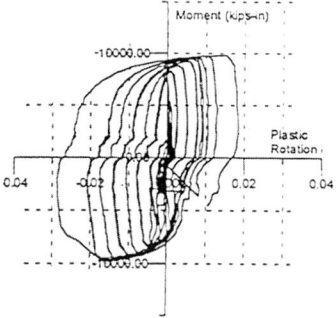

Fig. 5. Moment vs. Plastic Rotation

Fig. 6. Plastic Hinge, Weld Replacement

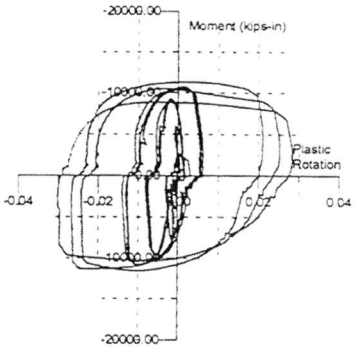

Fig. 7. Moment vs. Plastic Rotation

Fig. 8. Plastic Hinge and Buckle, Cover Plate

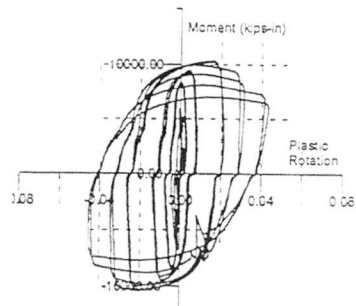

Fig. 9. Moment vs. Plastic Rotation

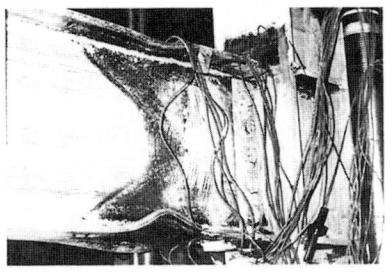

Fig. 10. Plastic Hinge and Buckle, Mini Cover Plate

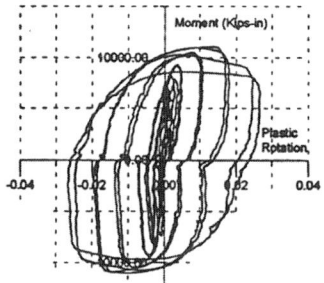

Fig. 11. Moment vs. Plastic Rotation

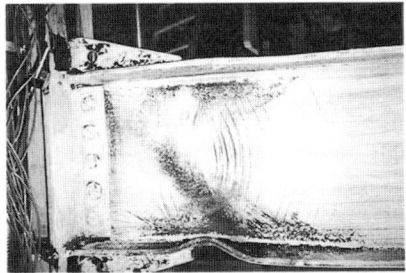

Fig. 12. Plastic Hinge and Buckle, Fin with hole

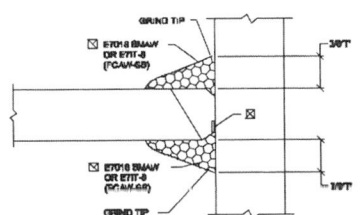

Fig. 13. Minimum Overlay

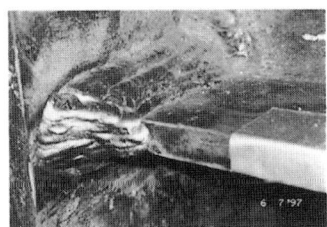

Fig. 14. Overlay Weld, Bottom Flange

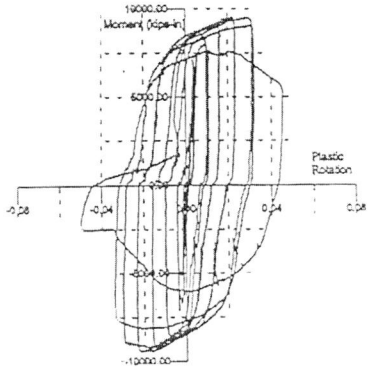 

Fig. 15. Moment vs. Plastic Rotation   Fig. 16 Plastic Hinge. Weld

ULTIMATE BEARING CAPACITY OF WELDED HOLLOW SPHERICAL JOINTS IN SPATIAL RETICULATED STRUCTURES

Q.H. HAN and X.L. LIU

Department of Civil Engineering, Tianjin University, Tianjin 300072, P.R.China

ABSTRACT

The bearing capacity of the welded hollow spherical joints with the diameter limit of 500mm has been presented in the *Chinese Specification for Design and Construction of Reticulated Structures (JGJ7-91)*. In this specification, the collapse properties of the joint under tension loads are strength problem and those under compression loads are stability problem. The ultimate tension capacity of the joint is related to the tension strength of the material, but the ultimate compression capacity of the joint is only related to the geometry of the joints, not the compression strength of the material. The ultimate bearing capacity was calculated using three-dimensional degenerated curved shell element. The tangent stiffness matrix and the elasto-plastic matrix of D_{ep} were derived. In this article, based on the numerical analysis of the experimental data for the six tension joints and six compression joints using the multi-linear isotropic hardening model and Von-Mises yield criterion, the strength collapse criterion and ultimate criterion were put forward. The load-displacement curve under axial force was traced using sub-incremental cylindrical arc-length technology. The collapse was strength problem under axial tension load but it was elasto-plastic buckling collapse under axial compression load. Both of them were related to the design strength of the material. The formulas for the bearing capacity of the welded hollow balls are obtained on the basis of the regression of 64 different calculated data and the diameter limit has been developed to 900*mm*.

KEYWORDS

Welded hollow spherical joint, three-dimensional degenerated curved shell element, collapse criteria, collapse properties, ultimate tension capacity, and ultimate compression capacity

INTRODUCTION

Grid shell and reticulated structures are formed by a large number of tension and compression bars connected with joints, which determine the application prospect of the spatial structures. Up to now, more than one hundred types of joints have been invented. Although some of them are applied in the current actual project, most node forms are discarded because of their complicated construction or expensive cost. In 1965, Prof. LIU X.L. developed the welded hollow spherical joint, and first applied it in the project of Science & Technology Hall in Tianjin. After that, the welded hollow spherical joint has been extensively used as a reasonable node pattern for grid structures in China.

The current Chinese *Specification for Design and Construction of Reticulated Structures (JGJ7-91)* gives the calculating formulas for welded hollow spherical joints as follows,

$$N_t \leq 0.55 \eta_t t d \pi f \tag{1}$$

$$N_c \leq \eta_c (400 t d - 13.3 \frac{t^2 d^2}{D}) \tag{2}$$

Where, N_t is tension-bearing capacity and N_c is compression bearing capacity. However, the above formulas are only applied to joints with diameter less than 500mm and assume that the failure mode of joints under tension load is strength collapse but the failure mode under compression load is stability collapse. So, the tension bearing capacity is related to the tensile strength of steels but the compression bearing capacity is only determined by the geometrical dimension, such as the external diameter of bar d, the external diameter of joint ball D and the thickness of joint ball t.

Based on the ideal elasto-plastic mechanical model and the Von-Mises collapse criterion, the authors (YAO, 2000) analyze the welded hollow spherical joints with different diameter by using the solid element with four nodes and present the regression formula for ultimate bearing capacity of hollow balls with diameter more than 500mm as follows,

$$N_{t,c} \leq \eta(0.32 + 0.6 \frac{d}{D}) t d \pi f \tag{3}$$

Where, $N_{t,c}$ is the tension and compression resistance of the welded hollow ball. The above formula can be applied to balls with diameter less than 900mm. The authors hold that if the diameter-to-thickness ratio $D/t<35$, the failure modes under tension and compression load are all strength problem and are all related to the design strength of f.

In this paper, based on the numerical analysis of the experimental data of six tension joints and six compression joints, using multi-linear isotropic hardening model and Von-Mises yield criterion, the strength collapse criterion and ultimate criterion are put forward. The load-displacement curves under axial force are traced and the failure modes for different construction are analyzed. The collapse is strength problem under axial tension load while it is elasto-plastic buckling collapse under axial compression load and both failure modes are related to design strength of the material. Additionally, the authors calculate 64 different welded hollow balls under axial tension and compression load and present the formulas for the bearing capacity of welded hollow balls with diameter less than 900mm.

THREE-DIMENSIONAL DEGENERATED CURVED SHELL ELEMENT

Element Tangent Stiffness Matrix under U.L. Formulation

The element sub-incremental stiffness equation can be expressed as follows,

$$[K_T]^e \{\Delta u\}^e = \{R\}^e - \{r\}^e \tag{4}$$

Where, $[K_T]$ is the element tangent stiffness matrix, which is express as follows,

$$[K_T] = [K_L] + [K_\sigma] \tag{5}$$

Where, $[K_L]$ is the linear elastic stiffness matrix and $[K_\sigma]$ is the initial stress stiffness matrix.

Yield Criterion

For metal materials, the Von-Mises yield criterion is usually used, that is

$$F = f - k = 0 \tag{6}$$

Where, F is the yield function, f and k can be calculated by the following equations,

$$f = \frac{1}{2}\sigma'_{ij} \cdot \sigma'_{ij} = \frac{1}{2}(\sigma'^2_x + \sigma'^2_y + \sigma'^2_z + 2\tau'^2_{xy} + 2\tau'^2_{yz} + 2\tau'^2_{zx}) \tag{7}$$

$$k = \frac{1}{3}\sigma^2_{x0} \tag{8}$$

Where, σ_{x0} is the initial yield stress.

The multi-linear isotropic hardening model is selected in this paper. It holds that if the material has become yielding, the loading curved surface would uniformly expand in all directions but its shape, center and direction in the stress space would not change. The post yield function would be the same as Equation (6), except for

$$k = \frac{1}{3}\sigma^2_s(\bar{\varepsilon}^p) \tag{9}$$

Where, σ_s is the elasto-plastic stress, which is a function of equivalent plastic strain $\bar{\varepsilon}^p$. The stress-strain curve used in this paper is shown in Fig.2 and all of the parameters are displayed in Table 1.

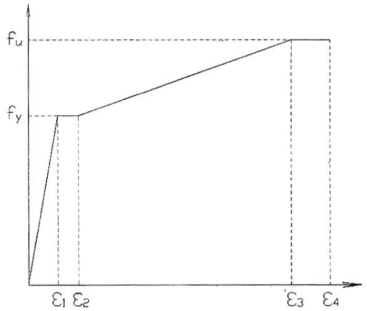

Figure 1: Stress-strain curve of steel Q235 and Q345

TABLE 1
DATA FOR THE PARAMETERS OF STEEL Q235 AND Q345

Steel type	f_y (Mpa)	f_u (Mpa)	ε_1 (%)	ε_2 (%)	ε_3 (%)	ε_4 (%)
Q235	235	375	0.114	2	20	25
Q345	345	510	0.170	2	20	25

Constitutive Equation

Prandtl-Reuss constitutive equation in incremental form can be expressed by matrices as follows,

$$\{dS\} = [D_{ep}]\{d\varepsilon\} \quad (10)$$

Where, $\{dS\}$ is the 2nd Piola-Kirchhoff stress tensor increment and $\{d\varepsilon\}$ is the Green strain increment. $[D_{ep}]$ is the elasto-plastic matrix, which can be determined by the following equation

$$[D_{ep}] = [D] - [D_p] \quad (11)$$

Where, $[D]$ is the elastic matrix in global coordinate and $[D_p]$ is the plastic matrix,

$$[D_p] = \frac{\{d_D\}\{d_D\}^Y}{\{a\}^Y\{d_D\} + (4/9)\sigma_s^2 E''} \quad (12)$$

Where

$$\{d_D\} = [D]\{a\} \quad (13)$$

$$\{a\}^Y = \{\sigma'_x \quad \sigma'_y \quad \sigma'_z \quad 2\tau'_{xy} \quad 2\tau'_{yz} \quad 2\tau'_{zx}\} \quad (14)$$

$$E'' = \frac{E E'}{E - E'} \quad (15)$$

NUMERICAL ANALYSIS OF THE EXPERIMENTAL DATA

Collapse criterion and ultimate criterion

Under the axial tension load, the deformation and the internal strain of welded hollow ball increase gradually with the increasing of load. Based on the monotonic tension test of steels, the strain criterion is taken as the strength collapse criterion (HAN, 2001),

$$\|\varepsilon_i\| \leq 0.25 \quad (16)$$

Under the axial compression load, the probability that the collapse is strength or stability problem will vary according to different construction of joints. But for the commonly used diameter-to-thickness ratio, the failure mode is usually compression buckling collapse. As a result, the following ultimate criterion is adopted in this paper (HAN, 2001),

$$\left.\begin{array}{l}P_{i-1} \leq P_i \\ P_{i+1} \leq P_i\end{array}\right\} \quad (17)$$

The corresponding load P_i is regarded as the ultimate compression capacity of welded hollow spherical joint.

The tension capacity of welded hollow spherical joints

The experimental data of six welded hollow balls are summarized in Table 2.

TABLE 2
ULTIMATE BEARING CAPACITY OF WELDED HOLLOW BALLS UNDER AXIAL TENSION LOAD (Q235)

No.	D (mm)	t (mm)	d (mm)	Test data (kN)	F.E.M.result (kN)	Error (%)	Strain (%)	Disp. (mm)
1	250	8	108	769	722	-6.1	25	9
2	400	18	130	1852	1947	5.1	25	17
3	450	12	159	1571	1466	-6.7	25	15
4	450	16	159	2020	1968	-2.6	25	14
5	500	14	160	1684	1627	-3.4	25	15
6	500	16	160	1964	1869	-4.8	25	15

Note: D-diameter of the ball, t-thickness of the ball, d- outer diameter of the pipe

From Table 2, it can be found that the numerical results agree well with the test results. The load-displacement curves of six joints are shown in Figure 2.

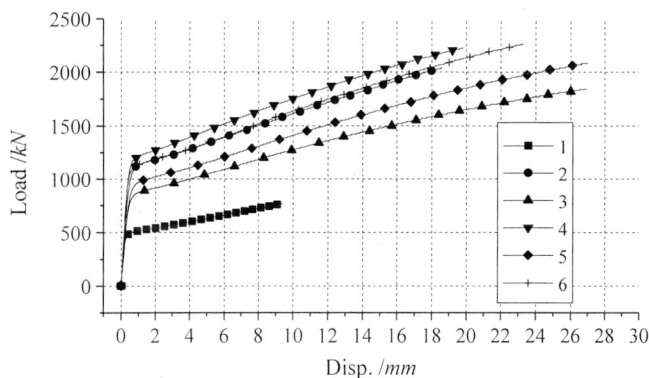

Figure 2: Load-displacement curves of welded hollow balls under axial tension load

Under axial tension load, stress in the hollow ball is compression stress in the circumferential direction but tension stress in the radial direction. After linear buckling analysis, it can be found that the failure mode of welded hollow balls under axial tension load could not be buckling collapse, but strength collapse.

The compression capacity of welded hollow spherical joint

The experimental data of six welded hollow balls are summarized in Table 3. From Table 3, it can be found that the numerical results agree well with the test results. Besides, the buckling load factors of every joint are listed in Table 3. Because the factors are relatively large, the probability of elastic buckling collapse is rather small. The load-displacement curves of six joints are shown in Figure 3.

TABLE 3
ULTIMATE BEARING CAPACITY OF WELDED BALLS UNDER AXIAL COMPRESSION LOAD (1:Q235,2~6:Q345)

No.	D (mm)	t (mm)	d (mm)	Test data (kN)	F.E.M.result (kN)	Error (%)	Disp. (mm)	Buckling factor
1	250	8	108	562	480	-14.6	-0.8	38
2	500	16	219	3140	2525	-19.6	-1.6	21
3	500	20	219	3700	3169	-14.4	-1.7	34
4	550	25	219	4000	4239	6.0	-2.2	52
5	550	25	180	3400	3217	-5.4	-2.3	36
6	650	25	219	4000	3905	-2.4	-2.5	39

Note: D-diameter of the ball, t-thickness of the ball, d-outer diameter of the pipe

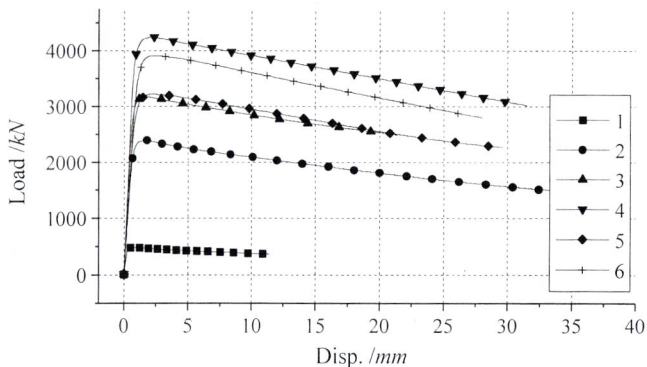

Figure 3: Load-displacement curves of welded hollow balls under axial compression load

Additionally, the failure mode will be different according to different diameter-to-thickness ratios D/t. The larger the D/t is, the more probability of the asymmetry the collapse would have but the smaller the D/t is, the more probability of the symmetry the collapse would have. After the compression load bearing capacity of those six test joints made of different steels are calculated, it can be concluded that the bearing capacity is directly related to the design strength of steels. From Q345 to Q235, the decreasing extent of capacity is 31%, which is perfectly equal to the decreasing extent of the design strength.

REGRESSION ANALYSIS OF ULTIMATE BEARING CAPACITY

Regression formula of axial tension bearing capacity

The tension bearing capacity of 32 joints is calculated with the external diameter of spherical joints varying from 160mm to 900mm. Then the calculation equation of tension bearing capacity for welded hollow spherical joints with the assurance rate of 95 percent is obtained as follows,

$$N_t \le 0.56\eta_t td\pi f \qquad (18)$$

Where, N_t is the design tensile resistance of hollow balls under axial load (unit: N), t is the thickness of hollow balls (unit: mm), d is the external diameter of the pipe, f is the design strength of steels (unit: N/mm^2), η_t is the enhancing coefficient for the balls with stiffeners which is equal to 1.1 or 1.0 according to if or not reinforcing the ball. Although Equation (18) has the same form of Equation (1), the applicable scope is widened.

The bearing capacity error curves of 32 welded hollow spherical joints under tension load are shown in Figure 4. Obviously, the error of Equation (18) presented in this paper is between that of Equation (1) and Equation (3). It means that Equation (18) is more reasonable.

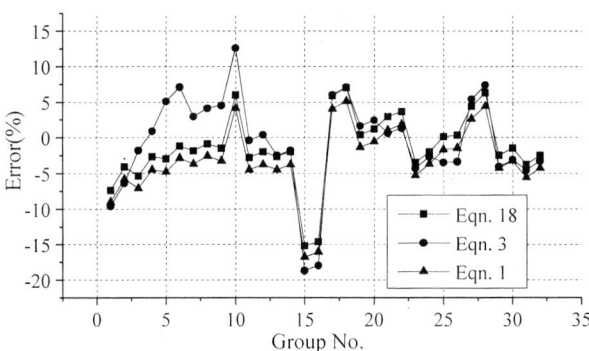

Figure 4: Error curves of tension bearing capacity of 32 welded hollow spherical joints

Regression formula of axial compression bearing capacity

The compression bearing capacity of 32 joints is calculated with the external diameters of spherical joints varying from 160mm to 900mm. Through calculation, it can be found that the compression bearing capacity increases with td increasing and decreases with D increasing. That is contrary to the results of Equation (2). Taking $\frac{N_c}{td}$ as the dependent variable and $\frac{1}{dD}$ as the independent variable, the calculation equation of compression bearing capacity for welded hollow spherical joints can be obtained as follows,

$$N_c \le \eta_c (260td + 70000\frac{t}{D})\frac{f}{215} \qquad (19)$$

Where, N_c is the design compressive resistance of hollow balls under axial load (unit: N), D is the external diameter of hollow balls (unit: mm), t is the thickness of hollow balls (unit: mm), d is the external diameter of pipe (unit: mm), f is the design strength of steels (unit: N/mm^2), and η_c is the enhancing coefficient for the balls with stiffeners which is equal to 1.4 or 1.0 according to if or not reinforcing the ball.

The bearing capacity error curves of 32 welded hollow spherical joints under compression load are shown in Figure 5. It can be found that the calculated values with Equation (19) are very close to the theoretical values and the errors are kept between those of Equation (2) and Equation (3). The calculated values of Equation (2) are generally higher than the theoretical values and the maximum error is 60%. For Equation (3), when $D \leq 500$, the calculated values are generally higher than theoretical values and the maximum error can also be up to 50%; when $D > 500$, the calculated values

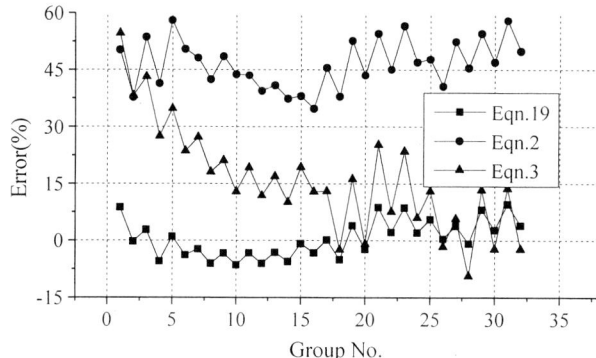

are close to the theoretical values and the errors are as small as that of Equation (19).

Figure 5: Error curves of compression bearing capacity of 32 welded hollow spherical joints

CONCLUSIONS

1. The test results can be simulated by using the Von-Mises yield criterion, the multi-linear isotropic hardening model, the strain collapse criterion and ultimate criterion put forward in this paper.
2. The failure mode is strength collapse under axial tension load while it is elasto-plastic buckling collapse under axial compression load. Both collapse are related to the design strength of the material;
3. Based on the calculation of the bearing capacity of 32 welded hollow spherical joints, the Equation (18) for tension bearing capacity is obtained with the assurance rate of 95% and the applicable external diameter of joint can be from 160mm to 900mm.
4. The compression bearing capacity of welded hollow balls increases when td increasing and decreases with D increasing which is contrary to Equation (2). Equation (19) is obtained for the compression bearing capacity and the applicable external diameter of joint can be from 160mm to 900mm.

REFERENCES

CHEN Z.H. (1990). Analysis of Collapse Mechanism and Experimental Study on the Load Capacity of Welded Hollow Spherical Joints in Space Structures. *Master Thesis of Tianjin University*, Tianjin, P.R.China

HAN Q. H. (2001). Behavior of the Single-Double Layer Reticulated Domes and Ultimate Bearing Capacity of the Welded Hollow Spherical Joints. *Post-Researching Report of the Institute of Engineering Mechanics of Chinese Earthquake Bureau*, Harbin, P.R.China

YAO N. L., DONG M., etc. (2000). Bearing Capacity Analysis of Welded Hollow Spherical Joints. *Building Structures* **30:4**,36-38.

ULTIMATE STRENGTH OF WELDED THIN-WALLED SHS-CHS T-JOINTS UNDER IN-PLANE BENDING

F.R. Mashiri[1], X.L. Zhao[1], L.W. Tong[2] & P. Grundy[1]

[1]Department of Civil Engineering, Monash University, Clayton, VIC 3800, AUSTRALIA
[2]Department of Building Engineering, Tongji University, Shanghai, 200092, P.R. China

ABSTRACT

Welded thin-walled T-joints made up of square hollow section (SHS) chords and circular hollow (CHS) section braces are tested under static in-plane bending load. The hollow sections are cold-formed and have thicknesses less than 4 mm. The SHS-CHS T-joints are used in building the undercarriages and structural supports of equipment and structural systems used in the road transport and agricultural industries. Failure in the SHS-CHS T-joints was observed to occur as a result of chord-face yielding. Chord cracking was also observed after large deformations, resulting in a peak load being attained in these joints. In this paper, load versus chord flange indentation graphs, for the SHS-CHS T-joints are used to determine the deformation limit that can be used in defining the ultimate strength of the joints. The deformed shape of the chord observed from experimental tests is used to create a yield line model. A formula is derived, for the ultimate strength of the SHS chord and CHS brace vierendeel connections based on a plastic mechanism analysis using yield line theory. The ultimate strength determined through the use of the deformation limit criteria is compared to the ultimate strength calculated using the formula obtained from yield line theory.

KEYWORDS

Ultimate strength, deformation limit, plastic mechanism, yield line theory, static strength, in-plane bending, thin-walled sections, steel hollow sections

INTRODUCTION

Vierendeel connections made up of circular hollow section (CHS) chords and as well as circular hollow section braces require brace end preparation. The profile cutting at the end of the brace that allows the brace to sit as well as match the profile on the circular hollow section chord prior to welding, significantly adds to the cost of the joint. Alternatively, the circular hollow section chord can be replaced by a square hollow section (SHS) chord. This enables the circular hollow section brace to be welded to the square hollow section chord without the need for profile cutting thereby reducing the cost of manufacture of the structure. The connection between SHS chords and CHS braces i.e. SHS-CHS T-joints are being used in the manufacture of equipment and structural systems in the road transport and agricultural industry. The tube thicknesses are typically less than 4mm. A review of the static strength of vierendeel connections has shown that formulae have been developed for the moment capacity of connections where both the brace and the chord are made up of either square hollow sections (Packer et al 1992) or circular hollow sections (Wardenier et al 1991). However there is no formula for determining the static strength of vierendeel connections made up of SHS chords and CHS braces.

This paper describes a series of static tests on SHS-CHS T-joints with different brace diameter (d_1) to chord width (b_0) ratio (β), brace wall thickness (t_1) to chord wall thickness (t_0) ratio (τ), and chord width (b_0) to chord wall thickness (t_0) ratio (2γ). The β values range between 0.34 and 0.64. The τ values range between 0.67 and 0.97. The connections have 2γ values of 25 and 33. During the static tests of SHS-CHS T-joints, failure was observed to be due to chord-face yielding followed by cracking at the weld toes in the chord after a large deformation in the joints due to chord-face yielding. Moment versus chord-flange-indentation graphs are plotted to determine the static response of the SHS-CHS T-joints under in-plane bending. The moment versus chord-flange-indentation graphs are used to determine the loads at 1%b_0 deformation and 3%b_0 deformation. The load at 1%b_0 deformation is generally used as the serviceability deformation limit (IIW 1989). The load at 3%b_0 deformation was suggested as the ultimate deformation limit by Lu et al (1994). Lu et al (1994) also proposed a method to determine the ultimate strength of a connection failing through the local plastification of the chord by defining deformation limit criteria based on the local deformation of the chord face at the intersection between the brace and the chord. Lu et al (1994)'s proposals are mainly based on hot rolled section connections. Zhao (2000), has however verified Lu et al (1994)'s proposals for welded cold-formed RHS section T-joints under axial load. The ultimate strength of the SHS-CHS T-joints under in-plane bending will thus be determined using Lu et al (1994)'s deformation limit criteria. This paper will also develop a formula for ultimate strength of SHS chord and CHS brace vierendeel connections based on a plastic mechanism analysis using yield line theory. The yield line model is based on observed deformation of the chord in the tested SHS-CHS T-joints under in-plane bending. The ultimate strength of the connections as determined by the formula from plastic mechanism analysis is compared to the ultimate strength of the SHS-CHS T-joints under in-plane bending as determined by the deformation limit criteria.

EXPERIMENTAL INVESTIGATION

The parameters of the six SHS-CHS T-joint specimens tested in this investigation are given in Table 1. Both the square and circular hollow sections used in the manufacture of the T-joints were of grade C350LO and conform to AS1163-1991 (SAA 1991). The C350LO tubes have a specified minimum yield stress of 350MPa and a specified minimum tensile strength of 430MPa. The square hollow section tubes and the circular hollow section tubes have a specified minimum elongation of 16% and 20% respectively. The gas-metal arc-welding process, also known as MIG was used in fillet-welding the circular hollow section braces to the square hollow section chord flange. The measured mean yield stresses for the tubes used in the manufacture of the SHS-CHS T-joints are shown in Table 1.

The setup of the static tests for the thin-walled SHS-CHS T-joints is shown in Figure 1(a). The tests were performed in a 500kN capacity Baldwin Universal testing machine. Some of the failed specimens are shown in Figure 1(b). Failure of the thin-walled SHS-CHS T-joints was observed to be predominantly due to chord face yielding as shown in Figure 1(b). Cracking of the thin-walled SHS-CHS T-joints, as shown in Figure 2(a), was also observed close to the peak load of the joints. Cracks developed at the weld toes in the chord on the tension side under in-plane bending in all the specimens tested.

TABLE 1: SHS-CHS T-joint specimens

Joint	Chord Member		Brace Member		β $\left(=\dfrac{d_1}{b_0}\right)$	τ $\left(=\dfrac{t_1}{t_0}\right)$	2γ $\left(=\dfrac{b_0}{t_0}\right)$
	Size $h_0 \times b_0 \times t_0$ (mm)	Measured Yield Stress, f_y (MPa)	Size $d_1 \times t_1$ (mm)	Measured Yield Stress, f_y (MPa)			
S3C1	100x100x3	432.5	48.3x2.9	393.5	0.48	0.97	33
S3C2	100x100x3	432.5	48.3x2.3	384.6	0.48	0.77	33
S3C4	100x100x3	432.5	33.7x2.6	436.2	0.34	0.87	33
S3C5	100x100x3	432.5	33.7x2.0	399.8	0.34	0.67	33
S6C1	75x75x3	395.0	48.3x2.9	393.5	0.64	0.97	25
S6C2	75x75x3	395.0	48.3x2.3	384.6	0.64	0.77	25

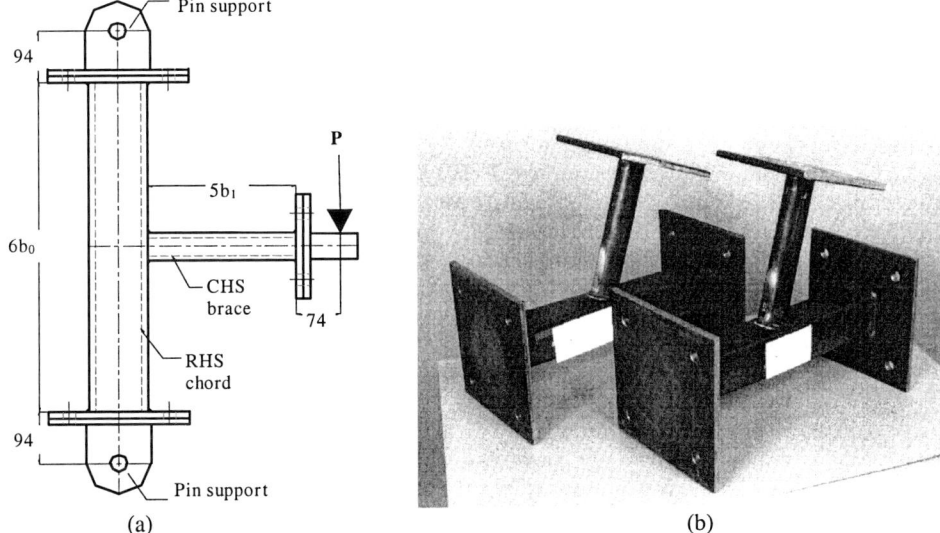

Figure 1: (a) Schematic diagram for thin-walled SHS-CHS T-joint static test, (b) Failed thin-walled SHS-CHS T-joint after static test

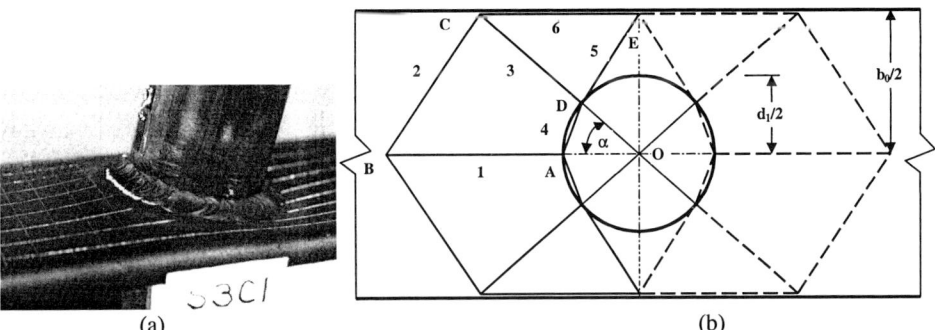

Figure 2: (a) Chord-face yielding and chord-cracking in thin-walled SHS-CHS T-joint, (b) Yield line model adopted

ULTIMATE STRENGTH BASED ON DEFORMATION LIMIT

Lu et al (1994) used the concept of chord flange indentation to propose criteria that can be used to determine ultimate strength of different sections welded to CHS and RHS chords. Lu et al (1994)'s proposals as summarized by Zhao (2000) are given here and are used as a guide in identifying the deformation limit used in determining the ultimate strength of SHS-CHS T-joints under in-plane bending: (i) For a joint which has an obvious peak load at a deformation around $3\%b_0$, the peak load or the load at $3\%b_0$ deformation is considered to be the ultimate load, where b_0 is the width of the chord member; (ii) For a joint which does not have a pronounced peak load, the ultimate deformation limit depends on the ratio of the load at $3\%b_0$ to the load at $1\%b_0$. If the ratio is greater than 1.5, the deformation limit is $1\%b_0$, i.e. serviceability is in control. The ultimate strength is taken as 1.5 times the load at $1\%b_0$. If the ratio is less than 1.5, the deformation is $3\%b_0$, i.e. strength is in control. The ultimate strength is taken as the load at $3\%b_0$; and (iii) A validity range of β ($=d_1/b_0$) and 2γ ($=b_0/t_0$) is given to determine whether the design is governed by serviceability or by strength.

The load-deformation graphs for the thin-walled SHS-CHS T-joints tested in this investigation were plotted to determine the loads at $1\%b_0$ deformation, $3\%b_0$ deformation as well as the peak load. A typical load-deformation curve is shown in Figure 3. Figure 3 shows that cracking in the chord was first observed after the $3\%b_0$ deformation load had been reached. Chord cracking ultimately causes a peak load to be reached. As chord cracking progresses this results in a reduction in area used for transferring applied load from the brace to the chord, thus reducing the load carrying capacity of the connection. In thin-walled tubular joints, susceptibility to cracking is increased by the undercuts at the weld toes of these joints, which become significant as the tube wall thickness becomes smaller.

Table 2 shows the values of the loads at $1\%b_0$ deformation ($M_{1\%b0}$), $3\%b_0$ deformation ($M_{3\%b0}$), the load at which chord cracking is first visually observed ($M_{visualcrack}$), as well as the peak load (M_{peak}). The ratio of the load at $3\%b_0$ to the load at $1\%b_0$ in all the connections ranges between 1.8 and 2.3. Since the ratio of the load at $3\%b_0$ to the load at $1\%b_0$ is greater than 1.5, this means that from the proposals of Lu et al (1994), the deformation limit used for determining the ultimate strength of the thin-walled SHS-CHS T-joints under in-plane bending is $1\%b_0$. Therefore serviceability is in control. The ultimate strength, M_{ult}, is therefore taken as 1.5 times the load at $1\%b_0$ as shown in Table 2. The T-joints have β values ranging from 0.34 to 0.64 and 2γ values ranging from 25 to 33. The validity ranges given by Lu et al (1994) to predict whether strength or serviceability control occurs in connections with RHS chords, show that strength control is expected for a connection with β and 2γ values of 0.64 and 25 respectively. Specimens S6C1 and S6C2, whose β and 2γ values are 0.64 and 25 respectively, show that a serviceability control is observed for SHS-CHS T-joints rather than the strength control predicted by Lu et al (1994)'s validity range. For SHS-CHS and SHS-SHS T-joints, with the same β value, the contact area between CHS and SHS is smaller than that for an SHS-SHS T-joint. A larger cut off value for serviceability control is thus expected in SHS-CHS T-joints. A peak load was reached in these connections due to chord cracking. The peak load reached is significantly higher than the load at $3\%b_0$ deformation. The ratio of the peak load to the load at $3\%b_0$ deformation ranges between 1.7 and 1.9. The significant difference between the peak load and the load at $3\%b_0$ deformation confirms the fact that serviceability rather than strength controls the determination of ultimate strength in thin-walled SHS-CHS T-joints under in-plane bending. Chord cracking was also observed to start occurring after the $3\%b_0$ deformation load. The ratio of the load at first visual chord crack to the load at $3\%b_0$ deformation load ranges between 1.4 and 1.7. Although the determination of the load at first visual crack is subjective and depends on the lighting during testing, it however gives an idea of the start of the occurrence of cracks in welded thin-walled SHS-CHS T-joints under in-plane bending.

TABLE 2: Loads at 1%b$_0$ deformation (M$_{1\%b0}$), 3%b$_0$ deformation (M$_{3\%b0}$), the load at which chord cracking is first visually observed (M$_{visualcrack}$), peak load (M$_{peak}$) and ultimate strength (M$_{ult}$).

Connection	M$_{1\%b0}$ (kNm)	M$_{3\%b0}$ (kNm)	M$_{visualcrack}$ (kNm)	M$_{peak}$ (kNm)	$\dfrac{M_{3\%b0}}{M_{1\%b0}}$	$\dfrac{M_{peak}}{M_{3\%b0}}$	$\dfrac{M_{visualcrack}}{M_{3\%b0}}$	M$_{ult}$ (kNm)
S3C1	0.479	0.998	1.681	1.74	2.1	1.7	1.7	0.719
S3C2	0.454	0.973	1.689	1.69	2.1	1.7	1.7	0.681
S3C4	0.254	0.538	0.738	0.99	2.1	1.8	1.4	0.381
S3C5	0.230	0.522	0.759	0.90	2.3	1.7	1.5	0.345
S6C1	0.578	1.040	1.599	1.92	1.8	1.8	1.5	0.867
S6C2	0.570	1.024	1.558	1.97	1.8	1.9	1.5	0.855

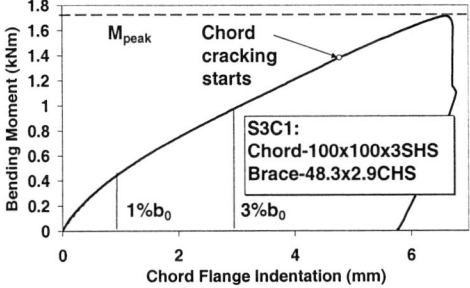

Figure 3: A typical load-deformation curve for thin-walled SHS-CHS T-joints under in-plane bending.

ULTIMATE STRENGTH BASED ON YIELD LINE MODEL

Yield line analysis has been used successfully to determine the strength of connections made up of hollow sections (Wardenier 1982, Zhao and Hancock 1991, Cao et al 1998a, Cao et al 1998b, Zhao and Hancock 1993). In these studies, plastic mechanism analysis using yield line theory has been used to develop formulae for the strength of connections made up of square or rectangular section chords with either square or rectangular section braces or longitudinal plates welded onto the face of the chord member.

The thin-walled SHS-CHS T-joints tested in this investigation were observed to fail due to chord face yielding, see Figure 2(a). The yield line model shown in Figure 2(b) was adopted for analysis. From experimental observation, the length AB in Figure 2(b) was found to have a mean value of 0.87b$_0$. Plastic mechanism analysis of the yield line model shown in Figure 2(b) will enable the development of a formula for determining the capacity of SHS-CHS T-joints under in-plane bending. A simplified approach is adopted in this yield line analysis. Membrane action and strain hardening effects are ignored in this analysis, which means that these models are based on small deflections. The material in this model is assumed to be a rigid perfectly plastic material (Wardenier 1982). The effect of the main member wall thickness and weld size are also ignored so that the plastic hinges are around the brace member and main member surfaces (Cao et al 1998b). Wardenier 1982 reported that by ignoring the influence of the weld, a yield line model generally becomes valid for both butt and fillet welded connections. In order to determine the formula for the strength of SHS-CHS T-joints under in-plane bending, the energy due to the deformation of yield lines is equated to the energy due to the external moment, M$_{ip}$, resulting from the externally applied load. The energy due to the deformation of yield lines (E$_d$) is given by:

$$E_d = \sum n_i \cdot l_i \cdot \phi_i \cdot m_p = m_p \sum n_i \cdot l_i \cdot \phi_i \qquad (1)$$

where n_i is the number of yield lines for ith yield line type, l_i is the length of a yield line, ϕ_i is the rotation of the yield line and m_p is the plastic moment per unit length of the yield line. The plastic moment per unit length of the yield line is given by (Bakker 1990):

$$m_p = f_{y0} t_0^2 / 4 \qquad (2)$$

where f_{y0} is the yield stress of the chord and t_0 is the tube wall thickness of the chord. The energy due to an external moment, M_{ip}, applied through a circular brace of diameter, d_1, and causing a chord flange indentation, δ, at location A in the yield line model (Figure 2(b)) is:

$$E_{ext} = M_{ip} \cdot 2\delta / d_1 \qquad (3)$$

A summary of the yield lines in Figure 2(b) and their properties is given in Table 3.

TABLE 3: Properties of yield lines in plastic mechanism shown in Figure 2(b).

Yield Line i	No. of Yield Lines n_i	Yield Line Rotation ϕ_i	Product of length and rotation of yield line for ith yield line type $n_i \cdot l_i \cdot \phi_i$
1	2	$\dfrac{2\delta\{l_1 - [(b_0/2)/\tan\alpha - d_1/2]\}}{(b_0/2) \cdot l_1}$; where $l_1 = 0.87 b_0$	$2[2\delta\{1.74 - \cot\alpha + \beta\}] = F_1$
2	4	$\dfrac{\delta \cdot l_2}{0.87 b_0 \cdot (b_0/2)}$; where $l_2 = \sqrt{\begin{array}{l}(0.5 b_0)^2 + \\ (0.87 b_0 - 0.5\cot\alpha - 0.5 d_1)^2\end{array}}$	$4\left[2.3\delta\left(\begin{array}{l}1.0069 - 0.87\cot\alpha - \\ 0.5\beta\cot\alpha + 0.25\cot^2\alpha + \\ 0.87\beta + 0.25\beta^2\end{array}\right)\right] = F_2$
3	4	$\dfrac{\delta \cos\alpha}{l_3 \tan\alpha}\left\{1 - \dfrac{(1.74 - \cot\alpha + \beta - \tan\alpha)}{1.74 + \beta}\right\}$	$4 \cdot \dfrac{\delta}{1.74 + \beta}\left[\dfrac{\cot\alpha}{\sin\alpha}\right] = F_3$
4	4	$\dfrac{\delta(1 - \cos\alpha)}{l_4}[\tan(\alpha/2) + \cot(\alpha/2)]$	$4[2\delta(\csc\alpha - \cot\alpha)] = F_4$
5	4	$\dfrac{\delta}{l_5}\left[\dfrac{\beta \cos^2\alpha}{1 - \beta \sin\alpha} + \dfrac{1 - \beta \sin\alpha}{\beta}\right]$	$4 \cdot \delta\left[\dfrac{\beta^2 + 1 - 2\beta \sin\alpha}{\beta(1 - \beta \sin\alpha)}\right] = F_5$
6	4	$\dfrac{\delta \cos\alpha}{(b_0/2) - (d_1/2)\tan\alpha}$; and $l_6 = 0.5\left(d_1 \cos\alpha + \dfrac{b_0 - d_1 \sin\alpha}{\tan\alpha}\right)$	$4 \cdot \delta\left[\dfrac{\cos\alpha}{\tan\alpha - \beta \sin\alpha}\right] = F_6$

Since the energy due to external moment is equal to the energy due to deformation of yield lines, equations 1 and 3 as well as the data in Table 3 yields the following:

$$M_{ip} = \frac{d_1}{2\delta} \cdot \frac{f_{y0} t_0^2 d_1}{4}[F_1 + F_2 + F_3 + F_4 + F_5 + F_6] = \frac{f_{y0} t_0^2 d_1}{2}[f(\alpha, \beta)] \qquad (4)$$

where $f(\alpha, \beta)$ is a function expressed in terms of the angle, α, in the yield line model shown in Figure 2(b) and the brace diameter to chord width ratio, β. The yield strength of the connection should be the minimum value of M_{ip}. In order to minimize the right hand side of equation 4, graphs of $f(\alpha, \beta)$ versus angle, α, are plotted for different brace diameter to brace width ratios, β, and shown in Figures 5(a) and 5(b). Figures 5(a) and 5(b) show that for a given value of β, a minimum value of $f(\alpha, \beta)$

occurs at a given angle. The angle at which $f(\alpha,\beta)$ is a minimum, α_{min}, however varies with β as shown in Figure 6(a). Since the minimum values of $f(\alpha,\beta)$ have been defined for various β values, a plot of $\min[f(\alpha,\beta)]$ versus β can be defined and a trend derived for determining $\min[f(\alpha,\beta)]$ values at different β values as shown in Figure 6(b). An equation relating $\min[f(\alpha,\beta)]$ to β is shown in Figure 6(b). The formula for determining the capacity of welded thin-walled SHS-CHS T-joints under in-plane bending can therefore be given by:

$$M_{ip} = \frac{f_{y0}t_0^2 d_1}{2} \cdot \min[f(\alpha,\beta)] = \frac{f_{y0}t_0^2 d_1}{2}[9.95\beta^2 - 10.427\beta + 9.6434] \quad (5)$$

The results of the ultimate strength of welded thin-walled SHS-CHS T-joints tested in this investigation, as determined by equation 5, are given in Table 4. A comparison of the capacities calculated using equation 5 and the capacities of the SHS-CHS T-joints determined from the deformation limit criteria shows that equation 5 generally underestimates the ultimate strength of welded thin-walled SHS-CHS T-joints as shown in Table 4. The ratio of the the ultimate strength from the formula derived from yield line theory to the ultimate strength predicted from the deformation limit criteria has a mean value of 0.983 for the tested specimens.

CONCLUSIONS

Welded thin-walled SHS-CHS T-joints tested in this investigation under in-plane bending fail primarily through chord face yielding, although chord cracking has been observed at large deformations. Using the deformation limit criteria proposed by Lu et al (1994), failure in thin-walled SHS-CHS T-joints is found to be controlled by serviceability, i.e. the deformation limit is $1\%b_0$, for $0.34 \leq \beta \leq 0.64$ and $25 \leq 2\gamma \leq 33$. A formula derived from a plastic mechanism analysis using yield line theory, in general, underestimates the ultimate strength of the thin-walled SHS-CHS T-joints, likely due to the fact that membrane action and strain hardening effects are ignored in the analysis.

ACKNOWLEDGMENTS

The authors wish to thank the Monash University, Civil Engineering laboratory staff, Mr. Roger Doulis, Mr. Graham Rundle, Mr. Roy Goswell and Mr. Don McCarthy, for their help and support with the testing. Thanks to OneSteel Market Mills, Australia, for providing the tubes used in these tests. This project was funded by CIDECT.

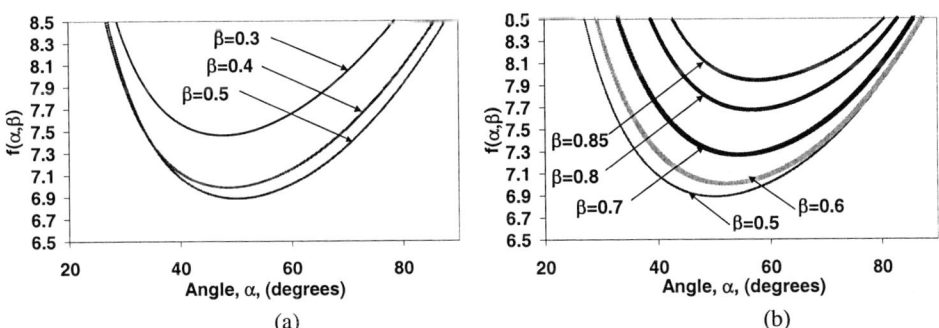

Figure 5: Graphs of $f(\alpha,\beta)$ versus angle, α, are plotted for different β values.

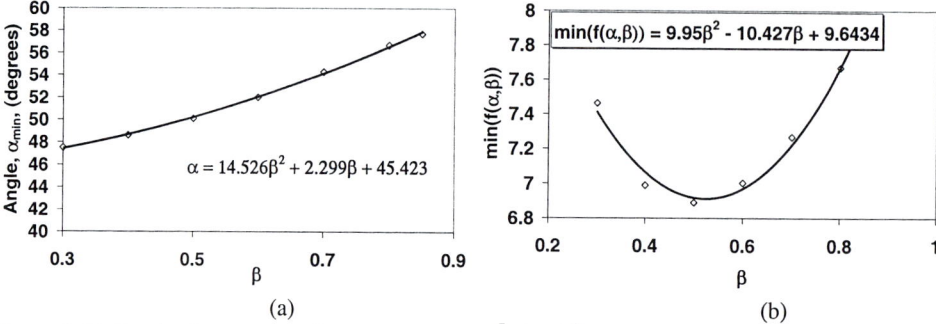

Figure 6: (a) Graph of α_{min} versus β, (b) Graph of $\min[f(\alpha,\beta)]$ versus β.

TABLE 4: Comparison of capacities from the deformation limit and the yield line theory

Connection	$M_{1\%b0}$ (kNm)	M_{ult} (kNm)	M_{ip}(Eqn. 5) (kNm)	$M_{ip}(Eqn.5)/M_{ult}$
S3C1	0.479	0.719	0.651	0.906
S3C2	0.454	0.681	0.651	0.956
S3C4	0.254	0.381	0.476	1.250
S3C5	0.230	0.345	0.476	1.380
S6C1	0.578	0.867	0.606	0.699
S6C2	0.570	0.855	0.606	0.708
			MEAN	0.983
			COV	0.285

REFERENCES

1. Bakker M.C.M. 1990, *Heron*, Vol. 35, No.3, 1990
2. Cao J.J., Packer J.A. and Kosteski N. 1998a, *J. of Constr. Steel Res.*, Vol. 46:1-3, Paper No. 134
3. Cao J.J., Packer J.A. and Yang G.J. 1998b, *J. of Constr. Steel Res.*, Vol. 48, pp. 1-25
4. IIW 1989, IIW Doc. XV-701-89, 2nd Edition, International Institute of Welding, 1989.
5. Lu L.H., de Winkel G.D., Yu Y. and Wardenier J. 1994, *Tubular Structures VI*, Editors: Grundy, Holgate & Wong, Balkema, Rotterdam, 1994, pp. 341-347
6. Packer J.A., Wardenier J., Kurobane Y., Dutta D. and Yeomans N. 1992, "Design Guide for Rectangular Hollow Section (RHS) Joints under Predominantly Static Loading" Construction with Hollow Steel Sections, CIDECT, Series 3, Verlag TUV Rheinland GmbH, Cologne, Germany.
7. SAA 1991: Structural Steel Hollow Sections, AS1163-1991, Standards Association of Australia
8. Wardenier J. 1982: *Hollow Section Joints*, Delft University Press, Delft, The Netherlands.
9. Wardenier J., Kurobane Y., Packer J.A., Dutta D. and Yeomans N. 1991, "Design Guide for Circular Hollow Section (CHS) Joints under Predominantly Static Loading" Construction with Hollow Steel Sections, CIDECT, Series 1, Verlag TUV Rheinland GmbH, Cologne, Germany.
10. Zhao X.L. 2000, *Journal of Constructional Steel Research*, Vol. 53, pp. 149-165
11. Zhao X.L. and Hancock G.J. 1991, "Plastic Mechanism Analysis of T-joints in RHS under Concentrated Force", Steel Structures, Journal of Singapore Structural Steel Society, 2(1), 31-43
12. Zhao X.L. and Hancock G.J. 1993, "Plastic Mechanism Analysis of T-joints in RHS under Combined Actions", In: Tubular Structures V, Coutie M.G. and Davies G. (eds), London: E &FN Spon, 345-352

TESTS AND DESIGN OF LONGITUDINAL FILLET WELDS IN VERY HIGH STRENGTH (VHS) STEEL CIRCULAR TUBES

Tong-Wei Ling, Xiao-Ling Zhao and Riadh Al-Mahaidi

Department of Civil Engineering, Monash University, Clayton, Vic 3800, Australia

ABSTRACT

Very high strength (VHS) tubes offer superior strength and enormous reduction in weight. Test carried out at Monash University showed that the average yield stress of the VHS tubes is around 1,350 MPa with an ultimate tensile strength of 1,500 MPa. There is no information on the behaviour of longitudinal fillet welded connections in VHS tubes. Existing research on this topic was limited to square and rectangular hollow sections (SHS and RHS) with a yield stress up to 450 MPa. Existing design rules may not be appropriate for VHS circular tubes. This paper describes a series of tests on longitudinal fillet welds in VHS steel circular tubes. The thickness of VHS tubes varies from 1.6mm to 2.0mm. The weld length (L_w) of longitudinal fillet welds varies from 15mm to 65mm. Two failure modes were observed, namely failure along the weld and tube failure. The test results were compared with the formulae developed for C450 SHS. The effect of strength reduction in the heat-affected-zone (HAZ) on design was included in the proposed design formulae. Critical weld length formula was derived, which can be used to predict the failure mode.

KEYWORDS

Circular Tubes, Longitudinal Fillet Welds, Thin-Walled Sections, High Strength Steel

INTRODUCTION

VHS tubes offer superior strength and enormous reduction in weight. The section capacity of VHS circular tubes has been studied and reported by Zhao (2000), Jiao and Zhao (2001, 2002a). The tests carried out at Monash University showed that the average yield stress of the VHS tubes is around 1,350 MPa with an ultimate tensile strength of 1,500 MPa. Due to the high strength and lightweight properties, VHS steel tubes have been used in automotive industry, and possibly in mechanical and structural applications. Extensive research has been performed on welded connections in C350 and C450 square and rectangular hollow sections (SHS and RHS) [Zhao and Hancock (1995a, 1995b, 1996), Zhao et al (1999)]. It was found that different failure modes might occur as the yield stress of steel tubes increase. However, existing research on this topic was limited to SHS and RHS with a yield stress up to 450 MPa. Therefore, existing design rules may not be appropriate for VHS circular tubes for the following reasons:

(a) The material properties of steel tubes in the HAZ (heat affected zone) may change significantly after welding.
(b) The failure mechanism may change due to the significant difference between the strength of the base metal and that of the weld metal.
(c) The different shape of cross section (circular tubes rather than SHS and RHS) may change the stress distribution around the weld especially for longitudinal fillet welds.

Tests were performed on butt-welded and transverse fillet welded connections in VHS tubes [Jiao and Zhao (2002b)]. Significant reduction in strength in the HAZ was observed. Design formulae were proposed by Jiao and Zhao (2002c).

This paper describes a series of tests on longitudinal fillet welds in VHS steel circular tubes. The thickness of VHS tubes varies from 1.6mm to 2.0mm. Two different failure modes were observed, namely failure along the weld and tubes failure. The existing design formula developed for C450 RHS was checked against the current test data with failure along the weld. A reduction factor of 0.628 was found appropriate for designing VHS tubes. A new design formula was derived for tube failure mode. The critical weld length was also given, which can be used to predict the failure mode.

MATERIAL PROPERTIES

Tensile coupons were used to measure the material properties (yield stress f_y, ultimate tensile strength F_u and Young's modulus E) in accordance with AS1391 (SAA 1991). The 0.2% proof stress was adopted as the yield stress. Coupon test results for the specimens are shown in Table 1. The measured material properties are used in predicting the capacities of welded connections later in the paper.

TABLE 1
COUPON TENSILE TEST RESULTS

Section ID	F_u (MPa)	f_y (MPa)	F_u/f_y	E (Gpa)
T1	1538	1380	1.11	194
T6	1524	1370	1.11	207
T8	1496	1325	1.13	195
MEAN	1534	1372	1.12	201
COV	0.022	0.018	0.010	0.022

WELDING PROCEDURES

The welding procedures comply with AS/NZS 1554.4:1995(SAA 1995). The electro gas welding (EGW) method was adopted. The high strength consumable wires were used and comply with the specification of AWS A5.28 (AWS 1996). The nominal tensile strength of the weld metal was 760MPa.

TESTS ON LONGITUDINAL FILLET WELDS

Specimens and Test Se up

A total of 4 specimens were designed for tensile testing. The label had a designation of ***Tabc*** where the first two symbols ***Ta*** represent section identification number as listed in Table 1. The symbol ***b*** specified as P or W refers to the types of specimen set up, P for plug in type and W for welding type. Last symbol c refers to the weld length (L_w shown in Figures 1 and 2) of the test region. The specimen labels and measured section dimensions are listed in Table 2.

TABLE 2

SPECIMEN DIMENSION

Specimen Label	Diameter D (mm)	Thickness t (mm)	Weld Length L_w (mm)
T1P15	31.7	1.6	15
T6P65	38.2	2.0	65
T8W18	75.2	1.6	18
T8W38	75.1	1.6	38

Two types of methods were used to apply the loading. One is called plug in type as shown in Figure 1, which was used for tubes with diameters less than 50 mm. In this method, a steel rod was plugged in the tube at the end without welding. Therefore the machine can grab the tube during the tensile testing. For plug in type specimen, the length of the tube was 250mm. This method was used by Zhao (2000), Jiao and Zhao (2001). The other method is called welding type as shown in Figure 2, which was used for tubes with diameters larger than 50 mm. The diameter of such tubes is too large to be grabbed by the testing machine. For welding type specimen, the length of the tube was 400mm. This method of applying loads was similar to that used by Zhao and Hancock (1995b), Zhao et al (1999). For all the connections welds were applied continuously around the end of the 10mm plate. The connecting region between the end return welds and the longitudinal welds was referred to as the "transition region".

Tensile tests were carried out in a 500 kN capacity Baldwin Universal Testing machine at Monash University. Displacement control was used with a loading speed of 2.0mm/min. Two Linear Variable Displacement Transducers (LVDTs) were used to measure the displacement.

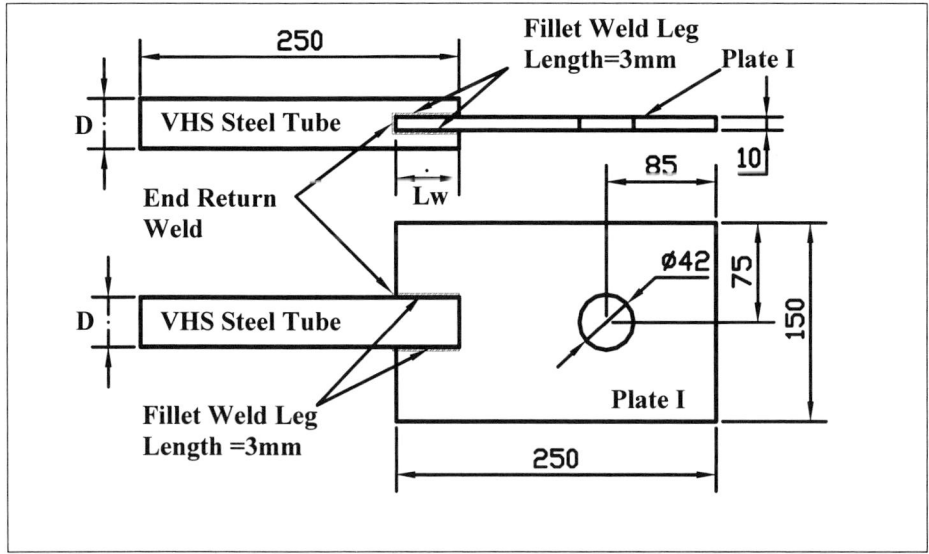

Figure 1 Loading method 1: Plug In Type (Dimensions in mm)

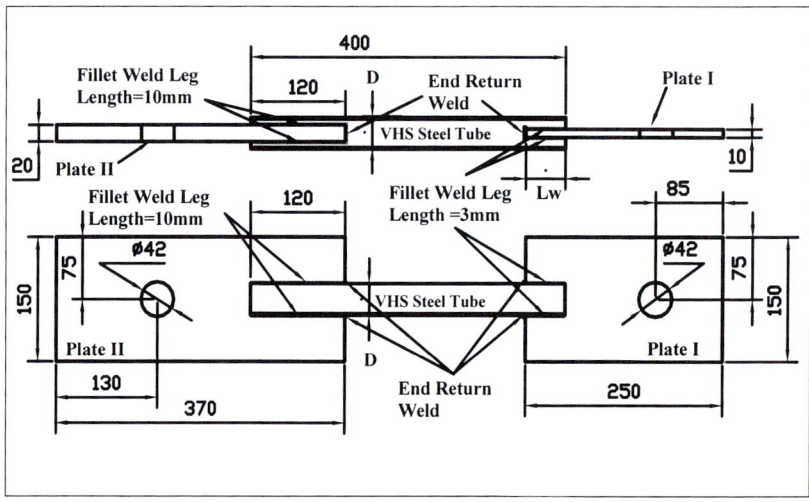

Figure 2 Loading method 2: Welding Type (Dimensions in mm)

Failure Modes

All failures of the specimens happened in the test region. Two kinds of failure modes were observed. The failure of T1P15, T8W18 and T8W38 can be grouped as failure mode 1. In the failure mode 1 (Figure 3, Figure 4), initial cracking occurred at the end return welds or transition region. Then the crack was tore along the welded line until it reached the end of the tubes. Zhao et al (1999) observed the same failure mode in C450 rectangular hollow section under the same kind of testing (Figure 5). In Zhao et al (1999), the weld length (L_w) was 40mm for all specimens.

However, different failure mode was observed on T6P65, which referred as failure mode 2. Figure 6 shows the failure mode 2 of T6P65. Although this kind of failure has the same initial cracking point as failure mode 1, but the cracking was propagated to the middle of the tube, instead of along the welded line. This may be because that a much longer welding was applied in T6P65, which made the capacity of the welding larger than that of the tube itself.

Figure 3 Failure Mode 1 of T8W18 (left) and T8W38 (right)

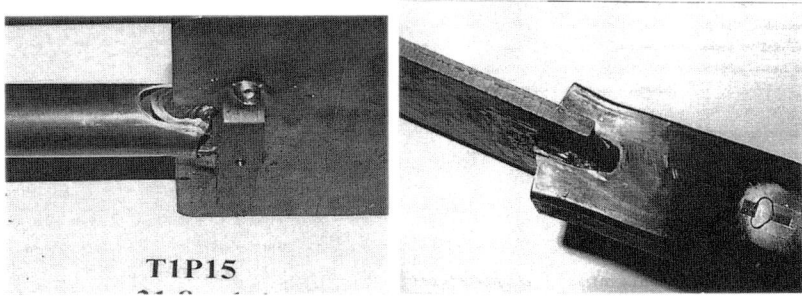

Figure 4 Failure Mode 1 of T1P15   Figure 5 Failure Mode of C450 SHS

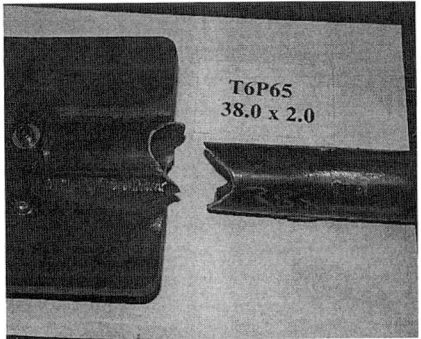

Figure 6 Failure Mode 2 on T6P65

Testing Results

Typical load versus displacement curves are shown in Figure 7 for Failure Mode 2 (T6P65) and Failure Mode 1 (T8W38). It can be seen that failure mode 2 has a period of transition during which displacement increases while the load maintains almost the same before the failure occurs. This somehow indicates the yielding of the tubes. For failure mode 1, the graph shows a slightly different behaviour. There is a drop in load before a transition period occurs. The maximum load (P_{exp}) obtained in each test is listed in Table 3.

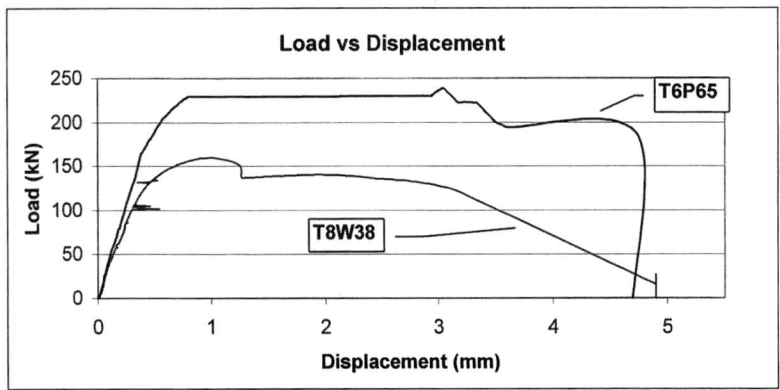

Figure 7 Typical load versus displacement curves

DESIGN OF LONGITUDINAL FILLET WELDS

Failure Mode 1

For the failure mode 1, Zhao et al (1999) proposed a formula to obtain the nominal strength (P_n):

$$P_n = 0.75 F_u tL \qquad (1)$$

where F_u is the tensile strength of parent (base) metal, L is the total weld length and t is the tube thickness. The experimental maximum load (P_{exp}) obtained from the tensile testing is compared with Equation 1 in Table 3.

TABLE 3

COMPARISON OF EXPERIMENT RESULTS WITH EQUATION 1

Specimen	L_w/t	L (mm)	P_{exp} (kN)	P_n (kN)	P_{exp}/P_n
T1P15	9.5	80	101	148	0.682
T8W18	11.1	92	113	165	0.685
T8W38	23.2	172	159	309	0.515
Mean					0.628
COV					0.127

From Table 3, the reduction in shear strength was due to the softening of heat-affected zone (HAZ). For butt-welded VHS connections the reduction factor was found about 0.550 on F_u, for transverse fillet welds the reduction factor was about 0.474 on F_u [Jiao and Zhao (2002c)]. It seems that the reduction factor for longitudinal fillet welds is about 0.628 from Table 3. This value may change slightly when more test results are obtained. If the reduction factor is called φ, then the proposed formula for failure mode 1 becomes:

$$P_{m1} = 0.75 \cdot \varphi \cdot F_u tL = 0.75 \cdot \varphi \cdot F_u t \cdot [4 \cdot L_w + 2T] \qquad (2)$$

where L_w is the weld length and T is the plate thickness. New comparison using Equation 2 was shown in Table 4 where a mean ratio of 0.998 is obtained.

TABLE 4

COMPARISON OF EXPERIMENT RESULTS WITH EQUATION 2

Specimen	P_{exp} (kN)	P_{m1} (kN)	P_{exp}/P_{m1}
T1P15	101	92.9	1.087
T8W18	113	104	1.087
T8W38	159	194	0.820
Mean			0.998
COV			0.126

Failure Mode 2

The failure in mode 2 is similar to tension failure of tube. According to the Australian Standards AS4100 (SAA, 1998), the nominal section capacity shall be taken as the lesser of (Af_y) and ($0.85AF_u$) where A is the cross-sectional area, f_y is the yield stress and F_u is the ultimate tensile strength. For very high strength (VHS) steel, f_y is always larger than $0.85F_u$. There are some reductions due to HAZ softening around the welds that need to be considered. The HAZ length can be estimated as (T+4s) for each weld on top and bottom, where T is the plate thickness and s is the weld leg length. The strength along this length is φF_u while the strength along the rest of section is F_u. Therefore the proposed capacity for failure mode 2 can be written as Equation (3):

$$P_{m2} = 0.85 \cdot [(A - A_{HAZ}) \cdot F_u + A_{HAZ} \cdot \varphi \cdot F_u] = 0.85 \cdot A \cdot F_u \cdot [1 - \frac{A_{HAZ}}{A} \cdot (1 - \varphi)] \qquad (3)$$

in which A is the cross-sectional area of the tube, A_{HAZ} is the HAZ area for both top and bottom welds, i.e.

$$A = \frac{\pi}{4} \cdot [D^2 - (D - 2 \cdot t)^2] \qquad (4)$$

$$A_{HAZ} = 2 \cdot \frac{\arcsin(\frac{T + 4 \cdot s}{D})}{4} \cdot [D^2 - (D - 2 \cdot t)^2] \qquad (5)$$

By substituting Equation (4) and Equation (5) into Equation (3), P_{m2} becomes:

$$P_{m2} = 0.85 \cdot F_u \cdot \frac{\pi}{4} \cdot [D^2 - (D - 2 \cdot t)^2] \cdot [1 - \frac{2 \cdot \arcsin(\frac{T + 4s}{D})}{\pi} \cdot (1 - \varphi)] \qquad (6)$$

For T6P65, T = 10mm, s = 3mm, D = 38.2mm, t = 2.0mm, F_u = 1524MPa and φ is taken as 0.628. The calculated P_{m2} according to Equation (6) is 252 kN which is about 5% larger than the experimental value of 239 kN.

Critical Weld Length

Critical weld length ($L_{w,cr}$), beyond which tube failure will occur, can be obtained from the condition of $P_{m1} = P_{m2}$, i.e.

$$0.75 \cdot \varphi \cdot F_u t \cdot [4 \cdot L_{w,cr} + 2T] = 0.85 \cdot F_u \cdot \frac{\pi}{4} \cdot [D^2 - (D - 2 \cdot t)^2] \cdot [1 - \frac{2 \cdot \arcsin(\frac{T + 4s}{D})}{\pi} \cdot (1 - \varphi)]$$

$$L_{w,cr} = \frac{0.85 \cdot \frac{\pi}{4} \cdot [D^2 - (D - 2 \cdot t)^2] \cdot [1 - \frac{2 \cdot \arcsin(\frac{T + 4s}{D})}{\pi} \cdot (1 - \varphi)]}{3 \cdot \varphi \cdot t} - \frac{T}{2} \qquad (7)$$

The calculated $L_{w,cr}$ and predicted failure mode for the 4 specimens described in this paper are listed in Table 5 together with the observed failure mode. It can be seen that the critical weld length given in Equation (7) predicts the correct failure mode.

TABLE 5
PREDICTED CRITICAL WELD LENGTH AND PREDICTED FAILURE MODES

Specimen Label	Observed Failure Mode	L_w (mm)	Predicted $L_{w,cr}$ (mm)	Predicted Failure Mode
T1P15	Failure mode 1	15	30	Failure mode 1
T6P65	Failure mode 2	65	39	Failure mode 2
T8W18	Failure mode 1	18	92	Failure mode 1
T8W38	Failure mode 1	38	92	Failure mode 1

CONCLUSIONS

The following observations and conclusions are made based on the limited test results:
- There were two failure modes in longitudinal fillet welds in VHS tubes, namely failure along the weld and tube failure.
- For failure mode 1, a reduction factor, $\varphi = 0.628$, on F_u was proposed to modify the formula derived for longitudinal fillet welds in C450 SHS. This is due to the HAZ softening around the welds.
- For failure mode 2, a design formula was derived which considered the reduction in the HAZ.
- A formula was derived for the critical weld length, beyond which tube failure will occur. This critical weld length can be used to predict failure modes.

More tests are planned to verify the above-proposed formulae using reliability analysis approach.

ACKNOWLEDGEMENTS

The authors are grateful to Monash University and OneSteel Market Mills for financial support. Thanks are given to Mr Graeme Rundle and Mr. Roger Doulis for assistance of performing the tests on VHS tubes. Comments from Mr. Hayden Dagg and Mr. Tony Gunn are appreciated.

REFERENCES

AWS (1996). "Low Alloy Steel Electrodes for Gas Shielded Metal Arc Welding." American Welding Society Welding Code, AWS A5.28, Miami, USA

Jiao, H. and Zhao, X.L. (2001), Material Ductility of Very High Strength (VHS) Circular Steel Tubes in Tension. *Thin-Walled Structures,* **39(11)**, 887-906.

Jiao, H. and Zhao, X.L. (2002a), Imperfection, Residual Stress and Yield Slenderness Limit of Very High Strength (BHS) Circular Steel Tubes, *Journal of Constructional Steel Research,* accepted for publication

Jiao, H. and Zhao, X.L. (2002b), Strength of Butt Welds and Transverse Fillet Welds in Very High Strength (VHS) Circular Steel Tubes, *Sixth Pacific Structural Steel Conference,* Beijing, China, 775-780

Jiao, H. and Zhao, X.L. (2002c), Butt Welds and Transverse Fillet Welds in Very High Strength (VHS) Circular Steel Tubes, *Journal of Structural Engineering,* ASCE, under review

SAA (1991), Methods for Tensile Testing of Metals, *Australian Standard AS1391*, Standards Association of Australia, Sydney

SAA (1995). "Structural steel welding Part 4: welding of high strength quenched and temped steels." *Australian Standard AS/NZS1554.4*, Standards Australia, Sydney

SAA (1998), Steel Structures, *Australian Standard AS4100*, Standards Association of Australia, Sydney

Zhao, X.L. and Hancock, G.J. (1995a). Butt Welds and Transverse Fillet Welds in Thin Cold-Formed RHS Members. *Journal of Structural Engineering,* ASCE, **121(11)**, 1674-1682.

Zhao, X.L. and Hancock, G.J. (1995b). Longitudinal Fillet Welds in Thin Cold-Formed RHS Member. *Journal of Structural Engineering,* ASCE, **121(11)**, 1683-1690.

Zhao, X.L. and Hancock, G.J. (1996). Welded Connections in Thin Cold-Formed Rectangular Hollow Sections. *Connections in Steel Structures III*, Eds. Bjorhovde, R. et al., Oxford: Pergamon, 89-98.

Zhao, X.L., Al-Mahaidi, R. and Kiew, K.P. (1999). Longitudinal Fillet Welds in Thin-Walled C450 RHS Members, *Journal of Structural Engineering*, ASCE, **125(8)**, 821-828.

Zhao, X.L. (2000). Section Capacity of Very High Strength (VHS) Circular Tubes under Compression. *Thin-Walled Structures,* **37(3)**, 223-240.

EXPERIMENTAL BEHAVIOUR OF END PLATE I-BEAM TO CONCRETE-FILLED RECTANGULAR HOLLOW SECTION COLUMN JOINTS

Luís Costa Neves[1], Luís Simões da Silva[2] and Pedro C.G. da S. Vellasco[3]

[1] Assistant - Civil Engineering Department, University of Coimbra, Portugal
[2] Associate Professor - Civil Engineering Department, University of Coimbra, Portugal
[3] Associate Professor - Structural Engineering Department, UERJ - State University of Rio de Janeiro, Brazil

ABSTRACT

In the framework of the component method, adopted by Eurocode 3 to assess the moment-rotation characteristics of the connections, any joint is decomposed into a set of springs known as components. These components are assembled in a mechanical model, that is able to predict the behaviour of any joint-geometry, provided that the response of the components is fully characterized. The behaviour of joints in steel rectangular hollow sections filled with concrete is frequently governed by the deformability of the column face loaded in bending. Therefore a good estimate of the moment-rotation response of the joint relies on the knowledge of the behaviour of this component, yet to be fully characterized. This paper describes an experimental study performed at the Civil Engineering Department of the University of Coimbra, that consists of a series of six tests on extended and flush end plate connections, with one or two bolt rows in tension, loaded under monotonic conditions. A description of the tested geometries is given, along with the test setup, load application method and the specimens instrumentation. Finally, some of the available results are presented, focusing on the joint moment-rotation response.

KEYWORDS

Experimental, concrete-filled, hollow section, joints, component method, eurocode 3, extended end plate, flush end plate, moment-rotation behaviour, ductility.

1. INTRODUCTION

Eurocode 3 (1998) adopts a methodology for evaluating the properties of the connections that is known as the component method. This method is based on a mechanical model with a set of springs, representing the contribution of each source of deformability in the connection, Figure 1. The rotational behaviour of any joint may then be derived, as long as the behaviour of each joint component (spring) is a priori characterised. In the case of connections where the beam is connected to the face of a rectangular hollow section filled with concrete (Figure 2), the governing component, in the great majority of cases, is the column loaded face.

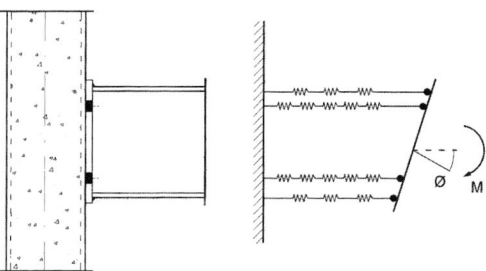

Figure 1: Spring model for the evaluation of the rotational behaviour

The behaviour of connections between concrete-filled rectangular hollow sections and I-beams has been studied by several authors: Lu, Puthli & Wardenier (1993); Lu & Wardenier (1998); Matsui (1986), Matsui et al. (1997). In the compressed zone the concrete supports the deformation and in the tension zone the column face is bent out of the plan. Vandegans (1996) showed that the deformability of the column is due almost exclusively to the deformability of this chord face loaded with the tensile force.

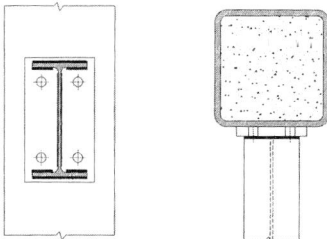

Figure 2 – Flush end plate connection between an I-beam and a RHS filled with concrete

The typical rotational behaviour of the loaded chord face is illustrated in Figure 3: M is the moment transmitted to the column, and ϕ is the rotation of the joint resulting from the deformation of the loaded column face. This moment-rotation $(M - \phi)$ curve may, in the initial elastic range, be characterized by the initial stiffness $S_{j,ini}$. Due to the large out-of-plane deformations δ of the loaded chord for higher values of bending moment, often greater than the chord thickness t_{wc}, the typical post-limit stiffening plate behaviour is observed. It is worth noting that neglecting the overstrength of the loaded chord may result in unforeseen overstressing of other joint components (bolts or welds, for example), which may fail suddenly due to their brittle behaviour, Simões da Silva (2001). This brittle behaviour was noticed in some of the tests performed in the present investigation.

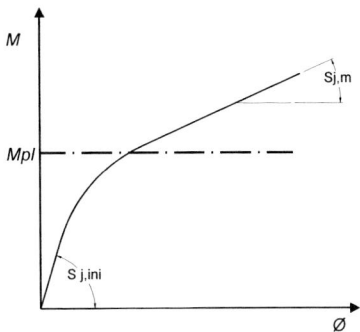

Figure 3 – Typical behaviour of the loaded column face

DESCRIPTION OF THE EXPERIMENTAL WORK

Test Setup

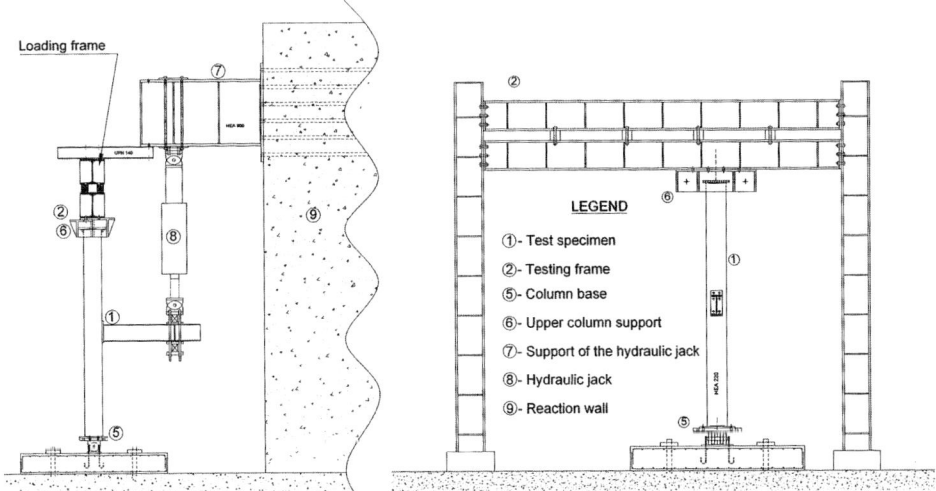

Figure 4: Test layout

Figure 4 shows the test layout, with the loading frame that supports the specimens. The load was applied by mean of a hydraulic jack with capacity of 200 kN and a stroke of 200 mm. It is worth noting that this stroke may be a limiting parameter, as the tested geometry may accommodate large rotations in the post-limit range. The adopted test layout could cope with rotations of about 200 mrad in monotonic tests and 100 mrad in cyclic tests, far above the rotations attained in these tests, due to brittle failure of the studs.

The column base was supported by the hinge shown in Figure 5b, fixed to a cast in place concrete foundation. At the top, the column was fixed with a device shown in Figure 6c. A general view of the test layout may be observed in Figure 5a.

Figure 5 – General view and details of the test setup

Description of the experimental tests

Table 1 presents a general overview of the performed tests: a total of six static monotonic tests in which two (E13, E14) connecting a IPE 240 beam to a RHS 300*6 column by an extended end-plate with a single bolt row in tension in the extended plate. Two other tests (E16, E17) had a similar geometry, but a second bolt row in the tension zone was added inside the beam flange. These two geometries were tested under positive and negative moments. Two other tests (E19, E21) connecting a IPE 300 beam to a SHS 300 column with 6 mm and 16 mm thickness, respectively, (by a flush end plate) complemented this test series.

TABLE 1
DESCRIPTION OF THE EXPERIMENTAL PROGRAMME

Test	Column	Steel	Beam	Steel	Type	Loading
E13	SHS 200*6	S 355	IPE 240	S 275	1 bolt row in tension	Monotonic M +
E14						Monotonic M -
E16	SHS 200*6	S 355	IPE 240	S 275	2 bolt rows in tension	Monotonic M +
E17						Monotonic M -
E19	SHS 300*6	S 355	IPE 300	S 275	flush	Monotonic M +
E21	SHS 300*16	S 355	IPE 300	S 275	flush	Monotonic M +

The connectors were studs welded to the column face and the end plate was machined to accommodate the welds.

Equipment and instrumentation

Figures 5 to 8 show the instrumentation of the test specimens. Data acquisition was made by a data logger TML TDS 602 with an extension box, allowing the monitoring of up to 80 channels. Measurement of the forces was by 200 kN load cells in the upper and in the lower beam faces in the line of the hydraulic jack. Linear transducers (LVDT) with strokes ranging from 10 mm to 200 mm evaluated the displacements and were placed in the beam and in the column. Strains were measured in the most relevant points of the column face and of the beam, using linear strain gages (TML FLK-6-11) or rosettes (TML FRA-5-11). To evaluate the forces in each bolt row in tension, special (TML BTM-6-C) strain gages were placed inside each bolt.

In Figure 6, LVDT no. 1 measures the displacement of the beam under the load application point, and no. 2 at mid-span. These two measurements give an approximation of the beam rotation. The two LVDT's no. 3 enable the evaluation the rotation of the beam in the section close to the joint, that is the rotation of the joint plus the rotation of the column (as a whole and by shear at the joint). This column rotation is then measured by the LVDT's no. 4 and 5 for the rotation as a rigid body, and by the no. 6 that includes shear in the joint. In addition to these measurements, no. 7 controls the extension of the column side faces.

The rotation of the joint may then be calculated isolating the relevant sources of deformability. It should be noted that the deformation of the loaded column face may be derived with the help of the readings of LVDT's no. 8-10.

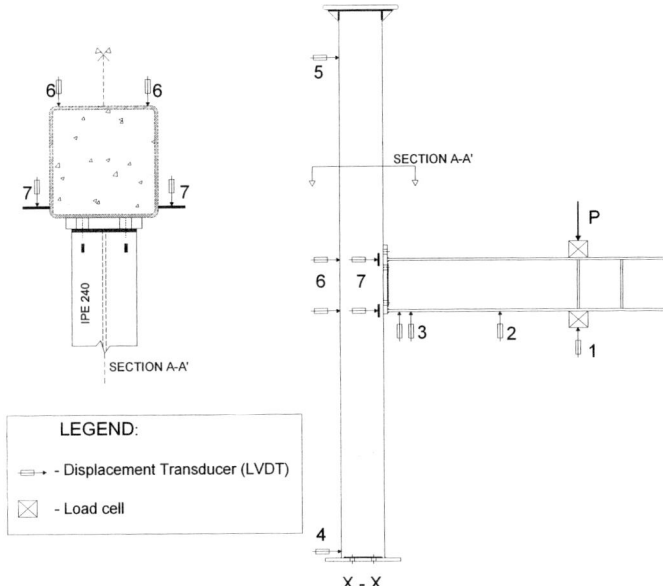

Figure 6: Location of LVDT's and load cells in the test specimens

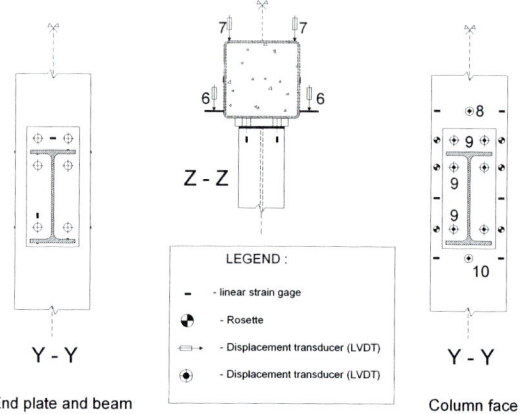

Figure 7: Strain gages and LVDT's in the beam and column face

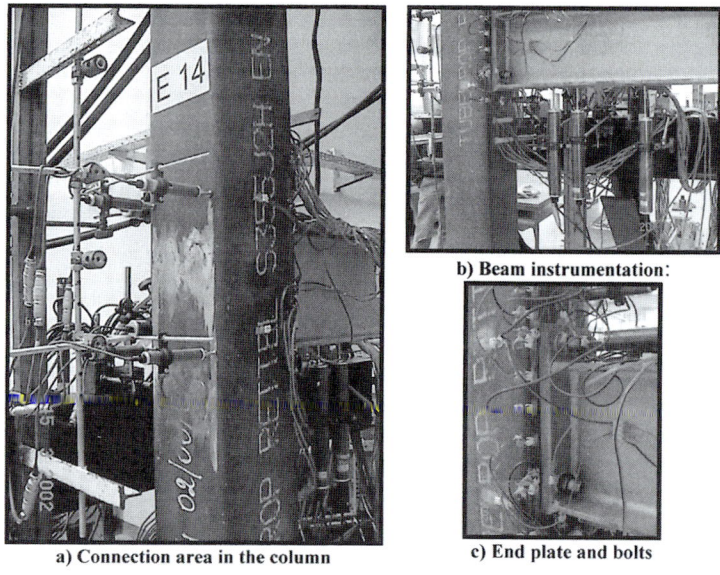

Figure 8: Details of the instrumentation

RESULTS AND CONCLUSIONS

Moment-Rotation $(M - \phi)$ curves

Complete yielding of the loaded column face is generally associated to a plastic mechanism, and corresponds to the plastic moment M_{pl}. This value may be calculated, for example, from Gomes (1996). Yielding, however, starts much earlier than that, especially for thin plates, Neves (1996). The behaviour of this curve after the plastic moment is strongly dependent of the slenderness of the loaded column face. For thick plates (with a ratio width/thickness around 10), the post plastic stiffness tends to vanish, while for very thin plates (with a ratio width/thickness around 50) the curve after *Mpl* does not exhibit significant decrease of stiffness. In the present test programme, this ratio was of 31 for tests E13 to E17, 48 for test E19, and 17 for test E21.

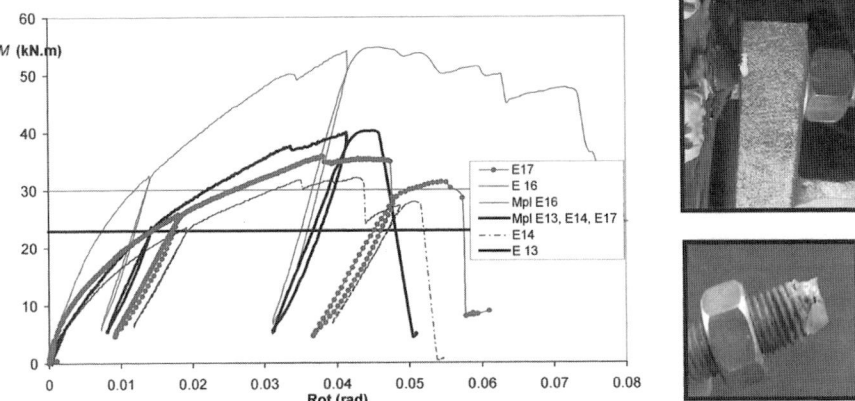

Figure 9: Moment rotation curves and details for tests no. 13 and 16

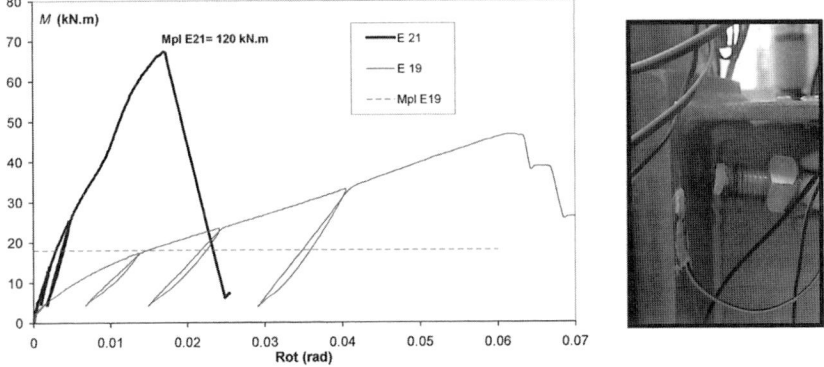

Figure 10: Moment rotation curves and details for tests no. 19 and 21

In tests no. E13 to E19 the governing component was the loaded column face that attained the plastic moment, being responsible for most of the joint deformation. The welded studs, designed for a plastic moment higher than the value of the critical component failed suddenly (brittle failure) as the ductile column face developed an overstrength by membrane action. This behaviour is illustrated in Figure 9, for the tests E13 to E17. Test E16 exhibits increased stiffness and resistance when compared to test E13 (Figure 9), a clear consequence of the extra row of bolts. The plastic moment is attained for a rotation between 10 and 20 mrad, with continous increase of moment and rotation until sudden failure of the studs in the weld zone. The maximum rotation capacity was about 45 mrad, that is clearly a limiting factor when comparing to other connection systems. Tests E14 and E17 with negative moment and one bolt row in tension, differ from test E13 by the lever arm, that influences both the plastic moment and the stiffness.

Moment rotation curves for tests no. 19 and 21 are shown in Figure 10. These two tests, with flush end plate, present different thicknesses of the column face: 6 mm in test E19 and 16 mm in test E21. The first test exhibits large ductility, with brittle failure of the studs for a rotation greater than 60 mrad, and a moment greater than 2.5 times the plastic resistance of the critical component. For the second test poor results were observed, with brittle failure (Figure 10) for a moment of 60 % of the column face plastic moment and a rotation of 20 mrad; contribution from other sources of deformability, like the beam in bending, was also noted.

Finally, we should mention that these tests are part of a study that aims to characterize the behaviour of this joint geometry in the scope of the component method and to derive simple formulae to be implemented in Eurocode 3

REFERENCES

Eurocode 3 (1998). ENV - 1993-1-1:1992/A2, Annex J, Design of Steel Structures – Joints in Building Frames. *CEN, European Committee for Standardisation,* Document **CEN/TC 250/SC 3**, Brussels.

Gomes, F.C.T.(1996). Moment capacity of beam-to-column minor-axis joints, *Proceedings of the IABSE International Colloquium on Semi-Rigid Structural Connections, Turkey*

Kawano, A. Matsui, C. (1997). New connections using vertical stiffeners between H-shaped beams and hollow or concrete filled square tubular columns, *Proceedings of the 1996 Engineering Foundation Conference on Composite Construction in Steel and Concrete III*, **ASCE, N.Y.**,172-185.

Lu, L. H., Puthli, R.S., Wardenier, J. (1993). Semi-Rigid Connections Between Plates and Rectangular Hollow Section Columns *Proceedings of the Fifth International Symposium on Tubular Connections* held at Nottingham, UK, **E & FN Spon, London**, 723-731.

Lu, L. H., Wardenier, J. (1998). The ultimate strength of I-beam to RHS column connections, *J. Constructional Steel Research,* **46: 1-3** paper no.139.

Matsui, Chiaki (1986). Strength and deformation capacity of frames composed of wide flange beams and concrete filled square steel tubular columns, *Proceedings of the 1986 Pacific Structural Steel Conference, Auckland NZ*, New Zealand Heavy Engineering Research Assoc, 169-181.

Neves, L.F.C. (1996). "Semi-rigid connections in steel structures. Assessment of stiffness for minor-axis geometries". (in Portuguese), *MSc. thesis*, University of Coimbra, Coimbra, Portugal.

Simões da Silva, L., Gervásio, H., Rebelo, C. and Girão Coelho, A. (2001). Assessment of overstrength effects in steel and composite connections using Monte Carlo methods, in *Proceedings of IABSE International Conference on Safety, Risk and Reliability – Trends in Engineering*, Malta, March 21-23.

Vandegans, D. (1996). "Application de la méthode des composantes selon l'Eurocode 3 aux assemblages par goujons filetés dans le cas de profils creux remplis de béton", *Construction Métallique*, **no. 3**, 25-37.

COMPOSITE CONNECTIONS AT PERIMETER LOCATIONS IN UNPROPPED COMPOSITE FLOORS

M. Dhanalakshmi[1], M.P. Byfield[1] and G.H. Couchman[2]

1. Cranfield University, RMCS Shrivenham, Swindon, SN6 8LA, UK
2. Steel Construction Institute, Silwood Park, Ascot, SL5 7QN, UK

ABSTRACT

Of all new multi-storey buildings in the UK, approximately 40% use composite floor construction. This type of construction is structurally efficient because it exploits the tensile resistance of the steel beams and the compressive resistance of the concrete slabs. This composite action allows shallower steel beams to be used because of the increased flexural strength and stiffness. Research has shown that more savings, approximately 25% on weight or depth of individual beams, can be achieved if composite connections are adopted. With such connections, increased moment of resistance can be achieved by introducing dedicated slab reinforcement, which acts like an additional row of bolts in an extended end plate. However, the use of composite connections is not widespread. This is due in part to two problems addressed herein. Firstly, the existing composite connections design and detailing rules can currently only be used with beams that are propped during construction, whereas unpropped construction is generally a preferred and more economic construction method. For applications where it is important to minimise beam depth, even if this is at the expense of heavier perimeter columns, practical details are also required for single sided moment resisting connections. In order to investigate these problems, a test was carried out at the Building Research Establishment, UK, on a full-scale unpropped sub-frame that incorporated a novel exterior column connection. The results of this test, with an emphasis on the exterior column connection, are presented in this paper.

KEYWORDS

Composite connections, ductility, composite construction, rotation capacity, steel structures, steel beams, unpropped construction.

INTRODUCTION

Full strength rigid connections may not offer the most economical solution for the construction of multi-storey steel framed buildings because of high fabrication costs. In order to improve design efficiency, Eurocodes 3 and 4 permit the use of 'semi-continuous construction', in which connections exhibit characteristics of partial strength, ductility and full or semi-rigidity. By limiting the strength of connections, fabrication costs can be reduced and problems related to fully restrained connections, such as local buckling and brittle failures by fracture can be avoided, Anderson & Najafi (1994).

Semi-rigid composite connections can result in savings on weight and depth of individual beams, BCSA/SCI (1998). This can facilitate service integration and allow a reduction in the overall building height, together with a reduced cost of cladding. However, the concern amongst steelwork designers and fabricators is that these benefits may remain largely unrealised because: (a) composite action is not currently recommended at perimeter column connections, due to problems anchoring the reinforcement; and (b) unpropped construction has become a popular practice of construction. Due to the lack of test data, the composite connections design guide, BCSA/SCI (1998), does not adequately model unpropped construction.

To take advantage of composite behaviour it is essential to examine the interactions between composite beams and their composite connections through to failure of the complete system. Full-scale testing is the only method of obtaining accurate and believable data on these interactions. Numerical modelling techniques can be developed, Ahmed and Nethercot (1995), but to ensure total confidence they should be calibrated against test data. Very few tests have been undertaken on unpropped construction and the test reported herein was carried out to validate or otherwise the design procedure for composite connections in the unpropped situation. The test provides data on the following:

(a) Ductility. Composite connections are less ductile than their bare steel counterparts. This is because of the large strains that need to be accommodated by the reinforcing bars as the connection rotates. When the beam is unpropped the bare steel rather than the composite section supports the dead load and this has two conflicting effects on the performance of the beam and connection system. Firstly, greater strains and curvatures are induced in the steel when unpropped construction is adopted. The rotation required from the bare steel connection is therefore greater in unpropped construction. Secondly, because the slab only behaves structurally as the concrete gains strength the reinforcement is not subject to any strain due to dead load. The available rotation capacity of the connection will therefore be greater when it is used on a beam that is unpropped during construction. Importantly however, substantially more curvature and end rotation is required to generate the plastic hinge in the beam, Anderson and Najafi (1997). Almost all previous tests on composite connections have been undertaken on isolated cruciform or single-sided composite connections, such as those carried out by Davidson, Lam and Nethercot (1990). These tests represent a connection to a beam that is propped during construction and therefore the data from these tests cannot be used to establish the performance of the more commonly occurring unpropped situation.

(b) Exterior connections. The few tests undertaken to simulate external column connections show that the problem of anchoring reinforcement restricts the moments that can be sustained, Tschemmernegg, Huber, and Pavlov (1995). The Composite Connection Design Guide, BCSA/SCI (1998), recommends that connections to external columns should be non-composite to avoid these problems. The choice of non-composite connections at external locations generally results in an increase in the size of the connected beam and this can erode the benefits associated with composite connections. This work reports a test that utilised a new exterior connection detail that mitigates some of the problems of anchoring the reinforcement, Figure 1. In this new detail the reinforcement is passed through the holes predrilled into the flanges of the column, with the reinforcement ends bent through 90°. The improved anchorage and continuity resulting from the new detail leads to improved moment rotation characteristics. Furthermore, the predrilled flange holes were of the same diameter at those holes drilled for bolts and did not therefore add significantly to fabrication costs.

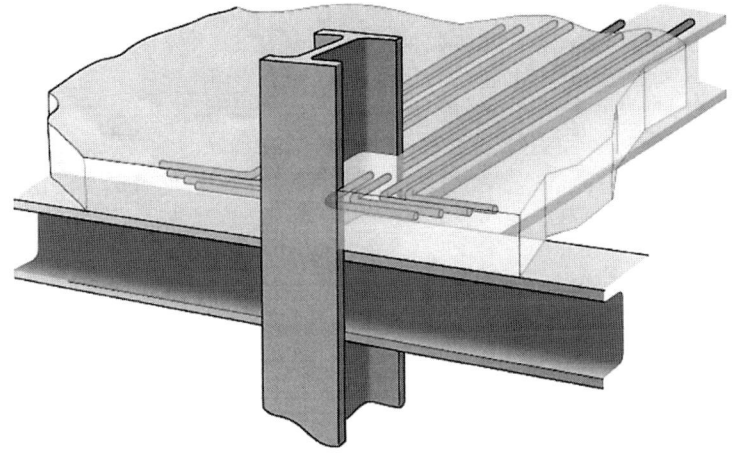

Figure 1: Novel Perimeter Column Connection

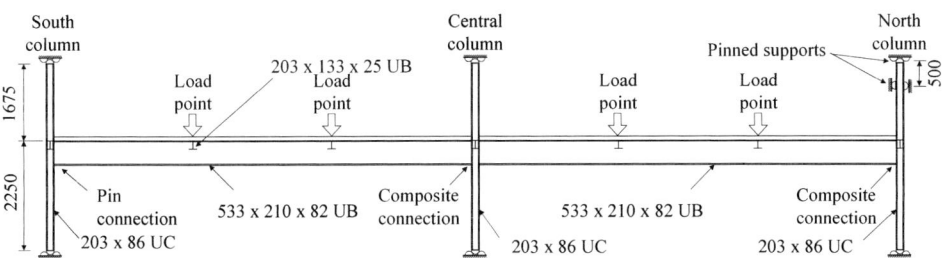

Figure 2: Side Elevation of BRE test frame

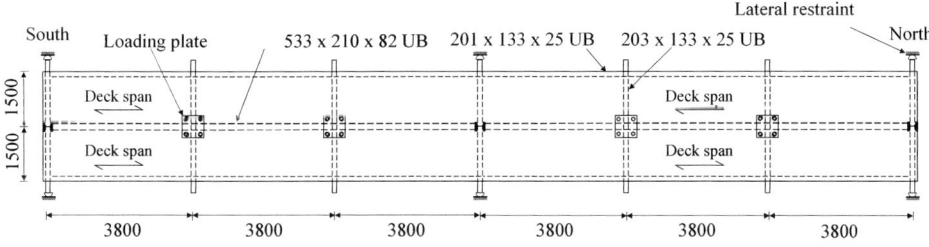

Figure 3: Plan view of test frame

EXPERIMENTAL TESTING

In order to investigate the above-mentioned problems, a full-scale test has been carried out at the Building Research Establishment on a 3.0m wide frame, comprising two bays, each of which were 11.4m long, see Figures 2 and 3. The length of the supporting columns was selected to reflect the span between the points of contra-flexure that occur at mid-height in the columns of multi-storey buildings.

Therefore, the column height reflects a half-storey below and above the slab. The slab width was chosen to be the effective width for the span of the beam i.e. span/4. The reaction frame was designed to provide a simple support to the top and bottom of each of the three supporting columns and consisted of a steel portal frame braced out of plane by two sets of diagonal tension cables.

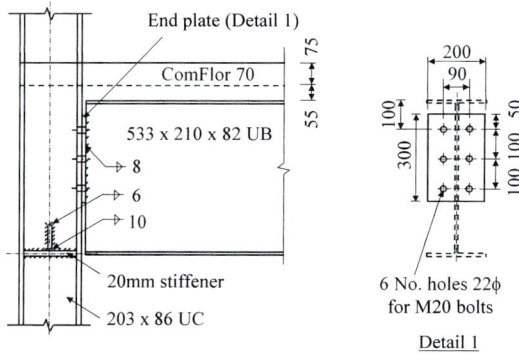

Figure 4: South (pinned) column connection detail

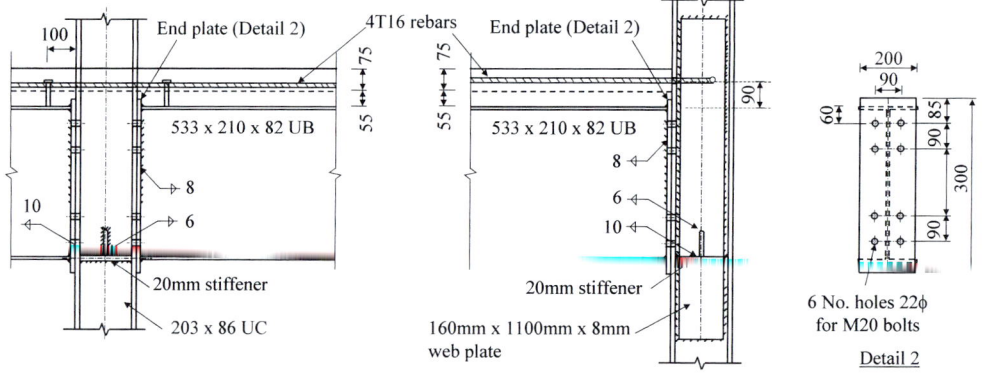

Figure 5: North and central column details

This frame employed standard composite connections at the interior joints and a new form of composite connection at the exterior joint located at the "North" column, see Figure 2. A standard 'green book', BCSA/SCI (1995), pin connection was used at the other exterior joint connecting to the "South" column, see Figure 4. The internal composite connections were designed in accordance with the SCI/BCSA composite connection design guide (1998), with 4No. 16mm ϕ high yield steel reinforcement bars, Figure 5. A 20mm web stiffener was used to prevent buckling of the column. The composite connection at the exterior joint detail, see Figure 5, was similar to the central composite connection, comprising the identical endplate and bolt arrangement together with a total of 4No. 16mm ϕ high yield steel reinforcement bars, which passed through 22mm ϕ holes drilled through the interior flange of the column. The design of the rebar anchorage for the North Column was in accordance with the BS8110 requirements and involved cranking the bars at two points through a total of 90°. Since the connection was unbalanced it was necessary to stiffen the column web to prevent a shear failure. This was achieved with a web plate stiffener applied to one side of the column web, in accordance with the recommended details listed in the SCI/BCSA moment connections design guide (1995). The props were removed before the casting of the slab, which allow the test to simulate the unpropped situation. The arrangement of the reinforcement for all the connections and the slab is shown in detail in Figure 6.

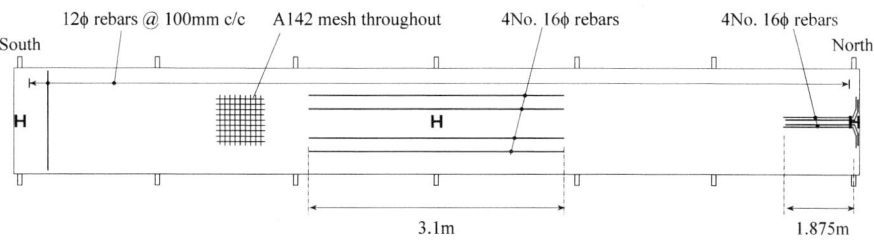

Note, high tensile reinforcement throughout ($f_y = 460N/mm^2$)

Figure 6: Rebar layout

All the experimentally recorded material properties of the test frame are listed in Tables 1, 2 and 3. Grade S275 structural steel was used throughout and ordinary high yield reinforcement bars were used as slab reinforcement. Normal weight C30 concrete was used for the slab. Three steel coupons were taken from the top flange, bottom flange and web of the north beam, south beam and the column. The tensile mill tests were carried out at the Corus (British Steel) Lackenby Mill and were in accordance with the British Specification BS EN 10025. Three of the concrete cubes were water tank cured and six cubes were site cured.

Location	Yield strength (N/mm²)			Ultimate strength (N/mm²)		
	Top	Bottom	Average	Top	Bottom	Average
south beam flange	298	290	294	477	477	477
south beam web	343	-	343	495	-	495
north beam flange	305	300	302	480	480	480
north beam web	303	-	303	499	-	499
column flange	266	268	267	466	466	466
column web	280	-	280	462	-	462

Table 1: Material properties of the beams and columns

Coupons	Yield strength (N/mm²)					Ultimate strength (N/mm²)					% Elongation			
	1	2	3	4	Avg.	1	2	3	4	Avg.	1	2	3	4
Rebar T10	482	488	486	481	484	557	559	558	558	558	27.1	30.9	30.9	28.8
Rebar T16	556	556	556	-	556	644	642	642	-	643	25.6	25.6	23.3	-

Table 2: Material properties of the reinforcement bars

Load number	Mean strength (N/mm^2)	
	Standard cured	Site cured
Mix 1	41	35
Mix 2	43	33

Mix 1: Three cubes standard cured; Six cubes site cured
Mix 2: Three cubes standard cured, Six cubes site cured
Site curing: under wet hessian and polythene sheet adjacent to the composite frame for 28 days
Standard curing: in water at 20°C for 28 days.

Table 3: Concrete cube strengths at 28 days

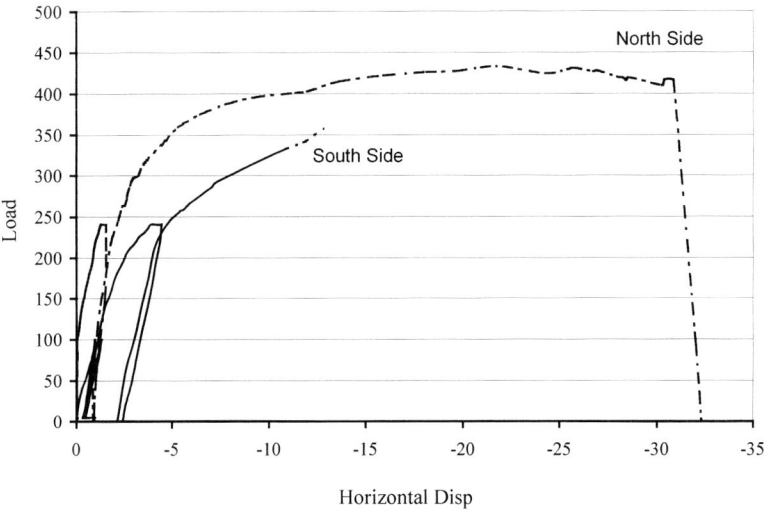

Figure 7: Load vs. horizontal displacement (final displacement recorded for south side was 26mm)

RESULTS

The horizontal displacements recorded at the perimeter beam column connections are shown in Figure 7. The figure demonstrates that significant sideways movement was produced as the bottom flanges of the beams reacted against the columns, causing significant bending moments in the perimeter columns. The moment vs. rotation performance of the connections are shown in Figure 8 and Figure 9, where the rotation is the net rotation of the connection, which does not include the column rotation. The observed behaviour of the connections will now be considered in detail.

The 'nominally pinned' south column connection. Figure 8 shows that this connection behaved as a nominally pinned connection during the construction phase, but subsequently stiffened considerably after the concrete slab was cast. The bending moments generated during rotations of up to 40mrads exceeded those predicted by the conventional design rules of BS5950-1. At rotations of greater than 45mrads the connection strength increased up to bending moments in excess of 300kN.m, as beam rotation caused the bottom flange of the beam began to react against the face of the column. This effect caused considerable sideways movement in the column as shown in Figure 7.

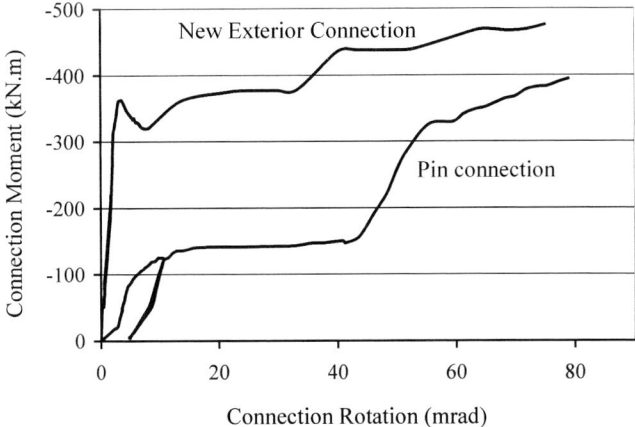

Figure 8: Moment vs. rotation for exterior beam column connections

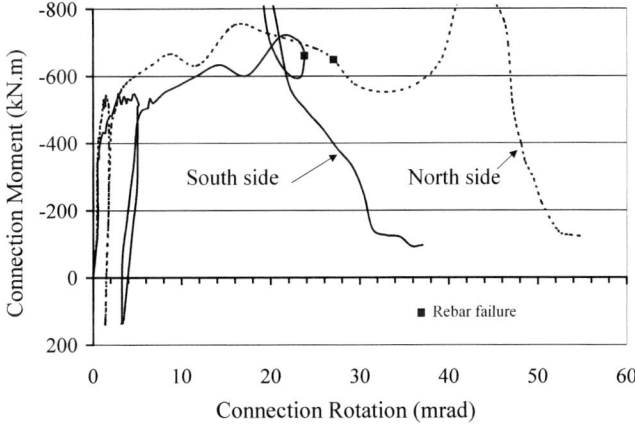

Figure 9: Moment vs. rotation for interior beam column connections

The north column composite connection: Figure 8 shows the strength, ductility and stiffness characteristics of this novel connection detail. The stiffness was lower than the internal connections (Figure 9) although the ductility was improved. The design moment capacity of this connection (calculated in accordance with the composite connections design guide, BCSA/SCI (1998), using the measured material and geometric properties recorded for the test frame) was 406kN.m. Inspection of Figure 8 shows that the test specimen failed to reach this design moment within an acceptable degree of rotation. In addition, end rotation of the beam caused considerable sideways movement of the column, see Figure 7.

The interior composite connections: Figure 9 shows the moment vs. rotation behaviour of the central connections, which clearly demonstrates the strength and stiffness characteristics of the connection.

The connection strength exceeded the design strength of 406kN.m and the south side connection began to fail after a post construction rotation of 23mrads. However, the identical north side connection achieved considerably more rotation before failure and the reason for this difference is the focus of on going consideration. The failure of a rebar initiated the decline in strength.

CONCLUSION

Data has been presented on the strength and ductility of standard composite connections using unpropped construction. In addition, the strength and ductility of a novel form of edge detail has been assessed. The test results show that the novel composite connection located at the perimeter column could generate significant moments, although the strength fell short of the expected design moment. The data presented can be used to assess the applicability of standard composite connections for use with unpropped construction, it can also be used to formulate design formulae for connections utilising the novel form of perimeter connection detail.

ACKNOWLEDGEMENTS

The authors would like to acknowledge the contributions of Dr David Moore of the Building Research Establishment in the UK, and Professor David Anderson of Warwick University, who were collaborators on the research presented herein, which was financed by the Engineering and Physical Sciences Research Council and the DETR.

REFERENCES

Ahmed, B. and Nethercot, D. (1995). Numerical modelling of composite flush end-plate connections. *Journal of Singapore Structural Steel Society* 6:1, 87-102.
Anderson, D.; Najafi, A.A(1994). Performance of composite connections: major axis end plate joints. *Journal of Constructional Steel Research* 31:1, 31-57.
Anderson, D. & Najafi, A. A (1997). Ductile steel-concrete composite joints. *Composite construction - Conventional and Innovative*, Innsbruck, Conference Report.
BCSA and SCI (1991). Joints in simple construction, Vol. 1: design methods. The Steel Construction Institute, Publication No. 205.
BCSA and SCI (1995). Joints in steel construction – moment connections. The Steel Construction Institute, Publication No. 207.
BCSA and SCI (1998). Joints in steel construction – composite connections. The Steel Construction Institute. Publication No. 213.
Brugger, R. (1993). Zur Schubtragfahigkeit von Verbundknoten, *Ph.D Thesis*. University of Innsbruck, Austria.
Davidson, J., Lam, D., Nethercot, D. (1990). Semi-rigid action of composite joints. *The Structural Engineer* 68:24.
Tschemmernegg, F., Huber, G., Pavlov (1995). Tension Region in the Panel Zone of a Composite Joint. *Paper T4, COST-C1/ECCS TC11 Drafting Group for Composite Connections*, University of Innsbruck, Austria.

ANALYSIS OF STEEL AND COMPOSITE BRACED FRAMES WITH SEMI-RIGID JOINTS

A. Kozlowski[1,2]

[1] Department of Civil Engineering, University of the Beira Interior,
Covilha, PT 6200, PORTUGAL
[2] Rzeszow University of Technology,
Rzeszow, PL 35959, POLAND

ABSTRACT

One of the recommended structural systems for multi-storey buildings is such, where composite technologies are used, e.g. steel skeleton combined with concrete slabs. Such a system has the following advantages: it allows to reduce the construction time, composite members can be made slender and at the same time stiff and strong, semi-rigid joints can be used which leads to cheaper connections and reduces steel consumption. The objective of the paper is to present an analysis of savings in steel consumptions, resulting from the usage of semi-rigid joints. To analyze influence of joint parameters on frame economy it is not enough to focus on moment distribution, but also on the results of frame elements dimensioning, according to limit states. In the paper, results of analysis of few steel and composite braced frames are presented. During analysis, frame structural elements (beam and columns) sections were changed according to changes of internal forced caused by joint stiffness changes. Efficiency of composite steel-concrete floors in the framework of buildings as well as economy confirmation of using semi-rigid joints in braced steel and composite structures have been demonstrated. Comparison of the bare steel frames with the frames with composite beams of the same joint stiffness shows that composite frames are more economical of 15 – 25 %. Other advantage of composite floor is lower beam sections, and consequently shorter column and smaller global volume of the building. Use of semi-rigid joint leads to additional saving in steel from 5 to 12 %.

KEYWORDS

Semi-rigid joint, structural analysis, braced frame, optimal design, composite floor.

INTRODUCTION

It is well known, that internal forces, especially bending moments distribution in frames, change due to change of joint stiffness. Application of joints with certain stiffness, instead of "pinned" connection in braced frames leads to lower values of the beam mid-span moments (which govern member section size), and to increase of the end beam moment. This may results in profitable equalizing the values of bending moments along the beams. Many results of semi-continuous frames analysis were published, for example, Christopher & Bjorhovde (1998), Hasan & Kishi & Chen & Komuro (1997), but most of them consider only internal forces distribution change. Design process consists not only of global analysis but also of member dimensioning according to codes regulation. Not always change in moment values in the frame involves also change in member section size, because cross-sections of standard steel members have discrete values. To analyze influence of joint stiffness on frame economy it is not enough to focus on moment distribution, but also on the results of frame elements dimensioning, according to limit states. When, after checking of limit states, member sizes have to be changed, also the ratio of the joint to member stiffness changes, what requires repeating the global analysis.

Framed structures are used mainly for residential and office buildings, where concrete slabs must be applied. One of the recommended structural systems for multi-storey buildings is such, where composite technologies are used, e.g. steel skeleton combined with the concrete slabs. Composite steel-concrete construction is very effective and attractive to designers because of its greater stiffness and resistance capacity compared to non-composite construction. Table 1 presents the comparison of the moment resistance and stiffness of steel and composite beams.

TABLE 1

COMPARISON OF THE MOMENT RESISTANCE AND STIFFNESS OF THE STEEL AND COMPOSITE BEAMS

Steel beam IPE	Moment resistance of the steel beam M_S [kNm]	Stiffness of the steel beam $E_S J_S$ [kNm^2]	Moment resistance of the composite beam * M_C [kNm]	Stiffness of the composite beam * B_C [kNm^2]	$\dfrac{M_C}{M_S}$	$\dfrac{B_C}{E_S J_S}$	Steel substitute of the composite beam IPE	Steel saving [%]
240	69,7	7970	209,9	35500	3,01	4,45	360	46,0
270	92,3	11800	257,7	46700	2,81	4,18	400	45,5
300	119,7	17100	314,7	67600	2,63	3,95	450	45,4
360	194,4	33300	456,8	111000	2,35	3,33	500	37,3
400	249,4	47400	556,1	130000	2,21	2,87	550	37,4
450	322,5	69200	687,0	176000	2,13	2,54	600	36,4
* concrete plate of the 150 mm thickness of concrete C20; profile steel grade: S 235								

Usage of composite structure enables to reduce the height of beams and the depth of floor structure. Further decrease in composite beam section can be obtained by appropriate design of beam to column connections. In so-called "composite connection", resistance to hogging moment is provided by properly anchored tension reinforcement, placed in concrete slab, together with steel part of beam-to-column joints.

The aim of the paper is to present results of the analysis of two steel and composite braced frames and on this basis to point out economical solution of such a structures. During analysis, frame structural elements (beam and columns) sections were changed according to changes of internal forced caused by joint stiffness changes.

FRAMES ANALYSIS

Methods of analysis

As a matter of fact, the presented analysis is an optimization task in which the objective function is the frame members volume (mass). The constraints are: stiffness of the joints, type of beam and column profiles and code requirements. Analyses were conducted on the example of two frames shown in Figure 1. Results of other frames analysis are included in Kozlowski (1999).

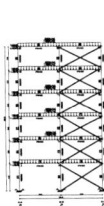
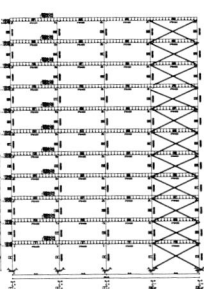

Figure 1: Analyzed frames

During analysis of each frame the stiffness of the semi-rigid beam-to-column connections were changed. The following values of the joint initial stiffness $S_{j.ini}$ were applied:
- 0; ideally pinned joint – conventional approach,
- few values from $0{,}5\dfrac{EJ_b}{L_b}$ to $10\dfrac{EJ_b}{L_b}$.

Analyses were conducted using linear spring joint model. Non-linear character of M-φ curves was taken into account by the method of equivalent secant stiffness. Secant joint stiffness S_{wc} was calculated using formula:

$$S_{wc} = \frac{S_{j.ini}}{\eta} \qquad (1)$$

where:
    E – Young modulus,
    J_b – moment of inertia of the beam section,
    L_b – length of the beam span,
    η - reduction factor, proposed by Kozlowski (2000).

Also "real" values of the column bases stiffness were adapted in the analysis. These column bases stiffness was taken from Wald (1995). Beam spans of 6,0 m, first story height of 4,2 m and a height of 3,6 m for a remaining stories were chosen.

Only one load combination was considered, involving vertical gravity loading and wind. The dead load was 30 kN/m, live load 12 kN/m, wind load was represented by horizontal concentrated forces of magnitude 11,8 kN, applied at each floor level, and of magnitude 5,9 kN applied at the roof level.

Steel grade S235 (Fe 360) was used for each structural members. Beams were designed using IPE section and columns using HEB profiles. Moment resistance and stiffness of composite floor beams were calculated according to the Polish Standard PN-82/B-03300, for the 150 mm t concrete plate with the reinforcement ratio 0,9 % in the negative moment zone.

During global analysis of the framed structures with composite steel-concrete floors the major difficulty is in proper evaluation of the composite beam member stiffness. As a matter of fact, composite beam stiffness is not constant along beam length. In the zone of positive span moment, this stiffness has the biggest value calculated according to "effective width" method. In the zone of negative moment, beam stiffness is dramatically reduced due to concrete slab cracking. In the present analysis, equivalent beam stiffness $S_{cp.eq}$ was applied, Leon & Hoffman (1996):

$$S_{cp.eq} = 0,6 S_{b.span} + 0,4 S_{b.neg} \qquad (2)$$

where:
$S_{b.span}$ – stiffness of the composite beam in the positive moment zone,
$S_{b.neg}$ – stiffness of the composite beam subjected to negative moment.

During analysis, frame structural elements (beam and columns) sections were changed according to changes of internal forced caused by joint stiffness changes. To satisfy design practice, the beam members had the same section in the whole frame, sections of column were changed every three storey.

Members of the steel frame were dimensioned according to the Polish Code PN-90/B-03200, (in nature very similar to EC3). Lateral - torsional buckling of the beams was neglected. Buckling length of the column was calculated with the buckling coefficients: $\mu_y = 0,8$ (in the frame plane) and $\mu_z = 1,0$ (out of plane). Limit for the top-story lateral drift was taken as H/500, where H is the total height of the frame.

The analyses were conducted using ROBOT V6 package. "Compatibility joint" option was applied, which allows linear joint model to be implemented. Second order P-Δ analysis was applied. At each stage of the analysis, the ultimate limit state (ULS) and serviceability limit state (SLS) for whole structure and each member groups were checked. In the case when any of the limit state was not fulfilled, member sections and also joint stiffness related to them were changed and the global analysis was repeated. These required many recalculation of each frame.

Results of the frames analysis

Results of the analysis of two braced frames are presented graphically in Figures 2 and 3, which show variation of the span moment M_p, moments in beam ends M_w, masses of the all steel frame m_s and composite frame m_z members, versus the secant joint stiffness S_{wc} variation. The figures are accompanied by a table, which presents:
- sections of the beam and column obtained from the ULS check,
- saving of the steel (in %) in comparison to traditional solution, i.e. frame with pinned connections,
- joint label, whose coordinates refer to their secant stiffness.
In the rows concerning composite frames, additional values of steel saving are presented, taken from the comparison to steel frame with the same joint stiffness (B) and to steel frame with pinned joints (marked as *).
Typical steel and composite joints used in the analysis are shown in Figure 4 and 5 respectively. Three values for each joint stiffness were considered: lower (D), middle (S) and upper (G). Joint labels used in table below frame analysis result have the following meaning: digit is the number of joint in Figures 4 or 5, letter is the joint parameter. For example, 7D means unstiffened flush end plate of lower parameters. In this way, it is possible to point out such joint, which gives the best economical solution.

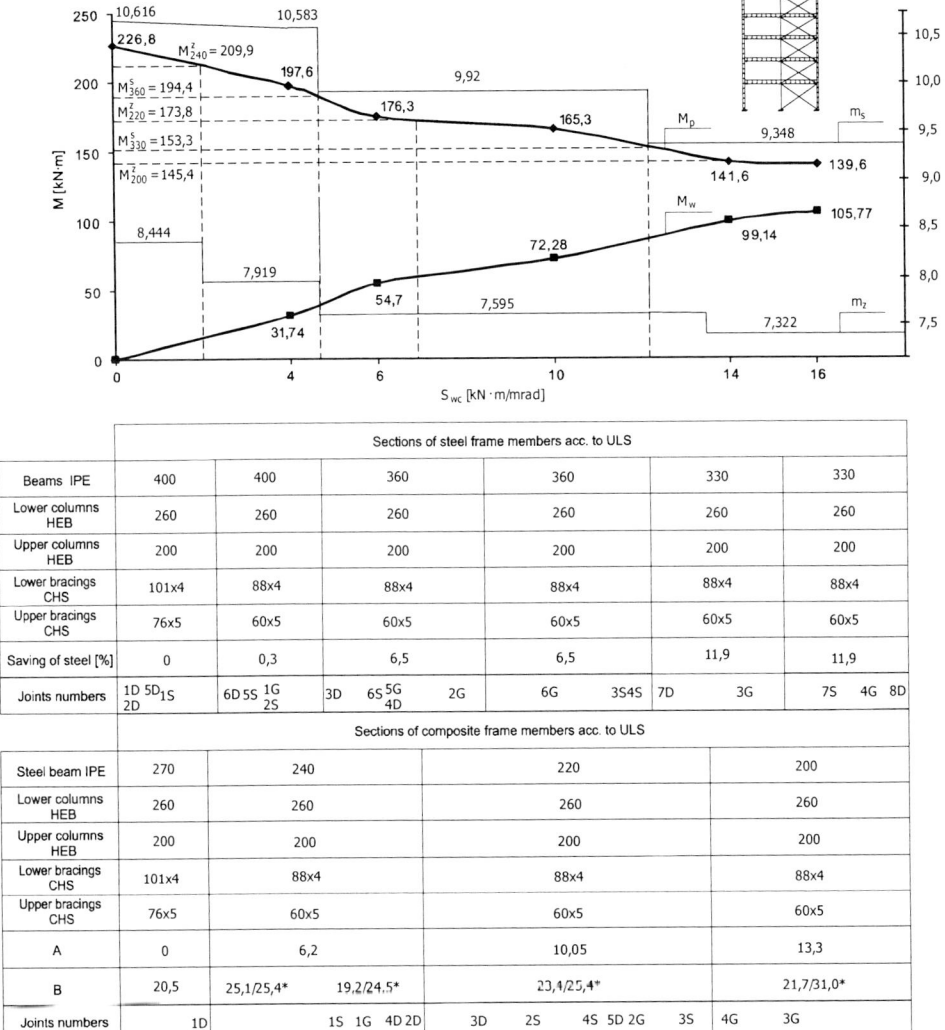

Figure 2: Results of two-bay six-storey frame analysis

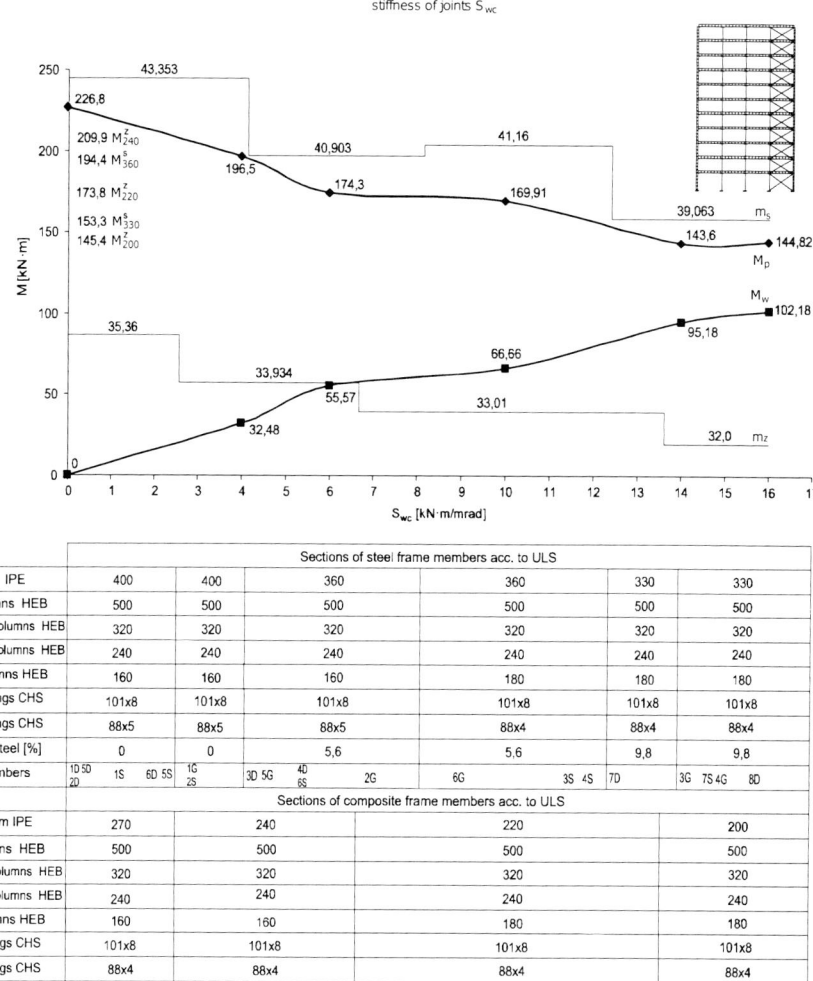

Figure 3: Results of four-bay eleven-storey frame

The following conclusions can be drawn from braced frame analysis:
a) increasing joint stiffness leads to smaller beam section, due to smaller value of the maximum moment (in span), governing the beam section sizing. This is the main source of the steel saving. Simultaneously, moments in the beam ends increase and are transmitted to the columns, but it does not

affect column sections because governing case for column dimensioning is out of plane buckling of internal column,

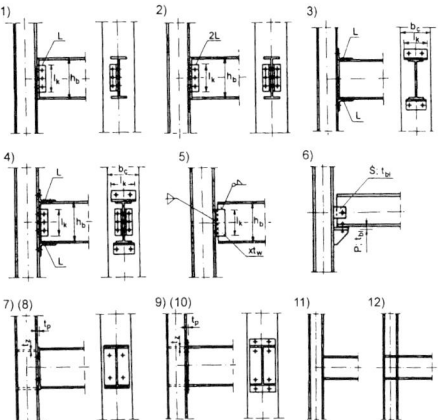

Figure 4: Steel joints applied in the analysis

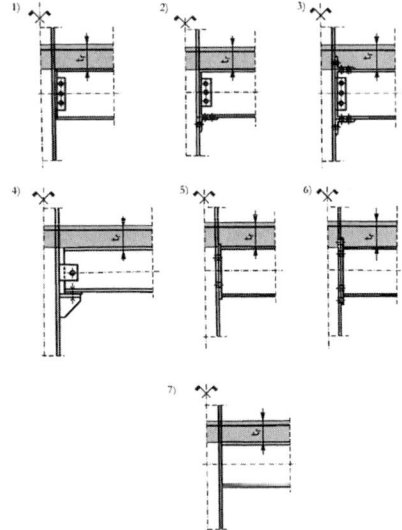

Figure 5: Composite joints used in the analysis

b) it is possible to equalize moments in the mid-span and beam ends, what can give the smallest beam section. This requires higher joint stiffness and consequently, higher cost of the joint. This can reduce profit from smaller beam section,
c) the higher joint stiffness the smaller sections of bracing,
d) in steel frames, the highest economy is achieved (steel saving up to 11,9 %), for the following joints: unstiffened flush end plate (type 7), top and seat angle (type 3) and top and seat with web angles (type 4). In this case, beam section is reduced from IPE 400 obtained in classical design (pinned joints) to IPE

330. Even when "simple" joint is used, for example web angle (type 2), beam section is reduced to IPE 360,
- in composite frames, saving of steel obtained in semi-rigid design is from 4,0 to 13,3 %, in comparison to pinned joint solution,
- comparison of the steel and composite frame analysis results shows that composite frames are much more efficient. Saving of steel is from 15,5 to 25,1 % when comparison is made between steel and composite frames subjected to the same loading and of the same joint stiffness. When one compares composite semi-rigid frame with bare steel frame with pinned connections (conventional solution) this steel saving can be even 31 %.

CONCLUSIONS

Efficiency of composite steel-concrete floors in the framework of buildings as well as economy confirmation of using semi-rigid joints in braced steel and composite structures have been demonstrated. Comparison of the bare steel frames with frames with composite beams of the same joint stiffness shows that composite frames are more economical by 15,5 to 25,1 %. Another advantage is lower beam sections in composite floor, shorter column and smaller global volume of the whole building. Use of semi-rigid joint leads to additional saving in steel from 5 to 12 %.

In this analysis only steel consumption was considered. Next step is to take into account the cost of joints and member fabrication and erection, like in Brognoli & Gelfi & Zandonini & Zanella (1998). Other possibility of including the joint costs in economy analysis is to use the sensitivity analysis, Xu Lei (2000).

The presented results can encourage designers to apply the semi-rigid design concept. Theoretical solution of such a design and software are now available and presented, for example, Chen & Toma (1994) or Chan & Chui (2000). The option of the semi-rigid joints is also started to be included in the increasing number of commercial packages for global analysis.

REFERENCES

Brognoli M., Gelfi P., Zandonini R. and Zanella M. (1998). Optimal Design of Semi-Rigid Braced Frames Via Knowledge-Based Approach. *Journal of the Constructional Steel Research*, **46:1-3.**
Chan S.L. and Chui P.P.T. (2000). *Non-linear Static and Cyclic Analysis of Steel Frames with Semi-Rigid Connections,* Elsevier, Oxford, UK.
Chen W.F. and Toma S. (1994). *Advanced Analysis of Steel Frames, Theory, Software and Application,* CRS Press, London, UK.
Christopher J.E and Bjorhovde R. (1998). Response Characteristics of Frames with Semi--Rigid Connections. *Journal of the Constructional Steel Research*, **46:1-3.**
Hasan R., Kishi N., Chen W.F. and Komuro M. (1997). Evaluation of Rigidity of Extended End-Plate Connections. *Journal of Structural Engineering*, **123:2.**
Kozlowski A. (1999). *Shaping of the steel and composite skeletons with semi-rigid joints.* Monograph (in Polish). Rzeszow University of Technology, Rzeszow, Poland.
Kozlowski A. (2000). Secant stiffness of semi-rigid joints for braced frames. *19th Czech and Slovak International Conference.* Strbskie Pleso, Slovakia.
Leon R.T. and Hoffman J.J. (1996). Plastic Design of Semi-Rigid Frames. *in: Connections in Steel Structures III: Behaviour, Strength and Design,* Elsevier, London, UK.
Wald F. (1995): *Column Bases.* eske Vysok U eni Technick , Praha, Czech Rep.
Xu Lei. (2000). Design Optimization of Semi-Rigid Steel Frames, *in: Practical Analysis for Semi-Rigid Frame Design,* World Scientific, Singapore.

NUMERICAL EVALUATION OF THE DUCTILITY OF A BOLTED T-STUB CONNECTION

Ana M. Girão Coelho[1] and Luís Simões da Silva[2]

[1] Delft University of Technology, Civil Engineering – Steel and Timber Structures,
PO Box 5048, 2600 GA Delft, THE NETHERLANDS (a.m.girao@citg.tudelft.nl)
[2] Department of Civil Engineering, Universidade de Coimbra,
Polo II, Pinhal de Marrocos, P-3030 Coimbra, PORTUGAL (luisss@dec.uc.pt)

ABSTRACT

This paper describes a numerical model for the assessment of the nonlinear force-deformation response of bolted T-stub connections and its ductility in particular. The results of existing experimental work are used to calibrate this model. A parametric study is also presented to determine the effect of the variation of some geometrical connection parameters on the overall behaviour.

KEYWORDS

Component method, Deformation capacity, Ductility, Finite element method, Resistance, Stiffness, T-stub model

INTRODUCTION

The prediction of the overall moment-rotation response of bare steel and steel-concrete composite joints by means of the component method requires the evaluation of the full force-deformation (F-Δ) curve of each active joint-component. In the particular case of bolted connections, many joint-components are modelled with equivalent T-stubs. Thus, the assessment of the F-Δ behaviour of bolted T-stub connections up to failure is of utmost importance.

The F-Δ curve is intrinsically nonlinear. The characterization of such complex behaviour is not easily open to simple analytical formulations, usually requiring a numerical finite element (FE) analysis. This paper presents a three-dimensional FE model that accounts for all the geometrical and material nonlinearities. Contact phenomena and fracture of the flange or bolt material are also assessed. The calibration of the FE model is based on experimental test programs available in the literature.

Because ductility is such an important characteristic of connection performance, a parametric study is also introduced to determine the effect of the variation of some geometrical connection parameters, such as the pitch of the bolts, the gauge and the bolt diameter, on the overall response of the T-stub.

CALIBRATION OF A FE MODEL FOR THE EVALUATION OF THE NONLINEAR FORCE-DEFORMATION RESPONSE OF A BOLTED T-STUB CONNECTION

Geometry of the Model

The calibration of the FE model is based on the experimental test program carried by Bursi and Jaspart (1997). The specimen T1 is obtained from an IPE300 beam profile and corresponds to two T-elements connected through the flanges by means of two snug-tightened M12 bolt-rows. Owing to the geometrical symmetry only one eighth of the T-stub is modelled, adopting the adequate boundary conditions (Fig. 1). Whilst the xy and yz planes are geometrical planes of symmetry, in the xz plane there is no symmetry since the bolt elongation behaviour is not symmetrical along the y direction. To comply with the requirements for symmetry in this plane, an 'equivalent bolt' is then defined in such a way that its geometrical stiffness is identical to that of the actual bolt (Bursi and Jaspart, 1997; Wanzek and Gebbeken, 1999).

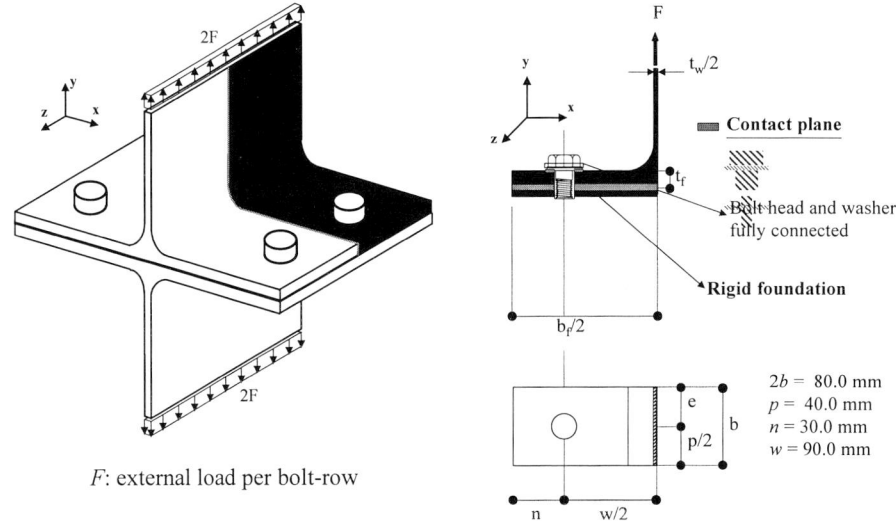

a) T-stub representation   b) T-stub model: one eighth of the specimen
Figure 1: Finite element geometry model assuming symmetry in the xy, xz and yz planes

Mesh Description

The modelling of the T-stub connection is performed with the commercial FE package LUSAS (2000) by means of three-dimensional elements, solids and compatible joints. In particular, the solid elements are hexahedral eight-noded 'enhanced' strain bricks and are used to model the continuum (Fig. 2) (LUSAS, 2000; Simo and Rifai, 1990). These elements exhibit improved accuracy in nonlinear problems involving plasticity and contact phenomenon when compared to higher order elements since they allow for a better representation of the discontinuities at element edges and of the strain field. The joint elements are employed in the simulation of element contact (LUSAS, 2000). These connect two adjacent nodes by means of extensional springs with adequate properties. Both element types have three degrees-of-freedom per node (u, v and w) and are numerically integrated.

The kinematic description of the solid elements in nonlinear geometrical analysis is based on the updated Lagrangian formulation that accounts for large displacements and moderately large strains.

For the material nonlinearity, an elastoplastic constitutive law based on the Von Mises yield criterion is adopted. A plastic potential defines the flow rule. The constitutive model is integrated by means of the explicit forward Euler algorithm (Owen and Hinton, 1980). For this algorithm the hardening data and direction of plastic flow are evaluated at the point at which the elastic stress increment crosses the yield surface.

With respect to the joint elements, the local element stiffness matrix is formulated directly from user input stiffness coefficients and is subsequently transformed to the global Cartesian system. These elements possess no geometrically nonlinear terms in their formulation. However, they may be used in second-order nonlinear analysis but will remain geometrically linear.

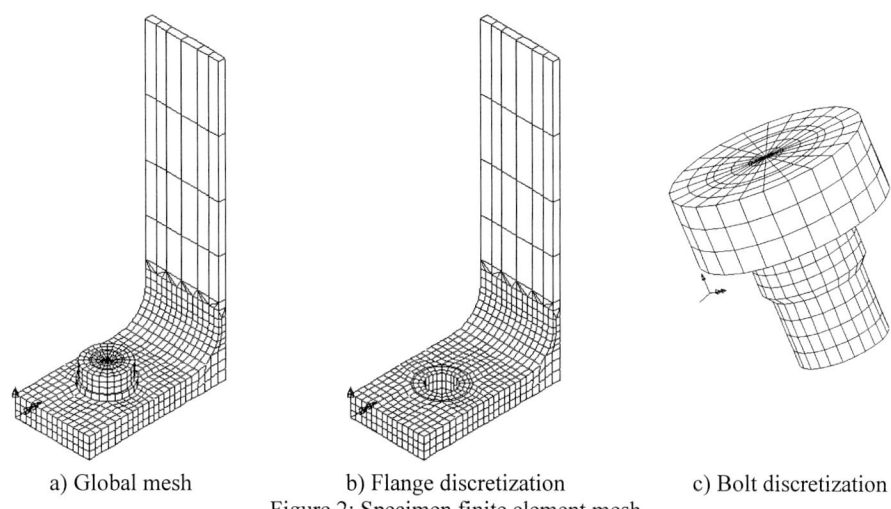

a) Global mesh   b) Flange discretization   c) Bolt discretization
Figure 2: Specimen finite element mesh

Loading and Boundary Conditions

The nodes in the symmetry planes xy and yz are fixed with symmetric geometrical boundary conditions: in plane xy, the nodes are fixed in the z direction on one side and in plane yz in the x direction, along the back of the half of the web. The 'symmetry' plane xz between the two flanges is modelled by contact elements on a rigid foundation (Fig. 1b). The nodes on this rigid base are fully restrained. Also, complying with geometrical symmetry, the bottom bolt nodes are fixed in the y direction. The interface boundary between the flange and washer is represented by means of contact elements (Fig. 1b). In order to reduce the number of contact planes the bolt head or nut and the washer are assumed fully connected.

Regarding the interface boundary conditions, a nonlinear frictional model is implemented. The sliding and sticking conditions are reproduced with the classic isotropic Coulomb friction law. No friction has been assumed between the flanges interface because of the T-elements symmetric behaviour. For the flange-washer interfaces a friction coefficient, μ, of 0.25 is adopted.

Loads are applied to the specimen in a displacement-control fashion that enforces a better conditioning of the tangent stiffness matrix when compared to the classical load-control procedure. A uniform total prescribed displacement of 0.1 mm is hence applied at the top of the upper T-element in the positive y direction. In the nonlinear analysis, the total load factor is increased from 1.0 to collapse.

Material Properties

For good correlation with experimental results, the full actual stress-strain relationship of materials must be adopted in the numerical simulation. For both models a rate and temperature independent plasticity law with hardening is used for the T-elements and the high strength bolt. The material behaviour of the rigid foundation is described by a linear elastic curve, with $E = 1.0 \times 10^{15}$ MPa and $\upsilon = 0.45$. The onset of plastic deformation is specified under the Von Mises yield criterion, as already explained. The constitutive laws are reproduced with a piecewise linear model (Bursi and Jaspart, 1997). The nominal stress-conventional strain (σ-ε) relationship is illustrated in Fig. 3a. These quantities are measured in common tensile coupon tests and are defined as the axial force per unit initial area and the change in length per unit initial length, respectively. The modelling of the large strain behaviour, however, requires the use of a true stress-natural strain measure for the definition of the uniaxial material response, instead of the classic nominal constitutive law (Bathe and Wilson, 1976). The natural strain, ε_N, and the true stress, σ_N, are defined with respect to the current length and cross-sectional area of the coupon. They are related to the conventional quantities by means of the following relationships:

$$\varepsilon_N = \ln(1+\varepsilon) \qquad \sigma_N = \sigma(1+\varepsilon) \qquad (1)$$

which allow for the conversion of the nominal law into the σ_N - ε_N law (Fig. 3b).

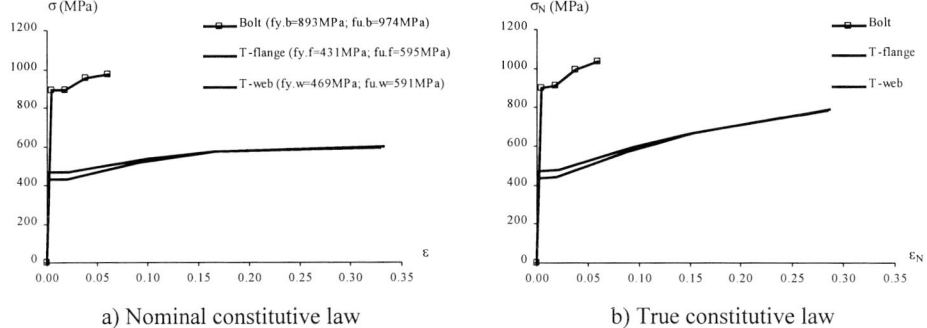

a) Nominal constitutive law b) True constitutive law
Figure 3: Material characteristics for specimen T1

Nonlinear Analysis

The T-stub response is traced with a load stepping procedure. Within each loading step the procedure is based on the standard Newton-Raphson method. This iterative procedure stops when the convergence criteria are met. The displacement-based criteria as well as the work norm are used as convergence criteria. Their choice has to be careful, as they have to be tight enough to guarantee sufficient accuracy but slack enough to reduce computation time. For predominantly materially nonlinear problems, in which high local residuals have to be tolerated, slack convergence criteria are usually more effective. For this particular problem, relatively slack tolerance norms have shown to be accurate enough with significantly reduced time computation. The termination criterion is the maximum load factor.

Numerical Results and Failure Criterion

The most significant characteristic describing the overall behaviour of the model is the F-Δ curve. The numerical response is compared with experimental results in Fig. 4a. The proposed model behaves close to the test results for the whole range. The experimental curve includes the web deformation that increases the flange total deformation. The real F-Δ response of the T-stub does not include the web contribution (Fig. 4a). Fig. 4b illustrates the evolution of the ratios of prying and bolt forces with the total applied load, per bolt row, showing an increase of such ratios with plastic straining in the flange.

The deformation capacity of a bolted T-stub connection primarily depends on the plate strength/bolt strength ratio and eventually is governed by bolt fracture or cracking of the plate material. In both situations, the modelling of the failure condition can be ascertained by assuming that cracking occurs when the ultimate strain, ε_u, is attained, either at the bolt or at the T-element critical sections (Fig. 3b). ε_u is then compared with the maximum principal strain, ε_{11}, at the relevant sections (Girão Coelho et al., 2002).

For specimen T1 and in line with the experimental observations, the collapse mode for this specimen involves significant flange yielding and inelastic phenomena in the bolts (Bursi and Jaspart, 1997). Under the above failure criterion, the ultimate conditions are governed by bolt fracture and occur for a global deformation of 7.91mm (Fig. 4a).

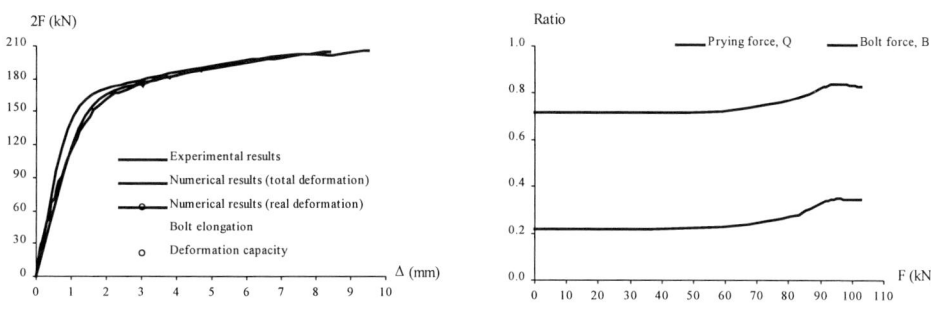

a) Load-deformation behaviour   b) Bolt and prying force
Figure 4: Numerical results for specimen T1

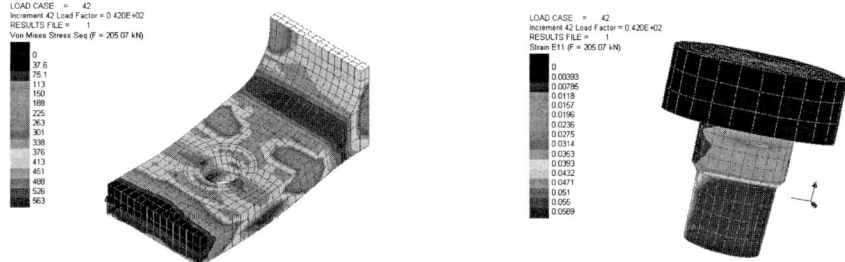

a) Von Mises equivalent stresses in the T-element flange ($f_{y,f}$ = 431 MPa)   b) Maximum principal strain ε_{11} in the bolt at collapse (ε_u = 5.89%)

Figure 5: Stress and strain contours in the T-stub at collapse (2F = 205.07 kN; Δ_{max} = 7.91 mm)

PARAMETRIC STUDY

Generality

The parametric study presented below aims at the analysis of the influence of two major parameters over the T-stub F-Δ characteristics: the connection geometry and the bolt.

Influence of Geometric Parameters

For the assessment of the influence of some key geometrical parameters over the F-Δ behaviour, a parametric study is now conducted. The main parameters accounted for in this analysis are the gauge (w), the pitch (p) and edge (e) distances. Table 1 summarizes the geometry for the tested specimens.

TABLE 1
GEOMETRICAL CHARACTERISTICS OF THE VARIOUS SPECIMENS

	Profile	Bolt	t_f (mm)	w (mm)	p (mm)	e (mm)	n (mm)
T1	IPE300	M12	10.7	90.0	40.0	20.0	30.0
P2	IPE300	M12	10.7	100.0	40.0	20.0	25.0
P3	IPE300	M12	10.7	80.0	40.0	20.0	35.0
P4	IPE300	M12	10.7	90.0	30.0	20.0	30.0
P8	IPE300	M12	10.7	90.0	80.0	20.0	30.0
E2_M12	IPE300	M12	10.7	90.0	70.0	35.0	30.0

Gauge distance (w)

Fig. 6a compares the numerical results for specimens T1, P2 and P3. In terms of the F-Δ response, the main conclusion to be drawn is that the increase of the gauge distance leads to smaller axial resistance and stiffness but improves the ductility of the connection, as also corroborated with experimental evidence (Swanson and Leon, 2000). For the three cases, the plastic condition is governed by flange yielding whilst the collapse condition is determined by bolt failure. Regarding the bolt behaviour, the three specimens yield identical results, as illustrated in Fig. 6a.

Pitch of the bolts (p)

The bolt-pitch variation is now analysed (specimens P4 and P8). As the pitch increases, the connection resistance also increases but the deformation capacity can be significantly reduced (Fig. 6b). The plastic collapse mechanism for specimens T1 and P4 is characterized by the formation of four plastic hinges: two hinges are located at the bolt axes, due to the bending moment caused by the prying forces and the other two hinges are located at the flange-to-web connection. Specimen P8 is governed by a type-2 plastic collapse mode, which involves the formation of two plastic hinges at the flange-to-web connection and the yielding of the bolts. The three specimens ultimate conditions are determined by bolt failure. The influence of the bolt on the overall response is similar in all cases (Fig. 6b).

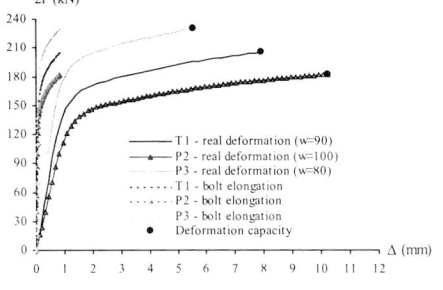
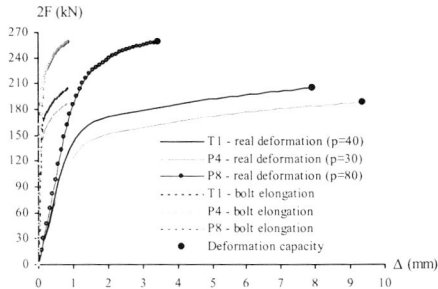

a) Gauge distance effect   b) Pitch distance effect

Figure 6: Assessment of the gauge and pitch distances effect over the global F-Δ response

Edge distance (e)

For specimen E2_M12 both edge and pitch distances are different from specimen T1. This example is designed according to Eurocode 3 (2000) to accommodate different bolt diameters (M16 and M20). The deformation capacity of this connection is rather limited and is governed by bolt fracture (Fig. 7).

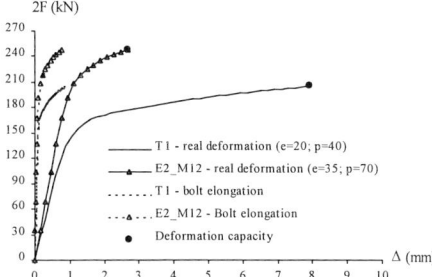

Figure 7: Assessment of the pitch and edge distance effect over the global F-Δ response

Influence of the Bolt

M12 bolts are used to fasten the T-stub connection T1. These bolts meet the requirements of DIN 6914 (short thread) and exhibit a nonlinear behaviour. In order to assess the effect of the bolt on the overall F-Δ response, the same T-stub configuration is analysed assuming elastic bolt behaviour (Fig. 8a). Up to the knee-range the bolt behaviour does not influence the response. The main effect of the bolt is evidenced in the post-limit F-Δ behaviour. Naturally, the ductility of the connection is now governed by fracture of the flange since the bolt behaves elastically until collapse. By assuming a long-threaded bolt M12 (DIN 933), the F-Δ characteristics barely change but show improved deformation capacity when compared to the short-threaded example (Fig. 8b). For geometry E2 (Table 1), where the bolt M12 is replaced with a bolt M16, the deformation capacity is greatly improved, as shown in Fig. 9. Also, both resistance and stiffness increase. Yet, even for this case, the bolt determines the collapse.

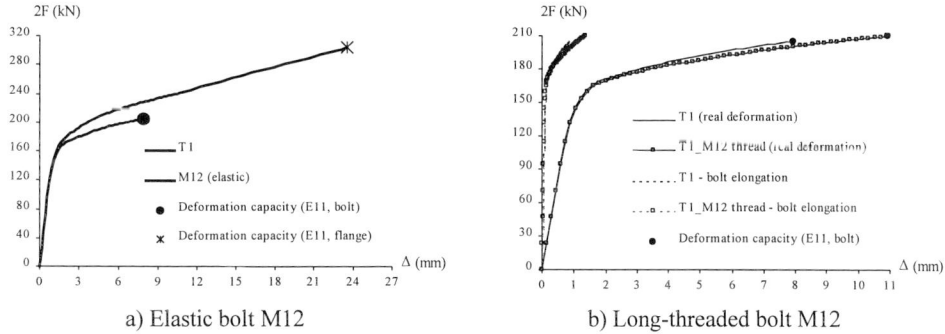

a) Elastic bolt M12   b) Long-threaded bolt M12
Figure 8: Influence of the bolt on the global F-Δ response

CONCLUSIONS

A reliable FE model for the assessment of the F-Δ response of a bolted T-stub has been proposed in this work. This model has led to some general conclusions to understand the influence of some key connection parameters on the overall behaviour. These are summarized in Table 2.

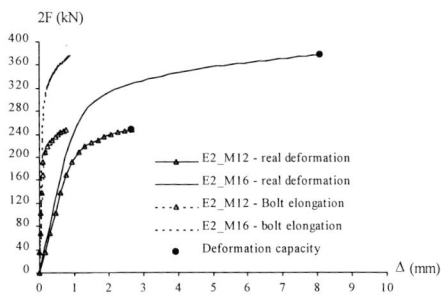

Figure 9: Effect of the bolt diameter on the global F-Δ response

TABLE 2
GENERAL TENDENCIES OF THE F-Δ BEHAVIOUR

	Axial strength		Axial stiffness		Deformation capacity, Δ_{max}
	Plastic resistance, F_{Rd}	Ultimate resistance, F_u	Elastic stiffness, $k_{e.0}$	Post-limit stiffness, $k_{pl.0}$	
	Connection geometry				
Gauge, w	$w\uparrow \Rightarrow F_{Rd}\downarrow$	$w\uparrow \Rightarrow F_u\downarrow$	$w\uparrow \Rightarrow k_{e.0}\downarrow$	$w\uparrow \Rightarrow k_{pl.0}\downarrow$	$w\uparrow \Rightarrow \Delta_{max}\uparrow$
Pitch, p	$p\uparrow \Rightarrow F_{Rd}\uparrow$	$p\uparrow \Rightarrow F_u\uparrow$	$p\uparrow \Rightarrow k_{e.0}\uparrow$	$p\uparrow \Rightarrow k_{pl.0}\uparrow$	$p\uparrow \Rightarrow \Delta_{max}\downarrow$
Edge distance, e	$e\uparrow \Rightarrow F_{Rd}\uparrow$	$e\uparrow \Rightarrow F_u\uparrow$	$e\uparrow \Rightarrow k_{e.0}\uparrow$	$e\uparrow \Rightarrow k_{pl.0}\uparrow$	$e\uparrow \Rightarrow \Delta_{max}\downarrow$
	Bolt characteristics				
Diameter, ϕ	$\phi\uparrow \Rightarrow F_{Rd}\uparrow$	$\phi\uparrow \Rightarrow F_u\uparrow$	$\phi\uparrow \Rightarrow k_{e.0}\uparrow$	$\phi\uparrow \Rightarrow k_{pl.0}\downarrow$	$\phi\uparrow \Rightarrow \Delta_{max}\uparrow$
Bolt threaded length, L_t	No influence				$L_t\uparrow \Rightarrow \Delta_{max}\uparrow$

REFERENCES

Bathe KJ and Wilson EL (1976). *Numerical Methods in Finite Element Analysis*. Prentice-Hall, Englewood Cliffs, New Jersey, USA.

Bursi OS and Jaspart JP (1997). Benchmarks for Finite Element Modelling of Bolted Steel Connections. *Journal of Constructional Steel Research*, **43:1**, 17-42.

CEN EUROCODE 3, prEN 1993-1-8 (2001). 200x, Part 1.8: Design of Joints, Eurocode 3: Design of Steel Structures, Draft 3, October 2001. CEN, European Committee for Standardization, Brussels.

Girão Coelho AM, Bijlaard F and Simões da Silva L (2002). On the Deformation Capacity of Beam-to-Column Bolted Connections. ECCS, Document ECCS-TWG 10.2-02-003.

LUSAS 13.3 (2000). *Theory Manual*. FE Analysis Ltd, Version 13.3, Surrey, UK.

Owen DRJ and Hinton E (1980). *Finite Elements in Plasticity, Theory and Practice*. Pineridge Press Limited, Swansea, UK.

Simo JC and Rifai MS (1990). A Class of Mixed Assumed Strain Methods and the Method of Incompatible Modes. *International Journal for Numerical Methods in Engineering*, **29**, 1595-1638.

Swanson JA and Leon RT (2000). Bolted T-stub Connections: Tests on T-stub Components. *Journal of Structural Engineering*, **126:1**, 50-56.

Wanzek T and Gebbeken N (1999). Numerical Aspects for the Simulation of End Plate Connections. In: *Numerical Simulation of Semi-Rigid Connections by the Finite Element Method* (Ed. Kuldeep S. Virdi). COST C1, Report of working group 6 – Numerical simulation, Brussels.

STRENGTH AND STRESS ANALYSIS OF STEEL BEAM-COLUMN CONNECTIONS USING FINITE ELEMENT METHOD

Hong Chen, Yongjiu Shi, Yuanqing Wang, Zhaofan Li

Department of Civil Engineering, Tsinghua University, Beijing 100084, China

ABSTRACT

Following the Northridge and Kobe earthquakes, many studies, both experimental and analytical, have been carried out in order to get a better understanding of the seismic resistance of steel structures. Central to above issues is the seismic behavior of column-beam connection region. In order to investigate the performance of existing and improved connection configurations, four types of steel web-bolted and flange-welded (WBFW) beam-column connections with different details were analyzed by using nonlinear finite element methods. The main objective of this analysis is to determine whether the improved connection detail can contribute to the seismic resistance of column-beam joints. The ultimate strength, the stress distribution and the development process of plastic zones of connections with the raising of applied loading are calculated. The mechanical behaviors of the different detail connections are investigated and compared. This paper also briefly reviews the modeling techniques used in the nonlinear finite element analysis of steel connections for obtaining information on stress fields, the loading capacity and ultimate strength. The analytical results indicate that the seismic performance of connections can be significantly improved with alternative and modified connection details. The comparison between the results of the analysis and those of the experiment is satisfactory. The connection with modified details is proposed on the basis of analytical and experimental outcomes.

KEYWORDS

Strength, stress, steel beam-column connection, finite element analysis

INTRODUCTION

Steel WBFW beam-column connections are used extensively in tall buildings in seismic areas. These type connections were once believed to exhibit excellent behavior under earthquake action. However

---

*This project is supported by Natural Science Foundation of China through Grant No. 59878026

the premature brittle fracture of connections in 1994 Northridge earthquake and 1995 Kobe earthquake changed these minds about these connections. Since then, many researches on the cause of the observed damages, the major affected factors and retrofitting measures of connection configurations have been initiated and extensively conducted in USA, Japan and Europe (Roeder and Foutch, 1996; Mazzolani and Piluso, 1996; Miller, 1998; Nakashima et al., 1998; Uang et al., 2000; Simoes and Girao, 2001).

Up to now, a lot of the experimental and analytical results have been shown that the effect of connection details on strength, stress fields and seismic behavior is significant (Mao et al., 2001). Therefore many strategies to improve the connection behavior through modifying connection configurations are introduced to practical design on basis of research results (Mele, 2002). Of all these strategies, the method by weakening the beam section at interface of beam-column to shift the high stresses and strains sections away from the interface of the column-beam, such as Reduced Beam-Section Connection ("dog-bone"), web-slotted connection and "enlarge access hole" connection, is typically proposed very much. These methods are simple, practical and efficient.

To help get a better understanding of the ultimate strength, stress fields and ductility of the WFBW connections, eight specimens with four different details have been fabricated and tested to failure under cyclic loading in Tsinghua University (Chen et al., 2001). They are called the standard type connection (Specimen TS1), "dog-bone" type connection (Specimen TS2), "slot-web" connection (TS3) and "Enlarge access hole" type connection (TS4), as shown in Fig.1. In this paper, the elastic-plastic nonlinear finite analysis is used in order to acquire the stress distribution, plastic zone development and ultimate strength of connections. The mechanical behavior of connections with different details is assessed and compared. The good correspondence between analytical results and experimental results is verified.

THE NUMERICAL MODELS

The behavior of steel column-beam connections is mainly studied through experimental and numerical analysis methods. Whereas experimental methods are generally considered to be reliable ones, experimental results were universally erratic due to inherent randomness and uncertainties. Numerical analysis method is now commonplace in both research and design environments. It is mainly because the sophisticated and powerful finite element analytical software becomes widely available and computational costs fall rapidly.

Numerical simulation by using finite element, combining with reliable experimental calibration, has been proved to be a reliable and cost effective method to evaluate the behavioral characteristics of steel column-beam connections, including stress and stain fields, ultimate strength, hysteretic behavior and stress intensity factor. Several commercial general-purpose finite element programs, such as ANSYS (ANSYS, 1999), ABAQUS (ABAQUS, 1997)and MARC (MARC, 1998), can be used for comprehensive research of steel column-beam connections as numerical simulation tools. Of the three, ANSYS is perhaps relatively more flexible and easy to use.

In this paper, the behavioral characteristics of four type connection details are evaluated through numerical simulation by using ANSYS software package. For these specimens, the same cross sections of beam and column are used, as well as loading and constraints conditions. The beam flanges are welded to the column using full penetration welds and beam web is connected to the shear tab with high strength bolts. The specimens are made of grade Q235 structural steel with a yield strength of 270MPa, Poisson's ratio of 0.3 and a Young's elasticity modulus of 2.07×10^5 MPa.

It is assumed that the material would obey von Mises' yield criterion with the associated Prandtl-Reuss plastic flow rule. The stress-strain curve coupled with bilinear kinematic hardening criteria is assumed to be elastic-perfectly plastic in nonlinear analysis. For strength and stress analyses, shell elements are preferred to solid elements from the view of computational expense and efficiency. The specimens are modeled on four-node shell element in the numerical simulation for steel connections. Fig.2 shows the typical finite element mesh. To highlight the effect of connection details on behavior of connections, numerical simulation is carried out without considering other related factors contributed to damage to the steel connections, which mainly include weld defects, the stress relaxation of high strength bolts and slip between the shear tab and beam web.

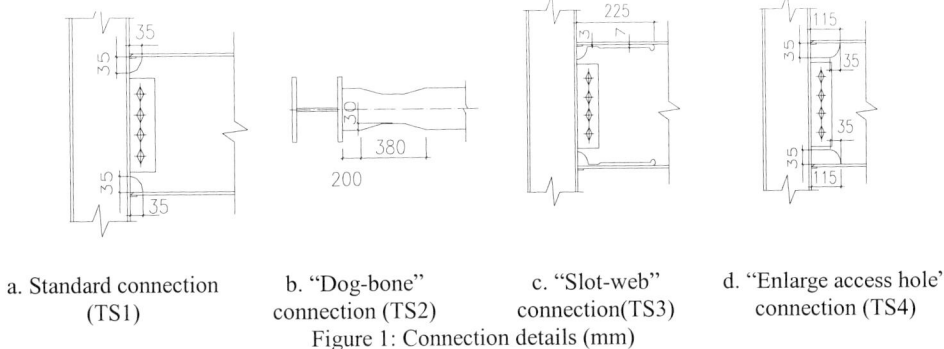

a. Standard connection (TS1)   b. "Dog-bone" connection (TS2)   c. "Slot-web" connection (TS3)   d. "Enlarge access hole" connection (TS4)

Figure 1: Connection details (mm)

For elastic states analysis, loading of 100kN which is lower than the elastic limit strength, is applied as a concentrated load acted in the beam tip. The displacement is applied in the beam tip in plastic analysis to acquire ultimate strength and load-deformation response. In a nonlinear analysis, load, whether by force or by displacement, should be applied in a number of small increments in order to gain convergent solution. Loading through increments of displacements is therefore preferable.

THE ELASTIC STRESS DISTRIBUTION ANALYSES

One of the most powerful and useful features of ANSYS software is its ability to map virtually any results data onto an arbitrary path through your model. This enables you to make clear, in the form of a graph or a tabular listing, how a result item varies along the path. In order to obtain the stress fields of connections, two paths, as shown in Fig. 3, are defined in the upper beam flange and the stress results are investigated on them. They are, path 1, on the upper face of the beam flange, along the longitudinal beam axis, and path 2, on the lower face, perpendicular to the longitudinal beam axis. The stress distributions along the above two paths and at interface of column-beam are analyzed.

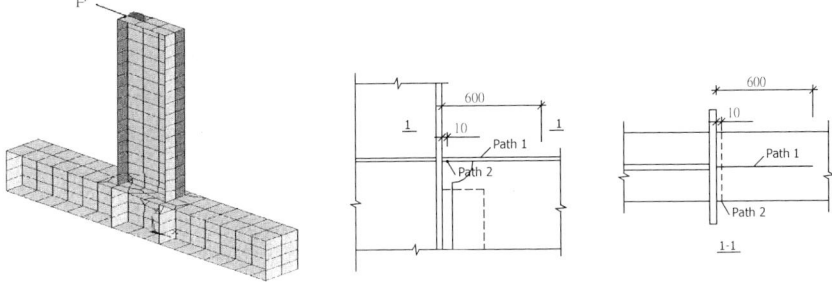

Figure 2: Finite element mesh   Figure 3: Two paths defined in the analysis

The Von Mises stress along path 1 is shown in Figure 4, from which we can find that the maximum stress in Specimens TS1 occurs at the end of the weld access hole, which is 232.31MPa and 55mm away from the column face, and the maximum stress in other Specimens is shifted from the column-beam interface. The maximum stress along path 1 is 268.63MPa for Specimen TS2, 236.19MPa for Specimen TS3, 228.70MPa for Specimen TS4. The maximum Von Mises stress in TS4 is smaller than those of other specimens The distances from the interface to the point of the maximum stress are respectively 390mm in Specimen TS2, 202mm in TS3 and 115mm in TS4. The Von Mises stress of Specimen TS2 is almost constant among the reduced section, forming an obvious platform in the curve near the column-beam interface. The above stress distributions show that the high stress and strain are shifted away from the column-beam interface, which is the result of reduced beam cross section. The analytical results correspond well with the experimental results. Nevertheless in places 200mm or more away from the column-beam interface, stress distribution is almost the same for all the specimens, therefore the effect of connection details on stress distribution is limited in the vicinity of column-beam interface.

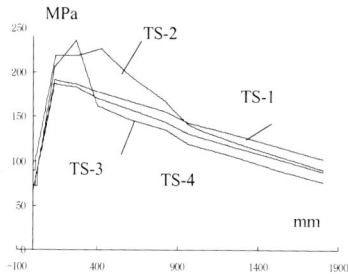
Figure 4: Von Mises stress on path 1

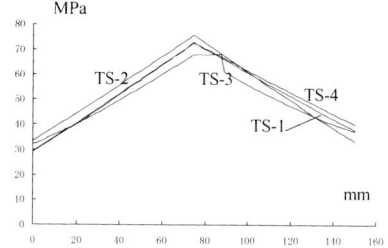
Figure 5: Von Mises stress on path 2

TABLE 1
PRINCIPAL STRESS IN BEAMS AT INTERFACE   MPa

Specimen No.	σ_1	σ_2	σ_3
TS-1	196.17	18.49	-0.67
TS-2	246.54	31.44	-0.38
TS-3	186.42	19.82	-0.01
TS-4	195.10	16.45	-0.16

The tri-axial principal stress in beams at column-beam interface is investigated, which is presented in Table 1. The tri-axial principal stress state and weld residual stress at column-beam interface would reduce lower the fracture toughness of welds and base steel, and leads to the premature brittle fracture of connections.

The Von Mises stress distribution on path 2 is shown in Figure 5, which is evidently different from one obtained on the basis of the beam theory, in which the constant stress is assumed in beam across section. Of all the specimens, the maximum stress is at the center of beam cross section, the minimum stress is at the edges of beam cross sections. The stress concentration is similar among Specimens TS-1, TS-3 and TS-4. The stress concentration in Specimen TS-3, which have a platform near the center of beam cross section, is less severe than one in other specimens, This difference may result from the slot-web, which weaken the constraints between the web and flange of beam.

The welds at column-beam interface, which are subjected to maximum stress, are the most weak because of welding defects and discontinuity. Under high stress state the welds with initial defects will crack and lead to brittle fracture of connections. Therefore the most efficient and practical measure to improve the seismic performance of connections is to reduce welds defects and lessen stress concentration.

The numerical analytical results show that by retrofitting the connection details, the high stress and strain can be shifted away from column-beam interface, the stress values at interface can be decreased. But the contribution of this improvement of connection details to the reduction of tri-axial stress and stress concentration of column-beam interface is not greater than expected.

THE STRENGTH AND LOAD-DISPLACEMENT RESPONSE ANALYSES

For plastic analysis, the loading is applied through the displacement, ANSYS employs the "Newton-Raphson" approach to solve nonlinear problems by use of Newton-Raphson equilibrium iteration. The elastic limit and ultimate strength are acquired and given in Table 2. The comparison made between the analytical and experimental results corresponds very well to each other. And this correspondence demonstrates that the numerical simulation approaches and results are very much reliable. From the Table 2 it can also be concluded drawn that the elastic limit and ultimate strength of connections with reduced beam section is smaller than that of standard connection. The strength of Specimen TS2 and TS3 decreases slightly larger than that of Specimen TS4 does.

TABLE 2
ELASTIC LIMIT AND ULTIMATE STRENGTH OF SPECIMENS

Specimen No.	Elastic limit strength (P$_y$: KN)		Ultimate strength (P$_u$: KN)		P^a_y / P^e_y	P^a_u / P^e_u
	P^a_y	P^e_y	P^a_u	P^e_u		
TS-1	123.11	123.08	155.67	173.95	1.00	0.89
TS-2	113.86	105.41	139.12	140.64	1.08	0.99
TS-3	120.08	104.97	139.88	132.82	1.14	1.05
TS-4	126.41	116.57	145.58	160.53	1.08	0.91

Note: 1. P^a_y and P^e_y stand for analytical and experimental results of elastic limit strength respectively.
2. P^a_u and P^e_u stand for analytical and experimental results of ultimate strength respectively.

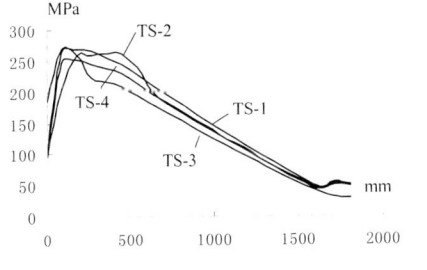
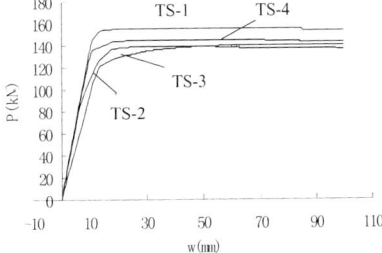

Figure 6: Von Mises stress on path 1   Figure 7: Load-displacement curves

The experimental results show that stress distributions along cross and longitudinal sections of beam in plastic stage are significantly different from that in elastic stage. The Von Mises stress of specimens at plastic stage on path 1 defined in Figure 3 is obtained through nonlinear finite analysis and shown in Figure 6. The Von Mises stress in plastic hinge range goes to constant for each specimen due to the plastic progress, thereby, a flat cure is formed in Von Mises stress distribution.

The stress platform of Von Mises stress distribution curves was shifted away from column-beam interface except for specimen TS-1, which is shown in figure 6. This analytical outcome also corresponds to experimental one. Besides, in plastic range, the Von Mises stresses along path 2 become constant for all specimens, accordingly the stress concentration is less than that in elastic stage. This is necessary result of plastic progress of beam cross sections.

The load-displacement curves of specimens obtained by nonlinear analysis are shown in Figure 7. The progress of plastic zones of connections is different from all specimens because of their unlike details, which can also be investigated through the nonlinear analysis. For specimen TS1, the plasticity occurs first at the interface of the beam and column flange, and then the plastic zone develops along the whole cross section of the beam flange along with the applied loading increases. The plastic zone gradually extends into the beam web and the panel zone of joint and finally leads to the failure of the connection.

For Specimen TS2, the plasticity happens at the beam flange in the first instance, where the cross sections is most smallest, and the plastic zone extends rapidly along the whole cross section of the beam flange with applied loading increases. The plastic zone subsequently extends to the whole beam cross section, at last the plastic hinge forms and results in the failure of connection, at this time the panel zone usually doesn't exhibit the plasticity. While for Specimen TS3, the plastic zone first appears at the flange near slot-web tip, and it develops along the flange section to column-beam interface along with applied loading augments. The plastic zone is limited in the beam flange by the slot-web. The failure of connections generally results from the local bucking of the compressed beam flange.

The plasticity occurs firstly at the beam flange near the access hole tip for Specimen TS4, whose extension of plastic zones is similar with one of the specimens TS1. The plastic hinge forms near the column-beam interface, while its location is slightly farer away from the column-beam interface than that of specimen TS1. And also the plastic zone in panel zone of Specimen TS4 is smaller than that of specimen TS1 at ultimate state.

CONCLUSIONS AND SUGGESTIONS

The strength, stress field and load-displacement response can be accurately investigated by using nonlinear finite analysis, the analytical results correspond well to experimental ones. The numerical simulation as a credible tool can be extensively applied in researches and practical design.

There is complicated tri-axial stress state and high strain and stress concentration at the column-beam interface, which weakens the fracture toughness of welds and base steel, as a result, the connections become susceptible to premature brittle fracture. So decreasing the weld defects and high stress concentration at the column-beam interface is the important approach to improve the seismic performance of connections.

The connection details have a great effect on behavior of connections. By modifying the connection details, such as Reduced Beam-Section Connection, Web-Slotted Connection and Enlarge Access Hole connection, the high stresses and strains can be shifted away from the interface of the column-beam. While this method is simple, practical and efficient, whose contribution to the reduction of tri-axial stress and stress concentration of column-beam interface is not greater than expected.

The analytical and experimental results show that Specimen TS4 with an enlarged web access hole exhibited an excellent seismic performance in terms of strength and ductility. Therefore, Specimen TS4 as an ideal modified connection detail is strongly recommended, while extensive researches must be conducted in order to further improve the seismic performance of this connection detail.

While the elastic plastic finite element analysis presented in this paper is basic and straightforward, the load displacement relations and stress and strain distributions of column-beam connections are correctly described and these help better understand the performance of column-beam connections. Nevertheless in this paper some key factors aren't considered, the analytical results do not completely explain the actual failure process of a steel beam to column connection. The key issue is how to deal with material details such as heat affected zone, material properties in different directions, notch and corner effects, and fracture criterion adopted. Theses issues should be added as critical parameters for further investigations.

REFERENCE

ABAQUS. (1997) Users' Manual I and II. Hibbit, Karlsson and Sorensen Inc.

ANSYS. (1999) Users' Manual. Swanson Analysis Systems Inc.

Chen H., Shi Y.J et al. (2001). Experimental research on beam-column connections under low-cycle cyclic loading. *Proceedings of Sixth Pacific Structural Steel Conference.* Beijing, China, October, Seismological Press,787-792.

Mao C., Ricles J., Lu L.W. and Fisher J. (2001). Effect of local details on Ductility of welded moment connections. *Journal of Structural Engineering.* **127:9**, 1036-1044.

MARC. (1998) User Information Manual. MARC Analysis Research Corporation.

Mazzolani F.M. and Piluso V. (1996) Theory and Design of Seismic Resistant Steel Frames. Champmann and Hall, E&FN Spon.

Mele E. (2002). Moment Resisting Welded Connections: an Extensive Review of Design Practice and Experimental Research in USA, Japan and Europe. *Journal of Earthquake Engineering.* **6:1**, 111-145.

Miller D.K. (1998). Lessons Learned from the Northridge Earthquake. *Engineering Structures* **20:4-6**, 249-260.

Nakashima M., Suita K., Morisako K. and Maruoka Y. (1998). Tests on Welded Beam-Column Subasemblies. I: Global Behavior. II: Detailed Behavior. *Journal of Structural Engineering.* **124:11**, 1236-1252.

Roeder C.W. and Foutch D.A. (1996). Experimental Results for Seismic Resistant Steel Moment Frame Connections. *Journal of Structural Engineering.* **122:6**, 581-588.

Uang C.M., Yu Q.S., Noel S. and Gross J. (2000). Cyclic Testing of Steel Connections Rehabilitated with RBS or Welded Haunch. *Journal of Structural Engineering.* **126:1**, 57-68.

SCAFFOLDS AND SLENDER STRUCTURES

GEOMETRIC NON-LINEAR ANALYSIS OF FLEXIBLE SUPPORTING SYSTEM

Z. Wang, Y.Q. Wang, Y.J. Shi

Department of Civil Engineering, Tsinghua University, Beijing, 100084, China

ABSTRACT

In recent years a new type of glass curtain wall which is supported by flexible system has been widely used. The flexible supporting system is usually constructed with steel bars or cables. After having been prestressed, most of the structural members will become tension member under working load. So that smaller cross-section is enough to resist the loading. At the same time the pretension can also significantly increase the structural stability and stiffness. Normally large deformation and displacement will occur in flexible supported system on duty. The geometric nonlinearity should be considered when analyzing the structural behavior. A finite element model was proposed in this paper to simulate the large deformation and pretension. The element stiffness matrix for pretension member is derived and the fundamental equations under global coordinate system are established. Based on the numerical model, some typical examples of flexible supporting system are given. The findings are useful for the reference of the designer in using the efficient and elegant structural form of this system.

KEYWORDS

Flexible supporting system   Point supported glass curtain wall   Geometric non-linear   Large deformation   Steel structure   Incremental Method

INTRODUCTION

Point supported glass buildings usually use flexible supporting system as its support system. There are many factors that affect its loading capacity such as geometrical non-linear. Based on 2D-beam element and 2D-bar element, a computer program is designed in this paper and two typical flexible supporting systems are analyzed by the program.

These typical flexible supporting systems are shown in Fig1.

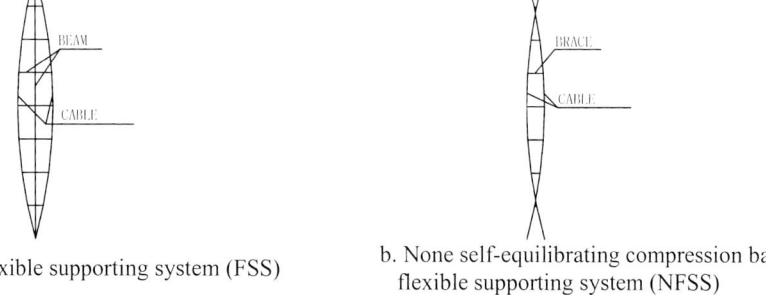

a. Normal flexible supporting system (FSS)

b. None self-equilibrating compression bar flexible supporting system (NFSS)

Figure 1: Common flexible supporting system

GEOMETRICAL NON-LINEAR ANALYSIS

Geometrical Non-Linear Analysis

In normal working, the condition wind load on glasses is transferred to the node of flexible supporting system. The deformation of the supporting system is a key factor to keep the glass serviceability.

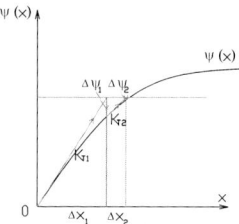

Figure 2: Diagrammatic sketch of Incremental Method

As can be seen from Fig2, tangent stiffness K_T and unbalanced force $\Delta\psi$ are needed in Incremental Method.

Tangent Stiffness K_T and Unbalanced Force $\Delta\psi$ of 2D-bar Element (2D-cable Element)

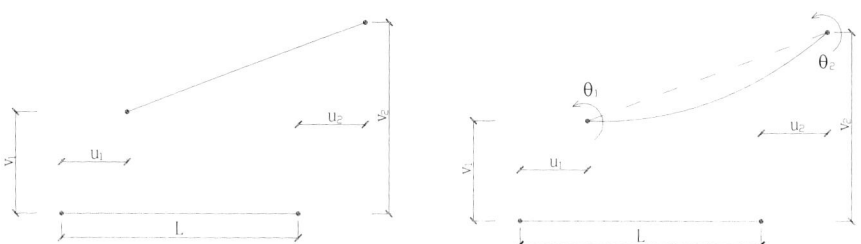

Figure 3: Diagrammatic sketch of 2D-barelement   Figure 4: Diagrammatic sketch of 2D-beamelement

The possible displacements of 2D-bar element δ^e is:

$$\delta^e = [u_1 \quad v_1 \quad u_2 \quad v_2]^T \tag{1}$$

The axis strain ε_x is:

$$\varepsilon_x = \frac{du}{dx} + \frac{1}{2}(\frac{dv}{dx})^2 \tag{2}$$

The shape function of the element is N_1、N_2:

$$N_1 = 1-\xi$$
$$N_2 = \xi \tag{3}$$
$$\xi = \frac{x}{L}$$

Use Equ.(1)- Equ.(3), we get $d\varepsilon_x$:

$$d\varepsilon_x = \frac{1}{L}[-1 \ 0 \ 1 \ 0]d\delta^e + \frac{1}{L^2}[0 \ -1 \ 0 \ 1]\delta^e[0 \ -1 \ 0 \ 1]d\delta^e \tag{4}$$

The first part of the Equ.(4) is strain-deformation linear transformation matrix and the second part is strain-deformation non- linear transformation matrix.

At last, we get:

$$K_0 = \frac{EA}{L}\begin{bmatrix} 1 & 0 & -1 & 0 \\ 0 & 0 & 0 & 0 \\ -1 & 0 & 1 & 0 \\ 0 & 0 & 0 & 0 \end{bmatrix}$$

$$K_\sigma = \frac{N}{L}\begin{bmatrix} 0 & 0 & 0 & 0 \\ 0 & 1 & 0 & -1 \\ 0 & 0 & 0 & 0 \\ 0 & -1 & 0 & 1 \end{bmatrix} \tag{5}$$

$$K_L = \frac{EA}{L}\begin{bmatrix} 0 & \theta & 0 & -\theta \\ \theta & \theta^2 & -\theta & -\theta^2 \\ 0 & -\theta & 0 & \theta \\ -\theta & -\theta^2 & \theta & \theta^2 \end{bmatrix}$$

Because the rotation of bar is small, so:

$$\theta = \frac{1}{L}[0 \ -1 \ 0 \ 1]\delta^e = \frac{v_2 - v_1}{L} \tag{6}$$

In fact, the unbalanced force is the initial element bar force vector minus element bar force vector when the deformation is δ^e.

$$\Delta\psi = \begin{bmatrix} -\cos\theta \\ -\sin\theta \\ \cos\theta \\ \sin\theta \end{bmatrix} \cdot (EA\varepsilon_x) - P \tag{7}$$

Tangent Stiffness K_T and Unbalanced Force $\Delta\psi$ of 2D-beam Element

The possible displacements of 2D-beam element δ^e is:

$$\delta^e = [u_1 \quad v_1 \quad \theta_1 \quad u_2 \quad v_2 \quad \theta_2]^T \tag{8}$$

The strain vector of the element is:

$$\varepsilon = \begin{bmatrix} \dfrac{du}{dx} + 0.5(\dfrac{dv}{dx})^2 \\ -y\dfrac{d^2v}{dx^2} \end{bmatrix} \tag{9}$$

The shape function of the element is N_1、N_2:

$$N_1 = [0 \quad 1 - 3\xi^2 + 2\xi^3 \quad l(\xi - 2\xi^2 + \xi^3) \quad 0 \quad 3\xi^2 - 2\xi^3 \quad l(-\xi^2 + \xi^3)]$$
$$N_2 = [1 - \xi \quad \dfrac{-y}{L}(-6\xi + 6\xi^2) \quad -y(1 - 4\xi + 3\xi^2) \quad \xi \quad \dfrac{-y}{L}(6\xi - 6\xi^2) \quad -y(-2\xi + 3\xi^2)] \tag{10}$$

Use Equ.(8) - Equ.(10), we get $d\varepsilon_x$:

$$d\varepsilon = \begin{bmatrix} H_1 \\ -yH_3 \end{bmatrix} d\{U\}^e + \begin{bmatrix} H_2\{U\}^e H_2 \\ 0 \end{bmatrix} d\{U\}^e$$

$$H_1 = \begin{bmatrix} -\dfrac{1}{l} & 0 & 0 & \dfrac{1}{l} & 0 & 0 \end{bmatrix}$$

$$H_2 = \begin{bmatrix} 0 & -\dfrac{6x}{l^2} + \dfrac{6x^2}{l^3} & 1 - \dfrac{4x}{l} + \dfrac{3x^2}{l^2} & 0 & \dfrac{6x}{l^2} - \dfrac{6x^2}{l^3} & -\dfrac{2x}{l} + \dfrac{3x^2}{l^2} \end{bmatrix} \tag{11}$$

$$H_3 = \begin{bmatrix} 0 & -\dfrac{6}{l^2} + \dfrac{12x}{l^3} & -\dfrac{4}{l} + \dfrac{6x}{l^2} & 0 & \dfrac{6}{l^2} - \dfrac{12x}{l^3} & -\dfrac{2}{l} + \dfrac{6x}{l^2} \end{bmatrix}$$

Be Similar to 2D-bar element, the first part of the Equ.(11) is strain-deformation linear transformation matrix and the second part is strain-deformation non-linear transformation matrix.
At last, we get:

$$K_0 = G_0$$

$$G_0 = \begin{bmatrix} \dfrac{EA}{l} & 0 & 0 & -\dfrac{EA}{l} & 0 & 0 \\ 0 & \dfrac{12EI}{l^3} & \dfrac{6EI}{l^2} & 0 & -\dfrac{12EI}{l^3} & \dfrac{6EI}{l^2} \\ 0 & \dfrac{6EI}{l^2} & \dfrac{4EI}{l} & 0 & -\dfrac{6EI}{l^2} & \dfrac{2EI}{l} \\ -\dfrac{EA}{l} & 0 & 0 & \dfrac{EA}{l} & 0 & 0 \\ 0 & -\dfrac{12EI}{l^3} & -\dfrac{6EI}{l^2} & 0 & \dfrac{12EI}{l^3} & -\dfrac{6EI}{l^2} \\ 0 & \dfrac{6EI}{l^2} & \dfrac{2EI}{l} & 0 & -\dfrac{6EI}{l^2} & \dfrac{4EI}{l} \end{bmatrix} \tag{12}$$

$$K_\sigma = \int_l \begin{bmatrix} 0 & 0 & 0 & 0 & 0 & 0 \\ 0 & A_1^2 & -A_1B_1 & 0 & -A_1^2 & A_1C_1 \\ 0 & -A_1B_1 & B_1^2 & 0 & A_1B_1 & -B_1C_1 \\ 0 & 0 & 0 & 0 & 0 & 0 \\ 0 & -A_1^2 & A_1B_1 & 0 & A_1^2 & -A_1C_1 \\ 0 & A_1C_1 & -B_1C_1 & 0 & -A_1C_1 & C_1^2 \end{bmatrix} dx \cdot N$$

$$\int_l A_1^2 dx = \dfrac{6}{5l} \quad \int_l A_1B_1 dx = \dfrac{1}{10} \quad \int_l A_1C_1 dx = \dfrac{1}{10} \quad \int_l B_1^2 dx = \dfrac{2l}{15} \quad \int_l B_1C_1 dx = \dfrac{l}{30} \quad \int_l C_1^2 dx = \dfrac{2l}{15}$$

$$K_L = G_{0L} + G_{0L}{}^T + G_L$$

$$G_{0L} = \begin{bmatrix} 0 & G_{12} & G_{13} & 0 & G_{15} & G_{16} \\ 0 & 0 & 0 & 0 & 0 & 0 \\ 0 & 0 & 0 & 0 & 0 & 0 \\ 0 & G_{42} & G_{43} & 0 & G_{45} & G_{46} \\ 0 & 0 & 0 & 0 & 0 & 0 \\ 0 & 0 & 0 & 0 & 0 & 0 \end{bmatrix}$$

$$G_{12} = \frac{6E}{5l^2}(v_j - v_i) - \frac{E}{10l}\theta_i + \frac{E}{10l}\theta_j$$

$$G_{13} = \frac{E}{10l}(v_j - v_i) - \frac{2E}{15}\theta_i + \frac{E}{30}\theta_j$$

$$G_{15} = -G_{12} \quad G_{42} = -G_{12} \quad G_{43} = -G_{13}$$

$$G_{16} = \frac{E}{10l}(v_j - v_i) - \frac{E}{30}\theta_i + \frac{2E}{15}\theta_j$$

$$G_{45} = G_{12} \quad G_{46} = -G_{16}$$

$$G_L = \begin{bmatrix} 0 & 0 & 0 & 0 & 0 & 0 \\ 0 & H_{22} & H_{23} & 0 & H_{25} & H_{26} \\ 0 & H_{32} & H_{33} & 0 & H_{35} & H_{36} \\ 0 & 0 & 0 & 0 & 0 & 0 \\ 0 & H_{52} & H_{53} & 0 & H_{55} & H_{56} \\ 0 & H_{62} & H_{63} & 0 & H_{65} & H_{66} \end{bmatrix}$$

$$H_{22} = \frac{72E}{35l^3}(v_j - v_i)^2 + \frac{3E}{35l}\theta_i^2 + \frac{3E}{35l}\theta_j^2 - \frac{18E}{35l^2}(v_j - v_i)\theta_i - \frac{18E}{35l^2}(v_j - v_i)\theta_j$$

$$H_{23} = \frac{9E}{35l^2}(v_j - v_i)^2 - \frac{E}{140}\theta_i^2 + \frac{E}{140}\theta_j^2 - \frac{3E}{35l}(v_j - v_i)\theta_i + \frac{E}{70}\theta_i\theta_j$$

$$H_{26} = \frac{9E}{35l^2}(v_j - v_i)^2 + \frac{E}{140}\theta_i^2 + \frac{E}{140}\theta_j^2 - \frac{6E}{35l}(v_j - v_i)\theta_i + \frac{E}{70}\theta_i\theta_j$$

$$H_{33} = \frac{3E}{35l}(v_j - v_i)^2 + \frac{2El}{35}\theta_i^2 + \frac{El}{210}\theta_j^2 + \frac{E}{70}(v_j - v_i)\theta_i - \frac{E}{70}(v_j - v_i)\theta_j - \frac{E}{70}\theta_i\theta_j$$

$$H_{36} = \frac{-El}{140}\theta_i^2 - \frac{El}{140}\theta_j^2 - \frac{E}{70}(v_j - v_i)\theta_i - \frac{E}{70}(v_j - v_i)\theta_j + \frac{El}{105}\theta_i\theta_j$$

$$H_{66} = \frac{3E}{35l}(v_j - v_i)^2 + \frac{El}{210}\theta_i^2 + \frac{2l}{35}\theta_j^2 - \frac{E}{70}(v_j - v_i)\theta_i + \frac{E}{70}(v_j - v_i)\theta_j - \frac{E}{70}\theta_i\theta_j$$

$$H_{25} = -H_{22} \quad H_{32} = H_{23} \quad H_{35} = -H_{23} \quad H_{52} = H_{22} \quad H_{53} = H_{23} \quad H_{55} = H_{22} \quad H_{56} = -H_{26} \quad H_{62} = H_{26}$$

$$H_{63} = H_{36} \quad H_{65} = -H_{26}$$

The unbalanced force is:

$$\Delta \psi = \left[G_0 + \frac{1}{2}G_{0L} + G_{0L}{}^T + \frac{1}{2}G_L \right] \begin{bmatrix} u_1 \\ v_1 \\ \theta_1 \\ u_2 \\ v_2 \\ \theta_2 \end{bmatrix} - P \tag{13}$$

About Cable (Tension bar) Element

Similar to bar element, the cable element use the tangent stiffness matrix of 2D-bar element. Cable element can only be tensioned, so we set the stiffness matrix to zero in the program when it will be compressed.

Solution Steps

(1) Estimate a displacement vector of the structure
(2) Use Equ. (7) and Equ. (13) to get unbalanced force
(3) Use Equ. (5) and Equ. (12) to get global K
(4)
$$\Delta\delta = -K_T^{-1} \cdot \Delta\psi \tag{14}$$
(5)
$$\delta = \delta + \Delta\delta \tag{15}$$
Go to Equ(2) until $\Delta\delta$ is small enough to be ignored.

EXAMPLES

In this section, two typical flexible supporting systems which are shown in Fig.1, is analyzed by the program based above solution steps.

The Governing Factors of Loading Capacity

The governing factors of the loading capacity include the cross-section area of cable (or tension bar), the value of prestress, the ratio of depth to span of supporting system and the shape of cable (generally Para-curves).

For architectural beauty, the diameter of cable can not be greater than 50mm and the ratio of depth-to-span of flexible supporting should be as small as possible. The prestress cannot be greater than 80Mpa otherwise it is difficult to construct.

Factors of Model

The material of rigid frame is Q235 steel and yield strength is 235MPa and Young's modulus is 2.06×10^5MPa. The material of tension bar is Cr18Ni9 and yield strength is 235Mpa and Young's modulus is 1.95×10^5MPa. The size of glass panel is 2m × 2m. The basic wind pressure is 0.35MPa. So, the design wind pressure is

$$\omega_k = 2.25\times 1\times(0.8+0.5)\times 1.4\times\omega_0 = 1.43 kN/m^2 \tag{16}$$

Every nodal force is
$$F = \omega_k \times 2\times 2 = 5720N \tag{17}$$

The variables of model are list at table1.

Table 1 VARIABLES OF MODEL

	Diameter D(mm)	Prestress P(Mpa)	Depth-to-Span K
Changing diameter of cable (Fig.5 and Fig.6)	Changing from 10 to 40	50	1:8
Changing Prestress of cable (Fig.7 and Fig.8)	20	Changing from 10 to 100	1:8
Changing Depth-to-Span (Fig.9 and Fig.10)	20	50	Changing from 0.05 to 1

CALCULATION RESULT

The Effect of Changing Cable Diameter (tension bar)

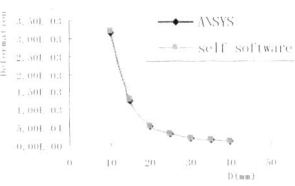

Figure 5: The cable diameter of FSS VS. Deformation

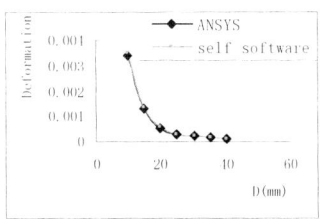

Figure 6: The cable diameter of NFSS VS. Deformation

The curve in Fig.5 and Fig.6 shows the effect of changing the diameter of tension bar on two typical flexible supporting systems.

As can been seen in these figures, the deformation reduces while the diameter of tension bar increases. We can conclude that increasing diameter can increase side stiffness significantly. When the diameter is too large, the effect to increase side stiffness is reduced.

The Effect of Changing Value of Prestress of Cable (tension bar)

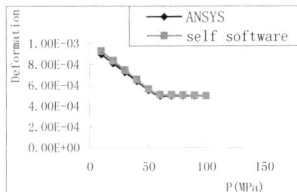

Figure 7: The prestress of FSS VS. Deformation

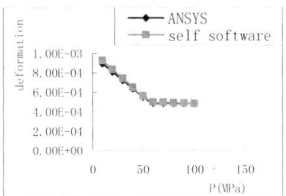

Figure 8: The prestress of NFSS VS. Deformation

The curve in Fig.7 and Fig.8 shows the effect of changing of prestress of tension bar in two type of flexible supporting system.

Similar to case one, the deformation reduces while the value of prestress of tension bar increases. But when the diameter is larger than some limit value, the effect of increasing side stiffness is reduced suddenly.

The Effect of Changing the Depth-To-Span of Flexible Supporting System

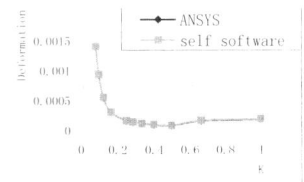

Figure 9: The Depth-to-Span of FSS VS. Deformation

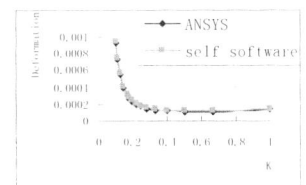

Figure 10: The Depth-to-Span of NFSS VS. Deformation

The curve in Fig.9 and Fig.10 shows the effect of changing the Depth-to-Span ratio of flexible supporting system on two typical flexible supporting systems.

The curves in two figures are similar. When the Depth-to-Span ratio increased, the deformation reduces to some value then increases. So, there is an optimum value.

CONCLUSIONS

A nonlinear numerical method has been developed and utilized to flexible supporting system. The geometric non-linear analysis in pretension and external load stages has been performed. The influence of various design parameters which include pretension, the area of cable and The Depth-to-Width of the system is studied. The conclusions of this study can be summarized as follows.

Incremental Method can solve geometric non-linear problem easily. The results based on this theory accord with ANSYS software. The software approach can be applied easily to any flexible supporting system.

Side stiffness of the structure will be increased significantly and the deformation of structure will be deduced by using pre-tensioned or increasing the diameter of the cable (tension bar). The larger diameter or the value of the prestress the cable has, the less of curvature the curve has (fig.5 to fig.8).

There is a least deformation when a particular Depth-to-Span ratio of supporting system is taken. That is the optimum ratio for design.

REFERENCE

S.L.Chan. G.P.Shu. Z.Y.Lv.(2002) Stability analysis and parametric study of pre-stressed stayed columns Engineering Structures, **Vol 24**,115-124

K.Kebiche. M.N.Kazi-Aoual. R.Motro.(1999). Geometrical non-linear analysis of tensegrity systems Engineering Structures, **Vol 21**,864-876

D.S.Wakefield(1999) Engineering analysis of tension structures theory and practice Engineering Structures, **Vol 21**,680-690

Blandford GE.(1996)   Large deformation analysis of inelastic space truss structures. J Struct Eng ;**122:4** ,407–15.

Yang YB, Chiou HT.(1987)   Rigid body motion test for nonlinear analysis with beam elements. J Eng Mech ASCE;**113:9**, 1404–19.

Argyris JH, Sharps DW.(1972)   Large deflection analysis of prestressed networks. J Struct Div Proc ASCE ;**98:ST3**, 633–54.

DETERMINATION OF THE FACTORS OF SAFETY OF STANDARD SCAFFOLD STRUCTURES

B Milojkovic[1], R G Beale[2] and M H R Godley[2]

[1]Halcrow Group, 44, Brook Green, Hammersmith, London, W6 7BY, formerly Oxford Brookes University, UK
[2]Department of Civil Engineering & Construction Management, Oxford Brookes University, Oxford, OX3 OBP UK

ABSTRACT

A three-dimensional, non-linear, elasto-plastic, finite element model of a scaffold structure typical of small domestic construction was developed with the objective of determining the factors of safety against collapse. All connections between standards and ledgers and between standards and transoms were considered to be semi-rigid. The influence of different types of fault in the structure on factors of safety was determined. The solution obtained from the elasto-plastic analysis was compared with an alternative approach which combined a linear elastic analysis and eigenvalue buckling with the use of member buckling interaction formulae given in the European Standard ENV 1993-1-1. This alternative analysis was used to generate tables of factors of safety. In a standard finite element program elastic buckling load factors are normally calculated by increasing all the applied loads simultaneously. For the purposes of this research, however, it was necessary to determine buckling load factors for imposed load only, dead and wind loads being kept constant. A regression curve was fitted to a series of buckling load factors for different imposed loads. This curve was used to enable the factors of safety against collapse of the structure to be calculated. The performance of scaffold structures is affected by faults in both the design and site erection stage and in the operational stage. A table was constructed which showed the effects of combinations of faults enabling an understanding to be achieved of those faults which may make a small scaffold collapse under normal wind and imposed load combinations.

KEYWORDS

Steel, scaffold, factors of safety, semi-rigid, structures, non-linear analysis

INTRODUCTION

During the six years from 1986-1993 the UK Health and Safety Executive (HSE) investigated 1091 safety related incidents (MARCODE HSE database). In addition, during the same period there were 471 reported collapses out of an estimated 7.5 million scaffold erections giving a failure rate of 6 collapses per 10000 erections. Scaffold collapses were primarily caused by missing ties (28%), overload (25%), faulty components (13%) and missing bracing (9%). The objectives of this research were to examine the effects of faults in design and construction on the safety factors of a scaffold which represented typical domestic construction (see Figure 1). The model scaffold was designed according to BS 5973 (BSI 1994).

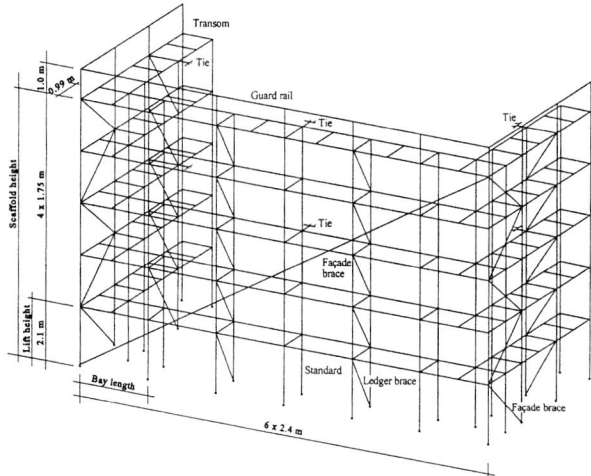

Figure 1: Model scaffold

Traditionally engineers have used simple methods, such as linear analyses, to analyse scaffolds for practical design. However, because linear analyses do not include geometrically non-linear effects the results obtained are not accurate even in the range of working loads (Chan et al, 1995; Chu et al, 1996; Huang et al, 2000). The authors have previously demonstrated the importance of the semi-rigid connections between ledgers and standards, and between transoms and standards. (Godley and Beale, 1997 and 2001) on the behaviour of the structure. However the normal eccentricity of the connection in traditional tube and fitting scaffolds has been shown both experimentally and theoretically to have little effect (Milojkovic et al, 1996).

STRUCTURAL ANALYSIS

The scaffold shown in Figure 1 was analysed using the Lusas finite element program (Lusas 1998). A full elasto-plastic analysis was initially conducted of the structure. Convergence difficulties caused by the wide variation of stiffnesses within the model and long processing times (each analysis was taking several days) occurred with this model. In all the analyses the diagonal brace failed in buckling before the rest of the structure had achieved its maximum capacity. As the structure was still capable of carrying load a reduced model without the brace was used to validate an alternative procedure. The full details of the analysis are given in Milojkovic (1999).

The failure of the structure was determined using the interaction formulae for member buckling proposed in ENV 1933-1-1 (1993). In this context collapse was defined as the load factor at which the structure first failed to satisfy the ultimate limit state requirement.

Buckling Analysis

This approach was applied member by member and every member was treated in the same way by combining a linear elastic analysis with an elastic buckling analysis. The axial forces and the bending moments in the members were taken from the results of linear elastic analyses, while the non-dimensional slenderness ratio was calculated by combining results from the linear elastic and elastic buckling analyses. Normally an interaction formula is used to check a member for which the end moments and axial forces, as well as the elastic buckling load are known. If this expression is less or equal to unity, the check succeeds. At failure, this expression is equal to unity. In this approach the load factor was adjusted until the interaction formula equalled unity. This load factor was taken to be the load factor at failure.

The finite element program Lusas calculated buckling load factors by uniformly increasing all loads together at the same time; dead load, wind load and imposed load where appropriate. However, these elastic buckling load factors were not appropriate in this case because the purpose of this project was to determine the safety factors against collapse when the imposed load alone was increased. This could not be done directly so the following method was adopted:

If the imposed load only is increased by a factor γ, the force in any member F, is given by:

$$F = F_D + F_W + \gamma F_I \qquad (1)$$

where F_D is the force in a member due to dead load, F_W is the force due to the wind load and F_I is the force due to the imposed load.

If a buckling analysis is made for any value of γ then buckling occurs when the value of the elastic buckling load F_{cr} is

$$F_{cr} = \lambda_{cr}(F_D + F_W + F_I) \qquad (2)$$

Naturally, λ_{cr} will depend on the value of γ chosen. We need to find the value of γ called γ_{cr} for which

$$F_{cr} = F_D + F_W + F_I \qquad (3)$$

and this was done by calculating values of λ_{cr} for different values of γ, and noting the value of $\gamma = \gamma_{cr}$ at which at $\lambda_{cr} = 1$. The calculated γ_{cr} is the elastic buckling load factor when only the imposed load is increased, while the dead load and the wind are kept constant. Values of λ_{cr} were calculated for a range of values of γ and for a load combination of dead load and a working load of 1.5kN/m² in every bay on the top level are shown in Figure 2. From this curve, the function relating γ and the critical load factors was given by, using regression analysis:

$$\lambda_{cr}(\gamma) = 8.823 - 2.649\gamma + 0.4185\gamma^2 + 0.00309\gamma^3 + 0.0008519\gamma^4 \qquad (4)$$

From this polynomial the critical load was determined when $\lambda_{cr} = 1$. In this case $\gamma_{cr} = 12.25$. A different regression curve was found for all structures analysed. Values of the elastic buckling load factors γ_{cr} are given for all combinations in Table 1.

Interaction formula

When an axially loaded member is also subjected to bending moments, the total effect cannot simply be computed by the direct addition of the corresponding values of the component cases. The reason for this is that the deflections caused by the bending moment increase or "amplify" the secondary effects produced by the axial compression, leading to a non-linear interaction. To account for this effect an interaction formula was used to estimate the overall factor of safety, λ, against collapse for this structure (members in compression) in accordance with ENV 1993-1-1, clause 5.5.4:

$$\frac{N_{SD}}{\chi A f_y} + \frac{k_y M_{y,SD}}{W_{PL} f_y} + \frac{k_z M_{z,SD}}{W_{PL} f_y} \leq 1 \qquad (5)$$

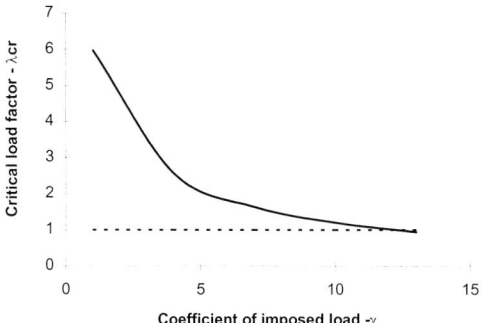

Figure 2: Graph of the variation of the critical load against the coefficient of imposed load

where N_{SD} is the axial force, $M_{y,SD}$ and $M_{z,SD}$ are the bending moments about the y and z axes in a scaffold element. $\chi A f_y$ is the characteristic resistance load of the element in compression, $k_y/(W_{PL}f_y)$ and $k_z/(W_{PL}f_y)$ are the characteristic resistance of the element in compression about the y and z axes. The forces were taken from linear analyses of scaffolds and amplified by the buckling loads determined above.

The non-linear analysis of the structure showed that local failure of the top ledger elements occurred before overall collapse. In order to investigate overall collapse the moment of resistance against bending of ledger elements was arbitrarily increased 1000 times. Failure of the structure then occurred by buckling failure parallel to the facade. The interaction formula equation (5) was then applied to the twelve central bottom elements on the front of the scaffold.

LOAD CASES AND COMBINATIONS

The dead load was considered with three alternative imposed load conditions to give the following load cases:

Load Case 1: Dead load + Working load of 1.5 kN/m² in every bay at the top level.
Load Case 2: Dead load + Working load of 1.5 kN/m² in alternate bays at the top level.
Load Case 3: Dead load + Overload on the top platform comprising two 300 kg loads acting on an area of 0.96m × 0.46m at the centre of each and every span.

In addition to this, wind loading was considered acting normal to and parallel with the façade. Thus each of the load cases was considered in the following combinations:

Combination 1: vertical load condition with no wind.
Combination 2: vertical load condition with wind parallel to the façade, blowing from left side to right.
Combination 3: vertical load condition with wind perpendicular to the façade - sucking, assuming that the façade itself offered no wind resistance.

There were therefore initially a total of nine load case combinations.

However, when Load Case 2 was analysed it was discovered that all standards had higher overall load factors against collapse than occurred in Load Case 1. This is because at the top level the load applied in Load Case 2 is only half (applied on alternate bays) of that in Load Case 1. The ledgers, however, have lower overall load factors in Load Case 2, because bending moments are greater. This increase in bending moment caused local failure of the structure but did not lead to an overall collapse. In addition, the faults which are introduced to the structure in this research do not have any influence on this local failure of the top ledgers. As the aims of this research were to investigate overall failure and not local failure Load Case 2 was neglected. This reduced the number of combinations to be analysed to six.

STRUCTURAL FAULTS

Inspection by the HSE has identified the most common faults which can exist in scaffold structures (Maitra, 1997). The faults were:

1   A partial settlement caused by the base-plate of the scaffold not being level. This fault induces a bending moment into the bottom leg of the scaffold. This was modelled by inclining the bottom leg of the scaffold by 2%.

2   A gross settlement caused by a member not being supported by the ground. For maximum effect this was modelled by removing the support at the most heavily loaded base.

3   Initially the standards were regularly placed at 2.4m centres. In this case the overall width of the scaffold was unaltered but spacing of the central two standards was changed to 2.1m and 2.7m. The effect of this fault was to increase the strength of the scaffold. This is due to the fact that for regularly spaced scaffolds the flexural rigidity provided by the ledgers is a minimum.

4-7 The middle standards were assumed to be out-of-plumb by 1% (fault 4), 2% (fault 5) parallel to the facade and by 1% (fault 6) and 2% (fault 7) normal to the facade.

8-10 Curvature was applied to the standards below the bottom lift of a maximum of 3mm (fault 8), 6mm (fault 9) and 12mm (fault 10).

11 The height to the first lift was increased to 2.7m. In order to keep the whole scaffold to the same height the top lift was reduced to 1.15m.

12 To gauge the effects of corrosion one standard was reduced from 4mm thickness to 3.25mm.

13 Connections between transoms and ledges adjacent to standards were made with right-angle couplers. In order to model the common fault of using putlog connections, pinned joints were inserted at all these positions.

14 Ledger bracing was initially placed as close to standards as possible. In practice this is not possible and in faulty structures the ledger bracing was placed 300mm vertically away from the correct position. In addition, swivel connectors instead of right-angled couplers were also used.

15 The perfect structure had ledger braces at every level. Commonly, however, the bottom diagonal is omitted. This was taken to be fault 15.

16-18 The most common cause of failure is inadequate tying. The perfect structure had two ties on the middle standard. Fault 16 – top tie omitted; fault 17 – bottom tie omitted, fault 18 – no ties.

TABLE 1
LOAD FACTORS FOR THE WHOLE STRUCTURE

Fault	Description	γ_{cr}		Overall Load Factors for Whole Structure (λ)						%age reduction in (λ)
		Load Case 1	Load Case 3	Load Case 1 Combination			Load Case 3 Combination			
				LC1	LC2	LC3	LC1	LC2	LC3	
-	Perfect Structure	11.79	6.45	10.91	9.88	9.88	6.02	5.39	5.25	-
1	Partial settlement	11.82	6.49	10.34	10.32	9.19	5.66	5.98	4.78	8.5%
2	Gross settlement	11.54	6.37	7.00	7.00	6.63	3.36	3.10	3.24	41.0%
3	Irregular standards	11.78	6.46	10.81	9.87	9.83	6.03	5.45	5.27	-0.5%
4	1% out-of-plumb parallel	11.92	6.44	10.41	9.35	9.31	5.60	5.10	4.80	8.0%
5	2% out-of-plumb parallel	11.72	6.43	9.76	8.90	9.23	5.39	4.93	5.05	6.0%
6	1% out-of-plumb normal	11.92	6.49	9.96	8.95	8.74	5.30	4.84	4.47	15.0%
7	2% out-of-plumb normal	11.80	6.42	9.24	8.45	8.85	4.90	4.40	4.60	16.0%
8	3mm curvature	11.71	6.34	10.24	9.37	9.11	5.52	4.96	4.56	13.0%
9	6mm curvature	11.72	6.37	9.61	9.15	8.03	5.28	4.94	4.13	21.5%
10	12mm curvature	11.74	6.39	8.61	8.23	6.74	4.71	4.45	3.35	36.0%
11	2.7m bottom lift	10.26	5.52	9.61	8.69	8.60	5.32	4.79	4.48	14.5%
12	Corrosion	11.73	6.37	10.84	9.83	9.83	5.97	5.41	5.19	1.0%
13	Putlog couplers	9.68	5.01	9.23	8.19	7.98	4.87	4.35	4.00	24.0%
14	Eccentric bracing	11.42	6.10	10.55	9.55	9.15	5.59	5.12	4.76	9.5%
15	No bottom brace	11.63	6.21	10.48	9.49	9.29	5.40	4.91	5.02	6.0%
16	No top tie	11.21	5.89	10.45	9.37	9.11	5.52	4.96	4.56	13.0%
17	No bottom tie	9.69	5.16	9.07	7.66	8.03	4.92	4.16	4.14	21.0%
18	No ties	9.48	5.06	9.02	7.62	7.87	4.86	3.53	3.70	29.5%

RESULTS OF ANALYSES WITH SINGLE FAULTS

Table 1 is a summary of all the analyses. The lowest buckling load factors for Load Cases 1 and 3 are given. These always occurred for Load Combination 3 (called LC3). The buckling load factors for Load Combinations 1 (LC1) and 2 (LC2) were less than 5.6% higher for Load Case 1 and 9% higher for Load Case 3. Full details are found in Miljokovic (1999). It can clearly be seen that the most significant reductions in load factor occurred for the cases of Gross Settlement (Fault 2), excessive curvature in the lowest standards (Fault 10), inadequate ties (Faults 17 and 18) and the incorrect use of putlog couplers instead of right angled couplers (Fault 13).

COMBINED FAULTS

In practice when scaffolds are designed and erected single faults do not arise on their own. Analyses were therefore firstly undertaken with combinations of faults. In most cases the reduction in capacity of a set of faults could approximately be estimated by adding the reductions in load factor given by the single faults in Table 1 above. For example, a combination of putlog couplers (instead of right angled couplers) in combination with no ties (i.e. Faults 13+18) gave a reduction of 59%. The sum of the two faults is 54.5%. The major exception occurred in the case of putlog couplers in combination with gross settlement. In this case the analysis produced a reduction of 28% instead of the combined total of 65%. The effect of putlog couplers is to change the structure from one containing semi-rigid connections to a structure with pinned joints only. This therefore prevents moment transfer between adjacent elements in a frame. When the interaction formulae are used to determine collapse the additional moments transferred from the frame farthest from the façade precipitate failure in the most heavily loaded member. When putlog couplers are used these moments are not transferred. The most heavily loaded members in this case are those adjacent to the 'settled' support in the same plane frame. They have higher axial loads than members with right angle couplers but much lower moments. The use of the interaction formulae produces a higher collapse load.

A typical scaffold is influenced by two important stages in its lifetime. The first of these stages is the design and construction stage. In this stage, decisions are first made on paper about the specification required to meet a defined set of purposes. Then the scaffold is erected in accordance (or not) with this specification. The design and erection stage of the process may be carried out by experts, and would then perform as intended, or it may be carried out less well, so that if the full demands specified for the structure were actually asked of it, the scaffold would not perform to specification. The second stage in the life of the scaffold occurs once it has been erected on site, when its performance is influenced by the quality of the site control. Site control is good if the scaffold is correctly used, so that the imposed loading never exceeds that specified for it, and if it is conscientiously inspected and serviced at regular intervals so that no significant deterioration takes places during its life.

Both stages can be subdivided into three categories: good, average and poor according to the faults which are present. Some faults may be generated in both stages of the life of the structure while others may be generated in one or the other. For instance, ties may be missed by the designer or the erector, or they may be removed while the scaffold is in use and, because of poor site control, not replaced. This is a fault possible in both stages. Table 2 summarises the combinations of faults assigned to each category. Note that overload, Load Case 3, should not occur in practice and has hence only been assigned to faults in site control. Common geometric faults such as standards out-of-plumb by 1% or with 6mm curvature and partial settlement have been assigned to average categories. Gross settlement, or extreme errors of curvature (12mm), or large out-of-plumb (2%) have been assigned into the category of poor site control. The use of putlog couplers was considered to be an element of poor design or poor construction. Joint eccentricity usually arises due to poor design. The absence of ledger braces was considered to belong to poor erection/site control as it should be noticed and corrected.

TABLE 2
FAULT COMBINATIONS

	Site Control		
Design/Erection	Good	Average	Poor
Good	-	17+LC3	15+18+LC3
Average	1+4/5+9+11+13+15	1+4/5+9+11+13+17+LC3	1+4/5+9+11+13+15+18+LC3
Poor	2+3+6/7+10+11+12+13+14+18	2+3+6/7+10+11+12+13+14+18+LC3	2+3+6/7+10+11+12+13+14+15+18+LC3

The resulting load factors are given in Table 3. The table indicates poor site control is potentially more detrimental to the safety of scaffolds than poor design. Each combination has been analysed in the same way as the single faults. It is interesting to note that the extreme case of poor design in conjunction with poor site control led to a negative safety factor. In this case, if all the faults were present, it would not be possible to erect the scaffold. In practice, not all the faults occur and so such scaffolds are constructed but with very little margin of safety. Although the analyses undertaken in this paper have dealt with only one scaffold it is thought that the principles and results obtained are applicable to scaffolds in general; particularly, the absence of ties and use of putlog connectors.

TABLE 3
LOAD FACTORS FOR DIFFERENT SITE DESIGN/SITE CONTROL COMBINATIONS

Design/Erection	Site Control		
	Good	Average	Poor
Good	9.88	4.14	0.98
Average	3.88	1.79	0.01
Poor	1.19	0.51	-0.18

CONCLUSIONS

The paper has described the analysis of a scaffold system representing typical domestic construction in the UK. The derivation of the regression equation enabling the buckling load factor of a scaffold against imposed loads only when static and wind loads are first applied is given and the use of the interaction formula given in Eurocode 3 elaborated.

The effects of individual faults on the safety of the scaffold have been determined. A table of combined faults has been drawn up to illustrate the relationship between design/erection and site control has been evaluated. It can be seen that site control has a greater influence on safety than the design of scaffolds.

REFERENCES

BS 5973: (1994). Code of Practice for Access and Working Scaffold and Special Scaffold Structures in Steel, BSI, London

Chan W.F., Zhou, Z.H., Chen W.F., Peng, J.L. and Pan A.D. (1995), Stability Analysis of Semi-rigid Steel Scaffolding, *Engineering Structures*, 17,568-574

Chu A.Y.T, Zhou Z.H., Koon, Z.H., Chan S.L., Peng, J.L. and Pan A.D. (1996), Design of Steel Scaffolding using an Integrated Design and Analysis Approach, in *Advances in Steel Structures*, Hong Kong 245-250

Godley M.H.R and Beale R.G. (1997), Sway Stability of scaffolding structures, *The Struct. Eng.* 75:1:4-12

ENV 1993-1-1: 1992. (1993). Eurocode 3. Design of Steel Structures: Part 1.1: General Rules for Buildings, BSI, London

Godley M.H.R and Beale R.G. (2001), Analysis of Large Proprietary Scaffold structures, *Struct and Buildings.* 146:1:31-39

Huang, Y.L, Chen, H.J, Rosowsky, D.V and Kao, Y.G. (2000), Load-carrying capacities and failure modes of scaffold shoring systems, Part 1: Modelling and experiments, *Struct. Eng & Mech.*10:53-66

Lusas User Manual Version 12.2 (1998). FEA Ltd, UK

Maitra, A. (1997). Accidents Associated with Scaffolding in the UK, Technical seminar, Design and use of Temporary Structures, Institution of Civil Engineers, London

Milojkovic, B. (1999), Factors of Safety for standard scaffold structures, PhD Thesis, Oxford Brookes University

Milojkovic, B., Beale, R.G. and Godley, M.H.R. (1996), Modelling Scaffold Connections, in Proc 4th ACME UK An Conf, Glasgow, 85-88

SWAY STABILITY OF STEEL SCAFFOLDING AND FORMWORK SYSTEMS

S.Vaux[1], C.Wong[2] & G. Hancock[3]

[1] SMEC Australia Pty Ltd, Level 5, 118 Walker Street, North Sydney, Australia, 2060
[2] Robert Bird & Partners International Consulting Engineers, Level 10, 30 Clarence Street, Sydney, Australia, 2000
[3] Centre for Advanced Structural Engineering, Department of Civil Engineering, University of Sydney, Australia, 2006

ABSTRACT

This work was undertaken with the objective of examining overall stability and non-linear behaviour of a proprietary high strength steel scaffolding and formwork system. The investigation incorporates both prototype testing of five sub-frames and computer analysis in order to record and then predict the load carrying capacity of the system. Two computer packages were utilised to develop models of the formwork. Analysis assuming linear behaviour was first applied in order to estimate buckling capacities prior to testing. Once testing was completed, non-linear analysis was then used to investigate the stability of the system further. References to current Australian Standards and previous research assisted assumptions and testing methodology. Conclusions about the critical failure mechanisms are drawn and commentary on the overall stability are made.

KEYWORDS

Formwork, scaffold, falsework, non-linear analysis, sway frame, stability, Cuplok

INTRODUCTION

Scaffolding and formwork are temporary structures widely used to support loads on construction sites. Both are typically composed of slender, structural steel members connected together by a variety of proprietary systems. Scaffolding is generally a light simple frame used to carry working loads such as labourers, construction equipment and material. Formwork however is a more robust frame designed to carry larger loads such as wet concrete when poured in-situ.

There has been little development in academic research of scaffolding and formwork systems until recent years, probably due to the temporary nature of the structures. Because of this lack of research, relatively few standard documents exist specifically for the design of falsework systems on construction sites. This is unfortunate since surveys in both Australia and overseas point to scaffolding and formwork as a main cause of failures during construction.

Super Cuplok is a relatively new and innovative modular formwork system developed by Boral Building Services. Based on their prior scaffolding system Cuplok, Super Cuplok is an enhancement, manufactured with higher strength 450 MPa steel. Cuplok and Super Cuplok share the same coupler connection design. This design innovatively requires no nuts, bolts or fixing devices. The ledgers (beams) feature forged blades on both ends of the members. The blades slot into the steel cups that are built into the standards (columns) and are held in place by a locking ring. This patented design proves extremely effective, as it is simple to use and quick to install.

The research carried out in this paper examines the stability of a Super Cuplok frame as a complete sub-assembly. The first stage is identification of the variables that may influence and introduce instability into the system. The focus of this paper is on three variables in particular.

1. Length of jacks
2. Influence of support eccentricities
3. Flexibility of Cuplok connections

Full details are given in Vaux (2000) and Wong (2000).

THEORETICAL DEVELOPMENT

The research described in this paper examines the Super Cuplok formwork system operating under the assumption of a sway frame. A sway frame may be defined as a structure whose upper levels are free to displace laterally. Although Super Cuplok uses cross bracing to provide lateral resistance between beam (ledger) levels, the base jacks may sway producing an overall sway of the frame. The Super Cuplok frame is typically loaded axially into the standards and through the jacks. If load is applied eccentrically, then the standards and jacks may undergo bending, which in turn causes in-plane rotation at the connections. This will therefore place the ledgers in bending and the standards in a combined action of bending and compression. It is the combined action of bending and compression in the vertical members, along with the frames ability to sway that is believed to cause the weakness of the system. Sources of influence on sway in the system are represented in Figure 1.

Adjustable Base Jacks

Common to most scaffolding and formwork, the Super Cuplok system uses circular threaded sections as the supporting members of the structure. The jacks are fitted inside the standard tubes and load is transferred through an adjustable wing nut. These allow the falsework to stand vertically regardless of the foundation level. Super Cuplok adjustable jacks are 38 mm in diameter and may vary between 85 and 600 mm in length set by the height of the wing nut as shown in Figure 1. As the inner diameter of the standards is approximately 40 mm, the difference between sections allows the jacks to be installed simply by hand, prior to erection of the uprights.

The design of the adjustable jacks permits the potential for movement of one vertical element relative to the other. This means that at the point of load transfer, the jack-standard connection may experience lateral movement when subjected to minor load. Theory suggests that this inbuilt eccentricity will be maximised when jack extension is at its greatest. It is predicted that the eccentricity will lead to second-order moments due to the P-Δ effect, and that the additional P-Δ bending moment will reduce the axial capacity of the frame.

Eccentricity of Load

Unintentionally, it is common for scaffolding and formwork to be loaded eccentrically. This may be caused by uneven ground beneath the structure as shown in Figure 1, or by irregular timber formwork on top. Eccentricity of this nature has the effect of introducing moments into the structure upon loading. This will place the standards in a combined action of compression and bending, causing a reduced capacity for vertical loading alone.

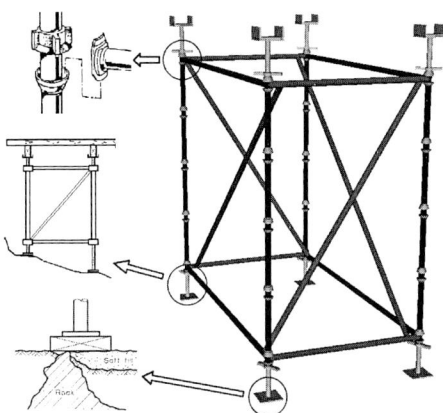

Figure 1: Research Parameters of Instability

The experimental tests described in this paper incorporate an eccentricity of the base jacks in order to replicate and study the base plate support shown in Figure 1. The test conditions of the structure incorporated an eccentricity in the base jack supports as specified in the Australian Formwork Standard AS 3610-1995 Formwork for Concrete, Appendix A3. For sample evaluation of any formwork assembly, for an assembly of height L, the standard recommends a load offset B of:

$$B = \frac{L}{200} \qquad (1)$$

To better understand the significance of this parameter, one of the five prototype tests was carried out without base eccentricity and the remainder with eccentricity.

Rotational Stiffness of the Super Cuplok Connection

Steel connections are commonly assumed and designed as either pinned or rigid. The Super Cuplok nodal connection however is a unique device that exhibits a non-linear semi-rigid nature. By design, the ledgers are not rigidly attached to the standards in the Cuplok connections, but are capable of small, unrestricted rotations. Under load, initially the ledgers possess a degree of flexibility. This rotation continues until prevented by the Cuplok cup. At this point, the connection locks up and begins to act rigidly. To observe how the connection rotational stiffness affects stability, one of the five prototype tests was set apart from the others by the state of its connections. The Cuplok connections of four of the prototypes each held only two elements due to the arrangement of the frame. Each connection in the fifth frame engaged four elements (two ledgers and two ledger ends) to stiffen the connections.

LINEAR-ELASTIC ANALYSIS BASED ON AS 4100-1998 STEEL STRUCTURES

The initial investigation of the Super Cuplok system involved analysing the ultimate limit states of a single bay of the formwork structure as specified in the Australian Steel Structures Standard AS 4100-1998. This analysis was then used to form an estimation of the prototype failure loads during the testing phase.

Under the assumption of a sway frame, the Frame Buckling Load Method (AS 4100-1998 4.4.2.3 (a) (iii)) was selected to calculate the moment amplifications for the frame. The moment amplification factor δ_s in this method is calculated from the equation:

$$\delta_s = \frac{1}{\left(1-\left(\frac{1}{\lambda_c}\right)\right)} \qquad (2)$$

in which λ_c is the elastic buckling load factor, determined from a rational buckling load analysis of the whole frame. The software selected to complete this task (Program PRFSA) was developed by the University of Sydney Centre for Advanced Structural Engineering for the purpose of predicting the in-plane behaviour of plane rigid frames.

Seven ultimate limit states were analysed using AS 4100-1998. Column and beam-column capacities were calculated for both the standard and jacks sections in the uprights, as illustrated in Figure 2.

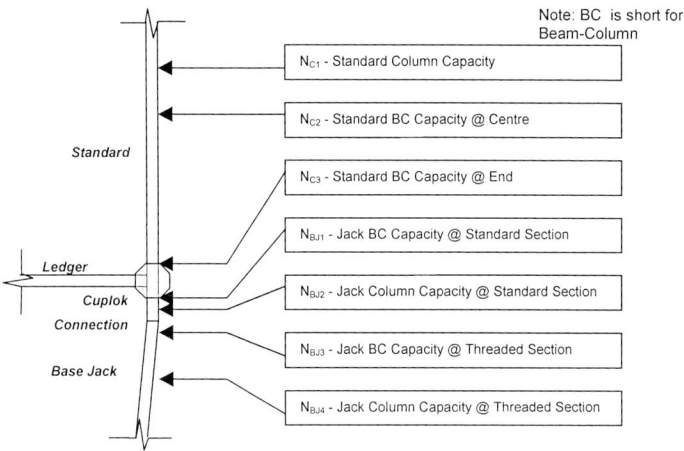

Figure 2: Identified Ultimate Limit States

The ultimate load N^* was calculated for each of the seven limit states with respect to the base jack extension, ranging between 0.1 and 0.6m. The beam-column capacity of the standard at the centre (N_{c2}) is critical for most jack extensions, until the jacks reach close to full extension. At approximately 0.5m extension, the jack column capacity at the standard section (N_{BJ2}) becomes very close to N_{c2}, with only a negligible difference between them.

PROTOTYPE TESTING ON SUPER CUPLOK FRAMES

Five Super Cuplok single bay assemblies were tested in accordance with AS 3610-1995 Formwork for Concrete, Appendix A4 – Destructive Testing. The frames were similar to the configuration in Figure 2 and were loaded axially and lateral deflections measured. The height of the standards between ledgers was 2.0m. The five test frame configurations were constructed to effectively investigate the three variables listed in the Introduction. The five frames are differentiated as follows.

Frame	Configuration
BSF1	Top and Bottom Jacks at 600mm, Eccentric Loading
BSF2	Top and Bottom Jacks at 300mm, Eccentric Loading
BSF3	Top and Bottom Jacks at 300mm, Concentric Loading
BSF4	Top Jack at 300mm, Bottom Jack at 400mm, Eccentric loading
BSF5	Top and Bottom Jacks at 300mm, Eccentric loading, Cuplok connections stiffened

Mechanical theodolites were used to record the plane lateral deflections of the prototypes with one theodolite observing the lower left ledger-standard Cuplok connection (Gauge 1) and the other observing the upper right connection (Gauge 2). Loading on the frames was applied through four hydraulic jacks, one on each standard. The vertical load was increased gradually in increments equivalent to 2.225 kN per standard. The load was increased until structural failure, recognised by rapid lateral deformations in the uprights. The load-deflection plots from the tests are provided in Figure 3.

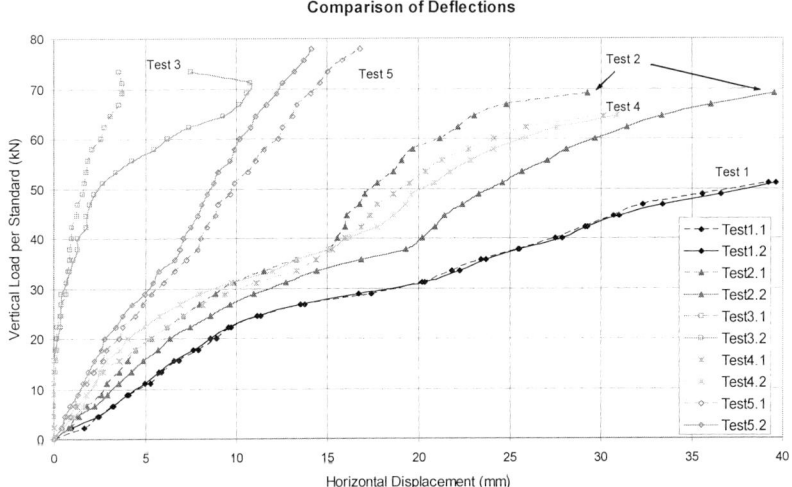

Figure 3: Prototype Test Load-Deflection Results

Observations

The first observation to note from the results is the large range of both ultimate failure and deflections of the five frames. The recorded ultimate capacities range from approximately 51 kN to 78 kN per standard, and the ultimate sway deflections range from approximately 3.7 mm to 39.6 mm.

Although there is a broad variation in results, there are also several trends recognised in this data. An interesting characteristic of the Super Cuplok system is that in all four eccentrically loaded frames, the frames stiffen at loads between 30 and 40 kN. This corresponds to the eccentrically loaded base plate reaching its travel limit as specified in AS 3610.

Failure Load Comparison

The failure loads are given in Table 1 where the percent difference is from BSF2 which is regarded as the reference frame. Raising the jacks to the maximum extension (600mm) caused the greatest reduction in capacity. This confirms the expected influence jack extension holds on structural stability.

TABLE 1: EXPERIMENTAL ULTIMATE TEST

Frame	Ultimate Load (kN)	Difference (%)	Point of Failure
BSF1	51.18	-25.8	Top Jack
BSF2	68.98	-	Standard
BSF3	73.43	6.5	Standard
BSF4	64.53	-6.5	Top Jack
BSF5	77.88	12.9	Bottom Jack

The configuration of BSF4 was loaded concentrically to compare against the eccentrically loaded BSF2 and show a rise of almost 10% when loaded without the eccentric base plates. Another interesting outcome is the substantial capacity rise in BSF5. The results show how stiffening the Super Cuplok connections is an effective method for increasing the strength of the structure.

ADVANCED ELASTIC-PLASTIC ANALYSIS USING SOFTWARE NIFA

The objective of the advanced analysis was to develop a theoretical model that simulated the Super Cuplok formwork system. This form of analysis is a plastic-zone method of finite element analysis which accurately determining forces within the structure to a finer degree of precision than both first and second-order elastic analyses.

NIFA (Nonlinear Inelastic Frame Analysis) was developed at the University of Sydney (Centre for Advanced Structural Engineering) for the purpose of elastic and inelastic non-linear analysis of two-dimensional frames. The 2D NIFA frame model is shown in Figure 4.

As NIFA by default assumes rigid frame member connections, the joints in the Super Cuplok system were modelled using infinitesimal virtual elements with adjusted flexural stiffness. By altering the Young's Modulus (E) of a very small element adjacent to a connection, the reduced rotational stiffness of the connections could be simulated.

In order to simulate the test frame results, the flexural stiffness of the small elements were used as variables. Seven NIFA models were constructed with different combinations of connections as listed in Table 2.

By design, the standards sit on the wing nut attached to the jacks. No pins or connection devices are used at this joint and the standard is free to lift off the jack. The Jack-Standard (J-S) connections consequently permits a degree of movement between the two members and is therefore a variable in this model.

The first stage of the advanced analysis was to develop a NIFA model based on the configuration of the prototype frame BSF2. A comparison with the NIFA models 1-7 with the test BSF2 is shown in Figure 5.

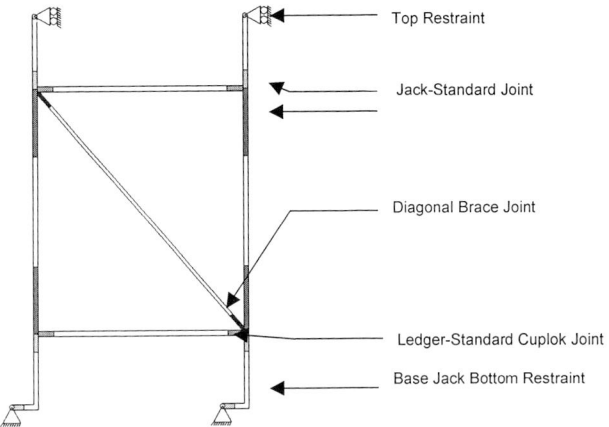

Figure 4: Effective NIFA Model (Not to Scale)

TABLE 2: NIFA TEST VARIABLES

Test	E Values			Test
	Jack-Standard	Cuplok	Base Jack	
NIFA1	Rigid	Measured	Rigid	Initial Values
NIFA2	Rigid	E/1000	Rigid	Pinned Cuplok
NIFA3	Rigid	Rigid	Rigid	Rigid Frame
NIFA4	E/50	Rigid	Rigid	Rigid Cuplok / Flexible J-S
NIFA5	E/50	E/1000	Rigid	Pinned Cuplok / Flexible J-S
NIFA6	E/50	Rigid	E/10	Rigid Cuplok / Flexible J-S & Base
NIFA7	E/50	E/1000	E/10	Pinned Cuplok / Flexible J-S & Base

Note: E = 200,000 MPa

Discussion of Results

The advanced analysis results, when compared against each other, show significant and conclusive trends. The most accurate predictions are typically produced by analyses NIFA2 and NIFA7 shown in Table 3.

The conditions applied in NIFA2 assumed that all joints are rigid except the Cuplok connections, which are pinned. These assumptions proved extremely effective at predicting the failure load of the eccentric test frames, except the concentrically loaded frame. The test conditions of NIFA7 incorporated flexibilities in the model at the Jack-Standard and Base-Jack connections in order to promote further deflections than those produced in NIFA2. As shown in Table 3, predictions are generally much more conservative than the tests except for the concentrically loaded frame which failed in a 3D mode and could not be predicted by NIFA. The results from the two model conditions both confirm that for stability predictions of the Super Cuplok system, the Cuplok connections are best represented with pin joints. This conclusion reinforces the initial theories and clarifies how the connection mechanism operates in a complete frame.

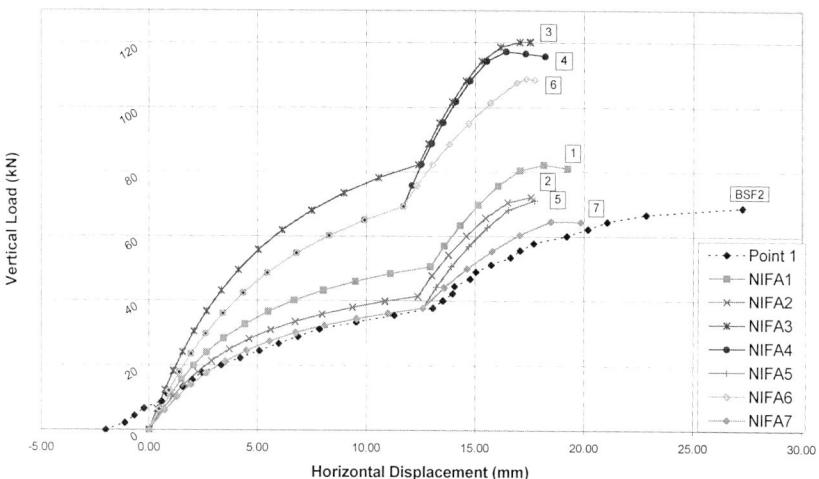

Figure 5: NIFA Output for Frame BSF2

TABLE 3: COMPARISON OF TESTS WITH NIFA2 AND NIFA7 MODELS

TEST	Failure Load (kN)		Accuracy (%)		Accuracy (%)
	Actual	NIFA2		NIFA7	
1	51.18	48.17	-5.87	41.65	-18.61
2	68.98	72.28	4.79	64.45	-6.56
3	73.43	100.50	36.87	88.02	19.87
4	64.53	64.66	0.20	56.39	-12.61
5	77.88	77.36	-0.66	67.58	-13.22

LINEAR DESIGN LOADS TO AS 4100-1998

Linear elastic analysis was used to form initial predictions on the test frame ultimate capacities. Column and beam-column capacities were calculated at specific sections along the uprights as shown in Figure 2. The accuracy of the AS 4100-1998 predictions can now be assessed by comparing with the recorded test data as given in Table 5. The estimates are approximately 10 – 23 % below the test failure loads, demonstrating the conservative nature of the AS 4100-1998 capacity calculations. This result is to be expected due to the safe assumptions incorporated throughout the analysis.

TABLE 4: COMPARISON OF AS 4100-1998 AGAINST TEST RESULTS

Test	AS 4100-1998 Predictions		Test Results		Accuracy
	Failure Load (kN)	Point of Failure	Failure Load (kN)	Point of Failure	(%)
BSF1	45.81	Jack	51.18	Top Jack	-10.48
BSF2	59.77	Standard	68.98	Standard	-13.35
BSF3	-	-	73.43	Standard	-
BSF4	56.39	Standard	64.53	Top Jack	-12.61
BSF5	59.77	Standard	77.88	Bottom Jack	-23.25

Comparing the frames with no additional ledger blades (i.e. BSF1, BSF2 and BSF4), the AS 4100-1998 predictions with capacity factor $\phi=1.0$ prove conservative by approximately 10 – 13 %. This level of accuracy is ideal for design purposes as it forms an estimate with a satisfactory margin of safety. The frame BSF5 incorporated extra ledger blades in each Cuplok joint to strengthen up the connections. In comparison of the results, the AS 4100-1998 predict was conservative by 23 %. From these comparative results it is concluded that the AS 4100-1998 analysis produced conservative predictions for all eccentric frames, regardless of the Cuplok connection flexibility. The linear elastic analysis overall developed safe under-estimates of strength, which is suitable for design.

CONCLUSIONS

This research investigated the ultimate limit states that influence sway stability of the Super Cuplok Formwork Support System. Five complete sub-assemblies were tested in order to observe and record the effects of permissible variations in frame configurations. Ultimate load predictions were developed with both linear and non-linear techniques and the accuracy of the calculations appraised. It may be concluded from these investigations that the structural stability of the Super Cuplok formwork system is complex to predict. By design, the strength of the formwork may be substantially altered by permissible adjustments to configuration. Both forms of analysis produced satisfactory results. The conservative nature of the AS 4100-1998 is beneficial for preliminary design techniques, producing safe and effective capacity estimates. The NIFA analysis achieved a much higher level of precision, appropriate for more advanced stability investigations.

REFERENCES

Vaux, S (2000). *Sway Stability of Steel Scaffolding and Formwork Systems*, Honours Thesis, BE (Civil) Degree, University of Sydney.

Wong, C. (2000). *Sway Stability of Steel Scaffolding and Formwork Systems*, Honours Thesis, BE (Civil) Degree, University of Sydney.

Standards Australia (1998). *Steel Structures*, AS 4100-1998.

Standards Australia (1995). *Formwork for Concrete*, AS 3610-1995.

ACKNOWLEDGEMENTS

The provision of test specimens by Boral Building Services is appreciated.

SECOND-ORDER ANALYSIS AND DESIGN OF STEEL SCAFFOLD USING MULTIPLE EIGEN-IMPERFECTION MODES

S.L. Chan[1], C. Dymiotis[2] and Z.H. Zhou[1]

1   Department of Civil and Structural Engineering
Hong Kong Polytechnic University
Email : ceslchan@polyu.edu.hk

2   Department of Civil & Offshore Engineering
Heriot-Watt University
Edinburgh EH14 4AS Scotland, UK

ABSTRACT

Casualties due to scaffold collapse are common in many countries. However, current design of scaffolds still mainly relies on experimental results and judgement, which are non-scientific, uncertain and unreliable. From experience, lack of fit at joints between scaffold members greatly affects their buckling strength, implying that initial imperfection is an important factor influencing the overall buckling strength of scaffolds. This paper proposes a design approach that uses multiple eigen-buckling modes as initial imperfections in order to simulate imperfection due to lack of fit or crooked members.

KEYWORDS

Design, second-order analysis, nonlinear analysis, buckling, steel structures, frames, scaffolding

INTRODUCTION

There are two main types of scaffolding, namely access scaffolds and the support scaffolds. The former are subjected to moderate loads and are therefore aimed for light to medium duty, whereas the latter take much higher loads from, for instance, concrete weight and workers. Even though collapse frequently occurs in both types of scaffolding systems, this paper addresses the analysis and design of support scaffolds for taking heavy vertical loads.

The main difficulty for designing support scaffolds is the assumption of effective length. Recommendations in design codes are mainly for permanent structures such as portal frames, angle struts in trusses and steel buildings. Separate codes are normally required for the design of scaffolds, as these are temporary structures and exhibit some significantly different characteristics. However, recommendations in design codes or design manuals are normally applicable only to short scaffolds since high and large-scale scaffold systems were seldom tested. The determination of buckling strength of relatively high scaffolds by available methods, such as the eigenvalue type of elastic Euler's buckling load and the effective length method, is inaccurate.

EIGENVALUE METHOD

Conceptually this method assumes that Euler's buckling load is equal to the failure load. No material yielding is assumed and the buckling equation can be written as follows:

$$| K_L + \lambda K_G | = 0 \tag{1}$$

where K_L and K_G are the linear and the geometric stiffness matrices, respectively and λ is the eigen-buckling load factor.

As all practical members deform when load is applied, this method does not consider the large deflection or the P-Δ effect, hence it is not accurate. The calculated buckling load is an upper bound solution to the actual failure load, as shown in Figure 1.

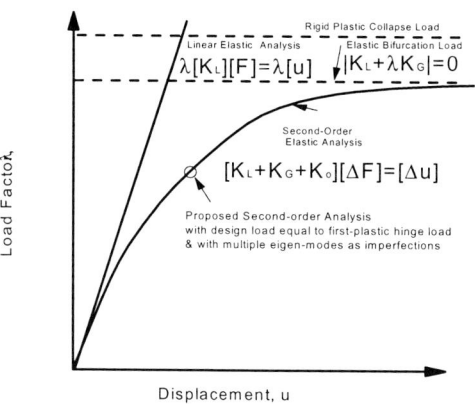

Figure 1 General Analysis Types
for Framed Structures

EFFECTIVE LENGTH METHOD

This approach is normally based on visual inspection of the effective length and the use of Perry Robertson formula for determination of member strength. As an error in assuming an effective length can lead to a very substantial error in buckling strength, the method is considered rather unreliable.

COMBINED EIGENVALUE BUCKLING ANALYSIS AND EFFECTIVE LENGTH METHOD

A more scientific method is to determine the effective buckling load by an eigenvalue bucking analysis as follows:

$$L_e = \sqrt{\frac{\pi^2 EI}{\lambda F}} \qquad (2)$$

where L_e is the effective length, E is the Young's modulus of elasticity, I is the second-moment of area and F is the design load.

This method determines the buckling strength of the most critical member, whereas the effective length of other members subjected to lower loads cannot be reliably determined. For example, in the case of a post under a relatively small axial load, the effective length becomes excessively long and the results cannot be used for sizing the majority of the remaining less critical members.

PROPOSED DESIGN AND ANALYSIS METHOD BY MULTIPLE EIGEN-IMPERFECTION MODES

In the proposed design method for steel frames, an eigenvalue analysis is first carried out by determining the elastic bifurcation load as in Equation 1.

The next question is the selection of the number of eigen-modes to be imposed onto the initial geometry of the structure. Obviously, the mode for the lowest eigenvalue buckling load needs to be included. In some cases, however, the second eigen buckling mode may also need to be considered. For example, when a sway buckling load is close to a non-sway mode, the simple ignorance of the second buckling mode may be dangerous. In the proposed method, the lowest few buckling modes are chosen for imperfections.

With the solution of eigen-vectors and modes, the frequencies are then compared sequentially as,

$$\frac{\lambda_i}{\lambda_{i+1}} = \phi_i \qquad (3)$$

When the ratio of buckling load factors, ϕ, is larger than 1.2, the next mode will be included in the imperfection calculation. This process is terminated once ϕ_i is larger than 1.2, indicating the next eigen-buckling load is sufficiently larger than the lower few eigen-loads and can be considered non-critical.

The eigen-buckling mode does not have an absolute displacement and its displacement vector can only be fixed by assuming a displacement at a particular degree of freedom. Assigning a magnitude for the largest nodal displacement in the eigen-vector, the eigen-imperfection displacement matrix can be determined as,

$$\delta_i = \Phi_i \Big|_{\overline{\Phi}_{max} = \delta_0} \qquad (4)$$

where δ_i is the imperfection displacement vector of the ith mode, Φ_i is the eigen-vector for the same mode and $\overline{\Phi}_{max}$ is the degree of freedom with the maximum relative displacement set equal to δ_0, which is the assumed imperfection displacement assigned by the user.

The complete imperfection displacement vector is determined by adding all the selected eigen-displacement vectors in Equation 4 so that

$$x_i = x_o + \Sigma \delta_i \qquad (5)$$

where x_i and x_o are the imperfect and original coordinates of the structure, respectively.

The second-order analysis procedure containing Equations 3 to 5 above have been incorporated into the computer program Nida for second-order analysis and design by section capacity check without assuming an effective length. A complete description of the method of second-order analysis with section capacity checks is available in the manual of Nida and related publications. Nonetheless, the concept of the formulation is given below for clarity.

Imperfect Element

A member containing imperfection can be written by using an imperfection v_0 in the energy equation as follows:

$$EI\ddot{v} = P(v+v_o) + \frac{M_1+M_2}{L}(\frac{L}{2}+x) - M_1 \qquad (6)$$

$$EI\dddot{v} = P\dot{v} + \frac{M_1+M_2}{L}$$

$v = v_1$ and $\dot{v} = \theta_1$ at $x = 0$
$v = v_2$ and $\dot{v} = \theta_2$ at $x = L$

where E is the Young's modulus of elasticity, I is the second moment of area, v is the lateral deflection, P is the axial force and L is the member length. A super dot represents a first derivative with respect to x. M_1 and M_2 are the nodal moments in a member. It is noted that a single element is adequate for modelling a member using the above formulation.

Using a fifth order polynomial function, the element stiffness matrices can be derived from the six constraints in Equation 6 by a standard procedure with careful consideration of the various important second-order terms. It is of interest to note that the coupling between the axial force and bending moment must be considered, otherwise the element is expected to experience a convergence difficulty during the iterative process.

DESIGN METHOD

The Newton-Raphson method improved by an minimum residual displacement iterative scheme (Chan, 1988) and the arc-length incremental scheme is used to trace the equilibrium path. This has a better convergence than the conventional Newton Raphson method and also benefits from the capacity to traverse the limit point.

The deflection in Equation 3 can be obtained for every member in each load cycle after satisfaction of equilibrium condition. Unlike the conventional cubic Hermite function, the displacement function here is a function of axial force, as well as nodal rotations. In order to obtain the location for maximum curvature, the third derivative of the function is evaluated and the zero slope condition is sought from

$$\dddot{v} = 0 \text{ at } x = x_m \qquad (7)$$

The maximum bending moment can then be determined from

$$M_m = EI\ddot{v}_{x=x_m} \tag{8}$$

There exist at most two solutions for maximum moment and the larger moment will then be chosen for strength design. With these values of moment and axial force, the maximum bending stress in the element can be computed simply as,

$$\sigma_m = \frac{M_m}{Z_y} + \frac{M_m}{Z_y} + \frac{P}{A} \tag{9}$$

where M_m is the bending moment, Z is the section modulus, and A is the cross sectional area. Depending on whether the formation of the first plastic hinge is allowed or a purely elastic design is preferred, the elastic or plastic section modulus can be used in Equation 9.

EXAMPLE: SECOND-ORDER ANALYSIS AND DESIGN OF A 3-STOREY SCAFFOLD

The support scaffold shown in Figure 2 is analysed. The height of the structure is 5.37m and the other dimensions are a width of 1.79m and a depth of 1.22m. The yield stress of the material is 350 MPa and the assumed Young's modulus is 200 kN/mm^2. The cross-sectional areas of the vertical and horizontal members are 687mm^2 and 306mm^2, respectively, while the second moments of areas are 271,700mm^4 and 63,300mm^4, respectively. The cross bracing has an outer diameter of 25mm with a 2.2mm thick tubular section.

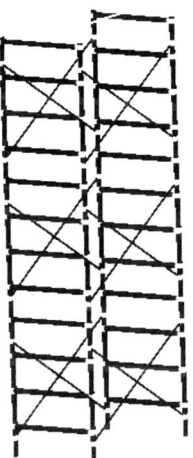

Figure 2. The configuration of the analysed scaffold (height = 5.37 m)

Two specimens of identical dimensions were tested to failure by Weesner and Jones (2001). These were loaded by a jack at the top, with lateral restraint at the top enabled by friction between the jack and the unit, which may not be achievable on-site. Using the proposed method with an imperfection of 1/500 of the scaffold height, the design strength is determined as 396 kN and the buckling load as 509 kN, which are less than the experimental loads. However, the elastic eigenvalue buckling load is 595 kN, which is significantly higher than the two tested failure loads. Using this buckling load to calculate the effective length and then BS5950(2000), the permissible design load is 505 kN, which is between the two tested loads.

From this example, it is found that the simpler eigenvalue buckling load cannot be taken as the design load capacity since it is much larger than the tested failure load. The proposed analysis using the eigen-mode as the initial imperfection is conservative and results in values that are lower than the tested failure load by 25-30%. The large deflection buckling load is close to the tested failure load, but it appears that the actual scaffolds on site may have a larger imperfection than the ones tested in the laboratory and the safety factor may not be adequate.

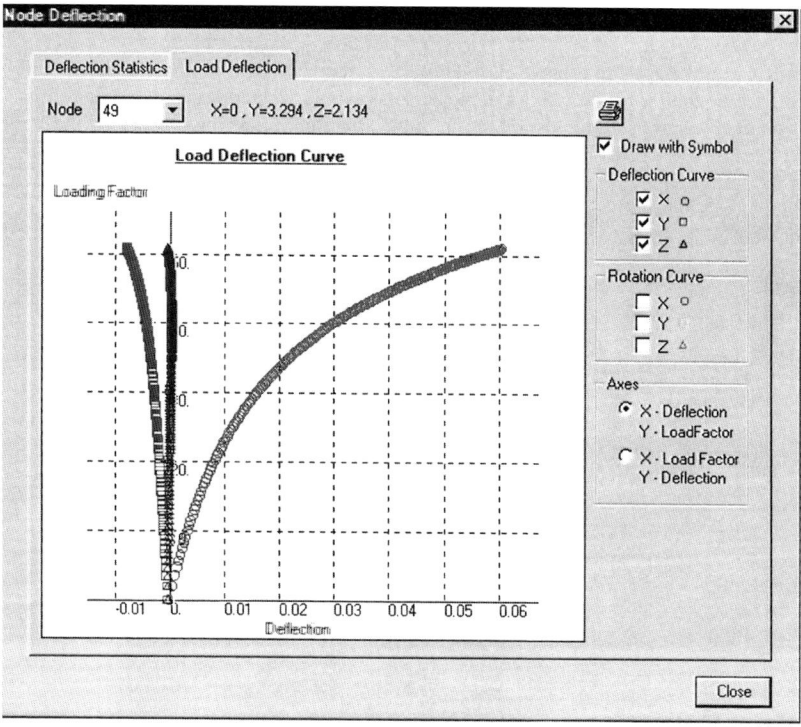

Figure 3. The load vs. deflection path

CONCLUSIONS

A second-order analysis method using the eigen-buckling mode as imperfection is proposed for design of practical scaffolds on site. The magnitude of the imperfection is taken as 1/500 of the scaffold height. Although the design load is smaller than the ones tested in the laboratory by 25-30%, the tested specimens were exposed to a more ideal condition than the actual scaffolds on site. Nevertheless, a design load closer to the tested load can be obtained by adjusting the imperfection to, say, 1/1000 of the scaffold height. It is therefore possible for engineers to assess more directly the condition of a new or used scaffold in-situ or in the laboratory by assuming the length of the global imperfection rather than a reduced effective length for an adverse condition. This method can be used in design of scaffolds under various conditions.

ACKNOWLEDGEMENT

The authors gratefully acknowledge that the work described in this paper was partially supported by a grant from the Research Grant Council of the Hong Kong Special Administrative Region on the project "Advanced analysis and design of steel frames allowing for beam-column inelastic buckling (Project No. B-Q465)".

REFERENCES
British Standards Institution, BS5950, Part 1 (2000), Structural Use of steel-work in buildings, BSI, London.
Chan, S.L.,"Geometric and Material Nonlinear Analysis of Beam-Columns and Frames using the Minimum Residual Displacement Method", *International Journal for Numerical Methods in Engineering,* vol. 26, 1988, pp.2657-2669.
Chan, S.L. and Zhou, Z.H., "Second order analysis of frame using a single imperfect element per member", *Journal of Structural Engineering, ASCE,* vol. 121, No. 6, June, 1995, pp.939-945.
NAF-Nida (2002), "Non-linear integrated design and analysis", Naf Series, User's Manual, version 5.
Weesner, L.B. and Jones, H.L., "Experimental and analytical capacity of frame scaffolding", Engineering Structures, 23, 2001, pp.592-599.

COLD-FORMED STEEL

ON THE DISTORTIONAL POST-BUCKLING BEHAVIOUR OF COLD-FORMED LIPPED CHANNEL STEEL BEAMS

Luís Carlos Prola[1]   Dinar Camotim[2]

[1] Civil Eng. Dept., SU Maringá, Av. Colombo, 87020-900 Maringá, Brazil
[2] Civil Eng. Dept., IST, TU Lisbon, Av. Rovisco Pais, 1049-001 Lisboa, Portugal

ABSTRACT

This paper presents and discusses the results of an investigation on the elastic distortional post-buckling behaviour of cold-formed lipped channel steel beams (members subjected to pure bending). Such results, obtained from spline finite strip analyses accounting for initial geometrical imperfections (critical buckling mode shape), consist of (i) equilibrium paths and (ii) curves describing the post-buckling evolution of the relevant (ii$_1$) displacement and stress longitudinal profiles and (ii$_2$) cross-section stress distributions. In particular, a non negligible post-buckling asymmetry, with respect to the cross-section distortion sign, is revealed and shown to stem from differences in the flange-stiffener warping behaviour. The results presented also concern the influence of the local end support conditions on the beam distortional post-buckling behaviour (stiffness and strength). Such end conditions involve the restraint to (i) the end section warping and/or to (ii) the flexural rotation of the wall element transverse edges.

KEYWORDS

Distortional post-buckling, Cold-formed steel beams, Lipped channel beams, Spline finite strip method, Warping effects, Warping restraint, Plate rotation restraint.

INTRODUCTION

The structural behaviour of cold-formed steel beams is often affected by an instability phenomenon known as *distortional buckling*, a designation stemming from the buckling mode (distortional mode – *DM*) nature: it involves cross-section *distortion*, *i.e.*, a number of internal wall elements experience "quasi rigid-body" membrane transverse displacements, which implies displacements and deformations at beam internal longitudinal edges (fold lines). In order to illustrate this phenomenon, Figure 1 schematically displays the distortionally buckled configuration of a lipped channel segment subjected to major axis pure bending. Notice the displacements taking place at the top (compression) flange-stiffener fold line and observe also that (i) the top flange and stiffener remain practically undeformed and (ii) only the upper half of the web exhibits significant flexural deformation. In physical terms, an analogy can be made between the beam distortional buckling behaviour and the torsional instability of an "elastically restrained sub-member subjected to uniform compression". In fact, one may say that the beam distortional buckling behaviour is

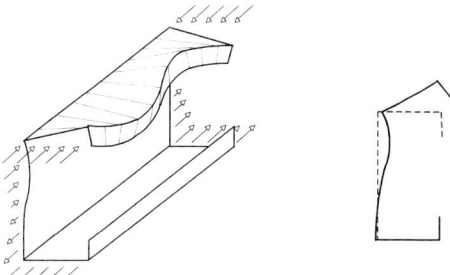

Figure 1: Deformed configuration of a distortionally buckled beam segment

"triggered" by the torsion of the "top flange-stiffener plate sub-assembly" about the web-flange edge, which is continuously restrained by the remaining walls (in order to ensure compatibility, the web exhibits flexural deformations). Therefore, such behaviour may be assimilated to the torsional instability of an "L-shaped" uniformly compressed column with one longitudinal edge elastically restrained against the rotation and the other free. This similarity was first detected by Lau & Hancock (1987), who used it to devise the column distortional buckling structural model depicted in Figure 2(b), later extended to beams by Hancock (1996). The model provides accurate distortional bifurcation stress estimates and has been adopted by the current Australia/New Zealand cold-formed steel design code (1996). Notice, however, that its application is restricted to beams with pinned and free-to-warp end sections.

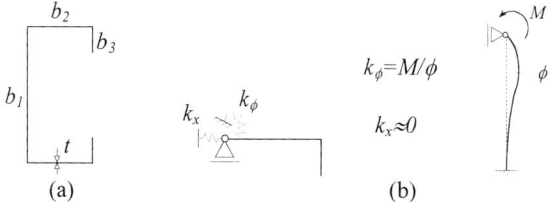

(a) (b)
Figure 2: (a) Cross-section geometry and (b) distortional buckling structural model

Sridharan (1982) carried out the first study on the thin-walled member distortional post-buckling behaviour, adopting a semi-analytical method to analyse lipped channel columns compressed between "fictitious rigid platens". About 10 years later, Kwon & Hancock (1992, 1993) resumed this topic and investigated, numerically (spline finite strip method) and experimentally, the behaviour of lipped channel columns with fixed and free-to-warp end sections. Quite recently, Prola (2001) and Prola & Camotim (2002) also used spline finite strip analyses to study the distortional post-buckling behaviour of lipped channel columns and to assess how the local end support conditions affect their post-buckling strength. Finally, one should mention the work of Schafer (2000), reporting that, regardless of the (elastic) critical buckling mode nature, cold-formed steel short members predominantly exhibit (elastic-plastic) distortional collapses.

The objective of this paper is to present and discuss the results of an investigation on the elastic distortional post-buckling behaviour of cold-formed lipped channel steel *beams*, *i.e.*, members subjected to major axis pure bending[1], thus extending the previous investigation. The results reported here were obtained from spline finite strip analyses, which account for critical-mode-shape initial geometrical imperfections, and consist of: (i) post-buckling equilibrium paths, (ii) displacement and stress longitudinal profiles and (iii) cross-section stress distributions. Such results have been chosen with the aim of illustrating several distortional post-buckling aspects, namely (i) the effect of the initial cross-section distortion sign, (ii) the displacement and stress evolution and (iii) the influence of the local end support conditions.

---

[1] Such members are commonly used as roof purlins.

GEOMETRY AND SUPPORT CONDITIONS

All the lipped channel beams analysed displayed the same cross-section geometry, defined by the following mid-line dimensions (see Figure 2(a)): $b_1=118.7\,mm$, $b_2=88.7\,mm$ ($b_2/b_1 \approx 0.75$), $b_3=5.46\,mm$ ($b_3/b_1 \approx 0.046$) and $t=1.08\,mm$ ($b_1/t \approx 110$). Together with the beam length values selected, such dimensions ensure that, for all the end support conditions dealt with, the critical local bifurcation stress corresponds to distortional buckling in a mode exhibiting the least possible half-wave number. The steel properties $E=205\,GPa$ (Young's modulus) and $\nu=0.3$ (Poisson's ratio) were adopted and all the post-buckling analyses performed assumed the beam to contain critical-mode-shape (DM) initial geometrical imperfections with an amplitude $v_{f\text{-}s0}/t=0.1$, where $v_{f\text{-}s0}$ is the maximum initial transverse displacement at the compressed (top) flange-stiffener longitudinal edges and t is the wall thickness.

An accurate assessment of the beam *local* end support conditions (*i.e.*, conditions involving the membrane and flexural displacements of the wall element transversal edges) is essential if one aims at meaningful comparisons between numerical and experimental results. In fact, performing numerical simulations of experimental tests without adequately modeling the end conditions provided by the testing set-up is bound to make any comparison virtually meaningless and may even lead to erroneous conclusions. First of all, one must clarify the meaning of *local* end conditions, as opposed to *global* end conditions: while the former involve the displacements of the wall element transverse edges, the latter are related only to the end section rigid-body motions. Concerning the local end conditions, it is still convenient to deal separately with the *longitudinal* (warping) and *transverse* (flexural) displacements.

As far as the end section warping is concerned, the two conditions considered here correspond to (i) *free* longitudinal displacements and (ii) longitudinal displacements *conditioned* in such a way that the end section is forced to remain plane, although it may experience rigid-body translations and/or (flexural) rotations. Such conditions lead to *free-to-warp* (*FW*) and *warping-free* (*WF*) end sections, respectively. Notice that, in physical terms, the last condition corresponds to an end section "glued" to a "rigid platen", a situation often arising in experimental tests – the specimen end section is in full direct contact with the testing machine end plate internal surface, thus being prevented from warping (the support condition of the end plate *external* surface then provides the member *global* boundary condition).

The transverse displacements of the wall element transverse edges are always assumed fully prevented (recall that, depending on the wall element orientation, such displacements may be either membrane or flexural). However, concerning the edge flexural rotations, two conditions are dealt with here, namely (i) *pinned edges* (*PE*) and (ii) *fixed edges* (*FE*). Physically, these conditions correspond to having the end section connected to a "rigid platen" by means of rollers and welds, respectively. In experimental tests, the "natural" local support condition (specimen end section simply resting in direct contact with the testing machine end plate internal surface) invariably leads to an intermediate situation, *i.e.*, to partially restrained edges. However, specific measures are often adopted to ensure fully restrained edges (Rasmussen, 2000).

WARPING EFFECTS

Before turning our attention to the numerical results, it is important to address a surprising phenomenon recently reported by Prola & Camotim (2002), which specifically concerns lipped channel members and is related to the influence of cross-section warping on the distortional post-buckling behaviour. It stems from the fact that the warping taking place at the compressed (top) stiffener and flange heavily depends on the cross-section distortion sign (*i.e.*, on whether the top flange "opens" or "closes"), as illustrated in Figures 3(a) and 3(b). Such figures show the longitudinal deformations appearing in an arbitrary centred stiffener segment of length a_0 ($a_0 \leq a$, where a is the beam length), due to warping displacements related to negative ($v_{f\text{-}s}>0$) and positive ($v_{f\text{-}s}<0$) cross-section distortion. The stiffener segment belongs to a beam with *FW-PE* end sections and the observation of the figures below leads to the following conclusions:

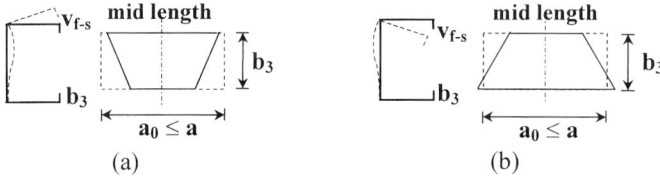

Figure 3: Characterisation of the compressed stiffener warping displacements: (a) $v_{f\text{-}s}>0$ and (b) $v_{f\text{-}s}<0$

(i) The (compressed) stiffener warping displacements are *qualitatively different* for $v_{f\text{-}s}>0$ and $v_{f\text{-}s}<0$. The same applies to the compressed flange (near the flange-stiffener corner), but to a smaller extent.

(ii) For $v_{f\text{-}s}>0$ (*negative* distortion), the stiffener deformed configuration due to warping exhibits larger inward displacements near the free end, *i.e.*, it is similar to the configuration of a beam acted by negative bending moments.

(iii) For $v_{f\text{-}s}<0$ (*positive* distortion), an opposite behaviour takes place, as larger inward displacements occur near the flange-stiffener corner, while the free end experiences small outward displacements. Thus, the stiffener deformed configuration is now similar to that of a beam acted by positive bending moments.

Due to the above cross-section warping asymmetry, the compressed stiffener (mostly) and flange contribute differently to the beam distortional post-buckling behaviour for buckling modes involving negative and positive cross-section distortion. This means that an accurate characterisation of such beam behaviour requires a distinction between the two cases, as will be amply demonstrated in the next section.

DISTORTIONAL POST-BUCKLING RESULTS

The numerical results presented and discussed here, obtained from spline finite strip analyses, comprise:
(i) *Post-buckling equilibrium paths*, *i.e.*, curves relating the average applied stress σ (normalised by the critical bifurcation stress σ_{cr}) and the maximum compressed flange-stiffener edge vertical displacement $v_{f\text{-}s}$ (normalised by the wall thickness t). As already mentioned, all post-buckling paths correspond to beams with DM-shape initial imperfection of amplitude $|v_{f\text{-}s0}|=0.1t$.

(ii) Sequences of *displacement longitudinal profiles*, *i.e.*, curves showing the evolution, as post-buckling progresses, of the variation of $v_{f\text{-}s}$ along the beam length ($v_{f\text{-}s}(x)$).

(iii) Sequences of *normal stress distributions* along (iii$_1$) the compressed flange-stiffener edge ($\sigma_{f\text{-}s}(x)$) and (iii$_2$) the most deformed cross-section mid-line ($\sigma(s)\equiv\sigma_s$), which make it possible to visualise the post-buckling stress evolution.

In addition, the beams analysed have the following three local support conditions (identical for both end sections): (i) FW-PE, (ii) FW-FE and (iii) WF-PE. In the first case, the critical buckling mode exhibits a single half-wave and results concerning both positive and negative cross-section distortion are presented. In the remaining two cases, on the other hand, the beams buckle in two half-wave modes and the results reported are associated only to negative $v_{f\text{-}s}$ values (*i.e.*, they correspond to just one of the half-waves).

Beams with Free-to-Warp End Sections and Pinned Edges (FW-PE)

Figure 4 depicts the post-buckling equilibrium paths ($v_{f\text{-}s}$ at mid-span) of two identical beams ($a=380\,mm$ – $a/b_1\approx 3.2$ and $\sigma_{cr}=55.6\,MPa$) containing positive and negative *DM* initial imperfections with amplitude $|v_{f\text{-}s0}|=0.1t$. First of all, one notices that, regardless of the $v_{f\text{-}s}$ sign, the beam distortional post-buckling strength is much smaller than its local-plate counterpart (*e.g.*, Prola, 2001). Concerning the difference between the two curves, the following remarks are appropriate:

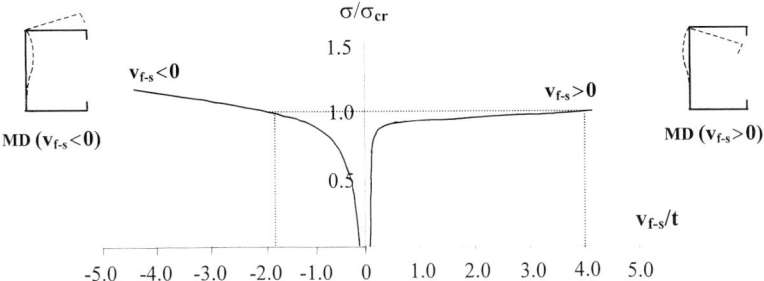

Figure 4: Distortional post-buckling equilibrium paths ($|v_{f-s0}|=0.1t$)

(i) The beam distortional post-buckling behaviour exhibits a non negligible dependence on the cross-section distortion sign and, therefore, is clearly *asymmetric*. For the same applied stress level, the negative v_{f-s} values are a bit smaller than the positive ones[1], thus indicating a higher post-buckling strength (*i.e.*, a larger warping restraint).

(ii) In addition, the beam pre-buckling behaviour appears to be also asymmetric with respect to v_{f-s}. In fact, the equilibrium paths related to $v_{f-s}>0$ and $v_{f-s}<0$ are qualitatively and quantitatively different, as the former (ii$_1$) is clearly stiffer and (ii$_2$) displays a more "abrupt" stiffness reduction (pronounced curvature) in the vicinity of the critical bifurcation applied stress level. Such difference, surprising at first glance, stems for the occurrence of *global flexural* deformations, which, for positive bending moments, correspond to *negative* v_{f-s} values. Therefore, for $v_{f-s0}>0$, the global flexural and distortional buckling v_{f-s} values have opposite signs and the latter only becomes predominant near the critical bifurcation stress ($\sigma/\sigma_{cr} \approx 1$). For $v_{f-s0}<0$, on the other hand, the two effects are additive, which leads to significantly higher v_{f-s} values and explains the more flexible and "smooth" behaviour.

Figures 5(a)–(b) display, both for $v_{f-s}>0$ and $v_{f-s}<0$, the post-buckling evolution ($0.8 \leq \sigma/\sigma_{cr} \leq 1.4$) of the flange-stiffener edge normal stress distribution ($\sigma_{f-s}(x)$). They clearly show a strong dependence of the post-buckling stress distribution on the distortion sign, which is due to the fact that such stresses are mostly generated by the warping restraint (the bending deformations are negligible). In addition, from the observation of the two sets of curves one concludes that:

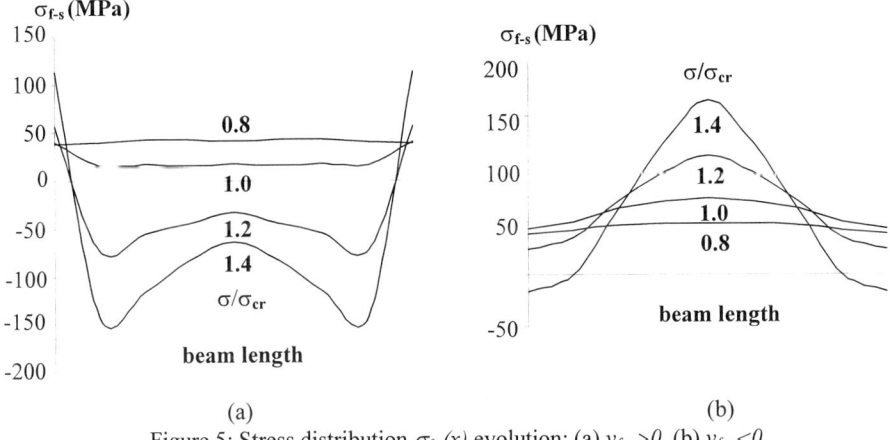

Figure 5: Stress distribution $\sigma_{f-s}(x)$ evolution: (a) $v_{f-s}>0$ (b) $v_{f-s}<0$

---

[1] The evolution of the v_{f-s} longitudinal profile, not shown here due to space limitations, also shows that the single half-wave deformed configuration remains unaltered throughout the whole post-buckling range (Prola, 2001).

(i) For $v_{f\text{-}s}>0$, the warping stresses are *tensile* along the whole length and they decrease as one travels from mid-span to the (free-to-warp) end sections. As a result, the post-buckling stresses are mostly tensile (compressive stresses only near the ends), with a maximum value at about 1/8-span.

(ii) For $v_{f\text{-}s}<0$, the warping stresses are *compressive* and they also decrease as one travels from mid-span to the end sections. Thus, the post-buckling stresses are mostly compressive (small tensile stresses appear near the ends, for $\sigma/\sigma_{cr}=1.4$), with a maximum value at mid-span.

Finally, Figures 6(a)–(b) show, once again for $v_{f\text{-}s}>0$ and $v_{f\text{-}s}<0$, the evolution of the stress distribution at the beam mid-span cross-section ($\sigma(s)\equiv\sigma_s$), for $0.8 \leq \sigma/\sigma_{cr} \leq 1.4$. The observation of these two sets of 4 diagrams shows that the stress distribution configuration remains practically unaltered in the bottom stiffener, flange and half-web, where the applied stresses are tensile[1]. On the other hand, the distortion sign strongly influences the post-buckling stress distribution on the compressed top stiffener (mostly) and flange (near the flange-stiffener corner). In particular, it is important to remark that:

(i) For $v_{f\text{-}s}>0$, the compressive stresses "move away" from the flange-stiffener corner and "accumulate" near the web-flange corner and, to a larger extent, near the stiffener free end. Notice that (small) tensile stresses develop around the flange-stiffener corner and that the stresses exceed *1000 MPa* (more than 17 times the critical stress) near the free end.

(ii) For $v_{f\text{-}s}<0$, the compressive stresses "accumulate" around the flange-stiffener corner and "move away" from the stiffener free end, where an early development of significant tensile stresses takes place. Such tensile stresses exceed *700 MPa* – more than 15 times the critical stress value.

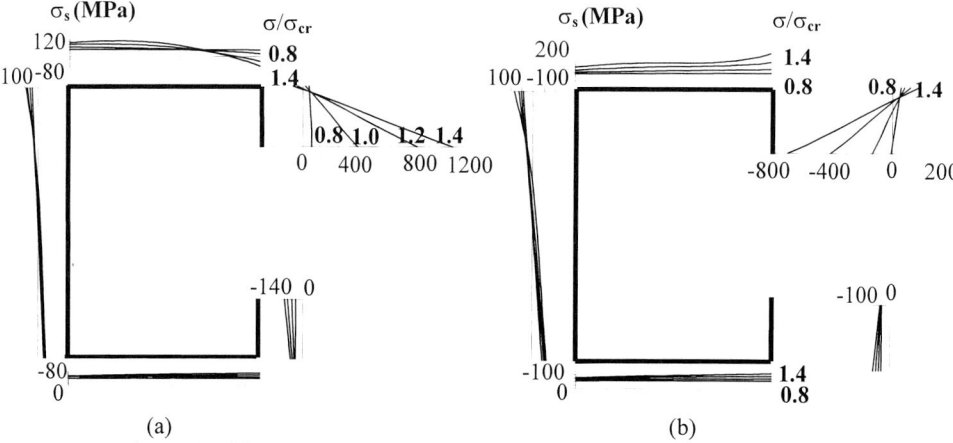

Figure 6: Mid-span stress distribution $\sigma(s)$ evolution: (a) $v_{f\text{-}s}>0$ (b) $v_{f\text{-}s}<0$

One last word to mention that both these stress distributions ($v_{f\text{-}s}>0$ and $v_{f\text{-}s}<0$) are radically different from the one associated to the local-plate post-buckling behaviour (*e.g.*, Prola, 2001), which implies that the classical "effective width" concept cannot be employed to estimate the ultimate loads of distortionally buckled beams (the compressed stiffener and flange stress diagram configurations are completely distinct).

Beams with Other End Support Conditions

In order to assess and compare the influence of preventing (i) the end section warping and/or (ii) the flexural rotation of the wall element transverse edges, we first determine the equilibrium paths of two additional beams. The lengths, minimum distortional bifurcation stresses and critical buckling mode half-wave numbers (*n*) of such beams are: (i) $a=800\,mm$, $\sigma_{cr}=71.3\,MPa$ and $n=2$, for the *FW-FE* beam, and (ii) $a=800\,mm$, $\sigma_{cr}=72.1\,MPa$ and $n=2$, for the *FW-PE* beam.

---

[1] Notice, however, that there are (small) differences in the bottom flange and stiffener stress distribution configurations.

Figure 7 displays the post-buckling equilibrium paths of the three beams, with the applied stress normalised with respect to the lowest critical stress ($\sigma_{crR}=55.6\,MPa$ – FW-PE beam). It is important to notice that (i) the *FW-PE* beam equilibrium path corresponds to $v_{f-s}<0$ and that (ii) the remaining two equilibrium paths correspond to the *maximum negative v_{f-s} value* (recall that $n=2$)[1].

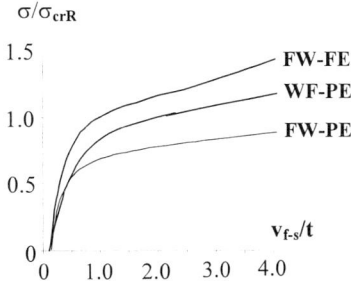

Figure 7: Influence of the local support conditions on the beam distortional post-buckling behaviour

A comparison between the three equilibrium paths shows that either fixing the wall element edges or preventing the end section warping significantly increases the beam post-buckling stiffness and strength, although the *FW–FE* and *WF–PE* curves exhibit different configurations. Notice that, due to the beam global flexural behaviour (obviously stiffer for fixed wall element edges) and in contrast with the column post-buckling behaviour (Prola & Camotim, 2002), the *WF–PE* curve always lies below the *FW–FE* one[2]. Just to provide an idea of how the local support conditions influence the beam post-buckling displacements and stresses, Figures 8(a)–(b), 9(a)–(b) and 10(a)–(b) display the evolution ($0.6 \leq \sigma/\sigma_{cr} \leq 1.4$) of (i) the longitudinal profile of v_{f-s}, (ii) the longitudinal profile of σ_{f-s} and (iii) the stress distribution σ_s at the cross-section with the highest *negative* v_{f-s} value, both for the *FW–FE* and *WF–PE* beams. These three pairs of figures provide a clear evidence that, in spite of a few quantitative differences, the two beams behave rather similarly, in qualitative terms. Moreover, the observation of such figures leads to the following remarks:

(i) The "negative" half-wave amplitudes are slightly larger than the "positive" ones. Together with the stress profiles presented in figure 9 (the longitudinal variation of σ_{f-s} is clearly distinct for $v_{f-s}>0$ and $v_{f-s}<0$), this result confirms the conclusions drawn from figure 4.

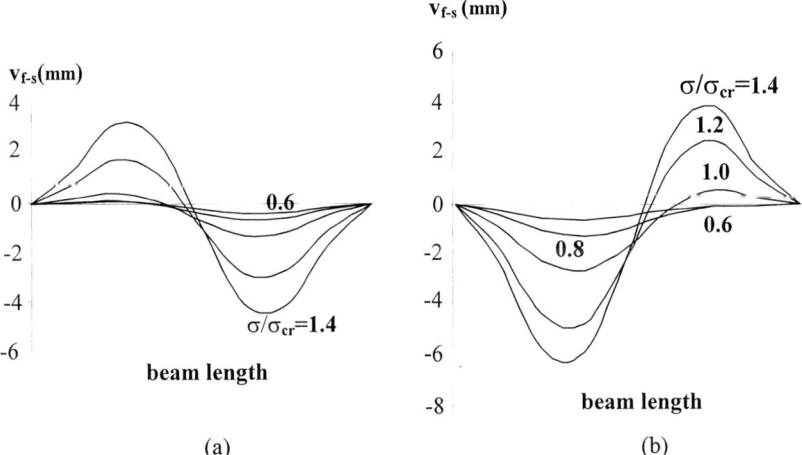

Figure 8: v_{f-s} longitudinal profile evolution: (a) *FW–FE* and (b) *WF–PE* beams

---

[1] The *FW-FE* and *WF-PE* beam maximum negative v_{f-s} values occur at the 3/4 and 1/4-span cross-sections (see Figure 8).
[2] Moreover, notice also that, for low σ/σ_{crR} values, the *FW-PE* and *WF-PE* curves practically coincide (global flexure only).

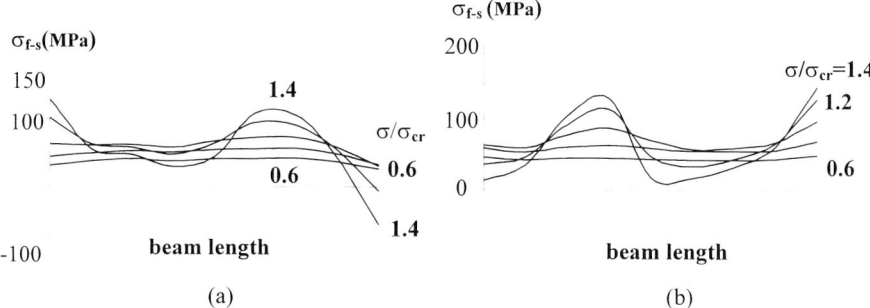

Figure 9: $\sigma_{f\text{-}s}$ longitudinal profile evolution: (a) *FW–FE* and (b) *WF–PE* beams

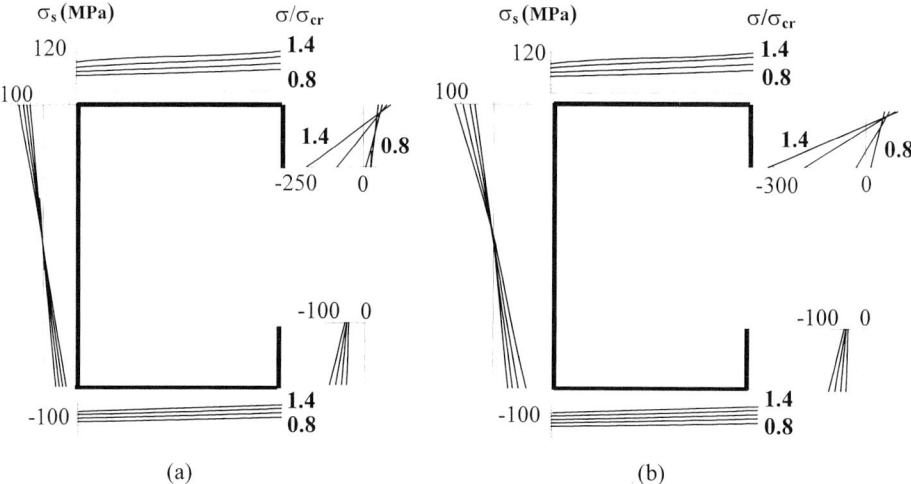

Figure 10: Stress distribution $\sigma(s)$ evolution: (a) *FW–FE* (3/4-span) and (b) *WF–PE* (1/4-span) beams

(ii) The stress distributions in the "most positively distorted" ($v_{f\text{-}s} < 0$) cross-section are qualitatively similar for the *FW–PE* (Figure 6(b)), *FW–FE* (Figure 10(a)) and *WF–PE* (Figure 10(b)) beams. However, the tensile stress values developing in the compressed (top) stiffener are significantly higher in the first case, a feature stemming from the more relevant role played by the warping effects.

(iii) The tensile stresses appearing in the vicinity of the compressed stiffener free end are somewhat (about 20%) higher in the *WF–PE* beam than in the *FW–FE* one. Moreover, for $\sigma/\sigma_{cr}=1.4$, such stresses reach values corresponding to 3-4 times the critical stress σ_{cr} (recall that, in the *FW–PE* beam, this ratio exceeded 15).

(iv) In the top (compressed) flange, the stress distribution remains fairly constant, as the "stress transfer" to the flange-stiffener corner is even smaller than the one observed in the *FW–PE* beam. This means that, for this particular cross-section geometry, the flange local-plate buckling effects are negligible.

(v) In the top (compressed) half of the web, virtually no "stress transfer" to the web-flange corner takes place, which indicates that, again for this particular cross-section geometry, web local-plate buckling effects are also absent (recall that a mild "stress transfer" was observed in the *FW–PE* beam).

(vi) As expected, the stress distribution configuration in the bottom cross-section half, which is subjected to linearly varying tensile stresses, remains practically unaltered throughout the whole (pre and post-buckling) applied stress range.

CONCLUDING REMARKS

The results of an investigation dealing with the distortional post-buckling behaviour of cold-formed lipped channel steel beams were reported and discussed. The beams analysed displayed a commonly used cross-section geometry and the results presented were obtained by means of spline finite strip analyses including initial geometrical imperfections (critical buckling mode shape). Such results consisted of post-buckling equilibrium paths and curves showing the evolution, along those paths, of the most relevant displacements and stresses: (i) the compressed flange-stiffener corner vertical displacement and normal stress longitudinal profiles and (ii) the stress distributions in the most distorted beam cross-section. This work also addressed the influence of the end section local support conditions on the beam distortional post-buckling behaviour and strength. The support conditions considered involved (i) the wall element transverse edge flexural rotation restraint and (ii) the end section warping restraint.

The results displayed provide a contribution to a better understanding of the distortional post-buckling behaviour of thin-walled lipped-channel beams. In particular, the following aspects deserve to be specially mentioned:

(i) Warping effects are responsible for a non negligible asymmetry, with respect to the cross-section distortion "sign", exhibited by displacements and stresses (influencing mostly the comprerssed stiffener stress distribution). Due to the global flexural behaviour, such asymmetry is a bit different from the one observed in uniformly compressed members (Prola & Camotim, 2002).

(ii) In qualitative terms, the beam distortional post-buckling behaviour was shown to be rather similarly affected by either fixing the wall element edges or preventing the end cross-section warping. In fact, in both cases (ii$_1$) the critical distortional mode exhibited two half-waves, instead of one, and (ii$_2$) the beam experienced significant stiffness and strength increases (higher in the first case).

REFERENCES

Hancock G.J. (1997). Design for Distortional Buckling of Flexural Members. *Thin-Walled Structures* **27:1**, 1063-1078.

Kwon Y.B. and Hancock G.J. (1992). Strength Tests of Cold-Formed Channel Sections Undergoing Local and Distortional Buckling. *Journal of Structural Engineering* (ASCE), **117:2**, 1786-1803.

Kwon Y.B. and Hancock G.J. (1993). Postbuckling Analysis of Thin-Walled Channel Sections Undergoing Local and Distortional Buckling. *Computers and Structures*, **49:3**, 507-516.

Lau S.C. and Hancock G.J. (1987). Distortional Buckling Formulas for Channel Columns. *Journal of Structural Engineering* (ASCE) **113:5**, 1063-1078.

Prola L.C. (2001). *Local and Global Stability of Cold-Formed Steel Members*. Ph.D. Thesis, Civil Eng. Dept., Technical University of Lisbon, Portugal. (in portuguese)

Prola L.C. and Camotim D. (2002). On the Distortional Post-Buckling Behavior of Cold-Formed Lipped Channel Steel Columns. *Proceedings of SSRC 2002 Annual Stability Conference*, 571-590.

Rasmussen K.J. (2000). Experimental Techniques in the Testing of Thin-Walled Structural Members. *Coupled Instabilities in Metal Structure* (eds. D. Camotim, D. Dubina and J. Rondal), ICPress, 225-239.

Schafer B. (2000). *Distortional Buckling of Cold-Formed Steel Columns*, AISI Report.

Sridharan S. (1982). A Semi-Analytical Method for the Post-Local-Torsional Buckling Analysis of Prismatic Plate Structures. *International Journal of Numerical Methods in Engineering*, **18:2**, 1685-1697.

Standards Association of Australia (1996). *The Australian/New Zealand Cold-Formed Steel Structures Standard*, AS/NZS 4600.

GBT-BASED DISTORTIONAL BUCKLING FORMULAE FOR THIN-WALLED RACK-SECTION COLUMNS AND BEAMS

N. Silvestre[1]   K. Nagahama[2]   D. Camotim[1]   E. Batista[2]

[1] Civil Eng. Dept., IST, TU Lisbon, Av. Rovisco Pais, 1049-001 Lisboa, Portugal
[2] Civil Eng. Prog., COPPE/UFRJ, P.O.B. 68.506, Rio de Janeiro, 21945-970 Brazil

ABSTRACT

The paper addresses the derivation of GBT-based analytical distortional buckling formulae for cold-formed steel rack-section columns and beams. Such formulae provide bifurcation stress estimates for members with arbitrarily inclined intermediate stiffeners and pinned/free-to-warp or fixed/warping-free end sections. The accuracy and validity of the GBT-based analytical estimates are assessed by means of a comparison with exact FEM results concerning several rack-section member geometries and, for some pinned/free-to-warp columns only, also with the values yielded by the formulae developed by Lau & Hancock.

KEYWORDS

Distortional buckling, Generalised beam theory (GBT), GBT distortional buckling formulae, Rack-section columns/beams, Cold-formed steel members, Pinned/free-to-warp members, Fixed/warping-free members.

INTRODUCTION

The structural behaviour and load-carrying capacity of cold-formed steel members displaying the cross-section configuration depicted in Figure 1(a) and commonly designated as *rack-section*[1] is often affected by *distortional buckling*, a local bifurcation instability phenomenon involving plate bending and fold line motions. Figure 1(b) displays the distortional buckling mode (*DM*) shapes exhibited by rack-section members subjected to uniform compression (columns) and major axis pure bending (beams). Since all the (few) available design rules dealing with the effect of distortional buckling on the member strength[2] require the knowledge of the corresponding bifurcation stress values, it is essential for designers to possess accurate and easy-to-use tools to determine such values. In spite of the growing availability of user-friendly finite strip (mostly) and/or finite element computer programs, (approximate) analytical formulae are still sought and regarded as the most popular and efficient design aids.

---

[1] This designation stems from the fact that such cold formed-steel profiles are often used in storage racks (see Hancock, 1985).
[2] It is worth mentioning that the Australian/New Zealand code (AS/NZS 4600) was the first to incorporate specific and rational provisions to account for the distortional buckling effect on member strength. A very thorough report on this subject recently made available by Schafer (2000) strongly suggests that the North American (AISI) Specification should soon follow this trend.

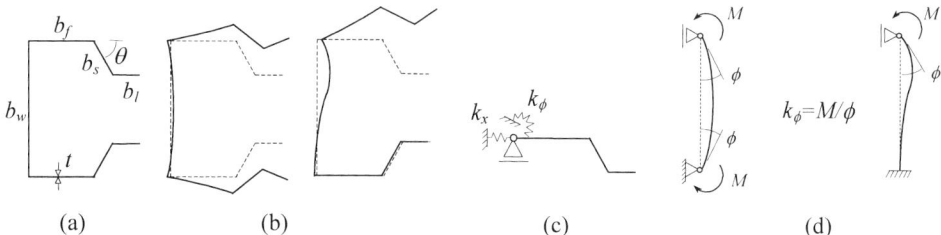

Figure 1: Rack-section (a) geometry, (b) column and beam distortional buckling mode shapes, (c) structural model and (d) column and beam web deformed configurations

Experimental and numerical results concerning distortionally buckled columns and beams provided clear evidence that (i) the compressed flange-stiffener-lip assemblies remain practically undeformed (they just "rigidly" rotate about the web-flange edge – see Figure 1(b)) and that (ii) meaningful flexural deformations appear only in the web. Such evidence led Lau & Hancock (1987) to develop column distortional buckling formulae[1], already included in the present Australian/New Zealand cold-formed steel code and which are based on the structural similarity between (i) the thin-walled member distortional buckling behaviour and (ii) the flexural-torsional buckling behaviour of the (fictitious) structural model shown in Figure 1(c) (uniformly compressed "flange-stiffener strut" elastically supported along the web-flange edge). Although the above approach is easy-to-use and provides accurate bifurcation stress estimates for commonly used cross-section dimensions, one must mention that it (i) applies only to columns[2] with pinned and free-to-warp end sections and (ii) does not yield accurate results in the presence of very slender webs, as recently pointed out by Schafer (2000). Finally, notice also that Generalised Beam Theory (GBT – Schardt, 1989, and Davies *et al.*, 1994) has also been used to develop an expression to estimate distortional bifurcation stresses in thin-walled pinned/free-to-warp columns (Davies & Jiang, 1998). However, as such expression involves a "distortional cross-section geometrical property", the evaluation of which requires a complex numerical procedure (first order GBT), it seems fair to say that its "analytical character" is quite debatable.

The objective of this work is to present the derivation and illustrate the application of analytical GBT-based formulae to estimate distortional bifurcation stresses in cold-formed steel rack-section columns and beams with arbitrarily inclined intermediate stiffeners (see Figure 1(a)). Two end support conditions are dealt with, namely members with (i) pinned and free-to-warp or (ii) fixed and warping-free end sections. In order to assess the accuracy and validity and show the potential of the derived analytical expressions, several numerical results are presented and discussed. In particular, the GBT-based analytical estimates are compared with (i) exact FEM results concerning several rack-section member geometries (length and cross-section dimensions) results and, only for pinned/free-to-warp columns with orthogonal intermediate stiffeners, also with (ii) the values yielded by the formulae developed by Lau & Hancock (1987).

GBT-BASED FORMULAE

Generalised Beam Theory, first developed by Schardt (1989), has been shown to provide a general and clarifying approach to investigate the stability behaviour of cold-formed steel members (Davies *et al.*, 1994, 1998). In fact, by decomposing the member buckling mode shape into a linear combination of cross-section *deformation modes* (including local modes, which involve the use of folded-plate theory), GBT analyses complement and compete with powerful numerical techniques, such as the finite element or finite strip methods. The distortional buckling formulae developed here provide a perfect illustration of this statement.

---

[1] Analytical *procedures* is probably a better way to describe the approach of Lau & Hancock.
[2] Hancock (1997) later developed similar distortional buckling formulae for beams. However, since such formulae only apply to single-lip channel sections, they are not valid for rack-section beams.

Due to space limitations and since the authors have recently published, in the context of single-lip channel members, a detailed account of the main aspects involved in the derivation of GBT-based distortional buckling formulae (Silvestre & Camotim, 2002a), only the expressions and procedures specifically related to rack-section members will be presented and discussed here. First, it is important to address the GBT modal decomposition of the column and beam *distortional* buckling mode shapes (depicted in Figure 1(b)) and its implications on the nature of the buckling formulae:

(i) In *columns*, the DM shape coincides with the GBT symmetric distortional (SD) deformation mode shown in Figure 2(a) and, therefore, the bifurcation axial load estimates are provided by

$$P_b = \frac{1}{X_{SD}} \left[EC_{SD}\mu_C \left(\frac{\pi}{L}\right)^2 + GD_{SD} + B_{SD}\mu_B \left(\frac{L}{\pi}\right)^2 \right] \qquad (1)$$

After solving $dP_b/dL=0$, one readily obtains the column *critical length* L_{cr} and *minimum* distortional bifurcation load $P_{b.min}$, which are given by

$$L_{cr} = \pi \sqrt[4]{\frac{EC_{SD}}{B_{SD}}} \sqrt[4]{\frac{\mu_C}{\mu_B}} \qquad P_{b.min} = \frac{2\sqrt{EC_{SD}B_{SD}}\sqrt{\mu_C\mu_B} + GD_{SD}}{X_{SD}} \qquad (2)$$

(ii) In *beams*, the DM shape combines the two GBT deformation modes depicted in Figure 2(b), namely (ii$_1$) a symmetric (SD) and (ii$_2$) an anti-symmetric (AD) distortional modes. The beam bifurcation bending moment estimates are then provided by

$$M_b = \frac{1}{X_{SD.AD}} \sqrt{EC_{SD}\mu_C\left(\frac{\pi}{L}\right)^2 + GD_{SD} + B_{SD}\mu_B\left(\frac{L}{\pi}\right)^2} \sqrt{EC_{AD}\mu_C\left(\frac{\pi}{L}\right)^2 + GD_{AD} + B_{AD}\mu_B\left(\frac{L}{\pi}\right)^2} \qquad (3)$$

Instead of solving $dM_b/dL=0$ (fourth order polynomial in L^2), it was found that expression

$$L_{cr} = \pi \sqrt[8]{\frac{E^2 C_{SD} C_{AD}}{B_{SD} B_{AD}}} \sqrt[4]{\frac{\mu_C}{\mu_B}} \qquad (4)$$

yields quite accurate L_{cr} estimates (errors never exceeding 1%). By incorporating (3) into (4), one is led to an expression which provides the beam minimum distortional bifurcation moment $M_{b.min}$.

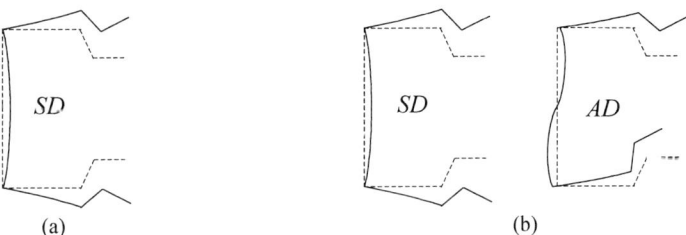

Figure 2: GBT deformation modes most relevant for (a) column and (b) beam distortional buckling

It is also important to point out that (i) C_k, D_k and B_k stand for cross-section modal properties (warping constant, torsion constant and transverse bending stiffness, respectively) and (ii) parameters μ_B and μ_C depend only on the functions adopted to describe (approximately) the modal amplitude variation along the member length, which vary with the member end support conditions and half-wave number n.[1] In regard to the member boundary conditions dealt with considered in this work, one should mention that:

---

[1] All approximation functions adopted here are *trigonometric* (*i.e.*, periodic with n half-waves).

(i) In members with *pinned* wall edges and *free-to-warp* end sections, the amplitude functions adopted (sinusoidal functions) constitute *exact* solutions of GBT equations and lead to

$$\mu_B = \frac{1}{n^2} \qquad \mu_C = n^2 \qquad (5)$$

(ii) In members with *fixed* wall edges and *warping-free* end sections, the amplitude functions adopted constitute *approximate* solutions of GBT equations (Silvestre & Camotim, 2002a) and they yield

$$\mu_B = \frac{3 \ (if \ n=1) \ or \ 2 \ (if \ n \geq 2)}{(n-1)^2 + (n+1)^2} \qquad \mu_C = \frac{(n-1)^4 + (n+1)^4}{(n-1)^2 + (n+1)^2} \qquad (6)$$

CROSS-SECTION MODAL PROPERTIES

The most relevant GBT feature, stemming from its folded-plate theory origin, consists of the fact that it is possible to define mechanical properties related to local (cross-section) deformation modes. In order to obtain such properties and also take full advantage of the GBT potential, one must (i) simultaneously diagonalise (symmetric) matrices *[C]* and *[B]* and (ii) identify a set of *orthogonal* warping functions u_k, procedures that require solving a standard auxiliary eigenvalue problem (Schardt, 1989). For single-lip channel and Z-section members, each exhibiting only 5 wall elements, such task could be performed analytically, by means of the symbolic manipulation program MAPLE (WMS, 2001), thus leading to fully analytical distortional buckling formulae (Silvestre & Camotim, 2002a, 2002b). Since this is no longer feasible for rack-section members (7 wall elements)[1], the above eigenvalue problem must be solved *numerically*, which constitutes a straightforward matter but somewhat "stains" the formulae analytical nature. Therefore, the formulae providing the modal properties of rack-sections with arbitrarily inclined intermediate stiffeners (see Figure 1(a)) are obtained following a sequential procedure which (i) includes a numerical eigenvalue problem solution[2] and (ii) is otherwise fully analytical (the components of a few selected eigenvectors are incorporated in formulae).

Auxiliary Eigenvalue Problem Definition and Solution

The definition of the aforementioned eigenvalue problem involves the following main steps:

(i) Determination of the *geometrical and mechanical parameters*

$$\alpha_1 = \frac{b_f}{b_w} \qquad \alpha_2 = \frac{b_s}{b_w} \qquad \alpha_3 = \frac{b_l}{b_w} \qquad \beta_1 = \alpha_1 \tan\theta \qquad \beta_2 = \alpha_2 \sin\theta \qquad K = \frac{Et^3}{12(1-\nu^2)}$$

(ii) Determination of matrices *[M]* and *[B]*, given by

$$[M] = -[F]^{-1}[\ddot{w}] \qquad [B] = -[\ddot{w}]^T [M] \quad ,$$

where the components of matrices *[F]* and *[ẅ]*, presented in annex, depend on the member cross-section dimensions and material properties.

(iii) Solution of the 8th-order eigenvalue problem defined by

$$([C]^{-1}[B] - \lambda[I])\{u\} = \{0\} \quad ,$$

where *[I]* is the identity matrix and the components of matrix *[C]* are also presented in annex. Four eigenvalues are *null*, as they correspond to "rigid-body" cross-section deformation modes.

---

[1] The eigenvalue problem analytical solution leads to very long, cumbersome and hard-to-handle expressions.
[2] Explicit expressions for the components of the matrices defining the eigenvalue problem are provided in annex.

(iv) Identification of the *smallest* (columns) or *two smallest* (beams) *non-null* eigenvalues ($0<\lambda_{SD}<\lambda_{AD}$) and evaluation of the respective eigenvectors $\{u\}_{SD}$ and $\{u\}_{AD}$. The symmetric and anti-symmetric components of such vectors are *nodal warping displacements* and must be written in the form

$$\{u\}_{SD} = \{1 \quad u_{3.SD} \quad u_{2.SD} \quad u_{1.SD} \quad u_{1.SD} \quad u_{2.SD} \quad u_{3.SD} \quad 1\}^T$$

$$\{u\}_{AD} = \{1 \quad u_{3.AD} \quad u_{2.AD} \quad u_{1.AD} \quad -u_{1.AD} \quad -u_{2.AD} \quad -u_{3.AD} \quad -1\}^T \quad .$$

(v) Determination of the *transverse bending moment* vectors $\{m\}_{SD}$ and $\{m\}_{AD}$, given by

$$\{m\}_{SD} = [M]\{u\}_{SD} = \{0 \quad 0 \quad m_{2.SD} \quad m_{1.SD} \quad m_{1.SD} \quad m_{2.SD} \quad 0 \quad 0\}^T$$

$$\{m\}_{AD} = [M]\{u\}_{AD} = \{0 \quad 0 \quad m_{2.AD} \quad m_{1.AD} \quad -m_{1.AD} \quad -m_{2.AD} \quad 0 \quad 0\}^T \quad .$$

Modal Properties Evaluation

Once the components of $\{u\}_{SD}$, $\{u\}_{AD}$, $\{m\}_{SD}$ and $\{m\}_{AD}$ are known, the evaluation of the cross-section modal properties is fully analytical and involves the following sequence of steps and expressions:

(i) Determination of *wall element chord rotations* (φ_0, φ_1, φ_2,) and *displacements* (w_0, w_1, w_2) – SD + AD

$$\varphi_0 = 2\gamma_3 \frac{u_2 - u_1}{\alpha_1 b_w^2} \qquad \varphi_1 = \frac{\beta_1(u_3 - u_2) + \beta_2(u_1 - u_2 + \gamma_2)}{\alpha_1 \beta_1 \beta_2 b_w^2} \qquad \varphi_2 = \frac{\alpha_1(u_3 - 1) + \alpha_3(u_2 - u_1)}{\alpha_1 \alpha_3 \beta_2 b_w^2}$$

$$\varphi_3 = \frac{\alpha_2 m_2 b_w}{6K} + \varphi_2 \qquad w_0 = \gamma_4 \frac{u_1 - u_2}{\alpha_1 b_w} \qquad w_1 = \frac{\beta_1(u_2 - u_3) + \beta_2(u_2 - u_1 + \gamma_2)}{2\beta_1 \beta_2 b_w}$$

$$w_2 = \frac{\alpha_2(u_2 - u_1)}{2\alpha_1 \beta_2 b_w} + \frac{\alpha_1(u_2 - u_3)}{\alpha_2 \beta_1 b_w} - \frac{\alpha_2(u_3 - 1)}{2\alpha_3 \beta_2 b_w} \qquad w_3 = -\frac{\alpha_3 \varphi_3 b_w}{2} + \frac{\alpha_3 \beta_1(u_2 - u_3) + \alpha_1 \beta_2(1 - u_3)}{\alpha_3 \beta_1 \beta_2 b_w}$$

Mode	γ_1	γ_2	γ_3	γ_4	γ_5
SD	3	0	0	1	15
AD	1	$2\beta_1 u_1$	1	0	1

(ii) Determination of *cross-section modal mechanical properties*:
C_{SD}, B_{SD} and D_{SD} (columns) + C_{SD}–C_{AD}, B_{SD}–B_{AD} and D_{SD}–D_{AD} (beams)

$$C = \frac{b_w t}{3}[\gamma_1 u_1^2 + 2\alpha_1(u_1^2 + u_1 u_2 + u_2^2) + 2\alpha_2(u_2^2 + u_2 u_3 + u_3^2) + 2\alpha_3(u_3^2 + u_3 + 1)]$$

$$B = \frac{b_w}{3K}[m_1^2(2\alpha_1 + \gamma_1) + 2\alpha_1 m_1 m_2 + 2m_2^2(\alpha_1 + \alpha_2)]$$

$$D = \frac{2b_w t^3}{3}(\alpha_3 \varphi_3^2 + \alpha_2 \varphi_2^2 + \alpha_1 \varphi_1^2 + \varphi_0^2/2) + \frac{b_w^3 t^3}{540K^2}[(8\alpha_1^3 + \gamma_5)m_1^2 + 14\alpha_1^3 m_1 m_2 + 8(\alpha_1^3 + \alpha_2^3)m_2^2]$$

(iii) Determination of *cross-section modal geometrical properties* : X_{SD} (columns) + $X_{SD.AD}$ (beams)

$$X_{SD} = \frac{t}{A}\left[\frac{2X_1}{\alpha_1 \alpha_2 \alpha_3 b_w} + \frac{(X_2 + X_3)b_w^3}{7560K^2}\right]$$

$A = b_w t(1 + 2\alpha_1 + 2\alpha_2 + 2\alpha_3)$

$X_1 = \alpha_1 \alpha_2(u_3 - 1)^2 + \alpha_2 \alpha_3(u_2 - u_1)^2 + \alpha_1 \alpha_3(u_3 - u_2)^2 + \alpha_1 \alpha_2 \alpha_3 b_w^2(w_0^2/2 + \alpha_1 w_1^2 + \alpha_2 w_2^2 + \alpha_3 w_3^2)$

$X_2 = b_w^2[(32\alpha_1^5 + 63)m_1^2 + 62\alpha_1^5 m_1 m_2 + 32(\alpha_1^5 + \alpha_2^5)m_2^2] + 42b_w K[\alpha_1^4 \varphi_1(m_1 - m_2) + \alpha_2^4 \varphi_2 m_2]$

$$X_3 = 1260K[w_0 m_1 + \alpha_1^3 w_1(m_1 + m_2) + \alpha_2^3 w_2 m_2 + K(\alpha_1^3 \varphi_1^2 + \alpha_2^3 \varphi_2^2 + \alpha_3^3 \varphi_3^2)]$$

$$X_{SD.AD} = \frac{t}{I}\left[\frac{X_1 + X_2}{\alpha_1 \alpha_2 \alpha_3} + \frac{b_w^3(X_3 + X_4 + X_5 + X_6 + X_7)}{15120 K^2}\right]$$

$$I = \frac{b_w^3 t}{12}[1 + 6(\alpha_1 + \alpha_2 + \alpha_3) + 4\alpha_2 \beta_2(2\beta_2 - 3) + 24\alpha_3 \beta_2(\beta_2 - 1)]$$

$$X_1 = \alpha_2 \alpha_3 (u_{1.SD} - u_{2.SD})(u_{1.AD} - u_{2.AD}) + \alpha_1 \alpha_2(1 - 2\beta_2)(u_{3.SD} - 1)(u_{3.AD} - 1) + \alpha_1 \alpha_3(1 - \beta_2)(u_{2.SD} - u_{3.SD})(u_{2.AD} - u_{3.AD})$$

$$X_2 = \alpha_1 \alpha_2 \alpha_3 b_w^2 [\alpha_1 w_{1.SD} w_{1.AD} + \alpha_2 w_{2.SD} w_{2.AD}(1 - \beta_2) + \alpha_3 w_{3.SD} w_{3.AD}(1 - 2\beta_2)]$$

$$X_3 = 1260 K^2 b_w [\alpha_1^3 \varphi_{1.SD} \varphi_{1.AD} + \alpha_2^3 \varphi_{2.SD} \varphi_{2.AD}(1 - \beta_2) + \alpha_3^3 \varphi_{3.SD} \varphi_{3.AD}(1 - 2\beta_2)]$$

$$X_4 = 1260 K^2 [-w_{0.SD} \varphi_{0.AD} + 2\alpha_2^2 \beta_2 (w_{2.AD} \varphi_{2.SD} + w_{2.SD} \varphi_{2.AD})]$$

$$X_5 = b_w^3 [(32\alpha_1^5 + 3) m_{1.SD} m_{1.AD} + (32\alpha_1^5 + (32 - 29\beta_2)\alpha_2^5) m_{2.SD} m_{2.AD} + 31\alpha_1^5 (m_{1.AD} m_{2.SD} + m_{1.SD} m_{2.AD})]$$

$$X_6 = 21 K b_w^2 \{a_1^4 [\varphi_{1.SD}(m_{1.AD} - m_{2.AD}) + \varphi_{1.AD}(m_{1.SD} - m_{2.SD})] + a_2^4 (1 + 2\beta_2)(\varphi_{2.SD} m_{2.AD} + \varphi_{2.AD} m_{2.SD}) - 3 m_{1.SD} \varphi_{0.AD}\}$$

$$X_7 = 21 K b_w \{30 a_1^3 [w_{1.SD}(m_{1.AD} + m_{2.AD}) + w_{1.AD}(m_{1.SD} + m_{2.SD})] + a_2^3 (30 - 28\beta_2)(w_{2.SD} m_{2.AD} + w_{2.AD} m_{2.SD}) + 2 m_{1.AD} w_{0.SD}\}$$

ILLUSTRATION, VALIDATION AND ACCURACY

In order to illustrate the application, validate and assess the accuracy of the derived GBT-based distortional buckling formulae, they are next applied to columns and beams displaying several geometries and end support conditions. The bifurcation stress estimates (σ_{gbt}) are compared with (i) exact FEM results (σ_{ex}), obtained using the commercial code ABAQUS (HKS, 1998), and, only for pinned/free-to-warp columns with orthogonal intermediate stiffeners, also with (ii) values provided by the formulae developed by Lau & Hancock (σ_{han}). Finally, one should mention that all the values concerning *pinned/free-to-warp* members are minimum values, associated to the member critical length L_{cr} corresponding to $n=1$ (single half-wave buckling mode) and yielded by expressions (2_1) (columns) or (4) (beams). In *fixed/warping-free* members, since the critical (distortional) buckling mode often exhibits more than one half-wave, it makes no sense to talk about "the minimum bifurcation stress value" (there is one such value per half-wave number and it may happen that neither is associated to the critical stress of a specific member). Thus, it is only meaningful to estimate distortional bifurcation stresses for (i) members with specified *lengths L* and (ii) buckling modes exhibiting given *half-wave numbers* n^1.

The cross-section geometry selected to illustrate the application of the formulae is defined by $b_w=90$ mm, $b_f=60$ mm, $b_s=10$ mm, $b_l=30$ mm, $t=1$ mm and $\theta=45°$ and the cold-formed steel properties adopted are $E=200$ GPa and $\nu=0.3$. The sequence of results is the following:

(I) Auxiliary eigenvalue problem solution: *nodal warping values* and *transverse moments*

$\lambda_{SD} = 0.000263$ N/mm^6 $u_{1.SD} = -0.318$ mm $u_{2.SD} = 1.640$ mm $u_{3.SD} = -2.430$ mm

$\lambda_{AD} = 0.000619$ N/mm^6 $u_{1.AD} = -0.424$ mm $u_{2.AD} = 0.990$ mm $u_{3.AD} = -1.632$ mm

$m_{1.SD} = -2.875$ N/mm $m_{2.SD} = 0.120$ N/mm $m_{1.AD} = -3.726$ N/mm $m_{1.AD} = -0.242$ N/mm

(II) Cross-section mechanical and geometrical properties

$C_{SD} = 220.11$ mm^4 $B_{SD} = 0.05795$ N/mm^2 $D_{SD} = 0.008238$ mm^2 $X_{SD} = 0.2415$

---

[1] The half-wave number leading to the *minimum* bifurcation stress value varies with the member geometry.

$$C_{AD} = 89.13 \ mm^4 \qquad B_{AD} = 0.05519 \ N/mm^2 \qquad D_{AD} = 0.004226 \ mm^2 \qquad X_{SD.AD} = 0.00427 \ mm^{-1}$$

As for the parameters related to the member end support conditions, given by (5) and (6), they depend on the value of n. For instance, if $n=1$, they read

$$\mu_B^{pin} = 1 \qquad \mu_C^{pin} = 1 \qquad \qquad \mu_B^{fix} = 0.75 \qquad \mu_C^{fix} = 4$$

Finally, introducing the above values in (2)–(4) leads to the following (i) critical length and bifurcation stress resultant estimates (pinned/free-to-warp column and beam) and (ii) bifurcation stress resultant estimates for $L=1000 \ mm$ and $n=1-3$ (fixed/warping-free column and beam). Such estimates are presented next, together with the exact FEM results (indicated between square brackets):

(i) Pinned/free-to-warp column: $L_{cr}=521 \ mm$ [526] and $P_{cr}=15.85 \ kN$ [16.32].
(ii) Pinned/free-to-warp beam: $L_{cr}=469 \ mm$ [475] and $M_{cr}=710 \ kNmm$ [701].
(iii) Fixed/warping-free column: $P_{cr}=28.1 \ (n=1); \ 22.2 \ (n=2); \ 29.5 \ kN \ (n=3) \ [22.0 - n=2]$.
(iv) Fixed/warping-free beam: $M_{cr}=1522 \ (n=1); \ 715 \ (n=2); \ 849kNmm \ (n=3) \ [707 - n=2]$.

Tables 1 and 2 displays distortional buckling stress values concerning, respectively, *pinned/free-to-warp* and *fixed/warping-free* columns and beams with steel properties $E=200GPa$, $v=0.3$. Several cross-section geometries are dealt with and one should recall that (i) Table 1 shows *minimum* flange distortional bifurcation stress values (in MPa) and (ii) Table 2 shows the *smallest* such value obtained for the indicated member length and $n=1, 2, ...$ (the n value leading to such value is also indicated). As alternative distortional buckling formulae are only available for pinned/free-to-warp columns, in the remaining three cases the GBT-based estimates are compared only with exact FEM results. From the observation of the results and comparisons presented, one may conclude that:

(i) For all the members considered here, the GBT-based formulae consistently yield accurate distortional bifurcation stress estimates (the error never exceeds 9%). In fact, the average and standard deviation values of the σ_{gbt}/σ_{ex} ratio are (i_1) *1.01* and *0.010* (pinned/free-to-warp columns), (i_2) *1.04* and *0.024* (pinned/free-to-warp beams), (i_3) *1.00* and *0.022* (fixed/warping-free columns) and (i_4) *0.97* and *0.025* (fixed/warping-free beams).

(ii) Concerning the *pinned/free-to-warp columns*, both the GBT-based ($\theta=45°–90°$) and Lau & Hancock ($\theta=90°$ only) formulae yield very accurate distortional bifurcation stress estimates. However, while all the GBT-based predictions (very marginally) overestimate the exact ones, the Lau & Hancock results lead to σ_{gbt}/σ_{ex} values rather "evenly distributed" around 1.0^1.

(iii) All the GBT-based estimates concerning *pinned/free-to-warp beams* are slightly unconservative (the maximum error is 9% and 6%, respectively for beams with orthogonal and inclined intermediate stiffeners). It is interesting to notice that the results tend to be comparatively less accurate for shorter webs and longer (more slender) flanges and/or stiffeners. The analysis of the exact FEM results showed that this tendency is due to compressed flange/stiffeners flexural (local) deformation effects, which are only partially accounted for by the GBT-based formulae.

(iv) All the GBT-based estimates concerning *fixed/warping-free columns* are very accurate (the error never exceeds 4%). Such estimates are (iv_1) mostly unconservative for $\theta=90°$ and (iv_2) always conservative for $\theta=45°$.

(v) Almost all the GBT-based estimates concerning *fixed/warping-free beams* are conservative, with the maximum error reaching 6%. It is clearly visible that the less accurate predictions correspond to the shorter beam lengths, i.e., to the occurrence of buckling modes with either two or three half-waves. This fact stems again from the partial accounting of the flexural (local) deformation effects.

---

[1] It should be mentioned that all the cross-section geometries considered in this work were taken from Lau (1988). It is conceivable that the conclusions drawn here do not remain valid for columns with different cross-section dimensions, as was shown to be the case for channel columns (Silvestre & Camotim, 2002a).

Table 1. Minimum (flange) distortional bifurcation stresses for pinned/free-to-warp members ($t=1\,mm$)

b_w (mm)	b_f (mm)	b_s (mm)	b_l (mm)	θ	\multicolumn{5}{c}{Columns}	\multicolumn{3}{c}{Beams}
					σ_{han}	σ_{gbt}	σ_{ex}	σ_{han}/σ_{ex}	σ_{gbt}/σ_{ex}	σ_{gbt}	σ_{ex}	σ_{gbt}/σ_{ex}
60	40	10	30	90°	140	139	137	1.02	1.01	241	232	1.04
90	40	10	30	90°	118	116	116	1.02	1.00	187	182	1.03
60	40	15	30	90°	179	183	179	1.00	1.02	413	379	1.09
90	40	15	30	90°	151	153	150	1.01	1.02	285	271	1.05
60	60	10	30	90°	99	99	97	1.02	1.02	168	156	1.08
90	60	10	30	90°	85	87	86	0.99	1.01	133	128	1.04
120	60	10	30	90°	74	76	76	0.97	1.00	115	112	1.02
60	60	15	30	90°	129	132	130	0.99	1.02	280	258	1.09
90	60	15	30	90°	111	116	114	0.97	1.02	206	193	1.07
120	60	15	30	90°	97	101	100	0.97	1.01	170	164	1.04
60	40	10	30	45°	-	82	82	-	1.00	124	121	1.03
90	40	10	30	45°	-	72	72	-	1.00	103	102	1.01
60	40	15	30	45°	-	100	99	-	1.01	175	167	1.05
90	40	15	30	45°	-	86	86	-	1.00	136	133	1.02
60	60	10	30	45°	-	61	60	-	1.01	89	86	1.03
90	60	10	30	45°	-	55	55	-	1.00	75	74	1.02
120	60	10	30	45°	-	49	49	-	1.00	67	67	1.01
60	60	15	30	45°	-	76	74	-	1.03	129	121	1.06
90	60	15	30	45°	-	68	66	-	1.03	102	98	1.04
120	60	15	30	45°	-	61	60	-	1.02	89	87	1.03
							Mean	1.00	1.01		Mean	1.04
							Sd.Dev.	0.020	0.010		Sd.Dev.	0.024

Table 2. Critical (flange) distortional bifurcation stresses for fixed/warping-free members

b_w (mm)	b_f (mm)	b_s (mm)	b_l (mm)	θ	t (mm)	L (mm)	\multicolumn{3}{c}{Columns}	\multicolumn{3}{c}{Beams}
							n	σ_{gbt}	σ_{ex}	σ_{gbt}/σ_{ex}	n	σ_{gbt}	σ_{ex}	σ_{gbt}/σ_{ex}
79	35.3	16.2	26.8	90°	1.652	800	2	479	499	0.96	2	707	750	0.94
79	35.3	16.2	26.8	90°	1.652	1300	3	410	409	1.00	3	714	729	0.98
79	35.3	16.2	26.8	90°	1.652	1500	3	391	385	1.02	4	712	719	0.99
79	35.3	16.2	26.8	90°	1.652	1700	4	385	379	1.02	4	711	710	1.00
79	35.3	16.2	26.8	90°	1.652	1900	4	374	368	1.02	5	710	708	1.00
78.6	35.3	16.4	27.05	90°	1.982	800	2	570	581	0.98	2	893	941	0.95
78.6	35.3	16.4	27.05	90°	1.982	1300	3	490	483	1.01	4	906	953	0.95
78.6	35.3	16.4	27.05	90°	1.982	1500	4	487	481	1.01	4	884	907	0.97
78.6	35.3	16.4	27.05	90°	1.982	1900	5	465	459	1.01	5	885	880	1.01
82.4	31.4	14.9	29.25	90°	2.395	800	2	672	660	1.02	3	1090	1121	0.97
82.4	31.4	14.9	29.25	90°	2.395	1100	3	620	608	1.02	3	1073	1115	0.96
82.4	31.4	14.9	29.25	90°	2.395	1500	4	587	576	1.02	5	1045	1065	0.98
82.4	31.4	14.9	29.25	90°	2.395	1700	5	580	569	1.02	5	1047	1040	1.01
78.6	35.3	16.4	27.05	45°	1.982	800	2	292	299	0.98	3	413	433	0.95
78.6	35.3	16.4	27.05	45°	1.982	1300	4	264	270	0.98	4	405	431	0.94
78.6	35.3	16.4	27.05	45°	1.982	1500	4	260	264	0.98	5	402	424	0.95
82.4	31.4	14.9	29.25	45°	2.395	800	3	372	386	0.96	3	500	536	0.93
82.4	31.4	14.9	29.25	45°	2.395	1100	4	345	356	0.97	4	500	532	0.94
82.4	31.4	14.9	29.25	45°	2.395	1500	5	330	339	0.97	6	501	527	0.95
									Mean	1.00			Mean	0.97
									Sd.Dev.	0.022			Sd.Dev.	0.025

CONCLUSION

The various concepts and steps involved in deriving GBT-based (approximate) analytical formulae to estimate distortional buckling stresses in thin-walled rack-section columns and beams were presented. Such formulae automatically incorporate folded-plate theory concepts, an important feature which is responsible for the fact that they directly account for cross-section distortion and (partially) flexural deformation effects. The derived formulae, which require the preliminary (numerical) solution of an auxiliary standard matrix eigenvalue problem and can be readily programmed even in a hand calculator, provide distortional critical lengths and bifurcation stress resultant estimates for rack-section columns and beams with arbitrarily inclined intermediate stiffeners and pinned/free-to-warp or fixed/warping-free end sections.

A detailed analysis of a set of four identical columns/beams with pinned/free-to-warp or fixed/warping-free end sections was presented, in order to illustrate the application of the proposed formulae. Moreover, an assessment of the accuracy of the GBT-based estimates was made by comparing them with (i) exact FEM results and, for pinned/free-to-warp columns with orthogonal intermediate stiffeners only, also with (ii) values yielded by the formulae developed by Lau & Hancock. Rack-section columns and beams with different dimensions and orthogonal or 45°-inclined intermediate stiffeners were considered and the analysis of the results obtained led to the following main conclusions:

(i) The GBT-based formulae consistently yielded quite accurate estimates for pinned/free-to-warp and fixed/warping-free columns and beams. In fact, the four sets of σ_{gbt}/σ_{ex} values exhibited averages and standard deviations varying from *0.97* to *1.04* and *0.010* to *0.025*, respectively. Moreover, a large number of estimates fell inside the 5% error range.

(ii) For pinned/free-to-warp columns with $\theta=90°$, the GBT-based estimates were shown to be as accurate as the values yielded by the formulae developed by Lau & Hancock.

(iii) For pinned/free-to-warp beams, the GBT-based estimates were found to slightly overestimate the exact bifurcation stress values, as indicated by the $\sigma_{dist}/\sigma_{d.ex}$ average value of *1.04*. However, such estimates exhibited a rather low scatter – the standard deviation was *0.024*.

(iv) For fixed/warping-free beams, the GBT-based estimates were found to be a bit conservative, as indicated by the $\sigma_{dist}/\sigma_{d.ex}$ average value of *0.97*. Once again, a low scatter was observed – standard deviation of *0.025*.

REFERENCES

Davies J., Leach P. and Heinz D. (1994). Second-Order Generalised Beam Theory. *Journal of Constructional Steel Research*, **31:2-3**, 221-241.

Davies J. and Jiang C. (1998). Design for Distortional Buckling. *Journal of Constructional Steel Research*, **46**:1-3, 174. (CD-ROM full paper – #104)

Hancock G. (1985). Distortional Buckling of Steel Storage Rack Columns. *Journal of Structural Engineering (ASCE)*, **111:12**, 2770-2783.

Hancock G. (1997). Design for Distortional Buckling of Flexural Members. *Thin-Walled Structures*, **27:1**, 3-12.

Hibbit, Karlsson and Sorensen Inc. (1998). *ABAQUS Standard* (version 5.8).

Lau S. (1988). *Distortional Buckling of Thin-Walled Columns*, Ph.D. Thesis, School of Civil and Mining Engineering, University of Sydney, Australia.

Lau S. and Hancock G. (1987). Distortional Buckling Formulas for Channel Columns. *Journal of Structural Engineering* (ASCE), **113:5**, 1063-1078.

Schardt R. (1989). *Verallgemeinerte Technische Biegetheorie*, Springer-Verlag, Berlin.

Schafer B. (2000). *Distortional Buckling of Cold-Formed Steel Columns*, AISI Report.

Silvestre N. and Camotim D. (2002a). GBT-Based Distortional Buckling Formulae for Thin-Walled Channel Columns and Beams. *Proceedings 16th International Specialty Conference on Cold-Formed Steel Structures*, Orlando, USA, October 17-18.

Silvestre N. and Camotim D. (2002b). GBT-Based Distortional Buckling Formulae for Thin-Walled Z-Section Columns and Beams. *Proceedings 3rd European Conference on Steel Structures (Eurosteel '02)*, Coimbra, Portugal, September 19-20.

Standards Association of Australia (1996). *The Australian/New Zealand Cold-Formed Steel Structures Standard*, AS/NZS 4600.

Waterloo Maple Software (2001). MAPLE *V* (release 7), University of Waterloo, Canada.

ANNEX: Matrices Involved in the Definition of the Auxiliary Eigenvalue Problem

In order to define the auxiliary 8th-order eigenvalue problem $([C]^{-1}[B] - \lambda[I])\{u\} = \{0\}$, it is necessary to know the expressions providing the *non-null* components of the matrices involved, *i.e.*, $[F]$, $[\ddot{w}]$ and $[C]$. Noticing that matrices $[F]$ and $[C]$ are *symmetric*, such components are given by:

$$F_{11} = F_{22} = F_{77} = F_{88} = 1 \qquad F_{33} = F_{66} = \frac{b_w}{3K}(\alpha_1 + \alpha_2) \qquad F_{44} = F_{55} = \frac{b_w}{3K}(\alpha_1 + 1)$$

$$F_{34} = F_{56} = \frac{b_w}{6K}\alpha_1 \qquad F_{45} = \frac{b_w}{6K}$$

$$\ddot{w}_{31} = \ddot{w}_{68} = \frac{1}{\alpha_3 \beta_2 b_w^2} \qquad \ddot{w}_{32} = \ddot{w}_{67} = \frac{\alpha_3 - \alpha_1}{\alpha_1 \alpha_3 \beta_2 b_w^2} \qquad \ddot{w}_{33} = \ddot{w}_{66} = -\frac{\beta_2 + 2\beta_1}{\alpha_1 \beta_1 \beta_2 b_w^2}$$

$$\ddot{w}_{34} = \ddot{w}_{65} = \ddot{w}_{43} = \ddot{w}_{56} = \frac{\beta_1 \beta_2 + \beta_1 + \beta_2}{\alpha_1 \beta_1 \beta_2 b_w^2} \qquad \ddot{w}_{35} = \ddot{w}_{64} = \ddot{w}_{46} = \ddot{w}_{53} = \frac{-1}{\alpha_1 b_w^2}$$

$$\ddot{w}_{42} = \ddot{w}_{57} = \frac{-1}{\alpha_1 \beta_2 b_w^2} \qquad \ddot{w}_{44} = \ddot{w}_{55} = -\frac{1 + 2\beta_1}{\alpha_1 \beta_1 b_w^2} \qquad \ddot{w}_{45} = \ddot{w}_{54} = \frac{2}{\alpha_1 b_w^2}$$

$$C_{11} = C_{88} = 2C_{12} = 2C_{78} = \frac{\alpha_3 b_w t}{3} \qquad C_{22} = C_{77} = \frac{(\alpha_2 + \alpha_3)b_w t}{3} \qquad C_{33} = C_{66} = \frac{(\alpha_1 + \alpha_2)b_w t}{3}$$

$$C_{44} = C_{55} = \frac{(\alpha_1 + 1)b_w t}{3} \qquad C_{23} = C_{67} = \frac{\alpha_2 b_w t}{6} \qquad C_{34} = C_{56} = \frac{\alpha_1 b_w t}{6} \qquad C_{45} = \frac{b_w t}{6}$$

TESTING AND NUMERICAL ANALYSIS OF COLD-FORMED C-SECTIONS SUBJECT TO PATCH LOAD

Robert Y Xiao[1], G P W Chin[1], K F Chung[2]

[1]Department of Civil Engineering, University of Wales, Swansea, UK, SA2 8PP
[2]Department of Civil and Structural Engineering, the Hong Kong Polytechnic University, Hong Kong, China

ABSTRACT

This paper aims to present the major findings of a research project on testing and numerical simulation on cold-formed C-sections under patch load. Testing on two lipped C-sections back-to-back beams subjected to patch load have been conducted. Based on measured physical and mechanical properties, a three-dimensional finite element analysis has been carried out; both material and geometrical non-linearity have been considered separately in the models. Moreover, the deformation and the stress distributions of the finite element models have been obtained from the numerical analysis for detailed discussion on the structural behaviour of web crippling in cold-formed steel C-sections.

KEYWORDS

Patch load, cold-formed steel C-section, contact elements, shell elements, non-linear analysis.

INTRODUCTION

Cold-formed steel sections are widely used in various applications and are more efficient in many lightly loaded structures compared with conventional hot-rolled members, which may be proven to be uneconomical (Hancock, 1998). Cold-formed structural steel can be manufactured into various shapes and sizes and even compound sections (Lawson et al 1998). In the past, cold-formed steel sections were commonly used as purlins but are now also applied in various structural applications. For structural use, cold-formed sections with edge lips are preferred for improved structural efficiency. That is why most cold-formed steel manufacturers have adopted section shapes with stiffened edges. The thickness typically ranges from 1.2 to 3.0 mm. However, recent advancement in cold formed steel technology enables sections with thickness of up to 12 mm can be produced and some works on these thick steel sections have been carried out by Young and Hancock (1999).

This paper aims to present the major findings of a research project on testing and numerical simulation on cold-formed C-sections under patch load. Testing on two lipped C-sections back-to-back beams subject to patch load have been conducted. A three dimensional finite element analysis was also established to give detailed information on both deformation and stress distributions of the cold-formed steel C-sections. The paper will also present comparisons between the experimental and the numerical results.

EXPERIMENTAL WORK

Patch load tests leading to web crippling in cold-formed steel C-sections with lip stiffeners have been carried out. Two identical C-sections with lips were tested under patch loads of 100 and 150mm wide respectively. The thickness of the C-sections is 1.6 mm and the dimensions of the C-sections are shown in Figure 1. Standard coupon tests of the C-sections were carried out, and an elastic modulus of 205GPa and a yield strength of 325GPa were obtained. The C-sections were rested on a flat surface, and no external restrains were provided. The patch load was applied through a solid steel block which was placed on the surface of the top flange at the mid-span of the C-sections. Force was applied onto the steel block by a hydraulic jack until failure.

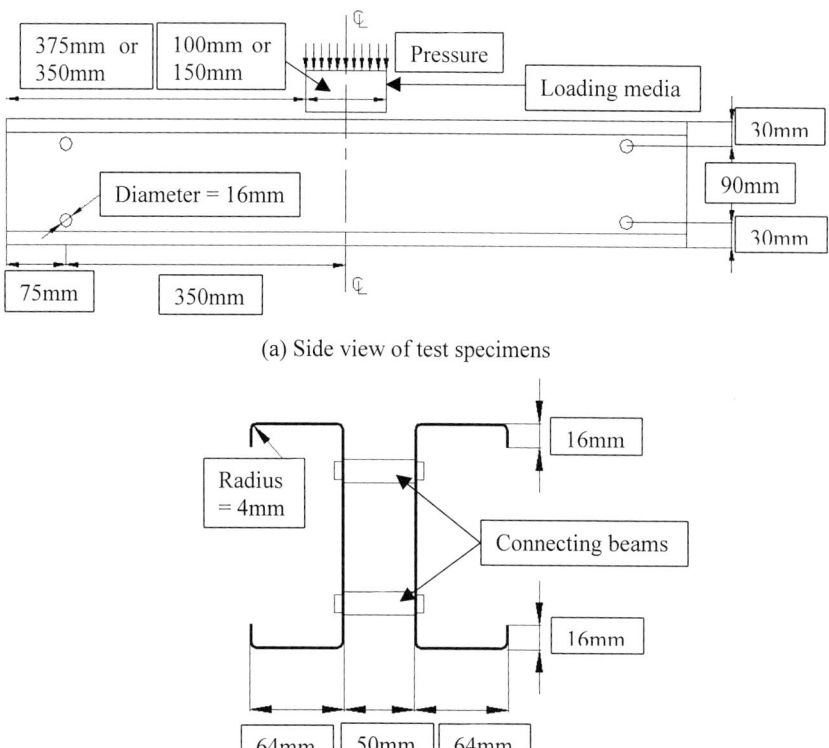

Figure 1 – Thin walled C-sections back-to-back with interconnections

FINITE ELEMENT MODELLING

A finite element package ANSYS (Version 5.6) was adopted in the numerical analysis. To model the C-sections, the shell element SHELL43 was selected which could deal with plasticity and various geometrical effects, such as large deflection, stress stiffening and large strain. For other parts of test specimens, such as the interconnections and the steel block, the volumetric element SOLID45 was used. To take advantage of the symmetrical condition, only a quarter size of the complete test specimens was modelled, as shown in Figure 2.

During the tests, the flange exhibited large deformation when the load was applied. As the load was applied through a steel block, the contact area with the flange reduced as the flange deformed or deflected. Two numerical techniques have been employed to tackle the problem. Firstly, a contact element is adopted to model the contact area between the flange and the steel block. Requiring the feature of a breaking contact condition, CONTAC52 is the only element that is capable of such behaviour. It is a two-node element which defines breaking of contact condition when a tensile force is developed within the element. Model A of Figure 2 shows the model that uses contact elements to simulate the loading behaviour. Secondly, a different approach of overcoming the problem with the changing of contact area is by applying pressure loading only on the inner strip of elements (approximately a width of 8 mm in total). The technique is to apply load on the flange that is in contact with the steel block at all times. During verification of this technique, a simple model with pressure loading acting over the contact area is used. It showed that even at a load of only 0.5 kN, the inner strip of elements exhibited significant deflection. The second model with the applied pressure loading is shown as Model B of Figure 2. For simplicity, the numerical model with contact elements is referred as Model A while the model with pressure loading is referred as Model B in subsequent discussion.

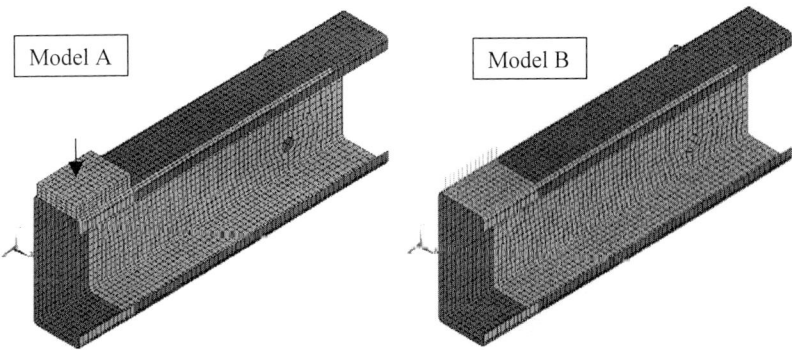

Figure 2 – Numerical models with contact elements as Model A, and with pressure load on a strip of elements as Model B

NUMERICAL RESULTS

In the numerical analysis, both material non-linearity and geometrical non-linearity have been considered. Since instability and plastic yielding are two distinctive failure modes for thin walled section, therefore the two non-linear analyses were carried out separately on each of the models with loading patch width of 100mm and 150mm, respectively. Figures 4 and 5 present the comparisons on the web deflection obtained from the experiments against the numerical models.

It is shown that results from Model B compare better than those from Model A against the numerical studies. In general, the numerical model considering only geometrical non-linearity give more accurate results. This indicates that buckling will be the dominant failure mode. From experimental observations, large deflection of the web was produced, which eventually led the webs of the sections to be in contact with each other. The numerical model using contact elements generally gives less accurate results due to severe breaking up of contact conditions. As the flange deflects downwards with significant rotation, most of the contact area over the width of the patch load is broken. This also causes higher distribution of stress along the boundary of patch load, as observed in both numerical models. Moreover, the numerical results of Model A under a 150 mm wide patch load is less accurate than that with a 100 mm wide patch load.

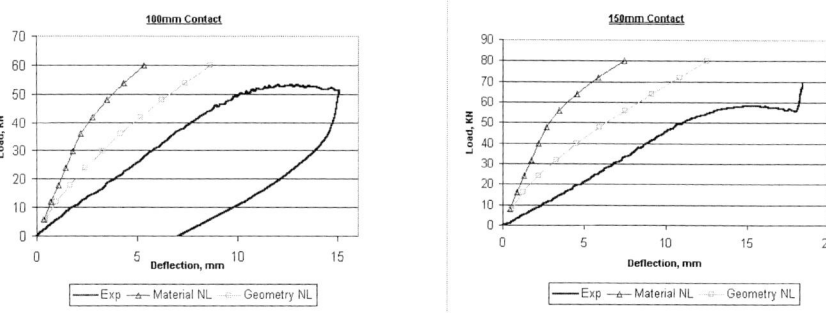

Figure 3 – Verification of the horizontal deflection of section web in Model A

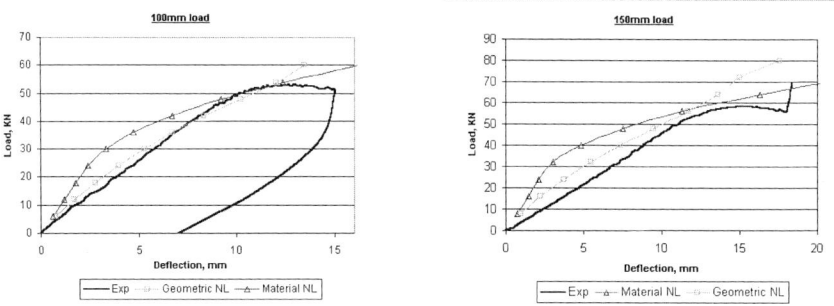

Figure 4 – Verification of the horizontal deflection of section web in Model B

For the results of Model B, it is observed that at the point where the two types of analyses meet, it also represents closely the failure load of the web section. This indicates that in the early stage of loading, the strength of web section is governed by the geometrical strength of the C-sections, and failure is more imminent after yielding commences. To verify the above assumption, the stress distribution of the model are examined. Observation is made to check the stress distribution when both numerical results of material non-linear and geometrical non-linear curves intersect. As anticipated, the stress distribution plots for both the models under patch loads of 100 and 150 mm width show similar patterns. Figure 5 shows the predicted yield stress plots that govern the onset of web failure. The web section is observed to be buckled before the flange was yielded locally. This shows that the flange does contribute in resisting the web crippling load.

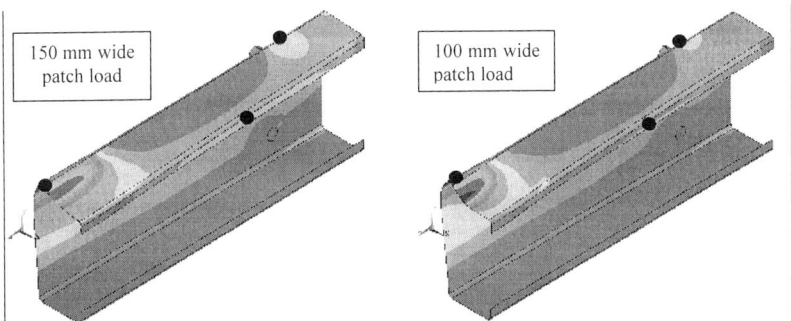

Figure 5 – Stress distribution plots of C-sections

To check reliability of the numerical results, the failure loads of the numerical models are compared to the experiment results and the design resistances to BS5950, as shown in Table 1; only the results from Model B were considered.

Table 1 – Comparison of failure loads

Width of patch load (mm)	Failure load (KN)				
	Experiment results	Finite element results with geometric non-linearity	Finite element results with material non-linearity	Design load to BS5950 with un-restrained web	Design load to BS5950 with restrained web
100	52	52	50	25.2	43.0
150	54.4	60	58	31.9	49.1

It can be seen that numerical modelling could assess the thin-walled section failure load accurately. For both tests, BS5950: Part 5 gives conservative resistances against web crippling. However, if the back-to-back C-sections are considered as two single unrestrained C-sections, then, this will lead to conservative strength prediction, and the predicted resistances are only half of the measured values. Further study is needed to investigate the structural behaviour of compound C-sections.

For the numerical results, good prediction of failure load for the 100 mm sample is achieved but failure load for the 150mm sample is marginally over predicted. Nonetheless, both the numerical results compare closely to the experimental failure loads.

CONCLUSIONS

Modelling the web crippling behaviour was achieved by applying pressure loading on specific elements which proves to be capable of producing sufficiently accurate results. The stress distributions from the finite element analyses show that the section flange with lip stiffener contributes to the load resistance against web crippling. In general, the presented finite element models with pressure loading can provide good approximations of the web crippling loads. The comparison with British Standard has also been presented. Further study is needed to look at the details of connecting members.

REFERENCES

1. Hancock, G.J., (1998). Design of cold-formed steel structures, 3rd Edition, *Australian Institute of Steel Construction*, Sydney, Australia.
2. Lawson, R.M., Chung, K.F. and Popo-ola, S.O. (2002). *Structural design to BS 5950-5:1998 section properties and load tables*, The Steel Construction Institute, UK.
3. Young, B. and Hancock, G.J., (1999). Section moment capacity of cold-formed unlipped channels, *Proceedings Of The Second International Conference on Advances in Steel Structures*, Vol.1,pp349.Hong Kong, China.
4. Young, B. and Hancock, G.J., (1999). Web crippling tests of high strength cold-formed channels, *Proceedings Of The Second International Conference on Advances in Steel Structures*, Vol.1, p357.
5. BSI, BS5950, Part 5, Structural Use of Steelwork in Building- Part 5, Code of Practice for Design of Cold Formed Thin Gauge Sections. 1998, London.

TORSIONAL BUCKLING EXPERIMENTS ON WIDE-FLANGE THIN-WALLED Z-SECTION COLUMNS

R.A.D Fish, M. Lee and K.J.R. Rasmussen

Department of Civil Engineering, University of Sydney, Australia

ABSTRACT

The paper addresses the problem of mode-switching in thin-walled point-symmetric sections. It is well-known that the torsional and flexural overall buckling modes are uncoupled in point symmetric section columns, and that the minor axis flexural mode generally is critical. Theoretically, this result remains true when the section is locally buckled at incipient overall buckling. However, local buckling may change the critical mode from the minor axis flexural mode to the torsional mode.

The paper presents a series of experiments conducted to substantiate this theoretical result. Fixed-ended column tests were performed over a range of lengths on wide-flange Z-sections brake-pressed from high strength steel. Most columns failed by interaction of local buckling with overall torsional buckling and/or overall flexural buckling. Comparisons are made between the test strengths and theoretical buckling curves obtained using the theory for the overall buckling of locally buckled members. Agreement is generally achieved.

KEYWORDS

Z-sections, Steel structures, Tests, Local buckling, Torsional buckling, Flexural buckling, Interaction buckling, Finite strip analysis, Bifurcation analysis.

INTRODUCTION

It is well-known that local buckling may influence the overall buckling behaviour of thin-walled sections. The influence depends on the end support conditions (pinned of fixed) and the symmetry characteristics. For a *doubly* symmetric section, such as an I-section, local buckling reduces the flexural rigidity and precipitates overall buckling at a reduced load but does not induce overall displacements. For a *singly* symmetric cross-section, such as a channel section, local buckling induces overall bending when the column is compressed between pinned ends but not when compressed between fixed ends (Young and Rasmussen 1997). For a *point*-symmetric cross-section, such as a Z-section column, theoretical results (Rasmussen 2000) have shown that local buckling does not induce overall displacements, as it does in pin-ended singly symmetric columns, but causes a coupling between the minor and major axis buckling displacements. In physical terms, this implies that the direction of overall buckling occurs about an axis rotated from the minor principal axis. This phenomenon was studied in detail in Rasmussen (2001).

According to the theory for the overall bifurcation of locally buckled sections, torsional and flexural overall buckling of point-symmetric sections are uncoupled. However, the critical overall buckling mode may switch from a flexural mode to a torsional mode in the case of Z-sections with wide flanges because local buckling reduces the warping rigidity (EI_ω) more severely than the minor axis flexural rigidity (EI_y). The purpose of this paper is to present tests and bifurcation analyses of fixed-ended Z-section columns to verify experimentally and numerically the behaviour predicted by the theory.

EXPERIMENTS

The test specimens were brake-pressed into section from nominally 1.5 mm thick G450 sheet steel. G450 is an Australian-produced steel to AS1397 (1993) with galvanized coating and nominal yield stress of 450 MPa. It has low tensile strength to yield stress ratio and limited ductility of the order of 10-15%.

The average measured cross-section dimensions are shown in Table 1 using the nomenclature defined in Fig. 1a. The coefficients of variation of the measured widths of the flanges (b_f) and web (b_w) were 0.013 and 0.023 respectively, indicating that a tight tolerance was achieved on the cross-section dimensions. The average base metal thickness (t) was 1.50 mm, measured after removing the galvanizing layer by etching. The b/t-ratios for the flanges and web were 26.7 and 26.3 respectively. The elastic local buckling stress (σ_l) and half-wavelength (l) were determined from a finite strip analysis (Hancock 1978), as also shown in Table 1. The local buckling mode, critical overall flexural buckling mode and overall torsional mode are shown in Figs 1b, 1c and 1d respectively.

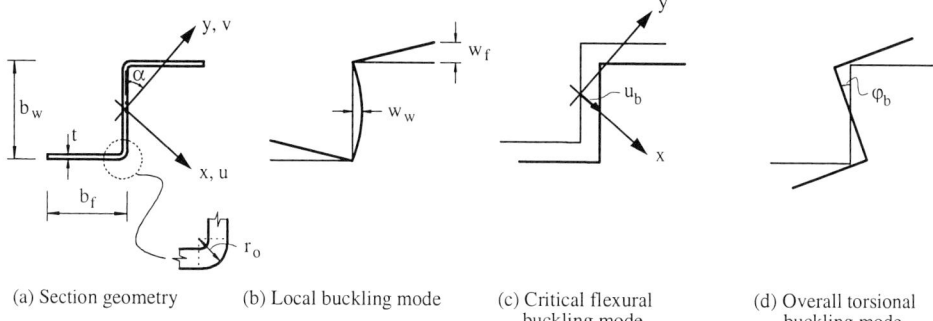

(a) Section geometry   (b) Local buckling mode   (c) Critical flexural buckling mode   (d) Overall torsional buckling mode

Fig. 1: Symbol definitions and buckling modes

b_f (mm)	b_w (mm)	t^* (mm)	r_o (mm)	A (mm^2)	I_y (mm^4)	I_x (mm^4)	I_ω (mm^6)	J (mm^4)	α	σ_l (MPa)	l (mm)
42.7	44.7	1.50	2.6	190.1	1.42×10^4	1.28×10^5	1.75×10^7	141	43°	212	95

* base metal thickness

Table 1: Average Measured Cross-section Dimensions and Section Constants

Two tensile and two compression coupons were cut from the steel sheets in the same direction as the longitudinal axis of the test specimens. Figures 2a and 2b show the initial part of the stress-strain curve and the complete stress-strain curve respectively, as obtained from one of the tension coupon tests. The average values of initial Young's modulus (E_0), yield stress (σ_y) and ultimate tensile strength (σ_u) are shown in Table 2. In the absence of a sharp yield point, the yield stress (σ_y) was obtained as the 0.2% proof stress. Significant softening was observed in the vicinity of yield, as shown in Fig. 2a. Based on the average values of 0.01% proof stress ($\sigma_{0.01}$) and 0.2% proof stress ($\sigma_{0.2}$), as determined from the stress-strain curves, the Ramberg-Osgood n-parameter shown in Table 2 was calculated using the equation $n=\ln(20)/\ln(\sigma_{0.2}/\sigma_{0.01})$.

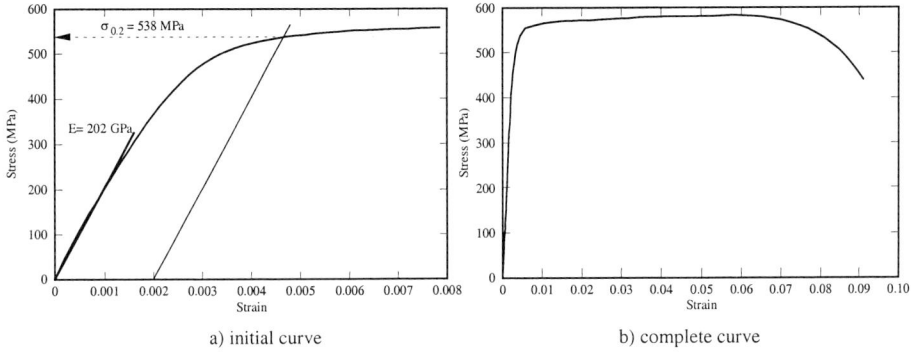

a) initial curve b) complete curve

Fig. 2: Stress-strain Curve Obtained from a Tensile Coupon Test

	E_0 (GPa)	$\sigma_{0.01}$ (MPa)	$\sigma_y = \sigma_{0.2}$ (MPa)	σ_u (MPa)	n
Tension	197	331	528	575	6.4
Compression	199	334	466	-	9.0

Table 2: Average Mechanical Properties

The test specimens were cut in lengths varying from 120 mm to 1600 mm. Subsequently, the ends were milled flat to ensure even loading. For each length, two nominally identical specimens were prepared. Local and overall geometric imperfections were measured on all specimens prior to testing. The overall geometric imperfections are shown as u_0 and φ_0 in Table 3 corresponding to the measured out-of straightness at midlength in the direction of the major x-axis, as shown in Fig. 1c, and the initial twist rotation at midlength, as shown in Fig. 1d. The out-of-flatness of the tip of the flanges (w_{f0}) measured at mid-length is also shown in Table 3. The values of w_{f0} were obtained as a mean of the values measured at each flange tip by adding or subtracting the values depending on whether their directions were sympathetic to the local buckling mode or not. The averages of the overall imperfection relative to the length (u_0/L) and the local imperfection of the flanges (w_{f0}) were 1/14400 and 0.20 mm respectively.

Specimen	Length (mm)	u_0/L	φ_0 (deg)	w_{f0} (mm)	N_{lE} (kN)	N_u (kN)	N_u/N_l
ztf120a	119.0	-	-	-	-	64.2	1.60
ztf120b	119.0	-	-	-	43.3	55.8	1.39
ztf200a	201.0	1/46500	0.12	0.05	39.2	62.2	1.55
ztf200b	199.6	1/42800	0.13	0.01	39.8	60.9	1.51
ztf400a	400.0	1/8500	0.20	0.17	42.3	59.1	1.47
ztf400b	399.0	1/6700	0.27	0.30	39.1	58.6	1.46
ztf600a	599.0	1/115000	0.49	0.09	41.5	54.5	1.36
ztf600b	599.0	1/248000	0.71	0.69	40.8	54.6	1.36
ztf800a	799.0	1/25000	0.98	0.11	40.9	50.2	1.25
ztf800b	798.0	1/21000	0.72	0.22	44.5	50.5	1.26
ztf1000a	999.0	1/34000	0.54	0.12	41.3	46.7	1.16
ztf1000b	999.0	1/22500	0.37	0.39	41.8	44.7	1.11
ztf1200a	1199.0	1/6700	0.07	0.36	40.0	41.6	1.03
ztf1200b	1199.0	1/14000	0.61	0.23	40.0	42.3	1.05
ztf1400a	1399.0	1/11700	0.19	0.49	-	41.9	1.04
ztf1400b	1399.0	1/5000	0.57	0.05	-	38.1	0.95
ztf1600a	1600.0	1/8400	0.64	0.08	-	33.7	0.84
ztf1600b	1599.0	1/1294000	0.03	0.19	-	36.5	0.91
Average	-	1/14400	-	0.20	41.1	-	-

Table 3: Geometric Imperfections and Ultimate Loads

The specimens were loaded between fixed ends in a vertical position. The top end platen was rigidly connected to the cross-head, thus preventing flexural and torsional rotations. At the base, a lockable spherical seat was used to ensure full contact between the end platen and the specimen during setup. Once contact was achieved, the seat was locked by tightening a bolt at each corner such that flexural rotations could no longer occur.

The specimens were uniformly compressed to failure using a 250 kN capacity MTS Sintec testing machine. The ultimate loads (N_u) are shown in Table 3. Readings were taken at regular intervals of local and overall deformations. The local deformations were measured using transducers mounted on an aluminium frame, which was attached to the corners of the specimen at midlength, as shown in Fig. 3. The frame followed the specimen during overall buckling and ensured that the local buckling deformations were measured at the same points in the cross-section throughout loading. Readings were taken at the centre of the web and near the free edge of the flanges. Two transducers were used at each of these location, spaced approximately a quarter-wave longitudinally to ensure non-zero local buckling readings from at least one of the two transducers. In addition, three transducers were used to record overall displacements at midlength.

The experimental local buckling load was estimated using the N vs w^2 method, according to which the local buckling load is the intersection of the load axis with the line fitted through the graph of the load (N) versus the square of the plate buckling deformation in the initial post buckling range. The experimental local buckling loads (N_{lE}) are shown in Table 3. They are generally close with an average of 41.1 kN and a COV of 0.039. The average experimental local buckling load (N_{lE}=41.1 kN) was close to the theoretical value (N_l=40.2 kN).

Figure 3: Specimen ztf1000b Undergoing Flexure and Torsion

BIFURCATION ANALYSIS

The general theory (Rasmussen 1997) for calculating the overall buckling loads of locally buckled members was applied to fixed-ended point-symmetric columns, such as Z-sections, in Rasmussen (2000). It was shown that the longitudinal buckling displacement w_b, the flexural buckling displacements u_b and v_b in the principal x- and y-axis directions respectively, and the buckling twist rotation φ_b are determined from the differential equations,

$$\left[(EA)_t \, w_b'\right]' - \left[(ES_\omega)_t \, \varphi_b''\right]' = 0 \tag{1}$$

$$\left[(EI_y)_t \, u_b''\right]'' + \left[(EI_{xy})_t \, v_b''\right]'' + \left(N_{cr} u_b'\right)' = 0 \tag{2}$$

$$\left[(EI_{xy})_t \, u_b''\right]'' + \left[(EI_x)_t \, v_b''\right]'' + \left(N_{cr} v_b'\right)' = 0 \tag{3}$$

$$-\left[(ES_\omega)_t \, w_b'\right]' + \left[(EI_\omega)_t \, \varphi_b''\right]'' - \left[(GJ)_t \, \varphi_b'\right]' + \left(N_{cr} \frac{\overline{W}}{\overline{N}} \varphi_b'\right)' = 0 \tag{4}$$

where N_{cr} is the overall buckling load, $(EA)_t$, $(ES_\omega)_t$, $(EI_x)_t$, $(EI_y)_t$, $(EI_{xy})_t$, $(EI_\omega)_t$, $(GJ)_t$ are tangent rigidities calculated at the buckling load, as described in detail in Rasmussen (1997), and

$$\frac{\overline{W}}{\overline{N}} = \frac{\int_A \sigma_0 (x^2 + y^2) dA}{\int_A \sigma_0 dA}. \tag{5}$$

In Eqn. 5, σ_0 is the longitudinal stress at incipient buckling. The tangent rigidities ($(EA)_t$, $(ES_\omega)_t$, $(EI_x)_t$, $(EI_y)_t$, $(EI_{xy})_t$, $(EI_\omega)_t$) can be found by subjecting a length of section equal to the local buckle half-wavelength to increasing levels of compression and then superimposing small increments of generalized strain (axial compression, curvature about the major and minor principal axes, and warping) at each compression level. The tangent rigidities are the ratios of the resulting stress resultant to the applied generalised strain, eg

$$(EI_y)_t = \frac{\Delta M_y}{\Delta \kappa_y} \quad (6) \qquad (EI_{xy})_t = -\frac{\Delta M_x}{\Delta \kappa_y} \quad (7) \qquad (EI_\omega)_t = -\frac{\Delta B}{\Delta \varphi''}. \quad (8)$$

The nonlinear inelastic finite strip local buckling analysis described and applied by Key and Hancock (1993) has been used in this paper to calculate the tangent rigidities.

For fixed-ended columns, the governing equations (1-4) and the boundary conditions are satisfied by the displacement field,

$$\frac{w_b}{C_w} = \sin(\frac{2\pi z}{L}) \quad (9) \qquad \frac{u_b}{C_u} = \frac{v_b}{C_v} = \frac{\theta_b}{C_\theta} = 1 - \cos(\frac{2\pi z}{L}) \quad (10)$$

which, upon substitution into Eqns (1-4), leads to an eigenvalue problem with the following non-trivial solutions,

$$N_{cr\varphi} = \left(\frac{4\pi^2 (EI_\omega)_t}{L^2} + (GJ)_t - \frac{4\pi^2}{L^2} \frac{(ES_\omega)_t^2}{(EA)_t} \frac{\overline{N}}{\overline{W}}\right) \tag{11}$$

$$N_{cr\,uv} = \frac{-B \pm \sqrt{B^2 - 4AC}}{2A} \tag{12}$$

where

$$A = 1 \quad (13) \qquad B = -(N_x + N_y) \quad (14) \qquad C = N_x N_y - \left(\frac{2\pi}{L}\right)^4 (EI_{xy})_t^2 \quad (15)$$

$$N_x = \frac{4\pi^2 (EI_x)_t}{L^2} \quad (16) \qquad N_y = \frac{4\pi^2 (EI_y)_t}{L^2}. \quad (17)$$

Figures 4a and 4b compare the test strengths shown in Table 3 with buckling curves determined from Eqns (11-17) for the magnitudes of local geometric imperfection of 0.02 mm and 0.2 mm respectively. The overall buckling loads and test strengths are nondimensionalised with respect to the theoretical local buckling load (N_l=40.2 kN). In all analyses, the local geometric imperfection was assumed to be in the shape of the elastic local buckling mode. The buckling curves are those corresponding to elastic material behaviour and elastic perfectly-plastic material behaviour with values of Young's modulus (E_0) and yield stress (σ_y) determined from the compression coupon tests, as shown in Table 2. The base metal thickness and average measured values of flange and web widths were used in the numerical calculations, as shown in Table 1, ignoring corner radii. One and two harmonics were used to model out-of-plane (flexural) and in-plane (membrane) buckling displacements respectively.

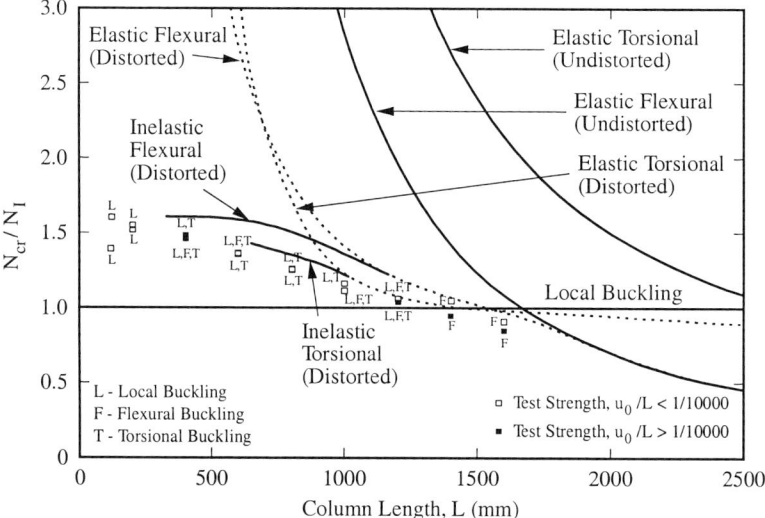

(a) Magnitude of local geometric imperfection of w_{f0}=0.02 mm in numerical analyses

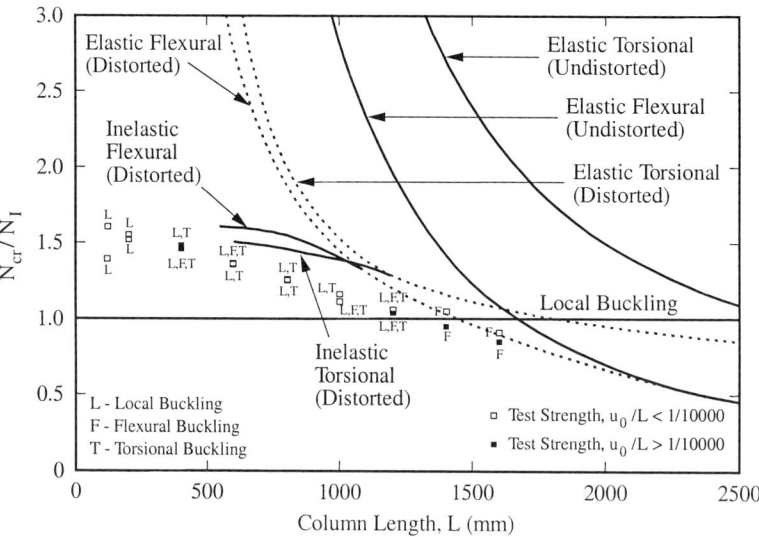

(b) Magnitude of local geometric imperfection of w_{f0}=0.2 mm in numerical analyses

Figure 4: Test Strengths and Bifurcation Curves

Figures 4a and 4b show the elastic and inelastic bifurcation curves of the locally buckled (distorted) cross-section, as well as the classical elastic overall flexural and torsional buckling curves assuming no local buckling (undistorted). According to classical theory, the critical overall mode is the flexural mode at all lengths. However, for small local geometric imperfections, the critical overall mode switches to a torsional mode in the short and intermediate length range, as demonstrated by the inelastic buckling curves shown in Fig. 4a. Torsional buckling becomes less critical for larger values of local geometric imperfection, as shown in Fig. 4b.

The experimental failure modes (local, torsional and/or flexural) are also shown in Figs 4a and 4b. The overall failure modes were determined by comparing the curvature about the minor axis (κ_u) with the rate of change of twist (κ_φ) at the ultimate load. Using the buckling functions given by Eqn. 10, the minor axis curvature and rate of change of twist were obtained as,

$$\kappa_u = -\left(\frac{2\pi}{L}\right)^2 u_b \cos\left(\frac{2\pi z}{L}\right) \qquad (18)$$

$$\kappa_\varphi = \left(\frac{2\pi}{L}\right) \varphi_b \sin\left(\frac{2\pi z}{L}\right) \qquad (19)$$

where u_b and φ_b are the values of minor axis displacement and twist rotation at mid-span measured at the ultimate load respectively. The overall buckling mode was determined using the following criterion:

$$\text{Overall failure mode} = \begin{cases} \text{Torsional (T)} & \text{when } |\kappa_\varphi(L/4)| > 2|\kappa_u(L/2)| \\ \text{Flexural and Torsional (F,T)} & \text{when } \tfrac{1}{2}|\kappa_u(L/4)| \leq |\kappa_\varphi(L/4)| \leq 2|\kappa_u(L/2)| \\ \text{Flexural (F)} & \text{when } |\kappa_\varphi(L/4)| < \tfrac{1}{2}|\kappa_u(L/2)| \end{cases}$$

The test strengths are shown with open or solid markers for overall minor axis geometric imperfections relative to the length (u_0/L) less than or greater than 1/10,000 respectively. The open markers thus represent the strength of nearly straight columns.

The bifurcation curves shown in Fig. 4a for $w_{f0}=0.02$ mm are generally in close agreement with the test strengths, particularly those with overall geometric imperfections less than $L/10,000$. The discrepancy is attributed mainly to the fact that a) the bifurcation analysis assumes perfect column geometry and so does not consider overall geometric imperfections (albeit small), b) the loss of stiffness at low stress values implied by the nonlinearity of the stress-strain curve was not considered in the nonlinear analysis used to calculate the inelastic tangent stiffness values, and c) the overall bifurcation loads are sensitive to local geometric imperfections. Overall geometric imperfections mainly affect the column strength in the intermediate to long length range, while the nonlinear stress-strain curve most significantly affects the column strength in the short to intermediate length range where the buckling stress exceeds the proportionality stress.

The predicted and observed overall failure modes generally agree except that in several tests, torsional buckling occurred in combination with flexural buckling and in the tests of the 1400 mm long columns the observed failure mode was flexural, while theory predicted failure by torsion, as shown in Fig. 4a. The discrepancy in predicted and observed failure modes at $L=1400$ mm is likely to be explained by reference to the magnitude of local geometric imperfection (w_{f0}). Notwithstanding that the values of w_{f0} shown in Table 3 are simply the measured mid-span values with no account given of the longitudinal variation of the local imperfection, (so that they are not directly comparable to the magnitude of local imperfection incorporated in the nonlinear local buckling analysis), they do indicate that the local imperfection varied significantly between the test specimens. To investigate the effect of local geometric imperfections, the imperfection magnitude was increased to $w_{f0}=0.2$ mm, resulting in the bifurcation curves shown in Fig. 4b. It can be seen that by increasing the local geometric imperfection, the critical mode becomes the flexural mode over a much wider range of lengths. This result implies that the flexural tangent rigidities (EI_y, EI_{xy}) are more sensitive to local imperfections than the warping tangent rigidity (EI_ω). It can also be seen that the effect of increasing the magnitude of local geometric imperfection is to decrease the strength for loads near and below the local buckling load, and increase the buckling strength at loads higher than the local buckling load.

The latter increase in buckling strength is consistent with the well-known result that the stiffness of a geometrically imperfect plate element increases with the magnitude of local imperfection in the post-buckling range. It appears from Figs 4a and 4b that better agreement between test and numerical analysis is achieved for the 1400 mm long columns by assuming $w_{f0}=0.2$ mm but the agreement is compromised at shorter lengths.

CONCLUSIONS

Tests have been presented on wide-flange thin-walled Z-sections uniformly compressed between fixed ends. Eighteen tests were conducted at nine lengths ranging from short to long columns, featuring failure by interaction of local and overall torsional and/or flexural buckling. The test specimens were brake-pressed from high strength steel with a nominal yield stress of 450 MPa. Coupon tests demonstrated significant material softening at stresses well below the yield stress.

The tests strengths have been compared with elastic and inelastic buckling loads predicted for a locally buckled section bifurcating in an overall mode. The bifurcation loads were generally in good agreement with the tests strengths, particularly for those columns with very small overall geometric imperfections. The main cause of the discrepancy between the experimental and numerical strengths is attributed to material softening which was not accounted for in the elastic-perfectly-plastic material model used in the numerical analysis.

The experimental failure modes were also in general agreement with those predicted by the overall bifurcation analysis. In particular, the switch in overall buckling mode from the flexural to the torsional mode predicted by the bifurcation theory was confirmed by the tests.

The overall bifurcation mode was shown to be sensitive to the magnitude of local geometric imperfection. The flexural mode proved more affected by local geometric imperfections causing the flexural mode to be critical over a wide range of lengths for a large local geometric imperfection compared to a small local geometric imperfection.

REFERENCES

AS1397, 1993. *Steel Sheet and Strip – Hot-dipped Zinc-coated or Aluminium/zinc-coated*, Standards Association of Australia, Sydney.

Hancock, GJ, 1978. Local, distortional and lateral buckling of I-beams, *Journal of the Structural Division, ASCE*, **104**(ST11), 1787-1798.

Hibbitt, Karlsson and Sorensen, Inc., 1995, 'ABAQUS Standard, Users Manual', Vols 1 and 2, Ver. 5.5, USA.

Key, PW and Hancock, GJ, 1993. A Finite Strip Method for the Elastic-plastic Large Displacement Analysis of Thin-Walled and Cold-Formed Steel Sections. *Thin-walled Structures*, **16**, 3-29.

Rasmussen, KJR, 1997. Bifurcation of Locally Buckled Columns, *Thin-walled Structures*, **28**(2), 117-154.

Rasmussen, KJR, 2000. Overall Bifurcation of Locally Buckled Point-symmetric Columns, *Proceedings*, Coupled Instabilities in Metal Structures, CIMS'2000, Lisbon, Portugal, 163-170.

Rasmussen, KJR, 2001. Bifurcation Experiments on Locally Buckled Z-section Columns, *Proceedings*, 3rd International Conference on Thin-walled Structures, Thin-walled Structures – Advances and Developments, Cracow, Poland, 217-224.

Young, B and Rasmussen, KJR, 1997. Bifurcation of Singly Symmetric Columns, *Thin-walled Structures*, **28**(2), 155-177.

STRUCTURAL STABILITY OF STAINLESS STEEL COMPRESSION MEMBERS

Y. Liu[1]   and   B. Young[2]

[1] School of Civil and Structural Engineering, Nanyang Technological University, Singapore 639798
[2] Department of Civil Engineering, Hong Kong University of Science and Technology, Clear Water Bay, Kowloon, Hong Kong (Formerly, School of Civil and Structural Engineering, Nanyang Technological University, Singapore 639798)

ABSTRACT

A test program on cold-formed stainless steel square hollow section subjected to pure axial compression is described in this paper. A series of tests was compressed between fixed ends. The specimens were cold-rolled from stainless steel sheets. The tests were performed over a range of column lengths, which involved local buckling and overall flexural buckling. Measurements of overall geometric imperfections and material properties of the specimens were conducted. The experimental column strengths were compared with the design strengths predicted by the American, Australian/New Zealand and European specifications for cold-formed stainless steel structures. Generally, the three specifications conservatively predicted the test strengths of the fixed-ended cold-formed stainless steel square hollow section columns. However, the design rules in the Australian/New Zealand Standard are slightly more reliable than the design rules in the American and European specifications.

KEYWORDS

Cold-formed steel, Columns, Experimental investigation, SHS, Stainless steel, Structural design, Structural stability, Tubular members.

INTRODUCTION

Cold-formed stainless steel members are used increasingly for structural applications. This is due to the aesthetic appearance, high corrosion resistance and ease of maintenance as well as ease of construction of stainless steel structural members. Design rules are available for cold-formed stainless steel structural members. These includes the American Society of Civil Engineers (ASCE, 1991) Specification for the Design of Cold-Formed Stainless Steel Structural Members, the Australian/New Zealand Standard (Aust/NZS, 2001) for Cold-Formed Stainless Steel Structures, and the European Code (Eurocode 3, 1996) Design of Steel Structures, Part 1.4: Supplementary Rules for Stainless Steels. The column design rules in the specifications are mainly based on the investigations of pin-ended columns. In practice, there is some degree of rotational restraint at the end supports, and the column is somewhere between fixed and pinned. Therefore, it is also important to obtain test data for the other limiting case of fixed supports. There are not many test data reported on fixed-ended cold-formed stainless steel tubular columns in the literature. Young and Hartono (2002) performed a series of tests on fixed-ended cold-formed stainless steel circular hollow section columns.

Cold-formed square hollow section is formed by cold-rolled with weld of annealed flat strip into a circular hollow section then further rolled into square hollow section. This process of forming by cold-working produces considerable enhancement to the material properties of the annealed steel. More economic design can be achieved by taking into account of the enhancement of the material properties due to cold-working. Hence, in this paper, the design strengths were calculated based on the material properties obtained from the finished specimens. The material properties were determined by tensile coupon tests as well as stub column tests. The purpose of this paper is to present a test program on fixed-ended cold-formed stainless steel square hollow section columns. In addition, the test strengths are compared with the design strengths predicted using the American (1991), Australian/New Zealand (2001) and European (1996) specifications for cold-formed stainless steel structures. Furthermore, the reliability of the design rules is evaluated using reliability analysis.

TEST SPECIMENS AND MATERIAL PROPERTIES

The tests were performed on square hollow sections (SHS) of austenitic stainless steel of type 304. The test specimens were cold-rolled from annealed flat strips. The specimens were supplied from the manufacturer in uncut lengths of 6000 mm. Each specimen was cut to a specified length ranging from 360 to 3600 mm, and both ends were welded to stainless steel end plates to ensure full contact between specimen and end bearings. The longest specimen lengths produced l_e/r_y ratios of 65 and 69 for Series S1 and S2 respectively, where l_e is the column effective length and r_y is the radius of gyration about the y-axis. Two series of SHS were tested, having nominal dimensions of 70 by 70 mm with thickness of either 2 or 5 mm. Each series was tested between fixed ends at various column lengths. The test specimens were labeled such that the test series and specimen length could be identified from the label. For example, the label "S2L0360R" defines the following specimen:
- The first two letters indicate that the specimen belongs to test Series S2, where the prefix letter "S" refers to square hollow section.
- The third letter "L" indicates the length of the specimen.
- The last four digits are the nominal length of the specimen in mm (360 mm).
- If a test was repeated, then the letter "R" indicates the repeated test.

The two test series were S1 and S2 of section sizes 70×70×2 and 70×70×5 respectively. Tables 1 and 2 show the measured cross-section dimensions of the test specimens using the nomenclature defined in Fig. 1. The cross-section dimensions shown in Tables 1 and 2 are the averages of measured values at both ends for each test specimen.

The material properties of each series of specimens were determined by tensile coupon tests as well as stub column tests. For tensile coupon tests, longitudinal coupons were taken from the finished specimens belonged to the same batch of specimens as the column tests. The location of the coupons is shown in Fig. 1, which taken from the centre of the column face at 90° angle from the weld. The coupon dimensions conformed to the Australian Standard AS 1391 (1991) for the tensile testing of metals using 12.5 mm wide coupons of gauge length 50 mm. The coupons were also tested according to AS 1391 in a 300 kN capacity Instron UTM displacement controlled testing machine using friction grips. A calibrated extensometer of 50 mm gauge length was used to measure the longitudinal strain. In addition, two linear strain gauges were attached to each coupon at the center of each face. The strain gauges readings were used to determinate the initial Young's modulus. A data acquisition system was used to record the load and the readings of strain at regular intervals during the tests. The static load was obtained by pausing the applied straining for 1.5 minutes near the 0.2% proof stress and the ultimate tensile strength. This allowed the stress relaxation associated with plastic straining to take place. The material properties obtained from the coupon tests are summarized in Table 3. The measured material properties are the static 0.2% ($\sigma_{0.2}$) and 0.5% ($\sigma_{0.5}$) proof stresses, the static tensile strength (σ_u), as well as the initial Young's modulus (E_o) and the elongation after fracture (ε_u) based on a gauge length of 50 mm.

TABLE 1
MEASURED SPECIMEN DIMENSIONS FOR SERIES S1

Specimen	Depth D (mm)	Width B (mm)	Thickness t (mm)	Radius r_i (mm)	Length L (mm)	Area A (mm^2)
S1L0360	69.9	70.1	1.91	1.9	360.0	512
S1L0360R	69.9	70.2	1.93	1.9	360.0	516
S1L1200	69.9	70.0	1.92	1.9	1199.0	513
S1L2000	69.9	70.1	1.95	1.9	2000.0	522
S1L2800	69.9	70.2	1.92	1.9	2800.5	515
S1L3600	70.0	70.1	1.93	1.9	3599.0	516
Mean	69.9	70.1	1.93	1.9	---	516
COV	0.001	0.001	0.007	0.000	---	0.007

Note: 1 in. = 25.4 mm.
COV = coefficient of variation.

TABLE 2
MEASURED SPECIMEN DIMENSIONS FOR SERIES S2

Specimen	Depth D (mm)	Width B (mm)	Thickness t (mm)	Radius r_i (mm)	Length L (mm)	Area A (mm^2)
S2L0360	70.0	70.0	4.86	4.1	358.5	1213
S2L0360R	70.0	69.9	4.91	4.1	358.5	1223
S2L1200	70.0	69.9	4.86	4.1	1199.0	1212
S2L2000	70.0	70.0	4.91	4.1	2000.0	1223
S2L2800	69.9	70.0	4.91	4.1	2799.0	1222
S2L3600	70.0	69.9	4.86	4.1	3600.0	1212
Mean	70.0	69.9	4.89	4.1	---	1217
COV	0.001	0.001	0.005	0.000	---	0.005

Note: 1 in. = 25.4 mm.
COV = coefficient of variation.

TABLE 3
MEASURED MATERIAL PROPERTIES FROM TENSILE COUPON TESTS

Series	$D \times B \times t$ (mm)	E_o (GPa)	$\sigma_{0.2}$ (MPa)	$\sigma_{0.5}$ (MPa)	σ_u (MPa)	ε_u (%)	n
S1	$70 \times 70 \times 2$	195	337	383	636	60	4
S2	$70 \times 70 \times 5$	194	444	500	688	61	5

Note: 1 ksi = 6.89 MPa.

For stub column tests, the material properties of the complete cross-section in the cold-worked state were obtained from the stub column tests. The shortest specimen lengths complied with the Structural Stability Research Council guidelines (Galambos, 1988) for stub column lengths. The measured cross-section dimensions and the measured specimen length of the stub columns are given in Tables 1 and 2. Two stub columns for each test series were tested, and the specimens were S1L0360, S1L0360R, S2L0360 and S2L0360R. Four longitudinal strain gauges were attached at mid-length of the stub columns. The strain gauges were located at the centre of each face of the columns. In addition, four displacement transducers were connected to the bottom end plate of the specimens, and the transducers measured the shortening of the specimens from the top end plate of the specimens. Similar to tensile coupon tests, the static load was obtained by pausing the applied straining for 1.5 minutes. Table 4 shows the measured initial Young's modulus (E_o) and the static 0.2% proof stress ($\sigma_{0.2}$), which were obtained from the strain gauges readings.

The measured stress-strain curves obtained from the tensile coupon tests and the stub column tests were used to determine the parameter n using the Ramberg-Osgood expression (Ramberg and Osgood, 1943). The parameter n is to describe the shape of the curve, which obtained from the measured 0.01% ($\sigma_{0.01}$) and 0.2% ($\sigma_{0.2}$) proof stresses. The values of n for tensile coupon tests and stub column tests are shown in Tables 3 and 4 respectively. In Tables 3 and 4, the 0.2% proof stresses of test series S1 and S2 differ by approximately 30%. This is probably due to the strain hardening of the materials.

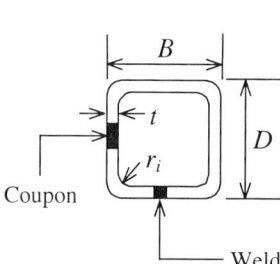

Figure 1: Definition of symbols and location of tensile coupon in cross-section

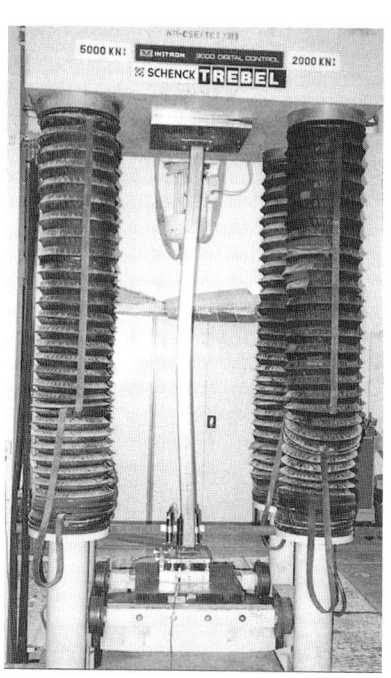

Figure 2: Overall flexural buckling of specimen S2L2000

TABLE 4
MEASURED MATERIAL PROPERTIES FROM STUB COLUMN TESTS

Series	$D \times B \times t$	E_o	$\sigma_{0.2}$	n
	(mm)	(GPa)	(MPa)	
S1		191	382	3
S1	$70 \times 70 \times 2$	192	379	3
Mean		192	381	3
S2		187	494	3
S2	$70 \times 70 \times 5$	189	499	3
Mean		188	497	3
Note: 1 ksi = 6.89 MPa.				

TEST RIG

Figure 2 shows the test rig and the test set-up of a typical column test. A servo-controlled hydraulic testing machine was used to apply compressive axial force to the specimen. Two stainless steel end plates were welded to the ends of the specimen. A moveable upper end support allowed tests to be conducted at various specimen lengths. A rigid flat bearing plate was connected to the upper end support, and the top end plate of the specimen was bolted to the rigid flat bearing plate, which was restrained against the minor and major axes rotations as well as twist rotations and warping. Hence, the top end of the column was fixed in position. The load was then applied at the lower end through a spherical bearing. Initially, the spherical bearing was free to rotate in any directions. The ram of the actuator was moved slowly toward the specimen until the spherical bearing was in full contact with the bottom end plate of the specimen having a small initial load of approximately 2 kN. This procedure eliminated any possible gaps between the spherical bearing and the bottom end plate of the specimen. The bottom end plate of the specimen was bolted to the spherical bearing. The spherical bearing was then restrained from rotations and twisting by using vertical and horizontal bolts respectively. The vertical and horizontal bolts of the spherical bearing were used to lock the bearing in position after full contact was achieved. Hence, the spherical bearing became a *fixed-ended* bearing. The fixed-ended bearing was considered to restrain both minor and major axes rotations as well as twist rotations and warping. Three displacement transducers were positioned on the fixed-ended bearing to measure the axial shortening of the specimen. Displacement control was used to drive the hydraulic actuator at a constant speed of 0.7 mm/min. The use of displacement control allowed the tests to be continued into the post-ultimate range. A data acquisition system was used to record the applied load and the readings of displacement transducers at regular intervals during the tests. The static load was recorded by pausing the applied straining for 1.5 minutes near the ultimate load. This allowed the stress relaxation associated with plastic straining to take place.

TABLE 5
MEASURED OVERALL GEOMETRIC IMPERFECTIONS AT MID-LENGTH

Specimen	δ_x/L	δ_y/L
S1L1200	1/4970	1/44100
S1L2000	1/2100	1/10500
S1L2800	1/4410	1/44100
S1L3600	1/950	1/5150
S2L1200	1/44100	1/3780
S2L2000	1/5250	1/1050
S2L2800	1/1100	1/44080
S2L3600	1/950	1/28350

GEOMETRIC IMPERFECTIONS AND TEST RESULTS

Initial overall geometric imperfections of the specimens were measured prior to testing. Geometric imperfections were measured for both x and y axes of the specimens. Two theodolites were used to obtain readings at mid-length and near both ends of the specimens. The overall geometric imperfections at mid-length for x-axis (δ_x) and y-axis (δ_y) over the specimen length (L) are shown in Table 5. The maximum overall geometric imperfections at mid-length was 1/950 of the specimen length for both Series S1 and S2.

The experimental ultimate loads (P_{Exp}) of the stub and long column tests are shown in Tables 6 and 7 for Series S1 and S2 respectively. The stub column tests were repeated and the test results are very close to the first test values, with a difference of 0.5% and 2.3% for Series S1 and S2 respectively. The small difference between the repeated tests demonstrated the reliability of the test results. Failure modes at ultimate load of the columns involved local buckling, overall flexural buckling and combined local and overall flexural buckling. Figure 2 shows the overall flexural buckling of specimen S2L2000 at ultimate load.

TABLE 6
COMPARISON OF TEST STRENGTHS WITH DESIGN STRENGTHS FOR SERIES S1

Specimen	Test	Comparison					
	P_{Exp}	$\dfrac{P_{Exp}}{P_{ASCE}}$	$\dfrac{P_{Exp}}{P_{Aust/NZS}}$	$\dfrac{P_{Exp}}{P_{EC3}}$	$\dfrac{P_{Exp}}{P^*_{ASCE}}$	$\dfrac{P_{Exp}}{P^*_{Aust/NZS}}$	$\dfrac{P_{Exp}}{P^*_{EC3}}$
	(kN)						
S1L0360	194.0	1.14	1.14	1.14	1.05	1.05	1.05
S1L0360R	193.1	1.14	1.14	1.14	1.04	1.04	1.04
S1L1200	189.8	1.12	1.12	1.12	1.03	1.03	1.03
S1L2000	188.1	1.17	1.39	1.15	1.03	1.28	1.07
S1L2800	159.2	1.22	1.44	1.11	1.10	1.34	1.02
S1L3600	115.0	1.05	1.25	0.97	1.01	1.17	0.92
Mean, P_m	---	1.14	1.25	1.10	1.04	1.15	1.02
COV, V_P	---	0.048	0.114	0.062	0.031	0.117	0.052
Reliability Index, β_o	---	3.29	3.04	2.67	2.99	2.73	2.40

Note: 1 kip = 4.45 kN.
COV = coefficient of variation.

TABLE 7
COMPARISON OF TEST STRENGTHS WITH DESIGN STRENGTHS FOR series S2

Specimen	Test	Comparison					
	P_{Exp}	$\dfrac{P_{Exp}}{P_{ASCE}}$	$\dfrac{P_{Exp}}{P_{Aust/NZS}}$	$\dfrac{P_{Exp}}{P_{EC3}}$	$\dfrac{P_{Exp}}{P^*_{ASCE}}$	$\dfrac{P_{Exp}}{P^*_{Aust/NZS}}$	$\dfrac{P_{Exp}}{P^*_{EC3}}$
	(kN)						
S2L0360	825.3	1.53	1.53	1.53	1.36	1.36	1.36
S2L0360R	843.9	1.56	1.56	1.56	1.39	1.39	1.39
S2L1200	669.3	1.24	1.32	1.24	1.11	1.23	1.11
S2L2000	510.4	1.09	1.36	1.06	0.93	1.27	0.98
S2L2800	407.2	1.06	1.36	1.04	1.02	1.29	0.99
S2L3600	280.7	0.87	1.17	0.94	0.92	1.12	0.92
Mean, P_m	---	1.23	1.38	1.23	1.12	1.28	1.13
COV, V_P	---	0.223	0.104	0.215	0.186	0.076	0.182
Reliability Index, β_o	---	2.36	3.48	2.13	2.35	3.39	2.07

Note: 1 kip = 4.45 kN.
COV = coefficient of variation.

DESIGN STRENGTHS

For the design strengths, the ASCE Specification adopts the Euler column strength while the Aust/NZS Standard allows the use of Euler column strength (identical to those in the ASCE Specification) or the Perry curve. The latter has been used for the purpose of comparison. The Eurocode 3 adopts the Perry curve. For the ASCE Specification, the tangent modulus (E_t) was determined using Equation (B-2) in Appendix B of the Specification. For the Aust/NZS Standard, the values of the required parameters α, β λ_o and λ_1 were obtained from Table 3.4.2 of the Standard, which depend on the type of stainless steel and these parameters are given as $\alpha = 1.59$, $\beta = 0.28$, $\lambda_o = 0.55$ and $\lambda_1 = 0.20$ for type 304. For the Eurocode 3, the values of imperfection factor and limiting slenderness were taken as 0.49 and 0.4 respectively, which were obtained from Table 5.2 of the Code. The three specifications require the determination of effective cross-section area (A_e) of the column. In the three specifications, the effective area was found to be equal to the gross area of cross-section (fully effective) for Series S2, whereas the effective area was found to be less than the gross area of cross-section at short column lengths for Series S1.

RELIABILITY ANALYSIS

Reliability analysis is detailed in the ASCE Specification (1991), and a target reliability index of 3.0 for structural members as a lower limit is recommended. A resistance factor (ϕ) of 0.85 for concentrically loaded compression members is given by the American and Australian/New Zealand specifications, while a ϕ factor of 1/1.1 is given by the Eurocode 3, and these factors are used in the reliability analysis. The load combinations of 1.2DL + 1.6LL, 1.25DL + 1.5LL and 1.35DL + 1.5LL are used in the analysis for American, Australian/New Zealand and European specifications respectively, where DL is the dead load and LL is the live load. The statistical parameters were obtained from Clause 6 of the ASCE Specification for structural members, where $M_m = 1.10$, $F_m = 1.00$, $V_M = 0.10$ and $V_F = 0.05$ which are the mean values and coefficients of variation of material and fabrication factors. The statistical parameters P_m and V_P are the mean value and coefficient of variation of tested-to-predicted load ratios respectively, as shown in Tables 6 and 7. The reliability indices (β_o) of the design rules were determined and the values are shown in Tables 6 and 7.

COMPARISON OF TEST STRENGTHS WITH DESIGN STRENGTHS

The fixed-ended test strengths (P_{Exp}) are compared with the unfactored design strengths predicted using the American (1991), Australian/New Zealand (2001) and European (1996) specifications for cold-formed stainless steel structures. The design strengths were calculated using the material properties obtained from both the tensile coupon tests as well as the stub column tests, in which the 0.2% proof stresses were used as the corresponding yield stresses. Tables 6 and 7 show the comparison of the test strengths with the design strengths for Series S1 and S2 respectively, where P_{ASCE}, $P_{Aust/NZS}$ and P_{EC3} are the design strengths calculated using the material properties obtained from tensile coupon tests for American, Australian/New Zealand and European specifications respectively. The P^*_{ASCE}, $P^*_{Aust/NZS}$ and P^*_{EC3} are the design strengths calculated using the material properties obtained from stub column tests. The test strengths are also compared with the column curves obtained from the three specifications as shown in Figs. 3 and 4 for Series S1 and S2 respectively. The theoretical elastic flexural buckling loads of the fixed-ended columns are also shown in Figs. 3 and 4. In calculating the design strengths and the theoretical buckling loads, the fixed-ended columns were designed as concentrically loaded compression members and the effective length (l_e) was assumed equal to one-half of the column length (L) for the fixed-ended columns ($l_e = L/2$) as recommended by Young and Rasmussen (1998). The design strengths and the theoretical buckling loads were calculated using the average measured cross-section dimensions and the measured material properties for each test series as detailed in Tables 1-4.

The design strengths predicted by the Aust/NZS Standard using the material properties obtained from tensile coupon tests and stub column tests are conservative for Series S1 and S2. The design strengths predicted by the ASCE Specification and Eurocode 3 are also conservative, except for some of the long columns, as shown in Tables 6 and 7. The reliability indices of the design rules for the Aust/NZS Standard are higher than the target reliability index of 3.0, except for the design strengths calculated using the material properties obtained from stub column tests with a reliability index of 2.73 for Series S1. The reliability indices of the design rules for the ASCE Specification and Eurocode 3 are lower than the target value, except for the ASCE design strengths calculated using the material properties obtained from tensile coupon tests for Series S1. Therefore, the design rules in the Aust/NZS Standard are slightly more reliable than the design rules in the ASCE Specification and Eurocode 3. It should be noted that the stub column test strengths are relatively higher than the design strengths for the more compact section Series S2. The stub column tests approximately reached the ultimate strength of the material, whereas the design strengths calculated based on the 0.2% proof stress.

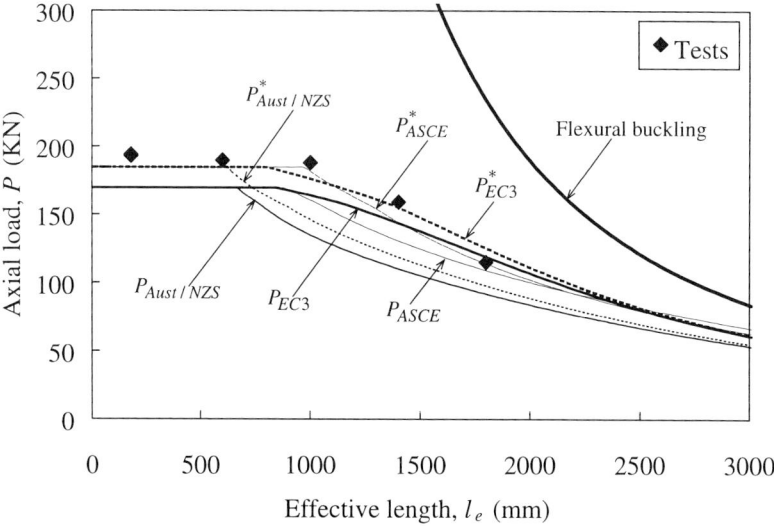

Figure 3: Fixed-ended column curves for Series S1

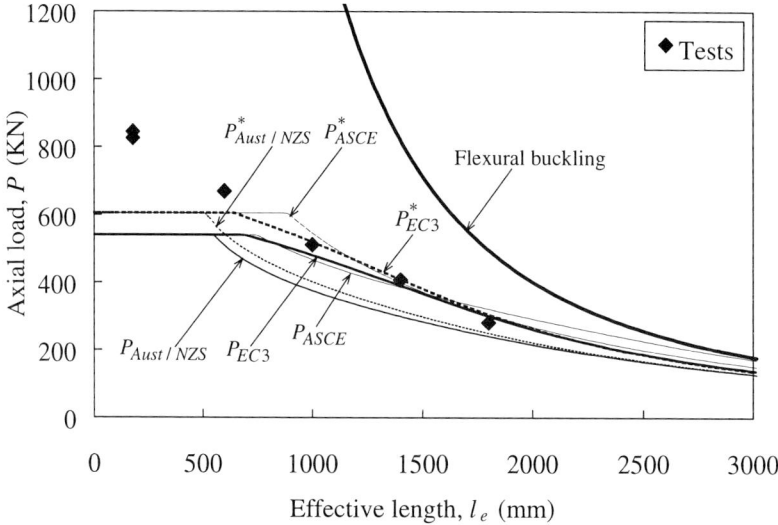

Figure 4: Fixed-ended column curves for Series S2

CONCLUSIONS

A test program on cold-formed stainless steel square hollow section columns has been described. A series of austenitic stainless steel of type 304 square hollow section were tested between fixed ends for stub and long columns. A comparison of the test strengths with the design strengths has been presented. The design strengths were predicted using the American, Australian/New Zealand and European specifications for cold-formed stainless steel structures. The design strengths were calculated based on the material properties obtained from the finished specimens, which takes into account of the enhancement of the material properties due to cold-working. Tensile coupon tests and stub column tests were conducted to determine the material properties. The design strengths were calculated based on an effective length of one-half of the column length. The reliability of the design rules has been evaluated using reliability analysis.

It is shown that the design strengths predicted by the three specifications are generally conservative for the tested fixed-ended cold-formed stainless steel square hollow section columns. However, the reliability analysis shown that the design strengths predicted by the Australian/New Zealand Standard are slightly more reliable than the design strengths predicted by the American and European specifications. Hence, it is recommended that a fixed-ended cold-formed stainless steel square hollow section column can be designed using the Australian/New Zealand Standard that calculated based on the material properties obtained from either stub column test or tensile coupon test.

REFERENCES

ASCE. (1991). *Specification for the Design of Cold-Formed Stainless Steel Structural Members*, American Society of Civil Engineers, ANSI/ASCE-8-90, New York.

Aust/NZS. (2001). *Cold-Formed Stainless Steel Structures*, Australian/New Zealand Standard, AS/NZS 4673:2001, Standards Australia, Sydney, Australia.

Australian Standard (1991). *Methods for Tensile Testing of Metals*, AS 1391, Standards Association of Australia, Sydney, Australia.

Eurocode 3 (1996). *Design of Steel Structures, Part 1.4: Supplementary Rules for Stainless Steels*, European Committee for Standardization, ENV 1993-1-4, CEN, Brussels.

Galambos T.V. (1988). *Guide to Stability Design Criteria for Metal Structures*, 4th ed., John Wiley & Sons, Inc., New York.

Ramberg W. and Osgood W.R. (1943). Description of Stress-Strain Curves by Three Parameters. *Technical Note* No. 902, National Advisory Committee for Aeronautics, Washington, D.C.

Young B. and Hartono W. (2002). Compression Tests of Stainless Steel Tubular Members. *Journal of Structural Engineering,* ASCE **128:6**.

Young B. and Rasmussen K.J.R. (1998). Tests of Fixed-ended Plain Channel Columns. *Journal of Structural Engineering,* ASCE **124:2**, 131-139.

MEMBRANE IMPERFECTIONS MEASURED IN COLD FORMED TUBES

Andrew Wheeler and Martin Pircher

Centre for Construction Technology & Research,
College of Science, Technology & Environment, University of Western Sydney,
Locked Bag 1797, South Penrith DC, NSW 1797, Australia

ABSTRACT

A number of studies have been carried out to determine the significance of initial imperfections on the local buckling of thin walled tubes subjected to flexural loading using numerical methods. However, to-date studies investigating the shape and magnitude of imperfections in real sections have been limited. Measurements taken from thin walled tube specimens are presented in this paper with particular emphasis on the shape and magnitude of the local imperfections.

The methods used to physically measure tubes, along with a numerical method derived to reduce the results are presented in this paper. The measured imperfections are assessed using a Fourier based analysis, producing the means to model the imperfections in a finite element analysis. Thus enabling assessment of the effect of real initial imperfections on the buckling capacity of the member when subjected to pure bending.

KEYWORDS

Thin Walled Steel Tubes, Initial Imperfections, Local Buckling, Flexural Loading, Imperfection Shape, Imperfection Size

INTRODUCTION

The use of concrete filled steel tubes in the construction industry continues to increase as an economical alternative to traditional construction methods. This composite system utilises the compressive strength of the concrete, with the steel tube providing tensile capacity for any tensile and/or flexural loading. The tube is also advantageous as it provides the necessary formwork during construction. Typically concrete filled tubes are used in members subjected to predominantly axial loads (columns). While concrete filled tubes have been used for flexural members in frames, generally comprising of rectangular hollow members, recent developments have seen circular hollow sections filled with concrete used as flexural members Nakamura (2002). However, circular concrete filled tubes have been used in predominantly axial loaded members, consequently extensive research into the behaviour of concrete filled tubes under axial loading has been published (O'Shea & Bridge, 2000). To enable better utilisation of these concrete filled tubes in beam column applications, an investigation into their flexural behaviour is currently being carried out at the Centre for Construction Technology and Research at the University of Western Sydney.

The first stage of this work endeavors to extend the current knowledge of the flexural behaviour of thin-walled tubes. Work conducted by Schilling (1965) and Sherman (1976) suggests that initial imperfections within the tube have a detrimental effect on the bending strength of circular hollow tubes. These suggestions have been reinforced by numerical studies into the flexural behaviour of thin walled tubes (Wheeler et al. 1999, Wheeler & Bridge 2000), in which effects such as magnitude and shape of the imperfections were studied. The publication of data on imperfections in thin-walled tubes, to the authors' knowledge is minimal. However, investigations on imperfections in thin-walled circular-cylindrical members and structures have been carried out in other contexts (Ding 1996, Bernard et al. 1999, Berry et al. 2000, O'Shea & Bridge 2000, Pircher et al. 2000).

In this paper, a new method for measuring and analysing the imperfections of cylindrical members is presented. The method is simple to implement in a laboratory environment yet very accurate. From the measurement results, three-dimensional imperfection maps for the measured tubes were produced. In producing these maps a method of analysis is applied to filter out any noise and the presence of any repetitive imperfections possibly caused by the manufacturing process assesed. To assist with future research the results are assessed using a Fourier analysis coupled with cross validation, to determine the dominant Fourier terms. This method has so far been applied to six cold-formed tubes with high diameter to wall-thickness ratios (D/t ratios).

SPECIMENS

The specimens used in the experimental work described in this paper are cold-formed circular hollow sections. Two tube sizes were examined, both with a nominal wall thickness of 6.4 mm and nominal outside diameters of 406.4 mm and 457.0 mm. According to AS 4100 (SA, 1998) these sections have a section slenderness (λ_s) of 89 and 100 respectively, classifying both groups of tubes as non-compact. For both tube sizes the imperfections of 3 different tubes lengths were measured, giving a net total of six specimens. The designated tube name and nominal diameter are listed in Table 1.

TABLE 1
Specimen Details

Tube	Wall-thickness (mm)	Diameter (mm)	Minimum (mm)	Maximum (mm)	Difference (mm)
t400a	6.4	406.4	-0.829	0.878	1.707
t400b	6.4	406.4	-0.828	1.224	2.052
t400c	6.4	406.4	-1.332	1.242	2.575
t457a	6.4	457.0	-1.539	1.681	3.220
t457b	6.4	457.0	-0.686	0.654	1.340
t457c	6.4	457.0	-0.992	1.254	2.246

It is recognised that the manufacturing process to form cold-rolled tubes generates tubes with a high degree of straightness. Consequently, to determine the imperfection characteristics of the tubular specimens, an accurate method of measurement was required. While the surface of the tubes is nominally coated with an epoxy resin to prevent surface corrosion, the tubes investigated in this report had incomplete placement of the protection coating, resulting in light rusting and runs of epoxy on the surface. These superficial variations could be considered "noise" in the context of the measured imperfections.

MEASUREMENT APPARATUS

The method of measurement is based on the assumption that a line of LVTD's placed perpendicularly to the circumference of a cylinder, and fixed in space, will measure any relative changes on the diameter of the tube as it is rotated about its longitudinal axis. Due to the length and size of the tubes, a purpose built rig was constructed in the structures laboratory at the Centre for Construction Technology and Research. The schematic diagram of the testing rig is shown in Figure 1, with the key components being the measurement Trolley and Rotating clamps shown in Figure 2a and 2b respectively.

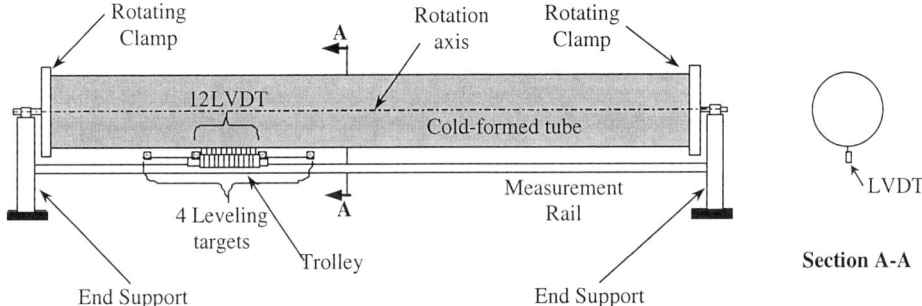

Figure 1 – Schematic Diagram of Measurement Rig

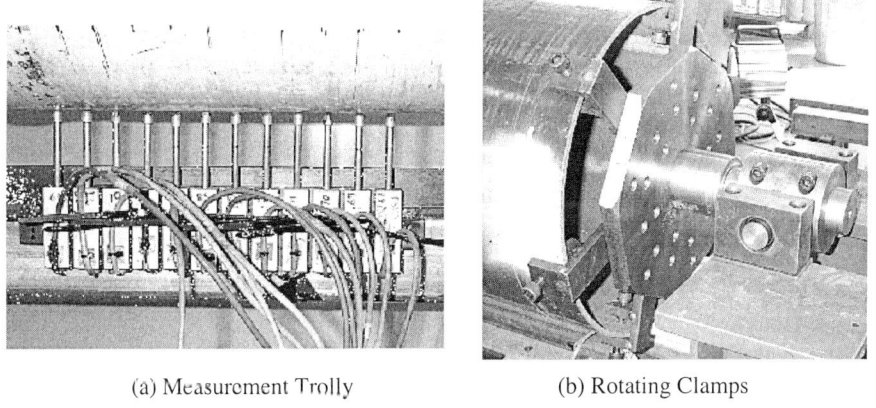

(a) Measurement Trolly   (b) Rotating Clamps

Figure 2 – Measurement Components

The specimens were to be tested in four point bending to determine their flexural capacity. Consequently, only imperfection measurements over the central two quarters were recorded, with an approximate length of two meters being measured for each specimen.

MEASUREMENT PROCEDURE

The specimens were placed on the end supports with purpose built clamps (Figure 2b) which allowed for the tube to be rotated about a fixed axis. The transducer trolley (Figure 2a) was then placed on the measurement rail parallel to the axis of rotation. Before the imperfections were measured a number of initial calibration measurements were made to establish a reference datum for each of the transducers on the trolley. The tube was then rotated to the starting meridian and measurements taken, after which the tube was rotated at 4° increments and measurements taken while the position of the trolley remained fixed. Upon completion of a full set of circumferential measurements (360°) the transducer trolley was moved to the adjacent region. Each individual set of measurement consisted of 12 measurements corresponding to the 12 LVDT's mounted on the measurement trolley. When the measurement trolley was displaced longitudinally, five LVDT's overlapped the previous measured region. This procedure was repeated twelve times to map the desired area. The effects of bending of the measurement rail were also recorded using a Wild NA2 automatic level coupled with the Wild GPM3 parallel plate micrometer. The data from the LVDT's was recorded digitally using a Solitron Data Recorder.

Additional information on the specimens, measurement apparatus and measurement procedure may be found in Wheeler Pricher (2002)

MEASUREMENT PROCESSING

The methods used in the measurement procedure resulted in twelve series of independent circumferential measurements, linked by overlapping regions. Initially numerical adjustments were made to correct for the different base values of each of the 12 LVDT's, these adjustments were determined from the datum established in the initial calibration. The mis-closure within circumferential sets and bending of the tube itself were also considered. Using the levels measured with the Wild NA2 the bending of the measurement rail is determined and corresponding corrections made to the LVDT's.

The adjacent circumferential sets were joined together by using the overlapping transducer measurements to initialise a Helmert transformation (eg. Methley, 1986). Corrections were also made for the errors introduced by the fact that the axis of rotation was not quite identical with axis of the best-fit perfect cylinder which in turn was not quite parallel with the measurement rail (O'Shea & Bridge, 2000).

All these mathematical operations resulted in imperfection maps which gave the deviations of the surfaces of the specimens from the best-fit perfect cylinders. Figure 3 shows a 3-dimensional flat representation of the imperfection map (Figure 3a) for the tube *T457c*. The same imperfections magnified by a factor of 20 is superimposed onto a ideal cylinder are also shown in Figure 3b. The extreme values of the measured imperfections for each tube are given in Table 1.

Measurement Analysis

A spectral analysis of the imperfection maps was performed using a least-square fit Fourier decomposition analysis in both the meridional and circumferential directions (Bracewell, 1978). The number of Fourier terms which can be analysed by this method is limited by the Nyquist frequency (eg. Hearn & Metcalfe 1995). The general expression used in the Fourier analysis is shown in Equation 1, with the shift δ, the meridional coordinate x and the length of the measured region L. The results generated are the amplitudes a_k for the various Fourier terms k. In the case of the measured tubes the highest computed Fourier term in the meridional direction was incidentally the same as for the circumferential direction ($m = 44$). For each measured meridian and circumference the wave spectra were generated and the individual amplitudes a_1 to a_{m+2} were recorded. The shift δ was also adjusted to ensure a best possible fit of the Fourier analysis.

$$w_i(x) = a_1 + a_2 \bar{x} + \sum_{k=1}^{m} a_{k+2} \cdot \sin(\pi k \bar{x}) \quad (1)$$

$$\text{with} \quad \bar{x} = \frac{x}{L} - \delta$$

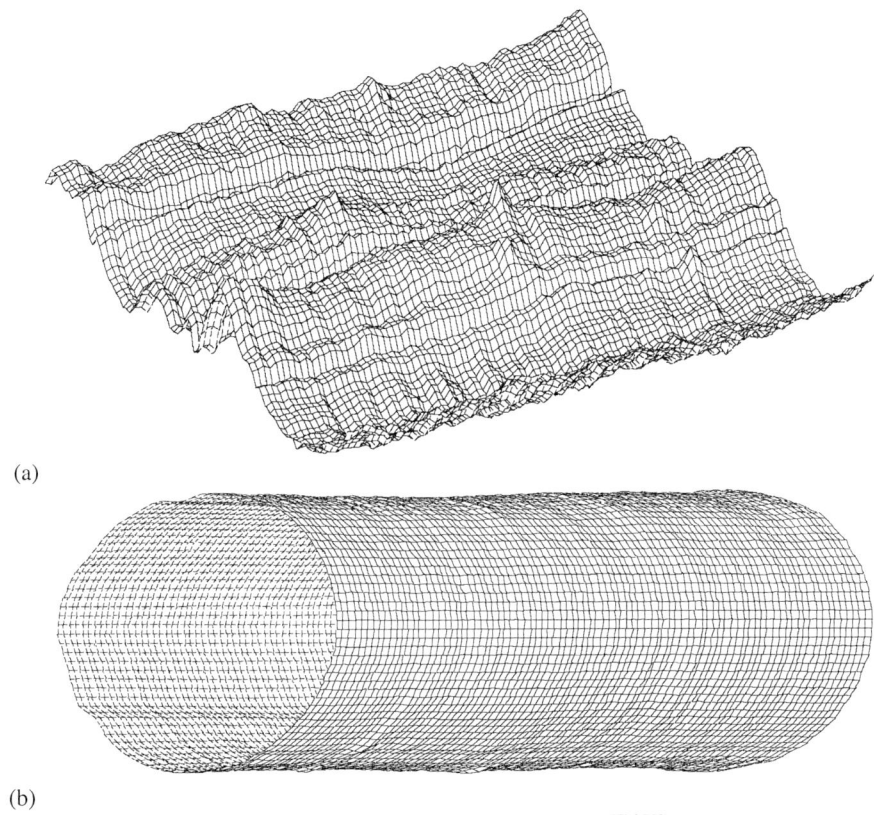

(a)

(b)

Figure 3 – Imperfections as measured (T457c)

Many of the computed Fourier terms can potentially be classified as noise and consequently be omitted from the analysis without loosing accuracy on the final result. To obtain an objective assessment of the significance of each term, the method of cross-validation is used (Thomas 1994, Kowalsky 1982). In the given context, cross-validation can also be used to help identify the optimal size for the Fourier spectra. The Prediction Error Sum of Squares (PRESS) has traditionally been used in the context of other types of spectral analyses (Haaland & Thomas, 1988) and was adopted for cross-validation of the Fourier spectra for the present study.

The method used for the cross-validation is described is detail in Wheeler and Pircher (2002). This method demonstrated that the number of Fourier terms required to accurately model the measured imperfections varies in both the circumferential and meridional directions. The significance of the Fourier terms are determined using the PRESS and the 15 most significant terms used to model the initial imperfections on the tube.

This procedure greatly reduced the data describing the imperfection maps, while filtering out noise introduced during the measurement process. Figure 4 gives a comparison between the imperfection maps before (Figure 4a) and after (Figure 4b) Fourier analysis and cross-validation. The 3-dimensional representations and tables of the final imperfection maps produced with only the significant Fourier terms for all tubes are given in Wheeler & Pircher (2002).

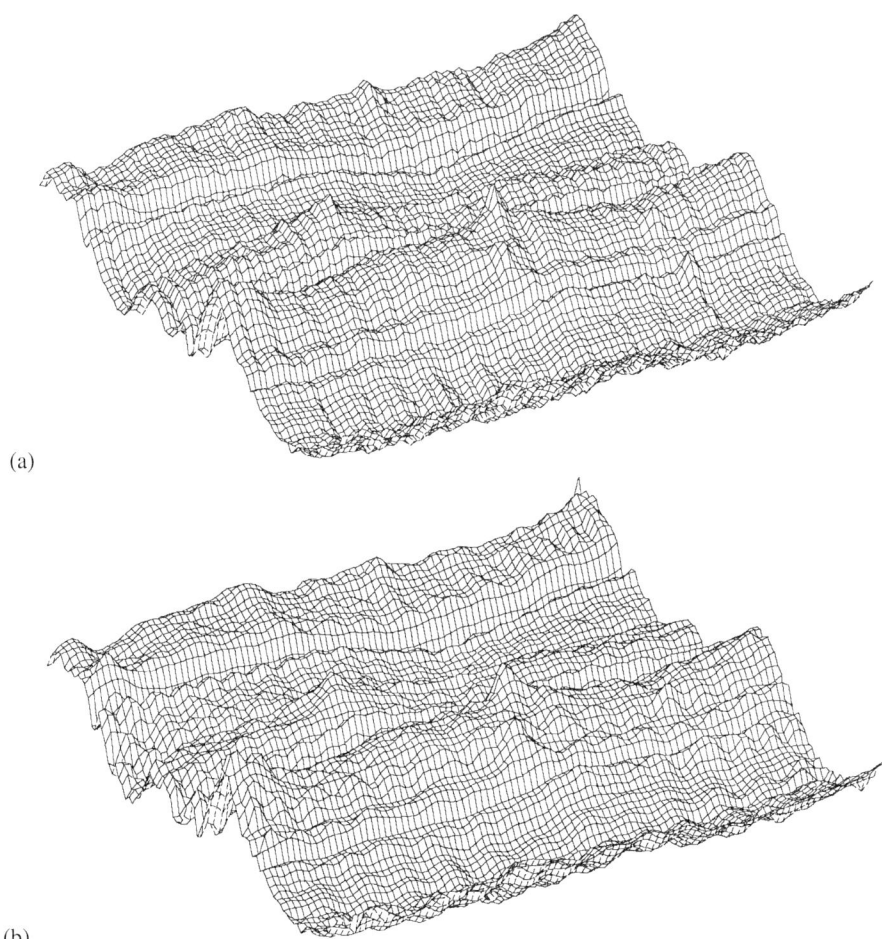

Figure 4 – Imperfection map of a tube (a) as measured, (b) after cross-validation

TABLE 2.
Fourier terms in meridional direction, ranked according to PRESS.

Tube	#1	#2	#3	#4	#5	#6	#7	#8	#9	#10	#11	#12	#13	#14	#15
T400a	1	2	3	4	5	7	6	9	8	10	11	13	12	17	15
T400b	1	2	3	4	5	8	7	9	10	11	6	13	14	15	12
T400c	1	14	16	15	13	12	2	17	11	10	8	18	9	6	7
T457a	1	2	3	6	4	5	7	9	8	12	11	10	13	14	16
T457b	1	2	3	4	5	6	12	11	8	9	7	13	14	15	10
T457c	1	6	2	7	12	3	14	13	8	4	9	5	15	17	10

THE FIFTEEN (15) MOST SIGNIFICANT FOURIER TERMS FOR EACH TUBE IN MERIDIONAL AND CIRCUMFERENTIAL DIRECTION ARE GIVEN IN

Table 2 and Table 3 respectively. In the meridional direction, the long wavelengths prevail. Only one of the tubes with a diameter of 406.4mm (*T400c*) also had significant shorter wavelength with great significance (Fourier terms 12 to 16 correspond to wavelengths of 133.9mm to 178.5mm). For the tubes with a diameter of 457mm (*T457a, T457b, T457c*) Fourier terms 6 and 12 consistently displayed increased significance. In the circumferential direction, Fourier term one (1) signifies ovalisation and is consistently ranked among the three most significant terms. Shorter wavelengths have also been found to have significance, as can be seen in Table 3.

TABLE 3.
Fourier terms in circumferential direction, ranked according to PRESS.

Tube	#1	#2	#3	#4	#5	#6	#7	#8	#9	#10	#11	#12	#13	#14	#15
T400a	1	3	5	13	7	21	11	27	15	29	23	19	24	4	9
T400b	3	24	1	6	8	28	2	26	14	34	32	30	37	17	4
T400c	9	27	1	3	21	11	17	25	7	29	13	31	5	41	33
T457a	1	7	15	18	11	32	20	28	23	12	6	9	30	27	22
T457b	11	1	3	6	26	5	19	24	14	30	20	25	18	21	17
T457c	7	3	1	22	24	26	14	5	28	36	18	12	11	17	32

CONCLUSIONS

A method for measuring the imperfections of thin walled tubes using has been developed using linear voltage displacement transducers and optical leveling. To ensure reliability of the measurements a number of overlapping data sets were produced, and using these redundancies numerical methods were developed to generate accurate measurements of the imperfections. Performing control measurements with an automatic level to ensure measurements based on a reference plane further increased the accuracy. Once implemented the method is simple and easy to use in a laboratory environment.

Three-dimensional imperfection maps where generated, with the deviations of the actual specimen from a perfect cylinder in each generated point. The map was constructed using various transformations and data reductions that account for mis-allignments within the measurement rig, bending of components of the measurement rig and the tube itself. Overlapping measurements are used to increase the precision of the resulting imperfection maps.

In order filter out any "noise" and to reduce the vast amounts of measurement data to a format which can be managed in subsequent analyses a spectral analysis using a least-square fit Fourier-decomposition was performed on the imperfection maps, in both meridional and circumferential directions. The significant Fourier terms were determined using the Prediction Error Sum of Squares (PRESS) as a cross-validation algorithm. PRESS was used to determine the significant Fourier terms for each measured meridian and circumference. It was shown that in the meridional direction short wavelengths are generally not among the 15 most significant Fourier terms. While ovalisation is the prevailing influence in the circumferential direction, short wavelengths were found to also play a role.

ACKNOWLEDGMENTS

The authors are grateful to OneSteel Pipe & Tube, Kembla Grange, for providing the specimens used in this investigation and to TDV GesmbH., Austria, for help on the implementation of the Helmert transformation. Many thanks also go to Bernhard Lechner for his help on the spectral analysis of the measurements.

REFERENCES

Bernard E.S., Coleman R., Bridge R.Q. (1999) "Measurement and Assessment of Geometric Imperfections in Thin-walled Panels", *Thin-Walled Structures*, v33, pp 103-126

Berry P.A., Rotter J.M., Bridge R.Q. (2000) "Compression Tests on Cylinders with Circumferential Weld Depressions", *Journal of Engineering Mechanics, ASCE*, v126, n4, pp 405-413

Bracewell R.N. (1978) *The Fourier Transform and Its Applications*, McGraw-Hill Book Company, New York

Ding X.L., Coleman R., Rotter J.M. (1996) "Surface Profiling System for Measurement of Engineering Structures", *Journal of Surveying Engineering, ASCE*, v122, n1, pp 3-13

Haaland D.M., Thomas E.V. (1988) "Partial Least-Squares Method for Spectral Analysis. Relation to Other Quantitative Calibration Methods and the Extraction of Qualitative Information", *Analytical Chemistry*, v60, pp 1208-1217

Hearn G.E., Metcalfe A.V. (1995) *Spectral Analysis in Engineering – Concepts and Cases*, Arnold, London

Kowalski B.R., Gerlach R., Wold H. (1982) "Systems under Indirect Observation", *Chemical Systems under Indirect Observation*, K. Joreskog & H Wold (Eds.), North-Holland, Amsterdam, pp 191-209

Methley B.D.F. (1986) *Computational Models in Surveying and Photogrammetry*, Blackie, Glasgow and London

Nakamura, S., Momiyama, Y., Hosaka t. and Homma K. (2002) "New technologies of steel/concrete composite bridges", *Journal of Constructional Steel Research*, Vol. 58, No. 1, January 2002, pp 99-130

O'Shea M.D. and Bridge, R.Q. (2000), "Design of Thin-walled Concrete Filled Tubes", *Journal of the Structural Division, ASCE*, v126, n11, pp 1295-1303.

Pircher M., Berry P.A., Ding X., Bridge R.Q. (2001) "The Shape of Circumferential Weld-Induced Imperfections in Thin-Walled Steel Silos and Tanks", *Thin-Walled Structures*, v39, n12, pp 999-1014

SA (1998), *AS 4100-1998: Steel Structures*, Standards Australia, Sydney.

Schilling, C.G. (1965), "Buckling Strength of Circular Tubes", *Journal of the Structural Division, ASCE*, v91, n5, pp 325-348

Sherman, D.R. (1976), "Tests of Circular Steel Tubes in Bending", *Journal of the Structural Division, ASCE*, v102, n11, pp 2181-2195

Thomas E.V. (1994) "A Primer on Multivariate Calibration", *Analytical Chemistry*, v60, n15, pp 795A-804A

Wheeler A., Pircher M. (2002) "Measured Imperfections in Six Cold-Formed Thin-Walled Steel Tubes", *Research Report CE20*, Centre of Construction Technology & Research, University of Western Sydney

Wheeler A.T., Pircher M., Bridge R.Q. (1999), "Modelling of Thin-walled Tubular Sections Subjected to Pure Bending", *Proceedings: The Thirteenth Australian Compumod Users' Conference*, Melbourne, Australia

Wheeler, A. and Bridge, R. (2000), "Thin-walled Steel Tubes Filled with High Strength Concrete in Bending", *Proceedings: Engineering Foundation Conference, Composite Construction IV*, Banff, Canada

FLEXURAL FAILURE OF COLD-FORMED SINGLE CHANNELS CONNECTED BACK-TO-BACK

M. Dundu and A.R. Kemp

School of Civil and Environmental Engineering, University of the Witwatersrand, P. Bag 3, WITS, 2050, South Africa

ABSTRACT

This paper describes the results of a series of experiments on the flexural behaviour of the eaves region of portal frames comprising cold-formed, single channels bolted back-to-back at the eaves joint. Observed modes of failure include local buckling of the compression zone of the flange and web of the channels, lateral torsional buckling of the channels between points of lateral support and bolts in bearing. Variables in the tests include the number of bolts in the connection, the lengths between lateral restraints, the width of the channel flanges and the strength of the channels. The tests demonstrate the considerable ductility that is achieved in back-to-back bolted connections, which should be sufficient in many cases to accommodate plastic analysis of the portal frames. The counter-balancing moments and forces in the back-to-back connections are shown to be important in enhancing the lateral buckling strength of the channels.

KEYWORDS

Flexural behaviour, eaves, strengths, plastic behaviour, local buckling, lateral torsional buckling, moment-rotation, moment-curvature.

INTRODUCTION

Cold-formed steel portal frames can be a viable alternative to conventional hot-rolled I-sections, which require heavy-duty engineering input in the pre-manufacturing of elements, and erection using skilled labour. The aim of this investigation is to develop a lightweight, bolted structure that can easily be fabricated and constructed without using sophisticated machinery. It is envisaged that the structure will be erected mainly by unskilled labour, using only site aids such as lightweight scaffolding, ladders and spanners. To accomplish this, material with low mass, i.e. thin walled section, is used so that channels may be manhandled into position. An investigation was initiated into the use of cold-formed channel sections in light, small span portal frames using back-to-back bolted connections in bearing at the eaves and apex. It is anticipated that these frames will find most use in small span industrial, single storey community and agricultural structures. It is proposed that the frame of the structure should be provided in a kit form and erected in "meccano" fashion, like the British Swagebeam system developed by Trebilcock (1994). The portal frame will be delivered to site in sections cut to length and with connection holes pre-punched at the factory. A comprehensive testing programme was therefore carried out to determine the performance of such structures. Full scale testing of the portal frames was abandoned because of the associated complexity and cost. Since the moments are greatest at the eaves, this region became the main focus of the investigation. Particular attention was paid to the moment resistances, modes of failure and possible plastic behaviour of the joints.

STRUCTURAL FORM AND EXPERIMENTAL MODEL

A complete portal frame structure, consisting of a span of 12m with an eaves height of 3m and a pitch of 10° was considered as a basis for the development of all the structures tested. A single 300x75x20x3 channel section was found to be suitable for this structure. The eaves connection is shown in Figure.1. Either a 100x75x20x2 cold-formed lipped channel or 125x65x20x3 lipped Z-section could be used for the purlins. The purlins are to be continuous over two spans of 4.5 m each and spaced at a maximum of 2.0m centres. The design of the frames was done in accordance with South African codes, SABS 0160 and SABS 0162: Part 2. For this particular structure purlins and rails are connected to the main frame through a 100x75x3 cold-formed lipped angle.

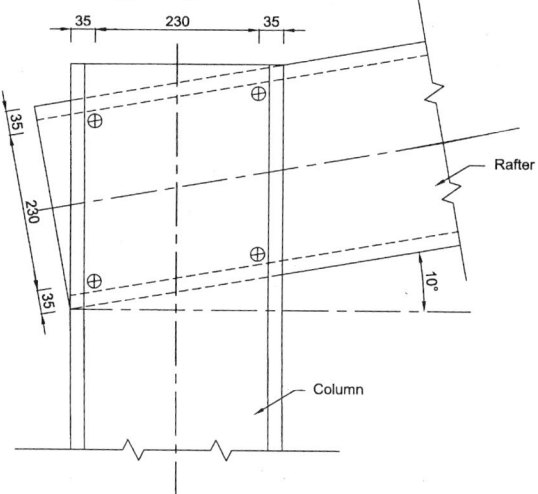

Figure 1: Eaves Connection

Two eaves frames were connected together and loaded equally in the experimental model to allow for interaction in the structure. This model involved extracting a portion of the two frames from the eaves joint to the points of contraflexure in the rafter and column respectively. Since the bending moment is at its maximum in this region, this represents the most highly stressed portion of the portal frame. K-bracings were used to stabilise the pair of frames. A schematic diagram of the general test set-up and the degrees of freedom at the supports are shown in Figure 2. All supports were free to rotate in any direction. Variables in the 4 tests include the number of bolts in the connection, the points of contraflexure, the width of the channel flanges and the strength of the channels. A list of the channels used in the four frames, the number and size of the bolts in the connections, the angle sections connecting the purlin to the frames, the distances from the eaves to the inflection points and the strength of the channels are given in Table 1.

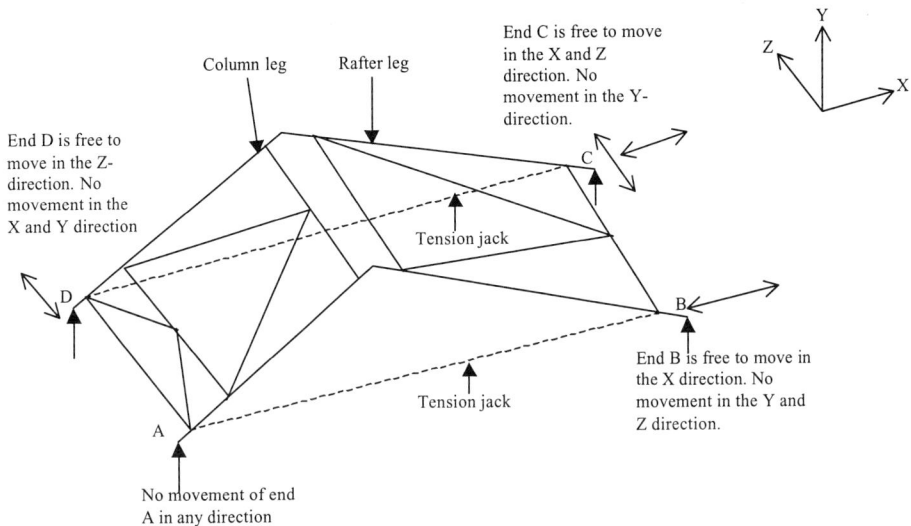

Figure 2: Schematic details of the test set-up

TABLE 1
VARIABLES IN THE STRUCTURES

Structure	Frame Section	No of Bolts	Size of Bolts	Angle Section	Eaves-contraflexure distance (mm)		Strength of Channels (MPa)	
					Column	Rafter	Yield	Ultimate
Structure 1	300x75x20x3	4	M20	100x75x6*	1500	1510	472	545
Structure 2	300x75x20x3	8	M20	100x75x3	1500	1845	472	545
Structure 3	300x65x20x3	4	M16	100x75x3	2000	1845	346	477
Structure 4	300x50x20x3	4	M16	100x75x3	2000	1845	264	349

* Mild steel angle section

The column and rafter legs were connected back-to-back at the eaves joint. The bolts at the eaves joint were increased to 8 in structure 2 to find out whether this would change the mode of failure of structure 1, which failed as a result of concentration of stress below the inside lower bolt. Each cold-formed angle was connected to the frame using two bolts of the same size as those at the joint, and to the 100x50x20x2mm cold-formed channel purlins using 2 or 4, 12mm diameter bolts. All structures were braced using 50 x 50x 2 mm K-bracing. The points of contraflexure for all structures tested are shown in Figure 3. The points of contraflexure in structure 1 (A-B) are for a 12m span frame with a rigid base. A-D represents the points of contraflexure in structure 2. The point of contraflexure in the rafter of this structure was moved 0.335m away from its position in structure 1, to facilitate a lateral-torsional buckling mode of failure. C-D represents the points of contraflexure in structures 3 and 4. The point of contraflexure in the column of these two structures was moved from its position in structures 1 and 2 by a further 0.5m away from the eaves joint. This new point of contraflexure in the column was established by assuming that the base of a complete portal frame is partially fixed. To stop the column from failing in lateral-torsional buckling the extra column length added had to be braced in a similar way to other points of lateral restraints.

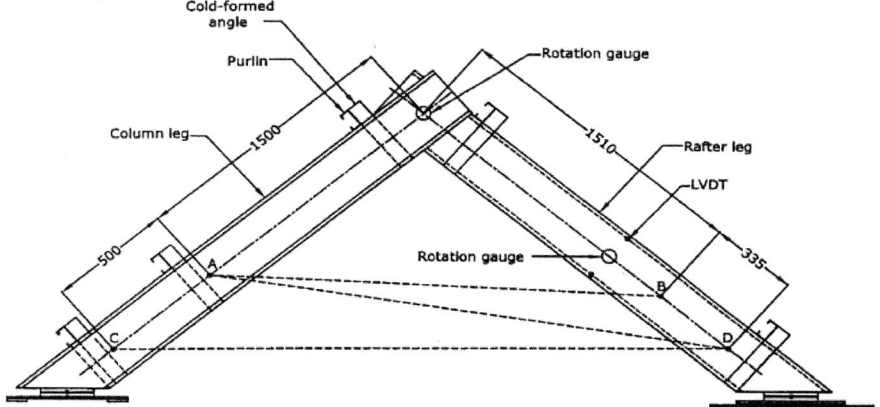

Figure 3: Points of contraflexure for all structures

EXPERIMENTAL PROGRAMME

The average yield and ultimate strength obtained from the coupons of the channels used in the four structures are given in Table 1 and were used to calculate the moment of resistance of channels and connections in Table 2 respectively. Young's modulus of elasticity (E = 207 GPa) was used in the buckling calculations. High strength structural bolts, size M16 and M20, all of Grade 8.8 steel were used for eaves connections and any other connection in the web of the cold-formed channels. Size M12, Grade 8.8 bolts were used for purlin connections. All these bolts are manufactured according to the South African Standard, SABS 136. Grade 8.8 bolts have a minimum tensile stress of $f_t = 800$ MPa. Standard washers were placed under both the head and the nut of the bolt to guard against deformation and possible buckling of the thin material adjacent to the bolt. The diameter of all bolt holes was made 1mm greater than the nominal diameter of the bolt to reduce slip in the connections.

The frames were fully instrumented so that the rotation of the eaves' joints, strain, and lateral torsional buckling of the frames could be measured. Data collection was fully automated by means of data-logging equipment. Scanning of results was performed using an Orion or Fluke data logger. In-plane rotation of eaves' joints was monitored using high-precision digital rotation transducers (also known as inclinometers). The centre of each joint was installed with two rotation transducers, with one transducer on each side of the joint. Strains and lateral torsional buckling of the longer laterally unsupported leg were measured using electric-wire strain gauges and two linear variable differential transformers (LVDT) respectively. The strain gauges were placed at anticipated points of high strain, just outside the joint to determine the moment-curvature behaviour of the frames.

In a typical portal frame building internal frames are subjected to similar loading and it is also logical to assume that when the frames are acted upon by the load they deflect downwards by the same amount. Based on this argument the two portal frames were subjected to equal loads applied through pins at the points of contraflexure using two 16-ton, hand-operated jacks. The load was applied in line with the shear centre to reduce twisting arising from loading away from the shear centre. Load cells were connected to the two jacks and the same load was generated in each frame.

PRESENTATION AND ANALYSIS OF RESULTS

Capacity of Connections

The calculated moments of resistance and the results of the frame tests are summarised in Table 2. M_{rb} and M_{rj} are the unfactored lateral-torsional buckling moment of resistance of the longer laterally unsupported leg and moment of resistance of the joint respectively, determined from SABS 0162-2-1993. This code is based on the Canadian code. M_{rb} was determined, based on modified section properties (effective width of compression elements) to control local buckling, an effective length factor of 0.85 (assuming a partially restrained member) and a moment-gradient factor of 1.67. M_{rj} was evaluated, based on the bearing resistance of the plate, which in all cases was much less than the shearing resistance of the bolts. Based on the design recommendation of Kemp (2001), a coefficient C of 1.8 for a standard washer was used in the bearing resistance calculations. This factor depends on the ratio of bolt diameter to member thickness. The unfactored moment of resistance of the channel at first yield, M_{ry} was obtained from an elastic stress-superposition approach, whereby the axial and bending stresses are combined. M_{ry} was based on the effective cross-sectional properties of the channels. The effect of the residual stresses formed as a result of the rolling process were not taken into account in determining this moment. No account was also taken of the strengthening effect at the corners due to cold-forming. M_{max1} and M_{max2} are the maximum first and second order moments respectively at the eaves connection and were determined from the product of the maximum applied load and the eccentricity e. M_{max1} is based on the initial eccentricity whilst M_{max2} included the effect of changing eccentricity at the joint. N_{max} is the maximum axial force.

TABLE 2
MOMENT AND COMPRESSIVE RESISTANCE, AND FRAME TEST RESULTS

Channel Section	Calculated Moment and Compressive Resistance				Frame Test Results				
	M_{rb} (kNm)	M_{rj} (kNm)	M_{ry} (kNm)	N_{rb} (kN)	M_{max1} (kNm)	M_{max2} (kNm)	N_{max} (kN) Column	Rafter	$(M_{max1}/M_{rb}) + (N_{max}/N_{rb})$
300x75x20x3	47.43	38.77	47.37	295.59	36.71	38.62	29.08	29.14	0.87
300x75x20x3	47.43	66.26	47.48	295.59	39.22	41.00	26.19	30.29	0.93
300x65x20x3	35.22	27.14	34.38	225.70	30.83	31.86	19.68	18.60	0.96
300x50x20x3	23.83	19.86	23.49	148.75	24.66	25.46	15.75	14.88	1.13

The value of the interaction equation in the last column of Table 2 is lower for high strength material (structures 1 and 2) than for the mild steel specimens. Stress concentrations at the bolts and shear lag in the connections of these structures contributed to the interaction values less than 1.0. Increasing the number of bolts from 4 to 8 in structure 2 had little effect on the resistance of the connection.

Failure Modes

Local buckling was the ultimate failure mode in all tests in virtually the same position. It occurred after considerable rotation of the channel section close to the eaves connection. A typical local buckling failure is shown in Figure 4(a) and is initiated in the bottom flange by stress concentrations and shear lag below the inside lower bolt, followed by buckling of the web. After local buckling, the applied load dropped slowly.

Lateral-torsional buckling was the second mode of failure of concern. Although the longer laterally unsupported leg (rafter) did not ultimately fail due to lateral-torsional buckling there was considerable lateral movement and twisting of the leg (see Table 3). This was more evident in frames where the points of inflection were moved further away from the eaves connection (structures 2, 3 and 4) and narrower flanges were used (structures 3 and 4) to promote lateral-torsional instability.

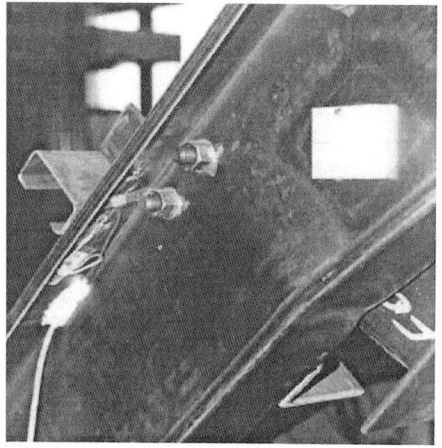 

(a) Local buckling failure   (b) Bearing distortion

Figure 4: Failure modes

TABLE 3
LATERAL DEFLECTIONS AND TORSIONAL ROTATIONS AT FAILURE

Structure	Top (mm)	Bottom (mm)	Rotation from LVDT ϕ (rad)
Structure 1	7.31	14.80	0.0326
Structure 2	6.69	16.09	0.0409
Structure 3	13.05	20.78	0.0336
Structure 4	9.08	17.36	0.0360

After testing, the frame members were disassembled for inspection. It was observed that the frames 1, 3 and 4 showed considerable bolt-bearing deformation around bolt-holes. This type of distortion was more pronounced at the inside lower bolt-hole as shown in Figure 4(b), where the bolt-holes were distorted significantly. Bearing distortion around bolt-holes is intended to provide the ductility required for moment redistribution.

Moment-Rotation Curves

The observed behaviour of the eaves connection for structures 1, 2, 3 and 4 is shown by moment-rotation curves in Figure 5(a). These moment-rotation curves represent the average curve for the two connections of each test structure. The moment-rotation relationship shows that there was virtually no rotation of the connections during the initial stages of loading. The applied load was carried by frictional resistance between the webs of the channels in the connection. Although this phase of load transfer was expected to be followed by a slip of the connection, there was not any significant slippage in structures 1, 3 and 4, implying that one or more bolts were initially in bearing in these tests. The response of the connections for structures 1, 3 and 4 became progressively more non-linear as ductile bearing distortions took place, in the presence of strain hardening, until failure occurred due to local buckling of the flange of the channel. Although the flange and web of the channel failed in local buckling there was also considerable deformation of the holes in bearing (Figure 4(b)). The moment-rotation response in test 2 with 8 bolts was different from that in structures 1, 3 and 4 with 4 bolts, as the moment-rotation relationship remained largely linear-elastic until failure. There was no apparent bearing failure of the holes. The higher initial rotation in structure 2 is due to lower bolt tension resulting in greater slip.

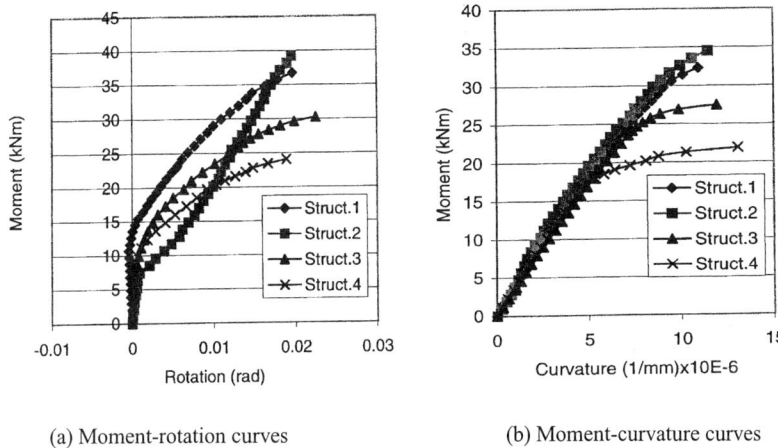

(a) Moment-rotation curves (b) Moment-curvature curves

Figure 5: Moment-rotation and curvature curves for all structures

The behaviour of all connections at low values of moment was influenced by the tightness of the bolts and whether bolts were in bearing or not. Since the bolts were hand tightened the frictional resistance between the webs of the channels varied as shown in the moment-rotation graphs. The non-linear response and decreasing connection stiffness exhibited late in the loading sequence is attributed, primarily, to local yielding, bolt-bearing deformations and eventual yielding at the flange below the inside bolt. The stiffness of the connection is generally dependent on the geometry of the connection, clearance between the size of the hole and bolt, size and number of bolts, thickness of the cold-formed steel plate, yield and ultimate stress of the cold-formed steel plate and whether the bolt shank is plain or threaded. Since all other parameters in structures 1, 3 and 4 are the same except the size of bolts, yield and ultimate stress of the cold-formed steel plate and the width of flanges, the difference in the moment-rotation relationship is a result of these parameters. All connections are able to develop continually increasing moments through the full range of rotations imposed during the tests. The maximum rotation for each connection of approximately 0.02 radians is shown in Figure 5(a) and reflects useful ductility. No significant advantage was gained in ultimate strength by using 8 bolts in the connections of test 2.

Moment Curvature Curves

The moment-curvature curves obtained from averaging the observed curvatures of the two frames are shown in Figure 5(b). The moment was derived by taking the product of the load and the initial eccentricity to the centre of the line of strain gauges. Curvature was calculated from the strain readings at the top and bottom flanges of the channels. The moment-curvature curves include a linear range followed by a non-linear range. It is evident from Table 2 that the yield moment was only achieved in structure 4.

CONCLUSION

The value of the interaction equation in Table 2 is lower for high strength material (structures 1 and 2) than for the mild steel specimens. Stress concentrations at the bolts and shear lag in the connections of these structures contributed to this lower value. Increasing the number of bolts to 8 in test 2 did not have any meaningful improvement in the moment capacity of the joint. After considerable rotation of the channel sections within the eaves connection, the source of failure in all structures was local buckling of the compression flange and web. Although the longer leg (rafter) did not ultimately fail due to lateral-torsional buckling there was considerable lateral movement and twisting of the leg. This was more evident in frames where narrower channels were adopted (structures 3 and 4) and the points of inflection were moved further away from the eaves connection (structures 2, 3 and 4) to promote lateral-torsional instability. Structures 1, 3 and 4 showed considerable bolt-bearing distortion around bolt holes. This type of failure was more pronounced at the inside lower bolt-hole.

The non-linear response exhibited late in the loading sequence of structures 1, 3 and 4 is attributed, primarily, to bolt-bearing deformation and yielding of the flange below the inside bolt. The moment-curvature relationships include a linear range followed by a non-linear range. The yield moment was only achieved in structure 4.

ACKNOWLEDGEMENTS

The authors wishes to thank the Southern Africa Institute of Steel Construction and THRIP for sponsoring this research, Robor and MacSteel for providing the materials and Tass Engineering for cutting and drilling holes of all items used in this research.

REFERENCES

Kemp A.R (2001). Bearing Capacities and Modes of Failure in Single-bolt Lap Joints. *Journal of the South African Institution of Civil Engineering*, **Vol. 43, No. 1,** 13-18.

SABS 0160 (1989). *South Africa Standard Code of Practice for the General Procedures and Loadings to be adopted in the Design of Buildings*, South African Bureau of Standards, Pretoria.

SABS 0162 (1993). *South Africa Standard Code of Practice for the Structural use of Steel, Part 2 - Limit States Design of Cold-formed Steelwork*, South African Bureau of Standards, Pretoria.

Southern African Institute of Steel Construction (1997). *South African Steel Construction Handbook, Limit State Design*, Southern African Institute of Steel Construction, Third Edition.

Trebilock P.J (1994). *Building Design using Cold-formed Steel Sections*, An Architect's guide, SCI Publication P130, The Steel Construction Institute, U.K.

Wezel van P, Kemp A.R and Trinchero P.E (1998). *Bolted Knee Joint Connection Tests of Portal Frame Constructed from Cold-Formed Steel Channel Sections*, Final Year Investigational Report, Department of Civil Engineering, University of the Witwatersrand.

ULTIMATE STRENGTH DESIGN OF BOLTED MOMENT-CONNECTIONS BETWEEN COLD-FORMED STEEL MEMBERS

J.B.P. Lim [1] and D.A. Nethercot [2]

[1] The Steel Construction Institute, Silwood Park, Ascot, Berkshire SL5 7QN, UK
[2] Department of Civil and Environmental Engineering, Imperial College, London SW7 2BU, UK

ABSTRACT

Bolted moment-connections between cold-formed steel members can be formed through brackets, bolted between the webs of the steel sections. Such brackets are easy to manufacture and the joints simple to assemble on site. However, owing to the thinness of the material, the joints will be susceptible to failure through buckling. In this paper, a combination of numerical analysis and full-scale testing are used to investigate the buckling failure of a common example of a bolted moment-connection: the eaves joint of a cold-formed steel portal frame. For such a joint, the following three modes of failure should be considered: (i) overall lateral-torsional buckling of the eaves joint, (ii) buckling of the stiffened free-edge of the eaves bracket into a one-half sine wave and (iii) local plate buckling of the exposed triangular area of the eaves bracket. Guidance for the design against all three buckling modes of failure is presented.

KEYWORDS

cold-formed steel, moment-connections, lateral-torsional buckling, stiffener buckling, local plate buckling

INTRODUCTION

A cold-formed steel portal framing system in which back-to-back cold-formed steel channel-sections were used for the column and rafter members was recently developed at the University of Nottingham (Lim (2001)). The dimensions of the back-to-back channel-sections used are shown in Figure 1; the nominal moment-capacity of the back-to-back channel-sections was 82.80kNm.

The general arrangement adopted for the eaves joint of the cold-formed steel portal framing system is shown in Figure 2. As can be seen, the most critical components are the back-to-back eaves brackets (Figure 3).

Using conventional design methods, the moment-capacity required from the eaves brackets is obviously equal to that of the back-to-back channel-sections (82.80kNm). Such a moment-capacity is substantially higher than that used in previous investigations of Baigent and Hancock (1982), Chung and Lau (1999) and Kirk (1986) of 9.19kNm, 17.88kNm and 32.00kNm, respectively.

However, the thickness of the brackets used for connecting the channel-sections in the proposed system of 3mm is similar to that used in these other investigations. The eaves joint can therefore be expected to be more susceptible to failure through buckling than the joints of the other frames described in the literature. Guidance for the design of the eaves bracket is therefore required against: (i) overall lateral-torsional buckling of the eaves joint, (ii) buckling of the stiffened free-edge of the eaves bracket into a one-half sine wave and (iii) local plate buckling of the exposed triangular area of the eaves bracket. In this paper, a combination of experimental tests and finite element analyses are used to provide guidance against all three buckling modes of failure.

LATERAL-TORSIONAL BUCKLING OF EAVES JOINT

Lateral-torsional buckling failure of the eaves joint is said to occur when the eaves bracket and the cold-formed steel sections bolted to the eaves bracket buckle and twist laterally out-of-plane. In order to prevent this mode of failure, sufficient lateral restraint should be provided around the eaves joint. To investigate this mode of failure, three laboratory tests were conducted on an eaves joint arrangement in which the position of the lateral restraint was altered.

Buckling of the free-edge of the eaves bracket was prevented by angle sections (30 x 30 x 4mm) bolted to both sides of the free-edge of the eaves bracket. Also, a single eaves bracket of thickness 6mm, as opposed to the 3mm of the back-to-back arrangement, was used; local plate buckling can therefore be expected not to be critical since flexural stiffness is proportional to the cube of the thickness. All premature modes of failure of the eaves bracket can therefore be considered to be prevented except for overall lateral-torsional buckling.

Figure 4 shows a photograph of the laboratory test arrangement. As can be seen, the eaves joint was tested horizontally on the laboratory floor..

Moment-capacity tests on the channel-sections were not conducted. Instead, the effective width approach of BS5950: Part 5, that takes into account local plate buckling, was used. The average yield and ultimate stresses of the back-to-back channel-sections, measured from three tensile coupons, were 300N/mm^2 and 371 N/mm^2, respectively. The average thickness was 3.19mm. Using these values, the moment-capacity was calculated as 102kNm. Also, in accordance with BS5950: Part 5 (1998), the maximum distance between points of lateral restraint should not exceed 1365mm otherwise the back-to-back channel-sections will fail through lateral-torsional buckling.

The lateral restraints used in Test 1 are shown in Figure 5(a). To ensure that the eaves joint did not fail through lateral-torsional buckling, three points of lateral restraint were applied around the eaves joint. The maximum distance between the points of lateral restraint was approximately 800mm, which is less than the unrestrained distance of 1365mm given by BS5950: Part 5 (1998). The eaves joint failed at a bending moment of 106kNm. This value is very close to the moment-capacity predicted by the British Standard. Failure was caused by buckling of the compression flanges of the column at the eaves joint. No lateral buckling of either the bracket or the back-to-back channel-sections was observed. Test 4-1 has therefore demonstrated that the proposed eaves joint, when fully restrained against lateral-torsional buckling is capable of carrying the full moment-capacity of the back-to-back channel-sections.

Test 2 was used to determine the effect on the moment-capacity of relaxing the lateral restraint located at the corner of the eaves bracket (see Figure 5(b)). The eaves joint failed at a bending moment of 75kNm or 74% of the moment-capacity of the back-to-back channel-sections. At the ultimate load, the observed mode of failure of the eaves joint was buckling of the compression flanges of the back-to-back channel-sections; lateral movement of the corner of the eaves brackets was not as pronounced. Lateral-torsional buckling of the eaves joint must therefore have resulted in the buckling of the channel-sections.

Test 3 was used to determine the moment-capacity of the eaves joint for the case of lateral restraint being provided only to the tension side of the eaves joint. Such a lateral restraint arrangement was consistent with lateral restraint being only provided to the tension flanges of the column and rafter members in a portal frame through purlins; torsional restraint was therefore not provided. The lateral restraint to the compression side of the column leg was arbitrarily removed (Figure 5(c)). The eaves joint failed at a bending moment of 92kNm or 90% of the moment-capacity of the back-to-back channel-sections. The observed mode of failure was again buckling of the compression flanges of the back-to-back channel-sections. However, unlike the case of Tests 1 and 2, the buckling of the compression flanges of the back-to-back channel-sections was also accompanied by twisting of both the eaves bracket and back-to-back channel-sections around the column leg where the lateral restraint had been released.

The three laboratory tests have therefore demonstrated that a lateral restraint at the corner of the eaves bracket is necessary in order to allow the eaves joint to carry the full moment-capacity of the back-to-back channel-sections. The removal of the lateral restraint at the corner of the bracket reduced the ultimate moment to 74% of the moment-capacity of the back-to-back channel-sections. Further details of the laboratory tests described in this Section can be found in Lim (2001).

The work described in the remainder of this paper will be based on the eaves bracket arrangement shown in Figure 3; the parameters used to describe the geometry of the eaves bracket are shown in Figure 6. It will be assumed that the eaves joint is sufficiently braced to prevent failure through overall lateral-torsional buckling.

BUCKLING OF STIFFENED FREE-EDGE OF EAVES BRACKET

The free-edge of the eaves bracket is susceptible to buckling into a one-half sine wave and in order to prevent this mode of failure, a stiffener should be provided along the free-edge (Figure 3). As can be seen, the stiffener is formed through a fold along the free-edge of the eaves bracket.

While the presence of the stiffener is necessary to prevent buckling of the free-edge, the material cost of the stiffener itself is small and so the depth of the stiffener should ideally be as large as is practically possible. Nevertheless, a simple design rule for the stiffener depth should be provided. This design rule will be based on the stiffener depth required for an idealised triangular plate under self-equilibrating bending tractions.

Figure 7 compares the type of load applied to the eaves bracket with that applied to an idealised triangular plate. As can be seen, the load applied to the triangular plate is more severe than that applied to the exposed triangular area of the eaves bracket. In this Section, a design rule for the depth of stiffener required for a triangular plate is presented; this design recommendation should be conservative when applied to the eaves bracket.

In the absence of more specific data, the maximum deviation from straightness of a stiffener formed through a fold as given in BS ENV 1090-2 (1998) has been adopted:

$$w_s = \frac{2.5}{1000} l_s \qquad \text{Equation 1}$$

Finite element analyses of a series of triangular plates of different widths a_t were conducted. For each triangular plate modelled, 16 elements were used along each side as shown in Figure 8; the number of elements used through the depth of the stiffener was chosen to maintain an aspect ratio close to unity for these elements. The stiffener depth d_s was increased until any further increase in d_s resulted in a negligible increase in the moment capacity of the triangular plate. The general purpose finite element program ABAQUS (Hibbit, Karlsson and Sorensen, Inc. (1995)) was used for the analysis, with the eight-noded thin shell element S8R5 being used for the eaves bracket.

Two design rules for d_s are suggested. The first is based on the elastic critical load, determined through elastic critical Eigen-value finite element analyses. The second is based on the results of non-linear large-displacement elasto-plastic finite element analyses, which incorporates both stiffener and local plate imperfections in the finite element model.

The elastic critical buckling moment M_{cr} was first determined for a triangular plate simply-supported along all three sides. This value of M_{cr} is designated as M_{crss}. The required depth of stiffener for the edge-stiffened triangular plate was then defined as the value of d_s that results in the value of M_{cr} for this plate being the same as M_{crss}. The proposed design rule for d_s is shown in Figure 9. Curve fitting the results gives

$$d_s/a_t = 2.26(t/a_t)^{0.67} \qquad \text{Equation 2}$$

where
$\quad d_s$ is the depth of the stiffener
$\quad a_t$ is the width of the triangular plate

The design recommendation for the stiffener depth determined through non-linear elasto-plastic finite element analyses adopted standard techniques for stiffener design in which the required depth of stiffener was defined as the value of d_s that allows the plate to carry 95% of the ultimate moment capacity of the triangular plate. The stress-strain relationship of the material forming the eaves bracket was assumed to be elastic-perfectly-plastic with a yield and ultimate stress of $280 N/mm^2$. The proposed design rule for d_s is also shown in Figure 9. Curve fitting the results gives

$$d_s/a_t = 1.95(t/a_t)^{0.543} \qquad \text{Equation 3}$$

Elastic critical Eigen-value analysis is not usually a sufficient basis for design of stiffeners as no account is taken of the reduction in strength caused by imperfections. Nevertheless, such analysis can illustrate the trends in behaviour and is still used in some Codes as the basis for design rules. On the other hand, non-linear large-displacement elasto-plastic finite element analysis is a sufficient basis for design as the effect of imperfections can be properly taken into account. It is not surprising therefore that the stiffener depth d_s designed on the basis of ultimate strength is larger than that designed on the basis of Eigen-value analysis. For this reason, Equation 3 will be adopted for the design recommendation of the stiffener depth d_s required for the eaves bracket.

LOCAL PLATE BUCKLING

To investigate local plate buckling, plate imperfections are modelled in the eaves bracket and the moment-capacity of the bracket determined. The stiffener along the free-edge is not modelled; instead the free-edge is assumed to be simply-supported, thus idealising the stiffener as a member having zero-axial rigidity and infinite flexural rigidity. With this approach, the plate is designed independently of the stiffener.

However, before investigating local plate buckling, non-linear large-displacement elasto-plastic finite element analyses are first used to determine the plastic moment-capacity M_p of the eaves bracket. This value represents the upper bound moment-capacity of the eaves bracket and is obtained on the assumption that the bracket has no geometric out-of-plane imperfections, does not buckle and therefore does not require any lateral restraints.

Figure 10 shows details of the idealised loading applied to the eaves bracket. As can be seen, the eaves bracket is loaded through the bolt forces of a 3x3 bolt-group. It should be noted that, although a 3 x 3 bolt-group of size $(4/5)a_e$ by $(2/3)b_e$ has been used, by St Venant's principle the results obtained should still be applicable to a different array of bolt-group of somewhat similar size.

A series of eaves brackets were modelled, having different values of a_e and b_e. For each eaves bracket considered, 16 elements were used along the side BC with the number of elements used along the side AB chosen to maintain an aspect ratio close to unity for these elements. Figure 11 shows the finite element mesh for a typical eaves bracket with $b_e/a_e = 0.85$. It can be seen from this figure that 14 elements are used along the side AB.

From geometric considerations of the eaves bracket, M_p is clearly a function of a_e, b_e/a_e and t as well as σ_y. To determine the relationship, finite element models of the eaves bracket were solved having $a_e = 500$mm, $b_e/a_e = 0.05$ to 0.9 in increments of 0.05, and t = 3mm. Although the analysis was restricted to this geometry of eaves bracket, it should be noted that, by virtue of dimensional analysis, the results will also be valid for different sizes and thicknesses of bracket. The stress-strain relationship of the material forming the eaves bracket was assumed to be elastic-perfectly-plastic with a yield and ultimate stress of 280N/mm². Figure 12 shows the variation of M_p against b_e/a_e obtained from the finite element analyses. By curve-fitting the results

$$M_p = \sigma_y a_e^2 t \{0.0675 + 0.4886(b_e/a_e) - 0.1466(b_e/a_e)^2\} \quad \text{Equation 4}$$

for the range $0.05 \leq b_e/a_e \leq 0.90$

Having determined the plastic moment-capacity M_p of the eaves bracket, the reduction in moment-capacity as a result of local plate buckling can be ascertained.

Figure 13 shows details of the idealised lateral restraint provided to the eaves bracket. The area of the bracket sandwiched between the back-to-back channel-sections in the eaves joint was considered to be fully laterally restrained and, as mentioned previously, the stiffened free-edge of the eaves bracket was assumed to be simply-supported.

Schafer et al (1996) proposed the following expression for local plate imperfections w_1 based on the thickness of the plate

$$w_1 = 6te^{-2t} \quad \text{Equation 5}$$

where
        t is in mm

Initial local plate imperfections are modelled only within the triangular area of the eaves bracket. The shape of these local plate imperfections was obtained from Eigen-value analyses of the eaves bracket. Figure 14 shows the typical mode shapes obtained from the Eigen-value analysis. Following standard practice, the first three mode shapes were summed and the resulting shape adopted for the initial imperfection.

The finite element method was used to determine M_u for eaves brackets with a_e = 400mm to 1200mm in increments of 200mm, b_e/a_e = 0.05 to 1 in increments of 0.1, t = 3mm, and a yield and ultimate stress of 280N/mm^2.

Figure 15 shows the variation of M_u against b_e for various values of a_e/t. M_u is non-dimensionalised by M_p to represent the reduction in moment-capacity of the eaves bracket owing to local plate buckling. It can be seen that as the value of the width-to-thickness ratio a_e/t increases (that is, as the plate becomes thinner) the plate becomes more susceptible to local plate buckling and so the ratio M_u/M_p decreases.

Application of Figure 15 can therefore be used to determine the reduction in the plastic moment-capacity (as obtained from Figure 12 or Equation 4) due to local plate buckling.

CONCLUSION

In this paper simple design recommendations have been proposed for the buckling failure of the eaves joint of a cold-formed steel portal framing system. Use of these design recommendations should help reduce the amount of laboratory testing required when undertaking the design of similar eaves joints.

REFERENCES

Baigent, A.H. and Hancock, G.J. (1982): 'The behaviour of portal frames composed of cold-formed members', *Thin-walled structures - Recent technical advances and trends in design, research and construction*, Oxford, Elsevier Applied Science, p209.

BS5950: Part 5 (1998): *Code of practice for design of cold-formed sections*, London, British Standards Institution.

Chung, K.F. and Lau, L. (1999): 'Experimental investigation on bolted moment connections among cold-formed steel members', *Engng Struct*, **21**, p898.

Hibbit, Karlsson and Sorensen, Inc. (1995): *ABAQUS Theory Manual*, Hibbitt, Karlsson and Sorensen, Inc.

Kirk, P. (1986): 'Design of a cold-formed section portal frame building system', *Proc. 8th International Speciality Conference on Cold-formed Steel Structures*, St. Louis, University of Missouri-Rolla, p295.

Lim, J.B.P. (2001): '*Joint effects in cold-formed steel portal frames*', University of Nottingham, PhD thesis.

Schafer, B.W., Grigoriu, M., Pekoz, T. (1996): 'A probabilistic examination of the ultimate strength of cold-formed steel elements', *Proc. 13th International Speciality Conference on Cold-Formed Steel Structures*, St Louis, University of Missouri-Rolla.

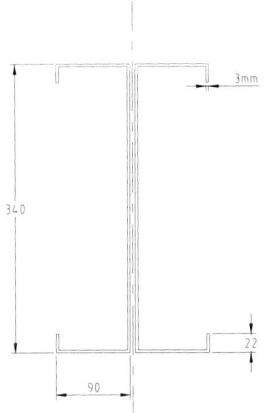

Figure 1: Dimensions of back-to-back channel-sections

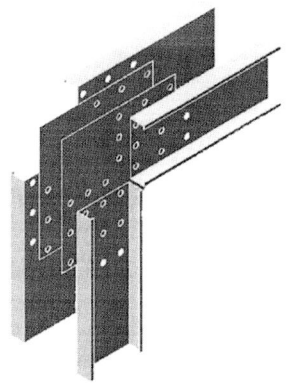

Figure 2: Details of the proposed arrangement for the eaves joint

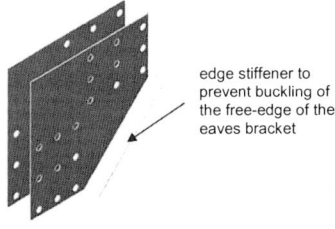

Figure 3: Details of back-to-back eaves brackets

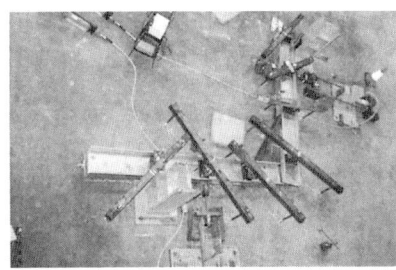

Figure 4: Photograph of laboratory test set-up of eaves joint

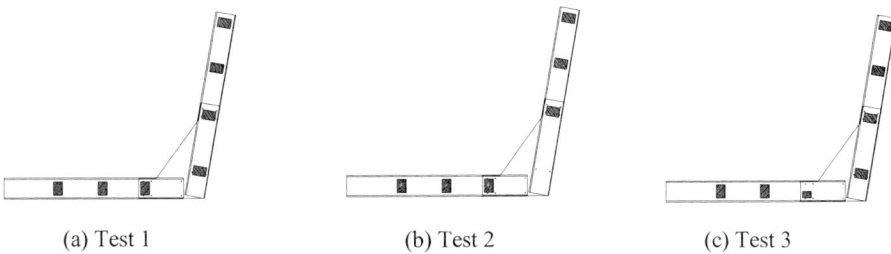

(a) Test 1 (b) Test 2 (c) Test 3

Figure 5: Details of lateral restraints in laboratory tests

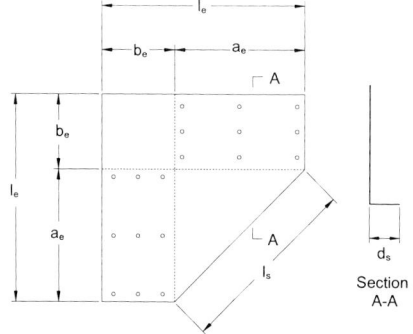

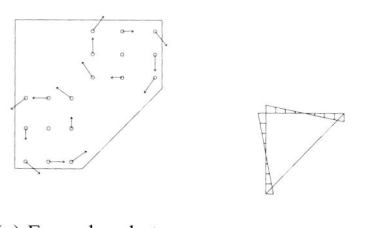

(a) Eaves bracket   (b) Triangular plate

Figure 7: Comparison between load applied to eaves bracket and triangular plate

Figure 6: Diagram showing parameters of eaves bracket

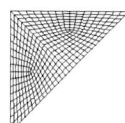

Figure 8: Details of finite element mesh of triangular plate

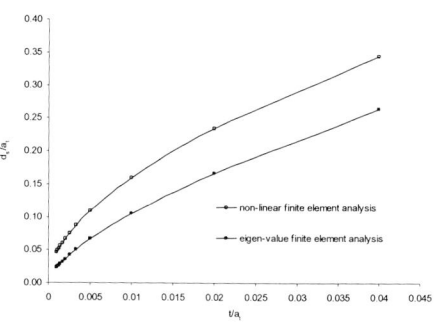

Figure 9: Design chart for variation of stiffener depth against plate thickness

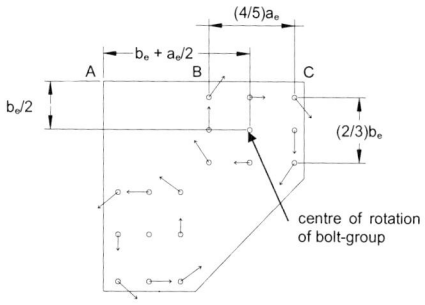

Figure 10: Details of idealised loading applied to eaves bracket

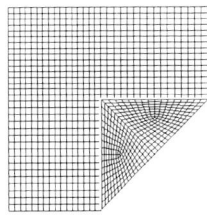

Figure 11: Details of finite element mesh of eaves bracket

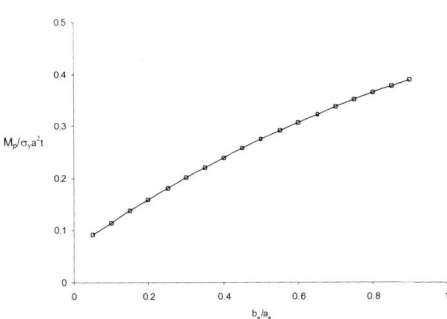

Figure 12: Variation of plastic moment-capacity M_p of eaves bracket against b_e/a_e

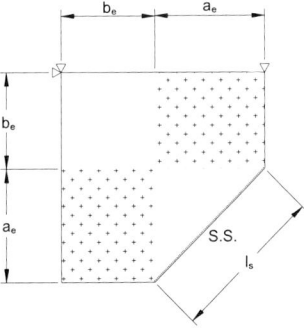

Figure 13: Details of idealised lateral restraints on eaves bracket used to determine M_u

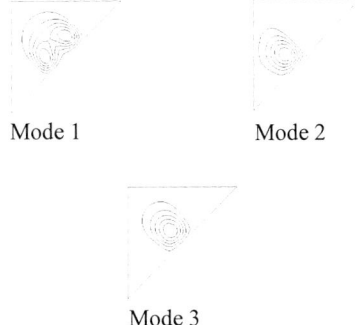

Figure 14: First three Eigen-mode shapes of exposed triangular area of eaves bracket

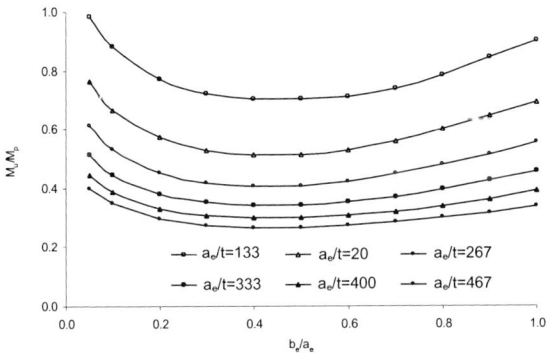

Figure 15: Variation of M_u against b_e for various values of a_e/t for eaves bracket

ANALYSIS OF CASSETTE SECTIONS IN COMPRESSION

P.A.VOUTAY, J.M.DAVIES

Manchester School of Engineering, University of Manchester, Oxford Road, Manchester, M13 9PL

ABSTRACT

Light gauge steel cassette sections offer an alternative form of load-bearing wall assembly for use in low rise steel-framed construction. The cassette sections utilised in this concept possess wide and slender flanges so that, by including intermediate stiffeners in the wide flanges, a significant increase in the ultimate load capacity may be achieved. However, the introduction of intermediate stiffeners in the wide flange also increases the number of buckling modes and therefore complicates the behaviour. This paper examines the behaviour of light gauge steel cassette sections in uniform compression and investigates their performance both with and without intermediate stiffeners. A detailed large deflection finite element analysis has been verified using experimental test results. This analysis, including both material and geometric non-linearity, was then used to assess the post buckling behaviour of these sections. "Generalised Beam Theory" (GBT), which allows the individual buckling modes to be considered separately and in predetermined combinations, used with the direct strength prediction method, is also presented as an alternative approach to design.

KEYWORDS

Cold formed steel cassette section, stiffener, compression, buckling, coupled instability, finite element, post buckling, GBT, direct strength approach.

INTRODUCTION

Cassette wall systems are an attractive form of construction for houses and low-rise commercial buildings (Davies 1998a, Davies 1998b, Davies 2000a) and this form of construction has been used successfully in a number of projects. In the preferred arrangement, cassettes span vertically in storey-height lengths between top and bottom tracks. In this configuration, the walls are subject to three separate force systems, namely compression, bending and shear. The effects of each of these forces have been the subject of recent research at the University of Manchester (eg Davies and Fragos 2001). This paper deals with compressive force and the influence of intermediate stiffeners in the wide and slender ($215 < w/t < 470$) flange.

Thomasson (1978) carried out a series of 46 tests on cassettes in compression with and without intermediate stiffeners. The authors have carried out a detailed analysis of these tests using a finite element model incorporating both geometric and material non-linearity. The behaviour of cassette sections without stiffeners is dominated by local buckling of the wide flange together with Euler buckling. Incorporating a single longitudinal stiffener in the wide flange and connecting the narrow flanges of the cassette together causes the behaviour of the cassette to become more complex. The number of critical buckling modes increases to five (local buckling of the plates between the edges of the cassette and the stiffener, web buckling, stiffener buckling, anti-symmetrical distortional buckling and Euler buckling). Additional longitudinal stiffeners further increase the complexity of the behaviour. This paper considers how such cassette sections may be designed.

As an alternative design solution, Generalised Beam Theory (GBT) used with the direct strength prediction method, is also presented here.

FINITE ELEMENT MODEL

Element type and mesh

Four different types of elements were used in the finite element model (FEM) in order to match the test conditions:

A **shell element** with four nodes and six degrees of freedom at each node was used as a basis. This element allows plasticity and large deflections and also uses reduced integration with five points of integration through the thickness of the shell. The element size was approximately 15×15 mm for specimens with stiffeners and 10×10 mm for specimens without stiffeners. These meshes were found to give a good prediction of the ultimate load, whereas coarser meshes gave inaccurate results and over-estimated the ultimate load.

Beam elements were used between the rigid end plates and the specimens. These beam elements were used in order to match the FEM with the test conditions. Their lengths were taken to be equal to the distance between the hinge centre and the top face of the support plate (95mm) as shown in Figure 1.

In order to connect the beam elements to the specimens, a three-dimensional **pin-joint element** was used. This element was free to rotate whilst the translations were fixed.

A **link element** is a three-dimensional spar element. This element is a uniaxial tension-compression element with three degrees of freedom at each node (translations). Due to the pin-jointed condition, no bending of the element is considered. This element was used to connect the narrow flanges of the panels, as realised during the tests by 30×3 mm flats at 300 mm centres, as shown in Figure 1.

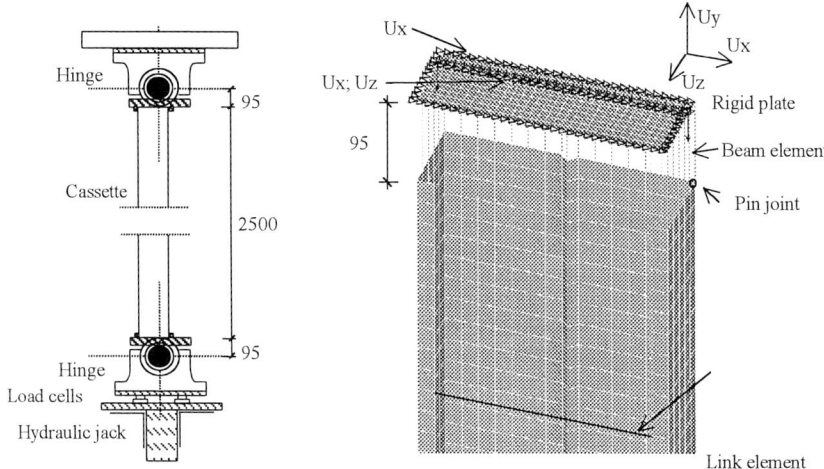

Figure 1: Test set up and Finite element model

Boundary conditions and loading

Full-length specimens were modelled with end conditions similar to the ones used during the full-scale tests. The hinges at the ends of the panels consisted of cylindrical rollers, which were mounted onto shafts of 110 mm diameter. 50 mm thick steel plates, through which the panel was loaded, were fixed to these shafts as shown in Figure 1. These end plates were modelled using rigid shell elements. For the bottom plate, the translations Ux, Uy and Uz were constrained on the line that coincided with the centroidal axis of the specimen. The translation Ux was constrained over the entire end plate. The same conditions were applied to the top plate except for the translation Uy in the axial direction, where the nodes on the centroidal axis were constrained to move together in the y direction as shown in Figure 1). The load was applied directly to the top plate.

Material properties

Thomasson determined the yield stress σ_y and the ultimate strength σ_{ult} for all the specimens tested. As proposed by Sivakumaran and Abdel-Rahman (1998), the multi-linear isotropic with strain hardening stress-strain curve and the Von Misses yield criterion have been used to model the non-linearity of the material, as shown in Figure 2. Young's modulus was taken to be 210 000 N/mm^2, Poisson's ratio to be 0.3 and the yield stress was in accordance with the values given by Thomasson (1978).

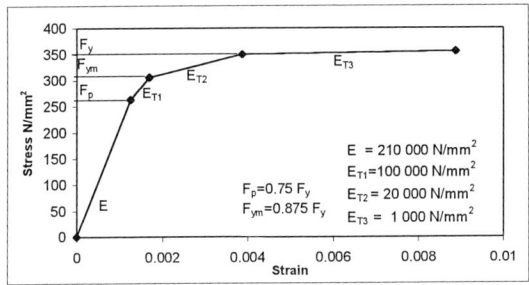

Figure 2: Stress-strain relationship for cold formed steel; $f_y = 350$ N/mm^2

Geometric imperfection and residual stress

Geometric imperfections were included in the FEM on the basis of the first eigenvalue buckling modes. Two distinct modes dominate the eigenvalue buckling analysis, the local mode and the stiffener mode, as shown in Figure 3.

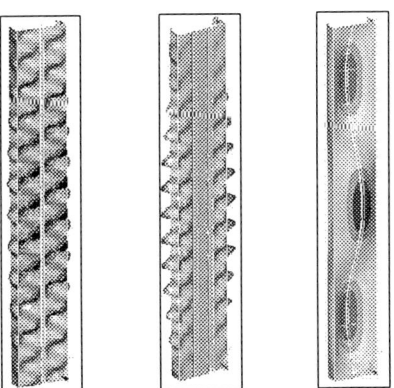

Figure 3: Local and stiffener buckling modes

In accordance with the imperfections measured by Thomasson, the amplitude of the local mode was limited to 0.5 mm for buckling of the plate elements between the stiffener and the edge and 0.25 mm when buckling of the web was predominant (cassettes with two stiffeners). The amplitude of the stiffener-buckling mode was limited to 1 mm. The residual stresses and the rounded corners of the specimen were not taken into account in the models. According to Thomasson the effect of axial residual stresses due to bending may be assumed to be of no consequence in this case.

Verification of the finite element model

The arc length procedure, allowing the load displacement path to be traced into the buckling and post buckling range, was used in this analysis. The basic shapes of the specimens are shown in Figure 4 and the geometry of their cross-sections is summarised in Table 1, together with the values of the yield stress and the ultimate load obtained from both the testing and finite element analysis. The values denoted 'd' indicate the depth of the stiffeners and 't' the thickness of the specimens. The stiffeners were of V shape with 45 degrees inclination.

The ultimate loads and lateral displacements taken at the middle of the specimens obtained with FEA are in good agreement with the test results. The FEA slightly over-estimated the experimental results, with a mean value of the ratio of experimental to FEA ultimate load of 0.97 and a standard deviation of 0.05. The difference between the FEA and test results may be explained by the fact that, in some of the specimens, the imperfection amplitudes measured were significantly higher (≈ 5 mm) than those used in the FEA. A similar influence can be observed in Thomasson's tests, where nominally identical tests with the same geometry, material properties etc, sometimes gave slightly different ultimate loads.

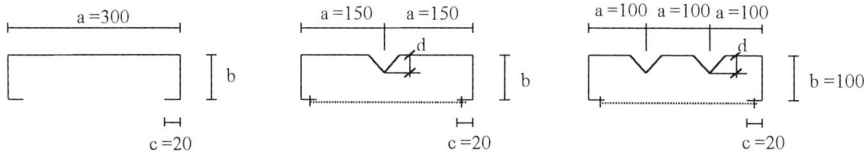

Figure 4: cross-sections and dimensions of specimens

Cassettes without stiffeners

Cassettes without stiffeners behave as typical cold formed steel columns with the predominant local buckling mode of the slender wide flange coupled to the Euler buckling mode. The distortional mode does not appear to be critical, so that the behaviour of these sections (deformations and ultimate loads) is the same, whether or not their narrow flanges are connected together.

Cassettes with one stiffener

Two buckling modes dominate the behaviour of cassettes with one intermediate stiffener, namely local buckling of the plate element between the stiffener and the edge of the cassette and buckling of the stiffener itself, as shown in Figure 3. Figure 5 shows typical load deflection curves for cassettes A101 (without stiffener), B101 (stiffener 10 mm depth), B102 (stiffener 14 mm depth), B103 (stiffener 16 mm depth). As also with Figure 6 which follows, it is the finite element results which are shown but these were very close to the test results. It can be see that, by including a 10 mm depth stiffener in cassette B101, the total area of the cassette is increased by 1.5% whereas the ultimate load is increased by 40%. A similar result is obtained with deeper stiffeners. The three cassettes with intermediate stiffeners of different depth have similar critical loads for local buckling but stiffener buckling appears after local buckling with different values of buckling load. In the post buckling range, the stiffener buckling mode is always present (i.e. the stiffener does not remain straight).

Specimen	t	d	σ_y	N_{TEST}	N_{FEA}	$\dfrac{N_{TEST}}{N_{FEA}}$
	mm	mm	N/mm²	KN	KN	
Cassette without stiffener						
A 7	0.64	∅	395.0	15.90	17.35	0.92
A 10	0.94	∅	460.0	35.88	37.59	0.95
A 15	1.39	∅	395.0	71.98	73.5	0.98
Cassette with one stiffener					Mean	0.95
B 71	0.63	10.0	350.0	23.00	22.98	1.00
B 72	0.64	14.0	350.0	23.50	23.07	1.02
B 73	0.63	16.0	350.0	25.25	24.70	1.02
B 101	0.95	10.0	435.0	45.00	49.85	0.90
B 102	0.95	14.0	400.0	54.00	51.96	1.04
B 103	0.96	16.0	400.0	53.50	54.33	0.98
B 151	1.40	10.0	400.0	96.90	96.90	0.96
B 152	1.40	14.0	400.0	108.58	108.58	1.00
B 153	1.44	16.0	400.0	120.00	117.58	1.02
Cassette with two stiffeners					Mean	0.99
C 71	0.63	10.0	350.0	24.50	24.37	1.01
C 72	0.63	14.0	340.0	25.75	26.97	0.95
C 73	0.63	18.0	350.0	26.75	28.59	0.94
C 101	0.95	10.0	410.0	54.50	56.59	0.96
C 102	0.95	14.0	410.0	56.80	64.61	0.88
C 103	0.95	18.0	410.0	70.00	69.01	1.01
C 151	1.42	10.0	405.0	118.00	125.29	0.94
C 152	1.43	14.0	390.0	146.50	138.35	1.06
C 153	1.44	18.0	370.0	150.00	152.71	0.98
					Mean	0.97

Table 1: Geometry of the specimens, FEA and test results

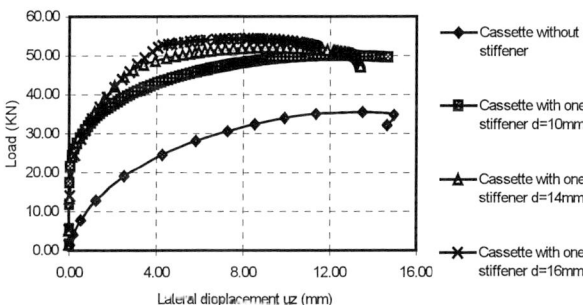

Figure 5: Comparison between lateral displacements for a cassette with one stiffener and a cassette without stiffeners

Cassettes with two stiffeners

The experimental tests carried out by Thomasson, and the FEA, show that the behaviour of a cassette with two intermediate stiffeners is dominated by three buckling modes namely local buckling of the web (and flanges), stiffener buckling, and the anti-symmetrical distortional mode. Figure 6 presents typical load deflection curves for cassettes A101 (without stiffeners), C101 (stiffeners 10 mm depth), C102 (stiffeners 14 mm depth), C103 (stiffeners 18 mm depth). For cassette C103, due to the higher stiffness of the wide flange and the predominance of web buckling, the lateral displacement in the Uz direction may change to the negative direction due to the shift of the centroidal axis. This phenomenon increased the stiffness of the cassette as well as the stress in the narrow flange. At some stage, the anti-symmetrical distortional mode becomes predominant, particularly if the local web buckling is anti-symmetrical, and this leads to sudden collapse of the specimen (see Figure 6).

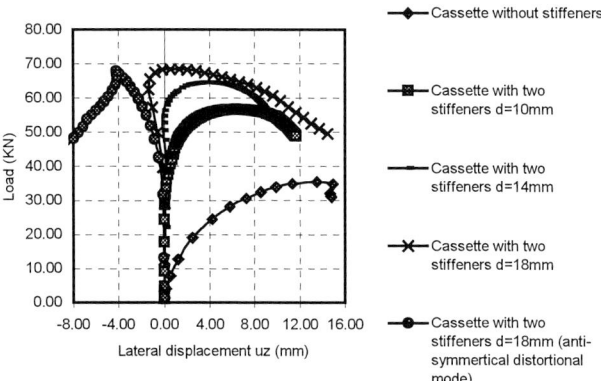

Figure 6: Comparison between lateral displacements for a cassette with two stiffeners and a cassette without stiffeners

The degree of fixity of the ends of the stiffeners can play a significant role in the performance of the cassettes. The above results were obtained on the basis that the stiffener end nodes were free to displace longitudinally relative to the ends of the wide flange, i.e. the ends of the stiffeners were pinned with respect to stiffener buckling. This is the conservative assumption and higher stiffener buckling loads are obtained if the ends of the stiffeners are constrained to have the same axial displacements as the ends of the wide flange.

GENERALISED BEAM THEORY

Generalised Beam Theory (GBT) is particularly powerful when used to analyse cold-formed steel sections in which distortion of the cross section is significant. GBT is an extensive and complex subject, which has been described by Davies and Leach (1994a and 1994b). One of its great advantages is its ability to identify the orthogonal modes of buckling and then isolate them so that they may be considered individually or in selected combinations. Due to the fact that the narrow flanges of the cassettes were attached together in Thomasson's tests, the symmetrical distortional mode was prevented. For this reason, the symmetrical distortional mode has been removed from the GBT analyses presented in this paper. Columns 1 and 2 of Table 2 give the elastic critical stresses for local and distortional buckling. The values given in column 2 are the minimum of the stiffener buckling mode and the anti-symmetrical distortional mode.

"DIRECT STRENGTH" APPROACH TO THE DESIGN OF COLD-FORMED SECTIONS

The "direct strength" method is analogous to the effective width approach for plate post-buckling. It uses a numerical elastic buckling solution for the entire member together with a strength curve. The method is presented in more detail in Shafer (2001) and some results obtained using this method are given Table 2. The nominal long column strength was calculated using the Eurocode 3 curve:

$$\overline{\lambda} = \left(\frac{\lambda}{\lambda_1}\right) \text{ with } \lambda = \frac{l}{i}, \; l = 2690 \text{ and } \lambda_1 = \pi \cdot \sqrt{\frac{E}{f_y}}$$

$$\phi = 0.5\left[1 + \alpha(\overline{\lambda} - 0.2) + \overline{\lambda}^2\right] \text{ and } \chi = \frac{1}{\phi + \sqrt{\phi^2 - \overline{\lambda}^2}} \text{ but } \chi \leq 1.0$$

$$N_E = \chi \cdot A \cdot f_y$$

Specimen	(1) Elastic GBT Local buckling N/mm²	(2) Elastic GBT Distortional buckling N/mm²	(3) Full Strength N_1 N_{FEA} N_1 = Local Schafer + Euler	(4) Full Strength N_2 N_{FEA} N_2 = Local EC3 + Euler	(5) Full Strength N_3 N_{FEA} N_3 = Distort. Hancock + Euler	(6) Full Strength Min (3) and (5)	(7) Full Strength Min (4) and (5)
Cassette without stiffener							
A7	5.00	32.03	1.02	0.69	1.36	1.02	0.69
A10	10.79	48.53	0.96	0.70	1.19	0.96	0.70
A 15	23.65	75.11	0.94	0.74	1.10	0.94	0.74
					Mean	0.98	0.71
Cassette with one stiffener							
B 71	23.71	43.66	1.34	1.06	1.18	1.18	1.06
B 72	25.34	76.24	1.37	1.09	1.58	1.37	1.09
B 73	25.04	75.26	1.28	1.02	1.47	1.28	1.02
B 101	54.30	66.52	1.35	1.14	1.07	1.07	1.07
B 102	55.95	115.49	1.28	1.09	1.34	1.28	1.09
B 103	57.02	115.97	1.24	1.05	1.29	1.24	1.05
B 151	119.65	100.85	1.33	1.19	0.99	0.99	0.99
B 152	125.61	177.39	1.22	1.11	1.18	1.18	1.11
B 153	131.45	180.40	1.16	1.06	1.11	1.11	1.06
					Mean	1.19	1.06
Cassette with two stiffeners							
C 71	36.95	38.40	1.48	1.22	1.03	1.03	1.03
C 72	37.13		1.36	1.12	1.35	1.35	1.12
C 73	37.40	51.00	1.30	1.07	1.05	1.05	1.05
C 101	83.98	80.09	1.37	1.20	1.04	1.04	1.04
C 102	84.52	117.51	1.21	1.07	1.11	1.11	1.07
C 103	85.11	120.95	1.14	1.00	1.06	1.06	1.00
C 151	182.29	121.12	1.22	1.13	0.87	0.87	0.87
C 152	187.28	175.18	1.13	1.05	0.95	0.95	0.95
C 153	195.91	188.67	1.02	0.95	0.87	0.87	0.87
					Mean	1.04	1.00

Table 2: GBT and direct strength approach results

The interaction between local buckling (or distortional buckling) and long column (Euler) buckling was taken into account using:

$$\lambda_L = \sqrt{\frac{N_E}{N_{crL}}} \text{ or } \lambda_D = \sqrt{\frac{N_E}{N_{crD}}}$$

$N_{ultimate}$ (for local buckling) was calculated using two alternative strength curves. The first is the one used in Eurocode 3: Part 1.3 that is derived from the Winter equation. Results obtained with this curve are given in column 4 of Table 2 in the form of the ratio N_{ult}/N_{FEA}.

If $\lambda_L \leq 0.673 \; N_{ultimate} = N_E$ if $\lambda_L > 0.673 \; N_{ultimate} = \left[1 - 0.2\left(\frac{N_{crL}}{N_E}\right)^{0.5}\right] \cdot \left(\frac{N_{crL}}{N_E}\right)^{0.5} \cdot N_E$

The second strength curve used is the one suggested by Schafer for local buckling:

$$\text{If } \lambda_L \leq 0.776 N_{ultimate} = N_E \text{ if } \lambda_L > 0.776 N_{ultimate} = \left[1 - 0.15\left(\frac{N_{crL}}{N_E}\right)^{0.4}\right]\cdot\left(\frac{N_{crL}}{N_E}\right)^{0.4} \cdot N_E$$

Results obtained with this curve are given in column 3 of Table 2 in the form of ratio N_{ult}/N_{FEA}. Distortional buckling (using the Hancock curve) and long column (Euler) interaction were taken into account as follows:

$$\text{If } \lambda_D \leq 0.561 N_{ultimate} = N_E \text{ if } \lambda_D > 0.561 N_{ultimate} = \left[1 - 0.25\left(\frac{N_{crD}}{N_E}\right)^{0.6}\right]\cdot\left(\frac{N_{crD}}{N_E}\right)^{0.6} \cdot N_E$$

Results obtained with this curve are given in column 5 of Table 2 in the form of ratio N_{ult}/N_{FEA}. Column 6 gives the minimum between columns 3 and 5 and column 7 gives the minimum between columns 4 and 5. The method leading to column 6 performed well for cassettes without stiffeners. The Schafer strength curve performed less well for cassettes with stiffeners and over estimated the ultimate load when local buckling was predominant. The method leading to column 7 (using the Winter strength curve as proposed in Eurocode 3) performed well for cassettes with stiffeners.

CONCLUSIONS

A finite element model has been developed which shows good agreement with full-scale test results for cassette sections subject to axial load. The effect of the stiffener in the wide flange has been found to be particularly favourable, even when the critical elastic buckling mode is stiffener buckling. Due to the fact that, in the tests, the narrow flanges were connected together, the elastic buckling analysis has been carried out by taking advantage of the facility offered by Generalised Beam Theory (GBT) whereby a selected combination of the buckling modes may be included in the analysis. The direct strength approach, used with GBT, offers an accurate method with which to predict the ultimate load.

References
Davies J. M. (1998a) "Light gauge steel cassette wall construction", *Nordic Steel Construction Conference 98*, Bergen, Sept. 14-16, pp 427-440.
Davies J. M. (1998b) "Light gauge steel framing for house construction", 2nd *Int. Conf. on Thin Walled Structures*, Singapore, Dec. 2-4, pp 17-28.
Davies J. M. (2000a) "Steel framed house construction", *The Structural Engineer*, Vol. 78, No.6, 21 March, pp 17-24.
Davies J. M. and Fragos A. S. (2001) "Shear strength of empty and infilled cassettes", 3rd *Int. Conf. on Thin Walled Structures*, Krakow, Poland, June 5-7, pp 3-18.
Davies J. M. and Leach P. (1994a) "First-order Generalised Beam Theory", *J. of Constructional Steel Research*, 31, pp 187-220.
Davies J. M., Leach P. and Heinz D. (1994b) "Second-order Generalised Beam Theory", *J. of Constructional Steel Research*, 31 pp 221-241.
Eurocode 3 (1996): *Design of Steel Structures - Part 1.3 :General rules – Supplementary rules for cold formed thin gauge members and sheeting*, CEN ENV 1993-1-3, February 1996.
Hancock, G. J., Rogers, C. A., Schuster, R. M. (1994) "Strength Design Curves for Thin-Walled Sections Undergoing Distortional Buckling". *J. of Constructional Steel Research,* 31(2-3), pp 169-186.
Shafer B. W. (2001) "Thin-walled column design considering local, distortional and Euler buckling" *Structural Stability Research Council Annual Technical Session and Meeting*.
Sivakumaran K. S. and Nabil Abdel-Rahman (1998) "A finite element analysis model for the behaviour of cold-formed steel members", *Thin-Walled Structures,* Vol. 31, pp 305-324.
Thomasson P. (1978). "Thin-walled C-shaped panels in axial compression", *Swedish Council for Building Research.* D1: 1978, Stockholm, Sweden.

PERFORMANCE OF WALL-STUD SHEAR WALLS UNDER MONOTONIC AND CYCLIC LOADING

Ludovic A. Fulop[1], Dan Dubina[1]

[1] Department of Steel Structures and Structural Mechanics, "Politehnica" University of Timisoara, 1. Ioan Curea st. 1900 Timisoara, RO

ABSTRACT

The ever-increasing need for housing generated the search for new and innovative building methods to increase speed, efficiency and enhance quality, one direction being the use of light steel profiles as load bearing elements and different materials for cladding. Wind and seismic behavior of these structures is influenced by the hysteretic characteristics of the shear wall panels. In this paper a review of actual research in the field and results of a full-scale shear test program on wall panels are presented. Based on tests, a numerical equivalent model for hysteretic behavior of wall panels working in shear was built to be used in 3D dynamic nonlinear analysis of cold-formed steel framed buildings.

KEYWORDS

Light gauge steel, houses, testing, shear, wall panels, cyclic, dynamic analysis.

INTRODUCTION

Steel-framed houses are usually built of light cold-formed load bearing structure and having different solutions of interior and exterior cladding. This technology is popular and accounts for an important and ever increasing market share in the US, Japan, Australia and Europe (Pekoz). The same method is used for buildings, of small dimensions, of other purposes (offices, schools, manufacturing premises, etc.), that are referred as small industrial buildings (SIB).

In such structures shear walls are the main structural elements to act against horizontal loads, e.g. wind and earthquake. Even if widely used in practice, the behaviour of shear walls subjected to earthquake is not fully understood and in recent years important effort has been made to clarify certain aspects related to shear wall strength, stiffness and ductility, as main parameters governing seismic behaviour.

LITERATURE REVIEW OF RECENT RESEARCH RESULTS

Research in the US (AISI 1997, Serette and Ogunfunmi 1996, Salenicovich et al. 2000) has been focused mainly towards experimental testing of shear walls typical to their home practice, in order to produce practical racking load values. Load bearing capacities are derived both from monotonic push-over curves, envelope and stabilised envelope curves from cyclic tests. Findings of these studies suggest a conventional elastic stiffness for a wall panel at 0.4 of the ultimate load (Serette 1998). Different frame typologies with various cladding materials were tested, studies being conducted to determine the influence of length/height ratios as well as the effect of openings. Even if very detailed, the majority of studies avoid addressing an important aspect of shear wall behaviour, energy dissipation capacity due to cyclic characteristics (Serette and Ogunfunmi 1996). The effect of gypsum wallboard was also studied, leading to the conclusion that both strength and stiffness are increased by the presence of gypsum wallboard, some results suggesting an increase in ultimate load of up to 30%, compared to the case of external sheeting.

Testing and numerical simulation was combined in order to account for hysteretic characteristics in an attempt to provide evidence on the possible values of response modification factors (q) (Kawai et al. 1999). Vibration tests of steel-framed houses were conducted and relatively large damping ratios were found due to interior and exterior finishes. According to the tests damping ratio of 6% was accepted for seismic analysis. A maximum 1/50 rad story drift angle limit is also suggested as acceptable during severe earthquakes. In FE analysis stage, a steel-framed house was subjected to two levels of seismic waves. The house exhibited good performance, reaching a maximum drift of 1/300 rad. Even when minimum required wall length was provided, the maximum drift did not exceed 1/60 rad.

The same issue is analysed by Gad (Gad and Duffield 2000, Gad et al. 1999), who proposes a new analytical approach to evaluate the ductility parameter (R_μ), and finds a value between 1.5 and 3.0 to be suitable. The same research briefly assesses inherent structural overstrength and finds it to be very important factor as far as earthquake resistance is concerned. The quantitative evaluation of overstrength is more difficult, but an empirical evaluation attempt is performed in the report. (Gad and Duffield 2000)

Experimental tests and FE modelling was employed by G. de Matteis (1998) to asses shear behaviour of sandwich panels both in single story and multi-story buildings. A number of six monotonic and six cyclic tests were performed on full-scale sandwich panel specimens of different configurations. In the final stage of the study, dynamic modelling on panels integrated in building structures, under real earthquake records is performed. According to the conclusions diaphragm action can replace classical bracing solutions only in low-rise buildings, and in areas of low seismicity. For multi-story frames cladding panels can only be used in an integrated system, sharing horizontal force with frame effect.

An important aspect of experimentation was to define acceptable damage levels and relate it to the performance objectives. Recent performance objective proposals are based on three or four generally stated goals (Rosowsky 2002): (1) serviceability under ordinary occupancy conditions: (2) immediate occupancy following moderate earthquakes; (3) life safety under design-basis events; (4)collapse prevention under maximum considered event.

SUMMARY OF THE EXPERIMENTAL PROGRAM

The experimental program was based on six series of full-scale wall tests with different cladding arrangements based on common practical solutions in both housing and SIB (TABLE 1). Each series consisted of identical wall panels, tested statically both monotonic and cyclic. The main frame of the wall panels were made of cold-formed steel elements, top and bottom tracks were U154/1.5, while studs were C150/1.5 profiles, fixed at each end to tracks with two pair SPEDEC SL4-F-4.8x16 (d=4.8 mm) self-drilling self-taping screws. In specimens using corrugated sheet as cladding the sheets were placed in horizontal position, with a useful width of 1035mm and one corrugation overlapping and tightened with seam fasteners SL2-T-A14- 4.8x20 (d=4.8 mm) at 200 mm intervals (TABLE 1). Corrugated sheet was fixed to the wall frame using SD3-T15-4.8-22 (d=4.8 mm) self-tapping screws, sheet ends being fixed in every corrugation, while on intermediate studs at every second corrugation. Additionally on the 'interior' side of specimens in Series II, 12.5mm thick gypsum panels (1200x2440mm) were placed vertically and fixed at 250mm intervals on each vertical stud.

TABLE 1
Description of Wall Specimens

Ser.		Exterior Cladding	Interior Cladding	Testing Method	Load Vel. (cm/min)	No. Test
O		-	-	Monotonic	1	1
I		Corr. Sheet LTP20/0.5	-	Monotonic	1	1
				Cyclic	6;3	2
II		Corr. Sheet LTP20/0.5	Gypsum Board	Monotonic	1	1
				Cyclic	6;3	2
III		-	-	Monotonic	1	1
				Cyclic	3	1
IV		Corr. Sheet LTP20/0.5	-	Monotonic	1	1
				Cyclic	6 ;3	2
OSB I		10 mm OSB	-	Monotonic	1	1
				Cyclic	3	1
OSB II		10 mm OSB	-	Monotonic	1	1
				Cyclic	3	1
Total Number of Specimens						15

Bracing was used in three specimens (TABLE 1), by means of 110x1.5 mm straps on both sides of the frame. Steel straps were fixed to the wall structure using SL4-F-4.8x16 and SD6-T16-6.3x25 self-drilling screws, the number of screws being determined to avoid failure at strap end fixings and facilitate yielding. 10 mm OSB panels (1200x2440mm) were placed in similar way as the gypsum panels in earlier specimens (TABLE 1), only on the 'external' side of the panel and fixed to the frame using bugle head self-drilling screws of d=4.2mm at 10.5 cm.

ANALYSIS OF EXPERIMENTAL RESULTS

The main outputs of the experiments were shear force versus horizontal displacement at the top of the wall-specimens. Furthermore, horizontal slip at the base of the wall and uplift displacement was measured in the two corners. As in case of the panels clad with corrugated sheet, the seams govern the failure, relative slip between two steel sheets was also recorded. Load versus lateral displacement curves are presented for all tested specimens (Figure 1), and in order to illustrate monotonic to cyclic results, stabilized envelope curves are being also presented for the cyclic curves.

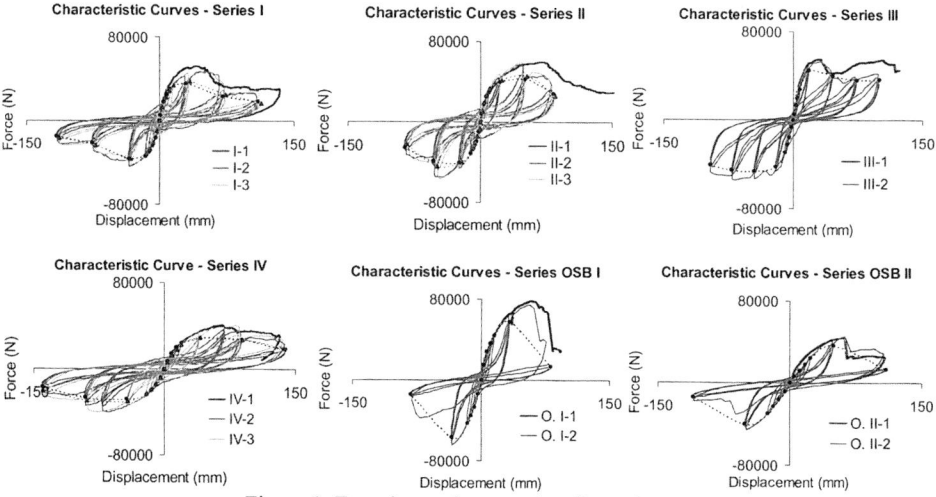

Figure 1. Experimental curves for all specimens

As seen wall-panels exhibited very complex, and highly non-linear behaviour. In order to evaluate specific properties like elastic modulus, ultimate force or ductility, curves have been interpreted according to an established procedure. The method has been reported by Kawai (Kawai & all. 1997). Initial stiffness is defied as the secant stiffness to the point of drift angle corresponding to 1/400 (D_{400}), while the yield line is chosen in a way that the hatched parts in Figure 2 have the same area. The allowable strength is referred as the minimum of the force at story drift angle 1/300 (F_{300}) and 2/3 F_{max}.

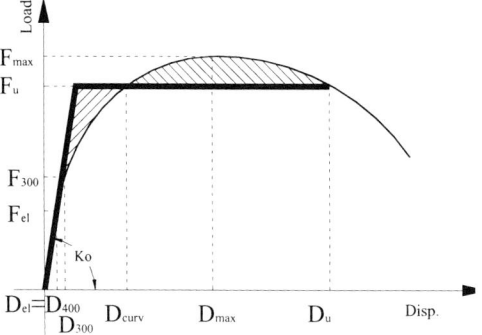

Figure 2. Method for determining equivalent elasto-plastic model

Results from monotonic and cyclic experiments are presented (TABLE 2); where two cyclic tests were performed the values are based on the mean value of the two. For cyclic tests, values are derived based on 1st envelope curve (envelope curve), and the 3rd envelope curve (stabilized envelope).

For design capacity, the minimum of 2/3F_{max} and F_{300} and F_{200} are relevant, since they represent the values accepted in the Japanese and US code respectively (Kawai et al. 1997).

Differences between monotonic and cyclic values can be observed as follows. Initial rigidity is not affected, values of cyclic and monotonic tests range within a difference of less than 20%. The same can be noted for ductility, exception being in case of OSB specimens where ductility is reduced by 10-25% for cyclic results. One important observation concerns ultimate load (F_u), where cyclic results are lower than monotonic ones by 5-10% even if we consider 1st envelope curve. If we take into account stabilized envelope curves, the difference can increase to 20-30%.

Based on the medium value of monotonic and cyclic results comparison of different series contribution of opening, gypsum board and other factors can be assessed, with the following conclusions:

Series I - Series II: Differences can be attributed to the effect of the gypsum board. There is an increase of in ultimate load of 16.2% and 17.8% respectively. As far as initial values are concerned (K_o, F_{400}, F_{300}, F_{200}) there seem to be no differences, but ductility is also improved slightly.

Series I – Series IV: There is significant decrease of initial rigidity (60.3% – 53.3%), for a lesser degree of ultimate load (16.4% – 21.0%), but ductility values are essentially unaffected

Series I – Series III: Comparison is more qualitative because of the different sheeting system. There are no differences as fare as initial rigidity is concerned; however an increase of ductility has been expected. This was not possible as failure mode for the strap-braced specimens was not the most advantageous one. Strap braced wall panels have the advantage of stable hysteretic loops, but also the disadvantage of higher pinching than the sheeted ones.

TABLE 2
SUMMARY OF EXPERIMENTAL RESULTS

Series/Curve		K_o(N/mm)	F_{400}(N)	F_{300}(N)	D_{curv}(mm)	F_u (N)	D_{uct}	2/3 F_{max}	F_{200} (N)
I	Mon	4088.1	24467.2	28690.95	24.8	47821.0	4.65	35250.91	35110.87
	1st Cycle	3446.7	20536.2	24086.55	22.5	39696.6	5.13	30430.85	29425.42
	3rd Cycle				14.9	33560.3	4.37		
II	Mon	3311.5	20088.5	24349.89	41.0	53801.3	5.03	39810.27	30508.56
	1st Cycle	3850.7	22904.4	26565.66	29.5	49029.8	5.05	38286.74	34024.80
	3rd Cycle				15.6	39818.6	5.54		
III	Mon	4187.5	25120.4	31980.04	19.2	51139.6	2.81	36765.07	40192.96
	1st Cycle	3626.8	21259.1	26956.51	21.5	48013.7	5.25	35682.62	36097.14
	3rd Cycle				14.8	42252.2	5.02		
IV	Mon	1598.3	9349.6	13723.80	41.7	35532.7	3.79	26813.45	18048.44
	1st Cycle	1766.3	10416.4	12869.81	37.2	33267.2	5.62	25163.27	17277.07
	3rd Cycle				23.8	26812.2	6.22		
OSB I	Mon	3909.6	23797.3	28470.15	37.2	68162.0	4.26	52517.67	37953.88
	1st Cycle	4197.3	24644.6	28942.46	21.4	54615.2	3.88	46562.66	37194.10
	3rd Cycle				17.5	48944.5	3.67		
OSB II	Mon	1814.9	10702.5	13779.63	36.4	37014.8	3.19	29586.53	18732.50
	1st Cycle	1610.5	9511.3	11850.46	32.3	37426.0	2.93	30539.23	16495.42
	3rd Cycle				27.8	33908.3	3.11		

Series I – Series OSB I: Comparison more qualitative, keeping in mind the different wall panel arrangements. Initial rigidity is of similar magnitude, with increase of ultimate load. Failure of OSB specimens under cyclic loading was more sudden than in case of corrugated sheet specimens, where degradation occurs gradually. This is also reflected by the reduced ductility for OSB specimens.

Series OSB I – Series OSB II: The effect of opening produced similar results as in cases of Series I – Series IV. Initial rigidity decreased with 64.6% – 59.1%, while ultimate load decreased with 32.5% - 36.9%. There is also an important decrease of ductility, probably highlighting the different failure modes of the two wall panels.

PERFORMANCE CRITERIA

In case of wall panels clad with corrugated sheeting damage is largely concentrated in seam fasteners. If plasticization in vicinity of fasteners increases the water-proof cladding layer looses its functionality and has to be replaced. For establishing global performance criteria the following acceptable deformations in the seam fasteners have been considered:

- If slip of the seams does not exceed the elastic limit, corresponding to $0.6F_{max}$ of the seam connection, damage is limited and can be considered negligible. In this case the cladding is still water-proof, no repairs are required and this would correspond to normal serviceability conditions.
- If slip is limited to the diameter of the screw (4.8mm) the cladding requires repair. There is damage, but not excessive and by minor interventions, like replacing screws with larger diameter ones the structure can be repaired. This could correspond to immediate occupancy criteria.
- In case of life safety criteria any kind of damage is acceptable, without endangering the safety of occupants. This criteria is not any more related to serviceability, but can correspond to the attainment of the ultimate force (F_{ult}) of the wall panel and the starting of the downwards slope.

Relative slip in seams has been measured for specimen I-3, II-2 and II-3 the first two criteria can easily be applied and relationship between slip and lateral deformation of the panel can be found (TABLE 3).

TABLE 3
PERFORMANCE CRITERIA

Specimen	Connection Deform. (mm)	Force (N)	Panel Top Disp. (mm)	Drift (%)
I-3	0.197	21423	6.71	0.274
	4.8	43885	29.22	1.197
IV-2	0.197	10106	7.96	0.326
	4.8	35613	44.13	1.808
IV-3	0.197	8849	8.11	0.332
	4.8	26332	42.22	1.730

Based on these assumptions the following performance criteria are suggested for wall panels clad with corrugated sheet: (1) fully operational (<0.003); (2) partially operational (<0.015); (3) safe but extensive repairs required (<0.025). Comparable design criteria can be established for other types of panels.

The first performance level does not provide ductility, because shear panel work is limited to elastic domain. This could be the design criteria for frequent, but low intensity earthquakes. In case of rare but severe earthquakes, the last two design criteria can be used and some ductility will be available.

SIMPLIFIED FINITE ELEMENT MODEL

In order to be able to provide suitable tool for time-history analysis of entire structures the global behavior of the wall panel has to be caught. A very detailed modeling of the panel, based on individual connection behavior could not be sufficiently simple to for this purpose, so modeling had to be simple and efficient on one hand, and accurate on the other. The basic idea is to replace the shear panel with equivalent cross bracing system, a technique often used in elastic calculations for design purposes.

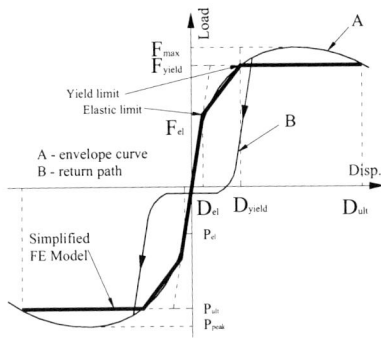
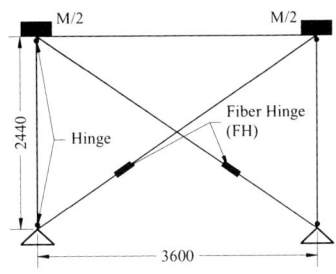

Figure 3. Scheme of hysteretic behavior   Figure 4. Simplified DRAIN-3DX model

A FE model based on DRAIN-3DX (Prakash and Powell 1993) computer code is proposed in order to get as accurate hysteretic behavior as possible. The model consists of a mechanism frame and a special bracing. As all column ends are hinged the frame itself is a mechanism and it does not contribute to load bearing capacity. Braces are modeled as 'TYPE 8' fiber hinge (FH) beam-column elements with FH to accommodate the hysteretic behavior (Figure 4). In order to calibrate the FE model experimental results from a full scale testing program has been used for comparison. Out of the six series five have been used as bases for the FE model calibration (TABLE 4).

TABLE 4
SERIES OF FE MODELS AND CALIBRATED VALUES BASED ON EXPERIMENTAL RESULTS

Series	I	II	IV	OSB I	OSB II
Sheeting	Corrugated Sheet	Corr. Sheet + Gypsum	Corrugated Sheet	OSB	OSB
Initial Rigidity (N/mm)	3446.6	3850.6	1766.3	4197.3	1610.5
Elastic Limit (F_{el}/D_{el})(N & mm)	24086/6.99	26566/6.90	128670/7.28	28942/6.89	11850/7.36
Yield Limit (F_{yield}/D_{yield})(N & mm)	33560/14.95	39819/15.58	26812/23.78	48944/17.49	33908/27.76
Ultimate Limit (D_{ult}/Duct) (mm)	42.61/4.37	57.29/5.54	94.35/6.22	42.85/3.67	65.57/3.11

Static analysis has been performed and the FE model was exposed to the same lateral displacement history as the experimental wall panels. Comparative experimental to FE curves are presented (for illustrative purposes) (Figure 5) and show good agreement with experimental results, the model being able to account for all important aspects of the experimental curves.

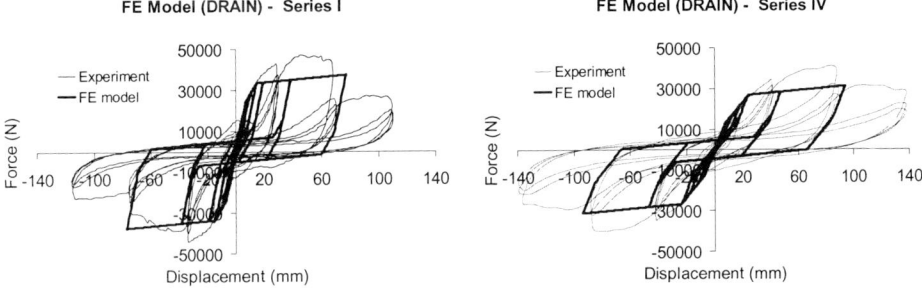

Figure 5. Modeled versus experimental hysteretic curve examples

FE ANALYSIS PROCEDURE AND RESULTS

Being tested trough static runs the developed hysteretic model could be used for dynamic time-history analysis. For this purpose five earthquake records have been selected from a group of seventeen historical records. All records have been scaled to peak ground acceleration (PGA) of 1cm/s². Three records (EL_{centro}, NE_{wall}, KO_{be}) re-assemble reasonably the proposed EC8 elastic spectra, while the other two (SH_{andon} and $VR_{ancea77}$) reflect extremely stiff and soft soil conditions respectively. Therefore, results from these two records representing very low and very high corner periods have to be treated with some precaution.

Using the single degree of freedom (SDOF) FE model described, time history analysis has been carried out using masses of 2000, 2500, 3000, 3500 and 4000kg, and records being scaled from 0.05g to 2g. Due to lack or reliable value, damping has not been considered even if values as big as 6% have been suggested at the level of an entire structure (Kawai et al. 1999). For second order effects a vertical force equal to 30% of the mass value has been also used in the model.

This procedure, known as incremental dynamic analysis (IDA) or dynamic pushover (DPO) is a common analysis method to more thoroughly estimate structural performance under seismic loads. It is important to note that IDA results are both structure and accelerogram sensitive, therefore large range of accelerograms is recommended. Outputs of the analysis consist in IDA curves, relating a performance parameter of the structure to an intensity measure (IM) of the record. The IM of the used IDA has to be scalable, PGA, spectral acceleration corresponding to the first mode period of the structure ($S_{a(T1,\ )}$) or effective peak acceleration (EPA) being the most common ones. Usual structural performance parameters are inter-story drift, maximum plastic rotation, accumulated plastic rotation and top story displacement, depending on the structural typology. Performance parameter can than be related to damage level of the structure and performance based criteria defined to describe the state of the structure after an event of certain intensity.

An important aspect of performance philosophy is to relate lateral displacement to damage and to define acceptable damage levels and relate it to the performance objectives of the panels. Recent performance objective proposals are based on three or four generally stated goals for buildings are (FEMA-273): Serviceability under ordinary occupancy conditions; Immediate occupancy following moderate earthquakes; Life safety under design-basis events; Collapse prevention under maximum considered event.

Such vague performance criteria can be translated into engineering practice by, for example, relating performance objective to deformations. For lateral loads (ie. wind, earthquake) this role can be played by lateral drift ( ) as measure. In these way performance criteria becomes related to a clearly understood parameter in practice. Based on the experimental investigation the lateral displacement limitations as defined in TABLE 4 has been identified for describing limit states of the panels.

DYNAMIC PERFORMANCE OF WALL PANELS

Based on the tolerable displacement values, corresponding earthquake IM levels have been identified for the different panel configurations and earthquake records. The three limit states, identified; D_{el} – or elastic limit of the panel up to which behavior can be considered elastic and it is the conventional capacity level to be used in design; D_{yield} - yield limit of the wall panel, where the panel lost its load bearing capacity, but it is still capable of deforming under the same load level, D_{ult} – ultimate state, the wall panel is not capable of sustaining a constant load level, and it's capacity is decreasing (Figure 3). Out of this three limit states the last two can be identified fairly accurately, and alternative methods of determination yield similar conclusions. Elastic limit is more a conventional value accepted in engineering practice and sometimes very different from code to code.

If elastic design is assumed, the limit of L_{el} is the basis of engineering calculations, even if the panels have important post elastic capacities. In seismic design this post-elastic behavior is accounted for by the behavior factor "q" – 'factor used for design purposes to reduce the forces obtained from linear analysis, in order to account for the non-linear response of the structure. (ENV 1998, Eurocode 8) Similar in US practice "R" is defined as 'response modification factor' or 'system performance factor' that intend to account for damping, energy dissipation capacity and over-strength, and is subdivided in period dependent strength factor R_s, period dependent ductility factor R , and redundancy factor R_R (A. Whittaker, G. Hart, C. Rojahn 1999). As the effect of over-strength has been identified (G. de Matteis et al. 1999), the idea of 'partial q factors' have also penetrated European thinking.

As panel behavior is highly non-linear and important strength reserve can be identified after the accepted design strength, it can be expected that over-strength plays an important role in the post-elastic performance. Based on previously defined limit states (D_{el}, D_{yield} and D_{ult}) partial behavior factors can be defined as ratios of the corresponding IM-s. Thus; q_1 has been defined as the ratio of Sa_{el} and Sa_{yield}, and it is primarily a measure of performance due to panel over-strength, while q_2, the ratio of Sa_{ult} and Sa_{yield} is to be understood as performance parameter due to ductility (TABLE 5).

TABLE 5
PERFORMANCE PARAMETERS Q_1, Q_2 AND Q_3

	Series I					Series II					Series IV					Series OSB I					Series OSB II				
	2t	2.5t	3t	3.5t	4t	2t	2.5t	3t	3.5t	4t	2t	2.5t	3t	3.5t	4t	2t	2.5t	3t	3.5t	4t	2t	2.5t	3t	3.5t	4t
q_1	2.84	2.57	2.42	2.25	2.39	1.89	2.35	2.17	2.38	2.26	3.04	3.12	3.80	3.09	3.37	2.14	2.97	3.04	2.70	2.47	3.46	3.43	3.80	3.95	4.29
q_2	1.25	1.49	1.59	1.53	1.45	1.51	1.44	1.74	1.72	1.83	1.92	2.12	2.40	2.54	2.82	1.42	1.21	1.44	1.43	1.37	1.73	1.85	2.02	1.65	2.15
$q_3 = q_1 \times q_2$	3.58	3.64	3.92	3.46	3.48	2.90	3.36	3.60	4.10	4.09	5.72	6.46	8.71	7.44	9.90	3.11	3.62	4.23	3.99	3.39	5.86	6.42	7.60	6.24	8.66

	Series I			Series II			Series IV			Series OSB I			Series OSB II		
Average	q_1	q_2	q_3	q_1	q_2	q_3	q_1	q_2	q_3	q_1	q_2	q_3	q_1	q_2	q_3
	2.50	1.46	3.62	2.21	1.65	3.61	3.28	2.36	7.65	2.66	1.38	3.67	3.78	1.88	6.96

It is important to mention that q_1 is highly dependant on the elastic limit defined (D_{el}), limit that is on the other hand conventional. As allowable strength definition was based on 1/300 (Kawai et al. 1997) story drift angle, for models with low initial rigidity (Series IV and Series OSB II) the criteria is very severe and a very low Sa_{el} is identified. This particularity yields to high values of q_1 and q_3 consequently. The value of q_2 instead is less dependent on conventional values, points corresponding to D_{yield} and D_{ult} being more readily definable.

CONCLUSIONS

It can be concluded that shear-resistance of wall panels is significant both in terms of rigidity and load bearing capacity, and can be effective against lateral load. The hysteretic behavior is characterized by very significant pinching, and therefore reduced energy dissipation. How this low dissipation affects earthquake performance of the building is to be evaluated.

Failure is starting at the bottom track in the anchor bolt region, therefore strengthening of the corner detail is very important. The ideal shape of corner detail is so that uplift force is directly transmitted from the brace (or corner stud) to the anchoring bolt, without inducing bending in the bottom track. Failing to strengthen wall panel corners has important effects on the initial rigidity of the system.

The seam fastener represented the most sensitive part of the corrugated sheeting specimens; damage is gradually increased in seam fasteners, until their failure causes the overall failure of the panel. Much of the post elastic deformation of the panel is in the region of seam fasteners, therefore increasing the load capacity and ductility of the seams will improve the behavior of the panels.

It is very important to underline that in case of all tested specimens the wall stud system proved a very good toughness. Even when damages were very important no collapse occurred. This is of real importance for buildings located in seismic areas. For corrugated sheet specimens, and similarly for others, performance design criteria can be suggested.

Based on experimental evidence a FE model has been calibrated to be used in earthquake modeling of shear wall-panels. The model is accurate enough to take into account all important aspects of the hysteretic behavior and simple enough to be incorporated in more complex structural schemes. A number of inelastic time-history runs have been performed using different wall panels, acting masses and earthquakes records, and the model has been found satisfactory for the purpose.

Using experimentally determined criteria three performance levels were associated with corresponding lateral displacement of the panels and 'partial behavior' factors have been identified for the panels based on time-history analysis results. The effect of over-strength is identified to be important in the post elastic behavior of panels and source of a possible design earthquake-force reduction. The resulting factor (2.2-2.6) is harmonizing reasonably with the value 1.5-5 suggested by Gad et al. (1999). The possibility of design force reduction due to ductility and energy dissipation seems to be more limited (1.4-1.6) probably due to low energy dissipation capacity of the hysteretic loops. This value is also in agreement with the findings of Gad et al. (1999).

Reference

AISI (1998). Shear Wall Design Guide, Publication RG-9804, AISI

AISI (1997). Monotonic Tests of Cold Formed Shear Walls with Openings, Prepared by NAHB research Center. Inc., The American Iron and Steel Institute

ECCS. (1985) Recommended Testing Procedure for Assessing the Behaviour of Structural Steel Elements under Cyclic Loads.

ECCS. (1995) European Recommendations for the Application of Metal Sheeting acting as a Diaphragm. Pub.88, European Convention for Constructional Steelwork, Brussels

Eurocode 8. (1998). Design provisions for earthquake resistance of structures

FEMA-273 (1997). NEHRP Guidelines for the Seismic Rehabilitation of Buildings, Prepared by the Building Seismic Safety Council for the Federal Emergency Management Agency, Washington.

E.F. Gad et al. (1999). Earthquake Ductility and Overstrength in Residential Structures, Structural Engineering and Mechanics, 8:4, 361-382

E.F. Gad and C.F. Duffield. (2000) Lateral Behaviour of Light Framed Walls in Residential structures, 12th World Conference on Earthquake Engineering

Gianfranco de Matteis. (1998) The Effect of Cladding in Steel Buildings under Seismic Actions, PhD Thesis, Universita degli Studi di Napoli Federico II

G. de Matteis et al. (1999). Seismic response of MR steel frames with different connection behaviours, Proceedings of the 6th International Colloquium SDSS'99, Elsevier, 409-420

Y. Kawai et al. (1997) Cyclic Shear Resistance of Light-Gauge Steel Framed Walls, ASCE Structures Congress, Poland

Y. Kawai et al. (1999) Seismic resistance and design of steel framed houses, Nippon Steel Technical report, No. 79

T. Pekoz, Building Design Using Cold Formed Sections, Publication of the Swedish Institute of Steel Construction no. 154

V. Prakash and G. H. Powell. (1994). Drain-3DX Base program description and user guide, Version 1.10, Department of Civil Engineering, University of California at Berkley

A. J. Salenicovich et al. (2000), Racking Performance of Long Steel-Frame Shear Walls, Fifteenth Int. Speciality Conference on Cold-Formed Steel Structures, St. Louis, Missouri, Oct. 19-20, 471-480

R.L. Serette and K. Ogunfunmi. (1996). Shear Resistance of Gypsum-Sheeted Light-gauge Steel Stud Walls, Journal of Structural Engineering, ASCE, 122:4

R.L. Serette et al. (1996) Shear Wall Values for Light Weight Steel Framing, AISI

R.L. Serette. (1998) Seismic Design of Light Gauge Steel Structures: A discussion, Fourteenth Int. Speciality Conference on Cold-Formed Steel Structures, St. Louis, Missouri, Oct.15-16

Wittaker, G. Hart, C. Rojahn. (1999). Seismic response modification factors, Journal of structural engineering 1999:2, 438-443

DIRECT STRENGTH METHOD FOR THE DESIGN OF PURLINS

Luis Quispe[1] and Gregory Hancock[2]

[1] Addicoat Hogarth Wilson, Level 12, South Tower, 1-5 Railway Street, Chatswood, NSW, 2067
[2] BHP Steel Professor of Steel Structures, Centre for Advanced Structural Engineering
Department of Civil Engineering, University of Sydney, Sydney, Australia, 2006

ABSTRACT

The Direct Strength Method (DSM) has recently been developed by Schafer and Peköz for the design of cold-formed steel structural members. What is now required is the calibration of the method against existing design methodologies for common structural systems such as roof and wall systems.

The paper firstly explains the application of the DSM for the design of simply supported and continuous purlins. Some generalizations, such as how to handle combined bending and shear at the ends of laps, have had to be made to implement the method for continuous purlin systems.

The method is then applied to study a range of section sizes in C- and Z-sections and a range of spans for simply supported, continuous and continuous lapped purlins. The results are compared with purlin design capacities to the Australian/New Zealand Standard AS/NZS 4600. This standard is similar to the AISI Specification except that it includes design rules for distortional buckling. Some modifications have had to be made to the strength equations in the DSM to achieve an accurate and reliable comparison. These modifications are included in the paper.

KEYWORDS

Cold-formed steel, design, stability, purlins, direct strength

INTRODUCTION

The Direct Strength Method (DSM) is a newly proposed approach by Schafer and Peköz (1998) for determining the strength of cold-formed members. Conventionally, the effective width method has been used as recognized in the current cold-formed steel design standards (eg. AISI Specification (1996), AS/NZS:4600 (1996). The DSM however uses full section properties with an appropriate strength design curve to give a direct strength. The purpose of this research is to compare the results of the DSM with the effective width method. To achieve this objective, a series of tables for purlin capacity have been created using the DSM for comparison with those based on the effective width method. The Lysaght limit state design capacity tables produced by BHP Building Products (2000) computed to AS/NZS:4600 were readily available and so were used. Both in (downwards) and out (uplift) load cases for single, double continuous, double lapped, triple continuous, and triple lapped spans were studied. In each of the ten cases, the ratio of the strength based on DSM to that based on effective width was calculated and the results illustrate the comparison of the two methods.

In order to refine the comparison, three different curves were used. These are:

(i) The AISI beam design curve (Section C3.1.2 AISI) equivalent to Clause 3.3.3.2(a) of AS/NZS 4600 including interaction of lateral and distortional buckling (Method 1).

(ii) Clause 3.3.3.2(b) AS/NZS 4600 including interaction of lateral and distortional buckling. This is the old permissible stress design curve method of AS 1538 which has a lower beam curve but which was used for the Lysaght load tables (BHP Building Products (2000)) (Method 2).

(iii) AISI beam curve (Section C3.1.2, AISI) equivalent to Clause 3.3.3.2(a) of AS/NZS 4600 excluding interaction of lateral and distortional buckling (Method 3).

BACKGROUND

This investigation is based on two source documents. The first source is Chapter 12 of "*Cold-Formed Steel Structures to the AISI Specification*" by G.J. Hancock, T.M. Murray and D.S. Ellifritt (2001) where the DSM is presented as a new approach for the design of cold-formed steel members. This new approach uses elastic buckling solutions for the entire cross section in lieu of the effective width method, which analyses each element of the cross section separately. The method makes full use of the readily available solutions from software that is detailed later in the paper. The local or distortional buckling strengths are combined with the overall (flexural, torsional or flexural torsional) buckling strength using the unified method of Schafer and Pek $\bar{z}$ (1998).

The second source used in this investigation is the "*Lysaght Zeds & Cees Purlin & Girt System Limit State Capacity Tables & Product Information*" Revised December 2000. These tables were computed using software developed at the University of Sydney for use by BHP Building Products and Stramit Industries. Primarily each capacity value in the tables has been calculated by following the effective width method, which is set out in AS/NZS 4600: 1996 "*Cold-formed Steel Structures*". Clause 3.3.3.2(b) of AS/NZS:4600 was used for the beam design curve. The section depths range from 100 mm to 350 mm in 50 mm increments, and the thicknesses range from 1.0 mm to 3.0 mm.

Finite element flexural-torsional buckling analyses (PRFELB, CASE (1997b)) were used to model the whole purlin system to compute the overall buckling load. The model considers both in-plane distributions of axial force, shear force and bending moments, as well as out-of-plane buckling modes. The analysis assumes that:

- All purlins bend about the axis which is perpendicular to the web;
- There is continuity at the laps;
- There is minor axis translation and twisting restraint at the bridging points;
- There is lateral stability in the plane of the roof at internal supports and the end of cantilevers; and
- Both screw fastened and concealed-fixed claddings provide diaphragm shear restraint.

Forces acting to hold cladding against a structure are called *inward* (in). Forces acting to remove cladding from a structure are called *outward* (out). Lap lengths are carefully chosen and range from 600 mm to 2400 mm; lap lengths depend on nominal section size and span. In order to cover a representative variety of sections for a useful comparison, the spans chosen have approximate span/depth values of 20, 30 and 40 and match with those in the Lysaght tables.

The location of the bridging was established in the Lysaght Tables, the options being zero, one, two, and three rows of bridging.

DIRECT STRENGTH METHOD

The DSM concept says that at plate failure the full width can be considered at the effective design stress instead of the effective width considered to be at yield. The starting point in the DSM is to calculate the elastic buckling solutions for local and distortional modes. There are three basic buckling modes: local, distortional and lateral (flexural-torsional).

Elastic Local and Distortional Buckling Stresses (F_{crl}, F_{crd})

Appropriate solutions are readily available for the local and distortional buckling stresses by means of the numerical finite strip method. These solutions are clearly presented by Hancock, Murray and Ellifritt (2001) in Chapter 3 and Hancock (1998) in Chapter 3. Cross-section analysis and finite strip buckling analysis can be obtained using a computer program THIN-WALL produced by The Centre for Advanced Structural Engineering at the University of Sydney (CASE, 1997a) or by the Cornell University Finite Strip Program CUFSM (Cornell University, 2001). The programs compute the local (F_{crl}) and distortional (F_{crd}) buckling stresses as shown in Fig. 1. From these stresses and the full section modulus (S_{xf}), the elastic buckling moments can be calculated (M_{crl}, M_{crd}) for both local and distortional buckling.

$$M_{crl} = F_{crl} \times S_{xf} \qquad (1)$$

$$M_{crd} = F_{crd} \times S_{xf} \qquad (2)$$

Since Australian Z-sections have unequal flanges to permit lapping, then properties for the Z-sections are calculated based on the equivalent C-section, where the Z-flanges are averaged and the top flange is reversed to produce an equivalent C-section.

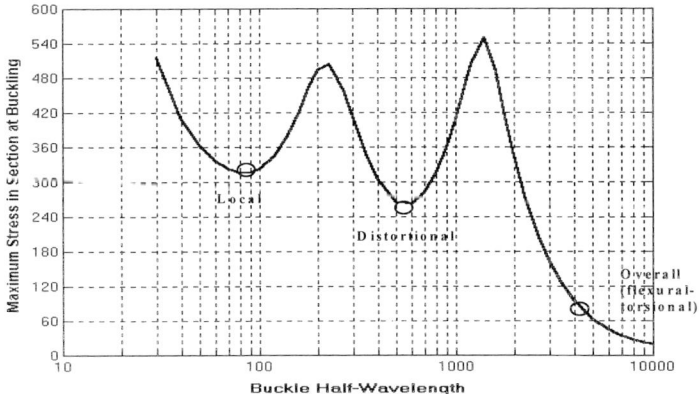

Figure 1: Elastic buckling from THIN-WALL

Elastic Lateral Buckling and Lateral Buckling Strength

The elastic lateral (flexural-torsional) buckling stress can be calculated for a continuous purlin system including laps using the program PRFELB (CASE, 1997b) developed at the University of Sydney and described in Chapter 5 of Hancock (1998) and Chapter 5 of Hancock, Murray, and Ellifritt (2001). Load factors from PRFELB (1997b) were used for each case to determine the elastic lateral buckling moment (M_e) as given by Eq. 3.

$$M_e = M_{max} \times \text{Load Factor} \tag{3}$$

M_{max} is the maximum moment in the bending moment diagram for the segment of the purlin under analysis.

The critical moment M_c is evaluated based on the limit state design procedure under the AISI 1996 beam design curve (Section C3.1.2, (1996)), which is described in Clause 3.3.3.2(a) of AS/NZS 4600 (1996) or Clause 3.3.3.2(b) of AS/NZS 4600 as appropriate. The following equations apply for singly, doubly and point symmetric sections.

$$M_c = M_y \quad \text{for} \quad M_e \geq 2.78 M_y \tag{4}$$

$$M_c = 1.1 M_y \left(1 - \frac{10 M_y}{36 M_e}\right) \quad \text{for} \quad 2.78 M_y > M_e > 0.56 M_y \tag{5}$$

$$M_c = M_e \quad \text{for} \quad M_e \leq 0.56 M_y \tag{6}$$

where M_y is the yield moment of the full section ($S_{xf}F_y$).

The value of the inelastic lateral buckling strength (M_{ne}) in the DSM is taken as the critical moment M_c given by the above Eqs 4, 5 and 6.

Direct Strength Computation

The computation of the local and distortional buckling strengths (M_{nl}, M_{nd}) is the next step in the calculations. These strengths account for the interaction of local buckling with lateral buckling and distortional buckling with lateral buckling by using the limiting moment M_{ne} instead of M_y in the calculations. Schafer and Peköz (1998) have developed local buckling strength equations and the following Eqs 7 to 9 define this buckling mode:

$$M_{nl} = M_{ne} \quad \text{for} \quad \lambda_l \leq 0.776 \tag{7}$$

$$M_{nl} = \left(1 - 0.15\left(\frac{M_{crl}}{M_{ne}}\right)^{0.4}\right)\left(\frac{M_{crl}}{M_{ne}}\right)^{0.4} M_{ne} \quad \text{for} \quad \lambda_l > 0.776 \tag{8}$$

where $\lambda_l = \sqrt{\dfrac{M_{ne}}{M_{crl}}}$ \hfill (9)

Distortional buckling strength can be derived similarly. Hancock, Kwon, and Bernard (1994) initially devised this method which was successfully adopted into AS/NZS 4600 (1996). Despite the fact that Schafer and Peköz suggest 0.25 as the coefficient and 0.6 as exponent as recommended in Hancock, Kwon and Bernard (1994), consideration has been in favour of the distortional buckling

Clause 3.3.3.3(a) of AS/NZS 4600 (1996), where 0.22 is the coefficient and 0.5 is the exponent for the distortional buckling strength equations, as follows:

$$M_{nd} = M_{ne} \qquad \text{for} \quad \lambda_d \leq 0.561 \qquad (10)$$

$$M_{nd} = \left[1 - 0.22\left(\frac{M_{crd}}{M_{ne}}\right)^{0.5}\right]\left(\frac{M_{crd}}{M_{ne}}\right)^{0.5} M_{ne} \qquad \text{for} \quad \lambda_d > 0.561 \qquad (11)$$

where: $\qquad \lambda_d = \sqrt{\dfrac{M_{ne}}{M_{crd}}} \qquad (12)$

These are Methods (1) and (2) in the introduction to this paper where the interaction of lateral and distortional buckling is considered. Method (3) ignores interaction of lateral and distortional buckling and replaces M_{ne} by the full section yield moment M_y.

From the above limiting strengths the nominal member capacity (M_n) is determined.

$$M_n = \text{The lesser of } (M_{nl}, M_{nd}) \qquad (13)$$

From the nominal member moment capacity (M_n) the design loads (w_u) are evaluated for each case and the current capacity resistance factor for flexure $\phi_b = 0.9$ still applies.

Shear, Bending and Combined Bending and Shear

Shear can become an important issue for the majority of cases studied for both inward and outward load configurations except when the configuration is a single span, where the maximum bending and shear are well separated. As stipulated in Clause 3.3.5 of the AS/NZS 4600 (Section C3.3 of AISI (1996), a combination of shear force and bending moment in the web produces a further reduction in the capacity of the web.

Combined Bending and Shear Capacity

The capacity factors adopted in this investigation are the same as in AS/NZS 4600 (1996) and the AISI Specification (AISI, 1996), namely for section moment capacity $N_b = 0.9$ and likewise for shear capacity $N_v = 0.9$; Eq. 14 is the combined bending and shear interaction equation:

$$\left(\frac{M_u}{\phi_b M_{nxo}}\right)^2 + \left(\frac{V_u}{\phi_v V_n}\right)^2 \leq 1 \qquad (14)$$

M_{nxo} is given by Eq. 13 when M_{ne} in Eqs (7) to (12) is taken as M_y.

When Eq. 14 is greater than 1.0 for a purlin design controlled previously by lateral buckling, then combined bending and shear controls the design. In order to obtain the reduced design load (w_u), the interaction equation is divided by a factor that will bring the right hand side of this equation to one.

DIRECT STRENGTH METHOD TABLES

The three tables (Tables 1, 2 and 3) presented in this paper cover a wide range of the equivalent Lysaght tables. In each of the three tables, the mean ratio of the DSM design capacity to the effective width design capacity based on the Lysaght tables $\left(\frac{w_u \, \text{DSM}}{w_u \, \text{Lysaght}}\right)$ is given along with the statistical variation. The full set of values can be found in the report by Quispe (2001). The average deviation (AVEDEV) is a measure of the variability in a data set and it is the average of the absolute derivations of the data points from their mean $\left(\frac{1}{n}\sum |x - \bar{x}|\right)$. The standard derivation (STDEV) is a measure of how widely the values are dispersed from the average value (mean).

Method 1 AISI beam curve and lateral distortional interaction

The correlation between DSM capacities and Lysaght capacities is very close to one.

Method 2 AS/NZS 4600 beam curve (Clause 3.3.3.2(b)) and lateral-distortional interaction

Method 2 gives a lower comparison average (0.98) than Method 1 since it is based on a lower beam curve (Clause 3.3.3.2(b) of AS/NZS:4600). It demonstrates that using the same beam lateral buckling and distortional buckling strength curves the DSM is slightly conservative as it accounts for interaction of lateral and distortional buckling not previously accounted for in AS/NZS:4600 and hence the Lysaght design capacity tables.

Method 3 AISI beam curve and no lateral-distortional interaction

Method 3 gives a higher comparison on average (1.01) than Methods 1 and 2 since it uses the higher AISI beam curve and ignores lateral-distortional interaction. All three methods have comparable average and standard derivations.

Method 3 gives a slight increase in capacity when lateral-distortional interaction is ignored whereas Method 2 gives a decrease in capacity when lateral-distortional interaction is included.

CONCLUSIONS

It was found that the Direct Strength Method performs very similarly to the Effective Width Method when applied to the design of simply supported, continuous and lapped purlins for both inward and outward loading. Method 3 is recommended for implementation in design standards. Further it allows for development of new web-stiffened and lip-stiffened sections since the analysis is done for the entire section in lieu of element by element.

REFERENCES

BHP Building Products (2000), Lysaght Zeds and Cees Purlin and Girt System Limit State Capacity Tables and Product Information.

Centre for Advanced Structural Engineering, (1997a), THIN-WALL - A Computer Program for Cross-Section Analysis of Thin-Walled Structures, User's Manual, Version 1.2, Department of Civil Engineering, The University of Sydney.

Centre for Advanced Structural Engineering (1997b), PRFELB- A Computer Program for Finite Element Flexural-Torsional Buckling Analysis of Plane Frames and Design of Beams to AS 4100, Design of purlins to AS/NZS 4600 and AS 1538, User's Manual Version 3.0. Department of Civil Engineering, The University of Sydney.

Cornell University Finite Strip Program CUFSM, (2001), Version 2.5, Cornell University.

Hancock, G.J., Kwon, Y.B., and Bernard, E.S. (1994), Strength Design Curves for Thin Walled Sections undergoing Distortional Buckling, J Constr. Steel Res. Vol 31, Nos. 2/3, 1994, pp. 169-186.

Hancock, G.J. (1998), Design of Cold-Formed Steel Structures (to Australian/New Zealand Standard AS/NZS 4600:1996, Australian Institute of Steel Construction, 3rd edition.

Hancock, G.J., Murray, T.M. and Ellifritt, D.S. (2001), Cold-formed Steel Structures to the AISI Specification, edited by Marcel Dekker, Inc. 375-391.

Quispe, L.A. (2001), Direct Strength Method for Purlin Design, Unpublished Report, Department of Civil Engineering, University of Sydney, November.

Schafer, B.W. and Pek$\bar{z}$, T. (1998), Direct Strength Prediction of Cold-Formed Steel Members using Numerical Elastic Buckling Solutions, Thin-Walled Structures, Research and Development, edited by Shanmugan, N.E., Liew, J.Y.R., and Thevendran, V., Elsevier, pp.137-144 (also in Fourteenth International Specialty Conference on Cold-Formed Steel Structures, St Louis MO, Oct.1998).

Standards Australia/Standards New Zealand (1996), Cold-Formed Steel Structures, (AS/NZS 4600:1996).

Timoshenko, SP, and Gere, J.M. (1959), Theory of Elastic Stability, McGraw-Hill Book Co. Inc. New York. N.Y.

TABLE 1

STATISTICAL RESULTS OF THE COMPARISON BETWEEN DSM AND EFFECTIVE WIDTH METHOD FOR METHOD 1 ASSUMPTIONS

CASES A TO J AND 1568 CASES INVESTIGATED	STATISTICAL RESULTS OF w_uDSM/w_uLysaght				
	Mean	AVEDEV	STDEV	Minimum	Maximum
Case A: Single span in	1.02	0.03	0.06	0.92	1.34
Case B: Single span out	1.04	0.07	0.09	0.86	1.26
Case C: Continuous double span in	0.95	0.04	0.05	0.83	1.03
Case D: Continuous double span out	0.96	0.05	0.07	0.83	1.21
Case E: Continuous triple span in	0.96	0.07	0.11	0.80	1.45
Case F: Continuous triple span out	0.98	0.08	0.12	0.80	1.45
Case G: Lapped double span in	0.97	0.06	0.09	0.84	1.38
Case H: Lapped double span out	1.01	0.10	0.13	0.84	1.40
Case I: Lapped triple span in	1.01	0.05	0.07	0.92	1.39
Case J: Lapped triple span out	1.03	0.07	0.08	0.86	1.27
STATISTICS OF THE 1568 CASES ANALYSED	**1.00**	**0.07**	**0.10**	**0.80**	**1.45**

TABLE 2
STATISTICAL RESULTS OF THE COMPASON BETWEEN DSM AND EFFECTIVE WIDTH METHOD FOR METHOD 2 ASSUMPTIONS

CASES A TO J AND 1573 CASES INVESTIGATED	STATISTICAL RESULTS OF $w_u DSM/w_u Lysaght$				
	Mean	AVEDEV	STDEV	Minimum	Maximum
Case A: Single span in	1.02	0.03	0.06	0.92	1.34
Case B: Single span out	0.99	0.04	0.05	0.83	1.06
Case C: Continuous double span in	0.95	0.04	0.05	0.83	1.03
Case D: Continuous double span out	0.95	0.05	0.05	0.83	1.09
Case E: Continuous triple span in	0.96	0.07	0.11	0.80	1.45
Case F: Continuous triple span out	0.96	0.07	0.11	0.80	1.45
Case G: Lapped double span in	0.97	0.06	0.09	0.84	1.38
Case H: Lapped double span out	1.00	0.09	0.10	0.84	1.39
Case I: Lapped triple span in	1.01	0.05	0.07	0.92	1.39
Case J: Lapped triple span out	0.99	0.05	0.05	0.85	1.13
STATISTICS OF THE 1573 CASES ANALYSED	**0.98**	**0.06**	**0.08**	**0.80**	**1.45**

TABLE 3
STATISTICAL RESULTS OF THE COMPARISON BETWEEN DSM AND EFFECTIVE WIDTH METHOD FOR METHOD 3 ASSUMPTIONS

CASES A TO J AND 1568 CASES INVESTIGATED	STATISTICAL RESULTS OF $w_u DSM/w_u Lysaght$				
	Mean	AVEDEV	STDEV	Minimum	Maximum
Case A: Single span in	1.02	0.03	0.06	0.92	1.34
Case B: Single span out	0.08	0.09	0.10	0.92	1.38
Case C: Continuous double span in	0.95	0.04	0.05	0.83	1.03
Case D: Continuous double span out	0.98	0.07	0.10	0.83	1.36
Case E: Continuous triple span in	0.96	0.07	0.11	0.80	1.45
Case F: Continuous triple span out	0.99	0.10	0.14	0.80	1.45
Case G: Lapped double span in	0.97	0.06	0.09	0.84	1.38
Case H: Lapped double span out	1.02	0.11	0.14	0.84	1.42
Case I: Lapped triple span in	1.01	0.05	0.07	0.92	1.39
Case J: Lapped triple span out	1.06	0.09	0.11	0.92	1.35
STATISTICS OF THE 1573 CASES ANALYSED	**1.01**	**0.08**	**0.08**	**0.80**	**1.45**

COLD-FORMED PURLIN-SHEETING SYSTEMS

F. ALBERMANI[1] and S. KITIPORNCHAI[2]

[1] Department of Civil Engineering, University of Queensland, Australia
[2] Department of Building and Construction, City University of Hong Kong, Hong Kong

ABSTRACT

Purlin-sheeting systems used for cladding in roofs and walls commonly consist of cold-formed channel or zed section purlins, screw-connected to corrugated sheeting. This paper presents a nonlinear finite element model for purlin-sheeting systems. The model incorporate both the sheeting and the purlin, and is able to account for cross-sectional distortion of the purlin, the flexural and membrane restraining effects of the sheeting, and failure of the purlin by local buckling or yielding. The validity of the model is shown by its good correlation with available experimental results.

KEYWORDS

Cold-formed, purlin-sheeting, cladding, nonlinear analysis

INTRODUCTION

Cold-formed steel zed and channel section members are widely used as purlins or grits. Purlins are usually connected to the sheeting by way of a screw through the crest of the corrugated sheeting and the purlin flange. Due to the restraining action of the sheeting, the purlin tends to fail by localised plastic collapse or local buckling rather than overall flexural-torsional buckling. The sheeting provides two restraining effects to the purlin; shear and rotational effects. Both shear and rotational stiffness cause a significant increase in the capacity of the attached purlin and their negligence result in over-conservative estimates of the purlin load-carrying capacity.

In this paper, a finite element model (Albermani and Kitipornchai, 1999) of the purlin will be used in which the restraining action of the sheeting is simulated by equivalent springs attached to the flange of the purlin as shown in Figure 1. These springs, K_{ry} and K_{rx}, represent, respectively, the shearing and rotational restraint provided by the sheeting. The spring stiffness depends on; sheeting profile and thickness, purlin profile and thickness, sheeting span and connection details between the sheeting and the purlin, among other things. In the present research two numerical tools, Double Beam Shear Test Model (DBSTM) and Rotational Restraint Model (RRM), have been developed for the determination of the shear and rotational stiffness of any particular purlin-sheeting system (Lucas et al (1997 a,b)). The validity of the proposed model is demonstrated by comparison with available experimental results for single, double and triple span purlins under both uplift and downwards loading (Rousch and Hancock, 1995).

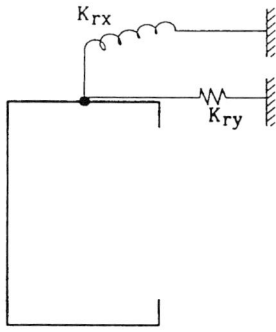

Figure 1: Modelling of sheeting restraint

SHEAR AND ROTATIONAL STIFFNESS

The shear restraint provided by the sheeting to the purlin is primarily independent of the purlin profile but does vary with both sheeting type and span. Using the DBSTM, the shear stiffness of various sheeting profiles can be determined. The shear stiffness for four standard corrugated sheeting profiles (Trimdek and Spandek Hi-Ten (Lysaght, 1991)) with 0.42 and 0.48mm thickness has been determined using the DBSTM for various sheeting span as shown in Figure 2. The shear stiffness of the sheeting increases with sheeting thickness and sheeting span.

The sensitivity of the purlin to the value of shear stiffness provided by the sheeting is investigated. It was found that for the practical range of shear stiffness for common sheeting profiles (Figure 2), the purlin, zed or channel section, is not significantly sensitive to the specific value used for the shear stiffness. For this reason a 'standard' value of 1000kN/rad is adopted to represent the shear stiffness of common sheeting profile.

The rotational restraint provided by the sheeting to the purlin is a complex parameter varying with each purlin-sheeting combination. Unlike the case for shear stiffness, the purlin is sensitive to the value of rotational restraint within the range of commonly used sheeting profiles. The RRM is used to determine the rotational stiffness of a number of different sheeting-purlin configurations. It was found that the effect of sheeting span on the rotational stiffness provided by the sheeting is minor. Similarly, the effect of sheeting profile and thickness (Spandek and Trimdek) on the rotational stiffness is in the range of 10%. The rotational stiffness provided by standard sheeting types to the purlin depends on the purlin profile as shown in Figure 3. If the purlin-sheeting system being analysed involves a significantly different sheeting type, purlin type or configuration, the RRM can be used to determine a system-specific value of rotational stiffness that can be used as an input in running the FE model of Figure 1.

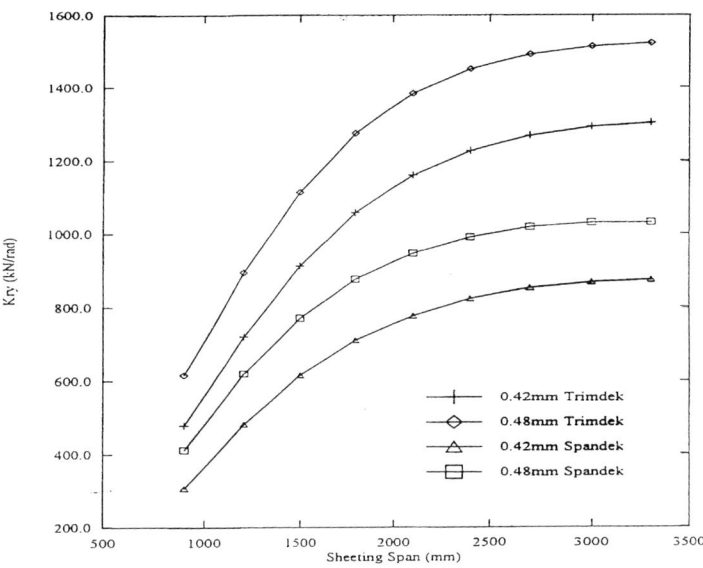

Figure 2: Shear stiffness for common sheeting profiles

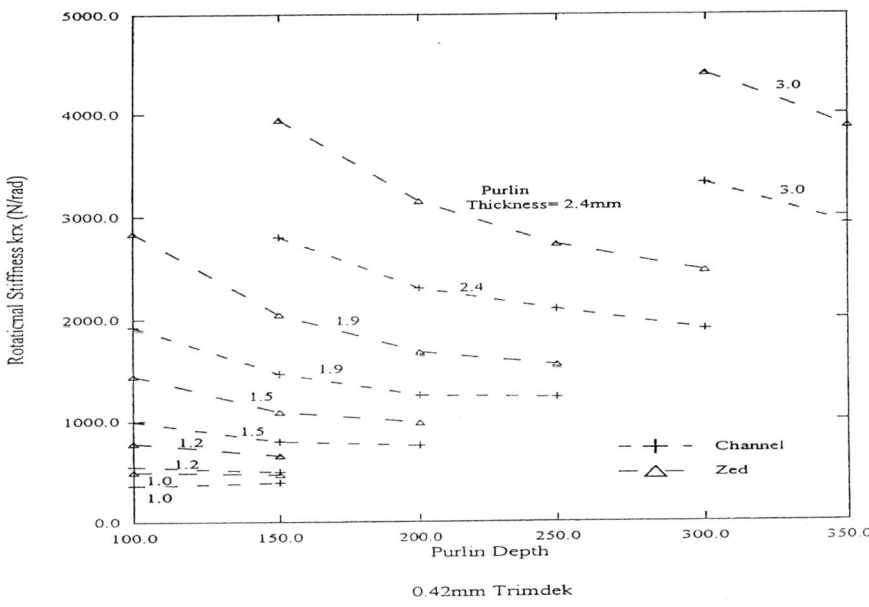

Figure 3: Rotational restraint for common purlin profiles

COMPARISON OF NUMERICAL AND EXPERIMENTAL RESULTS

Experimental results from vacuum test rig program reported by Rousch and Hancock (1995) were used to verify the accuracy of the proposed model. The shear and rotational stiffness values required for the running of the proposed model were determined as described in the previous section. Table 1 shows a comparison of the results obtained using the proposed model for single, double and triple span purlins with that from the vacuum test. All these tests were carried out under uplift loading except for test series S4T1-S4T6 was tested under downwards loading. The purlins presented in Table 1 include both channel and zed section purlins with zero, one and two rows of bridging and with single, lapped double and lapped triple span configurations. The proposed model predictions are within 10% of the experimental failure load. The proposed model shows that the dominant failure mode of the purlins is local plastic collapse of the free flange. The predicted location of this collapse mode along the purlin agrees with that obtained in the experiment. Figures 4 shows contour plots of the predicted deflected shape of the free flange for test S1T4 (triple span Z200-15 purlin, no bridging and under uplift loading) as obtained from the proposed model. Only one half of the flange length is shown in the figure. From the vacuum test experiment, this purlin was reported to has failed by local plastic collapse of the free flange approximately 2100mm from the simply supported end. Figure 4 clearly shows the development of instability in the free flange between 1000 and 2000mm from the simply supported end.

CONCLUSION

A numerical model for predicting the behaviour of purlin-sheeting systems has been presented in this paper. The validity of this model was shown by its good correlation with experimental results. The model requires the discretization of the purlin into an assembly of shell elements. The restraining action of the sheeting is obtained by augmenting the purlin finite element model by shear and rotational springs at the purlin-sheeting connection points. The stiffness of these springs could be determined numerically using two accompanied numerical tools, the DBSTM and RRM. These tools are modified simulations of physical experimental procedures and incorporate both the sheeting and the purlin.

From a study of sensitivity of the purlin-sheeting system to shear stiffness, it was found that a single value of shear stiffness could be used for all standard purlin-sheeting systems. A similar study on the sensitivity to rotational stiffness indicated that for each particular purlin section type, a particular value of rotational stiffness should be used. A chart was developed from which the value of rotational stiffness could be read for any standard purlin section.

The proposed model accounts for the distortion of the purlin's cross-section and for the restraining effect of the sheeting. The model is capable of predicting the ultimate capacity of purlins, whether is governed by yielding or buckling. The model together with the DBSTM and RRM tools could be easily used in design environment.

TABLE 1
PROPOSED MODEL AND TEST FAILURE LOADS

Test	Purlin Section	Span length (mm)	Rows of bridging/span	Test failure Load kN/m	Model failure load kN/m	Ratio Model/Test
(a) Single span purlin						
S7T1	Z200-15	7000	0	1.85	1.81	0.98
S7T2	C200-15	7000	0	1.70	1.70	1.00
S7T3	C200-15	7000	1	1.70	1.78	1.05
S7T5	C200-15	7000	2	1.95	1.93	0.99
S3T1	Z200-24	7000	0	3.28	3.59	1.09
S3T2	Z200-24	7000	1	3.69	3.60	0.98
S3T3	Z200-24	7000	2	4.76	4.53	0.95
S3T4	C200-24	7000	0	3.63	3.58	0.99
S3T5	C200-24	7000	1	3.63	3.36	0.93
S3T6	C200-24	7000	2	4.71	4.59	0.97
(b) Double span purlins						
S2T1	Z300-24	10500	0	4.33	4.47	1.03
S2T2	Z300-24	10500	1	4.93	5.02	1.02
S2T3	Z300-24	10500	2	5.77	5.75	1.00
(c) Triple span purlins						
S1T1	Z150-19	7000	0	2.31	2.53	1.10
S1T2	Z150-19	7000	1	2.63	2.82	1.07
S1T3	Z150-19	7000	2	2.98	3.07	1.03
S1T4	Z200-15	7000	0	2.58	2.60	1.01
S1T5	Z200-15	7000	1	2.94	2.82	0.96
S1T6	Z200-15	7000	2	3.87	3.91	1.01
S1T7	Z200-19	7000	0	3.51	3.77	1.07
S1T8	Z200-19	7000	1	4.28	3.95	0.92
S1T9	Z200-19	7000	2	4.55	4.30	0.95
S4T1/2	Z200-19	7000	1	3.97/4.42	4.16	0.99
S4T3/4	Z200-15	7000	0	2.90/2.94	2.89	0.99
S4T5	Z150-19	7000	0	2.92	2.94	1.01
S4T6	Z150-19	7000	1	2.69	2.92	1.09

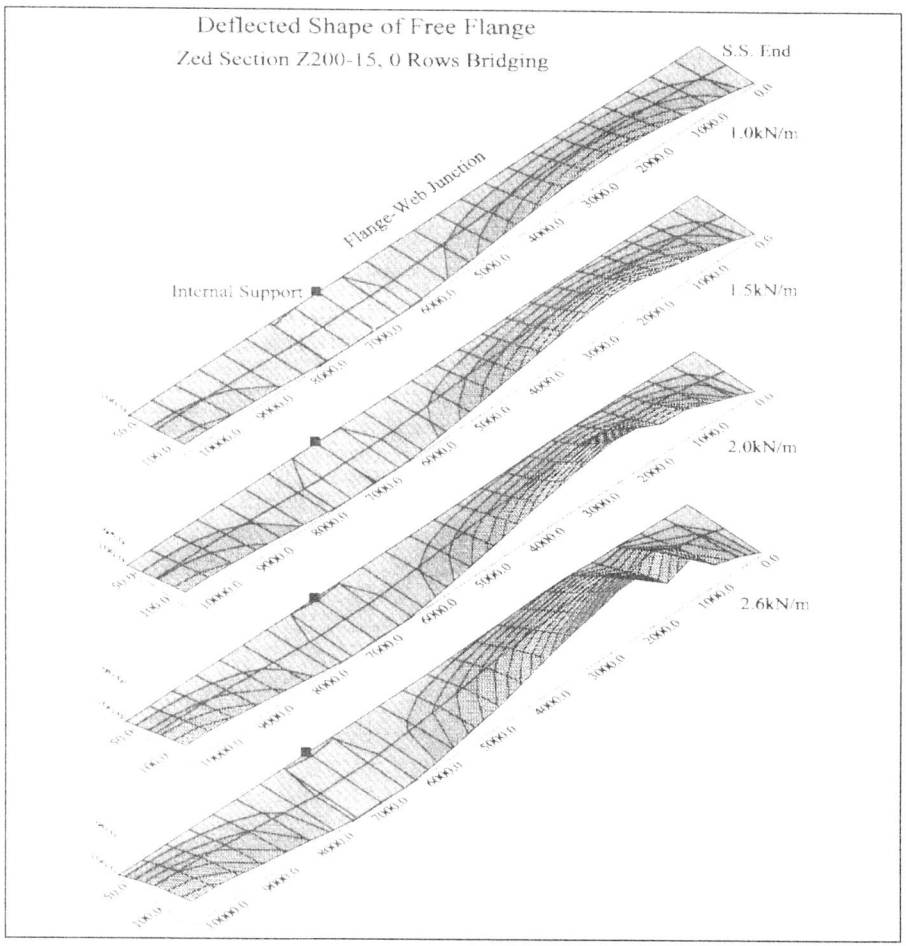

Figure 4: Predicted deflected shape of free flange for test S1T4

REFERENCES

Albermani F. and Kitipornchai S. (1999). Nonlinear Analysis of Thin-Walled Structures, *Chapter 6, Structural Dynamic Systems, Computational Techniques and Optimization*, Ed. C.T. Leondes, 239-87.

Lucas R.M., Albermani F. and Kitipornchai S. (1997a). Modelling of Cold-Formed Purlin-Sheeting Systems- Part 1: Full Model. *Thin-Walled Structures* **27:3**, 223-43.

Lucas R.M., Albermani F. and Kitipornchai S. (1997b). Modelling of Cold-Formed Purlin-Sheeting Systems- Part 2: Simplified Model. *Thin-Walled Structures* **27:4**, 263-86.

Lysaght Building Industries (1991), Design Manual: Steel Roofing and Walling.

Rousch C.J. and Hancock G.J. (1995). Tests of Channel and Z-section Purlins Undergoing Nonlinear Twisting. *Research Report R708*, School of Civil and Mining Engineering, University of Sydney.

AN EXPERIMENTAL INVESTIGATION INTO LAPPED MOMENT CONNECTIONS BETWEEN Z SECTIONS

H.C. Ho and K.F. Chung

Department of Civil and Structural Engineering,
The Hong Kong Polytechnic University, Hung Hom, Hong Kong, China.

ABSTRACT

The paper presents an experimental investigation on the structural performance of bolted moment connections between lapped cold-formed steel Z sections. A total of 12 one point load tests on generic lapped connections between Z sections with various lap lengths were carried out, and both the moment resistances and the effective flexural rigidities of the generic lapped connections between Z sections were compared. In all tests, flexural failure in the Z sections at the end of lap was found. Among all tests, the moment resistances of the generic lapped connections are found to range from 0.79 to 1.60 of the moment resistances of the connected sections. Thus, it is demonstrated that effective bolted moment connections in the generic lapped connections between cold-formed steel Z sections may be achieved. The experimental investigation provides test data for the development of design rules for bolted moment connections between generic lapped Z sections.

KEYWORDS

Bolted moment connections, Generic lapped sections, Roof systems, Moment resistance of connections, flexural failure.

INTRODUCTION

Cold-formed steel sections are economic building materials in construction, and they are cold-formed into sections of various shapes from steel sheets, strips, or plates. The most common shapes of the cold-formed steel sections are C sections and Z sections. The section depths generally range from 100 to 350 mm while the thicknesses range from 1.2 to 3.0 mm. Common yield strengths are 280 and 350 N/mm^2, but recently, steel with high yield strength up to 550 N/mm^2 may be found in strip products for improved structural economy. In industrial buildings, cold-formed steel sections are commonly used as secondary structural members such as purlins to support roof systems. Four different types of member configurations may be found in practical roof systems with different degrees of continuity: *(i) the single span system, (ii) the double span system, (iii) the continuous system with sleeves,* and *(iv) the continuous system with lapped sections*. The load carrying capacities of roof systems depend on many factors: section shapes, connection details and bracing conditions. Among all, the lapped system is the most common system due to simple and effective connection configurations between lapped sections at the purlin-rafter connections.

A number of research projects on the design development of modern roof systems have been reported by Fenske and Yener (1990) and Murray (1994), and a number of experimental investigations on the effects of lateral restraints on the structural performance of purlins were also reported by Willis and Wallace (1990), and Chung and St. Quinton (1996). Until recently, Ghosn and Sinno (1996) have investigated on the moment resistances of the bolted moment connections between Z sections, but no reduction in the flexural rigidities of the bolted moment connections was considered.

In general, design rules for section capacity and member resistance against buckling for cold-formed steel sections may be found in various codes of practice (AISI 1996; AS/NZ 4600 1996; Eurocode 3: Part 1.3 1996; and BS5950: Part 5 1998). However, for connection design, only the design rules for the load carrying capacity of individual fasteners are provided, and there is little technical guidance for engineers to assess the structural behaviour of bolted moment connections between lapped sections for general applications. Consequently, most modern roof systems with cold-formed steel purlins are developed through prolonged and expensive full-scale testing.

SCOPE OF WORK

In order to improve the buildability of cold-formed steel structures, a series of research and development projects was undertaken by the authors to study the structural behaviour of bolted moment connections between cold-formed steel sections. This paper aims to present the key findings of an experimental investigation on moment connections between lapped generic Z sections. A total of 12 tests on moment connections between lapped Z sections under one point load were executed to examine both the moment resistances and the effective flexural rigidities of the connections. The findings of the experimental investigation will facilitate the formulation of design rules for bolted moment connections between generic lapped Z sections. Due to severe cross-section distortion in Z sections at large deformation, the lapped connections were closely examined at large deformation in order to provide understanding of the structural performance of the lapped connections during unloading.

The research work aims to develop a set of design rules for bolted moment connections between cold-formed steel sections, in particular, for cold-formed steel purlins in roof construction. Moreover, the design rules will be complementary to existing codes of practice and suitably presented as a comprehensive procedure for design development. It is expected that the proposed procedure will shorten the conventional product development process significantly, and only limited full-scale system tests may be required for product validation.

BASIC CONFIGURATIONS OF BOLTED MOMENT CONNECTIONS

In order to enable effective bolted moment connections between lapped Z sections, a practical generic lapped connection between Z sections is proposed after considering ease of installation as follows:

- Only the webs of the Z sections are bolted together which in turn attach onto the rafters through hot rolled web cleats; section flanges are not connected.
- Six bolts per connection are used as a minimum configuration where four bolts are assigned to resist moment while the other two bolts are assigned to resist lateral load.

In general, the proposed moment connections are not able to develop full moment capacity of the connected sections due to discontinuity of load paths along section flanges in the sections.

TEST PROGRAM

In order to examine the structural performance of the lapped connections in Z sections under hogging moment in practical member and support conditions, a total of 12 generic lapped connections were carried out with two different section depths and five lap lengths. Generic lipped Z sections with two different section depths were employed, namely, Z15016 with 150 mm deep and 1.6 mm thick, and Z25025 with 250 mm deep and 2.5 mm thick; the design yield strength of both sections is 450 N/mm^2, which is designated as G450. Table 1 summarizes the test program while the section dimensions of the Z sections are illustrated in Figure 1. For each section size, a control test with continuous sections was carried out for reference.

In general, the points of inflection in continuous beams are commonly located at 0.20 to 0.25 of the span length between supports, depending on the number of spans and also the effectiveness of the connections over interior supports. Consequently, the member lengths of the test specimens, L_t, for sections Z15016 G450 and Z25025 G450 are selected to be 2.4 and 4.0 m respectively. For sections Z15016 G450 and Z25025 G450, the bolted moment connections are formed with six M8.8 bolts of 12 and 16 mm diameter respectively. A clearance of 2 mm is provided in each bolt hole for easy installation; all bolts are installed with a 50 Nm torque.

General Test Setup and Procedure

Typical set-up of the one point load tests is shown in Figure 2. All test specimens were tested in pairs with lateral restraints provided at regular intervals to avoid lateral buckling. Both the applied loads and the vertical deflections of the test specimens were recorded during loading, and the tests were terminated only after a mid-span deflection of 150 mm was recorded.

TEST RESULTS

In all tests, flexural failure of the connected section at the end of lap was critical. This failure mode was sudden, initiated by local buckling in the compression flange of the connected section, and then rapidly propagated to the section web, as shown in Figure 3. The load-deflection curves of all the test specimens are plotted in Figure 4. In general, it is shown that the deflections increased continuously even after the maximum applied load was reached. Unloading after attaining the maximum applied load is considered to be severe due to large cross-section distortion. Moreover, significant changes in slope in the load-deflections are also found in those tests with small lap lengths when compared with the section depths. Table 1 summarizes the results of the tests.

The moment resistance ratio γ_{max} is a parameter, which describes the effectiveness of a lapped connection against bending. It is defined as the ratio of the maximum moment resistance of the connection, M_{max}, to the moment capacity of the control specimen, M_{con}:

$$\gamma_{max} = \frac{M_{max}}{M_{con}}$$

For connections with sections Z15016 G450, the moment resistance ratios, γ_{max}, were found to lie between 0.88 and 1.60. For connections with sections Z25025 G450, the values of γ_{max} were found to lie between 0.79 and 1.49. Consequently, the proposed generic lapped connections between Z sections is shown to be structural efficient in providing continuity between lapped Z sections. However, it is shown in Figure 4 that after attaining the maximum applied loads, there is significant reduction in the moment resistances of the lapped sections at large deformation, i.e. the lapped connections cannot sustain their maximum moment resistances upon further deformation. Such behaviour is similar to the large deformation behaviour of a single Z section failed in flexure as the failure mode of the lapped connections is the flexural failure of the connected section at the end of lap. Consequently, the residual moment resistance ratio, γ_{res}, is also established to describe the effectiveness of a lapped connection against bending at large deformation, and it is defined as the ratio of the moment resistance of the connection, M_{res}, at large deformation to the moment capacity of the control specimen, M_{con}:

$$\gamma_{res} = \frac{M_{res}}{M_{con}}$$

The residual moment resistances of the lapped connections at large deformation may be evaluated from Figure 4. Among all tests, the residual moment resistance ratios, γ_{res}, are found to lie between 0.42 and 0.79 for connections with both sections *Z15016 G450* and *Z25025 G450*. These lapped connections are thus demonstrated to be unable to exhibit ductility at large deformation which is always required in conventional plastic hinge design.

The effective flexural rigidity ratio, α, is a parameter which describes the effective flexural rigidity of a lapped connection in comparison to that of a single section. According to the load levels, two effective flexural rigidity ratios are established as follows:

$$\alpha_i = \frac{(EI)_{test,i}}{(EI)_{con}} \quad ; \quad \alpha_f = \frac{(EI)_{test,f}}{(EI)_{con}}$$

where
$(EI)_{test,i}$ is the effective flexural rigidity of the connection at low load level;
$(EI)_{test,f}$ is the effective flexural rigidity of the connection under maximum applied load, P_{max}; and
$(EI)_{con}$ is the flexural rigidity of the control test specimen.

The effective flexural rigidities of the lapped connections are evaluated using the measured initial and final slopes of the load-deflection curves through basic structural analysis. For connections with sections *Z15016 G450*, the effective flexural rigidity ratios α_f were found to range from 0.103 to 1.212. For connections with sections *Z25025 G450*, the values of α_f were found to range from 0.223 to 1.415.

The variations in both the moment resistance ratios and the effective flexural rigidities of bolted moment connections obtained from the tests are plotted in Figure 5 for easy comparison. In general, the longer the lap length, the higher are the values of both the moment resistance ratio, γ, and the effective flexural rigidity ratio, α. It is interesting to note that for unity moment resistance ratio, i.e. '*full strength connection*', the lap length should be equal to 2 times the section depth. However, for unity effective flexural rigidity ratio, i.e. '*full stiffness connection*', the lap length should be about 5 times the section depth. This illustrates that it is easier to provide 'full strength connection' than to provide '*full stiffness connection*'.

CONCLUSIONS

In order to assess the structural performance of cold-formed steel sections with bolted moment connections for practical application on practical roof systems with lapped Z sections, a total of 12 tests were executed. In all tests, the test specimens failed in flexural failure of the connection sections at the end of lap. For test specimens with sections *Z15016 G450*, the moment resistances of the lapped connections were found to lie between 0.88 and 1.60 of the moment capacities of the control specimens. For test specimens with sections *Z25025 G450*, the moment resistances of the lapped sections were found to range from 0.79 to 1.49 of the moment capacities of the control specimens. Moreover, there is also significant enhancement in the effective flexural rigidities of the lapped sections. The test results provide important understanding and test data for the development of design rules for practical roof systems with lapped Z sections.

ACKNOWLEDGEMENTS

The research project leading to the publication of this paper is supported by the Research Grants Council of the Government of the Hong Kong Special Administrative Region (Project No. PolyU5040/99E), and the Research Committee of the Hong Kong Polytechnic University (Project No. G-W309).

References

AISI specification for the design of cold formed steel structural members. (1996). American Iron and Steel Institute.
Cold-formed steel structure code AS/NZ 4600:1996. (1996). Standards Australia / Standards New Zealand.
Eurocode 3: Design of steel structures: Part 1.3: General rules - Supplementary for cold-formed thin gauge members and sheetings, ENV 1993-1-3. (1996). European Committee for Standardization.
BS5950: Structural use of steelwork in buildings: Part 5: Code of practice for the design of cold-formed sections. (1998). British Standards Institution.
Fenske T. E. and Yener M. (1990). Analysis and design of light gage steel roof systems. *Thin-Walled Structures* **10:3**, 221-234.
Murray T. M. (1994). North American approach to the design of continuous Z- and C- purlins for gravity loading with experimental verification. *Engineering Structures* **16:5**, 337-341.
Willis C. T. and Wallace B. (1990). Behavior of cold-formed steel purlins under gravity loading. *Engineering Structures* **116:8**, 2061-2069.
Chung K.F. and St. Quinton D. (1996). Structural performance of modern roofs with thick over-purlin insulation – experimental investigation. *Journal of Constructional Steel Research.* **40:1**, 17-38.
Ghosn A. and Sinno R. (1996). Load capacity of nested, laterally braced, cold-formed steel Z-section beams. *Journal of Structural Engineering.* **122:8**, 968-971.

TABLE 1
SUMMARY OF TEST PROGRAMME AND TEST RESULTS

Section size	Test	Lap length, $2L_p$ (mm)	P_{max} (kN)	M_{max} (kNm)	γ_{max}	γ_{res}	α_i	α_f
Z15016 G450	CA018	180	27.98	16.78	0.88	0.53	0.396	0.103
	CA024	240	30.12	18.06	0.94	0.49	0.500	0.165
	CA030	300	33.92	20.36	1.06	0.51	0.729	0.316
	CA060	600	45.20	27.12	1.42	0.42	1.105	0.870
	CA090	900	51.00	30.60	1.60	0.79	1.400	1.212
	CA_con	---	31.94	19.16	1.00	0.29	1.000	1.000
Z25025 G450	CB030	300	51.90	51.90	0.79	0.43	0.334	0.223
	CB060	600	73.74	73.74	1.12	0.49	0.757	0.382
	CB090	900	86.96	86.96	1.32	0.44	1.130	0.807
	CB120	1200	99.02	99.02	1.49	0.51	1.286	1.097
	CB150	1500	95.10	95.10	1.44	0.49	1.516	1.415
	CB_con	---	66.04	66.04	1.00	0.31	1.000	1.000

All test specimens failed with flexural failure of connected section at the end of lap.

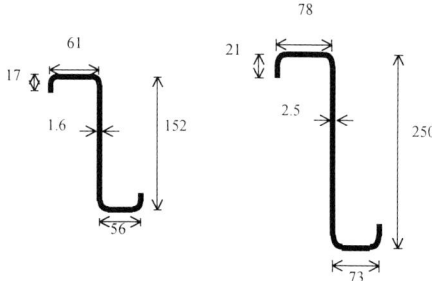

Figure 1: Section dimensions of *Z15016 G450* and *Z25025 G450*

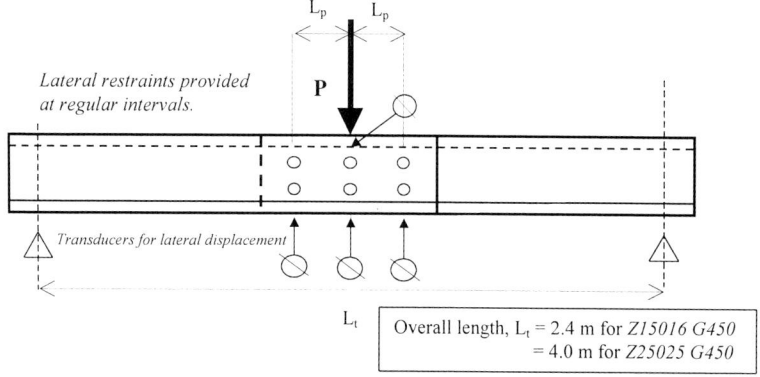

Figure 2: General set-up of one point load tests in lapped sections

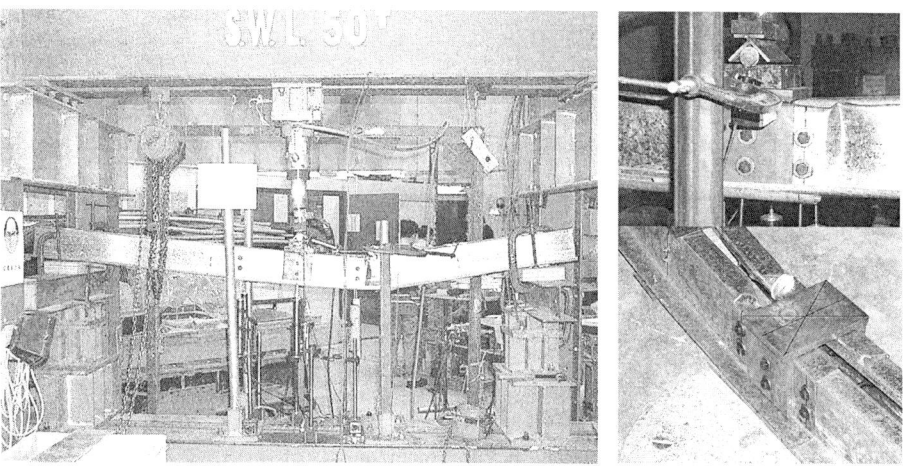

Figure 3: Typical mode of failure – *Flexural failure in connected section*

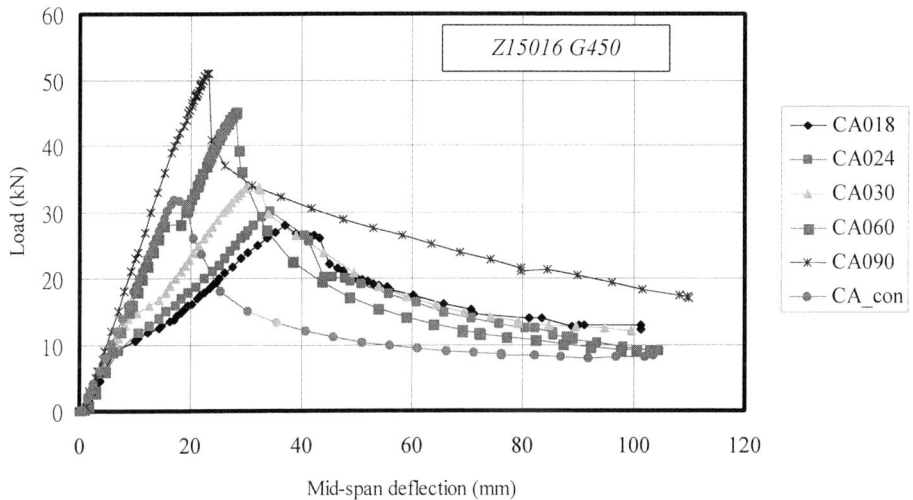

a) Test specimens with *Z15016 G450*

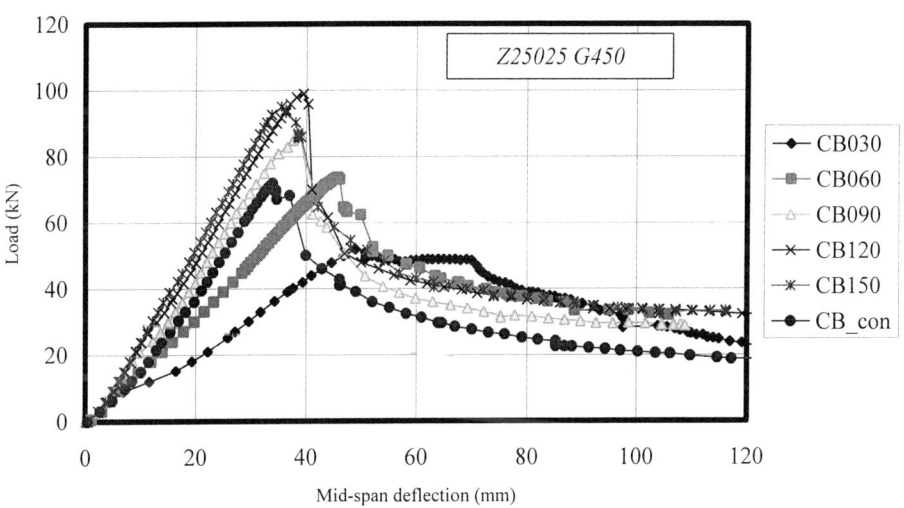

b) Test specimens with *Z25025 G450*

Figure 4 Load deflection curves of one point load tests

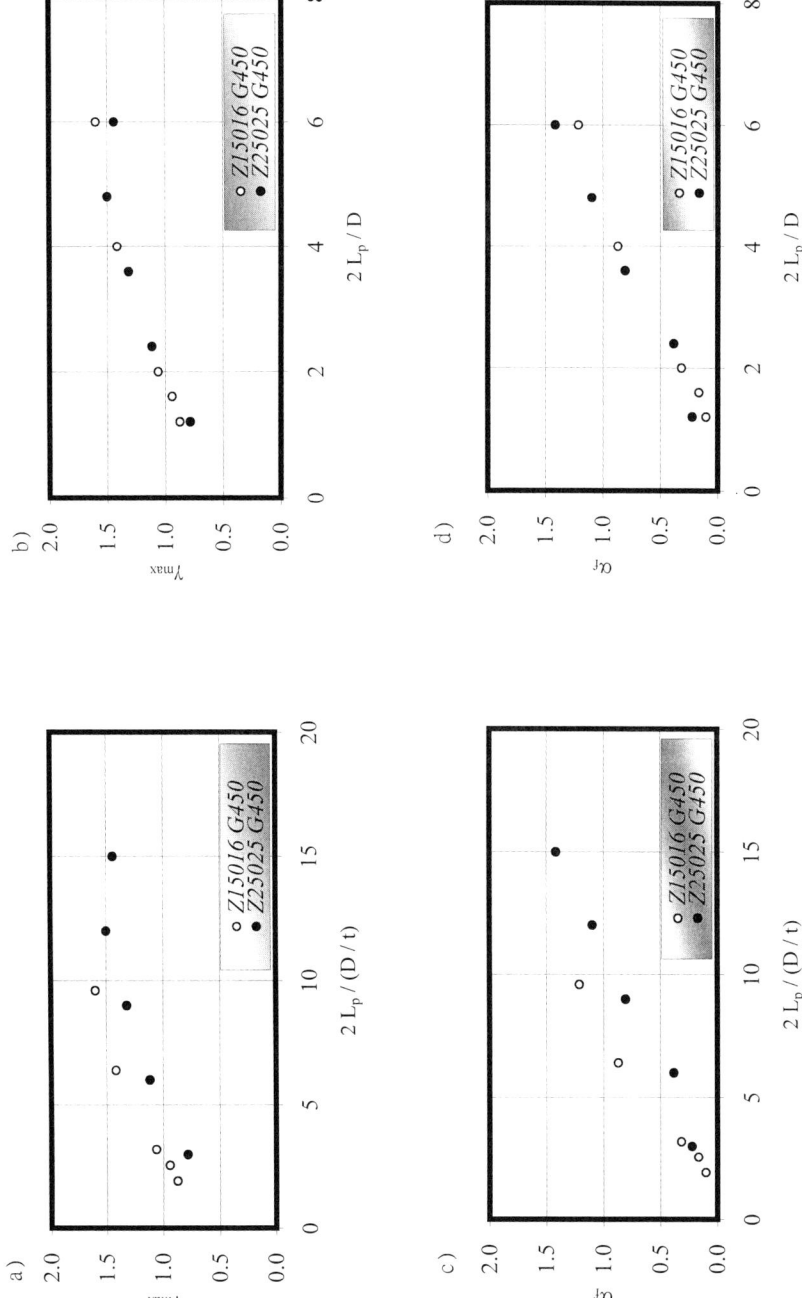

Figure 5: Variations in moment resistance ratios and effective flexural rigidities

PRACTICAL DESIGN OF COLD-FORMED STEEL Z SECTIONS WITH LAPPED CONNECTIONS

H.C. Ho and K.F. Chung

Department of Civil and Structural Engineering,
The Hong Kong Polytechnic University, Hung Hom, Hong Kong, China.

ABSTRACT

Based on the findings of an experimental investigation on generic lapped connections between cold-formed steel Z sections, it was established that due to discontinuity of load paths along section flanges of the sections, the lapped connections may not be able to develop full moment resistance nor the full flexural rigidity of the lapped sections. Consequently, reduction in both the strength and the stiffness of the lapped connections should be allowed for in practical design of cold-formed steel Z sections. This paper presents a number of design rules and expressions for multi-span cold-formed steel Z sections with lapped connections in modern roof systems under both gravity loads and wind up-lift.

KEYWORDS

Cold-formed steel purlins, Lapped connections, Effective flexural rigidity, Moment resistance ratios, Design development.

INTRODUCTION

Cold-formed steel sections are economic building materials in construction, and they are commonly used as secondary structural members such as purlins to support roof systems in industrial buildings. The most common shapes of the cold-formed steel purlins are Z sections, and the section depths typically range from 100 to 350 mm while the thicknesses range from 1.2 to 3.0 mm. Common yield strengths are 280 and 350 N/mm^2, but recently, steel with high yield strength up to 450 N/mm^2 may be found in some propriety purlins for improved structural economy. Typical spanning capacities of cold-formed steel Z sections range from 4.5 to 12 m, depending on section dimensions, steel grades, connection details and bracing configurations. As cold-formed steel sections are very weak against twisting, sag rods or bridgings are often provided during erection to prevent excessive member twisting before installation of roof claddings. Four different types of member configurations may be found in practical roof systems with different degrees of continuity: *(i) the single span system, (ii) the double span system, (iii) the continuous system with sleeves,* and *(iv) the continuous system with lapped sections.* Among all, the lapped system is the most common system due to simple and effective connection configurations between lapped sections at the purlin-rafter connections.

Although many research and development projects have been executed with extensive full-scale test programs for the product development of modern roof systems to achieve maximum structural efficiency, few test data was published. Nevertheless, a number of experimental investigations on the effects of lateral restraints on the structural performance of cold-formed steel purlins (Toma & Wittemann 1994; Chung & St. Quinton 1996; Laine 1998) were available in the literature. Design rules (Fenske & Yener 1990; Murray 1994; Epstein *et al* 1998) complementary to current codes of practice were also reported. In general, a number of design assumptions on both the effectiveness of lapped connections over supports and the effect of lateral restraints provided by roof systems and bracings are always made. In some cases, empirical expressions and data are also employed with limited ranges of applicability. It is generally considered that roof systems based on design to current codes of practice (AISI 1996; AS/NZ 4600:1996; Eurocode 3: Part 1.3 1996; BS5950: Part 5 1998) are structurally too conservative to acquire market competitiveness. A generalized and rational design method based on structural and design principles is highly desirable.

SCOPE OF WORK

In order to improve the buildability of cold-formed steel structures, a series of research and development projects was undertaken by the authors to study the structural behaviour of bolted moment connections between cold-formed steel sections. Based on the findings of an experimental investigation on generic lapped connections between cold-formed steel Z sections, it is established that due to discontinuity of load paths along section flanges of the sections, the lapped connections may not be able to develop full moment resistance nor full flexural rigidity of the lapped sections. Consequently, reduction in both strength and stiffness of the lapped connections should be allowed for in practical design of cold-formed steel Z sections. This paper presents a number of design rules and expressions for modern roof systems using multi-span cold-formed steel Z sections with lapped connections.

BASIC CONFIGURATIONS OF BOLTED MOMENT CONNECTIONS

In order to enable effective bolted moment connections between lapped Z sections, a practical generic lapped connection between Z sections is proposed after considering ease of installation as follows:

- Only the webs of Z sections are bolted together which in turn attach onto the rafters through hot rolled web cleats; section flanges are not connected.
- Six bolts per connection are used as a minimum configuration where four bolts are assigned to resist moment while the other two bolts are assigned to resist lateral load.

In general, the proposed moment connections are not able to develop full moment capacity of the connected sections due to discontinuity of load paths along section flanges in the sections.

DESIGN DEVELOPMENT

In conventional design of cold-formed steel purlins, the following assumptions are usually made:

a) Multi-span purlins are assumed to be continuous over internal supports, and thus, the moment coefficient, the shear force coefficient and the deflection coefficient derived from basic structural analysis on prismatic sections are applicable; these coefficients are taken as basic design parameters for load effects.
b) Lapped connections are assumed to be full effective, i.e. both the moment resistances and the flexural rigidities of the lapped sections are twice of those of the single sections. In some cases, the

structural enhancement due to lapping may be ignored conservatively, i.e. the lapped connections are considered merely as continuous sections without any increase in strength and stiffness.
c) The minimum lap lengths for full strength and stiffness connections are usually obtained from tests over a typical range of section depths and span lengths. The minimum lap lengths are then assumed to be applicable to all lapped connections with all span lengths.

It should be noted that while a long lap in a multi-span roof system will definitely increase the moment resistances of the lapped connections at internal supports, the increased flexural rigidity of the lapped connections, however, will inevitably attract more moments, i.e. both the moment resistances and the applied moments of the lapped connections are increased. Moreover, fixed length lapped connections will behave differently in multi-span roof systems of different spans. Consequently, it is important to quantify or even to 'engineer' the effect of lapped connections in multi-span roof systems for improved overall structural efficiency.

Effectiveness of Lapped Connections from Tests

Based upon the findings of an experimental investigation (Ho and Chung 2002) in lapped connections between cold-formed steel Z sections with different lap lengths, as shown in Figure 1, the moment resistances of the lapped connections were found to range from 0.79 to 1.60 of the moment capacities of the connected sections. Moreover, the effective flexural rigidities of the lapped connections were found to range from 0.103 to 1.415 of those of the connected sections. Thus, the lap lengths are shown to have significant effects on both the strength and the stiffness of the lapped connections, and hence, on the structural performance of the roof systems. It implies that in extreme situations, the assumptions in Items a) to c) mentioned above may under-estimate the design hogging moments while over-estimate the connection resistances at the same time, or vice versa.

Design Parameters for Lapped Span Beams under Gravity Loads

In order to allow for the presence of lapped connections in multi-span beams, a structural analysis using force method is carried out on a two lapped span beam and a four lapped span beam respectively with non-prismatic members. Figures 2 and 3 present the design coefficients for moment, shear and support reaction at critical positions of the two and the four lapped span beams respectively. Analytical expressions of the design coefficients are fully presented in Appendix. These design parameters are applicable to fully restrained beams under gravity loads with lapped connections similar to those tested in the experimental investigation. Figure 4 presents the variations of the moment coefficients for sagging and hogging moments against different lap lengths in two lapped and four lapped span beams. The moment resistances of the lapped connections obtained from the experimental investigation are also provided in Figure 4 for easy comparison. It is shown that the values of the moment coefficients differ significantly from those obtained for fully continuous beams. Moreover, for beams with different span lengths, the moment coefficients for both sagging and hogging moments differ significantly for short lap lengths although the differences diminish when the lap length increases. It should also be noted that for beams with short lap lengths, the hogging moment is reduced while the sagging moment is increased when compared with those in fully continuous beams. Depending on the support conditions and the connection configurations, there exist optimal lap lengths in beams where both the hogging and the sagging moments are equal, representing a highly desirable situation with high structural efficiency. However, additional checks against combined bending and shear on the cold-formed steel sections at the end of lap should be carried out to ensure structural adequacy of the beams.

Design Parameters for Purlins under Wind Up-lift

For purlins under gravity loads, the compression flanges are effectively restrained with roof cladding through fasteners. However, for purlins under wind up-lift, the compression flanges are usually unrestrained, and thus, lateral torsional buckling may be critical. While roof cladding is generally considered to be effective to provide partial restraints to the purlins, intermediate restraints such as sag rods and bridgings are always provided during erection to prevent excessive overall twisting during the installation of roof claddings. It is possible to utilize *solely* these intermediate restraints to reduce the member slenderness of the purlins to achieve full restraint condition. Intermediate restraints will reduce the effective lengths of the purlins, L_e, and at the same time, modify the associated bending moment envelopes for lateral torsional buckling check through the use of the design parameter C_b (AISI 1996). Both L_e and C_b in purlins under typical support conditions are presented in Table 1 for practical design. Normally, two intermediate restraints in purlins of practical section depths and span lengths will achieve full restraint condition, and the moment capacities of purlins are readily mobilized.

CONCLUSIONS

Lapped connections affect significantly the structural behaviour of multi-span cold-formed steel purlins in terms of both strength and stiffness. This paper presents a number of design rules and expressions for multi-span purlins with lapped connections in modern roof systems under both gravity loads and wind up-lift. The proposed design methods aim to improve the structural efficiency of design-based systems for increased market competitiveness at reduced product development efforts.

ACKNOWLEDGEMENTS

The project leading to the publication of this paper is supported by the Research Grants Council of the Government of the Hong Kong Special Administrative Region (Project No. PolyU5040/99E), and the Research Committee of the Hong Kong Polytechnic University (Project No. G-W309).

REFERENCES

AISI specification for the design of cold formed steel structural members. (1996). American Iron and Steel Institute.
BS5950: Structural use of steelwork in buildings: Part 5: Code of practice for the design of cold-formed sections. (1998). British Standards Institution.
Chung KF and St. Quinton D. (1996). Structural performance of modern roofs with thick over-purlin insulation – experimental investigation. *Journal of Constructional Steel Research.* **40:1,** 17-38.
Cold-formed steel structure code AS / NZ4600 : 1996. (1996). Standards Australia / Standards New Zealand.
Epstein HI, Murtha-smith E and Mitchell, JD. (1998). Analysis and design assumptions for continuous cold-formed purlins. *Practice Periodical on Structural Design and Construction.* **3:2**, 60- 67.
Eurocode 3: Design of steel structures: Part 1.3: General rules - Supplementary for cold-formed thin gauge members and sheetings, ENV 1993-1-3. (1996). European Committee for Standardization.
Fenske TE and Yener M. (1990). Analysis and design of light gage steel roof systems. *Thin-Walled Structures* **10:3**, 221-234.
Ho CH and Chung KF. (2002). An experimental investigation into lapped moment connections between Z sections. *Proceedings of the Third International Conference on Advances in Steel Structures.* December 2002, Hong Kong (in press).
Laine M. (1998). Design of steel purlins based on testing: Test methods and interpretation of test results. *Journal of Constructional Steel Research.* **46:1-3**, 189-190.
Murray TM. (1994). North American approach to the design of continuous Z- and C- purlins for gravity loading with experimental verification. *Engineering Structures* **16:5**, 337-341.
Toma T and Wittemann K. (1994). Design of cold-formed purlins and rails restrained by sheeting. *Journal of Constructional Steel Research.* **31:2-3**, 149-168.
Willis CT and Wallace B. (1990). Behavior of cold-formed steel purlins under gravity loading. *Engineering Structures* **116:8**, 2061-2069.

APPENDIX Expressions for Design Parameters

Two lapped span

$R_{2,0} = r_{2,0}\, qL$ where $r_{2,0} = \dfrac{(5+3\beta)(\alpha-1)(1-\beta)^3 + 5}{4\left[(\alpha-1)(1-\beta)^3 + 1\right]}$ (a)

$R_{2,1} = r_{2,1}\, qL$ where $r_{2,1} = \dfrac{3\left[(\alpha-1)(1-\beta)^4 + 1\right]}{8\left[(\alpha-1)(1-\beta)^3 + 1\right]}$ (b)

$M_{2,10} = m_{2,10}\, qL^2$ where $m_{2,10} = \dfrac{r_{2,1}^2}{2}$ (c) $M_{2,0} = m_{2,0}\, qL^2$ where $m_{2,0} = \left(r_{2,1} - \dfrac{1}{2}\right)$ (d)

$V_{2,1} = v_{2,1}\, qL$ where $v_{2,1} = r_{2,1}$ (e) $V_{2,0} = v_{2,0}\, qL$ where $v_{2,0} = 1 - r_{2,1}$ (f)

Four lapped span

$R_{4,1} = r_{4,1}\, qL$ where $r_{4,1} = \dfrac{1}{4} \cdot \left[\dfrac{c_{41}\alpha^2 + c_{42}\alpha + c_{43}}{c_{44}\alpha^2 + c_{45}\alpha + c_{46}}\right]$ (g)

where
$c_{41} = 32 - 204\beta + 516\beta^2 - 622\beta^3 + 297\beta^4$
$c_{42} = 204\beta - 804\beta^2 + 1126\beta^3 - 609\beta^4$
$c_{43} = 288\beta^2 - 504\beta^3 + 312\beta^4$
$c_{44} = 7 - 48\beta + 132\beta^2 - 180\beta^3 + 120\beta^4$
$c_{45} = 48\beta - 204\beta^2 + 324\beta^3 - 240\beta^4$
$c_{46} = 72\beta^2 - 144\beta^3 + 120\beta^4$

$R_{4,0} = r_{4,0}\, qL$ where $r_{4,0} = \dfrac{1}{2} \cdot \left[\dfrac{c_{47}\alpha^2 + c_{48}\alpha + c_{49}}{c_{44}\alpha^2 + c_{45}\alpha + c_{46}}\right]$ (h)

where
$c_{47} = 13 - 96\beta + 276\beta^2 - 378\beta^3 + 237\beta^4$
$c_{48} = 96\beta - 420\beta^2 + 666\beta^3 - 465\beta^4$
$c_{49} = 144\beta^2 - 288\beta^3 + 228\beta^4$

$R_{4,2} = r_{4,2}\, qL$ where $r_{4,2} = \left(2 - r_{4,1} - \dfrac{r_{4,0}}{2}\right) qL$ (i)

$M_{4,21} = m_{4,21}\, qL^2$ where $m_{4,21} = \dfrac{r_{4,2}^2}{2}$ (j) $M_{4,1} = m_{4,1}\, qL^2$ where $m_{4,1} = r_{4,2} - \dfrac{1}{2}$ (k)

$M_{4,10} = m_{4,10}\, qL^2$ where $m_{4,10} = \left(2 - \dfrac{r_{4,0}}{2}\right) \cdot r_{4,2} + r_{4,1}\left(1 - \dfrac{r_{4,0}}{2}\right) - \dfrac{(4-r_{4,0})^2}{8}$ (l)

$V_{4,2} = v_{4,2}\, qL$ where $v_{4,2} = r_{4,2}$ (m) $V_{4,1}^{\,-} = v_{4,1}^{\,-}\, qL$ where $v_{4,1}^{\,-} = 1 - r_{4,2}$ (n)

$V_{4,1}^{\,+} = v_{4,1}^{\,+}\, qL$ where $v_{4,1}^{\,+} = r_{4,1} - v_{4,1}^{\,-}$ (o) $V_{4,0} = v_{4,0}\, qL$ where $v_{4,0} = r_{4,0}/2$ (p)

α   is the effective flexural rigidity of lapped connection;
β   is the lap length coefficient where $2L_p = 2\beta L$; $2L_p$ is the total lap length and L is the span length
q   is the applied uniformly distributed load.

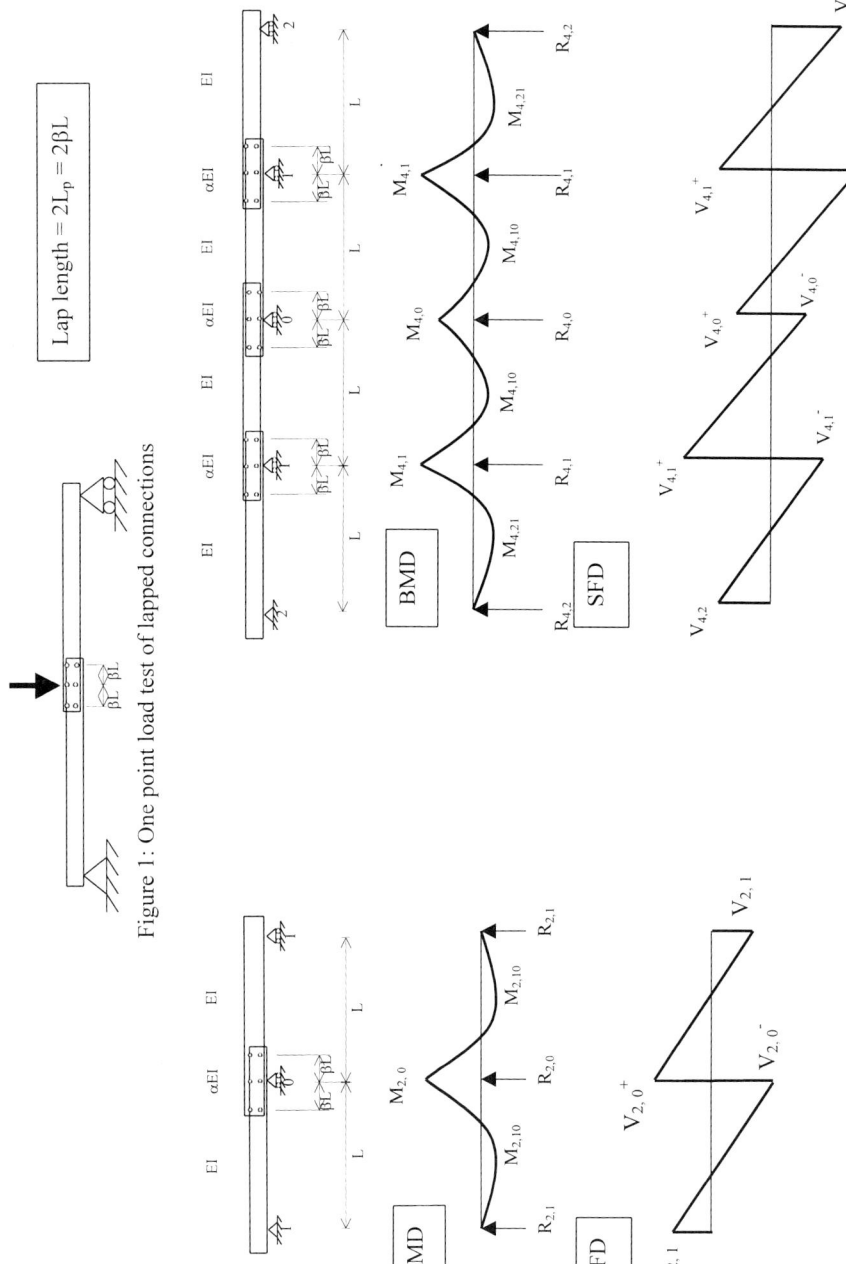

Figure 1: One point load test of lapped connections

Figure 2: Analysis of two lapped span purlin system

Figure 3: Analysis of four lapped span purlin system

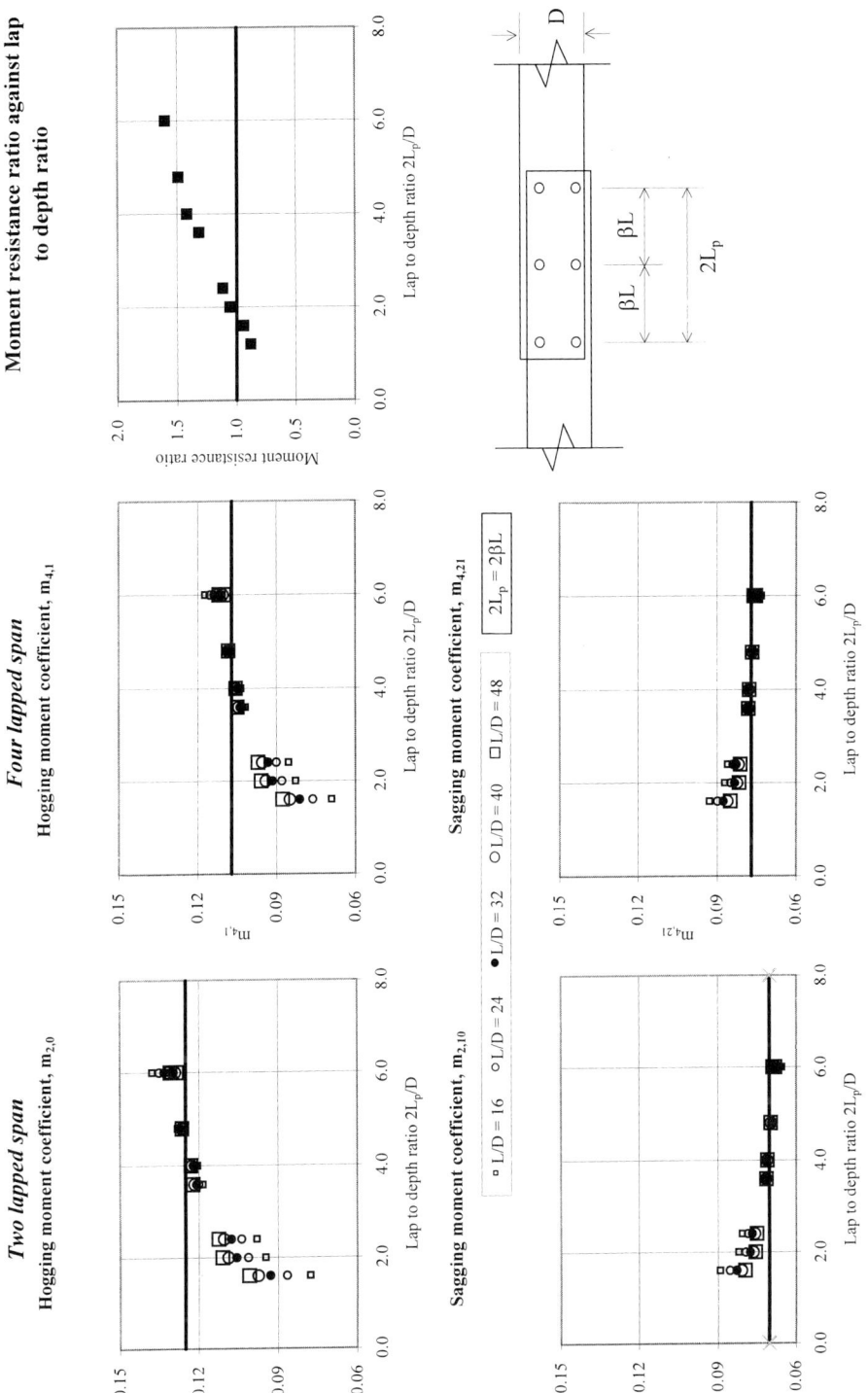

Figure 4: Variations of moment coefficients against lap length

Table 1: C_b FOR DIFFERENT CASES UNER WIND UPLIFT

Single span	Bending moment diagram	Four continuous span – End span	Bending moment diagram
1. No intermediate restraint $L_e = L$; $C_b = 1.13$		1. No intermediate restraint $L_e = 0.79L$; $C_b = 1.14$	
2. One intermediate restraint $L_e = 0.50L$; $C_b = 1.30$		1. One intermediate restraint $L_{e1} = 0.40L$; $C_b = 1.29$ $L_{e2} = 0.39L$; $C_b = 1.29$	
3. Two intermediate restraints $L_{e1} = 0.35L$; $C_b = 1.75$ $L_{e2} = 0.30L$; $C_b = 1.01$		2. Two intermediate restraints $L_{e1} = 0.34L$; $C_b = 1.37$ $L_{e2} = 0.26L$; $C_b = 1.05$ $L_{e3} = 0.19L$; $C_b = 1.53$	

Two continuous span	Bending moment diagram	Four continuous span – Internal span	Bending moment diagram
3. No intermediate restraint $L_e = 0.75L$; $C_b = 1.14$		1. No intermediate restraint $L_e = 0.54L$; $C_b = 1.14$	
4. One intermediate restraint $L_{e1} = 0.45L$; $C_b = 1.21$ $L_{e2} = 0.30L$; $C_b = 1.40$		2. One intermediate restraint $L_{e1} = 0.23L$; $C_b = 1.37$ $L_{e2} = 0.31L$; $C_b = 1.23$	
5. Two intermediate restraints $L_{e1} = 0.33L$; $C_b = 1.36$ $L_{e2} = 0.28L$; $C_b = 1.07$ $L_{e3} = 0.14L$; $C_b = 1.56$		3. Two intermediate restraints $L_{e1} = 0.10L$; $C_b = 1.56$ $L_{e2} = 0.26L$; $C_b = 1.04$ $L_{e3} = 0.18L$; $C_b = 1.47$	

DESTRUCTIVE MECHANISM OF LARGE SPAN COLD-FORMED SECTION ROOF TRUSS

Guo Yaojie, Li Kun and Du Xinxi

School of Civil Engineering, Wuhan University, Wuhan 430072, China

ABSTRACT

In order to obtain the information of ultimate bearing capacity and destructive characteristics of the cold-formed section roof truss, the test of a full-scale roof truss is carried out. The size and fabricated conditions of the testing truss are all as same as the actual one. On the basis of the test results, the destructive mechanism of the truss is analysed and the improved suggestions of design and fabrication for this type of truss are presented. Ultimate bearing capacity is larger than design load-carrying capacity. The enhanced strength is due partly to the effect of cold-forming. The strength should be utilized to save steel. The cold-forming residual stresses and the sub-stresses influence considerably on the local buckling of plate elements, which may not only cause the instability of members earlier but also damage the joints. The width-to-thickness ratio of the plate elements in compression chords should be strictly controlled. For the failure of joints is not prior to the collapse of members, the connection plates between the ends of web members and the welding surfaces of chords are necessary. The cold-forming and welding residual stresses exist simultaneously in the members, but the former plays an important part in local buckling of flanges of the compression chord. Fabricating eccentricity between the end sectional center of web members and the axis of chord should be small as far as possible to avoid the considerable torsional deformation in the chord.

KEYWORDS

steel roof truss, cold-formed section, destructive mechanism, ultimate bearing capacity, mechanical properties, strains distribution, local buckling, design suggestions

1 INTRODUCTION

There is an industrial building which overall length and width are 216m and 2×30m respectively. The distances between columns along the left and right side-row are 6m, along the mid-row columns 12m. Two roof trusses are symmetrically placed along the ridge line and simply supported on both the mid-row columns and left or right side-row columns. The top chord of each roof truss is inclined with a slope of 1:20 as shown in Fig.1. The span of roof truss is 30m. Purlins and desks are cold-formed steel shapes and profiled sheets respectively. The vertical design load applied at each loading joint of this roof truss is 20kN. These roof trusses are connected in the work site and each member is made of two cold-formed lipped channels connected by welding lip to lip.

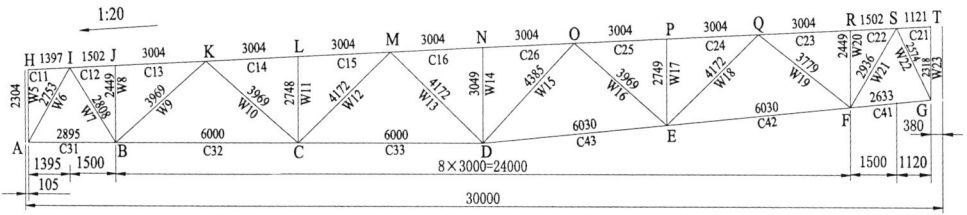

Fig.1 Serial Number and Length of Members

2 OUTLINE OF TEST

2.1 Basic Data of Members and Joints

The serial number of the members and their length are shown in Fig.1. There are four cross sections used in this truss. Their plate thickness and web depth are 4.5mm and 100mm respectively. But their flange widths are different, which of top and bottom chords are 160mm and 140mm respectively, and that of web members are 100mm for W5, W10, W18, W20 and W23 and 120mm for other web members. The lip depths of all cold-formed channels used to make above members are 20mm.

Every joint is marked by a letter as shown in Fig.1. In actual joints of this truss, the connection plates are placed between the welding surfaces of the chords and web members ends, but the ends of the W11, W13, and W17 are directly welded on the surfaces of the top chord at joints L, N and P respectively. In addition, there is a considerable eccentricity between the end sectional center of member W11 and the axis of the top chord at joint L.

2.2 Loading Programme and Lateral Braces

For the sake of convenience and safety, the reversal loading programme is adopted in the testing, i.e. the truss is reversed and its top chord is level as shown in Fig.2. Nine joint-loads are loaded by nine synchronous oil jacks and their loading directions are set with a slant of 1:20 for imitating the actual direction. Before loading, the self-weight of the truss is temporarily supported by two hand jacks set under the two end joints of the top chord. After loading, the supporting reactions of the truss are transmitted from the fix hinge and sliding hinge to bearing structures.

In order to assure both the lateral stability and only permissible the top chord instability in the plane of the truss, the lateral braces install at top chord joints expect both joints I and S. But the mid-joint D of the bottom chord is also braced laterally, which is helpful to the truss subjected to external loads during testing.

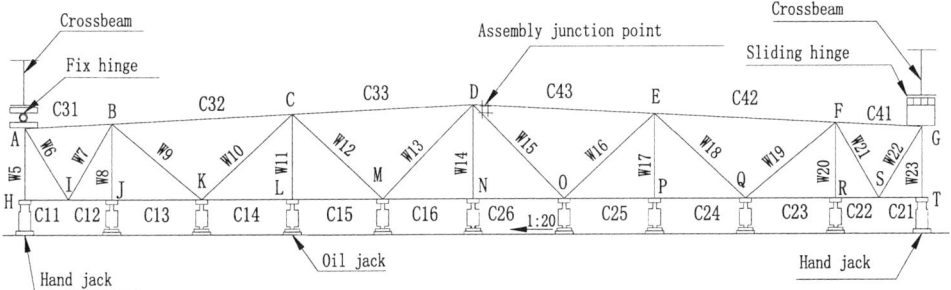

Fig.2　Reversal Loading Programme

3　TEST RESULTS

3.1 Steel Mechanical Properties

The coupons are cut from the web plates of cold-formed lipped channels used in the top chord of the truss. The average values of the main mechanical properties obtained from four standard tension tests are that the yield strength $F_y = 274 N/mm^2$, ultimate tension strength $F_u = 356 \, N/mm^2$, the elongation $\delta = 30\%$ and the modulus of elasticity $E = 226000 \, N/mm^2$. The measured value of modulus of elasticity is usually more 10% than the used value. The reasons are still not clear.

For making coupons, the cold-formed lipped channels are cut into strips and the bending occurs meanwhile in every strip. It shows that the cold-forming residual stresses locked into the lipped channels have been released [1]. Further more, residual stresses are tensile on the outside surface of the lipped channel section, and compressive on the inside surface. Their magnitudes are approximately equal, but they are considered to be varied linearly through the plate thickness, which is far from hot residual stresses and can influence on local and overall buckling as well as the safety of joints in the truss.

3.2 Strains

The strains are not uniformly distributed on the sections. As an example, the strain distribution of the section located at the right side of joint K at a distance 400mm is shown in Fig.4. When the joint load is no more than 20kN, average axial strains of members C16, C26, C33 and C43 in the chords and web member W6 are relatively close to theoretical values calculated by using the coefficients of axial forces as shown in Fig.3. When the joint load is exceeded 20kN, these measured values are greater than the theoretical values, which shows that inelastic strains have occurred in those sections.

When the joint load is increased to 55kN, average axial stains in above mentioned sections are 1989, 1958,1662 and $1926×10^{-6}$ for chord members C16, C26, C33 and C43 respectively, and $716×10^{-6}$ for web member W6.

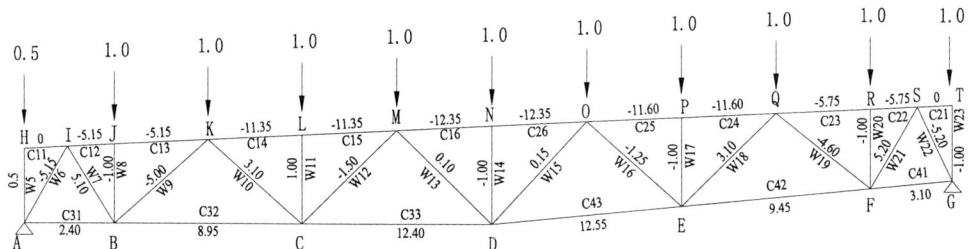

Fig.3 Coefficients of Axial Forces

When joint load is exceeded 55kN and closed to 60kN, some strains are growing up more and more, which shows that this truss was closed to be failed. Unfortunately, there are no resistance gauges bonded on the destructive member and joints.

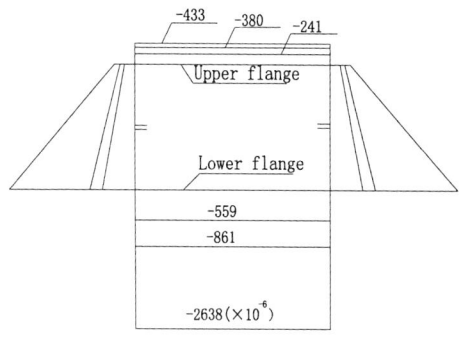

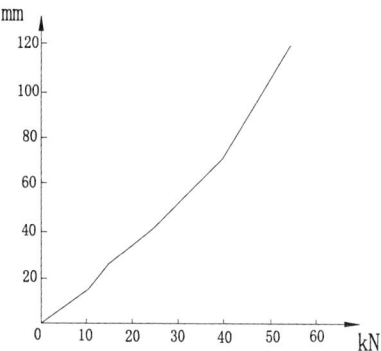

Fig.4 Distribution of Axial Strains

Fig.5 Mid-Span Deformation

3.3 Displacements

The vertical displacement of entire truss is shown in Fig.6. The deflection of the mid-joint N is approximately proportional to the joint load as shown in Fig.5. The deflection value is equivalent to one of a solid web beam simply supported on its ends and subjected to uniform distribution load and provided that the section stiffness is replaced by 0.75EI, where the moment of inertia I is computed by using top and bottom chord sectional areas with respect to the midpoint of their distance from center to center in the mid-span of this truss.

The relative displacement differences between adjacent joints are very little in the middle five joints but larger in two side eight joints, for example, the values between joints P and Q, and joints L and K are 16.9mm and 17mm respectively.

When joint load is increased to 50kN, the midpoints of top chord members C14 and C24 cave to 13mm and 16mm relative to the lines connected from their own ends respectively. Along with the joint load increase further the caving speed of member C24 becomes fast strikingly and destruction happens in this member.

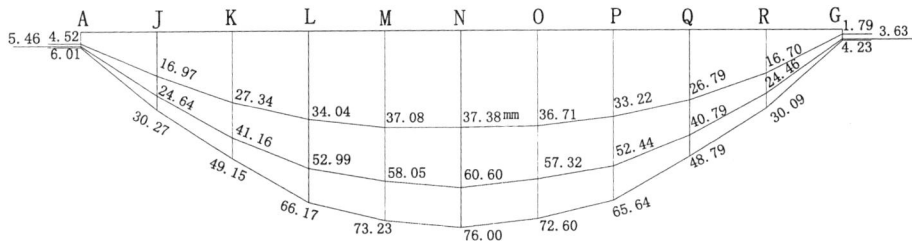

Fig.6 Vertical Displacements of Joints (Joint Load=20kN,30kN and 37kN)

Besides, because of the fabricated eccentricity between the end sectional center of web member W11 and the axis of top chord at joint L, both this joint load and above mentioned eccentricity generate a torsion moment, therefore, members C14 and C15 occur apparent torsional deformation which may affect its use and safety.

3.4 Destructive Characteristics and Ultimate Load-Carrying Capacity

When joint load is loaded to 60kN, only a few seconds, the truss is destroyed in the order of local buckling of the plate element nearby joint P, and the settlement of joint P or the end of web member W17 is compressed into the top chord flange, and the compression buckling of member C24 in the plane of the truss is found.

These events happen almost in the same time. It is certain that the destruction of this truss starts from the local buckling and failure of the joint rather than the buckling of the member. The damaged location is acted by all stresses including axial stress, residual stress and sub-stress (or the second order stress induced by the stiffness of joints in actual truss and the displacement difference between adjacent joints as the truss loading) rather than only axial stress.

Owing to the fact that the loading programme was reversal, the ultimate joint load 60kN involved the joint load 2.8kN induced by the truss self-weight, therefore, actual joint load is 2.85 times larger than the design load-carrying capacity in this test.

4 ANALYSIS OF DESTRUCTIVE MECHANISM

According to the popular truss design method, it is assumed that the joints are ideal hinges and the members are only subjected to axial stress. In fact, there are more or less sub-stress in the members as mentioned above. When the panel length to the sectional depth ratios in the chords are greater than 10, the sub-stresses are less and can be ignored in designing the truss of hot-rolled steel shapes. However, when designing the truss of cold-formed members, because the plate elements of cold-formed sections may be locally buckled and the maximum of cold-forming residual stresses may reach to $0.5F_y$[1], the whole stress including the sub-stress, cold-forming residual stress and axial stress may not only cause the buckling of the plate element prior to the member earlier but also damage the joint. This is the reason why the destruction of the testing truss starts from local buckling of the plate element and failure of the joint.

The design width-thickness ratio of the top and bottom flange in the top chord of the testing truss is 35.6. If cold-forming residual stress is $0.5F_y$, the elastic width-thickness ratio of the flange will be greater than 35.6 and the value is increased along with the increase of the joint load. Because of reaching critical value, the local buckling of the top chord flange nearby the joint P occurs first. Moreover, the connection plate between the end of web member W11 and the welding surface of top chord at joint P is omitted, the end of web member W11 may be suddenly compressed into the top chord flange, the joint P is equivalent to be settled abruptly in the same time, so that the displacement difference between joints P and Q is increased to a greater value.

Under the co-action of all compression stress including sub-stress, axial stress and cold-forming residual stress, the most pessimistic section nearby joint Q is squashed and the member C24, like a member hinged between joint P and the squashed section nearby joint Q, is buckled at once.

The axial stresses of the member C16 and C26 are slightly greater than that of the member C14 and C24, which can be seen from Fig.3, but the displacement difference between the ends of the former is less than that of the latter. Owing to the settlement of joint P, this value of C24 is greater than that of C14, and so do the sub-stress.

5 CONCLUSIONS AND SUGGESTIONS

Ultimate capacity is 2.85 times larger than design load-carrying capacity in the test and relative deflection of the mid span in the truss is 1/500 as the joint load is 30kN. It is obvious that the design seems too conservative. The enhanced strength resulted from the effect of cold-forming exceeds $0.1F_y$ in accordance with Chinese Standard(GBJ18-87)[2]. This strength should be utilized for steel saving.

The cold-forming residual stress and the sub-stress are of considerable influence on the local buckling of plate elements, which may not only cause the instability of members earlier but also damage the joints. If joints are assumed as ideal hinge in designing this type of trusses, the width-thickness ratio of the plate element in compression chord should be chosen carefully and controlled strictly.

It is hardly justifiable in design that the failure of the joint occurs before the member collapsed. For improving the stress states of the plate elements connected with web members, and preventing web member ends prone to be compressed into chord at joints and giving assurance that the failure of joints is not prior to the collapse of members in the truss, the connection plates between the ends of web members and the welding surfaces of chords are needed and justified.

The cold-forming and welding residual stresses exist simultaneously in the sections made of two cold-formed lipped channels connected by welding lip to lip, but the cold-forming residual stress plays an important part in the local buckling of flanges of compression chord, and the welding residual stress is higher tensile stress on the middle of web plates and lower compressive stress on top and bottom flanges of the sections. Especially, the cold-forming residual stress is worthy to be noted and to be studied further.

Fabricating eccentricity between the end sectional center of web members and the axis of chords should be small as far as possible to avoid the considerable torsional deformation of the chords. When the eccentricity exceeds the permissible error in fabrication, it should be done over again.

References

1 Weng. C. C. and Pekoz, T. (1988). Residual Stress in Cold-Formed Steel Members. Research Report, Cornell University, Ithaca, New York, 1-10.

2 Chinese Standard. (1987). Technical Code of Cold-Formed Thin-Wall Steel Structures (GBJ18-87), Beijing, China

Appendix —Notation

E = modulus of elasticity
F_y = yield strength
F_u = ultimate strength
I = moment of inertia
δ = elongation in a 2-inch gage length

SWAY BUCKLING OF DOWN-AISLE PALLET RACK STRUCTURES CONTAINING SPLICES

R G Beale and M H R Godley

Department of Civil Engineering & Construction Management, Oxford Brookes University,
Oxford, OX3 OBP UK

ABSTRACT

This paper presents an efficient approach to the determination of the buckling loads of down-aisle, spliced, unbraced, pallet rack structures subjected to vertical and horizontal loads. Pallet rack structures are analysed by considering an equivalent free-sway column and setting up the stability equations. The effects of semi-rigid beam-to-upright, splice-to-upright and base-plate-to-upright connections are fully incorporated into the analysis. Each section of upright between successive beam levels in the pallet rack is considered to be a single column element with two rotational degrees of freedom. A computer algebra package was used to determine modified stability equations for column elements containing splices. The influence of the position of splices in a pallet rack is clearly demonstrated. The results from the program are compared with a finite element analysis of a full frame with excellent agreement.

KEYWORDS

Cold-formed steel, pallet racks, design, stability, buckling

INTRODUCTION

Pallet rack structures are used in factories and warehouses for the storage of palletised goods. Such structures often have a large number of bays and beam levels. The connections between base-plates and uprights and between uprights and beams are usually semi-rigid. A typical example of such a structure is shown in Figure 1.

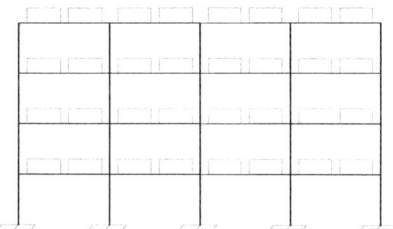

Figure 1: Typical beam and column arrangement

With increasing experience in use this type of structure is being constructed with more storeys and greater storey heights. As the rack gets higher it is often necessary to include splices in the uprights as the lengths of uprights required cannot be manufactured and processed in one piece and also for efficient design upright sections can be varied throughout the height of the rack.

In the design of such structures consideration must be given to the elastic stability of the racks. Previous investigations into rack stability have been made by several authors. Davies (1980 and 1992) analysed the down-aisle stability by considering a single upright model carrying both vertical and horizontal loads. His model took into account semi-rigid joints between beams and uprights and at the base of the rack. However, the model only allowed for column flexibility below the second beam level. As racks become more slender this approximation becomes less accurate. Lewis (1991) produced a simple rigid plastic model for a rack with pinned connections at the base which also assumed that bending distortion within the upright was negligible. The latest code by the Federation Europeanne de la Manutention (2000) requires a second order non-linear analysis to be undertaken in the design of the structures. The code also requires that splices are treated as semi-rigid joints.

The authors in a series of papers, Feng et al (1993), Godley et al (2000), Beale and Godley (2001), have developed a single column model which accurately predicts the behaviour of regular pallet racks subjected to horizontal and vertical loads. However, in common with all simplified models known to the authors, the original model is unable to analyse structures containing splices in the uprights. This paper derives the revised stability equations for this case and discusses the influence splices have upon the overall stability of the rack structure.

STRUCTURAL MODEL

The beam-upright pallet-rack system shown in Figure 1 can be treated as a free sway structure with uniform loads acting on the beams. Pallet rack systems normally consist of many bays, each identically loaded with the same beams and columns throughout. A single column structural model can therefore be used to analyse the rack. The loads from the beams can be equivalently applied to the centre of the line of the upright as a concentrated load. The rotational stiffness of the upright due to the beams can be represented by a rotational stiffness of the beam-to-upright joints. A single column structural model for this unbraced framework is given in Figure 2, where k_0 and k_i ($i=1,2,...,n$) are the rotational stiffnesses of the semi-rigid upright-base-plate and beam-to-upright joints respectively. k_{splice} represents the rotational stiffness of a splice. There can be several splices in an upright. The assumption is made that all uprights in a pallet rack are spliced at the same vertical heights.

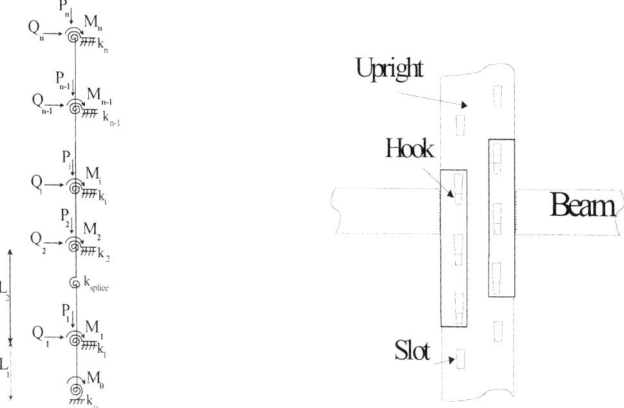

Figure 2: Structural Model   Figure 3: Typical Beam-to-Upright Connection

The rotational stiffness of a beam-upright connection can be shown to be (Godley et al 2000)

$$k_i = \frac{1}{\frac{l}{12EI_{bi}} + \frac{1}{4k_{li}} + \frac{1}{4k_{ri}}} \quad (1)$$

where l is the width of each bay in the rack, I_{bi} the second moment of area of the beam at level i, and k_{li} and k_{ri} the rotational stiffnesses of the beam-to-upright joints. As joints in pallet racks are often formed by hooks and slots (See Figure 3) they may have different rotational characteristics at the right and left hand ends of beams.

DERIVATION OF THE ELEMENT EQUATIONS

The general structural model shown in Figure 2 will apply to all regular pallet rack structures. To determine the buckling load of the sway structure the shear forces applied to the rack are zero. In this case the standard slope deflection equations for an element without a splice are

$$M_{ab} = i\theta_a \frac{v}{\tan v} - i\theta_b \frac{v}{\sin v} \quad (2)$$

and

$$M_{ba} = -i\theta_a \frac{v}{\sin v} + i\theta_b \frac{v}{\tan v} \quad (3)$$

(Horne & Merchant (1965)) where M_{ab} and M_{ba} are the bending moments at each end of the element and θ_a and θ_b the corresponding rotations at each end of the element. $i = \frac{EI}{L}$ and $v = \sqrt{\frac{P}{EI}}L$ where EI is the flexural rigidity of the element and P the axial load within the element.

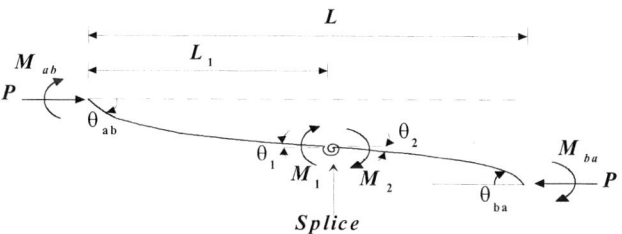

Figure 4: Column element containing splice

Figure 4 shows a single column element containing a splice distance L_1 from the left hand end. The rotational stiffness of the splice is k_{splice} and the moments on either side of the splice connection are respectively M_1 and M_2. The corresponding slope deflection equations for the two upright segments are

$$M_{ab} = i\theta_a \frac{v_1}{\tan v_1} - i\theta_1 \frac{v_1}{\sin v_1} \quad (4)$$

$$M_1 = -i\theta_a \frac{v_1}{\sin v_1} + i\theta_1 \frac{v_1}{\tan v_1} \quad (5)$$

for the sub-element below the splice and

$$M_2 = j\theta_2 \frac{v_2}{\tan_2} - j\theta_{ba} \frac{v_2}{\sin v_2} \qquad (6)$$

$$M_{ba} = -j\theta_2 \frac{v_2}{\sin v_2} + j\theta_{ba} \frac{v_2}{\tan v_2} \qquad (7)$$

for the sub-element above the splice where θ_1 and θ_2 are the rotations on either side of the splice. In this case $i = \frac{EI_1}{L_1}$, $j = \frac{EI_2}{L-L_1}$, $v_1 = \sqrt{\frac{P}{EI_1}} L_1$ and $v_2 = \sqrt{\frac{P}{EI_2}}(L-L_1)$ where I_1 and I_2 are the second moments of area on the column elements on the two sides of the splice.

At the splice continuity requires that

$$M_1 + M_2 = 0 \qquad (8)$$

$$\theta_1 - \theta_2 = \frac{M_2}{k_{splice}} = -\frac{M_1}{k_{splice}} \qquad (9)$$

Equations (4)-(9) contain the variables $M_{ab}, M_{ba}, M_1, M_2, \theta_{ab}, \theta_{ba}, \theta_1$ and θ_2. The standard slope deflection equations (2) and (3) contain $M_{ab}, M_{ba}, \theta_{ab}$ and θ_{ba}. Using the computer algebra package Mathcad (Mathsoft Inc 2001) we can eliminate M_1, M_2, θ_1 and θ_2 to produce the following equations:

$$M_{ab} = iv_1 \frac{\left(k_{splice}\left[\cos(v_1)\cos(v_2)jv_2 - \sin(v_1)\sin(v_2)iv_1\right] - v_1 v_2 ij \cos(v_2)\sin(v_1)\right)\theta_{ab} - k_{splice} jv_2 \theta_{ba}}{v_1 v_2 ij \cos(v_1)\cos(v_2) + k_{splice}\left[iv_1 \sin(v_2)\cos(v_1) + jv_2 \sin(v_1)\cos(v_2)\right]} \qquad (10)$$

$$M_{ba} = jv_2 \frac{\left(k_{splice}\left[\cos(v_1)\cos(v_2)iv_1 - \sin(v_1)\sin(v_2)jv_2\right] - v_1 v_2 ij \cos(v_1)\sin(v_2)\right)\theta_{ba} - k_{splice} iv_{12}\theta_{ab}}{v_1 v_2 ij \cos(v_1)\cos(v_2) + k_{splice}\left[iv_1 \sin(v_2)\cos(v_1) + jv_2 \sin(v_1)\cos(v_2)\right]} \qquad (11)$$

These modified equations are inserted into the column equations whenever an element using a splice is encountered.

Special cases of the equations can be derived in cases where the splice is adjacent to a beam-column intersection and when the splice has zero rotational stiffness. According to the FEM code (Federation Europeene de la Manutention (2000)) the latter case must be assumed whenever there is no experimental data for the splice. In the former case, using Mathcad we can simplify equations (10) and (11) to get the modified equations:

$$M_{ab} = k_{splice} jv_2 \frac{\cos(v_2)\theta_{ab} - \vartheta_{ba}}{jv_2 \cos(v_2) + k_{splice} \sin(v_2)} \qquad (12)$$

and

$$M_{ba} = jv_2 \frac{k_{splice}\left[\cos(v_2) - jv_1 \sin(v_1)\right]\theta_{ab} - k_{splice}\vartheta_{ba}}{jv_2 \cos(v_2) + k_{splice} \sin(v_2)} \qquad (13)$$

In this case we have given the equations for the case where the splice is considered to be at the bottom of the upper element. Length L_1 has been taken to be zero. A similar set can be derived for the case where the splice is considered to be at the top of the lower element. This latter case, however, yields lower buckling loads as can be seen in example 2 below.

The corresponding formulae for zero stiffness splices (effectively pinned connections in the middle of an element) are

$$M_{ab} = -iv_1 \tan(v_1)\theta_{ab} \qquad (14)$$

and

$$M_{ba} = -jv_2 \tan(v_2)\theta_{ba} \qquad (15)$$

GLOBAL EQUATIONS

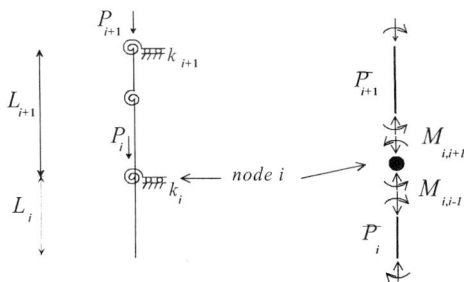

Figure 5: Compatibility and element equilibrium

Figure 5 shows the relationship between each element. The axial load acting on any element is the sum of the axial loads applied on each element above so that

$$\overline{P}_i = \sum_{j=i}^{n} P_j \qquad (16)$$

Moment equilibrium at a joint between a beam and upright implies that

$$M_{i,i+1} + M_{i,i-1} = -k_i\theta_i \qquad (17)$$

as the two column elements have the same rotation θ_i at the connection.

At the bottom (level 0) and top (level n) of the column the moment equilibrium equation reduces to

$$M_{0,1} = -k_0\theta_0 \qquad (18)$$

$$M_{n,n} = -k_n\theta_n \qquad (19)$$

Using equation (16) with the appropriate set of moment equations (2,3) or (10,11), or (12,13) or (14,15) depending upon the existence of a splice in a given column element and the type of splice we produce a tri-diagonal equation at node i of the form

$$c_{i,i-1}\theta_{i-1} + c_{i,i}\theta_i + c_{i,i+1}\theta_{i+1} = 0 \qquad (20)$$

Summing up equations (19)-(21) produce the system of equations:

$$[C]\{\theta\} = 0 \qquad (21)$$

where $\{\theta\}^T = \{\theta_0, \theta_1, \theta_2, \ldots, \theta_n\}$ and

$$\{C\} = \begin{bmatrix} c_{0,0} & c_{0,1} & & & & & \\ c_{1,0} & c_{1,1} & c_{1,2} & & & & \\ & & \vdots & & & & \\ & & c_{i,i-1} & c_{i,i} & c_{i,i+1} & & \\ & & & & \vdots & & \\ & & & & c_{n-1,n-2} & c_{n-1,n-1} & c_{n-1,n} \\ & & & & & c_{n,n-1} & c_{n,n} \end{bmatrix} \quad (22)$$

Equation (22) is a homogeneous linear matrix equation. The non-trivial solution for θ_i corresponds to the buckling equation. Hence the buckling load P_{cr} is given by

$$\det[C]_{P=P_{cr}} = 0 \quad (23)$$

BUCKLING LOAD ALGORITHM

The coefficients of equation (22) are transcendental functions of the load P. Expanding equation (23) by the Gauss elimination procedure gives

$$\prod_{i=0}^{n} c_{i,i}^* = 0 \quad (24)$$

where

$$c_{i,i}^* = c_{i,i} - c_{i-1,i} \frac{c_{i,i-1}}{c_{i-1,i-1}} \quad (i = 1, 2, \ldots, n) \quad (25)$$

The equation has $n+1$ roots representing the $n+1$ buckling loads. However, only the fundamental mode is required for structural design. The critical buckling load of a fixed ended column is given by $(2\pi)^2 EI/L^2$. The fundamental critical load of the column must therefore satisfy the inequality

$$0 < P_{cr} \leq \max\left\{\frac{(2\pi)^2 EI_i}{L_i^2}\right\} \quad (i = 1, 2, \ldots, n) \quad (26)$$

A fast algorithm to determine the buckling load is:

(i) Determine an upper bound to the critical load by finding the maximum value of equation (26).
(ii) Divide the interval between 0 and the maximum into 100 and evaluate the determinant, equation (23), increasing P from zero until a change in sign of the determinant is obtained.
(iii) Use the method of bisection to refine the buckling load to the required accuracy.

EXAMPLES

Example 1: A single column with three beam levels

The height to the first beam level was 1500mm and the two other storey heights were each 1200mm. $E = 210000$N/mm^2. Each storey carried a load of 4000N. The splice was placed 2m from the ground between the first and second beam levels. The base-plate/upright rotational stiffness was $1.2 \cdot 10^8$ Nmm/rad. The second moment of area of the upright below the splice was 600000mm^4 and above the splice 500000mm^4. The beam-to-upright rotational stiffness below the splice was 7.0·10^{8}Nmm/rad and above the splice was 6.0·10^{8}Nmm/rad. Note that although the same beam would normally be used throughout a pallet rack that as the beam is being connected to a smaller upright the structural characteristics of the beam end connector would be different in each case.

The rotational stiffness of the splice, called k_{splice} was varied from 0Nmm/rad to 1.0·10^{8}Nmm/rad covering the full range of splice conditions from a pinned to a rigid splice. The buckling load factors obtained from the analysis are presented in Table 1. The program is called Pallet. The results are compared with a finite element solution obtained using the Lusas program (FEA Ltd 2001) and show excellent agreement.

TABLE 1: COMPARISON OF BUCKLING LOAD FACTORS BETWEEN PALLET AND LUSAS

k_{splice} (Nmm/rad)	pinned	1.0·10^6	1.0·10^7	9.0·10^7	5.0·10^8	rigid
Pallet	7.087	7.135	7.396	7.683	7.737	7.750
Lusas	7.087	7.135	7.397	7.683	7.738	7.751

For a rotational stiffness less than 20000Nmm/rad the splice can be considered as pinned and for a rotational stiffness above 5.0·10^{8}Nmm/rad the connection is effectively a rigid connection.

Example 2: Effects of splice near to beam level

To investigate the effects on buckling load factors of splices near to beam levels the position of the splice in Example 1 was moved to be either just above, or just below, the first beam level. In this case the stiffness of the splice was 1.0·10^{6}Nmm/rad. The program was able to smoothly converge to the limiting values as the splice moved closer to the beam level. The values of the limiting buckling load factors are shown in Table 2.

TABLE 2: INFLUENCE OF SPLICE POSITION ON BUCKLING LOAD FACTOR

	Below beam level	Above beam level
Pallet	4.4915	6.1420
Lusas	4.4912	6.1419

Agreement between the two programs is again excellent with the results showing the influence of a weaker beam-column connection when the splice is below the beam level. The program assumes that splices and beam-column connections are infinitesimally small. In practice, however, they have a finite size, typically of the order of 200-300mm. For design purposes a splice would always be joined to the heavier section.

Example 3: Frame example

Figure 6 shows a typical pallet rack frame with a splice at mid-height.

The second moment of area of the upright below the splice is 7.0·10^{5}mm^4 and 6.0·10^{5}mm^4 above the splice. The second moment of area of the beams is 5.5·10^{5}mm^4. E = 2.1·10^{5}N/mm^2. The base-plate rotational stiffness is 9.0·10^{7}Nmm/rad. The beam-column rotational stiffness is 7.0·10^{7}Nmm/rad below the splice and is 5.0·10^{7}Nmm/rad above the splice. The rotational stiffness of the splice is 8.0·10^{7}Nmm/rad. Using equation (1) the equivalent beam-upright rotational stiffness below the splice is 1.1·10^{7}Nmm/rad and 8.3696·10^{7}Nmm/rad above the splice. The structure consists of five uprights with corresponding base-plates, forming 4 bays. Therefore the effective second moment of area of a single column in the model is given by (5/4)·7·10^{5}mm^4 = 8.75·10^{5}mm^4 below the splice and

$(5/4) \cdot 6 \cdot 10^5 \text{mm}^4 = 7.5 \cdot 10^5 \text{mm}^4$ above the splice. Similarly as there are 5 base-plates the effective base-plate stiffness is given by $(5/4) \cdot 9.0 \cdot 10^7 \text{Nmm/rad} = 1.125 \cdot 10^8 \text{Nmm/rad}$. For 5 splices the effective splice stiffness is $(5/4) \cdot 8.0 \cdot 10^7 \text{Nmm/rad} = 1.0 \cdot 10^8 \text{Nmm/rad}$.

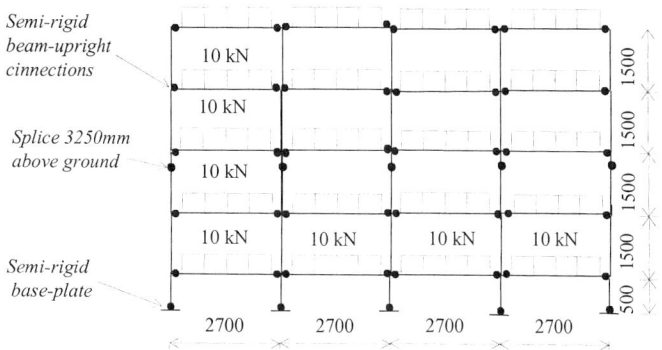

Figure 6: Example Frame

The buckling load factor from the Pallet program was 2.554 and from Lusas was 2.539. The difference is only 0.6% and is within the error margin of the finite element method.

CONCLUSIONS

The paper has presented an efficient procedure for the analysis of pallet racks containing splices. The position of splices close to beam-column intersections is shown to be critical in the buckling load of a frame.

The use of computer algebra packages has been shown to be an efficient procedure in the derivation of element stiffness matrices enabling complex algebraic manipulations to be easily carried out. It also enabled the limiting cases of zero stiffness splices and splices close to beam-column intersections to be determined.

REFERENCES

Beale R G and Godley M H R, (2001). "Problems arising with pallet rack semi-rigid base-plates", in *Proc. 1st Int. Conf. On Steel & Composite Structures*, Pusan, pp699-706

Davies J.M. (1980), "Stabilities of Unbraced Pallet Racks", in *Proc. 5th Int. Speciality Conf. On Cold-Formed Steel Structures*, St. Louis, 409-28

Davies J.M. (1992), "Down-aisle Stability of Rack Structures", in *Proc. 11th Int. Speciality Conference on Cold-Formed Steel Structures*, St. Louis, 417-435.

FEA Ltd., (2000), "Lusas User Guide Version 13"

Federation Europeenne de la Manutention, Section X, (2000*), Recommendations for the Design of Steel Static Pallet Racking and Shelving.*

Feng X., Godley M.H.R. & Beale R.G. (1993), "Elastic Buckling Analysis of Pallet Rack Structures", in *Developments in Structural Computing*, Edinburgh, 207-305.

Godley M.H.R., Beale R.G. and Feng X. (2000), "Analysis and Design of down-aisle pallet rack structures", *Comp. and Struct.*, 77(4), 391-401.

Horne M.R. and Merchant W. (1965), "The Stability of Frames", Pergamon

Lewis G.M. (1991), "Stability of Rack Structures", *Thin-Walled Struct.*, 27, 163-174.

Mathsoft Inc, (2001), "Mathcad User's Guide with Reference Manual"

COMPOSITE CONSTRUCTION

COMPOSITE ACTION IN NON-COMPOSITE BEAMS

R. Seracino and D.J. Oehlers

Department of Civil and Environmental Engineering, The University of Adelaide,
South Australia, 5005, Australia

ABSTRACT

Slab-on-girder bridges were a common form of bridge construction prior to the 1960s. In the design of these structures, it was assumed that there is no interaction between the steel beams and concrete deck as no mechanical form of shear connection was used. Due to the age of these structures, and recent increases in allowable live loads, load tests have been carried on several non-composite slab-on-girder bridges in order to assess their remaining strength. The load tests show that there is a measurable degree of interaction present and that the remaining strength is greater than anticipated. Several factors not accounted for in current assessment techniques contribute to the unintentional composite action observed. Due to the absence of mechanical shear connectors, the only mechanism available to transfer the longitudinal shear forces between the steel and concrete components is friction along the interface. The results of simply supported slab-on-girder beam tests were used to isolate interface friction from the other factors contributing to the reserve strength observed. This has lead to the development of a new mathematical assessment technique referred to as the Non-Composite Mixed Analysis Fatigue Approach that is used to quantify the effect of interface friction on the remaining fatigue strength or endurance of these structures.

KEYWORDS

partial interaction; non-composite; interface friction; slab-on-girder beams

INTRODUCTION

Non-composite slab-on-girder bridges have no mechanical shear connection at the steel-concrete interface. The original design therefore assumed that the steel and concrete components acted independently and often the strength of the section was taken as the strength of the steel girders alone. Ignoring the possibility of interaction between the two components allowed for a simple and conservative design approach. Fifty years later, the allowable load limits have increased and many slab-on-girder bridges are approaching the end of their anticipated design lives and hence, evaluation of these types of bridges is becoming increasingly common. Current analytical techniques still assume that there is no interaction between the steel and concrete components (Burdette & Goodpasture 1988) and in some cases, this leads to retrofitting or replacement of bridges that would be deemed adequate had unintentional composite action been taken into consideration. Although full scale bridge testing is sometimes employed to more accurately assess the behaviour of existing bridges, this option is very expensive, time consuming and disruptive. Even once a full-scale test has been performed, and a degree of composite action identified, there is not yet any precise way of dealing with the information since the characteristics of unintentional composite action have not been thoroughly investigated (Roberts et al 1997). Nowak & Tharmabala (1998) and Burdette & Goodpasture (1988) recommended that efforts should be made to improve analytical methods that account for the increase in strength of non-composite bridges due to unintentional composite action.

Full-scale bridge testing has consistently shown that there is a large amount of reserve strength in existing non-composite bridges (Nowak & Tharmabala 1998, Burdette & Goodpasture 1988). Research by Nowak & Tharmabala (1998) identified that this reserve strength is due to: 1) structural analysis; 2) strength of materials; and 3) interaction of components. Burdette & Goodspasture (1988) identified unintentional composite action as one of the primary reasons for the large reserve strength of slab-on-girder bridges and it was concluded that the unintentional composite action is caused by chemical bonding and friction at the steel-concrete interface. In addition, it was found that the chemical bond is broken early in the life of non-composite bridges and hence, will have little effect on its fatigue behaviour and need not be considered. Furthermore, bridge testing indicates that unintentional composite action has little effect on the flexural capacity of the bridge (Bakht & Jaeger 1992), but has significant influence under serviceability loads (Burdette & Goodpasture 1971). This paper deals with the increased fatigue strength or endurance of non-composite beams due to friction at the interface. Although frictional effects may be small from the original design point of view, in some cases non-composite bridges are only marginal with respect to structural adequacy and would only need a small increase in theoretical capacity to permit their continued use (Roberts et al 1997). As the fatigue life of a non-composite beam is a function of the range of stresses, it is necessary to understand the range of strain distributions that can exist.

STRAIN DISTRIBUTIONS

Figure 1 illustrates typical strain distributions for a steel-concrete beam subject to flexure depending on the analysis technique used. It is assumed that the materials remain linear-elastic under fatigue loading hence, the stresses are directly proportional to the strains. The no interaction strain distribution occurs when there is no shear connection and the steel and concrete components act independently. Consequently, two neutral axes occur, one each at the centroids of the steel and concrete components. Alternatively, a full interaction analysis assumes that there is no slip at the interface and hence, there is a continuous strain distribution through the section with only one neutral axis at the centroid of the composite section. Therefore, the no interaction and full interaction strain distributions define the range of possible strain distributions for a composite beam for any degree of interaction. Hence, the realistic partial interaction strain distribution must lie within the full and no interaction bounds (Seracino et al 2001) as shown in Figure 1. For any degree of interaction, except full interaction, there is a discontinuity in the strain distribution at the steel-concrete interface known as the slip-strain, ds/dx, which is a measure of the interface slip.

The presence of frictional forces at the interface of slab-on-girder beams provides the mechanism for interaction between the two components. Hence, the partial interaction slip-strain and curvature due to interface friction, ϕ_{fric}, is less than that given by no interaction theory. Therefore, the flexural stresses for a given applied bending moment are less than those predicted by a no interaction analysis which ultimately results in a reduced stress range in the steel and concrete and a corresponding increase in the residual fatigue strength or endurance. The effect of friction on the behaviour of a non-composite beam can therefore be evaluated by quantifying the partial interaction slip-strain or curvature. The proposed mathematical model is developed in the following sections.

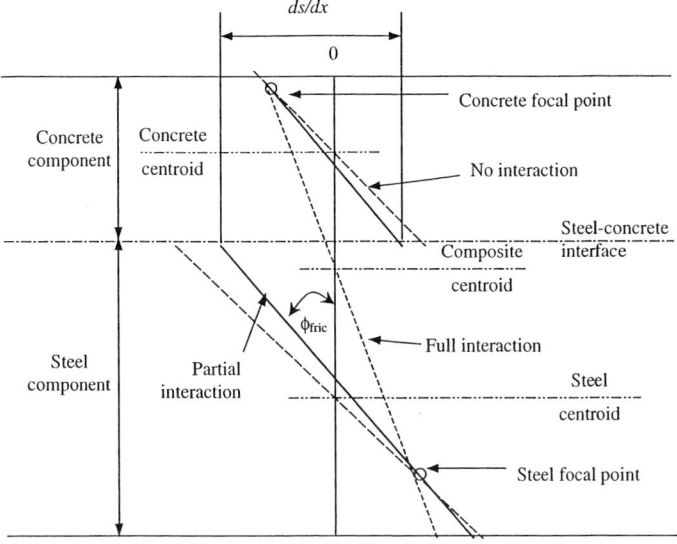

Figure 1: Strain distributions under flexure

PARTIAL INTERACTION SLIP STRAIN ALLOWING FOR INTERFACE FRICTION

If an assessment of an existing non-composite bridge is to be made, it is important that the concrete deck be inspected to determine the extent of cracking so that an appropriate concrete flexural rigidity can be utilised in theoretical calculations. Moses et al (1989) also recommends the use of site-specific data to improve predictions of the remaining life of existing bridges.

In developing the analytical procedure to estimate unintentional composite action, a model developed for limited-slip-capacity shear connectors by Oehlers & Sved (1995) was used. The limited slip capacity approach is referred to as a mixed analysis because it models the rigid plastic behaviour of the shear connectors, while the steel and concrete components remain linear elastic. This technique has been modified to suit the behaviour of non-composite beams at fatigue loads where the steel and concrete components behave elastically, while the frictional capacity at the interface is achieved. The modification involved replacing the total strength of the shear connectors in the shear span, P_{sh}, with the frictional capacity of the shear span, F_{fricsh}, which is a product of the coefficient of friction, μ, and the normal force across the interface, N

$$F_{fricsh} = \mu N \tag{1}$$

where μ may be taken as 0.7 based on experimental testing performed by Singleton (1985) to investigate cyclic friction between steel and concrete surfaces. The total normal force is defined as the sum of the normal force from the dead load of the concrete slab, N_{DL}, and the point load, N_{PL}.

In order to simplify the calculation of F_{fricsh}, it is assumed that the frictional resistance due to the point load is uniform throughout the shear span. This simplification, illustrated in Figure 2, was shown to be appropriate for the study of interface friction in composite beams (Oehlers et al 2000).

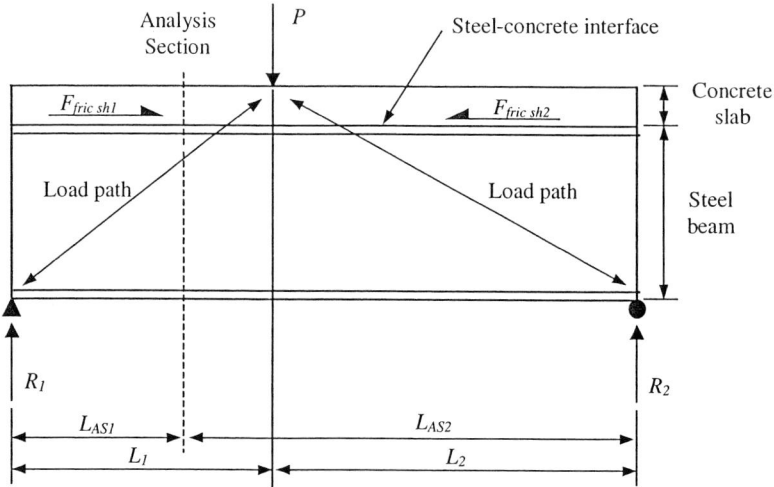

Figure 2: Assumed frictional force distribution for an analysis section

The normal force due to the point load for span i of a particular analysis section, N_{PLi}, is a proportion of the vertical shear acting within the shear span. For example, for spans 1 and 2 in Figure 2, the normal force is given by Equations (2) and (3) respectively

$$N_{PL1} = \frac{L_{AS1}}{L_1} R_1 \qquad (2)$$

$$N_{PL2} = R_2 - \frac{(L_1 - L_{AS1})}{L_1} R_1 \qquad (3)$$

where L_{ASi} = distance between the support and the analysis section for span i; L_i = length of shear span i; R_i = support reaction for shear span i.

The normal force due to the dead load of a particular analysis section for span i, N_{DLi}, is given by

$$N_{DLi} = w_{DL} L_{ASi} \qquad (4)$$

where w_{DL} = uniformly distributed load due to the concrete dead load.

As the frictional resistances of the spans to either side of the analysis section will in general be different, the frictional force for the particular analysis section is determined using the minimum of N_1 and N_2. In addition to the uniform distribution of normal force assumption, it is also assumed that the frictional resistance F_{fricsh} is constant regardless of the slip, which is consistent with the rigid plastic assumption used in the original mixed analysis approach.

The partial interaction strains, stresses and resultant forces at an analysis section allowing for interface friction is shown Figure 3 (Oehlers & Sved 1995). The stress profile illustrates that under fatigue loading, the steel and concrete elements are assumed to be linear elastic. Equations (5) to (7) define the mathematical model for the slip strain due to the effects of interface friction in non-composite beams. Equation (5) assumes that the steel and concrete components have the same curvature and hence, there is no separation at the steel-concrete interface, which is reasonable based on the experimental testing that was performed (Seracino & Hocking 2002)

$$\frac{M_c}{(EI)_c} = \frac{M_s}{(EI)_s} = \frac{1}{R} \qquad (5)$$

where M = bending moment; (EI) = flexural rigidity; R^{-1} = curvature; and the subscripts c and s represent the concrete and steel components respectively. The total applied moment M is resisted by three components given by

$$M = M_s + M_c + F_{fricsh}(h_s + h_c) \qquad (6)$$

where h_c = distance between the steel-concrete interface and the centroid of the concrete element; and h_s = distance between the steel-concrete interface and the centroid of the steel element. Hence, the slip strain can be calculated using the following

$$\frac{ds}{dx} = \left(\frac{h_c}{R} - \frac{F_{fricsh}}{(EA)_c}\right) - \left(\frac{F_{fricsh}}{(EA)_s} - \frac{h_s}{R}\right) \qquad (7)$$

where $(EA)_c$ = axial rigidity of the concrete element; and $(EA)_s$ = axial rigidity of the steel element.

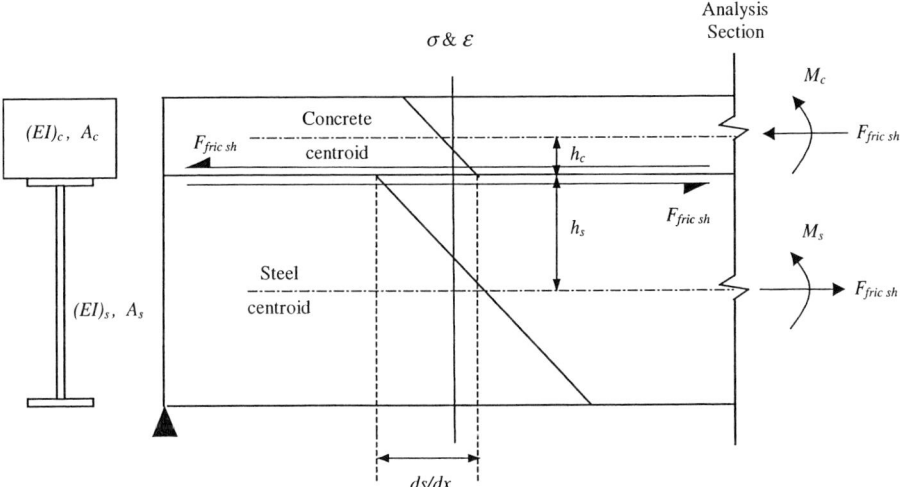

Figure 3: Analysis section of a non-composite beam

Validation with Experimental Results

The approach for determining the partial interaction slip strain presented in the previous section was validated by comparing with experimental data (Seracino & Hocking 2002). For reference, the theoretical no interaction slip strains including the concrete contribution are also compared.

Figure 4 shows typical results from these analyses where the slip-strain has been calculated using the proposed model (Eqs 5-7), no interaction theory and experimental results, for an analysis section assuming a fully cracked concrete section. The moment shown is that at the analysis section due to loading at various positions along the length of the beam based on the experimental testing. The proposed model shows an almost uniform reduction in slip-strain from the no interaction boundary condition of around 10% for a large range of moments and load locations. The conservative results are due to the fully cracked concrete assumption when in fact the slab exhibited only minor cracking.

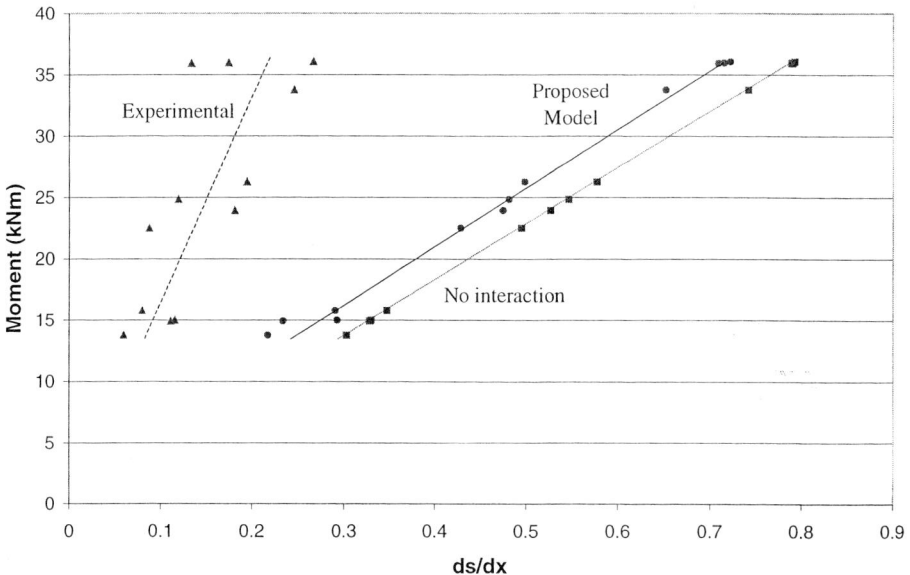

Figure 4: Variation of slip strain (assumed fully cracked $(EI)_c$)

THE NON-COMPOSITE MIXED ANALYSIS FATIGUE APPROACH

This section describes a procedure referred to as the Non-Composite Mixed Analysis Fatigue Approach that can be used to predict the more realistic partial interaction stress distribution in slab-on-girder beams allowing for interface friction so that a more accurate assessment of the remaining fatigue strength or endurance can be made.

The first step is to calculate the curvature, $\phi_{fric} = R^{-1}$ shown in Figure 1, which is found by solving Eqs (5) and (6) simultaneously. Then, in order to locate the strain distribution, so that the corresponding stresses may be determined, a point along the distribution must be known. The strain distribution can be located knowing the focal point in either the steel or concrete component (Seracino et al 2001), shown in Figure 1, which are the intersection points of the full and no interaction strain distributions. Seracino et al. (2001) showed that all partial interaction strain distributions pass through these focal points and that their location is constant for a specific cross-section, regardless of the degree of interaction.

As the strain, hence stress distribution can now be determined at an analysis section for any moment, an influence line diagram can be calculated from which the stress ranges can be found and the principle of superposition can be used to determine stress ranges for any combination of point loads representing a standard fatigue vehicle. At this point, the resulting increase in fatigue strength or endurance of a non-composite bridge due to interface friction will depend upon local design standards.

CONCLUSIONS

Load tests consistently show that slab-on-girder bridges exhibit composite action. The results from experimental testing confirm load test results and show that there is a definite reduction in slip-strain and curvature due to interface friction. It is important to note, however, that the material properties used in the fatigue assessment of an existing bridge should be determined carefully. In particular, the extent of concrete cracking should be determined so that an appropriate flexural rigidity can be used.

A new mathematical procedure referred to as the Non-Composite Mixed Analysis Fatigue Approach was developed. The basis of this approach was taken from a method developed to model the limited slip capacity of shear connectors in composite beams. In order to adapt the model to suit non-composite beams, the distribution of normal force at the steel-concrete interface is assumed to be uniform over the shear span and the strength of the shear connectors is replaced by the frictional resistance of the interface using a lower bound coefficient of friction. This approach was verified against experimental data and the theoretical boundary condition of no-interaction. Finally, the focal point concept is used to locate the strain distribution so that the stresses, stress ranges and remaining fatigue strength or endurance may be determined.

The Non-Composite Mixed Analysis Fatigue Approach provides engineers with a new assessment tool. If the remaining strength or endurance is still found to be insufficient after applying this new technique, a full scale load test could be undertaken to allow for the other factors contributing to the reserve strength, which are more difficult to quantify mathematically.

REFERENCES

Bakht B. and Jaeger L.G. (1992). Ultimate load test of slab-on-girder bridge. *Journal of Structural Engineering* **118:6,** 1608-1624.

Burdette E.G. and Goodpasture D.W. (1988). *Correlation of bridge load capacity estimates with test data*. National Cooperative Highway Research Report 306, Transportation Research Board, National Research Council, Washington, D.C.

Burdette E.G. and Goodpasture D.W. (1971). *Final report on full-scale bridge testing: An evaluation of bridge design criteria*. Office of Research and Planning Tennessee Highway Department and The Department of Transportation Federal Highway Administration.

Moses F., Schilling C.G. and Surya K.R. (1989). Reliability-based bridge life assessment. Structural Safety and Reliability: Proceedings of ICOSSAR '89, *The 5th International Conference of Structural Safety and Reliability*, New York, USA, International Association for Structural Safety & Reliability.

Nowak A.S. and Tharmabala T. (1998). Bridge reliability evaluation using load tests. *Journal of Structural Engineering* **114:10,** 2268-2279.

Oehlers DJ, Seracino R. and Yeo MF. (2000) Effect of friction on shear connection in composite bridge beams. *Journal of Bridge Engineering*, ASCE **5:2,** 91-98.

Oehlers D.J. and Sved G. (1995). Composite Beams with Limited-Slip-Capacity Shear Connectors. *Journal of Structural Engineering* **121:6,** 932-938.

Roberts W.S., Lake N.J. and Heywood R.J. (1997). *Investigation of the load capacity and degree of composite action of Salt Creek Bridge*. Report No. 97926a, Infratech Systems & Services, South Australia.

Seracino R. and Hocking P. (2002). The effect of interface friction on slab-on-girder beams. *17th Australasian Conference on the Mechanics of Structures and Materials*, Gold Coast, Australia, 12-14 June, Swets and Zeitlinger Publishers (Accepted for Publication).

Seracino R., Oehlers D.J. and Yeo M.F. (2001). Partial-interaction flexural stresses in composite steel and concrete bridge beams. *Engineering Structures* **23**, 1186-1193.

Singleton W.M. (1985). *The transfer of shear in simply supported composite beams subject to fatigue loading*. MSc Thesis, National University of Ireland, Department of Civil Engineering, University College, Cork, Ireland.

EFFECT OF CONCRETE INFILL ON NON-COMPACT TUBES SUBJECTED TO PURE BENDING

Andrew Wheeler and Russell Bridge

Centre for Construction Technology & Research,
College of Science, Technology & Environment, University of Western Sydney,
Locked Bag 1797, South Penrith DC, NSW 1797, Australia

ABSTRACT

The use of concrete filled thin walled steel tubes in structural applications is widespread particularly for axially loaded members. While extensive experimental studies have been carried out on tubes subjected to primarily axial loads, some with limited eccentricities, few studies have been carried out into the flexural behaviour of such members. In this paper the results from a testing program carried out at the University of Western Sydney will be presented. The program is designed to demonstrate the effect of the concrete in-fill on the flexural behaviour of the tubes, and will be used as a basis for developing design rules for concrete filled tubes subjected to flexure. Parameters investigated include the diameter to wall thickness ratio and concrete strength. Comparisons are made with identical tests carried out on bare steel tubes.

KEYWORDS

Concrete Filled Tubes, Bare Steel Tubes, Flexural Strength, Local Buckling, Experimental Study.

INTRODUCTION

The use of concrete filled steel tubes in flexural loading situations utilises the compressive strength of the concrete while the steel tube provides the tensile capacity. This method of construction provides an economical alternative to traditional construction methods and uses continue to increase. Traditionally concrete filled tubes have been used in columns where the loads are predominantly axial loads, resulting in extensive documented research into the axial behaviour of concrete filled tubes (O'Shea & Bridge, 2000). Some testing on flexural loading of concrete filled tubes has been presented (Lu, and Kennedy 1994). However these tests used rectangular sections. Nakamura et al, 2002 conducted a small number of tests on large diameter (600 mm) concrete filled circular tubes for use as flexural members for composite bridges. To increase the test data base for flexurally loaded large diameter (>300 mm) concrete filled tubes and enable better utilisation of concrete filled tubes, an investigation into the flexural behaviour of concrete filled tubes is currently being carried out at the Centre for Construction Technology and Research, University of Western Sydney. In this investigation thin-walled tubes filled with concrete are subjected to pure bending and observations made on the behaviour of the section.

In obtaining a full understanding of the behaviour of concrete filled tubes subjected to flexural loading, it is essential to have a reference datum of the bending capacity of the thin-walled tube. Previously Schilling (1965), Sherman (1976) and Elchalakani et al, (2002) have conducted flexural tests on thin-walled tubes. These studies indicated that initial imperfections within the tube have a detrimental effect on the bending strength of steel tubes. Numerical studies into the flexural behaviour of thin-walled tubes with initial imperfections have been conducted as part of the investigation at the University of Western Sydney (Wheeler et al. 1999, Wheeler & Bridge, 2000) in which the assumption that the initial imperfections have an effect on the flexural behaviour was verified. While a number of experimental studies into the flexural behaviour of circular hollow sections have been carried out (Elchalakani et al, 2002, Sherman, 1976 and Schilling, 1965), little data on the measured magnitude and exact shape of imperfections in steel tubes has been reported. Schilling (1965) machined the sections, prior to testing, to remove any significant imperfections, thus negating the issue of imperfection on bending strength. For the experimental work carried out by Sherman (1976) and Elchalakani et al, (2002), no details on measured initial imperfection of the tubes tested were reported.

In this paper, the results are reported of tests on a number of tubes with measured initial membrane imperfections (Wheeler and Pircher, 2002) subjected to flexural loading. These sections include both bare steel and concrete filled tubes. Further information regarding the test procedure and the results may be found in Wheeler (2002).

SPECIMEN DETAILS

This investigation examined the behaviour of concrete filled thin-walled circular hollow sections. The specimens were selected from commercially available circular hollow sections. Limitation on available section size and practical test sizes resulted in two tube sizes being selected for the test program. The nominal section sizes and D/t ratios are shown in Table 1. Further details on the sections can be found in the Design Capacity Tables for Structural Steel Hollow Sections (AISC, 1992). The nominal yield stress of the circular hollow sections is 350 MPa with the slenderness λ, calculated according to AS 4100 where $\lambda = (D/t)(f_y/250)$ also provided in Table 1. The sections were ccold-formed with seamless welds, manufactured to the requirements of AS 1163 (SA, 1991a).

TABLE 1
Nominal Section Sizes

Diameter (*D*) mm	Thickness (*t*) mm	D/t ratio	Slenderness ratio (λ)
406	6.4	63.4	88.8
456	6.4	71.3	100.0

A total of four (4) tests were performed on concrete filled circular hollow sections, two for each tube size. An additional two (2) tests were carried out on bare steel sections (no concrete infill). The bare steel tests were carried out to enable comparison of the bending capacity of the concrete filled tubes with that of the bare steel circular hollow sections.

The specimens tested in this series of tests were provided from the OneSteel Market Mills Kembla Grange manufacturing plant. For each tube size, three 4 m specimens were taken from the single 12 m tube supplied, ensuring constant material properties for constant tube size. The specimen numbers, size, filled/unfilled, and concrete strength are presented in Table 2.

TABLE 2
Specimen Details

Specimen No.	Diameter (D) (mm)	Concrete filled	Concrete Strength (MPa)
TPB001	406	No	–
TPB003	406	Yes	40
TPB004	406	Yes	55
TPB002	456	No	–
TPB005	456	Yes	48
TPB006	456	Yes	56

To ensure that the applied load was distributed at the loading and support points, a saddle system was used. The saddles consisted of a vertical loading plate, shaped to fit the cross-section. The saddles shown in Figure 1 consisted of a 20 mm plate profiled to fit half the cross-section positioned vertically. A 40 x 3 mm strap of steel was placed around the profile to minimise local effects during load transfer. The saddles were fastened to the tube using an additional strap placed around the tube and fastened firmly to the saddle as shown in Figure 1. This system was used to ensure an even loading pattern across the section and to prevent ovalisation at the loading and support points.

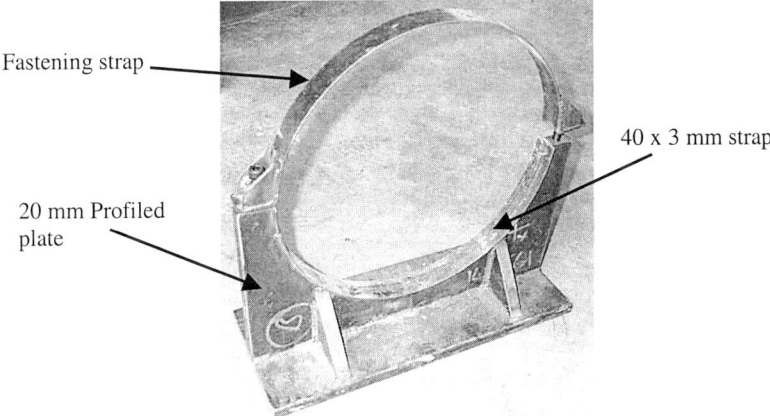

Figure 1 – Loading Saddles

While the first two tubes (TPB001, TPB002) were tested without concrete infill, the remaining tubes were filled with normal grade concrete. The tubes were filled by capping one end of the column then placing them at a one in three incline, then filling the tube using normal construction processes. A partial end cap was placed on the high end of the tubes and concrete placed to fill the tube totally.

MATERIAL PROPERTIES

Two tensile coupons were taken from each of the hollow sections, then prepared and tested in accordance with AS 1391 (SA, 1991b) to determine the yield stress (f_y), the tensile strength (f_u), and the percentage elongation (e_u) at the ultimate strength. A typical stress strain relationship for the tube material is shown in Figure 2, with a yield stress of 351 MPa and an ultimate strength of 446 MPa with the elongation at ultimate being approximately 18 percent. The initial elastic portion of the stress-strain curve was used to determine the elastic modulus (E) for the sections, with a value of $E = 210000$ MPa being determined.

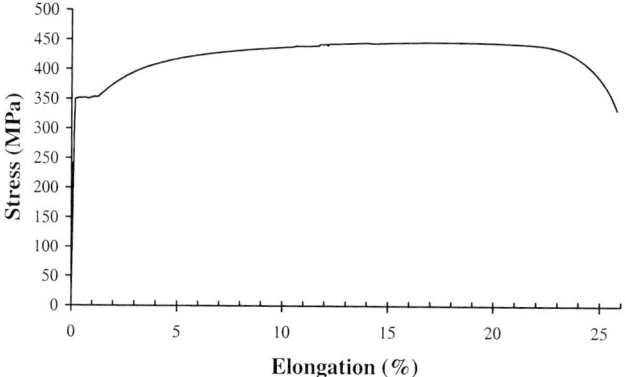

Figure 2 – Tube Material Properties

At the time of casting the concrete in the circular hollow sections a number of concrete cylinders where also cast. These cylinders were tested at a NATA Concrete Testing Laboratory and the age strength relationship was determined. The strength at time of testing was determined and these values are given in Table 2.

TEST SET UP AND PROCEDURE

A schematic view of the general test set-up, and a picture of a specimen being tested are shown in Figure 3 and Figure 4 respectively. Each test consisted of the specimen being positioned such that the saddle supports were placed on support bearings. The support bearings consisted of a half round located on roller bearings, giving an idealised roller pin support. Due to the fact that the beam specimens experienced high rotations during testing, the loading saddles were similarly set up using inverted support bearings as shown in Figure 3 and Figure 4.

Using this system of supports allowed both the support and loading points of the specimen to rotate and slide horizontally. This layout ensured the satisfactory modelling of the simply supported beam and avoided the introduction of a net tensile force into the specimen. Due to the high magnitudes of load required for testing, two 600 kN actuators were used in parallel, with a 1000 kN load cell placed at each actuator. A spreader beam was positioned between the two actuators for stability purposes only.

During the experiments a number of key quantities were measured to monitor the behaviour of the section. To measure the vertical deflections of the specimens, five vertical displacement transducers were placed. Two of the transducers were positioned to measure the displacement at the points of load application, and one transducer was placed at mid-span to obtain the maximum deflection. The remaining two transducers were placed evenly between the mid-span and load point transducers. The strains across the section at mid-span were measured using six 6 mm strain gauges spaced evenly around the section at mid-span.

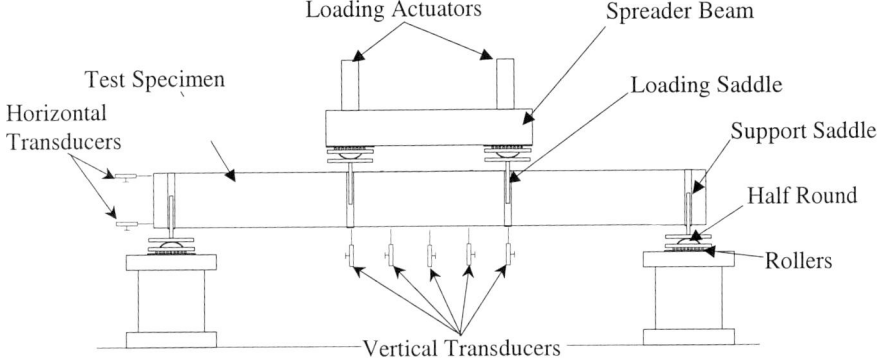

Figure 3 – Schematic Test Set-up

Figure 4 – Test Specimen in Place

The end slip of the concrete infill was measured using two horizontal transducers placed at each end of the cross section. The transducers were anchored to the cross sections at each end. This arrangement meant that only the relative movement between the concrete and the steel section was measured.

Each test specimen was placed in position in the testing machine. The instrumentation was then placed and the loading beam put in position. The two nominally identical hydraulic actuators connected in parallel applied the load under displacement control. The deformations where applied at a constant rate with the applied load, the transducers and strain gauges being read and recorded at a set interval using a Solatron data acquisition system.

The loading was continued until either a buckle formed in the tube with a significant loss in load as was observed in the bare steel tubes or when the actuators' travel limit was reached (in excess of 100mm) for the concrete filled tubes.

TEST RESULTS

The experimental moment-midspan displacement relationships for the tubes obtained from the experiments are shown in Figures 5 and Figure 6 for the 406 diameter and 457 diameter tubes respectively. Also shown in these figures is the Plastic Section moment for the sections and the predicted plastic capacity of the concrete filled tube as determined using advanced analysis software as detailed in Wheeler and Bridge (1993). For both tube sizes the difference in behaviour of the unfilled tube and the filled tube was significant with an increase in excess of 20 percent of the flexural capacity.

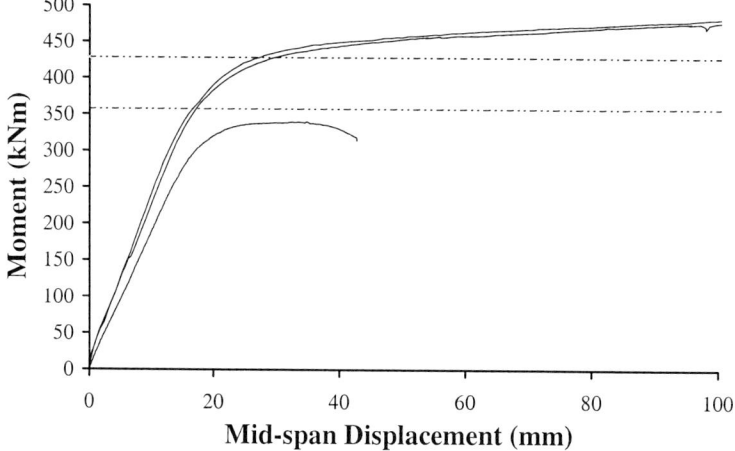

Figure 5 – Moment Displacement Curve for 406 tubes

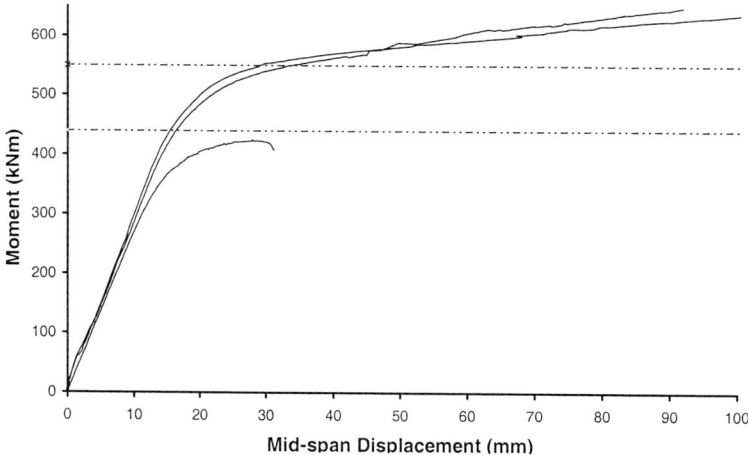

Figure 6 – Moment displacement curves fo 457 tubes

For the bare steel tubes the peak moment was reached when local buckles began to form in the section. These buckles occurred adjacent to the loading points as shown in Figure 7. It should be noted that while the location of these buckles were not ideal (at the supports rather than near to mid-span), it is assumed that the flexural-shear interaction at the loading points instigated the local buckling. Subsequent finite element analysis was able model identical buckles by slightly adjusting the conditions. The corresponding decrease in capacity of the numerical model was 0.5 percent.

According to AS 4100 both tubes are considered to be non-compact thus a reduction in the plastic section capacity is used for design, this is depicted in the results with both tests for the unfilled tube falling below the full plastic section capacity. At the ultimate flexural load, significant ovalisation of both tubes was observed. While the saddles were thought to be sufficient for restraining the ovalisation at the loading and support points, for the 457 diameter tube significant ovalisation occurred at the loading and support points, plastically deforming the saddle plates.

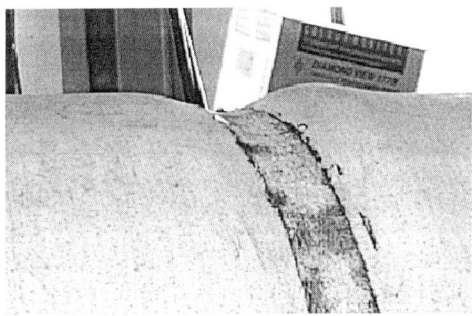

Figure 7 – Local Buckling of Bare Steel Tubes

In Figure 5 and Figure 6, the effect of the concrete infill is clearly shown for both size tubes. The behaviour becomes very ductile with mid-span deflections increasing four fold on those measured for the bare steel. In addition to the increase in ductility, the cross-sectional strength continues to increase to the actuator deformation limit of the tests.

With the inclusion of the concrete infill, ovalisation is prevented. Consequently the stiffness of the tubes is increased. Furthermore the concrete infill also prevents the inward buckles of the thin walled tube. This prevention of buckling was demonstrated in the extreme compressive strains of the tube with measured strains of the concrete filled tubes being in excess of five times larger than those measured for the bare steel tubes. End slip of the concrete infill from the steel tube was also observed, with approximately 0.5 mm observed at one end, indicating that the bond length of the concrete tube interface was sufficient to initiate full shear bonding after initial cracking of the concrete.

The section capacity for the concrete filled tubes shown in Figure 5 and Figure 6 are calculated using advanced analysis methods. The values were determined based on a 45 MPa concrete and a bi-linear elastic plastic stress strain relationship. The method used gives a good estimate for the capacity of the concrete filled tube. The additional strength is attributed to strain hardening of the steel tube as the strains exceed 10% and 20 % in the compressive and tensile regions respectively.

CONCLUSIONS

This paper has described a test program on the behaviour of both bare steel and concrete filled large diameter thin walled sections subjected to flexural loading. A detailed description of the test procedure and results has been presented. The tests clearly demonstrated that the behaviour of the tube, when filled with concrete, is enhanced. The ultimate strength of the sections is increased by over 20 percent. The ductility of the section is also increased, with deflections exceeding four times those measured for the bare steel tubes.

For the given tube sizes and D/t ratios, the tests demonstrated that under flexural loading the concrete in-fill enables full utilization of the thin-walled steel cross-section with the absence of local buckling.

REFERENCES

AISC (1992). *Design Capacity Tables for Structural Steel Hollow Sections*, Australian Institute of Steel Construction, Sydney, Australia.

Elchalakani M., Zhao, X. L. and Grzebieta R. "Bending tests to determine slenderness limits for cold-formed circular hollow sections," *Journal of Constructional Steel Research*, In Press

Lu, Y. and Kennedy, D. (1994), "Flexural Behaviour of Concrete filled Hollow Structural Sections", *Canadian Journal of Civil Engineering*, Vol. 21, No. 1, February, pp 11-130.

Nakamura, S., Momiyama, Y., Hosaka t. and Homma K. (2002) "New technologies of steel/concrete composite bridges", *Journal of Constructional Steel Research,* Vol. 58, No. 1, January 2002, pp 99-130

O'Shea M.D. and Bridge, R.Q. (2000), "Design of Thin-walled Concrete Filled Tubes", *Journal of the Structural Division, ASCE*, v126, n11, pp 1295-1303.

SA (1991), *AS 1163-1991: Structural Steel Hollow Sections*, Standards Australia, Sydney.

SA (1998), *AS 4100-1998: Steel Structures*, Standards Australia, Sydney.

Schilling, C.G. (1965), "Buckling Strength of Circular Tubes", *Journal of the Structural Division, ASCE*, v91, n5, pp 325-348

Sherman, D.R. (1976), "Tests of Circular Steel Tubes in Bending", *Journal of the Structural Division, ASCE*, v102, n11, pp 2181-2195

Wheeler A. T. (2002), "Bending Tests of Thin Walled Concrete Filled Circular Hollow Sections", Research Report CCTR003, Centre For construction Technology and Research, The University of Western Sydney, Sydney, Australia.

Wheeler A. T. and Bridge R. Q., (1993) "Analysis of Cross-sections in Composite Materials", *Proceedings*, Thirteenth Australasian Conference on the Mechanics of Structures and Materials, Wollongong, Australia, University of Wollongong, pp. 929- 937.

Wheeler A. T. and Pircher M (2002), "Measured Imperfections in Six Cold-Formed Thin Walled Steel Tubes", Research Report CCTR001, Centre For construction Technology and Research, The University of Western Sydney, Sydney, Australia.

Wheeler A.T., Pircher M., Bridge R.Q. (1999), "Modelling of Thin-walled Tubular Sections Subjected to Pure Bending", *Proceedings: The Thirteenth Australian Compumod Users' Conference*, Melbourne, Australia

Wheeler, A. and Bridge, R. (2000), "Thin-walled Steel Tubes Filled with High Strength Concrete in Bending", *Proceedings: Engineering Foundation Conference, Composite Construction IV*, Banff, Canada

SIMPLIFIED ELASTIC AND ELASTIC-PLASTIC ANALYSIS OF CONTINUOUS COMPOSITE BEAMS

P.A. Berry

Centre for Construction Technology & Research,
College of Science, Technology & Environment, University of Western
Sydney, Locked Bag 1797, Penrith South DC, NSW 1797, Australia

ABSTRACT

Continuous composite beams offer significant benefits over simply-supported composite beams: the load carrying capacity of otherwise identical beams is increased by 15-40% and the deflections are reduced by 60-70%. However, the elastic or elastic-plastic analysis of a continuous composite beam must satisfy compatibility as well as equilibrium. Due to cracking of the concrete slab in negative moment regions, the bending moment distribution and the variation of beam stiffness are interdependent, which in the past has required the use of iterative solution procedures.

This paper presents a simple method for the elastic or elastic-plastic analysis of continuous composite beams. By applying the moment area theorems, the point of contraflexure is obtained directly without the need for iteration and can be used either in hand calculations or as the starting point for a more sophisticated analysis. The method is applied to calculating the beam deflections, the level of moment redistribution, and the rotation demand at connections.

KEYWORDS

Composite beam, continuous composite beam, semi-continuous composite beam, contraflexure, elastic analysis, elastic-plastic analysis.

INTRODUCTION

Elastic Analysis of Continuous Composite Beams

Elastic analysis requires knowledge of the flexural stiffness at each cross-section, which presents some difficulties in the case of continuous composite beams. The positive moment stiffness is equal to that of a simply-supported beam with the same effective span, but the negative moment stiffness depends on the level of reinforcement and its extent is determined by the points of contraflexure, both of which are unknown in the initial stages of design.

Two simplified methods of analysis may be used to overcome this (British Standards Institution 1994):
- *Uncracked.* An uncracked analysis (i.e. in negative moment regions) assumes that the transformed second moment of area, I_t, based on the effective section in positive bending, may be applied uniformly to the entire length of the beam.
- *Cracked.* A cracked analysis assumes that the cracked second moment of area, I_{cr}, based on the effective section in negative bending with the concrete fully cracked, may be applied to a 15% length of the beam adjacent to the support and that the transformed second moment of area should be applied elsewhere.

This paper presents a more sophisticated method of analysis, which has been termed:
- *Contraflexure.* A contraflexure analysis uses the assumed second moments of area in positive and negative bending, typically I_t and I_{cr} as above, to determine the points of contraflexure for a given loading configuration, by solving for compatibility as well as equilibrium. Once the points of contraflexure are known, they can be replaced by internal hinges, and the structure becomes determinate. The structure may be represented as a simply-supported beam between the points of contraflexure, supported by cantilevers at the continuous supports. This approach is particularly useful for estimating the deflections at the serviceability limit state.

Elastic-Plastic Analysis of Continuous Composite Beams

Elastic-plastic global analysis of continuous composite beams is based on an idealised elastic, perfectly-plastic model of moment-curvature behaviour. The action effects may be determined from a superimposed series of elastic analyses. The initial moment-curvature behaviour is linear everywhere, so the first analysis can proceed elastically up to the load at which a plastic hinge starts to form. The cross-section at this plastic hinge now has zero flexural stiffness, so it may be replaced by a nominal pin and the elastic analysis recommenced. This is applied iteratively until a complete plastic collapse mechanism has developed. Unlike elastic and plastic analyses, it enables calculation of the required rotation capacity at plastic hinges, improving the rigour of ductility design for these cross-sections.

CONTRAFLEXURE ANALYSIS

General Principles

The elastic analysis of an indeterminate structure must satisfy equilibrium and compatibility. Satisfying equilibrium is straightforward, but satisfying compatibility is more difficult. The bending moment distribution depends on the variation of beam stiffness, and vice versa. In positive moment regions the stiffness can be taken as $I^+ = I_t$, while in negative moment regions it is $I^- = I_{cr}$. Adopting this distribution requires knowledge of the point of contraflexure. Much of the literature refers to the need for iteration (Oehlers & Bradford 1995, Johnson & Anderson 1993), but it is possible to obtain direct solutions by applying the moment area theorems.

Moment Area Theorems

Theorem of Slopes

The change in slope from point A to point B along a beam is given by

$$\theta_B - \theta_A = -\int_A^B \frac{M}{EI} dz \tag{1}$$

Theorem of Deflections

The distance measured from point A to the tangent from point B is given by

$$\delta_{AB} = \int_A^B \frac{M}{EI} z_A \, dz \qquad (2)$$

Symmetric Beams

For beams that are symmetric in all respects about their mid-span, the theorem of slopes can be applied using point A as the support ($z = 0$) and point B as mid-span ($z = L/2$). From symmetry, the slope at mid-span is zero, so the theorem of slopes can be reduced to

$$\theta_{z=0} = \int_0^{L/2} \frac{M}{EI} \, dz \qquad (3)$$

For continuous beams or semi-continuous beams with rigid connections, the slope at the support under elastic conditions must be zero, so the theorem of slopes can be further reduced to

$$0 = \int_0^{L/2} \frac{M}{EI} \, dz \qquad (4)$$

Assuming that the point of contraflexure occurs at $z = a$ and assigning the appropriate values of I,

$$0 = \int_0^a \frac{M}{EI^-} \, dz + \int_a^{L/2} \frac{M}{EI^+} \, dz \qquad (5)$$

Multiplying by EI^+ and assigning $\Gamma = I^+/I^-$, this compatibility condition can be expressed as

$$0 = \Gamma \int_0^a M \, dz + \int_a^{L/2} M \, dz \qquad (6)$$

Central Point Load, P

For a central point load, the moment can be expressed over the domain $z \in [0, L/2]$ as

$$M = M_A + \frac{P}{2} z \qquad (7)$$

If $z = a$ is the point of contraflexure, then $M_A = -Pa/2$ and

$$M = \frac{P}{2}(z - a) \qquad (8)$$

The compatibility integral reduces to the quadratic equation

$$0 = 4(\Gamma - 1)\left(\frac{a}{L}\right)^2 + 4\left(\frac{a}{L}\right) - 1 \qquad (9)$$

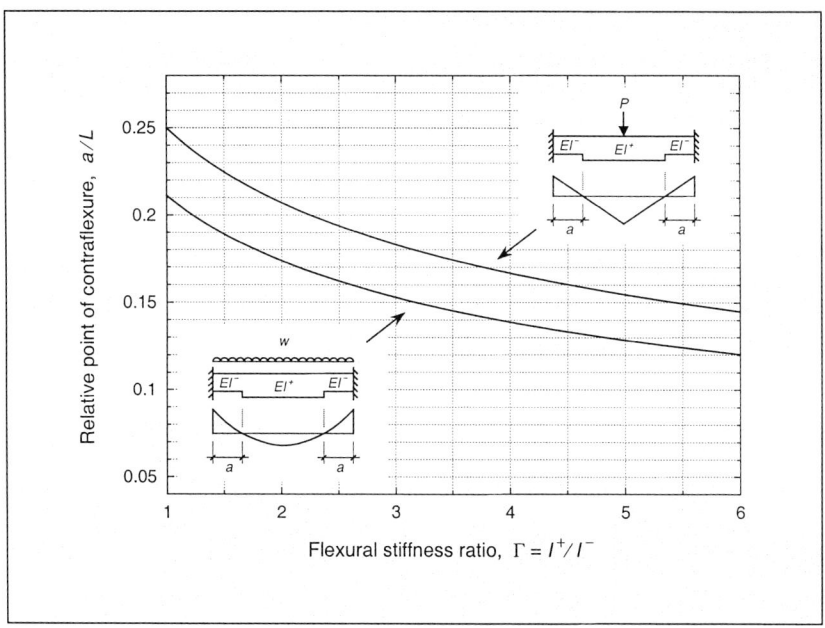

Figure 1: Point of Contraflexure for Fully Built-In Beams

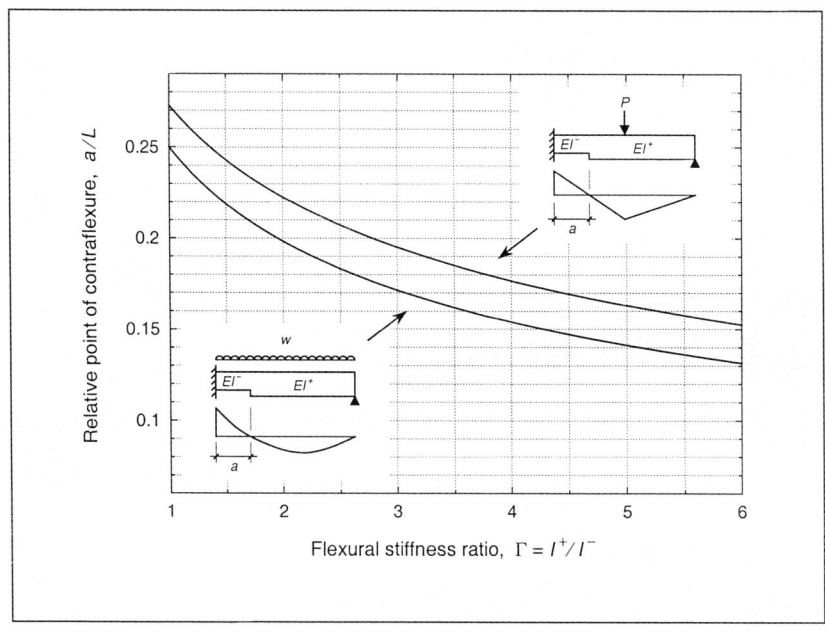

Figure 2: Point of Contraflexure for Propped Cantilevers

which has the desired solution as its positive root

$$\frac{a}{L} = \frac{\sqrt{\Gamma}-1}{2(\Gamma-1)} \tag{10}$$

Uniformly Distributed Load, w

For a uniformly distributed load, the moment can be expressed over the domain $z \in [0, L/2]$ as

$$M = M_A + \frac{wL}{2}z - \frac{w}{2}z^2 \tag{11}$$

If $z = a$ is the point of contraflexure, then

$$M_A = \frac{w}{2}a^2 - \frac{wL}{2}a \tag{12}$$

and the moment can be expressed as

$$M = \frac{w}{2}\left[\left(a - \frac{L}{2}\right)^2 - \left(z - \frac{L}{2}\right)^2\right] \tag{13}$$

The compatibility integral reduces to the cubic equation

$$0 = 8(\Gamma - 1)\left(\frac{a}{L}\right)^3 - 6(\Gamma - 2)\left(\frac{a}{L}\right)^2 - 6\left(\frac{a}{L}\right) + 1 \tag{14}$$

for which the solution is shown in graphical form in Fig. 1.

Propped Cantilevers

For propped cantilevers, the theorem of deflections can be applied using point A as the pinned support and point B as the fixed support. For continuous beams or semi-continuous beams with rigid connections, the slope at point B under elastic conditions must be zero, and hence the tangent at point B is horizontal. Assuming that point A does not undergo any settlement, the distance from point A to the tangent from point B, δ_{AB}, is zero, so the theorem of deflections can be reduced to

$$0 = \int_A^B \frac{M}{EI} z_A \, dz \tag{15}$$

Central Point Load, P

A similar solution procedure to that for symmetric beams leads to the following cubic equation, for which the graphical solution is shown in Fig. 2.

$$0 = 4(1-\Gamma)\left(\frac{a}{L}\right)^3 - 12(1-\Gamma)\left(\frac{a}{L}\right)^2 + 11\left(\frac{a}{L}\right) - 3 \tag{16}$$

Uniformly Distributed Load, w

A similar solution procedure to that for symmetric beams leads to the following quartic equation, for which the graphical solution is shown in Fig. 2.

$$0 = (1-\Gamma)\left(\frac{a}{L}\right)^4 - 4(1-\Gamma)\left(\frac{a}{L}\right)^3 + 6(1-\Gamma)\left(\frac{a}{L}\right)^2 - 4\left(\frac{a}{L}\right) + 1 \tag{17}$$

MOMENT REDISTRIBUTION

The effects of moment redistribution are best illustrated with reference to an example. Consider a fully built-in composite beam of span, L, subjected to a central design point load, P, as shown in Fig. 3. The relative values adopted in regions of positive and negative bending, although chosen for ease of calculation, are nevertheless typical: for strength, the design positive moment capacity, ϕM_{bv}^+, is taken to be three times the design negative moment capacity, ϕM_{bv}^-; and, for stiffness, the positive second moment of area, I_t, is taken to be four times the negative second moment of area, I_{cr}.

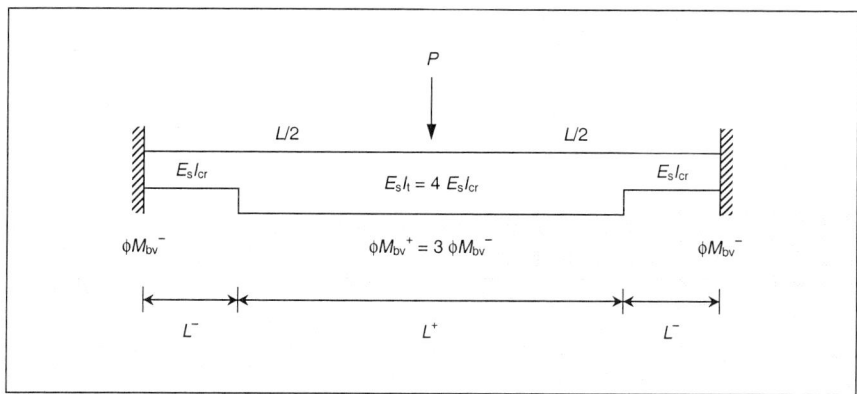

Figure 3: Beam Example with Fully Fixed Ends

ELASTIC CRACKED CONTRAFLEXURE ANALYSIS

The results of an elastic cracked contraflexure analysis with no moment redistribution are shown in Fig. 4(a). For a fully built-in beam with $I_t/I_{cr} = 4$ subjected to a central point load, the points of contraflexure are located at $L/6$ away from the supports (Eqn. 10, Fig. 1), which is very similar to the value of $0.15L$ used in a simplified cracked analysis. The critical cross-sections are at the supports, for which the required design moment capacity is $\phi M_{bv}^- = PL/12$, but there is considerable wasted capacity in the positive moment region, where the strength is 1.5 times the action effects. Fig. 4(b) shows the results of the same analysis, only this time after 25% moment redistribution: a moment of $PL/48$ equal to 25% of the original $PL/12$ is redistributed from the highly loaded negative moment region to the under-utilised positive moment region. Now, the support cross-sections and the mid-span cross-section are all critical. The required design negative moment capacity is $\phi M_{bv}^- = PL/16$, three-quarters of that required for the same analysis without moment redistribution, and there is no longer any wasted positive moment capacity. An equivalent uncracked analysis requires 50% moment redistribution to achieve the same result. In an attempt to maintain consistency, design rules specify different limits for the allowable level of moment redistribution depending on the type of analysis.

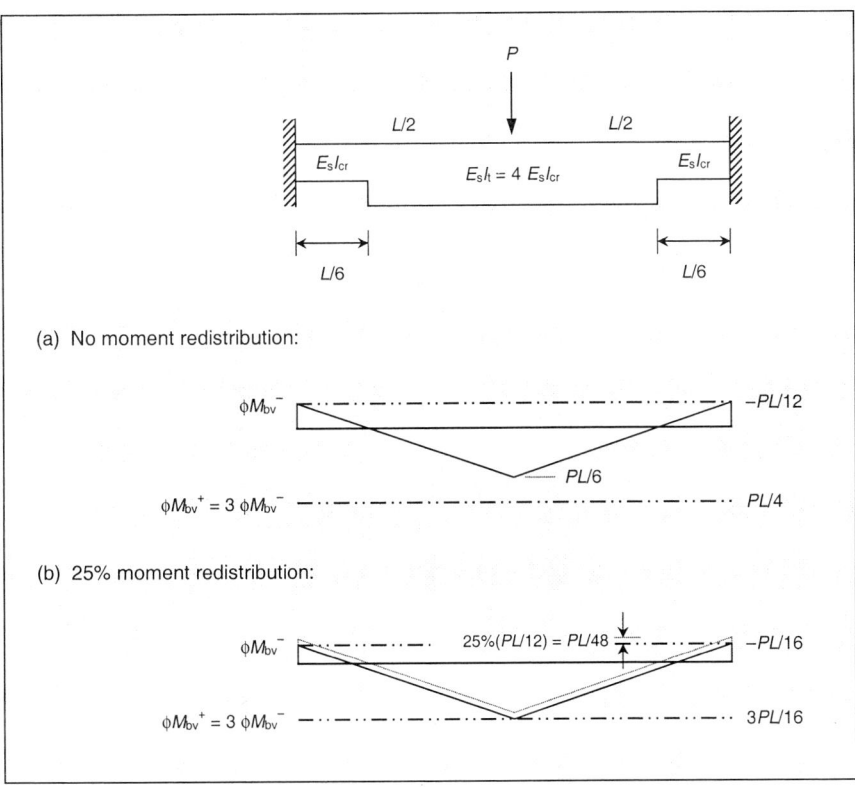

Figure 4: Elastic Cracked Contraflexure Analysis

ELASTIC-PLASTIC CRACKED CONTRAFLEXURE ANALYSIS

The results of an elastic-plastic cracked contraflexure analysis are shown in Fig. 5. The first phase of the analysis shown in Fig. 5(a) is elastic up to a load of $3P/4$, at which point plastic hinges form at the support cross-sections and the structure effectively becomes a simply-supported beam. The second phase of the analysis, shown in Fig. 5(b), ends after an additional load of $P/4$, when a hinge forms at the mid-span cross-section and the structure becomes a mechanism. The beam end rotations during this second phase represent the ductility required by the connections. For a beam of this geometry supporting a central point load, in this case $P/4$, the required rotation capacity may be calculated using the moment area theorem of slopes (Eqn. 1) as:

$$\theta = \frac{PL^2}{48 E_s I_t} \tag{18}$$

which is two-thirds of the value that results from an elastic-plastic uncracked analysis. Fig. 5(c) shows the final superimposed bending moment diagram.

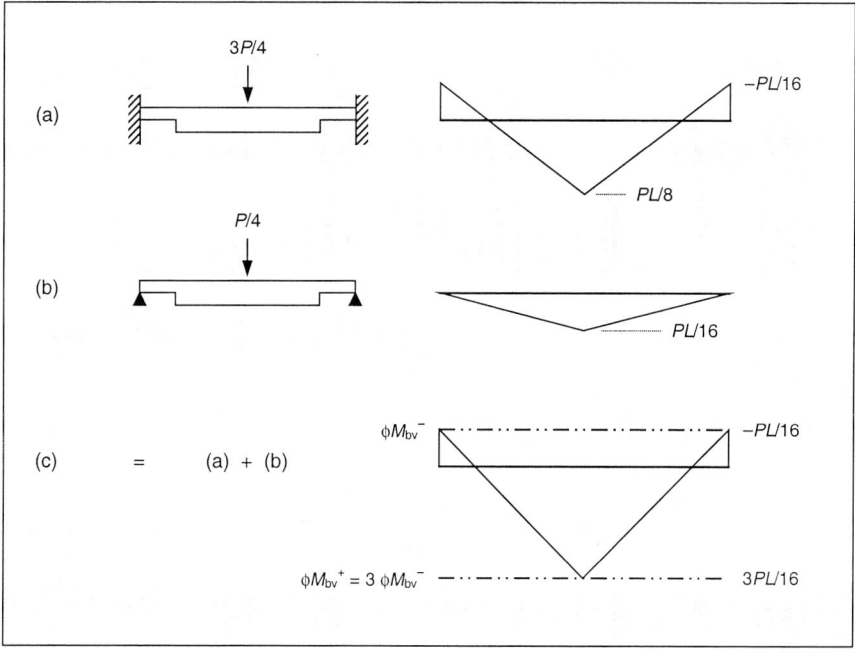

Figure 5: Elastic-Plastic Cracked Contraflexure Analysis

CONCLUSIONS

The contraflexure analysis presented in this paper is a simple method for the elastic or elastic-plastic analysis of continuous composite beams. The point of contraflexure is obtained directly without the need for iteration and can be used either in hand calculations or as the starting point for a more sophisticated analysis. Once the points of contraflexure are known, they can be replaced by internal hinges, and the structure becomes determinate. This approach is particularly useful for estimating the beam deflections, the level of moment redistribution, and the rotation demand at connections.

REFERENCES

British Standards Institution (1994). *Eurocode 4: Design of Composite Steel and Concrete Structures, Part 1.1. General Rules and Rules for Buildings (together with United Kingdom National Application Document)*, DD ENV 1994-1-1, London.

Johnson, R.P. and Anderson, D. (1993). *Designers' Handbook to Eurocode 4, Part 1.1: Design of Composite Steel and Concrete Structures*, Thomas Telford, London.

Oehlers, D.J. and Bradford, M.A. (1995). *Composite Steel and Concrete Structural Members: Fundamental Behaviour*, Pergamon, Oxford.

ELASTIC CROSS-SECTION ANALYSIS OF CONTINUOUS COMPOSITE BEAMS AFFECTED BY WEB SLENDERNESS

P.A. Berry

Centre for Construction Technology & Research,
College of Science, Technology & Environment, University of Western
Sydney, Locked Bag 1797, Penrith South DC, NSW 1797, Australia

ABSTRACT

Continuous composite beams offer significant benefits over simply-supported composite beams: the load carrying capacity of otherwise identical beams is increased by 15-40% and the deflections are reduced by 60-70%. However, the presence of tensile reinforcement in the concrete slab of negative moment regions may cause an otherwise non-compact web to be affected by local buckling. The Australian Standard AS 2327.1 on simply supported composite beams covers the analysis of cross-sections in positive bending, but there are currently no rules for designing cross-sections of continuous composite beams in negative bending.

This paper presents a method for the elastic analysis of cross-sections in negative bending, which eliminates the previous need for iterative solutions. It is based on the calculation of key levels of reinforcement that define the transition between different effective cross-sections. The method extends the 'hole in the web' approach used by AS 2327.1 and the European Standard Eurocode 4 for the plastic cross-section analysis of non-compact webs (EC4 Class 3) to cover the elastic cross-section analysis of slender webs (EC4 Class 4). The method is most applicable to welded beams, which are common in bridge construction and generally more slender than hot-rolled sections. The paper shows that whereas the use of a longitudinal web stiffener provides only minimal benefit, the use of an additional bottom flange plate extending for a short distance over the support can lead to substantial strength gains.

KEYWORDS

Composite beam, continuous composite beam, elastic analysis, welded beams, bridges.

INTRODUCTION

The 'hole in the web' approach for the plastic analysis of composite cross-sections originated in the British Standard BS 5950 (1990). It has been adopted for both positive and negative bending in the

European Standard Eurocode 4 (1992) and for positive bending in the Australian Standard for simply supported composite beams, AS 2327.1 (1996). However, the calculation procedure used to apply the method, which is elaborated by Johnson & Anderson (1993), has several disadvantages:
1. It requires an iterative solution procedure.
2. It produces a small discontinuity at the boundary between a Class 2 and Class 3 web.
3. It is not well suited to a piecewise-linear approximation.
4. Published formulae cover only a limited number of cases (Lawson 1990).

Berry, Patrick, Liang & Ng (2001) have developed a method for the plastic analysis of cross-sections in negative bending, based on the calculation of key reinforcement levels, which overcomes these disadvantages. The method does not require iteration, removes the discontinuity at the formation of a 'hole' in the web, is well suited to linear approximation, and is easily generalised. This paper extends that method to the elastic analysis of cross-sections in negative bending that are affected by local buckling of the web. The method is used to show the dramatic improvements to be gained from using an additional bottom flange plate that extends for a short distance over an internal support. The benefits of using a longitudinal web stiffener are also investigated and shown to minimal.

EFFECTIVE CROSS-SECTION

The 'hole in the web' approach given in AS 2327.1 for plastic cross-section analysis allows non-compact webs to be represented by an effective compact web, by assuming that the depth of the web that resists compression is limited to $15t_w\varepsilon$ adjacent to the compression flange and $15t_w\varepsilon$ adjacent to the new plastic neutral axis (PNA). This 'hole in the web' approach can be extended to elastic cross-section analysis by choosing an effective depth that models elastic local buckling. For this paper, an effective depth of $30t_w\varepsilon$ adjacent to supporting elements has been chosen to match the yield slenderness limits given in Table 5.1 of AS 2327.1. Thus, a slender web may be represented by an effective non-compact web, by assuming that the depth of the web that resists compression is limited to $30t_w\varepsilon$ adjacent to the compression flange and $30t_w\varepsilon$ adjacent to the new elastic neutral axis (ENA), as shown in Figure 1.

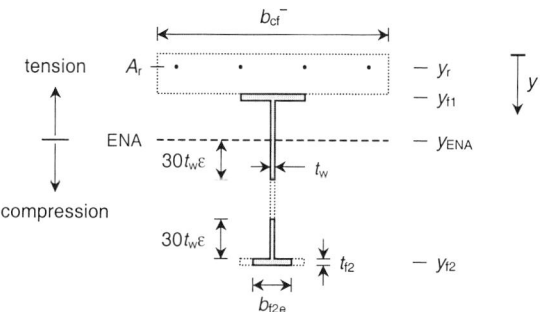

Figure 1: A composite cross-section reduced by local buckling

A general composite cross-section subject to negative bending, including a web stiffener and an additional bottom flange plate, is shown in Figure 2. The effective dimensions of each component and the coordinate from the top surface to its centroid are given in Table 1. The direct contribution of the web stiffener to the moment capacity of the member is conservatively ignored.

TABLE 1
EFFECTIVE SECTION DIMENSIONS

Component	Coordinate	Tension	Compression[1]
Flange 1	$y_{f1} = D_c + t_{f1}/2$	b_{f1}	$b_{f1e} = \min(b_{f1}, 2\lambda_{ey}t_{f1}\varepsilon_{f1} + t_w)$
Web	$y_w = D_c + t_{f1} + d_w/2$	$d_w = D_s - t_{f1} - t_{f2}$	$d_{we} = \min(d_w, 60t_w\varepsilon_w)$
Flange 2	$y_{f2} = D_c + t_{f1} + i_w + t_{f2}/2$	b_{f2}	$b_{f2e} = \min(b_{f2}, 2\lambda_{ey}t_{f2}\varepsilon_{f2} + t_w)$
Flange plate	$y_{fp} = D_c + D_s + t_{fp}/2$	b_{fp}	$b_{fpe} = \min(b_{fp}, \lambda_{ey}t_{fp}\varepsilon_{fp})$

[1] λ_{ey} is the appropriate value of yield slenderness limit taken from Table 5.1 of AS 2327.1.

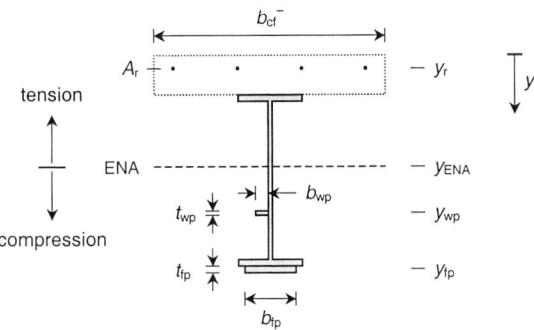

Figure 2: A general composite cross-section in negative bending

KEY LEVELS OF REINFORCEMENT

The first step in calculating the negative moment capacity of a slender composite member is to determine the location of the elastic neutral axis (ENA). Since the location of the ENA depends on the effective cross-section and vice versa, it would appear that an iterative process is necessary and that is the approach adopted in the literature (Johnson & Anderson, 1993). However, a direct solution technique is possible, based on the calculation of key levels of reinforcement that define the transition between different effective cross-sections. A negative result for a particular transition point indicates that it will not occur for the given steel section regardless of the level of reinforcement. The transition points and their applicable dimensions for key reinforcement calculations are given in Table 2.

TABLE 2
APPLICABLE DIMENSIONS FOR KEY REINFORCEMENT CALCULATIONS

Transition	k	A_{rk}	y_{ENA}	b_{f1a}	d_{wa}	b_{f2a}	b_{fpa}
Top of section	t	A_{rt}	$y_t = D_c$	b_{f1e}	d_{we}	b_{f2e}	b_{fpe}
Junction 1	j1	A_{rj1}	$y_{j1} = D_c + t_{f1}$	b_{f1}	d_{we}	b_{f2e}	b_{fpe}
Hole in web	h	A_{rh}	$y_h = D_c + d_w - d_{we}$	b_{f1}	d_w	b_{f2e}	b_{fpe}
Junction 2	j2	A_{rj2}	$y_{j2} = D_c + t_{f1} + d_w$	b_{f1}	d_w	b_{f2e}	b_{fpe}
Junction plate	jp	A_{rjp}	$y_{jp} = D_c + D_s$	b_{f1}	d_w	b_{f2}	b_{fpe}

The key areas of reinforcement can be calculated from the centroidal expression

$$A_{rk} = \sum_{i=f1}^{fp} \left(A_{ia} \frac{y_i - y_{ENA}}{y_{ENA} - y_r} \right) \quad \forall \text{(for all)} \, k \in \{t, j1, h, j2, jp\} \quad (1)$$

using the relevant values for each transition point given in Table 2.

KEY VALUES OF THE TOTAL CROSS-SECTION AREA

The key values of the total cross-section area, which are used in other calculations, are given by

$$A_k = A_{rk} + \sum_{i=f1}^{fp} A_{ia} \quad \forall k \in \{t, j1, h, j2, jp\} \quad (2)$$

using the relevant values for each transition point given in Table 2.

KEY VALUES OF THE SECOND MOMENT OF AREA

The key values of the second moment of area can be calculated from

$$I_k = \frac{A_{rk}(y_{ENA} - y_r)^2 + \sum_{i=f1}^{fp} A_{ia}(y_{ENA} - y_i)^2}{+ \frac{t_w d_w^3}{12} - \frac{t_w (d_w - d_{wa})^3}{12} + \sum_{i=f1,f2,fp} \frac{b_{ia} t_i^3}{12}} \quad \forall k \in \{t, j1, h, j2, jp\} \quad (3)$$

using the relevant values for each transition point given in Table 2.

GENERAL VALUES OF THE ELASTIC NEUTRAL AXIS COORDINATE

The elastic neutral axis coordinate, y_{ENA}, can be calculated for any given area of reinforcement, A_r, using an exact differential approach from the appropriate key level of reinforcement (Table 2).

$A_{rj1} \leq A_r \leq A_{rt}$

The elastic neutral axis is located in flange 1 and can be calculated from the key point, $k = t$, as

$$y_{ENA} = y_t + p(a,b,c) \quad (4)$$

where $p(a,b,c)$ is the most positive root of the quadratic equation $ax^2 + bx + c = 0$ with coefficients

$$\begin{aligned} a &= \frac{b_{f1} - b_{f1e}}{2} \\ b &= A_t - dA_r \quad ; dA_r = A_{rt} - A_r \\ c &= -dA_r(y_t - y_r) \end{aligned} \quad (5)$$

$A_{rh} \leq A_r < A_{rj1}$

The elastic neutral axis is located in the web, such that a 'hole' forms in the compressive region of the web below the ENA, and can be calculated from the key point, $k = j1$, as

$$y_{ENA} = y_{j1} + p(a,b,c) \tag{6}$$

where

$$a = \frac{t_w}{2}$$
$$b = A_{j1} - dA_r - \frac{A_{we}}{2} \quad ; dA_r = A_{rj1} - A_r \tag{7}$$
$$c = -dA_r(y_{j1} - y_r)$$

$A_{rh} \leq A_r < A_{rj1}$

The elastic neutral axis is located in the web, but in this case below the point at which a 'hole' forms in the compressive region, and can be calculated from the key point, $k = h$, as

$$y_{ENA} = y_h + p(a,b,c) \tag{8}$$

where

$$a = 0$$
$$b = A_h - dA_r \quad ; dA_r = A_{rh} - A_r \tag{9}$$
$$c = -dA_r(y_h - y_r)$$

$A_{rj2} \leq A_r < A_{rh}$

The elastic neutral axis is located in flange 2 and can be calculated from the key point, $k = j2$, as

$$y_{ENA} = y_{j2} + p(a,b,c) \tag{10}$$

where

$$a = \frac{b_{f2} - b_{f2e}}{2}$$
$$b = A_{j2} - dA_r \quad ; dA_r = A_{rj2} - A_r \tag{11}$$
$$c = -dA_r(y_{j2} - y_r)$$

GENERAL VALUES OF THE SECOND MOMENT OF AREA

The second moment of area, I, can be calculated for any given area of reinforcement, A_r, using an exact differential approach from the appropriate key value of the second moment of area (Equation 3) and key level of reinforcement (Table 2).

$A_{\text{rj1}} \leq A_{\text{r}} \leq A_{\text{rt}}$

Using the values of y_{ENA} and dA_{r} from Equations 4 and 5

$$I = I_{\text{t}} + A_{\text{t}}(y_{\text{ENA}} - y_{\text{t}})^2 - dA_{\text{r}}(y_{\text{ENA}} - y_{\text{r}})^2 + \frac{(b_{\text{f1}} - b_{\text{f1e}})(y_{\text{ENA}} - y_{\text{t}})^3}{3} \qquad (12)$$

$A_{\text{rh}} \leq A_{\text{r}} < A_{\text{rj1}}$

Using the values of y_{ENA} and dA_{r} from Equations 6 and 7

$$I = \frac{I_{\text{j1}} + A_{\text{j1}}(y_{\text{ENA}} - y_{\text{j1}})^2 - dA_{\text{r}}(y_{\text{ENA}} - y_{\text{r}})^2}{+ \frac{t_{\text{w}}(y_{\text{ENA}} - y_{\text{j1}})^3}{12} + \frac{t_{\text{w}}(y_{\text{ENA}} - y_{\text{j1}})(y_{\text{j1}} + d_{\text{we}} - y_{\text{ENA}})^2}{4}} \qquad (13)$$

$A_{\text{rh}} \leq A_{\text{r}} < A_{\text{rj1}}$

Using the values of y_{ENA} and dA_{r} from Equations 8 and 9

$$I = I_{\text{h}} + A_{\text{h}}(y_{\text{ENA}} - y_{\text{h}})^2 - dA_{\text{r}}(y_{\text{ENA}} - y_{\text{r}})^2 \qquad (14)$$

$A_{\text{rj2}} \leq A_{\text{r}} < A_{\text{rh}}$

Using the values of y_{ENA} and dA_{r} from Equations 10 and 11

$$I = I_{\text{j2}} + A_{\text{j2}}(y_{\text{ENA}} - y_{\text{j2}})^2 - dA_{\text{r}}(y_{\text{ENA}} - y_{\text{r}})^2 + \frac{(b_{\text{f2}} - b_{\text{f2e}})(y_{\text{ENA}} - y_{\text{j2}})^3}{3} \qquad (15)$$

NOMINAL NEGATIVE MOMENT CAPACITY

The nominal elastic negative moment capacity can be calculated at any point from

$$M_{\text{bv}}^- = \min\left(\frac{f_{\text{yr}}I}{y_{\text{ENA}} - y_{\text{r}}}, \frac{f_{\text{yf1}}I}{y_{\text{ENA}} - y_{\text{t}}}, \frac{f_{\text{yf2}}I}{y_{\text{jp}} - y_{\text{ENA}}}, \frac{f_{\text{yfp}}I}{y_{\text{b}} - y_{\text{ENA}}}\right) \qquad (16)$$

using the relevant values for each transition point given in Table 2.

RESULTS

The design elastic negative moment capacity, ϕM_{bv}^-, for an 800 WB 122 with a 200 mm slab is shown in Figure 3. The reinforcement is located at $y_{\text{r}} = 50$ mm and has a yield strength of $f_{\text{yr}} = 500$ MPa. The key levels of reinforcement are shown for the case of no moment-shear interaction, $\gamma \leq 0.5$.

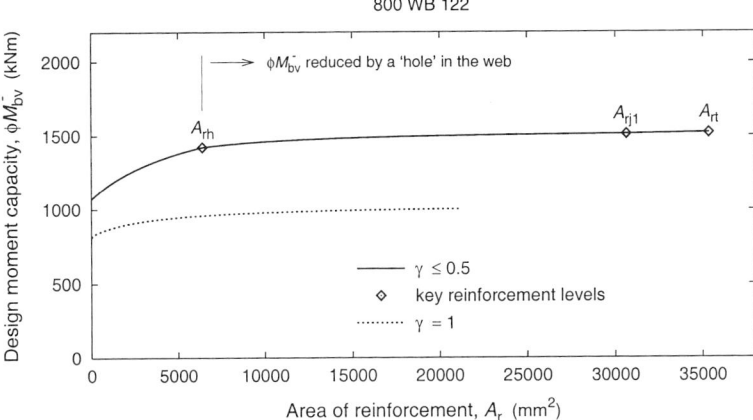

Figure 3: Design elastic negative moment capacity

Local buckling of the web limits the effectiveness of composite beams in negative bending. The design elastic positive moment capacity of an 800 WB 122, assuming an effective concrete slab width of 3000 mm, is 1670 kNm. For practical levels of reinforcement ($A_r = 4500$ mm^2), the design elastic negative moment capacity is 1320 kNm (Figure 3). Allowing for the different second moments of area in positive and negative moment regions, elastic analysis of an encastréd beam leads to negative bending moments at the supports that are 50% greater in magnitude than the positive bending moment at midspan. Therefore, this cross-section does not represent an efficient design.

Local buckling of the web can be overcome by the use of either a longitudinal web stiffener or an additional bottom flange plate. The latter prevents local buckling by lowering the ENA below the level at which a 'hole' develops in the web. The benefits of both approaches are shown in Figure 4, in which the web stiffener is conservatively assumed to make no direct contribution to the moment capacity of the cross-section and the additional bottom flange plate is taken to be the same size as flange 2.

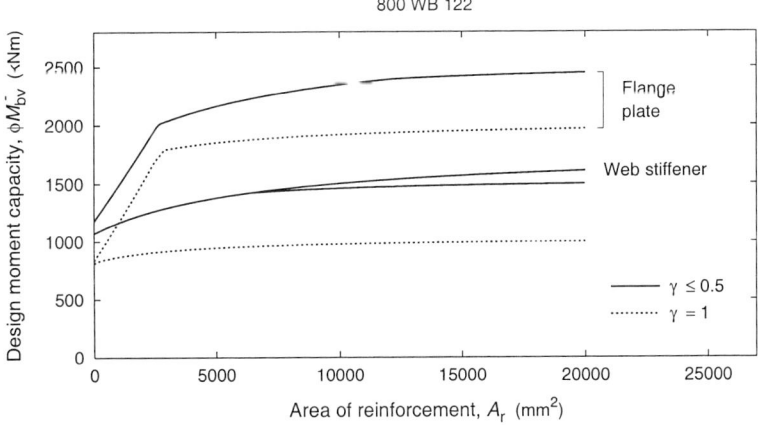

Figure 4: Benefits of adding a web stiffener or an additional bottom flange plate

The use of a web stiffener leads to only modest strength gains, which reduce to zero for full moment-shear interaction ($\gamma = 1$). The benefits of an additional bottom flange plate are much more substantial: for practical levels of reinforcement, the negative moment capacity is almost doubled, regardless of the shear ratio, γ. The design elastic negative moment capacity of an 800 WB 122 with an additional bottom flange plate and slab reinforcement of $A_r = 4500$ mm^2 is 2140 kNm (Figure 4), which is 30% greater than the positive moment capacity and represents a more efficient design.

CONCLUSIONS

This paper presents a new method for the elastic analysis of cracked cross-sections in negative bending. It follows the principles of AS 2327.1 and Eurocode 4, but has been formulated differently to overcome the disadvantages of the current calculation procedures. The new method does not require iteration, removes the discontinuity at the formation of a 'hole' in the web, is well suited to linear approximation, and is easily generalised.

The method is used to study the benefits of adding either a longitudinal web stiffener or an additional bottom flange plate to the cross-section in negative moment regions. The use of a web stiffener leads to only modest strength gains, but dramatic improvements can be achieved from using an additional bottom flange plate that extends for a short distance over an internal support. The negative moment capacity is almost doubled, which greatly improves the design efficiency of continuous composite beams.

REFERENCES

AS 2327.1 (1996). *Composite structures, Part 1: Simply supported beams*, Standards Australia, Sydney.

Berry, P.A., Patrick, M., Liang, Q.Q. and Ng, A. (2001). Cross-Section Design of Continuous Composite Beams. *Proceedings, ASEC 2001, Australasian Structural Engineering Conference*, Gold Coast, The Institution of Engineers, Australia, 491-498.

BS 5950 (1990). *The Structural Use of Steelwork in Buildings, Part 3: Design in Composite Construction, Section 3.1: Code of Practice for Design of Simple and Continuous Composite Beams*, British Standards Institution, London.

Eurocode 4 (1992). *Design of Composite Steel and Concrete Structures, Part 1.1: General Rules and Rules for Buildings*, European Committee for Standardization (CEN), Brussels.

Johnson, R.P. and Anderson, D. (1993). *Designers' Handbook to Eurocode 4, Part 1.1: Design of Composite Steel and Concrete Structures*, Thomas Telford, London.

Lawson, R.M. (1990). *Commentary on BS 5950: Part 3: Section 3.1: 'Composite Beams'*, Publication 078, Steel Construction Institute, Ascot, UK.

EFFECTS OF TRANSVERSE REINFORCEMENT ON COMPOSITE BEAMS WITH PRECAST HOLLOW CORE SLABS

D. Lam and T. F. Nip

School of Civil Engineering, University of Leeds, Leeds, LS2 9JT, UK

ABSTRACT

In composite steel beams with precast hollow core slabs, the amount of transverse reinforcement can have a significant effect on the shear and slip capacity of the mechanical shear connectors. The issue of connector ductility becomes especially important when partial shear connection is adopted, as premature failure of the shear connectors would lead to sudden failure of the composite beam. This paper presents its finding on the effect of transverse reinforcement on connector ductility and design equations are proposed.

KEYWORDS

Composite, transverse reinforcement, shear studs, precast, hollow core slab, steel, structural design

INTRODUCTION

Composite construction by use of headed shear studs to connect steel beams and precast floor slabs to generate composite action in building construction are now common in the UK. The precast slabs are seated directly on the top flange of the steel beam and the transverse reinforcement is placed on site into the slot made by opening the core of the precast floor units. The opening cores and the narrow gap between the precast units to facilitate the placement of shear connectors and transverse reinforcement are filled with in situ concrete as shown in Figure 1.

The behaviour of the shear connection in a composite beam depends mainly on the relationship between the shear capacity of the connectors and the longitudinal slip at the interface between the top flange of the steel section and the concrete slab. This load – slip behaviour, usually found from the push-off test, depends on the type of connectors, their sizes and dimensions, the amount of transverse reinforcement, their spacing and the gap and strength of the insitu concrete infill. The overall beam behaviour also depends on the span of the beam and the degree of shear connection. Early work by Lam (2000a) showed that for the beam with full shear connection, a slip of 2mm was observed in the full-scale beam test at the ultimate load and the effects of slip can be neglected. On the other hand, for beams with partial shear connection, the effects of slip cannot be ignored. The ability of the shear connectors to maintain the maximum capacity with slip, i.e. the ductility of the shear connector, became a very important issue.

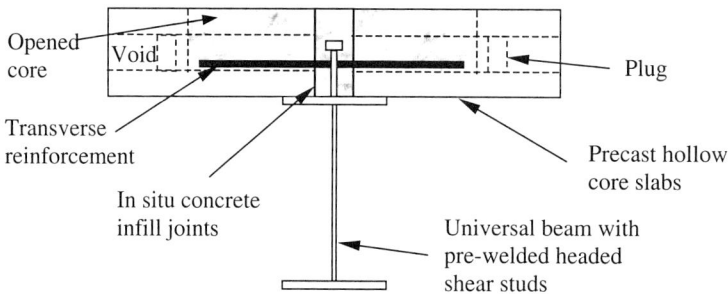

Figure 1: Composite beams with precast hollow core slabs

The current practice in designing composite steel beams with metal decking or solid RC slabs is to provide sufficient transverse reinforcement, in most of the cases, over-reinforced to avoid tensile splitting failure. The transverse bars are placed uniformly perpendicular to the line of shear connectors to ensure uniform shear flow. However, for composite beams with precast hollow core slabs, the cores for placing the transverse reinforcing bars are opened during casting and only alternative non-edge cores are opened to avoid the risk of core wall collapse. The typical spacing between two transverse bars can sometimes be bigger than 300mm across the transverse joint, which could affect the shear capacity of the connectors. This paper describes the functions of transverse reinforcement in the composite beams with hollow core floor slabs and presents design equations for calculate the shear capacity of studs based on the amount of the transverse reinforcement. The equations are compared with the experimental push off test results.

EFFECT OF TRANSVERSE REINFORCEMENT

Transverse reinforcement is used (a) to transfer shear force from the steel beam to the slabs and (b) to provide confinement against concrete splitting. The transverse reinforcement is the most influential parameter to the shear stud capacity for composite beams with precast hollow core slabs. Specimens with high reinforcement ratio showed significant increase in shear capacity and ductility over the specimens with low reinforcement ratio. This phenomenon was observed from all the push off tests when the mode of failure was concrete failure or combined stud shearing and concrete failure. An idealised load – slip relationship for shear connectors with high and low reinforcement ratio are shown in Figure 2.

The resistance against concrete splitting is provided by the tensile force of the concrete, F_{ct} and the tensile force of the transverse reinforcement, F_t. The shear capacity of the head shear stud of specimen with low reinforcement ratio, i.e. $F_t < F_{ct}$ will drop immediately after the concrete is cracked as the transverse reinforcement is already yielded. Whereas, the specimen with high transverse reinforcement ratio, i.e. $F_t > F_{ct}$ would able to maintain the load as redistribution of stresses take place between the concrete and transverse reinforcement with slip of at least 6mm. The specimen only failed when the ultimate strength of the transverse reinforcement is reached. This is particularly important when partial shear connection design is adopted.

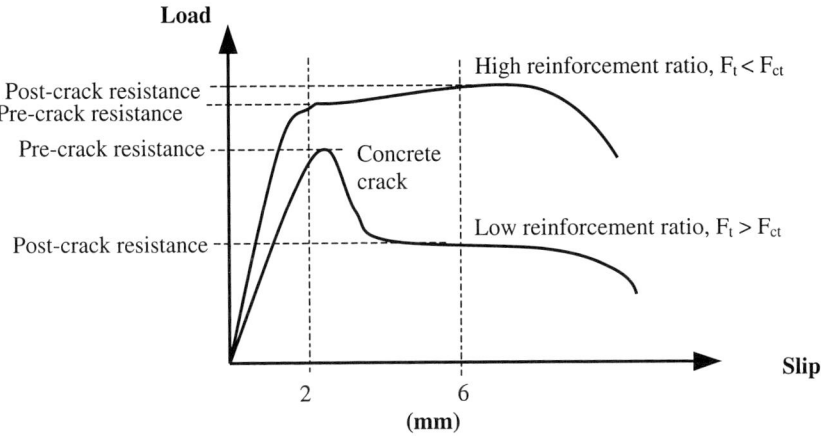

Figure 2: Idealised load – slip curves for specimen with high and low transverse reinforcement ratio

SPLITTING STRENGTH OF SHEAR STUD

Figure 3 shows the distribution of stresses when a concentrated load is applied, the dispersal of the concentrated load causes lateral tensile forces in front of the load and lateral compressive forces behind. Before concrete splitting occurs, both the concrete and the transverse reinforcement provide the resistance against the slab splitting. When the tensile strength of the concrete is reached, longitudinal crack forms first followed by the transverse cracks. As the longitudinal crack extends under increasing load, the transverse stresses became tensile and splitting failure occurs in this region.

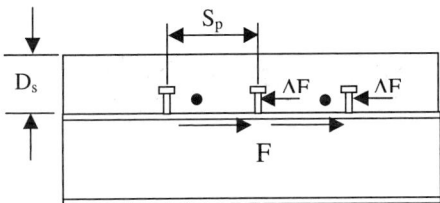

Figure 3: Longitudinal stress distribution of the composite beam

Therefore, the concrete splitting strength of a single stud is

$$P_{split} = f_{ct} \times D_s \times S_p \qquad (1)$$

Where f_{ct} = concrete splitting strength,
D_s = slab depth,
S_p = spacing between two studs

The tensile strength of the transverse reinforcement is

$$P_{rebar} = A_s \times f_y \qquad (2)$$

Where A_s = area of transverse reinforcement
f_y = yield strength of reinforcement

Post-splitting Strength

After the slab has split, the load is transferred from the slab to the stud through the cone of concrete by transverse reinforcement. Failure occurs when either the stud shear off or the transverse reinforcement yields. If the transverse reinforcement is not yielded before the concrete split, the load is transferred to the transverse reinforcement, therefore and no sudden drop of load capacity of the stud is observed. Since the plastic analysis allows the concrete cracking and stress redistribution in the floor slab to be considered. The tensile force exert by the transverse reinforcement can be taken as the ultimate strength of reinforcing bar.

If the post-splitting confinement strength by the transverse reinforcement is much lower than the pre-splitting strength, the shear resistance of the stud will drop and cause continuous slip across the interface. The corresponding slip will be less than the required slip of 6mm suggested for the partial interaction composition design (Johnson 1991). The shear connection will fail in concrete splitting. However, if the post-splitting confinement strength is higher than the pre-splitting strength, the shear connection will able to maintain the load until the ultimate strength of the transverse reinforcement is reached.

Proposed Equations

Two equations are proposed for estimate the shear capacity of headed shear connector for pre-splitting and post-splitting condition:

For pre-splitting strength, P_{pre}

$$P_{pre} = 1.1 f_{ct} \times h_s \times S_p + 0.67 A_s \times f_y \qquad (3)$$

Where f_{cu} = concrete cube strength of the insitu infill
h_s = height of headed stud

For post-splitting strength, P_{post}

$$P_{post} = f_{ct} \times h_s \times 3d + 0.67 A_s \times f_u \qquad (4)$$

Where f_u = ultimate strength of reinforcement
d = diameter of stud

PUSH OFF TEST RESULTS

Over one hundred push-off tests were conducted using the horizontal push-off test arrangement presented in Lam (2000b). The main parameters for the tests are transverse reinforcement, insitu infill concrete strength, stud diameter, insitu infill gap, and end details of hollow core unit. Test results of the 30 push-off tests related to the effect of transverse reinforcement are summarized in Table 1. In general, three mode of failure were observed; concrete failure; stud failure and combined concrete and shear stud failure. All tests listed in Table 1 failed in either concrete failure or combined concrete and stud yielding, no stud failure only was observed in these tests. Table 2 shows the yield and ultimate tensile strength of the reinforcing bars used in the tests.

TABLE 1
PUSH-OFF TEST RESULTS

Test No.	Slabs type and thickness [mm]	Transverse reinforcement	Stud Dia. [mm]	In-situ Concrete strength [N/mm^2]	Max. load per stud [kN]	Load at 6mm slip [kN]	P_{PRE} Eq. (3) [kN]	P_{POST} Eq. (4) [kN]
T10-C25-150-65SE	150 square-end	4 no. T10	19	23.4	57.33	54.12	64.75	48.19
T10-C35-150-80SE	150 square-end	4 no. T10	19	34.2	73.50	68.17	72.26	50.78
T10-C35-150-120SE	150 square-end	4 no. T10	19	38.8	83.67	72.48	75.09	51.76
T10-C50-150-120SE	150 square-end	4 no. T10	19	53.3	89.17	75.55	83.04	54.51
DT10-C40-150-80SE	150 square-end	8 no. T10	19	40.2	89.2	82.0	104.75	87.83
T12-C20-300-100SE	300 square-end	4 no. T12	19	19.9	91.0	81.3	72.53	57.67
T12-C40-300-100SE	300 square-end	4 no. T12	19	46.6	113.2	91.3	90.09	63.73
T12-C40-150-80SE	150 square-end	4 no. T12	19	40.3	81.33	74.62	86.54	62.51
DT12-C30-150-80SE	150 square-end	8 no. T12	19	29.0	101.1	90.2	118.79	106.26
T16-C25-200-80SE	200 square-end	4 no. T16	19	24.7	99.67	99.47	109.92	97.62
T16-C25-150-80SE	150 square-end	4 no. T16	19	23.4	97.33	96.0	108.93	97.28
T16-C25-150-80SE	150 square-end	4 no. T16	19	26.7	114.0	114.0	111.38	98.12
T16-C25-200-100SE	200 square-end	4 no. T16	19	28.4	100.17	99.15	112.58	98.54
T16-C40-150-80SE	150 square-end	4 no. T16	19	42.1	127.8	118.0	121.19	101.51
T8-C25-150-65TE	150 tapered-end	4 no. T8	19	23.5	56.7	50.1	54.05	32.74
T12-C25-150-65TE	150 tapered-end	4 no. T12	19	26.6	69.83	65.19	77.70	59.45
T16-C25-150-65TE	150 tapered-end	4 no. T16	19	26.6	81.18	81.17	111.31	98.10
T16-C40-150-65TE	150 tapered-end	4 no. T16	19	44.1	118	117.3	122.32	101.90
T16-C40-200-65TE	200 tapered-end	4 no. T16	19	44.1	120.0	118.0	122.32	101.90
T16-C40-250-65TE	250 tapered-end	4 no. T16	19	44.1	122.0	120.0	122.32	101.90
T16-C25-150-80SE22	150 square-end	4 no. T16	22	24.4	117.5	118.0	109.69	99.54
T16-C40-150-80SE22	150 square-end	4 no. T16	22	40.1	128.0	120.0	120.03	103.68
T20-C25-200-80SE22	200 square-end	4 no. T20	22	26.7	121.0	122.0	146.98	144.59
T20-C25-250-80SE22	250 square-end	4 no. T20	22	28.8	132.0	133.0	148.46	145.18
T16-C25-150-65TE22	150 tapered-end	4 no. T16	22	25.5	117.5	117.0	110.51	99.87
T16-C40-150-65TE22	150 tapered-end	4 no. T16	22	40.1	128.0	120.0	120.03	103.68
T16-C40-200-65TE22	150 tapered-end	4 no. T16	22	40.1	120.0	120.0	120.03	103.68
T16-C40-250-65TE22	150 tapered-end	4 no. T16	22	40.1	128.0	129.0	120.03	103.68
T20-C25-200-80TE22	200 tapered-end	4 no. T20	22	25.0	132.0	130.0	145.74	144.09
T20-C25-250-80TE22	250 tapered-end	4 no. T20	22	23.1	132.0	128.0	144.30	143.51

TABLE 2
STRENGTH OF TRANSVERSE REINFORCEMENT

Reinforcement	Yield Strength (N/mm^2)	Ultimate Tensile Strength (N/mm^2)
T8	525 – 545	589 – 617
T10	544 – 552	630 – 733
T12	514 – 525	601 – 619
T16	535 – 549	627 – 633
T20	505 – 526	604 – 623

In order to check the accuracy of the proposed equations, the results of push off tests were compared with the estimated pre-splitting and post-splitting shear capacity of the studs. The actual yield and ultimate tensile strength of the reinforcing bars were used. The calculated shear capacities are shown in Table 1. The estimate results are within 10% of the push-off test results, the discrepancy of the model are mainly due to the effect of the other parameters such as gap width, slab height, etc. Figure 4 shows the position of transverse reinforcement and end details of the push-off tests.

(a) Single reinforcing bar(T) per open core with square-end(SE)

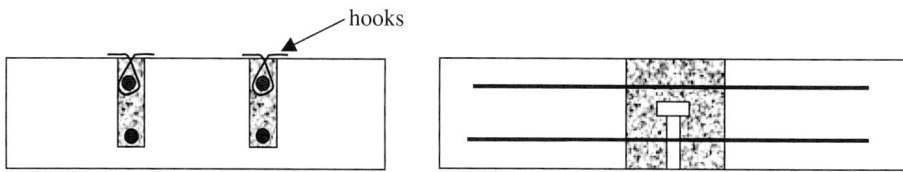

(b) Dual reinforcing bars(DT) per open core with square-end(SE)

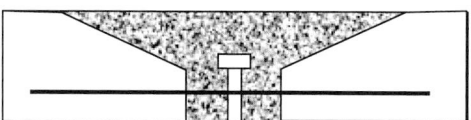

(c) Single reinforcing bar(T) per open core with tapered-end(TE)

Figure 4: Transverse reinforcement position and end details

For the push-off tests with dual traverse reinforcing bars, the purpose of the top bar is to control the concrete splitting. However, because of the position of the top bar being above the height of the headed stud, the shear capacity of the headed stud is not performed as well as the single transverse bar with equivalent cross section area, but it did allow some redistribution of load.

Figure 5 & 6 show the load – slip curve of T10-C35-150-80SE, specimen with low transverse reinforcement ratio and the load – slip curve of T16-C25-200-80SE, specimen with high transverse reinforcement ratio. The calculated pre-splitting and post-splitting load by the proposed equations were compared with the load – slip curves as well.

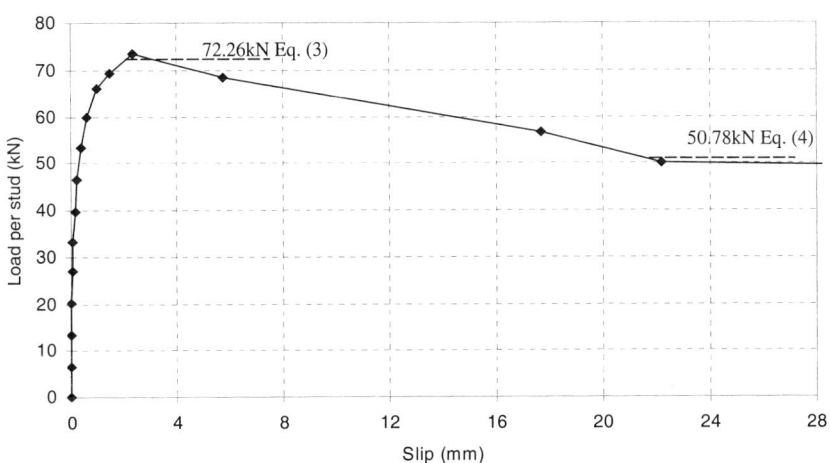

Figure 5: Load – slip curve of T10-C35-150-80SE

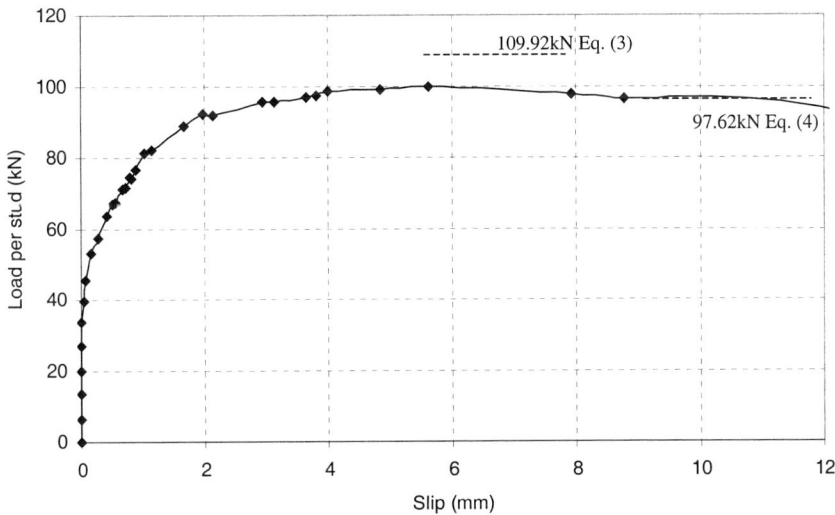

Figure 6: Load – slip curve of T16-C25-200-80SE

CONCLUSIONS

Transverse reinforcement is used to provide ties for the slabs and confined concrete from splitting. The ductility of the shear connector, i.e. slip capacity is directly affected by the amount of transverse reinforcement. Design equations presented in this paper for estimate the shear capacity of the headed shear stud showed a good correlation with the push-off test results. For full shear connection design, pre-splitting shear capacity of the headed stud can be used for the composite design, while for partial shear connection design; post-splitting shear capacity of the headed stud should be used. In general, a minimum transverse reinforcement of T16 bars should be used if partial shear connection design is used to ensure a minimum ductility of 6mm slip.

REFERENCES:

Johnson R. P. and Molenstra N. (1991) Partial shear connection in composite beams for buildings, *Proceeding of the Institution of Civil Engineers, Part 2*, Vol.91, pp 679-704.

Lam, D., Elliott, K. S. and Nethercot, D. A. (2000a) Experiments on composite steel beams with precast concrete hollow core floor slabs, *Proceedings of the Institution of Civil Engineers: Structures and Buildings*, Vol. 140, pp.127-138.

Lam, D. (2000b) New Test for Shear Connectors in Composite Construction, *United Engineering Foundation Conference, Composite Construction in Steel and Concrete IV*, Banff, Alberta, Canada.

ACKNOWLEDGEMENT

The authors would like to acknowledge the financial support provided by EPSRC, Bison Concrete Products Ltd. and Severfield-Reeve Plc. and the skilled assistance provided by the technical staff in the School of Civil Engineering at Leeds University.

SHEAR CONNECTION IN COMPOSITE BEAMS INCORPORATING PROFILED STEEL SHEETING WITH NARROW OPEN OR CLOSED STEEL RIBS

M. Patrick and R.Q. Bridge

Centre for Construction Technology & Research, College of Science,
Technology & Environment, University of Western Sydney, Kingswood,
NSW 2747, Australia

ABSTRACT

Most Australian steel deck profiles have a reentrant narrow open steel rib, a closed lap-joint rib, or a combination of both. A series of tests has been conducted to examine the influence of the narrow open steel rib on the performance of the shear connection in composite beams with the profiled steel sheeting ribs laid perpendicular to the beams. Where two or more studs are used at a cross-section, it was found that the minimum transverse spacing had to be increased to 80mm for the connection to behave in a similar manner to that in a solid slab. Rib punch-through of the narrow open steel rib is acceptable provided its effect on reducing the shear capacity of connectors is sufficiently small that it can be ignored in design. Another series of tests has been conducted to examine if the presence of closed lap ribs had any effect on the shear connection. It was found that studs, pairs or single, could be placed adjacent to the closed lap rib without any reduction in stud shear strength compared to that in a solid slab. For profiles consisting of only closed lap ribs at 300 mm centres, three stud pairs could be fitted to each pan at a nominal longitudinal 100 mm spacing which, allowing for a construction tolerance, means the studs can be placed within only 45 mm clear of the webs of the ribs.

For steel sheeting parallel to the beam, an additional requirement for the maximum spacing of the studs has been developed. The objectives of this are to prevent the occurrence of a brittle rib-shearing failure and to ensure that straight reinforcing bars placed transverse to the steel beam are effective as longitudinal shear reinforcement. If this limit is exceeded, additional waveform reinforcement is necessary to cross the horizontal shear surface passing at the level of the tops of the ribs that can form in the region between the shear connectors.

New design rules that account for all of these factors have recently been introduced into Australian Standard AS 2327.1, or are being considered for inclusion, and are described in the paper.

KEYWORDS

Closed steel ribs; composite beam; longitudinal shear; reentrant open steel ribs; reinforcing steel; rib punch-through failure; rib shearing failure; shear connector.

INTRODUCTION

Most Australian steel deck profiles have a reentrant narrow open steel rib, a closed lap-joint rib, or a combination of both as shown in Fig. 1.

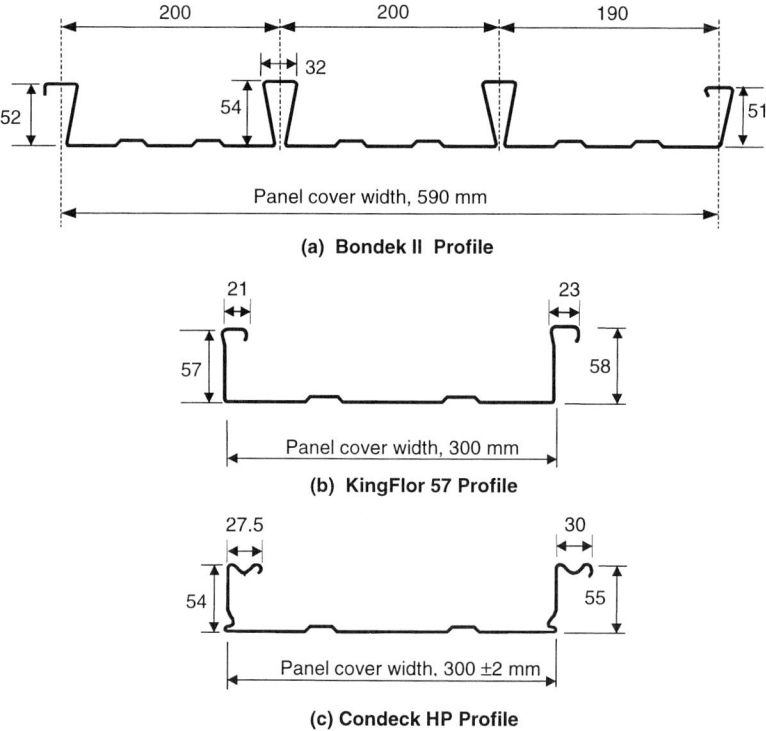

Figure 1: Typical Australian Steel Deck Profiles

The major design objectives for the shear connection in composite beams incorporating the Australian steel decks with the profiled steel sheeting ribs laid perpendicular to the beams is that it will not only be efficient and economical, but that it will not cause the composite beam to fail prematurely. Therefore, the shear connectors must exhibit sufficient ductility to allow the assumption to be made that all the connectors can effectively be loaded to their design shear capacity, as for a solid slab, between any adjacent pair of critical cross-sections or between a critical cross-section and the adjacent end of the beam. Secondary modes of failure, as detailed in Australian Standard AS2327.1, must be prevented. This is usually achieved by stringent detailing requirements: e.g. studs must be sufficiently far away from narrow open steel ribs; special types of reinforcement (OneSteel Reinforcing, 2001, Patrick and Bridge, 2002) must be used where necessary; and the transverse spacing of studs must be sufficiently wide. Rib punch-through of the narrow open steel rib is acceptable provided its effect on reducing the shear capacity of connectors below that for a solid slab is sufficiently small that it can be ignored in design. These details and further amendments have been determined from the results of extensive pushout and beam tests (Patrick, 2001).

For the shear connection in composite beams incorporating the Australian steel decks with the profiled steel sheeting ribs laid parallel to the beams, there are now new additional detailing requirements to prevent a form of premature longitudinal concrete rib shearing failure (Type 3 longitudinal shear failure) in which the failure surface forms horizontally between adjacent ribs and locally avoids the studs by passing over their heads (Patrick, 2000).

Therefore, a feature of the design rules in AS 2327.1 for determining the design shear capacity, f_{ds}, of welded studs placed through profiled steel sheeting is that the same values are used as for solid slabs and that the design shear capacity, f_{ds} is reached with very little slip (in practice, typically 2-4 mm) and maintained indefinitely (typically 8-10 mm). This strong ductile behaviour of the shear connection has been made possible, partly because it has not been permissible to use open-trough profile decks, and also because of the special detailing of the shear connection that is required.

DEFINITION OF NARROW OPEN RIBS AND CLOSED LAP RIBS

To meet the requirements of AS2327.1, a narrow open rib is defined as one where: the overall height of the steel rib is not greater than 80 mm, excluding any embossments; the width of the opening at the base of the steel rib is not greater than 20 mm; and the area of the voids formed by the steel ribs in the concrete shall not be greater than 20% of the area of the concrete within the depth of the steel ribs.

For a lap joint rib to be considered as closed, obviously some limitation must be placed on the gap between the nominal contact surfaces of the overlapping ribs. Based on observations of this form of overlapping rib for Australian profiles, it is proposed that the gap must not exceed 3 mm and that the vertical component of the overlapping ribs must be in nominal contact over more than 60% of its area. This is to account for non-contact regions in some deck profiles (see Fig. 1(c) and Fig. 4(a)) that may arise from fixing slots or embossments in these regions. These must be located more than 5mm above the pan of the sheeting.

SHEAR CONNECTION WITH THE PROFILED STEEL SHEETING PERPENDICULAR TO THE STEEL BEAM

Narrow Reentrant Open Steel Rib and Closed Lap joint Rib of BONDEK II

A panel of BONDEK II profiled steel sheeting comprises two narrow open ribs and a closed lap joint as can be seen at (a) in Fig. 1. The load-slip curves from two push-out tests performed on BONDEK II are shown in Fig. 2 (Patrick et al, 1995). In one of the tests (see Fig. 2(a)), a pair of 19 mm diameter studs with a transverse centre-to-centre spacing of 60 mm was located centrally in the middle pan between the narrow open ribs, and thrust against one of these open ribs.

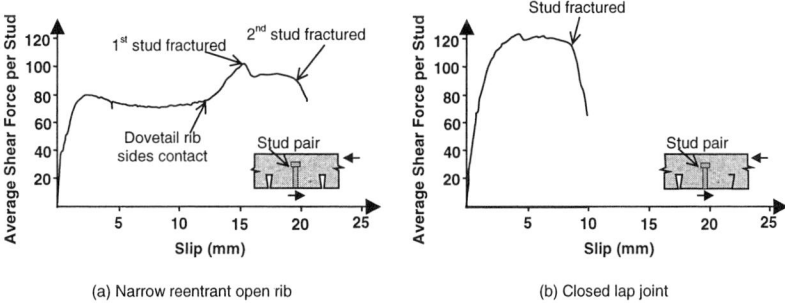

Fig. 2 Push-out test load-slip curves for open and closed ribs

In accordance with AS 2327.1, 60 mm is the minimum allowable centre-to-centre transverse spacing of 19 mm studs. In the other test (see Fig. 2(b)), a similar pair of 19 mm diameter studs, this time located centrally in an outer pan, thrust against the closed lap joint. The push-out specimens were both reinforced with DECKMESH™ (OneSteel Reinforcing, 2001) in order to prevent premature rib shearing failure, which could have led to a brittle load-slip response.

As seen from Fig. 2(a), the average shear force per connector reached a maximum value of approximately 80 kN at a slip of about 2.5 mm before falling away slightly. Rib punch-through occurred during this stage of the test. The shear force per connector then rose to about 105 kN before one of the studs fractured at its base at a slip of about 15 mm. At a slip of approximately 20 mm, the second stud broke and the specimen had then completely failed. During this test, a wedge of concrete formed in front of the studs, similar in shape to that seen in Fig. 3. The pressure from this wedge caused the open rib to close up adjacent to the studs. As indicated in Fig. 2(a), when the open rib sides contacted each other, the force on each stud began to rise again, which eventually caused both of the studs to fracture.

The load-slip curve in Fig. 2(b) closely resembles curves obtained from testing studs in solid slabs under otherwise similar conditions. It is apparent that the closed lap joint did not have a detrimental effect on the performance of the stud pair.

Tests performed on specimens similar to that corresponding to Fig. 2(a), but with only one 19 mm stud instead of a pair of studs per pan, produce a load-slip curve very similar to that in Fig. 2(b). It follows that if the transverse spacing of the studs in a stud pair located in the central pan of BONDEK II is increased beyond a certain limit, then their shear strength and load-slip curve should not be influenced by the presence of the narrow open rib.

Some tests have recently been performed to investigate the effect of increasing the transverse spacing between pairs of 19 mm diameter welded studs, placed in the middle of the pan between two open ribs of BONDEK II. The result of a typical test is shown in Fig. 3.

(a) Fractured studs and rib punch-through exposed by cutting away sheeting pan (b) Fractured concrete removed to expose side of open rib

Figure 3 Rib punch-through failure for narrow open steel rib of BONDEK II

A major objective was to determine for design purposes at what transverse spacing the studs could be considered to be acting in a solid slab. At 60 mm centres, the reduction in shear capacity, as seen in Fig. 2(a), was considered too great.

The tests were performed using a horizontal push-out test rig (Patrick, 2001) for economical one-sided test specimens with the transverse centre-to-centre spacing, s_t, equal to 80, 120 or 160 mm. A specimen with $s_t = 80$ mm is shown in Fig. 3. It can be seen that both of the studs fractured at their base. Cutting away the sheeting in the pan between the two open ribs exposed the crack pattern visible in Fig. 3(a). This pattern was observed in all the tests irrespective of the value of s_t. It is synonymous with the cone failure associated with fasteners embedded in concrete acted on by a horizontal force while close to a long free edge (Comite Euro-International du Beton, 1997). The distance that the cone extended past the centre of the stud was normally about 1.5 times the distance from the centre of the stud to the edge of the rib at its base (like the top diagonal crack in Fig. 3(a)). This value is normally assumed when designing fasteners (Comite Euro-International du Beton, 1997). As can be seen from Fig. 3(b), the depth of the cone that sheared off equalled the height of the open rib. This was shallower than what would have been expected had it been a deeper free edge. Therefore, some force was probably transmitted over the top of the open rib. The pan would also have restrained the concrete and increased the shear strength to some degree.

Despite rib punch-through occurring, the tests showed no significant difference in shear capacity for $s_t = 80$, 120 or 160 mm with the studs fracturing in all cases. According to Comite Euro-International du Beton (1997), treating the slab as only having a depth equal to the height of the open rib and also ignoring the presence of the sheeting and transverse reinforcement in the slab, the studs should have been weakened by 10 to 30 percent for the largest and smallest spacings, respectively. However, this was not observed in the tests, probably for the reasons given above. Therefore, it is recommended that the minimum centre-to-centre spacing of 19 mm welded studs be increased to 80 mm for BONDEK II. Then, no additional reduction factor is applied when calculating their design shear capacity to account for the presence of the open ribs.

Closed Lap joint Ribs of CONDECK HP

Push-out and beam tests were performed to investigate the effect of placing pairs of 19 mm diameter welded studs very close to the closed lap joint ribs of CONDECK HP according to the detail in Fig. 4(a). The ribs were perpendicular to the steel beam. This detail does not conform with Clause 8.4.1(c) of AS 2327.1, which requires a minimum of 60 mm of concrete between the face of a stud and the adjacent sheeting rib. Moreover, the stud pairs were spaced along the test specimens at 200 mm centres, noting that the ribs of CONDECK HP are at 300 mm centres (See Fig. 1(c)). Thus, a pair of studs was placed in the middle of the pans adjoining the pan shown in Fig. 4(a) with the studs next to the rib, and another pair of studs was placed in the same pan 200 mm away. Therefore, there were four studs in the pans with the studs close to the ribs.

From the push-out and beam tests, it was observed that many of the shear studs eventually fractured and that the shear force vs slip relationships were representative of shear connectors in a solid slab. Namely, the peak shear forces were well in excess of the nominal shear capacity given in Table 8.1 of AS 2327.1 for the appropriate concrete strength, and the slips measured at peak shear force were typical of tests on studs in a solid slab. It was concluded that CONDECK HP did not have a detrimental effect on the performance of the shear connectors compared with had they been in a solid slab. The tests also showed that DECKMESH™ reinforcement (OneSteel Reinforcing, 2001) is satisfactory for this intensity of shear connectors in edge beams, i.e. pairs of 19 mm studs at 200 centres longitudinally and 60 mm transversely.

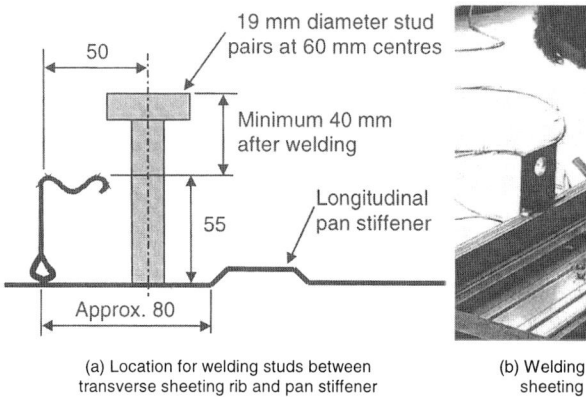

(a) Location for welding studs between transverse sheeting rib and pan stiffener

(b) Welding studs through CONDECK HP between sheeting rib and pan stiffener as shown in (a)

(c) Close-up of studs after welding adjacent to CONDECK HP rib, and with a 60 mm centre-to-centre spacing transverse. DECKMESH was used because it was an edge beam.

(d) Stud pairs along sheeting at 200 mm centre-to-centre spacing, resulting in 2-4-2-4-etc studs per pan, while AS 2327.1-2002 requires max. two studs per pan. It was an internal beam, so STUDMESH was placed directly on the sheeting ribs as Type 2 reinforcement.

Figure 4 Shear connection details for tests on push-out specimens and beams incorporating closed lap ribs of CONDECK HP

SHEAR CONNECTION WITH THE PROFILED STEEL SHEETING PARALLEL TO THE STEEL BEAM

Type 3 Longitudinal Shear Surface

A Type 3 longitudinal shear surface may form instead of a Type 2 surface if at least one steel sheeting rib runs parallel to the steel beam. It the rib is sufficiently close to the shear connectors, the failure surface first passes to the top of the rib instead of to the base of the slab, i.e. the failure surface will form along the path of least resistance. A possible shape of the failure surface in the region of a shear connector is shown in Fig. 5(a) passing over the top of the connector to the top of each adjacent rib. If the shear connectors are well spaced apart along the steel beam (which is parallel to the sheeting ribs), then it is possible that the shear surface will become approximately horizontal in the regions between the connectors, passing directly between the tops of the sheeting ribs as shown in Fig. 5(b). The narrower the width of the concrete rib, then the more likely this will occur. A horizontal shear surface has also been observed in tests when the shear connectors are short and may not even protrude above the tops of the sheeting ribs (Veldanda and Hosain, 1992).

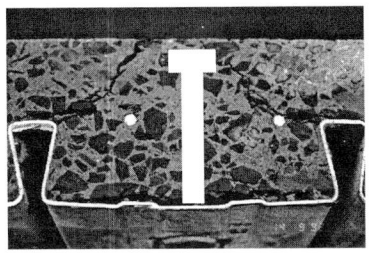
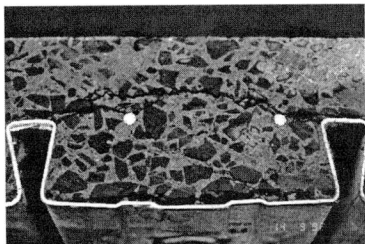

(a) Possible shape of shear surface in region of shear connector

(b) Possible shape of shear surface away from shear connector

Figure 5 Possible Type 3 longitudinal shear surface (photos altered by computer)

It follows from this discussion that various parameters can affect the shape of a Type 3 longitudinal shear surface. Limited account has been taken of this in AS 2327.1 to date, and a new rule on this matter is proposed as outlined in the following section. Moreover, if along the entire length of the beam the shear surface takes the form shown in Fig. 5(a), then straight reinforcing bars placed transverse to the steel beam will pass through the shear surface and therefore act as shear reinforcement. However, in regions where the shear surface is effectively horizontal, then some other form of reinforcement is required. OneSteel Reinforcing and the Centre for Construction Technology and Research have lodged a joint patent for a waveform reinforcement in the vertical plane which prevents longitudinal shear failure by passing through the horizontal regions of a Type 3 shear surface.

Additional requirement for detailing the shear connectors

From the discussion above, an additional requirement, as outlined in Fig. 6, is proposed concerning the maximum longitudinal spacing of shear connectors defined in Clause 8.4.1(b) of AS 2327.1. This applies if the profiled steel sheeting is deemed parallel to the steel beam and a Type 3 longitudinal shear failure is possible. The requirement is particularly relevant to primary beams. The objective of this rule is to ensure that conventional Type 1 or 3 shear surfaces can only form. Therefore, straight bars placed transverse to the steel beam will be effective as longitudinal shear reinforcement.

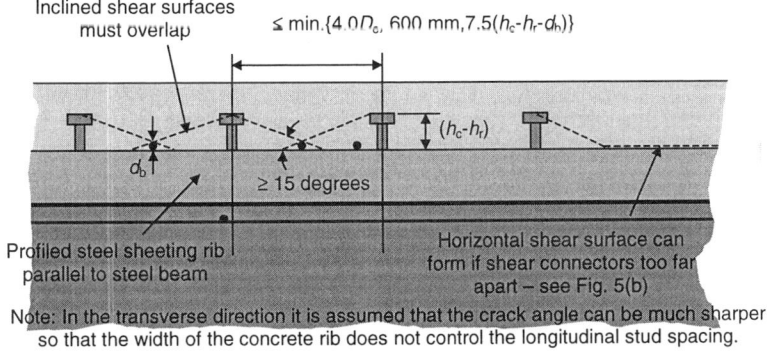

Fig. 6 Maximum longitudinal spacing of shear connector groups for possible Type 3 shear surface

If these spacing limitations are exceeded, vertical waveform reinforcement or special "handle-bar" reinforcement is required in order to cross the horizontal shear surface that can form in the region between shear connectors. Also, this rule may sometimes necessitate either using more-closely-spaced single shear connectors rather than pairs of connectors, or increasing the height of the connectors.

CONCLUSIONS

For composite beams with the profiled steel sheeting having open steel ribs laid perpendicular to the beams where two or more studs are used at a cross-section, it was found that the minimum transverse spacing had to be increased to 80mm for the shear connection to behave in a similar manner to that in a solid slab providing the distance of the studs from the ribs was greater than 60 mm. Rib punch-through of the narrow open steel rib could still occcur but its effect on reducing the shear capacity of connectors was sufficiently small that it could be ignored in design. The presence of closed lap ribs had little effect on the shear connection. It was found that studs, pairs or single, could be placed adjacent (within 45 mm) to the closed lap rib without any reduction in stud shear strength compared to that in a solid slab.

For steel sheeting parallel to the beam, the maximum longitudinal spacing of the studs has been limited to a value that prevents the occurrence of a brittle rib-shearing failure and ensures that straight reinforcing bars placed transverse to the steel beam are effective as longitudinal shear reinforcement. If this limit is exceeded, additional vertical waveform reinforcement or another form of reinforcement is necessary to cross the horizontal shear surface passing at the level of the tops of the ribs.

REFERENCES

Comite Euro-International du Beton (1997), *Design of Fastenings in Concrete – Design Guide*, Thomas Telford, London.

OneSteel Reinforcing (2001). *DECKMESH™ & STUDMESH® – Composite Slab Shear Reinforcement Solutions*, CD-ROM 3, Sydney, Australia.

Patrick, M. (2000). Experimental Investigation and Design of Longitudinal Shear Reinforcement in Composite Edge Beams, *Journal of Progress in Structural Engineering and Materials*, **2:2**, April-June, 196-217.

Patrick, M. (2001). *Design of the Shear Connection of Simply-Supported Composite Beams, Composite Structures Design Manual - Design Booklet DB2.1*, OneSteel Market Mills.

Patrick, M. and Bridge, R.Q. (2002). Novel New Reinforcing Components for Composite Beams, *Proceedings*, International Conference - Advances in Building Technology, Hong Kong, 4- 6 Dec.

Patrick, M., Dayawansa, P.H., Eadie, I., Watson, K.B. and van der Kreek, N. (1995). Australian Composite Structures Standard AS2327.1, Part 1: Simply-Supported Beams, *Steel Construction*, Australian Institute of Steel Construction, Vol. 29, No. 4, pp. 2-40.

Standards Australia (2002). *AS 2327.1-2002: Composite Structures - Simply Supported Beams*, Sydney.

Veldanda, M.R. and Hosain, M.U. (1992). Behaviour of Perfobond Rib Shear Connectors: Pushout Tests, *Canadian Journal of Civil Engineering*, Vol. 19, No. 1, February, pp. 1-10.

SHEAR CONNECTION IN COMPOSITE BEAMS INCORPORATING OPEN-TROUGH PROFILE DECKS

M. Patrick and R.Q. Bridge

Centre for Construction Technology & Research, College of Science,
Technology & Environment, University of Western Sydney, Kingswood,
NSW 2747, Australia

ABSTRACT

For composite steel-concrete beams, it is well known from tests that the shear capacity of welded-stud connectors can be reduced by the presence of profiled steel sheeting if it has an open-trough profile where the open-trough ribs, usually trapezoidal in shape, are laid transverse to the steel beam. This is compared with if the studs were located in a solid concrete slab (or adjacent to closed sheeting ribs).

The approach taken in overseas design Standards has been to apply a reduction factor, k_t (≤ 1.0), to the design shear capacity of connectors in a solid slab, to account for the presence of the open-trough profile deck. Values as low as 0.4 are not uncommon.

The failure mode can be of the type where a horizontal failure surface forms over the tops of the sheeting ribs, locally passing over the tops of the shear studs which extend above the ribs. This is referred to as rib shearing failure, which is a special type of longitudinal shear failure recognized in Australian Standard AS 2327.1. It is a brittle mode of failure that is not acceptable in AS 2327.1, which requires the shear connection to be ductile, all critical shear connectors being assumed to carry equal shear force at ultimate load. Rib shearing failures with open-trough profile decks can be prevented using special waveform reinforcement.

Rib punch-through is another possible mode of failure which is also brittle in nature. For this mode, a new, patented, low-cost small-scale component has been developed that, when used in conjunction with each stud connector, can significantly increase the strength and ductility of the shear connection resulting in improved values of the reduction factor k_t, in some cases the connection achieving the full capacity of connectors in a solid slab. An extensive series of tests is currently being undertaken to design the component, which takes the form of a steel ring positioned about the stud base, for a wide range of deck parameters and stud configurations.

KEYWORDS

Composite beam; shear connector; longitudinal shear; reinforcing steel; rib punch-through failure; rib shearing failure.

INTRODUCTION

The shear connection of a composite beam normally comprises the five components shown in Figure 1. They can all influence the behaviour of the shear connection, and can be considered to include the shear connectors, profiled steel sheeting, slab reinforcement (including longitudinal bars if beams are continuous), concrete slab and steel beam top flange.

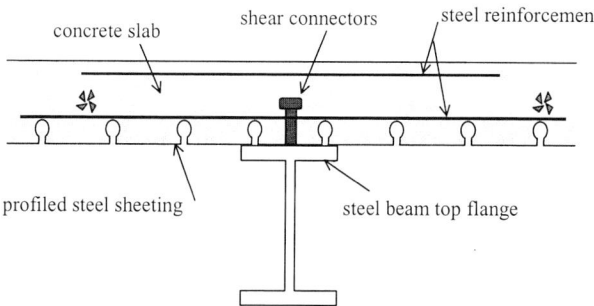

Figure 1: Components of the shear connection of a composite beam

The strength design method in AS 2327.1 (Standards Australia 2002) requires that the shear connection displays "ductile" behaviour, which can be assessed by knowing the shape of the load-slip curves from push-out tests, and also the magnitude of the slips required along typical composite beams at ultimate load. No specific guidance is given to designers on either of these matters in AS 2327.1. However, Eurocode 4, Part 1.1 (British Standards Institution 1994) requires ductile shear connectors to have a slip capacity of at least 6 mm. Ductile and brittle behaviour of the shear connection, as well as the assumed model for design in AS 2327.1, are shown diagrammatically in Figure 2. A feature of this model for the shear connection covered by AS 2327.1 is that the design shear capacity, f_{ds} is reached with very little slip (in practice, typically 2-4 mm) and maintained indefinitely (typically 8-10 mm).

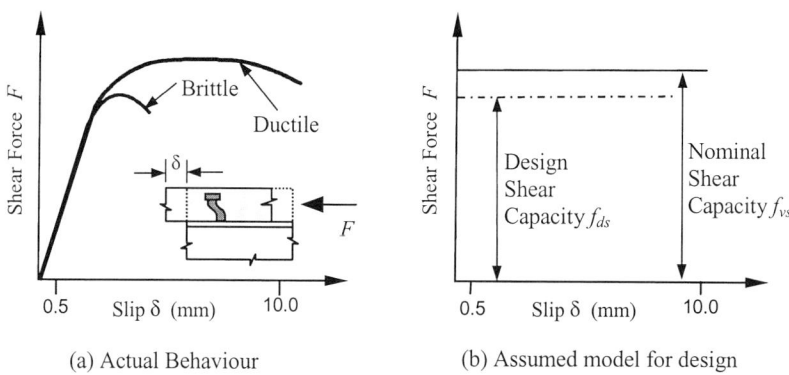

Figure 2: Load-slip behaviour of the shear connection

A feature of the design rules in AS 2327.1 for determining the design shear capacity, f_{ds}, of welded studs placed through profiled steel sheeting is that the same values are used as for solid slabs. This has been made possible, partly because it has not been permissible to use open-trough profile decks, and also because special detailing of the shear connection is required. For example: studs must be sufficiently far away from narrow open deck ribs; transverse spacing of studs may be increased; and special types of reinforcement must be used when necessary.

Three new types of patented reinforcing components have been developed in Australia to ensure that design engineers can detail economical longitudinal shear reinforcement in the presence of profiled steel sheeting with re-entrant or closed ribs (Patrick and Bridge 2002). Otherwise, it is possible that non-composite beams, or even other forms of construction, will be more economical. The new components were designed for use with automatically welded, headed studs, the only type of shear connector referred to in AS 2327.1 that may be attached directly through profiled steel sheeting.

One type of reinforcing component prevents rib shearing failure when narrow re-entrant or closed ribs are orientated perpendicular to the longitudinal axis of the steel beam. It is waveform in shape, and has been adapted for use with decks with an open-trough profile. It is described in the paper that once rib shearing failure is suppressed using this waveform reinforcement, however, another failure mode called rib punch-through can occur. This undesirable mode of failure causes the studs to have very low shear strength. Some of the initial work that has been undertaken to overcome the detrimental effects of rib punch-through failure is described.

REDUCTION IN SHEAR CAPACITY OF WELDED STUDS IN OPEN-TROUGH PROFILE DECKS

The approach taken in overseas design codes has been to apply a reduction factor, k_t (≤ 1.0), to the design shear capacity of connectors in a solid slab, to account for the presence of the sheeting. This approach of referencing solid-slab design values has been adopted in order to simplify design.

Johnson and Anderson (1993) explain the basis on which expressions in BS 5950:3.1 (British Standards Institution 1990), for ribs perpendicular to the steel beam with either one, two, or three or more studs per pan, were modified for Eurocode 4, Part 1.1 to give the following relationship (see Fig. 3 for definition of terms):

$$k_t = \left(\frac{0.7}{\sqrt{N_r}}\right)\left(\frac{b_{cr}}{h_r}\right)\left[\left(\frac{h_c}{h_r}-1\right)\right] \leq 1.0 \text{ (or 0.8 when } N_r=2) \tag{1}$$

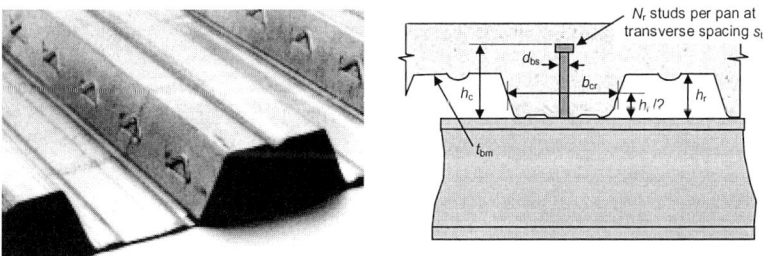

(a) Trapezoidal "3W" profile – $h_r=3"=76$ mm   (b) Reduction factor parameters

Figure 3: Open-trough deck and shear connector reduction factor parameters

Some restrictions that apply to the use of Eq. 1 are that $d_{bs} \leq 20$ mm, $t_{bm} > 1.0$ mm, N_r is not to exceed 2.0 in computations, $h_r \leq 85$ mm, $b_{cr}/h_r \geq 1$, $h_c \geq h_r+35$ mm, $s_t \geq 4d_{bs}$ and the studs must be welded through the steel sheeting. If the sheeting base metal thickness, $t_{bm} \leq 1.0$ mm then the upper limits for k_t in Eq. 1

become 0.85 and 0.7 for $N_r=1$ and 2, respectively. Similarly, if the sheeting is holed (which is not normal practice in Australia) then these limits are further reduced to 0.75 and 0.60, but studs with a diameter of up to 22 mm may then be used.

After analysing the results of 136 push-out tests with the sheeting ribs laid transverse to the steel beam, Johnson (2000) concluded that Eq. 1 can be unsafe, citing a case where the test strength was overestimated by 50 per cent. He explains that this is because two essential parameters are missing from the equation, viz.: the thickness of the steel sheeting; and the location of the stud/s relative to the webs of the profile. It appears that Eq. 1 is least satisfactory when rib punch-through failures can occur. This mode of failure is most likely when a central stiffener is located in each pan, and the studs are placed in the unfavourable location, such that the zone of concrete in compression in front of the studs is least. Despite these shortcomings, it appears that Eq. 1 will still be used in Eurocode 4 (CEN 2001). It is also worth noting that the real situation is likely to be worse than Johnson concluded because ductility was ignored in the study. The significance of this omission is partly described below.

The authors have also studied the accuracy of Eq. 1 and found it to be a very poor predictor. For example, for an open-trough profile deck like that shown in Fig. 3(a), with a single stud placed centrally in each pan, Eq. 1 can give $k_t=1.0$, and yet the strength in a test may only be about 50 to 60 percent that of the same stud in a solid slab. Moreover, the failure mode can be very brittle. For example, in tests using the decking shown in Fig. 3(a), a rib-shearing failure occurred (see Fig. 4(a)) and the load-slip behaviour was poor (see Fig. 4(b)) (Patrick 2001). Tests also show that when the failure is either by rib shearing (Fig. 4(a)) or rib punch-through (see Fig. 5), the shear strength is in fact hardly affected by the height of the stud, h_c, despite its apparent importance in Eq. 1.

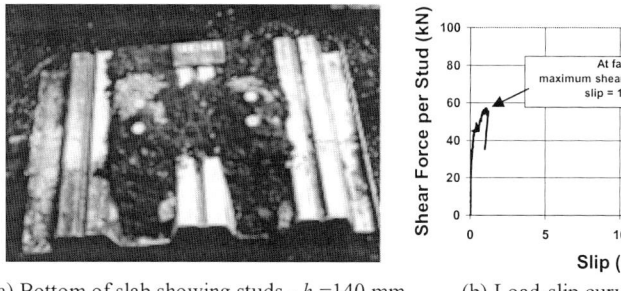

(a) Bottom of slab showing studs - $h_c=140$ mm   (b) Load-slip curve - $h_c=140$ mm

Figure 4: Brittle rib-shearing failure with an open-trough profile deck

WAVE-FORM REINFORCEMENT TO PREVENT RIB SHEARING FAILURE

Rib Shearing Failure Mode

A patented waveform reinforcing component, DECKMESH™ (OneSteel Reinforcing 2001), has been developed to prevent rib shearing failure in composite beams when profiled steel sheeting is laid perpendicular to the longitudinal axis of the steel beam, and welded-stud shear connectors are fastened directly through the sheeting. This brittle failure mode seen in Fig. 4(a) involves horizontal cracks that form in the concrete ribs between the tops of the steel sheeting ribs in pans where there are shear studs, while the failure surface locally avoids the studs by passing over their heads (Patrick 2000). It is now mandatory in AS 2327.1 to reinforce against rib shearing failure, in certain circumstances, which is referred to therein as Type 4 longitudinal shear failure. The product was initially developed for use with steel decks with re-entrant or closed ribs. A version has now been developed for use with open-trough profile decks.

Push-out Tests on Open-Trough Profile Decks with Waveform Reinforcement

A companion push-out specimen was tested with that shown in Fig. 4(a) that exhibited rib shearing failure at a slip of about 1 mm. As can be seen in Fig. 5(a), special, patented waveform reinforcement was added to the companion specimen to prevent rib shearing failure - the heavy top bars are not part of the waveform reinforcement, but provide flexural strength to the push-out specimen. Unlike the specimen without this waveform reinforcement, it exhibited significant shear strength at large slips (see Fig. 5(b)). After testing, no cracks were visible on the sides of the slab indicating that rib shearing failure did not eventuate, but rib punch-through occurred and was the cause of the noticeable drop-off in shear strength with slip.

Therefore, even with the waveform reinforcement present, the trapezoidal profile significantly reduced the shear strength of the welded studs. The results in Fig. 5(b) show that at a slip of 6 mm, the studs only had about 40 per cent of the shear strength they would have had in a solid slab with the same strength concrete. The results from similar tests also showed that while extending the studs some distance up into the cover slab improved their ductility, it only very slightly improved their shear strength at a slip of 6 mm.

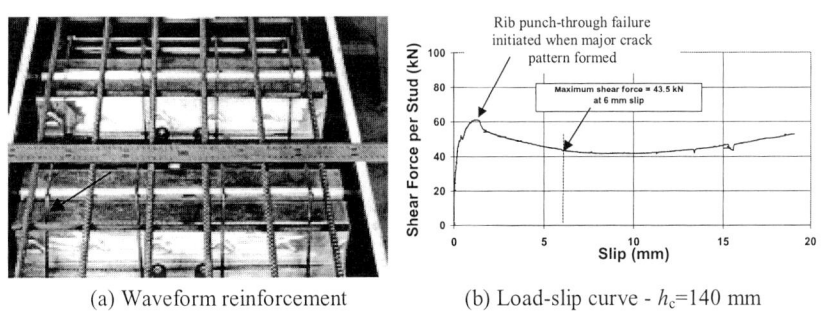

(a) Waveform reinforcement   (b) Load-slip curve - h_c=140 mm

Figure 5: Failure mode changed from rib shearing to rib punch-through with waveform reinforcement

Designing Against Rib Shearing Failure

Push-out tests like those described above and full-scale beam tests have shown that rib shearing failure is caused by a flexural crack that initiates in the concrete at the top of the sheeting rib on the opposite side to where the compressive force in the slab thrusts on the shear connectors (point A in Figure 6) (Patrick 2000). The crack propagates horizontally across the top of the concrete rib, locally passing over the tops of the welded studs, as shown in Figure 6. Tests have also shown that the slab must be reinforced over a width of at least 400 mm in order to suppress the failure mode completely. Clause 9.8 of AS 2327.1 requires four reinforcing wires or bars to be placed on each side of the concrete rib, with a maximum centre-to-centre spacing of 150 mm measured in the transverse direction. Correspondingly, it will be assumed that the effective width of the slab, b_{eff}, with respect to resisting rib shearing failure, normally equals 600 mm in edge beams, and is centred about the steel beam, except when the concrete flange outstand on one side of the beam is less than 300 mm measured from the vertical centreline of the steel beam.

It is proposed that reinforcement like that shown in Figure 5(a) is not required to control rib shearing failure provided the cracking moment of the concrete rib is not exceeded, viz.:

$$N_r f_{vs} h_r \le f'_{cf} b_{eff} b_r^2 / 6 \qquad (2)$$

where –

- N_r = number of welded studs per pan;
- f_{vs} = nominal shear capacity of a welded stud in accordance with AS 2327.1;
- h_r = overall height of profiled steel sheeting ribs;
- f'_{cf} = flexural tensile strength of concrete to AS 3600 (Standards Australia 2001);
- b_{eff} = effective width of concrete rib resisting rib shearing failure ($\le$ 600 mm); and
- b_r = width across top of concrete rib.

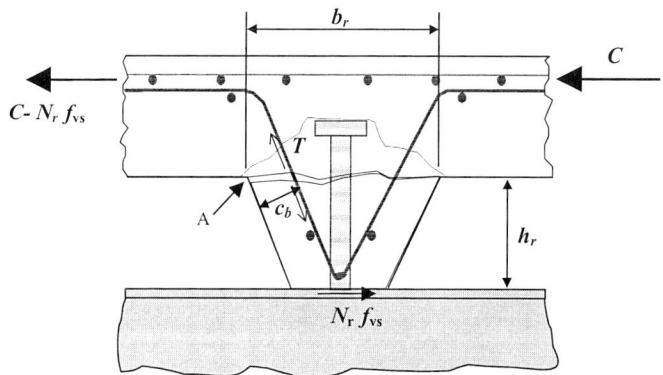

Figure 6: Design of wave-form reinforcement to prevent rib shearing failure

It follows that when Eq. 2 is not satisfied, rib shearing reinforcement is required. With four bars placed on each side of a concrete rib containing shear connectors, and ignoring any inclination of the reinforcement to the horizontal plane, it can be shown that the diameter of each wire or bar, d_b, must be such that:

$$d_b \ge \sqrt{\frac{N_r f_{vs} h_r}{2.8 \phi f_{sy} (b_r - c_b) k_b}} \qquad (3)$$

where –

- f_{sy} = design yield stress of steel reinforcement (=500 MPa in AS 3600);
- ϕ = capacity factor for bending of concrete rib (=0.8 in AS 3600);
- c_b = concrete cover measured from nearest edge of sheeting rib to centre of bar (see Figure 6); and
- k_b = bar anchorage factor, which equals 1.0 if the bars can develop f_{sy} across the horizontal failure surface, but is otherwise less than 1.0.

STEEL RING TO RESIST RIB PUNCH-THROUGH FAILURE

Rib Punch-Through Failure Mode

Rib punch-through causes the shear connectors to be weakened by the concrete punching through the side of an open steel rib (Patrick 2001). An example is shown in Fig. 7. A critical aspect of this failure mode is that the shear force falls off rapidly immediately the major crack pattern forms (see Fig. 5(b)). Beyond this point, the concrete no longer confines the base of the studs and the shear connection becomes much more flexible. At very high slips (>15 mm) the shear force may exceed the first peak, but this is not normally significant in design. As can be seen in Fig. 7(a), the crack pattern is synonymous with the cone failure associated with fasteners embedded in concrete acted on by a lateral force while close to a free edge. The loose concrete visible in Fig. 7(b) was easily removed.

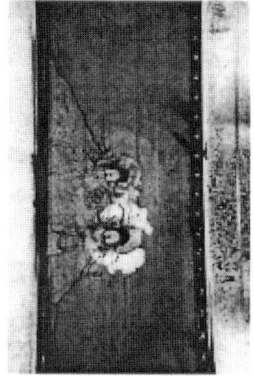
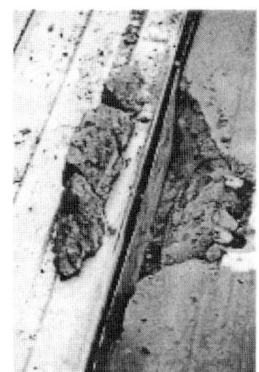

(a) Fractured studs and rib punch-through exposed by cutting away sheeting pan (b) Fractured concrete removed to expose side of dovetail rib

Figure 7: Example of rib punch-through failure

Eurocode 4: Part 1.1 Ductility Provisions for Shear Connectors

When ideal plastic behaviour of the shear connection is required in Eurocode 4, shear connectors must have a characteristic slip δ_{uk} of 6 mm. The slip capacity, δ_u, of a connector in a test is the larger slip measured at the characteristic load (the minimum failure load of three tests reduced by 10 percent). The characteristic slip capacity δ_{uk} is taken as the minimum test value reduced by 10 percent. According to this definition, even the response shown in Fig. 5(b) may not be considered ductile. A new, patented means for significantly improving the strength and ductility of shear connectors in open-trough profile decks that would otherwise fail by rib punch-through is now very briefly described.

Patented Steel Ring Confines Concrete around Stud Base

In the absence of a ring, a rib punch-through failure occurred leaving much of the stud unconfined by concrete at ultimate load causing the stud to bend significantly (see Fig. 8(a)(i)). In an otherwise identical test specimen, a short piece of steel tubing was cut and clipped into position over each stud. The rings successfully confined the concrete around the bases of the studs (see Fig. 8(a)(ii)). Rib punch-through still occurred (see Fig. 8(a)(iii)), but was of much less consequence. The effective diameter of the stud was increased over the height of the ring. The rings were 40 mm high while the height of the steel rib was 55 mm. This left a region 15 mm high above the ring in which the studs could bend and still cause their shear strength and stiffness to fall below that of the same stud in a solid concrete slab. Had higher rings been used then better performance would have resulted. The tests showed that having adequate embedment of the stud head into the solid concrete cover slab was also very important. In the tests the studs were only 95 mm high, leaving only 40 mm of the top of the stud to engage the slab after rib punch-through occurred.

The results of the two tests are presented in Fig. 8(b). The first two graphs show that the ring significantly improved the shear strength of the studs, i.e. at a slip of 6 mm, the shear force per stud was increased from 53 kN to 81 kN representing an improvement of over 50 per cent. Also, the slip capacity as defined in Eurocode 4 was greatly improved. Moreover, tests have shown that had a higher ring or stud been used, then the dip in the force-slip curve in Fig. 8(b) can be completely removed.

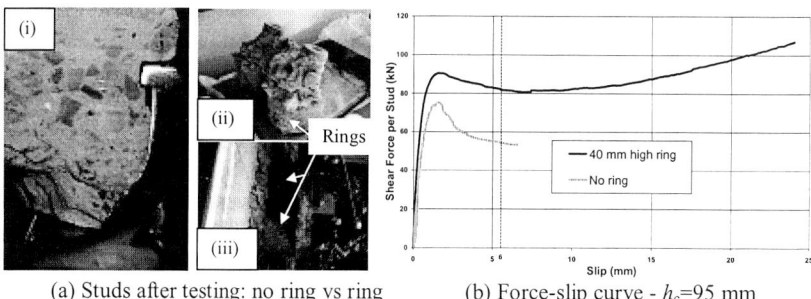

(a) Studs after testing: no ring vs ring   (b) Force-slip curve - h_c=95 mm

Figure 8: Steel ring greatly improves stud strength and ductility by confining concrete at stud base

CONCLUSIONS

Rib shearing and rib punch-through failures can cause major problems with the shear connection of composite beams incorporating open-trough profile decks layed transverse to the steel beam. Novel methods for overcoming the effects of these failure modes are under development and have been briefly described in the paper. These advances are being immediately put into practice in Australia where open-trough profile decks are now beginning to be used. The design rules under development will be considered for inclusion in the Australian Standard AS 2327.1 for composite beams.

REFERENCES

British Standards Institution (1990). *BS 5950: Part 3: Section 3.1: Code of Practice for Design of Simple and Continuous Composite Beams*, London.

British Standards Institution (1994). *Eurocode 4: Design of Composite Steel and Concrete Structures, Part 1.1. General Rules and Rules for Buildings (together with UK National Application Document)*, DD ENV 1994-1-1: 1994, London.

European Committee for Standardization (CEN) (2001). *Eurocode 4: Design of Composite Steel and Concrete Structures, Part 1.1. General Rules and Rules for Buildings,* Draft prEn 1994-1-1:2001.

Johnson, R.P. (2000). Shear Connection - Three Recent Studies, *Composite Construction IV, Engineering Foundation Conferences*, **T3: Shear Connectors**.

Johnson, R.P. and Anderson, D. (1993). Designers' Handbook to Eurocode 4, Part 1.1: Design of Composite Steel and Concrete Structures, Thomas Telford, London.

OneSteel Reinforcing (2001). *DECKMESH™ & STUDMESH® – Composite Slab Shear Reinforcement Solutions,* CD-ROM 3, Sydney, Australia.

Patrick, M. (2000). Experimental Investigation and Design of Longitudinal Shear Reinforcement in Composite Edge Beams, *Journal of Progress in Structural Engineering and Materials*, **2:2**, April-June, 196-217.

Patrick, M. (2001). *Design of the Shear Connection of Simply-Supported Composite Beams, Composite Structures Design Manual - Design Booklet DB2.1*, OneSteel Market Mills.

Patrick, M. and Bridge, R.Q. (2002). *Novel New Reinforcing Components for Composite Beams.* Proceedings, International Conference - Advances in Building Technology, Hong Kong, 4- 6 Dec.

Standards Australia (2001), *AS 3600-2001: Concrete Structures*, Sydney.

Standards Australia (2002). *AS 2327.1-2002: Composite Structures - Simply Supported Beams*, Sydney.

RESEARCH IN CANADA ON STEEL-CONCRETE COMPOSITE FLOOR SYSTEMS: AN UPDATE

M. U. Hosain[1] and A. Pashan[2]

[1]Professor [2]Graduate Student
Department of Civil Engineering, University of Saskatchewan
Saskatoon, SK, S7N 5A9 CANADA

ABSTRACT

This paper briefly summarizes the results of recent experimental research carried out in Canada on steel-concrete composite floor systems. Most of test programs were conducted at the University of Saskatchewan and involved the testing of full size composite beams and numerous push-out specimens. The first of the investigations, which involved 85 push-out specimens and 6 full size beams, led to the development of a new equation that can be used to calculate the shear capacity of a headed stud in narrow ribbed metal deck directly without having to use the current reduction factor formula. The new equation was found to provide much better correlation to test results than CSA and Eurocode 4 provisions. The experimentally determined ultimate flexural capacity of the six full size beam specimens agreed reasonably well with the predicted values based on the proposed equation. This equation has recently been incorporated in the latest edition of the Canadian Standard. In another investigation, a statistical analysis of 52 push-out specimens with solid slabs resulted in a new equation that was found to be more reliable compared to the current CSA and AISC provisions which are insensitive to important variables such as stud spacing and transverse reinforcement. The latest experimental program involves the testing of 60 push-out specimens and four full size composite beams with wide ribbed parallel metal deck. A regression analysis of the test data of 44 push-out specimens yielded a new equation which, once again, gives much better predictions than those obtained using the current provisions. The experimentally determined ultimate flexural capacity of the four full size beam specimens agreed very well with the predicted values based on the proposed equation. The better results may be attributed to the fact that the proposed equation takes into account the effects of longitudinal and transverse stud spacing and w_d/h_d ratio of the metal deck.

KEYWORDS

Composite floor systems, full size tests, push-out tests, headed stud shear connectors, regression analysis, wide ribbed metal deck, narrow ribbed metal deck, solid concrete slab.

INTRODUCTION

The University of Saskatchewan is currently the main Canadian centre for research in the area of steel-concrete composite floor systems. The experimental investigations, which involved the testing of full size composite beams and numerous push-out specimens, were divided into three categories:

Headed Studs in Solid Slab.
Headed Studs embedded in concrete slab with parallel Narrow Ribbed Metal Deck.
Headed studs embedded in concrete slab with parallel Wide Ribbed Metal Deck.

Only metal decks with ribs parallel to the beam were considered in the studies being reported in this paper. A considerable amount of research on composite beams with perpendicular metal deck was conducted earlier at the University of Saskatchewan by Jayas and Hosain (1987, 1988, 1989) and new provisions were included in the Canadian Standard [CSA (1989, 1994)] based on this research.

The overall objectives of the current research study were to acquire a better understanding of the behaviour of headed stud shear connectors and, if required, to develop new formulations that would provide better correlation to test results. This paper summarizes the results of some of the projects.

HEADED STUDS IN SOLID SLAB

Parametric Study with Push-out Specimens

In the current Canadian Standard CSA-S16-01 (2001) the factored resistance of a stud connector embedded in a solid concrete slab, q_{rs} is evaluated using Eq. 1, developed by Ollgaard et al. (1971).

$$q_{rs} = 0.5 \, \phi_{sc} \, A_{sc} \sqrt{f'_c \, E_c} \leq \phi_{sc} A_{sc} \, F_u \tag{1}$$

where, ϕ_{sc} is the resistance factor for shear connectors [0.8], A_{sc} is the area of steel shear connector [mm^2], f'_c is the compressive cylinder strength of concrete [MPa], E_c is the elastic modulus of concrete [MPa], F_u is the tensile strength of stud [MPa]. Eq. 1 is also included in AISC (1993) but without the resistance factor as shown below:

$$Q_n = 0.5 \, A_{sc} \sqrt{f'_c \, E_c} \leq A_{sc} F_u \tag{2}$$

Eurocode 4 (1994) provision for computing the factored resistance of a stud connector embedded in solid slab is more conservative than those of CSA and AISC as shown below:

$$q_{rs} = 0.369 \, \phi_{sc} \, A_{sc} \sqrt{f'_c \, E_c} \leq \phi_{sc} A_{sc} \, (0.8 \, F_u) \tag{3}$$

Unfortunately, all three equations [Eqs. 1 to 3] are insensitive to important variables such as stud spacing and transverse reinforcement. In order to resolve this issue, an experimental investigation involving 52 push-out specimens with solid slabs was conducted at the University of Saskatchewan by Gnanasambandam and Hosain (1996). The variables that were considered in this investigation included longitudinal and transverse stud spacing, transverse reinforcement, concrete strength and the size of stud connectors.

The experimental investigation consisted of 3 phases. Phase 1 involved 20 specimens and was directed towards evaluating the effects of longitudinal and transverse stud spacing on the stud capacity. Phase 2 involved 16 specimens with 150 mm slabs which were tested to study the effects of transverse reinforcement and concrete strength. Phase 3 had the same purpose as that of phase 2 but considered 16 specimens with 102 mm slabs. A statistical analysis of these 52 specimens resulted in Eq. 4.

$$q_u = 0.46 \, s_t \, h \, (f'_c)^{0.65} + 3.82 \, s_l \, d \, (f'_c)^{0.52} + 0.18 \, A_{tr} \, f_y$$

$$\leq 0.8 \, A_{sc} F_u \tag{4}$$

where s_l is the longitudinal stud spacing, s_t is the transverse stud spacing, h is the height of studs, d is the diameter of studs, A_{tr} is the area of the transverse reinforcement in mm^2, f_y is the yield stress of the transverse reinforcement in MPa.

Equation 4, which provides much better correlation to test results than those obtained using the current CSA and AISC provisions, is still under refinement. Those who would prefer to continue to use the reduction formula approach [AISC (1993), Eurocode 4 (1994)] to design composite beams with narrow and wide ribbed metal decks may find this equation useful.

HEADED STUDS IN NARROW RIBBED METAL DECK

In the nineties, North American [CSA (1994) and AISC (1993)] provisions specified that for parallel narrow ribbed metal deck, the nominal shear strength of a stud connector embedded in solid slab be multiplied by the following reduction factor suggested by Grant et al. (1977).

$$0.6 \frac{w_d}{h_d} \left[\frac{h}{h_d} - 1.0 \right] \leq 1.0 \tag{5}$$

where h is the height of stud connector after welding, w_d is the average width of the deck rib and h_d is the height of the deck.

Recent studies by Androutsos & Hosain (1993) have raised some doubts concerning the reliability of the reduction factor equation. Although this reduction factor has also been adopted by Eurocode 4 (1994) predicted values based on this reduction factor differ considerably from test results. A major drawback of the reduction factor approach is that the failure mechanism of a specimen with solid slabs could be different from that of a specimen with metal deck and thus the stud capacity cannot be arbitrarily adjusted. Moreover, the deficiencies of the parent equation, i.e. equation for stud connector embedded in solid slab, are inherited.

In order to resolve this issue, a comprehensive test program was started at the University of Saskatchewan in 1992. The main objective of this project was to develop an equation that can be used to calculate the shear capacity of headed studs in parallel narrow ribbed metal deck directly without having to use Eqs. 1 and 5.

Parametric Study with Push-out Specimens

The first phase of the experimental program involved the testing of 85 push-out specimens by Androutsos and Hosain (1994). Twenty six of the push-out specimens had solid slab and remaining specimens featured parallel narrow ribbed metal deck. The deck profile, w_d/h_d, varied from 0.78 to 2.0. The headed studs were either 16x76 mm or 19x125 mm depending upon the overall slab thickness of 102 mm and 150 mm, respectively. The longitudinal stud spacing was the principal experimental parameter. Concrete strength was also varied. For the push-out specimens with metal deck, the studs were welded through the decking. For those with solid slab, the studs were welded directly on the beam flange. A typical test setup for push-out specimens is shown in Figure 1.

A regression analysis of 85 push-out specimens resulted in the following new equation for predicting the capacity of headed studs in parallel narrow ribbed metal deck:

$$q_u = 0.92 \frac{w_d}{h_d} d h (f'_c)^{0.8} + 11.0 s d (f'_c)^{0.2} \leq 0.8 A_{sc} F_u \tag{6}$$

$$s \leq 120 \text{ mm and } w_d < 6d$$

where d and h are the diameter and height of the headed stud, s is the longitudinal stud spacing and f'_c is the compressive cylinder strength of concrete.

Eq. 6 was found to provide much better correlation to test results than CSA and Eurocode 4 provisions. The average absolute difference between the observed and those predicted by Eq. 6 was found to be approximately 7.34% compared to 40.72% and 63.85% for CSA and Eurocode, respectively. The standard deviation was estimated to be 0.0844. The better results provided by Eq. 6 may be attributed to the fact that, unlike CSA and Eurocode 4 provisions, it takes into account the influence of the stud spacing.

Figure 1: Test setup for push-out specimens

To provide a better appreciation of the degree of accuracy provided by CSA, Eurocode 4 and Eq. 6, Figures 2, 3, and 4 were prepared. In these figures, $q_{u(test)}$ and $q_{u(pred.)}$ are the observed and predicted values of the ultimate shear strength per stud, respectively. It is clear from the scattering of

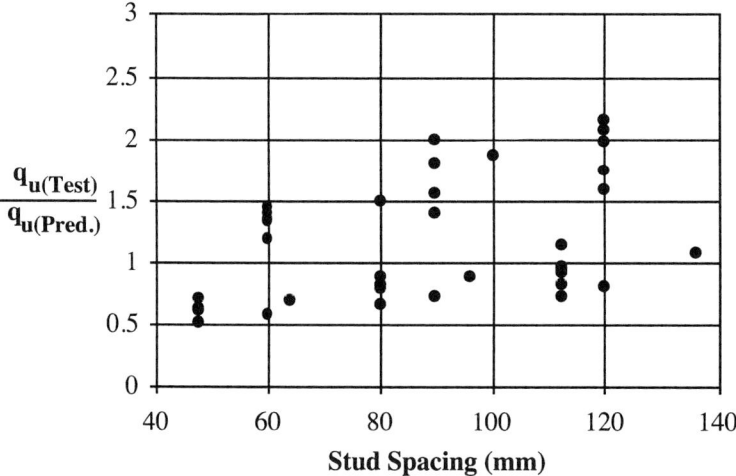

Figure 2: Distribution of the Test / Predicted Results Based on CSA and AISC Provisions

the results in Figures 2 and 3, that the CSA, AISC and Eurocode 4 provisions do not provide accurate predictions. On the other hand, Eq. 6 provides a much better correlation to tests results, as is apparent from the distribution of results in Figure 4.

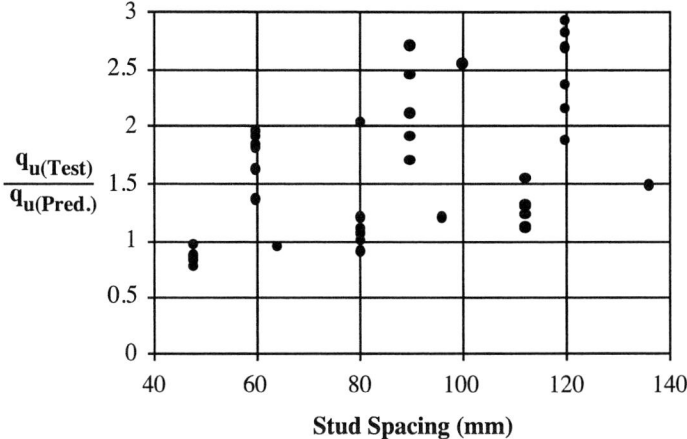

Figure 3: Distribution of the Test / Predicted Results Based on Eurocode 4 Provisions

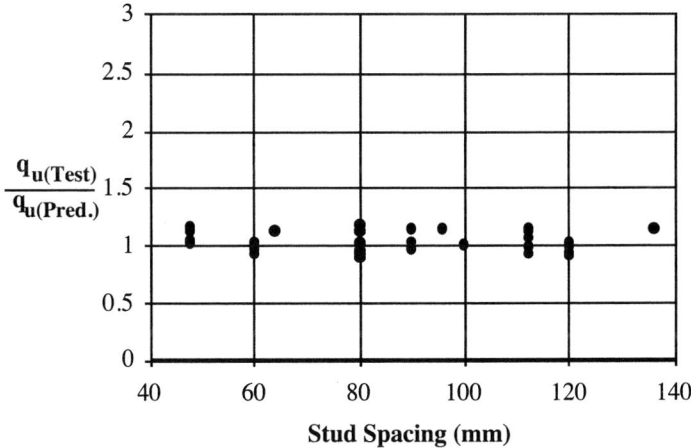

Figure 4: Distribution of the Test / Predicted Results Based on Eq. 6

Verification with Full Size Beam Tests

The second phase of this investigation involved the testing of six full size composite beams by Wu and Hosain (1997). The first three beams featured a 150 mm thick concrete slab with a 76 mm HB 308 type narrow-ribbed metal deck. 19x125 mm studs were welded on to the beam flange through the metal deck using a TR 2400 stud welder. The other three beams had a 102 mm thick concrete slab with a 38 mm HB 938 type narrow-ribbed metal deck The headed studs were 16x76.

The experimentally determined ultimate flexural capacity of the first three full size beam specimens agreed extremely well with the predicted values based on the proposed equation. However, for composite beams with 38 mm metal deck, the experimental values are somewhat higher than the predicted ones. This does not reflect on the accuracy of Eq. 6 since the same degree of discrepancy was also observed when the actual push-out test results were utilized to predict the moment capacity. Eq. 6 has recently been incorporated in the latest edition of the Canadian Standard.

HEADED STUDS IN WIDE RIBBED METAL DECK

Parametric Study with Push-out Specimens

Currently, the Canadian Standard CAN/CSA-S16.1-94 [CSA (1994)] specifies that Eq. 1, which is based on test results of push-out specimens with solid slabs can also be applied for calculating the stud capacity in wide ribbed metal decks, i.e., when the width to height ratio (w_d/h_d) of the metal deck exceeds 1.5. AISC and Eurocode 4 also provide the same specification. However, a recent study by Gnanasambandam and Hosain (1996) has raised some doubts concerning the reliability of this approach. Eq. 1 does not take into account the effects of stud spacing and transverse reinforcement. Moreover, the current approach ignores the influence of w_d/h_d ratio.

A parametric study was recently conducted to evaluate the effects of the aforementioned factors on the strength of headed studs in wide ribbed metal deck and to suggest an alternate formulation. After a long and arduous examination of the observed relationships between the ultimate load per stud, q_u, and the various experimental parameters, a general form of the proposed equation was first established by Wu and Hosain (1998) in terms of 11 different coefficients. A least square regression analysis of test results of the 44 push-out specimens with wide ribbed metal deck that were tested by the authors yielded the following long and complex preliminary expression that would, for sure, require considerable simplification.

$$q_u = \left[3.611 \left[\frac{S_l}{d}\right] - 0.209 \left[\frac{S_l}{d}\right]^2 \right] d^{0.589} h^{1.643} \sqrt{f'_c}$$

$$+ \left[37.870 \left[\frac{S_t}{w_d}\right] - 31.679 \left[\frac{S_t}{w_d}\right]^2 \right] d^{5.828} h^{-2.314} \sqrt{f'_c}$$

$$+ 2.628 \frac{w_d}{h_d} d^{-2.927} h^{3.163} \sqrt{f'_c} \leq 0.80 A_{sc} F_u$$

$$0.30 \leq \frac{S_t}{w_d} \leq 0.63 \ ; \quad 3 \leq \frac{S_l}{d} \leq 8$$

[7]

where,
q_u	=	Predicted ultimate load per stud (N)
S_l	=	Longitudinal stud spacing (mm)
S_t	=	Transverse stud spacing (mm)
f'_c	=	Concrete compressive strength (MPa)
w_d	=	Average width of metal deck flute (mm)
h_d	=	Height of metal deck (mm)
d	=	Studs diameter (mm)
h	=	Studs length (mm)
A_{sc}	=	Area of studs shear connector (mm^2)
F_u	=	Ultimate tensile strength of studs (MPa)

The average absolute difference between the observed values and those predicted by Eq. 7 was found to be approximately 2.90%, compared to 32.55% and 10.30% for CSA and Eurocode, respectively. The better results may be attributed to the fact that the complex equation takes into account not only the longitudinal stud spacing, concrete strength and w_d/h_d ratio of the metal deck, but also the transverse stud spacing.

The average arithmetic mean of the test/predicted ratio (μ), the standard deviation (σ) and the coefficient of variation (C.V) values for Eq. 7 as well as those for CSA and Eurocode provisions are given in Table 1. It is obvious that Eq. 7, although complex, provides much better predictions.

TABLE 1
STATISTICAL ANALYSIS OF PREDICTED VALUES

Statistics	CSA	Eurocode	Eq. 7
μ	0.66	0.90	1.002
σ	0.074	0.100	0.035
C.V	11.2%	11.2%	3.5%

Simplification of the Complex equation

Equation 7 gives very good prediction of test results. However, its major drawback is that it is too complicated to use without a computer. A series of regression analysis was carried out to obtain several simplified versions of Eq. 7 at the expense of some accuracy. Detailed computations are included in Wu and Hosain (1997). One of the simplified version is given by Eq. 8.

$$q_u = \left[0.264 \left[\frac{S_1}{d}\right] + 0.821\frac{w_d}{h_d} + 3.120 \right] d\,h\sqrt{f'_c}$$

$$0.30 \leq \frac{S_t}{w_d} \leq 0.63 \;;\; 3 \leq \frac{S_1}{d} \leq 8$$

[8]

The average absolute difference between the observed values and those predicted by Eqs. 7 and 8 were found to be approximately 2.90% and 5.50%, respectively. The average arithmetic mean of the test/predicted ratio (μ), the standard deviation (σ) and the coefficient of variation (C.V) for Eqs. 7 and 8 are listed in Table 2. Because of the simplicity of Eq. 8, with only a slightly lower accuracy, this equation is recommended for design purposes.

TABLE 2
STATISTICAL RESULTS OF PREDICTED VALUES

Statistics	Eq. 7	Eq. 8
μ	1.002	1.024
σ	0.027	0.018
C.V	2.7%	7.6%

Full Size Beam Tests

The proposed equation, i.e., Eq. 8, as well as the original complex equation (Eq. 7), were used to predict the ultimate moment values of the four full size beams. The average absolute difference between the observed ultimate moments and those predicted by Eqs. 7, 8 and by push-out test results were found to be approximately 2.77%, 2.36% and 3.86%, respectively. The average arithmetic mean of the test / predicted ratio (μ), the standard deviation (σ), the coefficient of variation (C.V) are given in Table 3.

TABLE 3

STATISTICAL RESULTS OF THE FULL-SIZE BEAMS

Statistics	Eq. 7	Eq. 8	Push-out
μ	1.018	1.019	1.033
$\sigma(kN)$	0.027	0.022	0.023
C.V	2.7%	2.2%	2.2%

This paper briefly summarizes the results of several experimental programs which involved the testing of numerous push-out specimens and full size composite beams. Regression analyses of the test data of the push-out specimens yielded three new equations, which provide much better correlation to test results than those obtained using the current provisions. The new equations take into account the effects of important parameters such as longitudinal and transverse stud spacing, concrete strength and metal deck ratio, etc. The experimentally determined ultimate flexural capacity of the full size composite beams agreed reasonably well with the predicted values based on the proposed equations.

References

American Institute of Steel Construction (1993). Load and resistance factor design specification for structural steel buildings, Chicago, Illinois.

Androutsos C. and Hosain M.U. (1994) Composite Beams with Headed Studs in Narrow Ribbed Metal Deck. Structural Engineering Research Report No. 42, August. Department of Civil Engineering, University of Saskatchewan, Saskatoon, Canada, S7N 5A9.

Androutsos, C. and Hosain, M.U. (1993). Composite Beams with Headed Studs in Narrow Ribbed Metal Deck. Composite Construction in Steel and Concrete II, Special Publication of the ASCE Structural Division. New York, N.Y., pp. 771-782.

Canadian Standard Association. (2001). Steel structures for buildings-limit states design: CSA-S16-01., Toronto, Ontario, Canada.

Canadian Standard Association. (1994). Steel structures for buildings-limit states design: CAN/CSA-S16.1-94, 1994, Rexdale, Ontario, Canada.

Canadian Standard Association. (1989). Steel structures for buildings-limit states design: CAN/CSA-S16.1-M89, 1989, Rexdale, Ontario, Canada.

Eurocode 4.(1994). Design of composite steel and concrete structures: General rules for buildings, EN 1994-1-1, CEN, Brussels.

Gnanasambandam, C. and Hosain, M.U.. (1996). Headed Stud Connectors in Solid Slabs and in Slabs with Wide Ribbed Metal Deck. Structural Engineering Research Report No. 43, August. Department of Civil Engineering, University of Saskatchewan, Saskatoon, Canada, S7N 5A9.

Grant, J.A., Fisher, J.W. and Slutter, R.G. (1977). Composite Beams with Formed Steel Deck. AISC Engineering Journal, Vol. 14, No. 1, pp. 24-43.

Jayas, B.S. and Hosain, M.U. (1989). Behaviour of Headed Studs in Composite Beams: Full Size Tests. Canadian Journal of Civil Engineering. Vol. 16, Number 5, pp. 712 - 724.

Jayas, B.S. and Hosain, M.U. (1987). Behaviour of Headed Shear Studs in Composite Beams: Push-out Tests. Canadian Journal of Civil Engineering,. Vol. 13, No. 1, pp. 106 - 115.

Jayas, B.S. and Hosain, M.U. (1988). Composite Beams with Perpendicular Ribbed Metal Deck. Composite Construction-I. ASCE/Engineering Foundation Special Publication, ASCE, New York, N.Y., pp. 511-526.

Ollgaard, J.G., Slutter, R.G. and Fisher, J.W. (1971). Shear Strength of Stud Connectors in Lightweight and Normal-Weight Concrete. AISC Engineering Journal, Vol. 8, No. 2 pp. 55-64.

Wu, H. and Hosain, M.U. (1997). Tests on Full Size Composite Beams with Wide Ribbed Metal Deck. CSCE Annual Conference, Sherbrooke, Canada, May 28-June 1, Vol. 6, pp. 249 - 256.

Wu, Hang and Hosain, M. U. (1999). Headed stud connectors in full-size composite beams with wide ribbed metal deck. Structural Engineering Research Report No. 45, University of Saskatchewan, Saskatoon, Canada, S7N 5A9, June.

EARLY AGE SHRINKAGE AND CASTING SEQUENCE EFFECTS IN COMPOSITE STEEL-CONCRETE GIRDERS

L. Dezi, G. Leoni and A. Vitali

Institute of Structural Engineering, University of Ancona, via Brecce Bianche 60131, Ancona, Italy

ABSTRACT

This paper analyses the problem of slab cracking in composite steel-concrete girders during the constructive phases due to early age concrete shrinkage and to casting sequence effects. The analysis is performed by means of a theoretical model, which includes connection deformability and creep effects. The numerical solution is obtained by means the finite element method and the step-by-step procedures. The proposed model is then applied to a real composite deck showing the advantages obtained by an optimised casting sequence in comparison to the continuous casting of the slab.

KEYWORDS

Composite steel-concrete girders, bridges, early age shrinkage, creep, fractionated casting, connection deformability.

INTRODUCTION

The problem of slab cracking in composite steel-concrete bridges during the constructive phases was evidenced by recent investigations on numerous newly constructed continuous decks in the US and France (Krauss and Rogalla, 1996; SETRA, 1995). On the basis of in situ measurements and laboratory tests, Ducret and Lebet (1999) demonstrated that high tensile stresses affect the concrete slab during construction leading to the development of transversal full depth cracks not only in the hogging regions but even in the sagging regions. All the authors are unanimous in stating that premature slab cracking is due to the casting sequences and to early concrete shrinkage, namely *endogenous* chemical shrinkage and *thermal shrinkage* produced by cooling after concrete end-setting.

Despite the considerable interest in this problem, very few papers devoted to the modelling of composite bridge decks with slab realised by sequential castings are available in technical literature. Marì (2000) and Kwak et al. (2000) proposed nonlinear models to predict the effects of the casting sequences on composite girders with rigid shear connection taking into account the long term behaviour of concrete but without focusing on concrete behaviour at early ages.

The authors have recently discussed the problem of the time dependent behaviour of continuous composite decks with fractionated slab casting taking into account the flexibility of the shear connection and drying shrinkage (Dezi et al., 2001).

In this paper, the proposed model is modified in order to analyse continuous composite decks with steel beams of variable depth, taking into account the effect of early shrinkage. The numerical solution is obtained by means the step-by-step procedures and the finite element method by using a 10 dof composite beam element. The model is applied to a real bridge deck showing the benefits obtained during the construction phases by adopting an optimised casting sequence.

ANALYTICAL MODEL AND NUMERICAL SOLUTION

Reference is made to the composite deck of Fig.1, having a steel beam of variable depth. In order to analyse the slab casting sequences, the girder is subdivided into n_{bs} sections each characterised by its own instant of concrete pouring (t_p). The reference frame $\{0;X,Y,Z\}$ of Fig. 1, with the co-ordinate plane XZ at the beam-slab interface, permits describing a generic composite deck with variable depth steel beam.

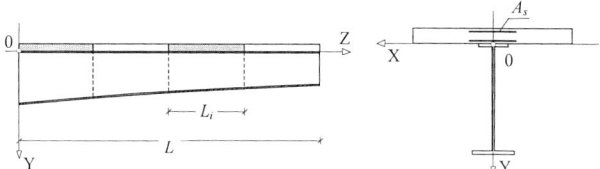

Fig. 1 Beam geometry and reference system frame

The kinematical model of composite beam with flexible shear connection, proposed by Newmark et al. (1951), is adopted. The displacements in X and Y directions are respectively

$$u(x,y,z;t)=0 \qquad v(x,y,z;t)=v_0(z;t) \qquad (1a,b)$$

where t is the generic instant of the analysis. The longitudinal displacements, assuming the preservation of the plane cross section for the steel beam and concrete slab considered separately, are expressed by

$$w_a(x,y,z;t)=w_{a0}(z;t)-y v_0'(z;t) \qquad w_c(x,y,z;t)=w_{c0}(z;t)-y v_0'(z;t) \qquad (2a,b)$$

where w_{a0} and w_{c0} are the longitudinal displacements of the steel beam and of the concrete slab measured at the points placed at beam-slab interface, respectively. Consequently, the interface slip Γ between the slab and the beam assumes the following expression:

$$\Gamma(z;t)=w_{a0}(z;t)-w_{c0}(z;t). \qquad (3)$$

It is important to note that the analytical description of the problem in question, characterized by concrete slab sections having different casting times, requires a geometrical domain including all the slab sections, even those which have not yet been poured and the slab displacement field is defined even where the steel beam is alone. Consequently, while the structural continuity of the steel beam implies the regularity of the two functions w_{a0} and v_0, w_{c0} is discontinuous at the interface cross sections between the adjacent slab segments. In these cross sections only the increments of all the functions are regular after concrete end-setting of the adjacent slab segments.

Linear elastic constitutive relationships are considered for the materials and for the shear connectors. The change in the structure scheme is introduced by defining the concrete constitutive law as follows:

$$\sigma_c(t) = H(t-t_e)\left\{\frac{[\varepsilon_c(t)-\varepsilon_c(t_e)]-\overline{\varepsilon}(t)}{J_p(t,t)} + \int_{t_e}^{t}\frac{\sigma_c(\tau)}{J_p(t,t)}\frac{\partial J_p(t,\tau)}{\partial \tau}d\tau\right\} \qquad (4)$$

where $H(t-t_e)$ is the Heavyside function (zero for $t<t_e$ and equal to one for $t\geq t_e$), $\varepsilon_c(t_e)$ is the concrete strain at the end-setting time t_e and thus $\varepsilon_c(t)- \varepsilon_c(t_e)$ is the actual concrete strain. Furthermore, J_p is the creep function of the concrete poured at time t_p, and $\overline{\varepsilon}$ is the cumulative strain due to endogenous, thermal and drying shrinkage. Analogously, under the assumption of perfect bond between concrete and reinforcing rebars, the stress in the reinforcing steel is given by

$$\sigma_s(t) = H(t-t_e)E_s[\varepsilon_c(t)-\varepsilon_c(t_e)] \qquad (5)$$

where E_s is the relevant Young modulus.

By assuming the displacements $\mathbf{s}^T = [w_{a0}, w_{s0}, v_0]$ as unknowns, the solving equilibrium condition is obtained by the Virtual Work Principle. By taking into account the displacement expressions and the constitutive relationships previously introduced, it can be written in the following form:

$$\sum_{n_{bs}}\int_{L_i}\left\{\mathbf{K}_a\mathscr{D}\mathbf{s}(t) + H(t-t_e)\left[\mathbf{K}_c(t)\mathscr{D}\mathbf{s}(t) - \mathbf{K}_c(t)\mathscr{D}\mathbf{s}(t_e) + \int_{t_e}^{t}\mathbf{f}_c(\tau)E_c(t)\frac{\partial J_p(t,\tau)}{\partial \tau}d\tau\right]\right\}\cdot\mathscr{D}\delta\mathbf{s}\,dz =$$
$$= \sum_{n_{bs}}\int_{L_i}[\mathbf{p}\cdot\mathscr{H}\delta\mathbf{s} + H(t-t_e)\mathbf{K}_c(t)\overline{\boldsymbol{\varepsilon}}(t)\cdot\mathscr{D}\delta\mathbf{s}]dz \quad \forall \delta\mathbf{s} \qquad (6)$$

where $E_c(t) = J_p(t, t)^{-1}$ is the concrete Young modulus at time t while $\mathbf{K}_a$ and $\mathbf{K}_c$ are stiffness matrix components (the former related to the steel beam and the shear connection and the latter related to the reinforced concrete slab) given by

$$\mathbf{K}_a = \begin{bmatrix} E_aA_a & 0 & E_aS_a & 0 & 0 \\ 0 & 0 & 0 & 0 & 0 \\ E_aS_a & 0 & E_aI_a & 0 & 0 \\ 0 & 0 & 0 & \rho & 0 \\ 0 & 0 & 0 & 0 & \rho \end{bmatrix} \quad \mathbf{K}_c(t)=\begin{bmatrix} 0 & 0 & 0 & 0 & 0 \\ 0 & E_c(t)A_c+E_sA_s & E_c(t)S_c+E_sS_s & 0 & 0 \\ 0 & E_c(t)S_c+E_sS_s & E_c(t)I_c+E_sI_s & 0 & 0 \\ 0 & 0 & 0 & 0 & 0 \\ 0 & 0 & 0 & 0 & 0 \end{bmatrix} \qquad (7a,b)$$

Furthermore, $\mathbf{f}_c^T = [0,N_c,M_c,0,0]$ is the vector grouping the stress resultants in the concrete, $\overline{\boldsymbol{\varepsilon}}^T = \overline{\varepsilon}[0, A_c, S_c, 0, 0]$ is the vector grouping terms related to the shrinkage and $\mathbf{p}^T = [p_{za}, p_{zc}, p_y, m]$ is the vector of the resultants of the applied external forces. Finally $\mathscr{D}$ and $\mathscr{H}$ are the formal differential operators defined as

$$\mathscr{D}^T = \begin{bmatrix} \partial & 0 & 0 & 1 & 0 \\ 0 & \partial & 0 & 0 & -1 \\ 0 & 0 & -\partial^2 & 0 & 0 \end{bmatrix} \qquad \mathscr{H}^T = \begin{bmatrix} 1 & 0 & 0 & 0 \\ 0 & 1 & 0 & 0 \\ 0 & 0 & 1 & -\partial \end{bmatrix} \qquad (8a,b)$$

In the previous expressions A, S and I denote the area and the first and second order inertia, calculated with respect to the X axis, of the concrete part (c), the reinforcing steel (s) and steel beam (a).

Eq.(6) is written as the summation of the contributions given by the different homogeneous beam sections. The Heavyside function permits catching the evolution of the structure during construction by eliminating the contributions of the slab for the instants before concrete end-setting.

The numerical solution of (6) can be obtained by introducing a double discretization, the first for the time domain and the second for the beam axis. Thanks to the time discretization, the integrals can be approximated with the trapezoidal rule according to the formula

$$\int_{t_e}^{t} G(\tau) \frac{\partial J(t,\tau)}{\partial \tau} d\tau \cong \frac{1}{2} \sum_{i=1}^{n} [G(t_i) + G(t_{i-1})][J(t,t_i) - J(t,t_{i-1})] \qquad (9)$$

where G is the generic function of t. The solution of Eq. (6) may be obtained for each time of the mesh, by solving a sequence of linear elastic-type problems in which the unknowns are the increments of the displacement functions in each sub-interval.

Beam axis discretization is required to numerically solve the elastic-type problems, at each calculus step, by means of the finite element method. In particular, in order to catch the discontinuity of the functions describing the axial slab displacements, double nodes are introduced at the cross sections separating the n_{bs} girder sections. The longitudinal displacements of the slab, measured at the two nodes, will not depend on each other for instants before end-setting of the concrete in both the adjacent slab sections. For the successive times, suitable constraints will impose the same displacement increments.

A 10 d.o.f. finite element is adopted in which the unknown displacements are the end vertical displacements and rotations of the element and the axial displacements of the steel beam and the concrete slab measured at the beam ends and at an intermediate node (Fig.2). A third order interpolating polynomial is used for the vertical displacement v_0, while second order interpolating polynomials are used for the longitudinal displacements w_{c0} and w_{a0}.

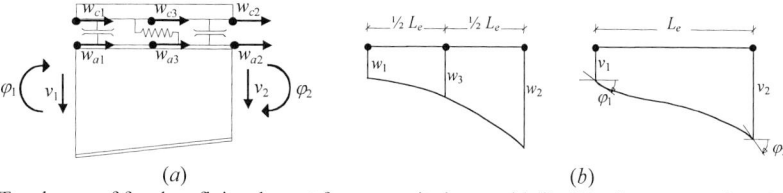

(a) (b)
Fig. 2 Ten degree of freedom finite element for composite beam with flexible shear connection: (a) nodal displacements, (b) shape functions for longitudinal and for vertical displacements.

It is important to underline that the choice of measuring the longitudinal displacements of the slab and of the steel beam at the beam-slab interface is particularly appropriate for analysing decks with steel beams variable geometry. In fact this element permits solving problems in which the steel beam undergoes sudden variations of the plate thickness and of the flange width without introducing artifices (i.e., the definition of rigid body constraints) in assembling the stiffness matrix of the entire structure.

The proposed numerical procedure makes it possible to catch the evolution of the complex static scheme due to the constructive phases and permits following the complete time evolution of displacements, stresses and shear flow at the beam-slab interface. The results can be organised in diagrams such as envelopes of maximum stresses obtained in the slab during construction or the stresses in each slab section for a fixed age of concrete. This allows a ready interpretation in order to establish the cracking tendency of the concrete slab.

APPLICATION TO A REAL DECK

The model presented is used to study early age shrinkage and casting sequence effects in a real five-span deck (35-65-90-65-35 m) with slab width of 13 m (Fig.3). The two steel beams have depth varying from a minimum of 2.00 m, at the external spans, to a maximum of 4.00 m over the two central supports, and 2.50 m at the central span. The web and flange thicknesses are shown in Fig.3b. The slab has a mean thickness of 0.29 m and a reinforcement geometric ratio of 1%, in the span sections, and 2% in the inner support sections. The analyses are performed with a constant value of the shear connection stiffness ($\rho = 3.0$ kN/mm^2).

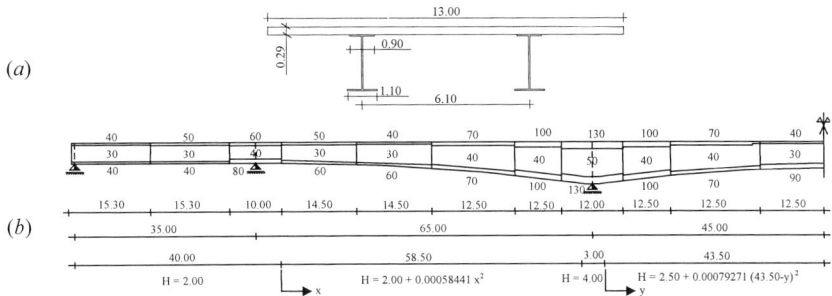

Fig. 3 Deck geometry: (*a*) cross section, (*b*) longitudinal view of the steel beam and sheet thickness

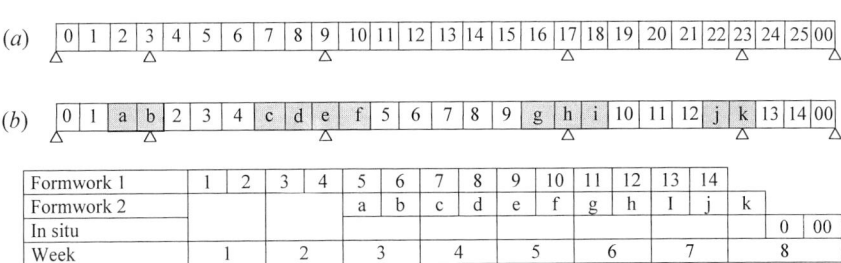

Fig. 4 Slab casting schemes: (*a*) continuous casting; (*b*) optimised sequential casting

The slab is poured by means of travelling formworks having a self-weight of 300 kN. Two casting schemes are considered: a continuous sequence using a single formwork travelling from one end to the other end of the deck (Fig.4a) and an optimised scheme performed by two travelling formworks following the casting sequences shown in Fig.4b.

For each slab section the following external actions are considered: at the instant of concrete pouring, the slab and the formwork weights are applied on the steel beam; subsequently, after 3 days, a negative load is applied on the composite cross section to simulate the removal of the formwork.

The time dependent analysis is performed by considering the creep and drying shrinkage functions suggested by the CEB-FIP model code 1990 (1988), assuming RH=75% and f_{ck}=35 MPa.

Since concrete of normal strength is used, endogenous shrinkage is disregarded. Thermal effects due to cement hydration are estimated on the basis of available laboratory test results (Ducret J.M. and Lebet, J.P., 1999). The effects of the brief initial heating period are neglected because they occur when the temperature reduction of 20°C is considered by approximating its time evolution with a linear law (Fig.5a).

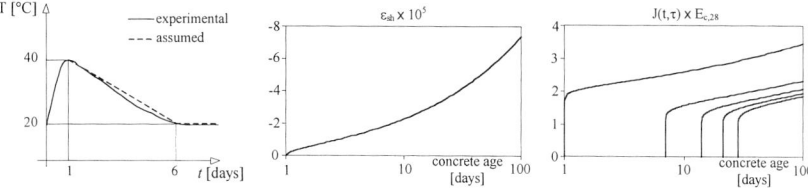

Fig. 5 Concrete behaviour: (a) thermal effects of hydration, (b) drying shrinkage, (c) creep functions

Fig.6 shows the concrete stresses produced by continuous casting after pouring of the slab sections 6 and 14. It is evident that all the slab sections are affected by tensile stresses that exceed the concrete tensile strength over the inner supports. Obviously, since the model is based on a linear analysis, the results are not realistic after concrete cracking; nevertheless it is possible to state that when continuous casting is adopted, concrete cracking occurs in large regions of the slab.

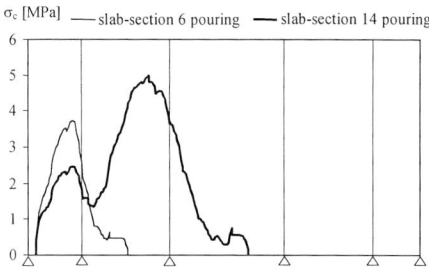

Fig. 6 Continuous casting of the slab: concrete stresses after pouring of slab-section 6 and 14

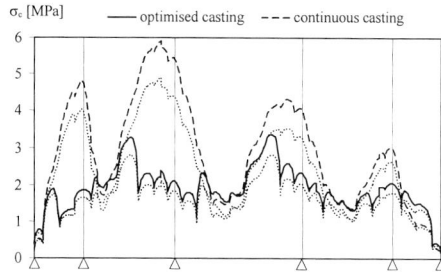

Fig. 7 Concrete stresses during constructive phases: envelopes of the maximum values

In Fig.7, the envelope of the maximum stress values, produced by the two casting techniques on concrete during construction, are compared. In both the cases the whole slab is characterised by a tensile stress state; nevertheless, optimised casting produces much lower tensile stresses than those produced by continuous casting especially at the hogging regions. Furthermore, notice that the maximum values obtained with the optimised sequence are lower than the average tensile strength of the concrete (f_{ctm}=2.9MPa) while, in the case of continuous casting, such a level is abundantly exceeded.

It is important to underline that since the whole concrete slab is under traction, creep analysis should be performed by using a compliance function validated for concrete in tension. In view of the fact that the functions available are valid for compressive concrete, the creep analysis performed leads to higher stress values than the real values. In fact, as it is well known, creep effects in tension are higher than those in compression. In a rough attempt to take into account the creep behaviour of concrete under traction, the analysis was repeated by doubling the creep coefficient. The dotted lines in Fig.7 show a substantial reduction of the traction values even if for the continuous casting they remain higher than the average concrete tensile strength.

Fig.8 shows the stresses in each slab section after six days from concreting and allows a comparison with the tensile strength of the young concrete. In the cases under examination, the optimisation of the casting sequences allows only a small reduction of the tensile stresses affecting the young concrete; nevertheless, if continuous casting is adopted, these tensile stresses increase significantly while small increases occur if optimised casting is chosen.

Fig.9 shows the time evolution of the concrete stresses produced in two cross sections, chosen in the second span and over the third support, by the two casting sequences considered.

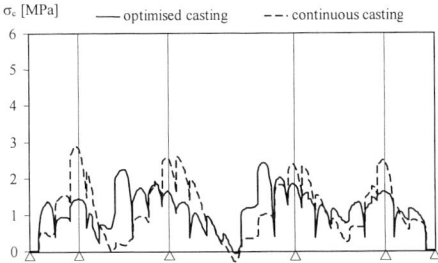

Fig. 8 Slab stresses in each section after six days from concreting

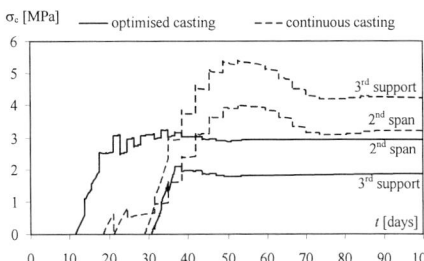

Fig. 9 Time evolution of concrete stresses for the optimised sequence

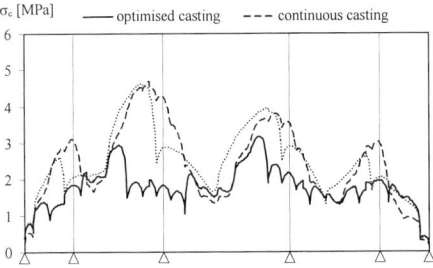

Fig. 10 Concrete stresses 3 months after the beginning of slab pouring

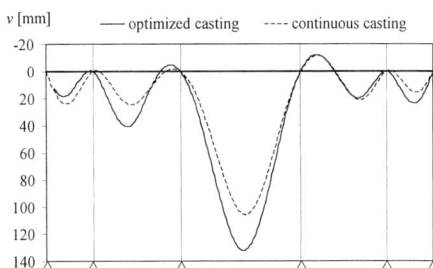

Fig. 11 Vertical displacements 3 months after the beginning of slab pouring

Fig.10 shows the stress diagram in the concrete slab a few weeks after complete slab pouring; on this residual stress state, additional stresses due to the pavement weight and the traffic loads will be superimposed. It is evident that the reduction of the tensile stresses over the support obtained with the optimised casting is particularly important in order to control cracking due to the external loads which induce negative moments just in these sections. In the same figure, the stress diagram produced by continuous casting with slab segments three times longer those considered in the previous cases, is also reported (dotted line). It is evident that the two continuous castings give almost the same results.

Finally, Fig.11 shows the vertical deck displacements after concreting of the whole slab. In this case, the optimised casting produces vertical displacements greater than those produced by continuous casting. Clearly, this is due to the fact that in the case of optimised casting sequences, the delay of slab casting over the supports implies a minor global stiffness of the deck. The higher displacements can be easily compensated with suitable shaping of the steel beam sections.

Once again, it is important to remember that the results achieved for continuous casting have only a qualitative significance because in the proposed model slab cracking is not taken into account; on the other hand, the results attained for optimised casting confirm the assumptions adopted. It must nevertheless be underlined that since neither creep functions for concrete subjected to tensile stresses nor functions describing the time evolution of the concrete tensile strength are available, fully realistic results cannot be achieved even by modelling the nonlinear slab behaviour.

CONCLUSIONS

A model for the time dependent analysis of continuous composite steel-concrete decks, accounting for early age shrinkage and slab casting sequences is presented. The numerical solution, achieved by the finite element method and by the *step-by-step* procedures, makes it possible to follow the complex time evolution of the stress state in all the structural elements. The proposed model has been applied to a real bridge deck, having variable depth steel beam, by considering two different casting schemes: a continuous sequence from one deck end to the other and an optimised scheme where the slab sections in the spans are poured before those over the inner supports. The following conclusions can be drawn:

- in both cases the whole slab is affected by a tensile stress state which develops just a few days after concreting and thus when the concrete tensile strength has not reached its maximum value; this result confirms the cracking tendency of concrete slabs during construction observed by Krauss and Rogalla (1996), SETRA (1995) and Ducret and Lebet (1999);

- the choice of the slab casting sequences assumes a fundamental role in containing slab cracking during the constructive phases, because it permits minimising the negative effects of the concrete and formwork weights;

- because the concrete slab is always in tension, the time dependent analysis should be performed by using a tension creep function, which is not available in technical literature; the slab stress values obtained with the creep coefficient given by the codes of practice, are then conservative;

- by using an optimised casting sequence, the tensile stresses on the concrete slab are lower than those produced by a continuous sequence; the stress reduction is more important over the inner supports where additional tensile stresses will be produced by service loads.

REFERENCES

CEB-FIP Model Code 1990 (1988). *CEB Bulletin d'Information n. 190*, CEB-FIP Comité Euro-International du Béton – Fédération International de la Précontrainte, Paris, France.

Dezi L., Leoni G. and Vitali A. (2001). Casting sequence analysis on steel-concrete composite bridges. *Proceedings of CONCREEP-6*, Boston, U.S.A., 767-772.

Ducret J.M. and Lebet, J.P. (1999). Behaviour of composite bridges during construction, *Structural Engineering International*, **3**, 212-218.

Gutsch A.W. (2001). Creep and Relaxation of Early-Age Concrete, *Proceedings of CONCREEP-6*, Boston, U.S.A. 619-624.

Marì A.R. (2000). Numerical simulation of the segmental construction of three dimensional concrete frames. *Engineering Structures* **22**, 585-596.

Newmark N.M., Siess C.P. and Viest I.M. (1951). Tests and analysis of composites beams with incomplete interaction, *Proc. Soc. Exp. Stress Anal.*, **9:1**, 75-92.

Krauss P.D. and Rogalla E.A. (1996). *Transverse Cracking in Newly Constructed Bridge Decks*, National Academy Press, Washington, USA

Kwak H.G., Seo Y.J. and Jung C.M. (2000). Effects of the slab casting sequences and the drying shrinkage of concrete slabs on the short-term and long-term behaviour of composite steel box girder bridges. *Engineering Structures* **23**, 1453-1480.

SETRA (1995), *Recommandations pour maîtriser la fissuration des dalles*, SETRA, Bagneux, France.

SHEAR STRENGTH OF PRESTRESSED CONCRETE ENCASED STEEL BEAMS WITH BONDED TENDONS

S.C. CHOY[1], Y.L. WONG[2] and S.L. CHAN[3]

[1] The Hong Kong Polytechnic University, Hung Hom, HONG KONG.
[2] The Hong Kong Polytechnic University, Hung Hom, HONG KONG.
[3] The Hong Kong Polytechnic University, Hung Hom, HONG KONG.

ABSTRACT:

This paper focuses on the shear transfer mechanism of prestressed concrete encased steel (PCES) beams- a hybrid structural member that has not been thoroughly studied. Although a PCES beam is a hybrid beam comprising the features of prestressed concrete beams and structural steel beams, it is doubtful that either Diagonal Compression Field Theory or Diagonal Tension Field Action mechanism or both can be used to model the shear resistance of the hybrid beam

In this paper, the results of shear tests on 6 PCES with bonded tendons are examined. The purpose of the test was to elucidate the shear performance of PCES with bonded tendons under monotonic loading. Evaluation and comparison of test results of PCES with the effect of various arrangements were made such as number of stirrups, shear span depth ratio and eccentricity of I beam.

The test results indicate that the shear resistance of PCES beams can be predicted by adding the contributions due to (i) concrete V_c, (ii) web of I section V_{web}, (iii) stirrups V_s, and (iv) vertical component of prestressed force V_p.

KEYWORDS

Steel, Prestressed, Bonded ,Tendons, AIJ-SRC, Shear Contribution

INTRODUCTION

There are lot of composite structures constructed with different combination of materials and mechanics. In reinforced concrete structures, medium grade steel reinforcement is placed in concrete to provide tensile or shear resistances. With the application of prestressing technique, the benefits of high strength steel tendons and pre-loaded concrete can be fully utilized in prestressed concrete structures. In fact, prestressing technique has been applied to structural steel members for strengthening purposes. It is also common to encase structural steel beams with concrete for better fire and strength resistances.

However, an exploration of a potentially super hybrid structural member- prestressed concrete encased steel beams (PCES) that compose four main components i.e. concrete, steel I section, stirrup and prestressing tendons, has not been reported. In this paper, we present our preliminary test results on the shear capacity, deformation, and failure modes of 6 PCES beams with bonded tendons, and propose an approach to predict the shear strength of PCES with bonded tendons.

EXPERIMENTAL PROGRAM

Test Specimens and Test Procedure

Test Specimens

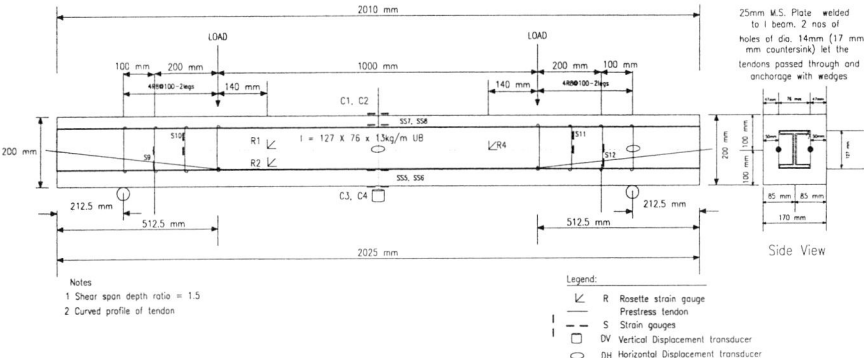

Figure 1: Dimensions of Typical Specimen and Test Layout

Figure 1 shows the dimensions of a typical specimen and test layout of this study. The details of specimens, with the identification codes, are listed in Table 1. Beam 3 have no prestressing strand, and they serve as reference beams for evaluating the enhancement in shear strength provided by the prestressing strands.

TABLE 1
SPECIMEN DETAILS

Beam Code	Dimension B x H (mm)	Shear Span depth ratio	Steel I Beam 127x76x13kg/m	Prestressing Strands		Stirrup		Concrete Grade
				Quantity	Total Prestressed Force (kN)	Quantity (2 legs)	Spacing	
1	170 x 200	1.5	Yes	2	194	-	-	40
2	170 x 200	1.5	Yes	2	200	4R8 in pair	100	40
3	170 x 200	1.5	Yes	0	0	-	-	40
4*	170 x 200	2	Yes	2	140	-	-	40
5	170 x 200	1.5	Yes	2	217	4R8	100	40
6	170 x 200	2	Yes	2	217	-	-	40

* eccentricity of tendons from centroid of I beam is 20mm more than the others

The measured material properties of reinforcement bars, steel I beam, prestressing tendon and concrete, and the concrete mix proportions are shown in Table 2.

TABLE 2
MEASURED MATERIAL PROPERTIES OF REINFORCEMENT BARS, STEEL I BEAM, PRESTRESSING TENDON AND CONCRETE

Material	Size of sample (mm)		Area (mm^2)	Yield Stress (MPa)	Ultimate Stress (MPa)	Young Modulus	Elongation (%)
	Width	Thickness					
Reinforcement	-	-	50.1	324	380	199	29
Web of I	25.1	4.62	-	317.5	445.3	185.4	31
Flange of I	25.1	6.88	-	323.9	458.9	199.9	30
Prestressing Tendon		12.9 dia.	99.6	171kN 0.1% proof load	1847 tensile	204	21.9
Concrete (kg/m^3)	Water 200	Cement 400	Coarse 950	Fine kg/m^3 700	F_c(MPa) 60	F_t(MPa) 4.2	E_c (N/mm^2) 34

Test Procedure and Instrumentation

All specimens, with same overall dimensions, were tested in a conventional beam-testing frame. Each test beam was simply supported, and monotonically loaded to failure, with two point loads, giving a shear span-to-depth ratio of 1.5 or 2.0. The loads were measured using load cells. Deflections at mid-span and load positions were measured using linear variable displacement transducers (LVDT). Rosette strain gauges and strain gauges were placed on the web and the flanges respectively of the steel I section. At the mid-span, strain gauges were bonded to the top and bottom surfaces of the concrete sections. A load cell was also attached to the anchorage of the prestressing strands at each end so that the actual prestressing force could be monitored. For beams containing stirrups in shear spans, strain gauges were also attached to the stirrups at four locations (two of each side). Details of instrumentation are also shown in Figure 1.

Test Results

Table 3 summaries the measured shear forces at stage 1 (occurrence of first crack), stage 2 (occurrence of diagonal crack), and stage 3 (maximum load) of the PCES beams, while Figure 2 shows typical total load - mid-span deflection responses of some test beams.

TABLE 3
SUMMARY OF TEST RESULTS OF PCES BEAMS AT VARIOUS STAGES

Shear Force(kN)	Stage	1	2	3	4	5	6
	1	43.8	51.5	4.55	37.25	54.3	20
	2	81.6	82	20.25	75.65	81.4	20
	3	143.4	141.5	55.6	140.5	149.2	94

Note : Stage 1-Formation of First Crack, Stage 2-Formation of Diagonal Crack Stage3 eak load

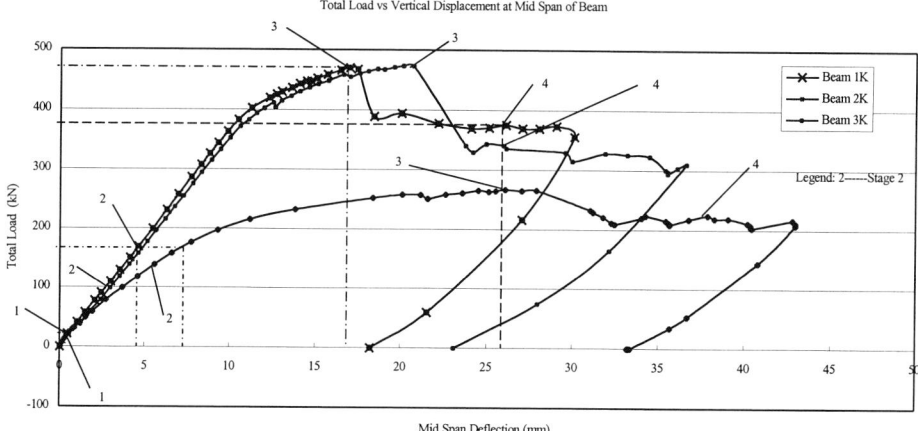

Figure 2 : Load- Mid-span Deflection Responses of Beams No. 1, Beam 2 and 3

It appeared that the load-deflection responses of test beams were linear prior to cracking. After the initiation of first crack (occurred in the flexure zone), the slope of load-deflection curve was slightly reduced. The reference beams Beam 3 (with steel web but no prestressing tendon) attained similar peak shear forces of 133.35kN with mid-span deflections of about 20mm, and thereafter exhibited insignificant strength reduction subjected to further deformation, till the mid-span deflection reached at least 43mm.

Other beams with steel web and prestressed forces (Beams 1, 2, 4, 5 and 6) developed higher shear resistances ranging from 163 to 243kN, with the corresponding mid-span deflections varied from 17 to 20mm. Their post-peak performance was characterized by strength degradation, at a rate comparable to that of the strength gain developed previously. Such degradation was accompany with diagonal crack widening and spalling of cover concrete at the compressive flange of the I section near the load application points, and eventually ceased when the residual strength approached the peak strength of the reference beams. Thereafter, their residual load-deflection curves became almost flat for the mid-span deflections exceeding 36mm.

The presence of 0.6% stirrup reinforcement at the shear spans for Beam 2 and 5 did not effectively improve the shear capacity (compared with Beam 1), but reduced the rate of strength degradation in the post-peak range of the loading history.

DISCUSSION OF TEST RESULTS

Outstanding Performance of PCES Beam System

As the design guidelines for PCES beams are not available, we studied the design approaches for a similar structural system: encased steel composite (ESC) beams, as recommended by some well-received design codes. It was found that many national codes, such as the American AISC-LRFD(1993), or the British BS5400(1979) recommended the steel web should resist total applied shear. Only the Japanese code AIJ-SRC (1987) considered the shear resistance contributed from the steel web, and concrete portion. In order to illustrate the beneficial contributions from the concrete and prestressing force, we performed an analytical simulation using ABAQUS software to determine the load-deflection response of the bare steel I beam used in this study under the load arrangement as for

Beam 3. The analytical results so obtained were compared with those of Beam 3 (ESC beam), and Beam 1 (PCES beam). It was found that measured shear capacities of Beam 3 and Beam 1 were 11% and 95% higher than the analytical shear capacity of the bare steel I beam. Under the total load of 100kN, the mid-span deflections of the bare I beam, Beam 3 and Beam 1 were 5.23mm, 3.7mm and 2.98mm respectively. Thus, the outstanding performance of the proposed PCES beams system, and the significant contribution due to concrete and prestressing force became obvious. The pre-tensioning force for Beam 4 was the least among all PCES beams. This is mainly due to the larger eccentricity between tendons and neutral axis of I beam. The prestressing force made Beam 4 distort laterally. Then the applied prestressing force was stopped at 140kN.

Shear Contribution of Each Component

We assume that the measured peak shear of a PCES beam V_{exp} is contributed by (i) the steel web of the I section V_{web}, (ii) the vertical component of prestressing force V_p, (iii) the steel stirrups V_s, and (iv) the concrete portion V_c. Since strain gauges and rosettes were installed in the stirrups and the steel web of I beams respectively, and load cells were used to measure the prestressing force of the tendons, the shear resisting components V_{web}, V_p, and V_s could be measured/calculated. Therefore, the term V_c could be considered as $V_c = V_{exp} - V_{web} - V_p - V_s$.

Figure 3 is a typical example showing the variation of each shear-resisting component at different loading stages. At Stage 1 (formation of first crack), concrete contributed more than 33% of the applied shear, whilst the contribution of steel web was less than 13%. At Stage 2 (formation of diagonal crack), the contributions of concrete and Steel web increased to 22% and 41% respectively. Upon further loading, the concrete contribution reduced but the steel web contribution increased. At Stage 3 (peak load), the concrete contribution became only 2%, while the steel web capacity was fully mobilized and contributed up to 58% of the shear capacity of the beam. The contribution of transverse reinforcement was also increased from 18% at Stage 2 to 33% at Stage 3. The shear contribution for prestressing tendon basically remained constant throughout the loading history. It was evident that shear contribution of each component varied at each stage. Concrete offered the most shear contribution before the formation of diagonal cracks. After the development of diagonal cracks, other components especially the steel web, gradually picked up more contribution. When the maximum load attained, the steel web yielded and contributed most in the shear resistance.

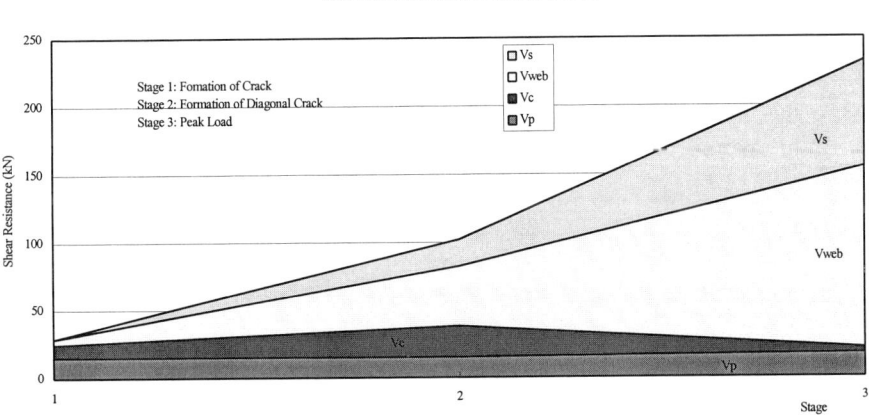

Figure 3: Shear Contribution of Each Component of Beam 5

Prediction of Shear Capacity of PCES Beams

Weng et al (1987) investigated the shear strength of encased steel beams, and found that the AIJ-SRC code gave good predictions for their beam series. Thus, we also tried to use the relevant basic equations from the AIR-SRC code, and added a new term accounting for the prestressing effects to formulate the proposed shear capacity of a PCES beam V_n as follows:

$$V_n = V_p + {}_uV_{web} + {}_uV_c \quad (1)$$

where V_p = shear carried by the vertical component of prestressing force, ${}_uV_{web}$ = shear carried by the steel web of I section, and ${}_uV_c$ = shear carried by concrete and stirrups.

$$V_p = N_p \cdot \sin \quad (2)$$

where N_p = prestressing force; = inclination of prestressing tendons

$${}_uV_{web} = \min\left(\frac{\sum {}_uM_s}{l'}, t_w \cdot d_w \cdot \frac{F_{ys}}{\sqrt{3}}\right) \quad (3) \qquad {}_uV_c = \min\left(\frac{\sum {}_uM_c}{l'}, {}_{su1}V_c, {}_{su2}V_c\right) \quad (4)$$

where ${}_uM_s$ = flexural capacity of steel I section; l' = shear span of beam; t_w = steel web thickness; d_w = depth of steel web; F_{ys} = yield stress of steel web; ${}_uM_c$ = flexural capacity of concrete section; ${}_{su1}V_c$ = shear carried by concrete and stirrups under diagonal shear failure of concrete section; ${}_{su2}V_c$ = shear carried by concrete and stirrups under shear bond failure.

$${}_{su1}V_c = B \cdot j_r \cdot (0.5 \cdot \alpha_r \cdot f_s + 0.5 \cdot \rho_w \cdot F_{yh}) \quad (5) \qquad {}_{su2}V_c = B \cdot j_r \cdot \left(\frac{b'}{B} \cdot f_s + \rho_w \cdot F_{yh}\right) \quad (6)$$

where B = gross width of composite beam; j_r = distance between centroids of tension and compression of concrete section under flexural; $_r$ = coefficient related to shear span ratio; f_s = shear stress of concrete($= F_c/k$); $_w$ = stirrup ratio; F_{yh} = yield stress of stirrups; and b' = effective width of concrete; F_c = concrete compressive strength; d_r = effective depth.

Table 4 compares the test results V_{exp} with the predicted shear strengths V_n evaluated using Equation 1. The table presents two different values of V_n that were calculated from different assumed shear stress of concrete or k values. In the AIJ-SRC code, k was taken as 20. In this study, we also tried k equal to 18. It was found that the average V_n/V_{exp} were 0.85 and 0.86 for k equal to 20 and 18 respectively. These indicate that the shear capacity of PCES beams can be satisfactorily predicted by our proposed design equations that are modified from the AIJ-SRC code.

TABLE 4
COMPARISONS OF PREDICTED SHEAR STRENGTHS WITH EXPERIMENTAL VALUES

Beam No.	V_{exp}	Vn, k=20	Vn, k=18	Vn/Vexp, k=20	Vn/Vexp, k=18
1	234.9	161.20	166.02	0.686	0.707
2	242.99	241.80	241.75	0.995	0.995
3	133.35	132.43	137.24	0.993	1.029
4	192	125.80	130.61	0.655	0.680
5	232.8	204.47	206.85	0.878	0.889
6	162.74	134.41	139.22	0.826	0.855
Average	213.09	173.54	176.89	0.85	0.86
S.D.	34.42	48.94	46.86	0.13	0.12

Note : Beam 3 and 4 are excluded in average and standard deviation calculation

SUMMARY AND CONCLUSIONS

This paper attempts to clarify the shear transfer mechanism of prestressed concrete encased steel (PCES) beams- a hybrid structural member that has not been thoroughly studied. It was found that the contribution of concrete section to the overall shear resistance was significant, particular at the onset of diagonal shear cracks. Upon further loading causing widening of he diagonal cracks, the concrete shear resisting ability reduced, while the web of the steel I section gradually played a more important role in resisting external shear. When the maximum shear capacity was developed, the steel web attained the Von Mises yield criteria or its maximum shear carrying ability. Shear reinforcement, in form of stirrups, did not appear to be effective in enhancing the shear resistance of PCES beams. However, the provision of stirrups apparently reduced the rate of strength degradation of the post-peak performance of PCES beams.

The contributions due to concrete, web of steel I-section, stirrups and vertical component of prestressed force were considered in the development of the new shear strength prediction that was modified from the AIJ-SRC Code (AIJ 1987). It was showed that our predictions matched well with the experimental results.

This paper also demonstrates that the new prestressed concrete encased steel (PCES) beam system has a good potential for practical applications because it exhibits high stiffness and strength at service and ultimate states respectively, and considerable dependable ductility in the post-peak performance. A further study on this area is therefore recommended.

REFERENCES

American Institute of Steel Construction (1993). *Load and Resistance Factor Design Specification for Structural Steel Buildings,* 2nd Ed. Chicago.
Architectural Institute Of Japan (1987). *Standards for Structural Calculation of Steel Reinforced Concrete Structures,* AIJ Code , Tokyo.
British Standards (1979). *Steel, Concrete and Composite Bridges: Part 5: Code of Practice for Design of Composite Bridge,* BS5400, London.
C.C. Weng, S.I. Yen and C.C. Chen (2001). Shear Strength of Concrete-Encased Composite Structural Members, *Journal of Structural Engineering*, ASCE Vol. 127, No.10, October, 2001, pp. 1190-1197.

INSTABILITY BEHAVIOR OF THE PRESTRESSED STEEL-CONCRETE COMPOSITE CONTINUOUS BEAM

Y. Han [1], Z.Z. Fang[2], Y.L. Guo[1]

[1]Department of Civil Engineering, Tsinghua University,
Beijing, 100084, P.R.China
[2]College of Civil Engineering and Architecture, Fuzhou University,
Fuzhou, 350002, P.R.China

ABSTRACT

In this paper, a finite element program for the nonlinear analysis of prestressed steel-concrete composite continuous beam is introduced. The program named NA-PSCB accounts for the nonlinear behavior of concrete, steel, prestressing tendons and shear connectors. Concrete is treated as an orthotropic nonlinear material. Steel is modeled as an elasto-plastic strain-hardening material, and the von Mises yield criterion with normality flow rule is applied. For shear connectors, an empirical nonlinear shear force-slip relationship is used. Also, the incremental displacement algorithm has been extended into the geometrically and materially nonlinear finite element for the structural stability analysis. The accuracy and reliability of the program are demonstrated by the analysis of an experimental beam throughout the entire loading range up to failure. Five simulated beams are presented to consider the effects of the main parameters --- the residual stress, the width-thickness ratio of web and flange. The results show that the onset of the elasto-plastic state of the structure will be earlier than expected for the existing residual stress, and the continuous beams will experience different buckle modeling from plastic, elasto-plastic to elastic which is connected directly with the section type from Class I section to Class III section (EC4).

KEYWORDS: prestressed, composite beam, stability, and finite element method

INTRODUCTION

The prestressed steel-concrete composite continuous beam is a new type of structural member that is developing rapidly in recent years. It fully utilizes the advantage of steel and concrete, enlarges the elastic range of structural materials and achieves a unity between good behavior and economy. The researchers around the world focus on it widely.

However, many previous works are limited to strength behavior of the prestressed steel-concrete composite beams. In fact, it is found from a large number of experiments and theoretical analyses that their failures are often induced by instability of steel beam in the region of interior bearing points. The further researches indicated that the prestress constrains the rotational capacity of the beam section. If the common plastic analysis method is directly adopted in this kind of structure, the Class I section prosperities will not provide the necessary ductility. Meanwhile, owing to neglecting the geometrical nonlinear effect in the numerical model, there were some differences between the previous FEM analytical results and corresponding experimental data.

An improved numerical analytical procedure based on the small elasto-plastic strains and updated Lagrangian formulation to account for the large displacements of the structure in this paper is therefore required to consider the material and geometrical nonlinear simultaneously and to simulate the full process behavior, especially stability.

FINITE ELEMENT PROGRAM

Assumption----- The prestressed steel-concrete composite continuous beam is a spatial thin-walled structure in a 3-dimensional stress field. However, because the normal stress of the steel and concrete top slab is smaller than any other directional stress, the dimensional problem can be simplified as a plan problem in the nonlinear analysis. The assumptions are included as follows: (1) a family of simplified elements to model directly the complex behavior of the concrete slab, steel beam, shear connection and prestressed tendons; (2) regarding the dividing line of internal elasto-plastic region along the boundary of the element to simplify the difficulty of determining the plastic region; (3) the material constitutive relationships of the concrete and steel under the simple loading condition; (4) neglecting the contribution of steel bar in concrete slab; (5) neglecting the friction between the interface of steel and concrete slab.

Incremental Displacement Algorithm---- The prestressed steel-concrete composite continuous beam is similar to a beam-column member, and its ultimate load will be emphasized. It is meaningful to trace the full-process curve and the incremental displacement algorithm is adopted from several available methods. The coordinates are built on the continuous deformed structure, i.e. the Updated Lagrange Formulation.

In order to stay as close as possible to the real response of the structure, a distinct displacement component will be controlled artificially. Regard the known component as Δu_2, the controlled component can be apart from the uncontrolled components in the stiffness matrix and displacement vector:

$$\begin{bmatrix} K_{11} & K_{12} \\ K_{21} & K_{22} \end{bmatrix}^i \begin{Bmatrix} \Delta u_1 \\ \Delta u_2 \end{Bmatrix}_n = \lambda_n^i \begin{Bmatrix} P_1 \\ P_2 \end{Bmatrix} + \begin{Bmatrix} R_1 \\ R_2 \end{Bmatrix}^i_{n-1} \tag{1}$$

In which K is the stiffness matrix; P_1 and P_2 are reference loads corresponding to the displacements Δu_1 and Δu_2; R is the residual load, λ is the load factor; i and n represent the iterative number and increment respectively.

By using the two-step solution method proposed by Batoz and Dhatt, condensing the equation (1) can give λ and the remaining $n-1$ unknown displacement after several iteration steps.

Element and material modelling

Element modelling

Based on the assumption, the steel slab and concrete slab are treated as shell elements with four nodes (Fig. 1(a)); the shear connection is treated as bar element (Fig. 1(b)) and the tension rod element (Fig. 1(c)) is used to simulate the prestressing tendon.

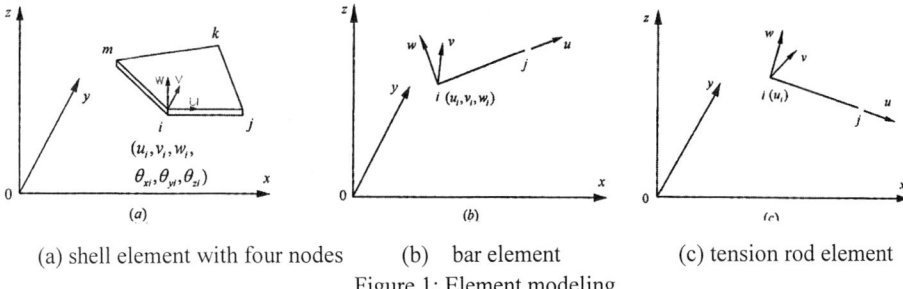

(a) shell element with four nodes   (b) bar element   (c) tension rod element

Figure 1: Element modeling

Material modelling

(1) The steel beam and the prestressed tendon are considered as the elasto-plastic strain hardening materials. The von Mises yield criterion with normality flow rule is used to represent the material plasticity as follows:

$$F([\sigma'],[\overline{\sigma}],H) = \sqrt{(\tfrac{2}{3}tr[\sigma']^2)} - \overline{\sigma}(H) = 0 \qquad (2)$$

where $[\sigma']$ is the deviatoric stress tensor, $\overline{\sigma}(H)$ the uniaxial yield stress and H the hardening parameter.

For subsequent yielding, the isotropic hardening rule is used. The plasticity matrix $[D_{ep}]$ is used for calculating $[K_T]$.

$$[D_{ep}] = [D_e] - \frac{1}{a}\overline{V}\overline{V}^T \qquad (3)$$

with $\overline{V}^T = \dfrac{\partial F^T}{\partial \sigma}[D_e]$ and $a = \overline{h} + \dfrac{\partial F^T}{\partial \sigma}\overline{V}$,

where $[D_e]$ is the elastic matrix and $\overline{h}$ is the slope of the uniaxial stress-strain curve.

(2) Concrete is assumed isotropic up to failure, either by crushing in compression or cracking in tension. The relation, as shown in Fig.2, is comprised of two parts: Part I is represented by the equation proposed by Saenz:

$$\sigma_i = \frac{E_0 \varepsilon_{iu}}{1+(\frac{E_0}{E_{sc}}-2)(\frac{\varepsilon_{iu}}{\varepsilon_{ic}})+(\frac{\varepsilon_{iu}}{\varepsilon_{ic}})^2} \quad (4)$$

where E_{sc} = secant modulus of elasticity at peak stress; E_0 = initial tangent modulus; σ_i = principal stress; ε_{iu} = equivalent uniaxial strain; ε_{ic} = strain at peak stress.

The tangent modulus E_t is given by

$$E_t = \frac{E_0[1-(\frac{\varepsilon_{iu}}{\varepsilon_{ic}})^2]}{[1+(\frac{E_0}{E_{sc}}-2)(\frac{\varepsilon_{iu}}{\varepsilon_{ic}})+(\frac{\varepsilon_{iu}}{\varepsilon_{ic}})^2]^2} \quad (5)$$

which can be used in biaxial compression and the compressive direction of the tension-compression. In the biaxial tension region, the initial tangent modulus is adopted. Part II accords to the expression:

$$\varepsilon_{ic} < \varepsilon < \varepsilon_{iu}, \quad \sigma_i = \sigma_{ic}[1-200(\varepsilon-\varepsilon_{ic})^2] \quad (6)$$

$$\varepsilon > \varepsilon_{iu}, \quad \sigma_i = 0.3\sigma_{ic} \quad (7)$$

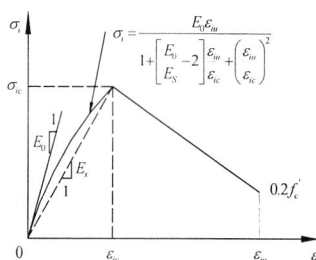
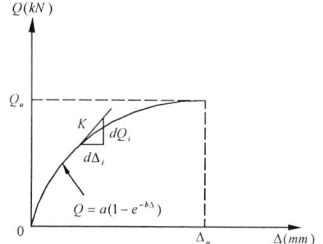

Figure 2: Equivalent uniaxial stress-strain law for concrete

Figure 3: Typical load-slip curve for flexible shear connector

The failure criterion is described by the updated formulation proposed by Kupfer and Gerstle. In order to reflect the influence of the high-stress state caused by the prestressing, an additional judging method is adopted which can be found in the reference 5.

(3) Two different constitutive laws are used separately to describe the normal and tangential behavior of shear connection. In the normal direction, the stiffness of a stud is evaluated as a usual steel bar element. On the tangent surface, the constitutive behavior is defined by the typical shear-slip function proposed by Yam and Chapman and shown in Fig.3. The following empirical load-slip equation reported by Ollgaard et al. has been used to calculate Q frequently:

$$\frac{Q}{Q_u} = (1-e^{-0.702\Delta})^{0.4} \quad (8)$$

with Q_u evaluated by the following equation adopted by the CSA S16-88 Code

$$Q_u = 0.5 A_d \sqrt{(f_c' E_c)} \le f_u A_d \qquad (9)$$

where f_u is the ultimate strength of steel and A_d is the cross-section area of the shear connector in mm², f_2' is the cylinder compressive strength of the concrete.

PROGRAM AND NUMERICAL EXAMPLES

The achievement of the FEM program

The instability mode of the negative moment region is complicated. During the loading process of the stability analysis, the arrays that describe the inplane and the out-of-plane displacements and the developing rates of the two displacements should be recorded. Since the instability of the structure can essentially be ascribed to the collapse of internal resistance, the neighboring displacement ratio adjacent to the limit point is suggested as the judgment of the instability.

However, the gobal deformation behavior of the structure including lateral deformation could not be represented by a point. The 2 order norm value is then calculated. Also, the average element strain of two adjacent increments is written down to record the onset of time of strain reversal in the steel flange and web elements of the negative moment region.

In theoretical analysis, if the stiffness is found to be negative or zero, the value of the last step load or the load related to the minimum stiffness before reloading is the ultimate load. In fact, there is another displacement failure criterion to the continuous prestressed composite beam. That is when the lateral displacement of the steel beam flange of the negative moment region is larger than twice the thickness of the web, the load will be defined as the ultimate load.

Numerical examples

Experimental example

The feasibility and reliability of the program NA-PSCB are demonstrated by an experimental example (Fig 4). The calculating process can be divided into two steps: the first step is to calculate the structure response when prestressing; the second step is to calculate the full-process response of applied load. Only half of the beam is considered in the analysis by taking advantage of the symmetric specimen with respect to the middle support. The finite element mesh is shown in Fig.5.

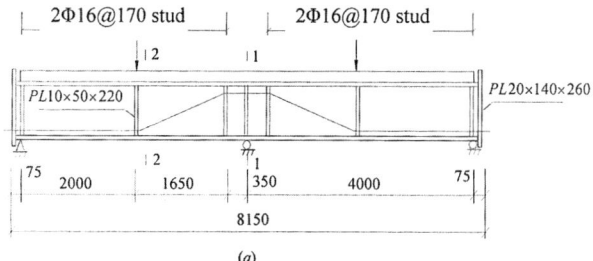

(a)

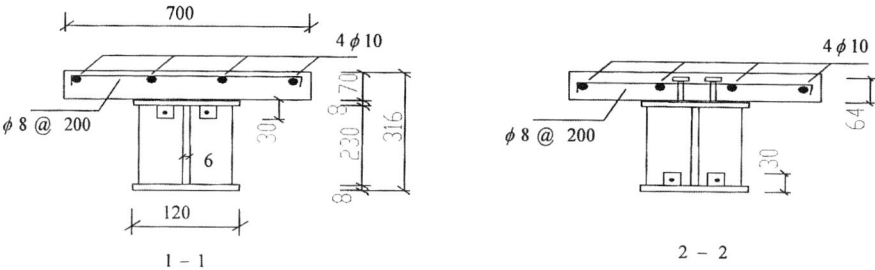

Figure 4: Detail of the specimen (unit:m)

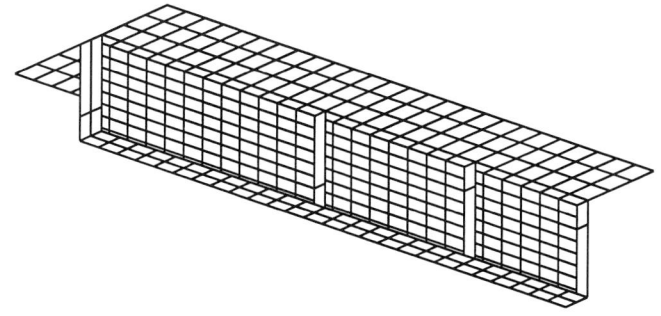

Figure 5: The finite element mesh

The load-midspan deflection, load-lateral deformation of the internal support region and the plastic region of the steel web are shown in Fig.6, 7 and 8, respectively. Good coincidence with the experimental results is observed.

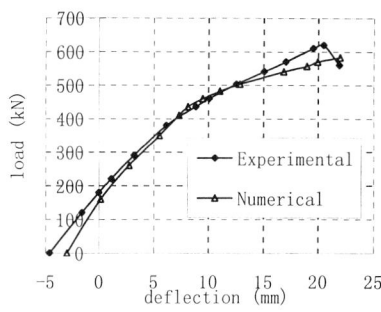
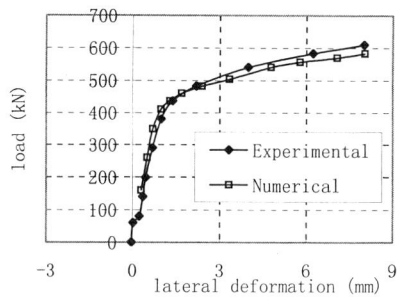

Figure 6: Load-deflection curve　　　　Figure 7: Load-lateral deformation curve

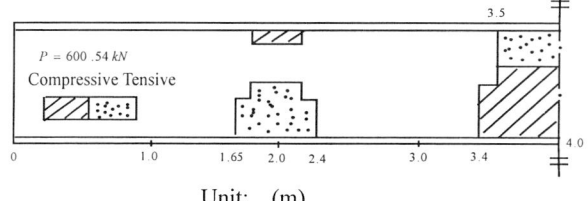

Unit: (m)

Figure 8: Plastic region

Effect of the main parameters

(1) *The width-thickness ratio of web*

Let $b_f/t_f=15$, $h_w = 240\sim300$mm varying from Class II to III sections of the EC4 code. The calculated results are shown in table 1. The final failure types belong to the distortional buckling.

TABLE 1
THE EFFECT OF THE WIDTH-THICKNESS RATIOS OF THE WEB

	b_f/t_f	h_w	t_w	h_w/t_w	Predominate failure type	Section type	P_{max}(kN)
B1	15	240	7	34.2	Local buckle	II	723.6
B2	15	320	8	40	Lateral buckle	III	533.7
B3	15	270	7	38.57	Local buckle	II	611.5
B4	15	310	7	44.27	Lateral buckle	III	460.2
B5	15	290	8	36.25	Local buckle	II	662.3

From table1, it is indicated that the load-carrying capability can increase with the decrease of the web width-thickness ratio. The distortional buckling model is the main failure type of a Class II section where the local buckle of web and flange predominate. With an increase of the width-thickness ratio of the web, lateral buckle gradually becomes obvious. The magnitude of the web width-thickness ratio in some degree decides onset of time of the buckling and the stress development of the section.

(2) *The width-thickness ratio of flange*

The values of $h_w/t_w=32\sim36$ vary from Class I to II sections of EC.4 code. As can be seen from the calculated results in table 2, the load-carrying capability will increase with a decrease of the flange width-thickness ratio. When the ratio increases further, the rigid constraint to web will decrease and the web presents the apparent buckling. The local buckle will not be considered as the ultimate failure of a Class II section.

TABLE 2
THE EFFECT OF THE WIDTH- THICKNESS RATIO OF FLANGE

	b_w/t_w	b_f	t_f	b_f/t_f	Section type	P_{max}(kN)	Predominate failure type
B6	32	90	10	9	I	788.6	/
B7	36	120	9	13.3	II	679.1	Web local buckle
B8	36	110	9	12.2	II	693.4	Bottom flange local buckle

(3) *The residual stress*

The distribution of residual stress is assumed to vary linearly as shown in Fig. 9 and the factors respectively are $\eta = 0.3$, $\xi = \dfrac{\eta}{1+\frac{2A_w}{A_f}} = 0.13$. In the process of solution, σ_r is considered as the initial stress superposition to the average element stress. As expected in Fig. 10, the residual stress decreases the bending stiffness of the element and the stable capability. The reason for this is that the section will enter the elasto-plastic state earlier and decrease value of inertia moment of elastic region I_e is higher than that of the situation without the residual stress.

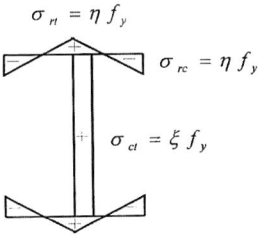
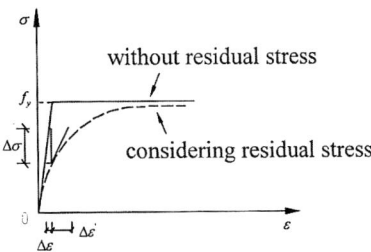

Figure 9: Distribution of the section residual stress   Figure 10: Effect of the residual stress

CONCLUSIONS

In view of the favorable comparisons between the finite element and experimental result, it can be stated that the finite element program NA-PSCB is capable of tracing the detailed response of prestressed composite continuous beam throughout the whole loading range from prestressing up to its failure. The program can simulate some detailed aspects of the behavior of such a beam, although this would in general require a finer mesh and more accurate modeling of the regions affected by such details. The program can be used to conduct extensive parametric studies in order better to understand the inelastic response. The mode of failure of the Class I, II, III sections classified by EC.4, is initiated by different buckling types from plastic, elasto-plastic to elastic. The existing design formula do not provide for the ascertained rotational capability and should be modified. A simplified design formula should be suggested after further analysis of parameters such as the influence of the extent of prestressing to the height of compressive region.

REFERENCE

[1] A. Ghani Razaqpur and Mostafa Nofal (1990). "Analytical Modeling of Nonlinear Behavior of Composite Bridges." J.of Structural Engineering, ASCE, vol116, No.6, 1715-1730
[2] J. J. Lin, M.Fafard. D.Beaulieut and B.Massicotte (1991). "Nonlinear Analysis of Composite Bridges By the Finite Element Method." Computers&Structures, Vol.40, No.5, 1151-1167
[3] A.R.Kemp (1996). "Inelastic Local And Lateral Buckling In Design Codes." J.of Structural Engineering, ASCE, Vol. 122, No.4, 374-381
[4]Z. H. Zong (1997). "The Research On the Satic and Dynamic Behavior of the Prestressed Steel-concrete Composite Beam." Dissertation of West-north Transportaion University
[5] Y. Han(2000). "The Research On the Stability Problem For Continuous Prestressed Steel-concrete Composite Beam." Graduate Thesis for Master Degree of FuZhou University

EVALUATION OF SIMPLIFIED SUPERPOSITION DESIGN METHOD FOR COMPOSITE COLUMNS

Jian-hai Zhong and Shui-fu Chen

Department of Civil Engineering, Zhejiang University,
Hangzhou 310027, China

ABSTRACT

In this paper, the accuracy and limitation of the simplified superposition design method for concrete encased composite columns subjected to biaxial bending recommended by the Chinese YB9082-97 Specification is evaluate using an accurate iterative computer-based method. Comparisons made between the two methods in the determination of load carrying capacities of rectangular cross sections with symmetrically placed H-shaped and box-shaped structural steel indicate that the simplified method appears to be conservative on all cross-sectional cases studied, but display different overall accuracies on cross sections with different structural steel shapes and different load eccentricities. Application of the simplified method to circular cross sections is also discussed. Based on the axial force and bending moment superposition concept, a new simplified design equation for such sections is proposed. Design examples are presented to demonstrate the simplicity and validity of the proposed design method.

KEYWORDS

Simplified Design Method, Steel-Concrete Columns, Structural Design, Composite Columns, Biaxial Bending

INTRODUCTION

Composite steel-concrete structural systems have been used in the construction of high and low-rise buildings throughout the world. With some promising characteristics, such as speeding up the construction process and protecting from fire damage, this kind of system is used more widely in China than ever. The concrete-encased composite column is one of the common composite structural elements. Many researchers, including Johnson and Smith (1980), Lachance (1982) and Roik and Bergmann (1984) proposed simple methods for analysis and design of rectangular composite columns under biaxial loading. Munoz and Hsu (1997) proposed a generalized interaction equation for design of biaxially loaded square and rectangular columns. In China, Specification for Design of Steel-Reinforced Concrete Structures, YB9082-97, declared on Nov.13, 1997, suggested a simplified superposition design method, which can be used to calculate rectangular cross section column with doubly symmetric structural steel under uniaxial/biaxial loading. In order to study the accuracy and error tendency of this method a general iterative computer-based design method proposed by Chen et al. (2001) is used. This computer-based method is suitable to arbitrary shaped composite columns with arbitrarily distributed structural steel and reinforcing bars and has been proved to be effective and accurate (Chen et al., 2001). Comparisons between the two methods in the determination of load capacities for rectangular cross sections with symmetrically placed H-shaped and box-shaped structural steel are first made. A modification to the simplified method is suggested so that the obtained results can match those by the computer method more closely. Application of the simplified method to the circular cross sections is then discussed. A simplified design equation for circular sections is suggested and its comparison to the computer-based method is presented in the paper.

BASIC ASSUMPTIONS AND DESIGN PROCEDURES

The simplified design method suggested in YB9082-97 Specification (1998) together with the computer-based method proposed by Chen et al. (2001) is based on the following basic assumptions:

(1) Plane sections before deformation remain plane after deformation.
(2) The cross section reaches its failure limit state when the strain of the extreme compression fiber of the concrete attains the maximum strain ε_{cu}.
(3) The stress-strain relationship of concrete in compression is represented by a parabola when the strain $\varepsilon_c < \varepsilon_0$ and a horizontal line when the strain $\varepsilon_0 \leq \varepsilon_c \leq \varepsilon_{cu}$. In YB9082-97 Specification (1998), ε_0 and ε_{cu} are set to 0.002 and 0.0033, respectively.
(4) The structural steel and the steel reinforcing bars are assumed to be elastic-perfectly plastic, i.e. the stresses are proportional to their strains up to the yield stresses and then remain the yield stresses constantly.
(5) Tensile strength of concrete is neglected.

In YB9082-97 Specification (1998), two cases are classified in the determination of the load carrying capacity of an existing rectangular cross-sectional column with structural steel and reinforcing bars being symmetrically placed. The final load capacity is the larger one calculated from the two cases. In the first case, when the design axial load (or assumed load capacity) $N \geq N_{c0}^{rc}$, the bending moment capacities M_x and M_y are determined by

$$N_c^{ss} = N - N_{c0}^{rc} \quad \text{and} \quad \frac{M_x}{M_{cy,x0}^{ss}(N_c^{ss})} + \frac{M_y}{M_{cy,y0}^{ss}(N_c^{ss})} \leq 1 \qquad (1)$$

and when $N < N_{c0}^{rc}$, the following equations are used,

$$N_c^{rc} = N \quad \text{and} \quad \frac{M_x}{M_{cu,x0}^{rc}(N_c^{rc}) + M_{cy,x0}^{ss}(0)} + \frac{M_y}{M_{cu,y0}^{rc}(N_c^{rc}) + M_{cy,y0}^{ss}(0)} \leq 1 \qquad (2)$$

In the second case, when $N \geq N_{c0}^{ss}$, the bending moment capacities M_x and M_y are determined by

$$N_c^{rc} = N - N_{c0}^{ss} \quad \text{and} \quad \frac{M_x}{M_{cu,x0}^{rc}(N_c^{rc})} + \frac{M_y}{M_{cu,y0}^{rc}(N_c^{rc})} \leq 1 \tag{3}$$

and when $N < N_{c0}^{ss}$, M_x and M_y are determined by

$$N_c^{ss} = N \quad \text{and} \quad \frac{M_x}{M_{cy,x0}^{ss}(N_c^{ss}) + M_{cu,x0}^{rc}(0)} + \frac{M_y}{M_{cy,y0}^{ss}(N_c^{ss}) + M_{cu,y0}^{rc}(0)} \leq 1 \tag{4}$$

In the above equations, N_{c0}^{ss} and N_{c0}^{rc} are the load carrying capacities of structural steel and reinforcing bars under pure compression, respectively; $M_{cy,x0}^{ss}(0)$, $M_{cy,y0}^{ss}(0)$, $M_{cu,x0}^{rc}(0)$ and $M_{cu,y0}^{rc}(0)$ are the moment carrying capacities of structural steel and reinforcing bars with respect to the x- and y-axes, respectively, when the axial load is zero. $M_{cy,x0}^{ss}(N_c^{ss})$ and $M_{cy,y0}^{ss}(N_c^{ss})$ are the moment carrying capacities of structural steel with respect to the x- and y-axes, respectively, when the axial load is equal to N_c^{ss}. $M_{cu,x0}^{rc}(N_c^{rc})$ and $M_{cu,y0}^{rc}(N_c^{rc})$ are the moment carrying capacities of reinforcing bars with respect to the x- and y-axes, respectively, when the axial load is equal to N_c^{rc}.

In the iterative computer-based method (Chen, et al., 2001), the stress resultants of the concrete are evaluated by integrating the concrete stress-strain curve over the compression zone, while those of the structural steel and reinforcement are obtained using the fiber element method (Mirza and Skrabek, 1991), in which the steel sections are discretized into small areas (fibers) and the reinforcing bars are treated as individual fibers. Using the iterative method, the depth and orientation of the neutral axis d_n and θ_n can be determined by the Quasi-Newton method within the Regula-Falsi numerical scheme (Yau et al., 1993) as follows

$$d_{n,k} = d_n + \frac{d_n' - d_n}{N' - N}(N_u - N) \tag{5}$$

where N' and N are the axial force capacities calculated with the neutral axis depth d_n' and d_n respectively, with N' being greater than the final load capacity N_u and N being smaller than N_u, and

$$\theta_{n,k} = \theta_n + \frac{\theta_n' - \theta_n}{\alpha_m' - \alpha_m}(\alpha_{md} - \alpha_m) \tag{6}$$

in which α_m' and α_m are the inclination of the resultant bending moment calculated with the neutral axis orientations θ_n' and θ_n respectively, with α_m' being greater than the design value α_{md}, and α_m being smaller than α_{md}. Here the angle α_m=arctg(M_y/M_x) and the design value α_{md}=arctg(M_{yd}/M_{xd}) or α_{md}=arctg(e_{xd}/e_{yd}), M_{xd} and M_{yd} are the given design bending moments and e_{xd} and e_{yd} are the given load eccentricities with respect to the x- and y-axes.

The final load-carrying capacity of the cross section at given eccentricities e_{xd} and e_{yd} can be determined by the following iterative equation

$$N_{u,k} = N_u + \frac{N_u' - N_u}{e_r' - e_r}(e_{rd} - e_r) \qquad (7)$$

in which N_u' and N_u are the axial load capacities calculated with the total load eccentricities at e_r' and e_r respectively, with e_r' being smaller than the design value e_{rd} and e_r being greater than e_{rd}. Here, $e_r = (e_x^2 + e_y^2)^{1/2}$ and $e_{rd} = (e_{xd}^2 + e_{yd}^2)^{1/2}$.

EVALUATION OF SIMPLIFIED DESIGN METHOD

In order to evaluate the accuracy and error tendency of the simplified superposition design method, the load-carrying capacities of two rectangular cross-sections, one with symmetrically placed H-shaped and the other with box-shaped structural steel are first determined using this method and are compared with those from the computer-based method. The layouts of the two cross-sections V1 and V2 are shown in Figure 1. The cylinder strength of concrete and yield stresses of structural steel and reinforcing bars are listed in Table 1. The calculated load carrying capacities and their comparison between the two methods are shown in Table 2 as well as Figure 2.

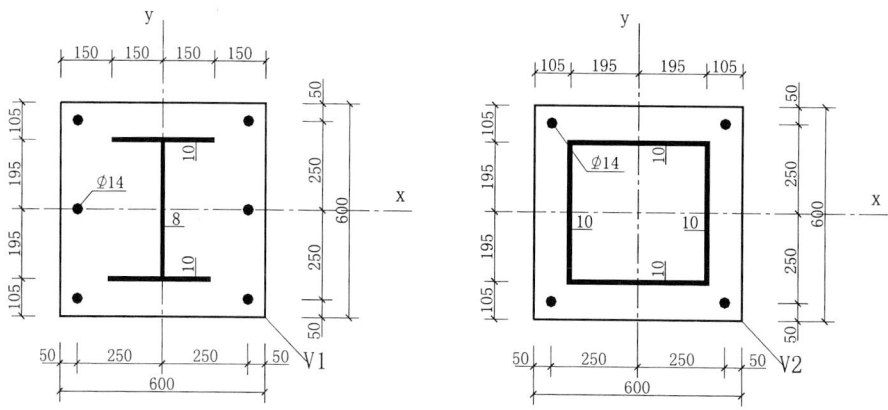

Figure 1:   Layout of V1 and V2

TABLE 1
MATERIAL PROPERTIES

Cross-section	f_c (N/mm^2)	f_s (N/mm^2)	f_y (N/mm^2)
V1 V2	12.5	345	315

f_c: characteristic strength of concrete in compression;
f_s: yield strength of structural steel;
f_y: yield strength of reinforcing bars.

TABLE 2
LOAD CARRYING CAPACITIES AND COMPARISON

Specimen	e_x (mm)	e_y (mm)	N_1 (kN)	N_2 (kN)	$(N_1-N_2)/N_2$
V11	0	0	7623	7450	2.3%
V12	15	15	5068	6800	-25.5%
V13	30	30	4197	6080	-30.9%
V21	0	0	9402	9335	0.7%
V22	15	15	7737	8535	-9.3%
V23	30	30	6574	7720	-14.8%

e_x, e_y: load eccentricities in the x- and y-direction with reference to the geometric centroid;
N_1, N_2: axial load carrying capacities calculated by the simplified method and the computer-based method, respectively.

It is clear from Table 2 and Figure 2 that the axial load carrying capacities determined by the simplified method are smaller than those by the computer-based method in all the cases presented, with the differences being larger while the load eccentricities e_x and e_y becoming larger. On the other hand, the differences between the two methods for cross sections with box-shaped structural steel are smaller than those with H-shaped structural steel. In addition these differences may vary according to the main parameters of the cross section, such as dimensions, material properties, amount and location of the structural steel and reinforcing bars.

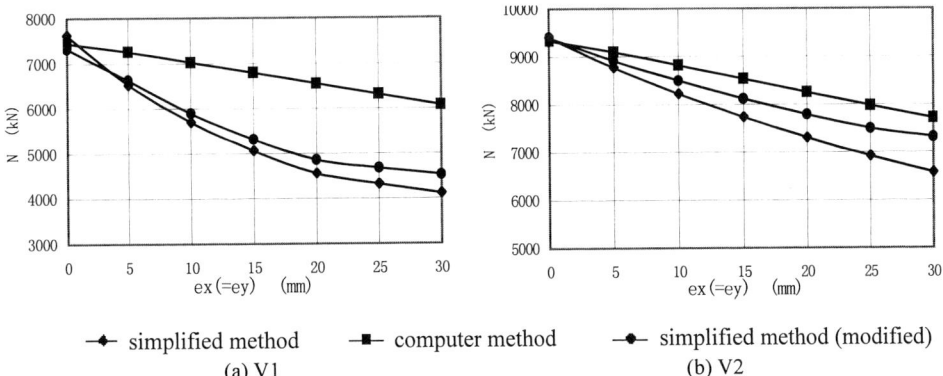

Figure 2: Comparison of determined load capacities of rectangular sections

Overall, it is found through the above analysis and comparison that the simplified superposition design method appears to be conservative in all studied conditions, with the degree of conservation being larger while the load eccentricity becoming larger. In order to reduce the overall error tendency of this method in comparison with the computer-based method, a modification factor is introduced and the final load carrying capacity then becomes

$$N_3 = k_1 N_1 \qquad (8)$$

where $k_1 = 1 + e_x/a + e_y/b$ is the modification factor, here a and b are the side length of the cross section along the x- and y-axes, respectively; N_1 is determined by Eqns. 1-4. The load capacities for cross sections V1 and V2 determined using Eqn.8 are shown in Figure 2. It is seen that the obtained results are more closely to those of the computer-based method.

DISCUSSION OF CIRCULAR CROSS-SECTIONS

The simplified design equations 1-4 are only recommended for the use of composite columns with rectangular cross sections (Metallurgy Ministry of China, 1998). For columns with circular cross sections, if this method is directly used, it is extremely difficult for designers to determine $M_{cy,x0}^{ss}(N_c^{ss})$, $M_{cy,y0}^{ss}(N_c^{ss})$, $M_{cu,x0}^{rc}(N_c^{rc})$ and $M_{cu,y0}^{rc}(N_c^{rc})$. Here based on the axial force and bending moment superposition concept, a new simplified design equation for such sections is proposed, i.e.

$$\frac{N - N_{c0}^{rc}}{N_0 - N_{c0}^{rc}} + \frac{M_x}{M_x^{ss}} + \frac{M_y}{M_y^{ss}} \leq 1 \qquad (9)$$

and

$$N_4 = k_2 N \qquad (10)$$

in which, N, M_x and M_y are the axial load and bending moment capacities of the cross section to be determined. N_0 is the axial load carrying capacity of the cross section under pure compression; M_x^{ss} and M_y^{ss} are the moment carrying capacities of structural steel with respect to the x- and y-axes, respectively; N_4 is the final load carrying capacity to be determined. $k_2=1+e_x/D+e_y/D$ is the magnification factor considering the effect of load eccentricities, here D is the diameter of the circular cross section.

Two examples, V3 and V4 as shown in Figure 3, are here presented to demonstrate the accuracy and validity of the proposed method. The material proprieties of the two sections are the same as those of V1 and V2 as listed in Table 1. The load carrying capacities determined by Eqns. 9 and 10 and their comparison with those of the computer method are shown in Table 3 as well as Figure 4.

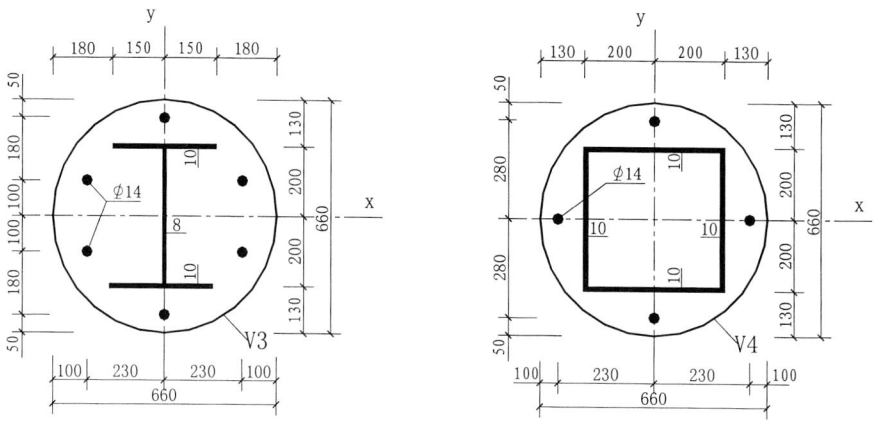

Figure 3: Layout of V3 and V4

TABLE 3
LOAD CARRYING CAPACITIES AND COMPARISON

Specimen	e_x (mm)	e_y (mm)	N_4 (kN)	N_2 (kN)	$(N_4-N_2)/N_2$
V31	0	0	7286	7225	0.84%
V32	10	10	5922	6800	-12.91%
V33	20	20	5033	6231	-19.23%
V34	30	30	4408	5850	-24.64%
V35	40	40	3945	5450	-27.60%
V36	50	50	3588	5050	-28.95%
V41	0	0	9273	9120	1.68%
V42	10	10	8326	8580	-2.95%
V43	20	20	7596	8026	-5.36%
V44	30	30	7014	7500	-6.47%
V45	40	40	6541	7020	-6.82%
V46	50	50	6147	6600	-6.85%

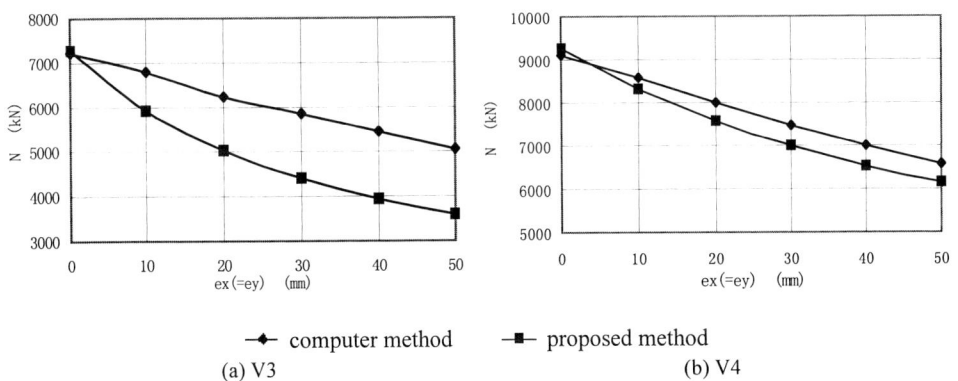

Figure 4: Comparison of determined load capacities for circular sections

It is seen that if the load eccentricities are not very large, the load carrying capacities determined by the proposed design equation agree closely with those by the computer-base method, although the former still appears conservative as a whole. Through further study it is known that the results of the proposed method would deviate increasingly to those of the computer method with the load eccentricities being larger.

CONCLUSIONS

The accuracy and error tendency of the simplified superposition design method for concrete encased composite columns subjected to biaxial bending recommended by the Chinese YB9082-97 Specification has been evaluated using an iterative computer-based analysis and design method. Comparisons between the two methods in the determination of load capacities of rectangular cross sections with symmetrically placed H-shaped and box-shaped structural steel were first made. It indicates that the simplified method appears to be conservative in all studied cases, with the degree of conservation being larger while the load eccentricities becoming larger. By introducing a modification factor considering the effect of load eccentricities, the overall conservation of the method is reduced. As the original simplified design equation is only recommended for rectangular cross sections, a new simplified equation for circular cross sections was then proposed. The obtained results by this method agree well with those of the computer-based method under low load eccentricities. For large eccentricity conditions the results may deviate to those of the computer-based method considerably. Therefore in these conditions, it is recommended that the obtained load capacity by the simplified superposition method should be checked by the computer-based method.

REFERENCES

Chen. S. F., Teng. J. G. and Chan. S. L. (2001). Design of Biaxially Loaded Short Composite Columns of Arbitrary Section. *J. Struct. Engrg.*, ASCE **127:6**, 678-685.

Johnson R. P. and Smith D. G. E. (1980). A Simple Design Method for Composite Columns. *Struct. Engrg.* **58A:3**, 85-93.

Lachance L. (1982). Ultimate Strength of Biaxially Loaded Composite Sections. *J. Struct. Div.*, ASCE **108:10**, 2313-2329.

Metallurgy Ministry of China (1998). *Specification for Design of Steel-Reinforced Concrete Structure*, Metallurgy Industry Press

Mirza S. A. and Skrabek B. W. (1991). Reliability of Short Composite Beam-Column Strength Interaction. *J. Struct. Engrg.*, ASCE **117:8**, 2320-2339.

Munoz P. R. and Hsu C. T. T. (1997). Bixially Loaded Concrete-Encased Composite Columns Design Equation, *J. Struct. Engrg.*, ASCE **123:12**, 1576-1585.

Roik K. and Bergmann R. (1990). Design Method for Composite Columns with Unsymmetrical Cross-Sections. *J. Construct. Steel Res.* **15**, 153-168.

Yau C. Y., Chan S. L. and So A. K. W. (1993). Biaxial Bending Design of Arbitrarily Shape Reinforced Concrete Column. *ACI Struct. J.* **90:3**, 269-278.

TESTS OF CONCRETE-FILLED DOUBLE SKIN (SHS OUTER AND CHS INNER) COMPOSITE STUB COLUMNS

X. L. Zhao, R. H. Grzebieta, A. Ukur and M. Elchalakani

Department of Civil Engineering, Monash University, Clayton, VIC 3168, Australia

ABSTRACT

This paper describes a series of compression tests carried out on concrete filled double skin tubes (CFDST). The outer tube is made of square hollow sections (SHS), while the inner skin is made of circular hollow sections (CHS). Four section sizes were chosen for the outer tubes with width-to-thickness ratio ranging from 17 to 50. Some sections are fully effective (i.e. full section yielding can be achieved in compression) while some sections are not fully effective (i.e. full section yielding cannot be achieved in compression due to local buckling) in order to investigate the effect of width-to-thickness ratio on the behaviour of double skin composite stub columns. Two section sizes are chosen for the inner tubes with diameter-to-thickness ratio of 17 and 21. The failure mode, strength, ductility and energy absorption of CFDST are compared with those of empty single skin tubes. Theoretical models are developed to predict the ultimate strength of CFDST stub columns. Increased ductility and energy absorption have been observed for CFDST stub columns especially for slender outer tubes.

KEYWORDS

Concrete-Filled Tubes, Double-Skin Composite Sections, Steel Hollow Sections, Stub Columns

INTRODUCTION

Concrete filled double skin tubes (CFDST) consist of two concentric steel cylinders or boxes with the annulus between them filled with concrete. This form of construction can be applied to sea-bed vessels, in the legs of offshore platforms in deep water, to large diameter columns and to structures subjected to ice loading [Montague 1978, Shakir-Khalil 1991, Wei et al 1995, Lin and Tsai 2001]. It can also be used as high-rise bridge piers [Sugimoto et al 1997, Yagishita et al 2000] to reduce the structure weight while still maintaining a large energy absorption capacity against earthquake loading.

There are four possible combinations in constructing CFDST, i.e. (1) SHS (square hollow section) as both outer and inner skins, (2) CHS (circular hollow section) as both outer and inner skins, (3) CHS as the outer skin and SHS as the inner skin and (4) SHS as the outer skin and CHS as the inner skin. The strength, ductility and energy absorption of stub columns with the first three combinations were studied by the authors and reported in Zhao and Grzebieta (2002), Zhao et al (2002), Elchalakani et al (2002). This paper describes a series of tests on stub columns with SHS as the outer skin and CHS as the inner skin. Four section sizes were chosen for the outer tubes with width-to-thickness ratio ranging from 17 to 50 in order to investigate the effect of width-to-thickness ratio on the behaviour of double skin composite stub columns. Two section sizes are chosen for the inner tubes with diameter-to-thickness ratio of 17 and 21. The failure mode, strength, ductility and energy absorption of CFDST are compared with those of empty single skin tubes. Theoretical models are developed to predict the ultimate strength of CFDST stub columns. The CFDST construction demonstrated significant increase in strength, ductility and energy absorption especially for slender outer tubes.

MATERIAL PROPERTIES

Cold-formed SHS and CHS were used in the construction of CFDST in this paper. The square hollow sections are in-line galvanised square tubes [Zhao and Mahendran 1998] with a nominal yield stress of 450 MPa. The circular hollow sections are manufactured to AS1163 (SAA 1991a) with a nominal yield stress of 350 MPa. The measured dimensions and yield stresses are given in Tables 1 and 2. Tensile coupons were used to measure the material properties in accordance with AS1391 (SAA 1991b). The 0.2% proof stress was adopted as the yield stress. The yield stresses of both flat faces (σ_{yfo}) and corners (σ_{yco}) for SHS outer tubes are listed in Table 1. The compressive strength of the concrete was determined by testing three concrete cylinders with a diameter of 100 mm and a height of 200 mm. The concrete cylinders were cured for 28 days and the average compressive strength (f_c) was 70 MPa.

TABLE 1
MEASURED SECTION DIMENSIONS AND YIELD STRESS OF SHS OUTER TUBES

Section ID No.	Depth D (mm)	Width B (mm)	Thickness t_o (mm)	σ_{yfo} (MPa)	σ_{yco} (MPa)
S1	100.2	100.2	6.12	500	559
S2	100.4	100.4	4.13	476	521
S3	100.4	100.4	3.15	491	594
S4	100.3	100.3	2.12	468	568
Mean	--	--	--	484	561
COV	--	--	--	0.030	0.054

TABLE 2
MEASURED SECTION DIMENSIONS AND YIELD STRESS OF CHS INNER TUBES

Section ID No.	Outer diameter d (mm)	Thickness t_i (mm)	σ_{yfi} (MPa)
C1	48.5	3.01	425
C2	60.6	3.08	400
Mean	--	--	413
COV	--	--	0.043

STUB COLUMN TESTS ON EMPTY TUBES

Four empty SHS stub columns and two empty CHS stub columns were tested in compression in a 500 kN capacity Baldwin testing machine. The length of the specimens was determined according to AS/NZS 4600 (SAA 1996). The ends of the stub columns were milled flat before testing so that they could properly seat on the rigid end platens of the testing machine. The specimens after testing are shown in Figure 1. The failure modes are the same as those observed by many researchers in the past for empty SHS and CHS [Zhao and Hancock 1991, Grzebieta 1990 for example only]. The ultimate load (P_u) obtained in the test is listed in Tables 3 and 4. The full section yield capacity (P_{yt}), which is calculated as a product of the measured cross-section area and measured yield stress (σ_{yfo} and σ_{yfi} in Tables 1 and 2), is compared with P_u in Tables 3 and 4. Similar comparison is also made between P_u and the full section capacity (P_{yn}) based on the nominal yield stress. The ratios of P_u/P_{yt} and P_u/P_{yn} for SHS are plotted in Figure 2 against the section slenderness. The results in Elchalakani et al (2002) are also plotted in Figure 2. The yield slenderness limit of 40 specified in AS 4100 (SAA 1998), below which a section can achieve full yielding, is also plotted. It seems that the yield slenderness limit specified in AS4100 is satisfactory for cold-formed in-line galvanised SHS stub columns.

TABLE 3
TEST RESULTS OF EMPTY SHS STUB COLUMNS

Specimen No.	Measured section slenderness $(\frac{B-2\cdot t_o}{t_o})\sqrt{\frac{\sigma_{yf}}{250}}$	P_u (kN)	P_u/P_{yt}	P_u/P_{yn}	Energy W_{SHS} (kNm)
ES1	20.3	1210	1.113	1.237	14.0
ES2	30.8	725	0.994	1.052	6.74
ES3	41.9	551	0.942	1.028	4.20
ES4	62.0	230	0.599	0.623	1.70

TABLE 4
TEST RESULTS OF EMPTY CHS STUB COLUMNS

Specimen No.	Measured section slenderness $(\frac{d}{t_i})\cdot(\frac{\sigma_{yf}}{250})$	P_u (kN)	P_u/P_{yt}	P_u/P_{yn}	Energy W_{CHS} (kNm)
EC1	27.4	195	1.067	1.295	2.63
EC2	31.5	245	1.100	1.258	3.20

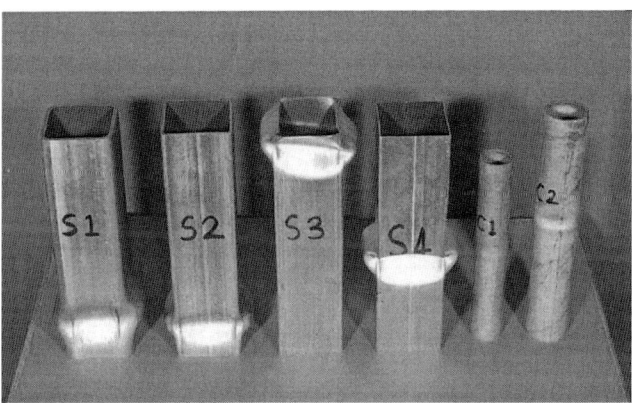

Figure 1 Empty stub columns after testing

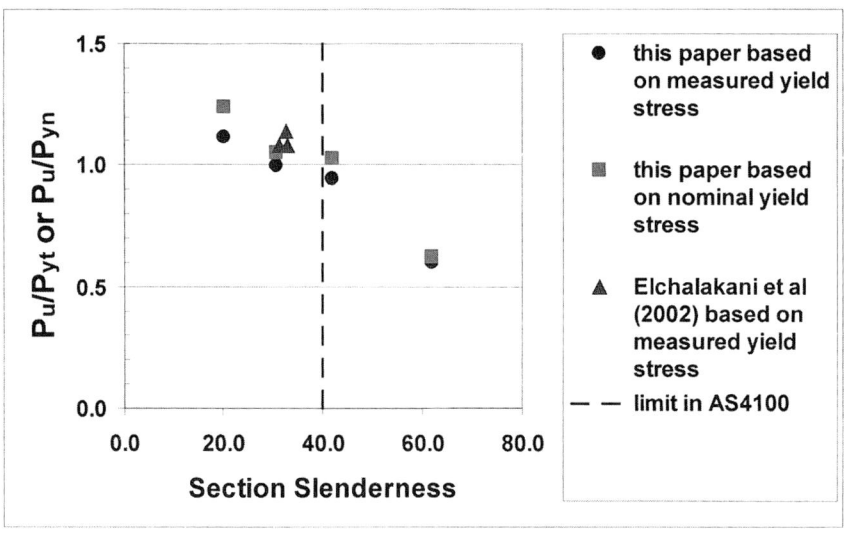

Figure 2 Section slenderness versus P_u/P_{yt} or P_u/P_{yn} for SHS

STUB COLUMN TESTS ON CFDST

Six CFDST stub column tests were carried out in a 5000 kN capacity Amsler machine. The test set up and procedures are the very much the same as those for empty stub column tests. A displacement control was used with a rate of 0.5mm/minute. The specimens after testing are shown in Figure 3. The outer SHS behaves the same way as a concrete-filled tube [e.g. Ge and Usami 1996], i.e. forming an outward folding mechanism. This is similar to that observed for CFDST stub columns with SHS as both outer and inner skins [Zhao and Grzebieta 2002]. A view of the CFDST after opening is shown in Figure 4. The failure mode of the inner CHS behaves differently from that for empty CHS in compression where an "elephant foot" occurs. The failure mode shown in Figure 3 is very much the same as that observed for the inner tube of CFDST with CHS as both outer and inner skins [Zhao et al 2002]. It was called "distorted diamond" mode in Zhao et al (2002). The axial load versus shortening curves are given in Figure 5. The maximum test load (P_{test}) for each specimen is listed in Table 5.

TABLE 5
TEST RESULTS OF CFDST STUB COLUMNS

Specimen No.	P_{test} (kN)	P_{theory} (kN)	P_{theory}/P_{test}	W_{CFDST} (kNm)	W_{CFDST}/W_{SHS}
S1C1	1677	1644	0.980	21.7	1.55
S2C1	1253	1315	1.049	14.88	2.21
S3C1	1065	1196	1.123	11.64	2.77
S4C1	935	892	0.954	9.60	5.65
S3C2	1040	1174	1.129	11.52	2.74
S4C2	870	870	1.000	8.68	5.11
Mean			1.039		
COV			0.0715		

Figure 3 CFDST stub columns after testing

Figure 4 Failure modes of the inner CHS tube

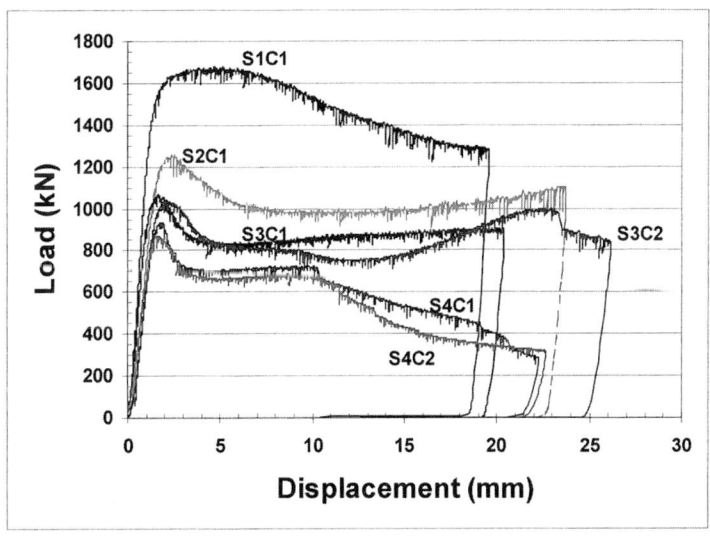

Figure 5 Load deflection curves for CFDST stub columns

ULTIMATE STRENGTH OF CFDST

The ultimate strength (P_{theory}) of CFDST can be estimated using the sum of the section capacities of the concrete, the outer steel tube and the inner steel tube, i.e.

$$P_{theory} = P_{concrete} + P_{outer} + P_{inner} \tag{1}$$

in which,

$$P_{concrete} = 0.85 \cdot f_c \cdot A_{concrete} \tag{2}$$

$$P_{inner} = \frac{\pi}{4} \cdot \sigma_{yfi} \cdot [d^2 - (d - 2 \cdot t_i)^2] \tag{3}$$

$$P_{outer} = P_{corner} + P_{flat} = \sigma_{yco} \cdot \pi \cdot (r_{ext}^2 - r_{int}^2) + 4 \cdot \sigma_{yfo} \cdot b_e \cdot t_o \tag{4}$$

and

$$A_{concrete} = (D - 2 \cdot t_o) \cdot (B - 2 \cdot t_o) - r_{int}^2 \cdot (4 - \pi) - \frac{\pi}{4} \cdot d^2$$

$$b_e = B - 2 \cdot r_{ext} \quad \text{if } \lambda_{so} \leq \lambda_{ey} = 40$$

$$b_e = (B - 2 \cdot r_{ext}) \cdot \frac{\lambda_{ey}}{\lambda_{so}} \quad \text{if } \lambda_{so} > \lambda_{ey} = 40$$

$$\lambda_{so} = (\frac{B - 2 \cdot r_{ext}}{t_o}) \cdot \sqrt{\frac{\sigma_{yfo}}{250}}$$

where the reduction factor 0.85 in Eq. (2) is defined in AS3600 (SAA 1994), f_c in Eq. (2) is the concrete cylinder compressive strength, σ_{yfi} is the yield stress of CHS listed in Table 2, σ_{yco} and σ_{yfo} are the yield stress of corners and flat faces of SHS respectively given in Table 1, d is the outside diameter of the inner CHS, D and B are the overall depth and width of the outer SHS, r_{ext} and r_{int} are the external and internal radii of the outer SHS, t_o and t_i are the thickness of the outer and inner tubes respectively.

The predicted ultimate strength (P_{theory}) is compared with the experimental value (P_{test}) in Table 5. A mean of 1.039 and a COV (coefficient of variation) of 0.0715 are obtained. It seems that Eq. (1) gives a good prediction of the ultimate strength of CFDST stub columns in compression.

DUCTILITY AND ENERGY ABSORPTION

The axial load versus axial shortening curve for specimen S1C1 is compared with that of the empty tube S1 with D/t of 16.4 in Figure 6 (a). Similar comparison is made in Figure 6 (b) for specimen S4C1 which has an outer tube S4 with a much larger D/t ratio of 47.3. It can be seen that the double-skin filling increases the ductility of CFDST specimens especially for slender outer tubes where the CFDST maintained higher loads for very large deformation compared to hollow sections.

Energy absorptions are calculated using the area under the load-deflection curve when axial shortening is up to 15 mm. The calculated values are listed in Tables 2, 3 and 4 for empty SHS, empty CHS and CFDST respectively. Comparisons are shown in Table 5 and Figure 7. There is a significant increase in energy absorption especially for more slender tubes, which emphasizes the efficiency of void-filling of thin SHS.

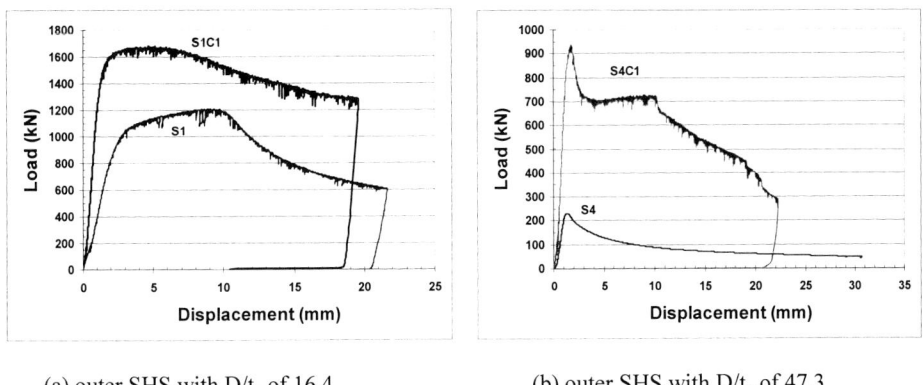

(a) outer SHS with D/t_o of 16.4 (b) outer SHS with D/t_o of 47.3

Figure 6 Comparison between CFDST and empty tubes

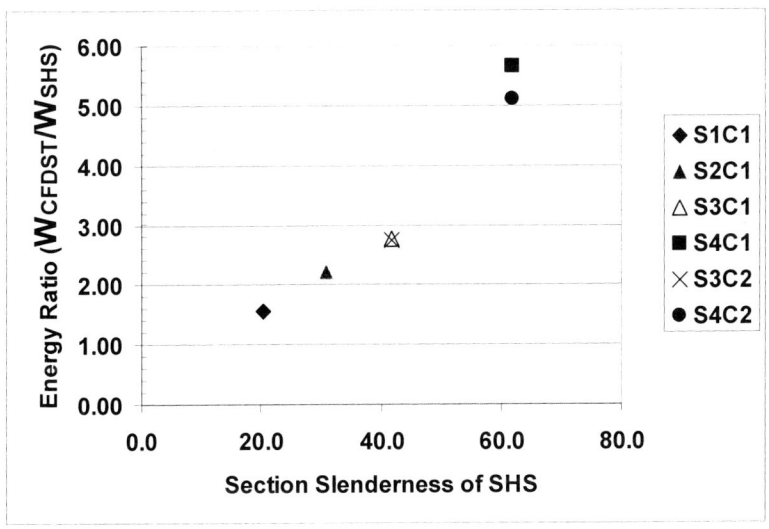

Figure 7 Energy absorption ratio

CONCLUSIONS

The following conclusions and observations are made based on the limited tests described in this paper, where the width-to-thickness ratio of the outer SHS ranged from 17 to 50 and from 17 to 21 for the inner CHS.

- The yield slenderness limit of 40 specified in AS 4100 seemed applicable to cold-formed in-line galvanised SHS stub columns.
- The failure mode of the outer SHS was similar to that observed for SHS fully filled with concrete. The failure mode for the inner CHS was found different from that for empty CHS.

- The proposed strength model predicted well the ultimate capacity of CFDST with SHS as the outer tube and CHS as the inner tube subjected to compression.
- Increased ductility and energy absorption were obtained especially for CFDST with thinner outer SHS.

ACKNOWLEDGEMENTS

The authors are grateful to the Australian Research Council for financial support. Thanks are given to Mr Graeme Rundle, Mr. Roger Doulis and Mr Geoff Doddrell for their assistance in conducting the tests.

REFERENCES

Elchalakani, M., Zhao, X.L., and Grzebieta, R.H. (2002). Tests on Concrete Filled Double-Skin (CHS Outer and SHS Inner) Composite Short Columns under Axial Compression, *Thin-Walled Structures*, **40:5**, 415-441

Ge, H.B. and Usami, T. (1996). Cyclic Tests of Concrete Filled Steel Box Columns, *Journal of Structural Engineering,* ASCE, **122:10**, 1169-1177

Grzebieta, R.H. (1990). On the Equilibrium Approach for Predicting the Crush Response of Thin-Walled Mild Steel Structures, *PhD Thesis*, Monash University, Australia

Lin, M.L. and Tsai, K.C. (2001), Behaviour of Double-Skinned Composite Steel Tubular Columns subjected to Combined Axial and Flexural Loads, *First International Conference on Steel and Composite Structures*, Pusan, Korea, 1145-1152

Montague, P. (1978). The Experimental Behaviour of Double Skinned, Composite, Circular Cylindrical Shells under External Pressure, *Journal of Mechanical Engineering Science,* **20:1**, 21-34

SAA (1991a), Structural Steel Hollow Sections, *Australian Standard AS1163*, Sydney

SAA (1991b), Methods for Tensile Testing of Metals, *Australian Standard AS1391*, Sydney

SAA (1994), Concrete Structures, *Australian Standard AS3600*, Sydney

SAA (1996), Cold-Formed Steel Structures, *Australian/New Zealand Standard AS/NZS4600*, Sydney

SAA (1998), Seel Structures, *Australian Standard AS/NZS4100*, Sydney

Shakir-Khalil, H. (1991), Composite columns of double-skinned shells, *Journal of Constructional Steel Research*, **19**, 133-152

Sugimoto, M., Yokota, S., Sonoda, K. and Yagishita, F. (1997). A basic consideration on double skin tube-concrete composite columns, Osaka City University and Monash University *Joints Seminar on Composite Tubular Structures*, Osaka City University, Osaka, July

Wei, S., Mau, S.T., Vipulanadan, C. and Mantrala, S.K. (1995). Performance of New Sandwich Tube under Axial Loading: Experimental, *Journal of Structural Engineering,* ASCE, **121:12**, 1806-1814

Yagishita, F., Kitoh, H., Sugimoto, M., Tanihira, T. and Sonoda, K. (2000). Double Skin Composite Tubular Columns subjected to Cyclic Horizontal Force and Constant Axial Force, *Proceedings of the 6th ASCCS Conference,* Los Angeles, USA, 22-24 March, 497-503

Zhao, X.L. and Hancock, G.J. (1991). Tests to Determine Plate Slenderness Limits for Cold-Formed Rectangular Hollow Sections of Grade C450. *Steel Construction*, AISC, **25:4**, 2-16.

Zhao, X.L. and Mahendran, M. (1998). Recent Innovations in Cold-Formed Tubular Sections, *Journal of Constructional Steel Research*, **46:1-3**, 472-473

Zhao, X.L. and Grzebieta, R.H. (2002). Strength and Ductility of Concrete-Filled Double Skin Square Hollow Sections, *Thin-Walled Structures*, **40:2**, 199-213

Zhao, X.L., Grzebieta, R.H. and Elchalakani, M. (2002). Tests of Concrete-Filled Double Skin CHS Composite Stub Columns, *Steel and Composite Structures – An International Journal,* **2:2**, in press

STRENGTH OF SLENDER CONCRETE FILLED COLUMNS FABRICATED WITH HIGH STRENGTH STRUCTURAL STEEL

B. Uy[1], M. Mursi[1] and H.B.A. Tan[1]

[1]School of Civil and Environmental Engineering, The University of New South Wales,
Sydney, NSW 2052, AUSTRALIA

ABSTRACT

The use of high strength steel in tall buildings has the ability to achieve tremendous benefits as the material provides a higher strength to weight ratio, hence allowing the reduction of cross-section dimensions. However as the strength of the steel is increased the buckling characteristics become more dominant with slenderness limits for both local and global buckling becoming more controlling. To arrest the problems associated with buckling of high strength steel, concrete filling can be utilised as it has the effect of changing the local buckling mode, which essentially increases the strength and stiffness of the overall member. This paper describes an experimental programme undertaken for concrete filled composite columns, which were designed to be slender in nature and thus fail by global buckling. A model for the axial strength that accounts for slenderness effects using an interaction approach is utilised for calibration and the column curve approach suggested in Eurocode 4 is also used to predict the experimental results. Each method is evaluated and discussed and recommendations for the design of high strength steel composite members is given. Conclusions and aspects of further research are also discussed.

KEYWORDS

buckling, columns, composite structures, high strength steel, steel structures, tall buildings

INTRODUCTION

Tall building gravity load systems can be significantly improved by reducing the cross-sectional dimensions of the vertical columns. Recent developments in the metallurgical qualities of high strength structural steel plate have seen it become extremely attractive for the design and construction of tall buildings throughout the world. The benefits of the use of high strength steel can be utilised in a braced frame where the external spandrel frame is used to resist gravity loads alone. High strength steel is most efficient when it is allowed to develop its' full yield stress and this can be achieved when local and overall buckling can be eliminated in a column design. This paper will summarise the previous applications of high strength steel in tall buildings throughout the world. The summary includes a description of the column type and grade of steel used to illustrate the methods in which high strength structural steel is being used. A detailed experimental program was conducted to evaluate the behaviour of slender high strength steel-concrete composite columns. The experiments were based on the behaviour of short and slender high strength steel-concrete composite columns loaded uniaxially. These experiments will be outlined herein and a numerical model for the calculation of the strength will be presented and shown to provide a conservative estimate of the column strengths. Comparisons with Eurocode 4 (British Standards Institution, 1994) are also made and the possible changes which are required to utilise this code with this variant material are highlighted. The paper will conclude by highlighting future research that is required to be conducted in order to promote the future use of these columns and for the development of international standards.

PREVIOUS RESEARCH

Rasmussen and Hancock (1992 and 1995) conducted tests on both high strength steel fabricated I-sections and box sections with a nominal yield stress of 690 MPa (N/mm^2). These tests established local buckling slenderness limits for high strength steel sections. Furthermore, slender columns were tested and the behaviour of these was compared with the slender column curves of the existing Australian Standard AS 4100-1998 (Standards Australia 1998). It was found that provided these local buckling slenderness limits were adhered to, then the slender column behaviour could be predicted using this standard developed specifically for mild structural steel.

Hagiwara et al. (1995) and Mochizuki et al. (1995) considered the behaviour of high strength structural steel for the application in super high-rise buildings in Japan. These studies considered the reliability inspection and the welding process for heavy gauge steel plate. These studies are pertinent to the application of the use of high strength steel in projects such as the Shimizu Super High Rise in Tokyo, Japan.

Sivakumaran and Yuan (1998) considered slenderness limits and ductility of steel sections fabricated with high strength steel with nominal yield stresses between 300 and 700 MPa (N/mm^2) respectively. The test programme involved testing 12 W shaped stub column sections with the objective being to determine the compression flange strength and strain ductility of sections of different steel grades.

Uy (1996) considered the behaviour of concrete filled steel box columns with a nominal yield stress of 690 MPa (N/mm^2). This study illustrated the advantages derived from filling the sections with concrete to increase the local buckling capacities. Furthermore, the members were considered under combined bending and compression to assess the strength of short columns. The results of these columns were compared with columns designed with normal strength structural steel, to show the reduced cross-sectional dimensions able to be achieved. Furthermore, comparisons of the cross-sectional ductility were made and showed that composite members composed of high strength structural steel still had a large degree of reserve strength after peak loading conditions were attained.

Uy (1999) presented the results of steel and composite sections using high strength structural steel of nominal yield stress 690 MPa (N/mm^2). These sections constructed as stubby columns were subjected to concentric axial compression. A theoretical model to predict the axial strength of these columns was provided and shown to be in good agreement with the models suggested by Eurocode 4 for encased and concrete filled sections.

Uy (2001) conducted an extensive experimental programme on short concrete filled steel box columns, which incorporated high strength structural steel of Grade 690 MPa (N/mm^2). The experiments were then used to calibrate a refined cross-sectional analysis method, which considered both the non-linear material properties of the steel and concrete coupled with the measured residual stress distributions in the steel. The model and experiments were then compared with the existing approach of Eurocode 4 and it was found that certain modifications were necessary. The Eurocode 4 approach, which employs the rigid plastic analysis method, was found to over predict the strength of the cross-sections. A modified technique known as a mixed analysis was therefore developed and found to be in good agreement with both the test results and the refined analysis procedure. This model considers the concrete to be plastic and the steel to be elastic-plastic and provides a much more realistic design approach for sections utilizing high strength structural steel, particularly when large flexural loads are present.

Bergmann and Puthli (2000) conducted an extensive experimental programme on short and slender high strength steel encased sections of 460 MPa (N/mm^2) grade steel subjected to combined compression and bending. These tests were then compared with the Eurocode 4 approach, which was found to be suitable for predicting the ultimate load for short columns. However, the results of the slender column tests proved to be inconclusive.

APPLICATIONS

Previous building projects, which have been completed and planned, are summarised in Table 1. This list identifies the type of projects and the potential benefits achieved from the use of high strength steel. In particular, this table reflects tall building projects in Australia where high strength steels have been used. In the design of Star City, Sydney the largest building project in Sydney since the Sydney Opera House, the major benefits derived from the use of high strength steel were in providing additional car space in the basement levels of the building. This was a mandatory requirement for the project specified by the Sydney City Council. The use of high strength steel in the other Australian buildings was justified in reducing column sizes and excavation costs and thus providing additional floor area and car park spaces in the building. This was also used on projects in Sydney, Melbourne and Perth in notable buildings such as Grosvenor Place, 300 Latrobe Street and Central Park. The Dai-Ichi building in Osaka, Japan, was designed utilizing high strength steel box columns in the perimeter frames. High strength steel was used to ensure that the structure remained in the elastic range under severe earthquake loading. The Shimizu Super High Rise (SSHR), which is a proposed project in Tokyo, Japan, will use high strength steel in box columns for the exterior spandrel frame.

TABLE 1
PROJECTS UTILISING HIGH STRENGTH STRUCTURAL STEEL

Building Name	City	Country	Year Completed	Number of Storeys	Column Type	Steel Grade MPa (N/mm^2)
Grosvenor Place	Sydney	Australia	1988	50	Encased	690
Central Park	Perth	Australia	1989	50	Encased	690
300 Latrobe St.	Melbourne	Australia	1990	30	Encased	690
Star City	Sydney	Australia	1997	20	Encased	690
Dai-Ichi	Osaka	Japan	Unknown	20	Filled	600
Shimizu SSHR	Tokyo	Japan	Proposed	120	Filled	600

DETAILS OF EXPERIMENTS

An experimental programme was conducted on columns of a short and slender nature fabricated with high strength steel plate, formed into boxes and filled with concrete. All columns were constructed with four equal width plates and joined at the vertices by means of a fillet weld as illustrated in Figure 1. The dimensions and other salient features from the testing are summarised for the experiments in Table 2. The columns were tested in a horizontal self-straining compression-testing machine. The slender columns were tested with a small eccentricity, e equivalent to 10% of the cross-sectional width of the specimen.

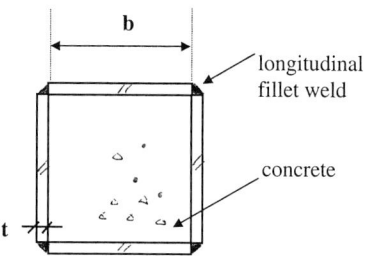

Figure 1: High strength steel square cross-sections

TABLE 2
DETAILS FOR THE TESTING OF SHORT AND SLENDER COLUMNS

Specimen Name	b (mm)	t (mm)	Length (mm)	e (mm)	Concrete Strength, f_c, MPa (N/mm^2)	Steel Strength, f_y, MPa (N/mm^2)
SH – C110	110	5	430	0	20	761
SH – C160	160	5	580	0	20	761
SH – C210	210	5	730	0	20	761
SL – C110	110	5	2800	10	20	761
SL – C160	160	5	2800	15	20	761
SL – C210	210	5	2800	20	20	761

STRENGTH MODELS AND COMPARISON WITH EXPERIMENTAL RESULTS

This section details the relevant models used to predict the short and slender column strength of the columns tested in this paper. The results of these analyses based on basic mechanics modelling and the procedures from Eurocode 4 is compared with the experimental results and recommendations for design are thereby suggested.

Model for Axial Strength of Short Columns

The short columns in this paper were tested in pure compression and in order to predict the strength of these members an axial compressive strength model was proposed which considers both the steel and concrete contributions to axial strength invoking the principle of superposition. The model shown in Equation 1 was used to provide a prediction of the axial strength of each of the columns

$$N_u = N_{uc} + N_{us} \tag{1}$$

where N_u is the ultimate axial strength of the composite column, $N_{uc}=f_c.A_c$ is the concrete contribution to axial strength and $N_{us}=f_y.A_{se}$ is the steel contribution to axial strength which incorporates the effects of local buckling as outlined by Uy (2000). This involves a very slight augmentation to the suggested model of Eurocode 4 (British Standards Institution 1994) to allow for slender plated sections.

Model for Axial Strength of Slender Columns

Interaction Approach

The model used to establish the slender column strength was based on the coupling of a cross-sectional strip analysis fully detailed by Uy (2000) and Uy (2001) with an interaction approach utilising a loading line. This strip analysis allows the determination of the locus of the strength interaction diagram for the cross-section incorporating high strength steel and concrete. Once the locus of the strength interaction diagram has been determined, a loading line is plotted. This method has been fully detailed by Uy (1998) and key features of the approach are outlined by Oehlers and Bradford (1995).

Column Curve Approach

The principle for checking sections under compression and uniaxial bending using the column curve approach is the approach inherent in Eurocode 4 and was also utilised herein. In this method an initial imperfection is incorporated, so that any additional consideration of geometrical imperfection is unnecessary in the calculations of moments within the column length. The axial resistance of the composite column in the absence of moment is given by $\chi N_{pl.Rd}$ as illustrated in Figure 2. Therefore, at the level $\chi = N_{Rd}/N_{pl.Rd}$, no additional bending moment can be applied to the column. The corresponding value for bending μ_k of the cross section is therefore the moment for imperfection of the column and the influence of this imperfection is assumed to decrease linearly to the value χ_n. For an axial load ratio less than χ_n, the effect of imperfections is neglected. It is important to recognize that the value χ_n accounts for the fact that the influence of the imperfections and that of the bending moment do not always act together unfavourably. For columns with end moments, χ_n may be obtained as follows:

$$\chi_n = \chi(1 - r)/4 \tag{2}$$

If transverse loads occur within the column height, then r is taken as unity and χ_n is thus equal to zero. With a design axial load of N_{Sd}, the axial load ratio χ_d is defined as:

$$\chi_d = N_{Sd} / N_{pl.Rd} \tag{3}$$

The horizontal distance from the interaction curve, μ (associated with $N.e$ in the case of the columns tested herein) defines the ultimate moment of resistance that is still available, having taken account of the influence of second order effects in the column. EC4 considers that the design is adequate when the following condition is satisfied:

$$M_{Sd} \leq 0.9 \, \mu \, M_{pl.Rd} \tag{4}$$

where M_{Sd} is the design bending moment, which may be factored to allow for second order effects, if any; μ is the moment resistance ratio obtained from the interaction curve and $M_{pl.Rd}$ is the plastic moment resistance of the composite cross section. The interaction curve has been determined without considering the strain limitations in the concrete. Hence the moments, including second order effects if necessary, are calculated using the effective elastic flexural stiffness, $(EI)_e$ and taking into account the entire uncracked concrete area of the cross section, (i.e. concrete is uncracked). Consequently, a reduction factor of 0.9 is applied to the moment resistance in Equation 4 to allow for the simplifications in the approach.

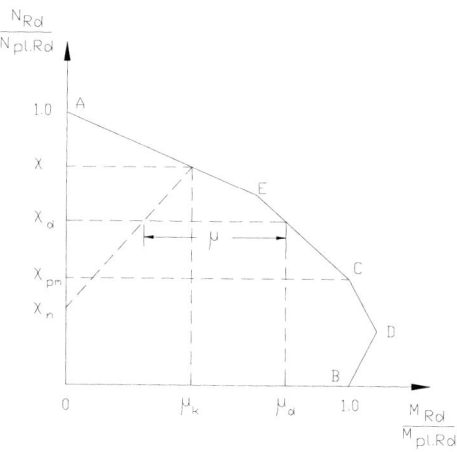

Figure 2: Interaction curve for compression and uniaxial bending using Eurocode 4 method

Comparisons of Short Columns

Results of the short column analysis are compared with the experimental results in Table 3. The Eurocode 4 model with appropriate modifications for plate slenderness is shown to be slightly unconservative in its' prediction of strength. However, this assumes that the steel stress is elastic-perfectly plastic which is not the case for high strength steel which has some strain hardening present.

TABLE 3
RESULTS AND COMPARISONS FOR SHORT COLUMNS

Specimen	Experimental Load, $N_{u.test}$ (kN)	Eurocode 4 Model, $N_{u.pred}$ (kN)	$N_{u.pred} / N_{u.test}$
SH – C110	1835	1954	1.06
SH – C160	2831	2985	1.05
SH – C210	3609	3381	0.94
Mean			1.02

Comparisons of Slender Columns

The experimental results of this study are compared with the slender column analyses and these are summarised in Table 4 and are shown to be conservative for all cases. The major reasons for the differences in the Eurocode 4 approach and the numerical model lie in the determination of the strength interaction envelope and the loading line used. The numerical model is based on a reinforced concrete approach, which does not account for imperfections and residual stresses whereas the Eurocode 4 approach is shown to give a much lower value for the predicted capacity of the columns due to the inclusion of imperfections. However, both methods are shown to provide a very conservative estimate of the strength and this may also be attributed to the columns in the tests having some partial rotational fixity. For design purposes it is suggested that the modified Eurocode 4 model be used for the determination of the strength envelope Uy (2001b) and the Eurocode 4 model for slenderness be used as this accounts for residual stresses and imperfections. Table 4 illustrates that the column curve approach become much more accurate when the steel contributing to the stiffness is more significant. Thus for Sl-C210, the concrete rigidity as a function of the overall column rigidity is about 3 %, and for this case the column curve approach underestimates the strength by only 7 %. For columns in real buildings where the stiffness of the steel is very significant, the column curve approach will prove to be much more accurate and is therefore recommended.

TABLE 4
RESULTS AND COMPARISONS FOR SLENDER COLUMNS

Specimen	Rigidity Ratio	Experimental Load, $N_{u.test, (kN)}$	Interaction Approach $N_{u.pred, (kN)}$	$N_{u.pred} / N_{u.test}$	Column Curves $N_{u.pred, (kN)}$	$N_{u.pred} / N_{u.test}$
SL – C110	0.06	1481	1060	0.72	878	0.60
SL – C160	0.04	2126	2080	0.98	1745	0.82
SL – C210	0.03	2939	3050	1.04	2725	0.93
Mean				0.91		0.78

CONCLUSIONS

The results of the experiments when compared with the analysis methods showed that both the numerical model and the Eurocode 4 model for short and slender column behaviour are fairly accurate. However it is suggested that for design the strength interaction diagram is determined using the modified Eurocode 4 method and that the slender column approach of Eurocode 4 be used for stability calculations as this includes the important effects of residual stresses and imperfections. Further research will require consideration of the effects of biaxial bending on both short and slender columns, as well as a more detailed look at the effects of ductility particularly on the behaviour of the welds.

ACKNOWLEDGEMENTS

This project was sponsored by a URSP Grant at The University of New South Wales and supported in kind by Bisalloy Steels. The authors would like to thank the staff of the Randwick Heavy Structures Laboratory at the University of New South Wales for assisting in the conduct of the experiments.

REFERENCES

Bergmann, R. and Puthli, R. (2000) Behaviour of composite columns using high strength steel sections, *Proceedings of Composite Construction IV, Engineering Foundation Conferences, May-June, Banff, Alberta.*

British Standards Institution (1994) *Eurocode 4, ENV 1994-1-1 1994. Design of composite steel and concrete structures, Part 1.1, General Rules and Rules for Buildings.*

Hagiwara, Y., Kadono, A., Suzuki, T. Kubodera, I. Fukasawa, T. and Tanuma, Y. (1995) Application of HT780 high strength steel plate to structural member of super high rise building: Part 2 Reliability inspection of the structure. *Proceedings of the Fifth East Asia - Pacific Conference on Structural Engineering and Construction, Building for the 21st Century,* Gold Coast, 2289-2294.

Mochuziki, H., Yamashita, T., Kanaya, K., and Fukasawa, T. (1995) Application of HT780 high strength steel plate to structural member of super high-rise building: Part 1 Development of high strength steel with heavy gauge and welding process. *Proceedings of the Fifth East Asia - Pacific Conference on Structural Engineering and Construction, Building for the 21st Century,* Gold Coast, 2283-2288.

Oehlers, D.J. and Bradford, M.A. (1995) *Composite steel and concrete structural members: fundamental behaviour*, Pergamon Press.

Rasmussen, K.J.R., and Hancock, G.J. (1992) Plate slenderness limits for high strength steel sections. *Journal of Constructional Steel Research*, **23**, 73-96.

Rasmussen, K.J.R., and Hancock, G.J. (1995) Tests of high strength steel columns. *Journal of Constructional Steel Research*, **34**, 27-52.

Sivakumaran, K.S. and Yuan, B. (1998) Slenderness limits and ductility of high strength steel sections. *Journal of Constructional Steel Research, Special Issue: 2nd World Conference on Steel Construction,* **46, (1-3),** 149-151, Full paper on CD-ROM.

Standards Australia (1998) *Australian Standard, Steel Structures, AS4100-1998*, Sydney, Australia.

Uy, B. (1996) Behaviour and design of high strength steel-concrete filled box columns. *Proceedings of the International Conference on Advances in Steel Structures,* Hong Kong, 455-460.

Uy, B. (1998) Strength, ductility and design of fabricated thin walled steel concrete filled box columns. *The International Journal of The Structural Design of Tall Buildings,* **7 (2),** 113-133.

Uy, B. (1999) Axial compressive strength of steel and composite columns fabricated with high strength steel plate. *Proceedings of the Second International Conference on Advances in Steel Structures,* Hong Kong, 421-428.

Uy, B. (2000) Strength of short concrete filled steel box columns incorporating local buckling, *Journal of Structural Engineering, ASCE,* **126 (3)**, 341-352

Uy, B. (2001) Strength of short concrete filled high strength steel box columns, *Journal of Constructional Steel Research,* **57,** 113-134.

CONCRETE-FILLED STEEL RHS COLUMNS SUBJECTED TO LONG-TERM LOADS

Lin-Hai Han[1]   Wei Liu[1] & You-Fu Yang[2]

[1]College of Civil Engineering and Architecture, Fuzhou University, Fuzhou, Fujian, 350002, China
[2]School of Civil Engineering, Harbin Institute of Technology, Haihe Road 202, Harbin, 150090, China

ABSTRACT

The present study is an attempt to predict the time-dependent behavior of concrete-filled steel RHS (rectangular hollow section) columns by means of a model that is proposed by the ACI Specifications (ACI Committee 209,1992). A theoretical model of analysis to account for shrinkage and creep effects on concrete-filled steel RHS columns under sustained loading is presented, and is a development of the analysis used for short-term loading (Han et al, 2001a). A comparison of results calculated using this model shows good agreement with test results. The main objectives of this paper were threefold: firstly, to report a series of tests on the behavior of concrete-filled steel RHS columns under long-term sustained loading, in addition to providing further tests on the static ultimate strength of the composite columns. Secondly, to analyze influence of several parameters, such as changing steel ratio, slenderness ratio, strength of the materials and load eccentricity ratio on the capacities of the concrete-filled steel RHS columns. Finally, to develop formulas for the calculation of the ultimate strength of the concrete-filled steel RHS columns with long-term sustained loading effects are considered, such formulas should be suitable for incorporation into building codes.

KEYWORDS

Concrete-Filled Steel RHS, column, long-term sustained load, creep, shrinkage, creep coefficient, stress-strain model, bearing capacity, parametric analysis, design method

1. INTRODUCTION

Concrete-filled steel columns at service loads in a building will suffer the effects of creep and shrinkage of the concrete. In the past, the behavior of concrete-filled HSS (hollow structural steel) columns under short-term loading has been the subject of many investigations. However, it appears that seldom attention has preciously been given to the influence of long-term sustained load on the behaviors of such members, this may be attributed to the fact that, except for exceptional cases, the structural strength is not compromised by the creep phenomenon. The lack of information on behavior of concrete-filled steel tubular columns under sustained loads indicates a need for further research in this area.

The only experimental observations of creep in concrete-filled steel tubular columns were reported by Furlong (1967), Han, et al (2001b), Nakai et al (1991), Terrey et al (1994), Morino et al (1996), Uy (2001) and Zhong (1994).

A series of tests on the behavior of concrete-filled steel RHS (rectangular hollow section) columns under long-term sustained loading, in additional to providing further tests on the static ultimate strength of the composite columns are presented in this paper. The differences of this test program compared with the similar studies carried out by other researchers mentioned above is that, the long-term sustained load ratio n ($=N_L/N_u$, where N_L is the long-term sustained load, N_u is the ultimate strength of the composite columns at short-term loading condition) is greater than 0.6. In the past, the specimens had been tested with the ratios of n less than 0.4.

There is attempt to predict the time-dependent behavior of concrete-filled steel RHS columns by means of a model proposed by the ACI Specifications (ACI, 1992) in this paper. A theoretical model of analysis to account for shrinkage and creep effects under sustained loading is presented, and is a development of the analysis used for short-term loading (Han et al, 2001a). A comparison of results calculated using this model shows good agreement with test results. Based on the theoretical model, influence of the changing steel ratio, slenderness ratio, strength of the materials and load eccentricity ratio on the capacities of the concrete-filled RHS columns are discussed. And finally, formulas for the calculation of the ultimate strength of the concrete-filled steel RHS columns with long-term sustained load effects considered are presented, which are suitable for incorporation into building codes.

2. EXPERIMENTAL PROGRAM

2.1 Specimens

To determine the steel material properties, tension coupons were cut from a randomly selected steel sheet. From these tests, the average yield strength (f_{sy}) was found to be 293.5 MPa, and the modulus of elasticity was 202×10^3 MPa.

The concrete mix was designed for a compressive cubic strength (f_{cu}) at 28 days of approximately 34.3 MPa. The modulus of elasticity (E_c) of concrete was found to be 29200 MPa. For each concrete mix batch used, three 150mm cubes were also cast and cured in conditions similar to the related specimens. The mix proportions were as follows: cement: 457 kg/m³; water: 206 kg/m³; sand: 608 kg/m³; coarse aggregate: 1129kg/m³. In all the concrete mixes, the fine aggregate used was silica-based sand, the coarse aggregate was carbonate stone.

Eight tests on concentrically loaded composite columns were carried out. A summary of the specimens is presented in Table 1, where L is the length of the specimen. λ ($=2\sqrt{3}L/B$) is determined as slenderness of the specimen, n is long-term sustained load ratio (n).

The tubes were all manufactured from mild steel sheet, with four plates were cut from the sheet. Each tube was welded to a rectangular steel base plate. The concrete was

TABLE 1
SPECIMEN LABELS AND MEMBER CAPACITIES

No.	Specimen	Section dimension $D \times B \times t_s$ (mm)	L (mm)	D/B	λ	Sustaining Load N_L (kN)	Load Ratio n
1	R-1	100×60×2.93	600	1.67	35	0	0.0
2	R-2	100×60×2.93	600	1.67	35	304	0.68
3	R-3	100×80×2.93	600	1.5	26	0	0.0
4	R-4	100×80×2.93	600	1.5	26	382	0.67
5	R-5	120×60×2.93	600	2	35	0	0.0
6	R-6	120×60×2.93	600	2	35	338	0.65
7	R-7	120×90×2.93	600	1.33	23	0	0.0
8	R-8	120×90×2.93	600	1.33	23	424	0.58

filled in layers and was vibrated by poker vibrator. The specimens were placed upright to air-dry until testing. During curing, a very small amount of longitudinal shrinkage of 0.6 to 0.8 mm or so occurred at the top of the columns. A high-strength epoxy was used to fill this longitudinal gap so that the concrete surface was flush with the steel tube at the top. The experimental program consisted of two stages, which are described below and designated as long-term service tests, static load versus deformation and ultimate load tests.

2.2 Long-term Service Load Tests

The long-term service load tests were carried out as soon as the age of 28 days after concrete casting. The experiment was carried out in a laboratory where the temperature and humidity are almost constant during the whole year, so that the influence of temperature gradient could be kept to minimum values.

The long-term sustained load (N_L) was applied by pre-stressing bars and controlled by means of a load cell. The load was kept constant with adjustments of these bars during the measurement. The values of the sustained loads as well as the long-term sustained load ratios (n) are shown in Table 1.

Strain measurements were obtained in the central part of the column for the steel. After observing the measured data of the axial strain (ε), it was found that for about 100 days, the process is tending to stabilize. Figure 1 shows the tested time-history of the axial strain (ε).

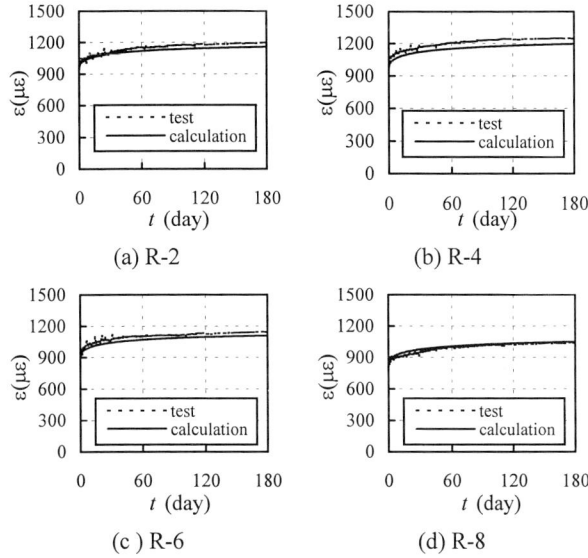

Figure 1: Tested time-history of the axial strain (ε).

2.3 Static Load versus Deformation and Ultimate Load Tests

At the completion of the long-term service load tests, all the specimens were removed from the creep-loading devices and tested to failure in a compression-testing device. In addition to this, specimens without long-term service load applied were also tested on the proposal of comparisons.

The compressive strength of concrete at the day of load versus deformation and ultimate load tests was 59.0 MPa (180 days).

All of the specimens were tested in pure compression. The tests were performed on a 5000 kN capacity testing machine. Eight strain gauges were used for each specimen to measure the longitudinal and transverse strains at the middle height. Two displacement transducers were used to measure the axial deformation. Three transducers were used to measure the lateral deflection.

It had been found that the deflections in the mid-height of the tested specimens were very small. Since the columns are very short, and the desired axially loads are achieved by accurately machining grooves mentioned above, and thus the failure load is governed by the axially compressive rigidity, one expects that this

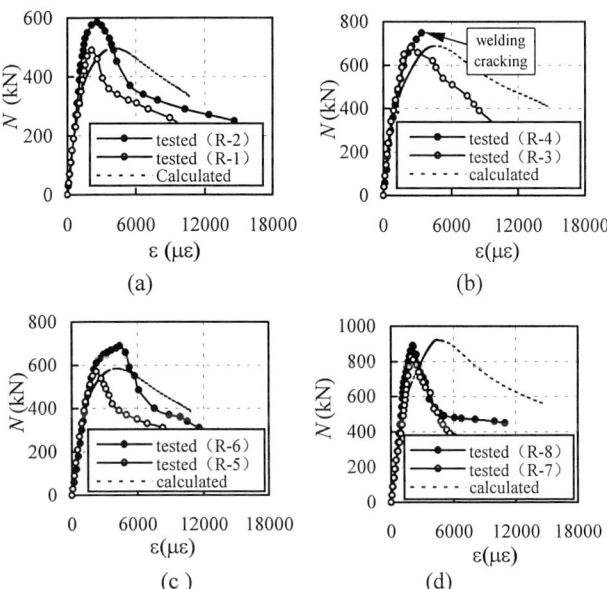

Figure 2: Tested axial load (N) versus axial strain (ε) relations

would be the case. Typical failure mode was overall buckling failure. The inward buckling of the steel tube is prevented by the concrete core, thus increasing the stability and strength of the column as a system. Total of the tested curves of load versus the longitudinal strain (ε) are shown in solid lines in Figure 2.

3. MODELS FOR LONG-TERM DEFORMATION

The main results, which were determined from the long-term tests, include both creep and shrinkage strain [$\varepsilon_{cr}(t)$ and $\varepsilon_{sh}(t)$]. The total strain was measured in the loaded specimens, and this is comprised of three components, which are highlighted in Equation (1).

$$\varepsilon_o(t) = \varepsilon_{elastic}(t) + \varepsilon_{cr}(t) + \varepsilon_{sh}(t) \tag{1}$$

As is well known, the total strain can be expressed as (Chu and Carreira,1986; Dezi, L., et al, 1993):

$$\varepsilon_o(t) = \frac{\sigma(\tau_0)}{E_c(\tau_0)} \cdot [1 + \varphi(t,\tau_0)] + \frac{[\sigma(t) - \sigma(\tau_0)]}{E_c(\tau_0)} \cdot [1 + \chi(t,\tau_0) \cdot \varphi(t,\tau_0)] + \varepsilon_{sh}(t) \tag{2}$$

where

$$\chi(t,\tau_0) = \frac{1}{1 - 0.91e^{-0.686\varphi(t,\tau_0)}} - \frac{1}{\varphi(t,\tau_0)} \tag{3}$$

is the aging coefficient (Song, 1993); $E_c(\tau_0)$ is the elasticity modulus of concrete at loading time (τ_0); $\sigma(\tau_0)$ and $\sigma(t)$ are the stress in concrete at time τ_0 and t respectively.

The creep fuction and shrinkage law recommended by the ACI Committee 209 (ACI, 1992) are calibrated herein with the test data with good agreement. The models express both the creep coefficient [$\varphi(t,\tau_0)$] and shrinkage strain [$(\varepsilon_{sh})_t$] as a function of time (t) given as follwings:

$$\varphi(t,\tau_0) = [\frac{(t-\tau_0)^{0.6}}{10+(t-\tau_0)^{0.6}}] \cdot \varphi_{max}(\tau_0) \tag{4}$$

$$(\varepsilon_{sh})_t = (\frac{t}{35+t}) \cdot (\varepsilon_{sh})_{max} \tag{5}$$

where $\varphi_{max}(\tau_0)$ represents the final creep coefficient, and $(\varepsilon_{sh})_{max}$ represents the total final shrinkage strain (ACI, 1992).

The analysis of axial deformation of concrete-filled steel RHS columns under long-term sustained load proceed as follows: (1). The stress and strain distributions in steel and in concrete under short-term loading are determined by using the model introduced in Han et al (2001a). (2). After a time of t of long-term sustained load, and due to creep effects, concrete stress is reduced while the steel stress is increased. The creep component of strain $\lfloor\varepsilon_{cr}(t)\rfloor_i$ is assumed. Then, knowing the creep coefficient and shrinkage after a time of t, the total strain of the concrete $\varepsilon_c(t)$ can be calculated by using Equation (1) and (5). Because of strain compatibility, the strain is equal to steel strain, i.e., $\varepsilon_s(t) = \varepsilon_c(t)$. The final stress in steel then may be computed by using equation of stress-strain relationship of the steel. Interal force N_s in steel tube can be determined. (3). Thus the internal force in concrete is $N_c = N_L - N_s$. The stress in the concrete can be calculated as $\sigma_c(t) = N_c/A_c$, the creep component of strain $\lfloor\varepsilon_{cr}(t)\rfloor$ in concrete is then can be calculated by using Equation (1) to (3). (4). Equivalence is then checked by verifying that $\lfloor\varepsilon_{cr}(t)\rfloor_i = \lfloor\varepsilon_{cr}(t)\rfloor_i$. If this condition is not satisfied, a new strain decrement $\lfloor\varepsilon_{cr}(t)\rfloor_i$ is assumed, as in Step (2), and Step (2) to (4) are repeated until the equivalence condition is guaranteed. The solution involves an iterative procedure and may require several trials. Procedures for assessing the convergence of the solution may be used but will not be discussed here. Repeat Step (1) to Step (4), the strains and internal forces in the steel tube and in the core concrete at any time t can be determined.

Based on the analytical results, it was found that the axial strain increases significantly in the earlier loading stage, it becomes moderate in the later stage. The force in the steel tube increases as the long-term loading time increases, the load in the concrete core decreases as the long-term loading time increases. It was because the axial creep deformation of the concrete core that the force in the concrete transfered to the steel tube gradually.

The average strains determined from the test program are compared with those obtained from the predictive model. Figure 2 shows the comparisons of the time-history of axial strain (ε) for the current tests. In order to check the accuracy of the theoretical analysis further, the results obtained using the method of analysis were also compared to the test results reported by other investigators, such as Nakai et al (1991), Morino et al (1997) and Uy(2001), a good agreement is obtained between the predicted and tested curves.

Based on the theoretical model mentioned above, the effects of the changing age time at loading (τ_0), time under sustained loading (t), level of sustained load (long-term sustained load ratio n), steel ratio (α) and strength of the materials on the long-term strain ε_L ($=\varepsilon_{cr}+\varepsilon_{sh}$, creep component and shrinkage component of strain respectively) due to creep and shrinkage of the column are analyzed. It was found that the strain of ε_L increases as the time immediately after loading (τ_0), the time under sustained loading (t), concrete strength, and the long-term load ratio n increases. The strain of ε_L decreases as the steel ratio (α) increases. The effects of the changing steel strength on the strain of ε_L are small.

4. EFFECTS OF THE SUSTAINED LOAD ON COLUMN STRENGTH

The load versus deformation relations of the columns subjected to sustained load can be established by using the method for short-term loading, which has been described in detail in Han et al (2001a). The behavior under sustained load is predicted by modifying the model under short-term loading condition. The modification consists of expressing the strain corresponding to the compressive stress for concrete, in terms of a function of time (t). Figure 3 shows the modified stress-strain diagram at any time ($t-\tau_0$) under a sustained stress(σ), applied at time τ_0. It is obtained by multiplying the strains in reference diagram by $\{1+\varphi(t,\tau_0)\}$, creep non-linearities are included in the creep coefficient [$\varphi(t,\tau_0)$] (Chu, et al, 1986). The calculation of the strength of the composite column with long-term sustained load effects considered involves the calculation of the deformation under long-term sustained load in the columns, as well as the ultimate strength of the composite columns with long-term effects are considered.

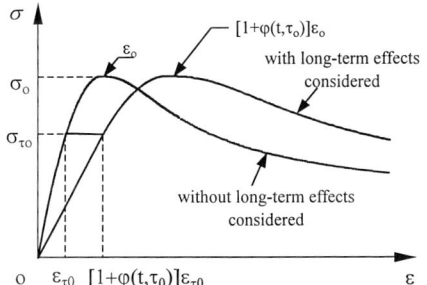
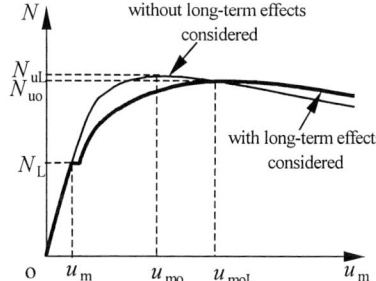

Figure 3: Modified stress (σ) versus strain (ε) diagram of concrete

Figure 4: Typical axial load (N) versus mid-height deflection (u_m) curves

After subjected to long-term sustained load, the strength of the composite column decreased with the duration of loading time (t). A numerical model was worked out for the analysis of the strength of the composite columns with long-term sustained load effects considered, and is a development of the analysis used for short-term loading condition (Han et al, 2001a). The model allows a differentiated consideration of all physical and geometrical non-linearity. In this method, For the calculation of the composite column with long-term sustained load effects considered, the following assumptions were made: (1). The cross sections remain plane during loading; (2). Perfect bond between concrete core and steel tube. (3). The deflection curve of the member is assumed as a sine wave. (4). The effect of shear force on deflection of members is omitted. (5). The contribution of concrete in tension is neglected. (6). The stress-strain relationship for steel, and for concrete in compression given in Han et al (2001a) is adopted. (7). Shrinkage stresses are very small compared with those from the applied load and are therefore neglected.

The load (N) versus mid-span deflection (u_m) relations can be established by using the method for short-term loading condition. The model has been described in detail in Han, et al (2001a). In order to obtain a basis of comparison for the effects of long-term sustained load on the strength of the composite columns, the aging time were set to be 50 years, i. e. The column subjected to a 50-year service load history was considered (Khor, et al, 2001). In the calculations, a small arbitrary load eccentricity of $L/1000$, reflecting a nearly ideal straight in axis of the column, has been selected for the initial eccentricity (Han, et al, 2001a).

Figure 4 gives schematic view of load versus mid-span lateral deflection curves. It can be found from this figure that, under long-term loading the increase in curvature with time t due to creep results in an increase in column deflection and reduced the column stiffness. In effect, the ultimate strength of a column under loading after a period of sustained load is reduced and is less than the corresponding value under short-term loading. The predicted curves of load versus lateral deflection (plotted in dashed lines) are compared in Figure 2 with experimental curves. Generally good agreement is obtained between the predicted and tested results.

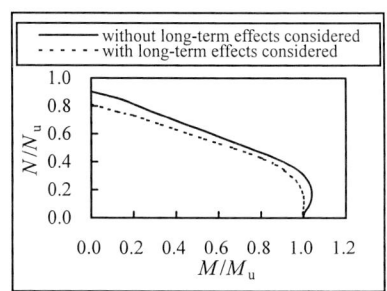

Figure 5: N/N_u versus M/M_u Interaction Curves

Figure 5 illustrates the typical calculated interaction relationship between compressive strength ratio (N/N_u) and bending strength ratio (M/M_u) of the composite columns. Figure 5 makes it clear that the critical load bearing strength decrease significantly because of the long-term sustained load effects. It was found that the ultimate strength of a column under quick loading after period of sustained load is reduced and is less than the corresponding value under short-term loading. Based on the theoretical model, influence of the changing slenderness ratio (λ), steel ratio (α), strength of steel and concrete, load eccentricity ratio (e/r) on the bearing capacity of the members was discussed. For convenience of analysis, strength index SI is defined to quantify the influencing of long-term sustained load on the concrete-filled steel RHS columns. It is expressed as:

$$SI = N_{uL}/N_u \qquad (6)$$

where N_{uL} is the ultimate load with long-term sustained load effects considered, and N_u is the ultimate load under short-term loading condition, which can be calculated using equations listed in GJB4142-2000 (2001), the equations were also listed in detail in Han et al (2001a).

Figure 6 shows the effects of different parameters on SI. (1). Slenderness ratio (λ) and steel ratio (α): It can be found that the ratio of SI of the composite columns with slenderness ratio λ less than 60 decreases as λ increases, except those columns with λ less than where the ultimate strength of the column is not altered by the sustained loads. The ratio of SI has no significant changes when the slenderness ratio λ is greater than

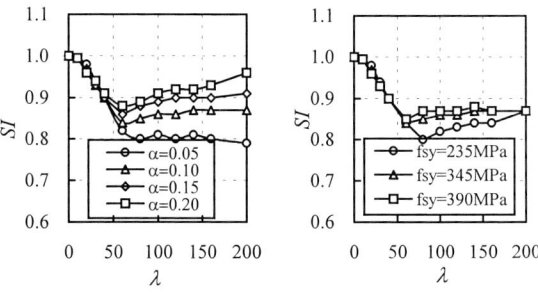

(a) Steel ratio and slenderness ratio (b) Steel strength

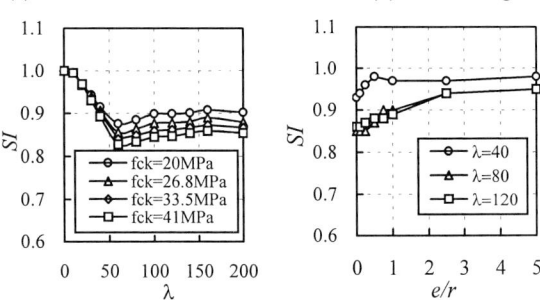

(c) Concrete strength (d) Load eccentricity ratio

Figure 6: Influence of different parameters on SI

60. It also can be found from Figure 6(a) that the ratio of *SI* is not altered by steel ratio α when λ is less than 40. However, the ratio of *SI* increases as α decreases as λ is greater than 40. (2) Yielding strength of steel (f_{sy}): It can be found from Figure 6(b) that generally, the ratio of *SI* decreases as f_{sy} decreases. (3) Strength of concrete (f_{ck}): in general, the ratio of *SI* decreases as f_{ck} increases (shown as in Figure 6 (c)). (4) Load eccentricity ratio (*e/r*): the consequence is a progressive loss of load bearing capacity of the composite columns by decreasing value of the ratio of *e/r* (show as in Figure 6 (d)). It can be found from the parametric analysis that, the maximum strength reduction of concrete-filled steel RHS columns to be roughly 20% of their strength under short-term loading conditions.

5. SIMPLIFIED MODEL

Using the relations between the strength index (*SI*) and the various parameters that determine it, the following formula for the strength index (*SI*) can be obtained by using regression analysis method, i. e.

$$SI = \begin{cases} (1 - 0.25 k_\lambda) \cdot \xi^{0.08 k_\lambda} \cdot [1 + 0.13 k_\lambda \cdot (1 - k_e)] & (k_\lambda \leq 0.4) \quad (7a) \\ (0.13 k_\lambda^2 - 0.3 k_\lambda + 1) \cdot \xi^{0.08 k_\lambda} \cdot (1 + \dfrac{1 - k_e}{15 + 25 k_\lambda^2}) & (0.4 < k_\lambda \leq 1.2) \quad (7b) \\ 0.83 k_\xi \cdot (1 + \dfrac{1 - k_e}{15 + 25 k_\lambda^2}) & (k_\lambda > 1.2) \quad (7c) \end{cases}$$

where, $k_\lambda = \lambda/100$; $k_e = (1 + e/r)^{-2}$; $k_\xi = \xi^{0.1}$, in which $\xi = A_s \cdot f_{sy} / A_c \cdot f_{ck} = \alpha \cdot f_{sy} / f_{ck}$ is defined as constraining factor (Han, et al, 2001a) to quantify the influences of steel ratio (α), yielding strength of steel (f_{sy}) and strength of concrete (f_{ck}).

The validity limits of Equation (7) are: *e/r*=0 to 3.0; α=0.04 to 0.2; λ=10 to 120; f_{sy} = 200 MPa to 500 MPa, and f_{ck} = 20 MPa to 60 MPa. To verify the validity of the formula, the ultimate strength as well as the strength index (*SI*) calculated with the formula were compared with those calculated with the mathematical model, which, as shown in Figure 7, predict the calculated strength index (*SI*) with reasonable accuracy. The ultimate strength calculated with the formula was compared with the experimental strength determined by Uy (2001) and the authors of this paper in Figure 8. It can be found that the accuracy with which the formula predicted the experimental strength is reasonable, and in general, the predictions are somewhat conservative.

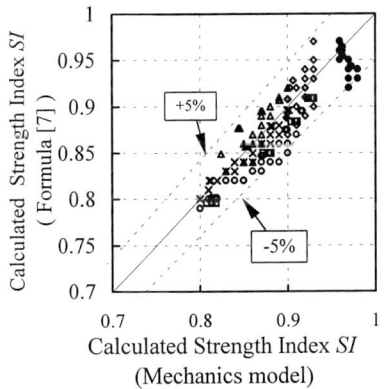

Figure 7: Comparison of calculated *SI* between formula [7] and mechanics model

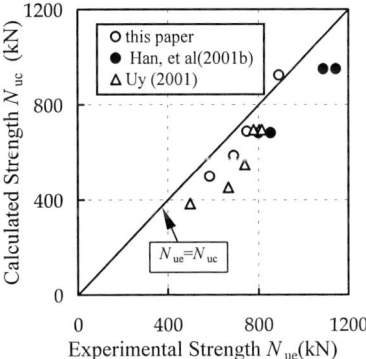

Figure 8: Comparison of ultimate strength between formula [7] and test results

6. CONCLUSIONS

The following conclusions can be made based on the limited research reported in the paper.

(1) After observing the measured data of the axial strain due to long-term sustained load effects in the current tests, it was found that for about 100 days, the process tends to stabilize.

(2) It was found that the long-term sustained load ratio (i.e. the ultimate load with long-term sustained load effects considered to the ultimate load under short-term loading condition) decreases as slenderness ratio increases when the slenderness is less than 60. The ratio decreases as the steel ratio and the steel strength decreases. The ratio decreases as the concrete strength increases.

(3) The load versus deformation relationship has been established for concrete-filled steel RHS columns with long-term sustained load effects considered.

(4) Sustained loads increase the deflections and decrease the strength of concrete-filled steel RHS slender columns. There is no strength reduction due to sustained load for a slenderness ratio λ less than 10. The maximum strength reduction due to long-term load effects can be expected to be roughly 20% of their strength under short-term loading.

(5) Formula should be suitable for incorporation into building codes, for the estimation of the strength for concrete-filled steel RHS columns with long-term load effects considered has been put forward.

REFERENCES

[1] ACI Committee 209 (1992). Prediction of Creep, Shrinkage and Temperature Effects in Concrete Structures. *Designing for Effects of Creep, Shrinkage and Temperature in Concrete Structures, ACI SP27-3*, Detroit, 51-93.
[2] Chu, K. H., Domingo, J. & Carreira, D. J. (1986). Time-Dependent Cyclic Deflections in R/C Beams. *Journal of structural engineering*, **112:5**, 943-959.
[3] Dezi, L., Ianni, C. & Tarantino, A. M. (1993). Simplified Creep Analysis of Composite Beams with Flexible Connectors. *Journal of Structural Engineering*, ASCE,**119:5**, 1484-1497.
[4] Furlong, R. W.(1967). Strength of Steel-Encased Concrete Beam-Columns. *Journal of Structural Division,* ASCE, **93: ST5**,113-124.
[5] GJB4142-2000 (2001). *Technical Specifications for Early-Strength Model Composite Structures.* Peking, China (in Chinese).
[6] Han, L. H., Zhao, X. L. & Tao, Z. (2001a). Tests and Mechanics Model of Concrete-Filled SHS Stub Columns, Columns and Beam-Columns. *Steel & Composite Structures-An International Journal*, **1:1**, 51-74.
[7] Han, L. H. Tao, Z. & Liu, W. (2001b). Concrete-Filled SHS Columns Subjected to Long-Term Loads. *Proceedings of The First International Conference on Steel & Composite Structures*, 14~16, June, Pusan, Korea, 1733-1740.
[8] Khor, E. H., Rosowsky, D. R. & Stewart, M. G. (2001). Probabilistic Analysis of Time-Dependent Deflections of R. C. Flexural Members. *Computer & Structures*, **79**, 1461-1472.
[9] Morino, S., Kswanguchi, J. & Cao, Z. S.(1996).Creep Behavior of Concrete Filled Steel Tubular Members. *Proc. of an Engineering Foundation Confer. on Steel-Concrete Composite Structure*, ASCE, Irsee, 514-525.
[10] Nakai, H., Kurita, A. & Ichinose, L. H. (1991). An Experimental Study on Creep of Concrete Filled Steel Pipes. *Proc. of 3rd Inter. Confer. of ASCCS*, Fukuoka, Japan, 55-60.
[11] Song, B. J. (1993). Creep theory of Concrete. *China Journal of Civil Engineering*, **3**, 66-68.
[12] Terrey, P. J., Bradford, M. A. & Gilbert, R.I.(1994). Creep and Shrinkage of Concrete in Concrete-Filled Circular Steel Tubes. *Proc. of 6th Inter. Symposium on Tubular Structures*, Melbourne, Australia, 293-298.
[13] Uy, B. (2001). Static Long-term Effects in Short Concrete-Filled Steel Box Columns under Sustained Loading." *ACI Structural Journal*, **98:1**, 96~104
[14] Zhong, S. T.(1994). *Concrete-Filled Steel Tubular Structures*, Heilongjiang Science and Technology Press, Harbin(in Chinese).

HYSTERETIC BEHAVIORS OF CONCRETE-FILLED STEEL SHS BEAM-COLUMNS

Z. Tao and L. H. Han

College of Civil Engineering and Architecture, Fuzhou University,
Gongye Road 523, Fuzhou, Fujian Province, 350002, P. R. China

ABSTRACT

The use of concrete filled composite columns has become increasingly popular in civil engineering. Only in China, it is reported that more than twenty-four high-rise buildings and two hundred arch bridges have been built due to their high strength and ductility, as well as large energy-absorption capacity. In the past, much progress had been made on the research of the static behaviors of concrete filled steel SHS (Square Hollow Section) columns, however, limited information is available on experimental behaviors of these columns when subjected to both axial and horizontal loading. In this paper, two series of experiments were conducted, the specimens of the first series were short columns subjected to cyclic shearing force as well as constant axial force, whereas the specimens of the second series were longer mainly subjected to a combination of constant axial load and cyclically flexural loads. The main parameters varied in the tests are axial load ratio and tube width to thickness ratio. Compared to conventional reinforced concrete columns, the SHS specimens exhibit richer energy-dissipation characteristics. The moment capacities of SHS beam-columns can be predicted with reasonable accuracy from simplified interaction curves. The experimental load versus hysteretic defection relations are compared to those predicted by integral method presented earlier, and satisfactorily agreements are gained.

KEYWORDS

Concrete filled steel tube, Beam-column, Hysteretic behavior, Strength, Ductility, Energy-dissipation

INTRODUCTION

Due to their high strength and ductility, as well as large energy-absorption capacity, the use of concrete filled steel tubular columns has become increasingly popular in civil engineering all around the world. It is reported that this kind of composite columns has been used in more than 24 high-rise buildings and almost 200 arch bridges only in China in recent years (Han 2000). The enhancement of structural behavior of the columns is in that the steel tube provides confinement for the concrete, and the concrete core prevents the inward buckling of the tube as well.

Because of the increased use of composite columns, a great deal of theoretical and experimental work has been carried out. In the past decades, much progress had been made on the research of monotonic behaviors, which accounting for the effects of local buckling, bond strength and confinement of concrete on the behavior of concrete filled steel SHS beam-columns (Shams and Saadeghvaziri 1997,

Han et al. 2001, Shanmugam and Lakshmi 2001). However, much less effort has been devoted to cyclic behaviors of such kinds of members. Nevertheless, a certain number of experimental programs have been performed (Sakino and Tomii 1981, Shiiba and Harada 1994, Boyd et al. 1995, Kang et al. 1998, Sakino et al. 1998, Varma 2000), and a few numerical models have been developed as well (Hajjar and Gourley 1997, Tsuiki et al. 1998). The lack of sufficient research works on hysteretic behavior on SHS beam-columns can be explained by the rather complicated behaviors of the composite columns, which require adequate experiments, sophisticated models and time-consuming calculations.

Two series of experiments were conducted, the purpose of the experimental study is to investigate the seismic behaviors of concrete filled SHS tubular columns. The effects of varying width to thickness ratio of the steel plate and axial load ratio were studied as well. The moment capacities of SHS beam-columns are compared between tests and simplified interaction curves, and satisfactorily agreements are gained. The experimental load versus defection relations are also compared to those predicted by integral method presented earlier.

OUTLINE OF EXPERIMENT

Test Specimens

The experiments were classified as two series, the difference between them was in that the specimens of series I were shorter with slenderness ratios (λ) of 20 or so, whereas specimens of series II were longer with slenderness ratios (λ) approaching 44 or 53 or so. The value of λ is defined as:

$$\lambda = 2\sqrt{3} L_e / B \quad (1)$$

where L_e is the effective length of a column, which is the same as the physical length of the column (L) with pin-ended supports, B is the tube width. The variable parameters in one series were axial load ratio and tube width to thickness ratio (B/t).

Material properties of steel tubes were determined from tensile coupon tests in accordance with the Chinese standard related to mental materials. The average yield strengths (f_{sy}) in each series are 274.5MPa and 340.0MPa respectively.

The concrete of one batch were used to fill the tubes for the specimens in the same series. Crushed coarse aggregate of 15 mm maximum size and sand of 2.5 mm maximum size were used to make the concrete. For each batch of concrete mixed, three 150mm cubes were also cast and cured in conditions similar to the related specimens. The composite columns were tested at an age of 28 days after concrete casting. The average compression cube strengths (f_{cu}) at that age were 25.2MPa and 20.1 MPa respectively. The modulus of elasticity (E_c) of concrete was measured in accordance with the Chinese standard related to concrete, the average values being 26,230 MPa and 25,306 MPa respectively.

The summaries of the specimens of the two series are presented in Table1 and Table2 respectively. The tube width to thickness ratio listed in Table 1 varied from 28.6 to 34.3, while that listed in Table 2 varied from 37.7 to 45.3. The range of load ratio (n) in Table 1 is between 0.03 to 0.35, while that in Table 2 is between 0.03 to 0.50. Load ratio is that of axial load (N) divided by ultimate strength of the specimen (N_u) which can be determined from equations presented by Han et al. (2001).

Test setup

Different test setup was used for series I and II. In series I, each specimen was fixed at the base and subjected to a constant axial load (N) and cyclically varying lateral load (P). A detailed description and schematic view of the test setup were shown elsewhere (Tao and Han 2001a). In series II, the testing specimens were laid horizontally which were simple supported at both ends. The vertical load (P) was applied to the middle span of the test specimen through a pair of loading plate. After the horizontal actuator had applied a specified axial load (N), then the vertical actuator applied cyclically increasing bending to the specimen. Details of the test setup were given elsewhere (You et al. 2002).

TABLE 1
SPECIMEN LABELS, MATERIAL PROPERTIES AND MEMBER CAPACITIES OF SERIES I

Specimen Label	B×t (mm)	B/t	L_e (mm)	λ	N (kN)	n	P_u (kN)	M_u^e (kN.m)	M_u^c (kN.m)	M_u^c/M_u^e
SHC1-1	100×3.5	28.6	600	21	20	0.03	57.5	17.3	15.4	0.89
SHC1-2	100×3.5	28.6	600	21	90	0.17	57.6	17.3	15.7	0.91
SHC1-3	100×3.5	28.6	600	21	190	0.35	50.2	15.1	14.0	0.93
SHC2-1	120×3.5	34.3	680	20	20	0.03	63.1	21.5	22.4	1.04
SHC2-2	120×3.5	34.3	680	20	120	0.17	69.4	23.6	23.0	0.97
SHC2-3	120×3.5	34.3	680	20	240	0.35	69.0	23.5	21.5	0.92

TABLE 2
SPECIMEN LABELS, MATERIAL PROPERTIES AND MEMBER CAPACITIES OF SERIES II

Specimen Label	B×t (mm)	B/t	L_e (mm)	λ	N (kN)	n	P_u (kN)	M_u^e (kN.m)	M_u^c (kN.m)	M_u^c/M_u^e
S100-1	100×2.65	37.7	1540	53	23	0.05	38.6	14.9	14.8	0.99
S100-2	100×2.65	37.7	1540	53	175	0.35	30.0	11.6	12.0	1.03
S100-3	100×2.65	37.7	1540	53	230	0.50	27.7	10.7	10.4	0.97
S120-1	120×2.65	45.3	1540	44	21	0.05	58.0	22.3	21.5	0.96
S120-2	120×2.65	45.3	1540	44	200	0.35	49.1	18.9	19.9	1.05
S120-3	120×2.65	45.3	1540	44	300	0.50	42.2	16.2	16.1	0.99
S120-4	120×2.65	45.3	1540	44	366	0.60	35.6	13.7	13.4	0.98
S120-5	120×2.65	45.3	1540	44	366	0.60	38.5	14.8	13.4	0.91
S120-6	120×2.65	45.3	1540	44	427	0.70	30.5	11.7	11.2	0.96

Lateral displacements (δ) of a test specimen were measured with displacement transducer at the top or middle of the test specimen in different series. Vertical-column displacements were also monitored by strain gauges.

For each specimen, yield displacement (δ_y) was experimentally defined firstly as the displacement when the measured maximum tensile stress of the steel tube attained the yield point (f_y). Then, the displacement was increased sequentially $n\delta_y$ ($n=2,3,\ldots$) with one cyclic loading in each $n\delta_y$ stage up to failure.

EXPERIMENTAL RESULTS AND DISCUSSIONS

Collapse Modes

Failure of all the specimens was apparently controlled by the elephant foot like buckling at the bottom end for series I and at the middle span for series II. There were no local buckling was observed at any other portion of all columns. The buckling was first observed in the flange plate when the lateral displacement attained $3\delta_y$ and subsequently, developed in the opposite flange plate during reversed loading. After that, the buckling behaved steadily up to failure. For specimens in series I or in series II, there were no apparent difference between their collapse modes. Thus, the failure of all the specimens was consequently not of shear but of bending with plastic hinge zone at the bottom or middle span of the column independent upon the various experimental parameters. Shown in Fig.1 is typical failure appearance of specimens in series I.

P-δ Hysteretic Curves

The lateral load versus lateral displacement hysteretic curves for the test specimens in series I and series II are shown in Figs 2 and 3 respectively (solid line). The maximum lateral load (P_u) and moment capacity (M_u^e) for each specimens are listed in Table 1 and Table 2, P_u and M_u^e are defined as the average of the peak lateral loads and peak moments in the two loading directions respectively. From Figs 2 and 3, it indicates that all specimens exhibited stable behavior and did not exist apparent pinch phenomenon despite the strength deteriorated or not.

Figure 1: Typical failure mode

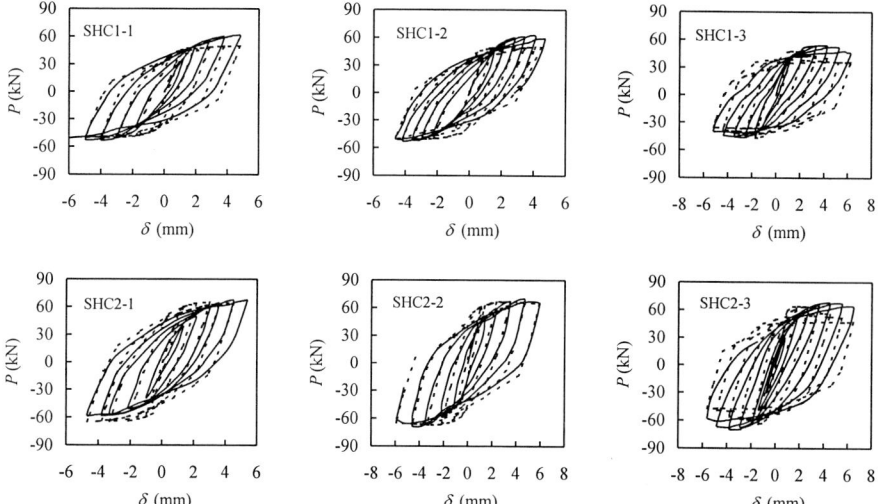

Figure 2: Horizontal load versus Horizontal displacement hysteretic curves (series I)

Effect of Width–Thickness Ratio

In order to make clear the effect of width to thickness ratio on the hysteretic curves, some envelop curves are nondimensionalized by P_u as shown in Fig.4. In previous research (Sakino and Tomii, 1981), the lateral loads have shown a tendency to decrease with increasing values of B/t. This trend was not observed in the tests reported here except the comparison of specimen S100-2 and S120-2, that is, no apparent decrease in flexural capacity was observed with the increase of B/t ratio, the reason may be is that the change of B/t ratio in the tests was too small to have any considerable effect on the hysteresis behavior.

Effect of Axial Compression Force Ratio

The P-δ envelope curves of specimens with different axial load ratio are shown in Fig.5. For specimens in series I, the P-δ envelope curves were not so greatly affected as those in series II. With the increasing of the axial load ratio, the specimens in series II exhibited less energy dissipation capacity as well as greater strength degradation. The different performances between specimens in the two series maybe generated from the influences of secondary moment which significantly influenced specimens in series II.

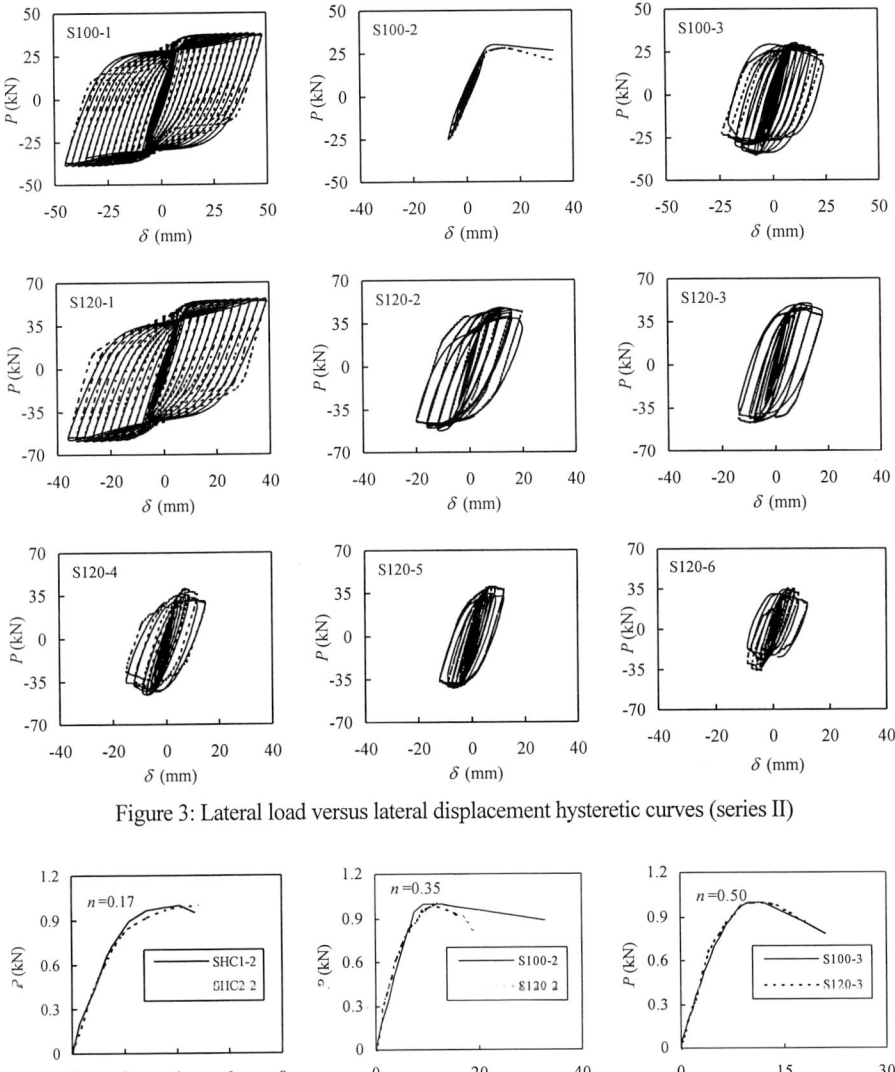

Figure 3: Lateral load versus lateral displacement hysteretic curves (series II)

Figure 4: Effect of width to thickness ratio on the hysteretic curves

COMPARISON OF EXPERIMENTAL RESULTS WITH STRENGTH PREDICTIONS

Axial load-bending moment capacity (P-M) interaction curves developed for concrete-filled SHS beam-columns have been presented by Han et al. (2001). The curves are established based on regression analysis. Table 1 and 2 list the experiment bending moments (M_u^e) and the corresponding predicted values (M_u^c). M_u^c were calculated when the columns were regarded as members subjected to both bending and axial loading. A mean ratio (M_u^c/M_u^e) of 0.967 is obtained with a COV of 0.047 for all 15 specimens. The predicted P-M interaction curves for different specimen types are compared in Fig.6 with the experimental values from the tests. Good agreement is achieved.

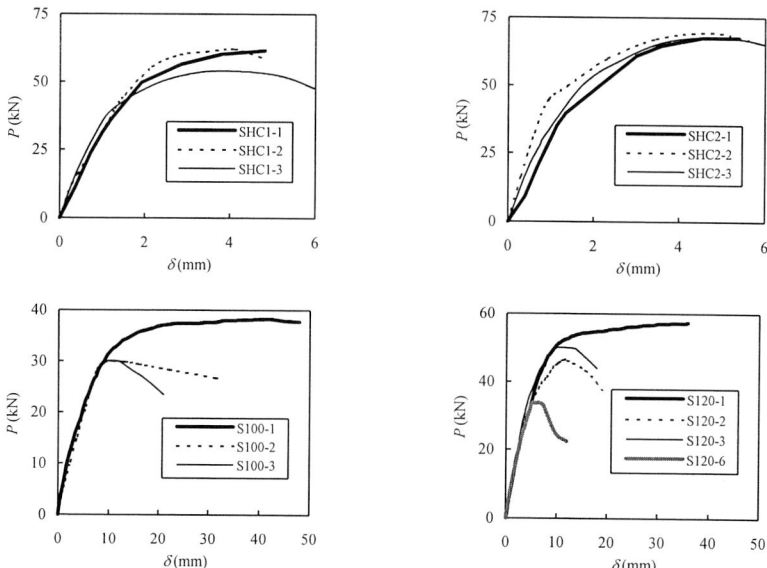

Figure 5: Effect of axial load ratio on the hysteretic curves

ANALYTICAL INVESTIGATIONS

In order to investigate the flexural behavior of concrete-filled SHS beam-column specimens when subjected to reversed cyclic lateral load and constant axial load, fiber-based models were developed by Tao et al. (2001b). In the reference paper, hysteretic models on the stress-strain of the steel and the concrete were presented, in which the envelope of the cyclic stress-strain curve and the hysteresis rules governing the cyclic behavior were specified. Fig.7 shows the hysteretic stress-strain behavior that assumed for the steel fibers. In the figure, the strain hardening and the Bauschinger effect of steel are taken into account. The skeleton curve consists of two stages: the elastic stage (from point o to point a) and the strain hardening stage (form point a to point b). The modulus of skeleton curve in strain hardening stage is taken as $0.01E_s$ in this paper, where E_s denotes the modulus of elasticity of steel. In the process of strain reversal, the stress decreases linearly with the modulus E_s until the stress level becomes equal to σ_d or $\sigma_{d'}$ (stress of steel at point d or d'). After that, the stress-strain curve exhibits the Bauschinger effect, which causes nonlinear stress-strain relationship and softening of the stress-strain curve before the stress reaches the yield stress in the opposite direction. The modulus in the softening stage is expressed as follows:

$$E_s' = (f_{sy} - |\sigma_d|)/(\varepsilon_y - |\varepsilon_d|) \qquad \varepsilon_y \leq |\varepsilon_d| < 6.11\varepsilon_y \qquad (2a)$$

$$E_s' = 0.1E_s \qquad |\varepsilon_d| \geq 6.11\varepsilon_y \qquad (2b)$$

where f_{sy} is the yield strength of steel; ε_y is the strain when steel yields and is equal to f_{sy}/E_s; ε_d is the strain at point d or d'.

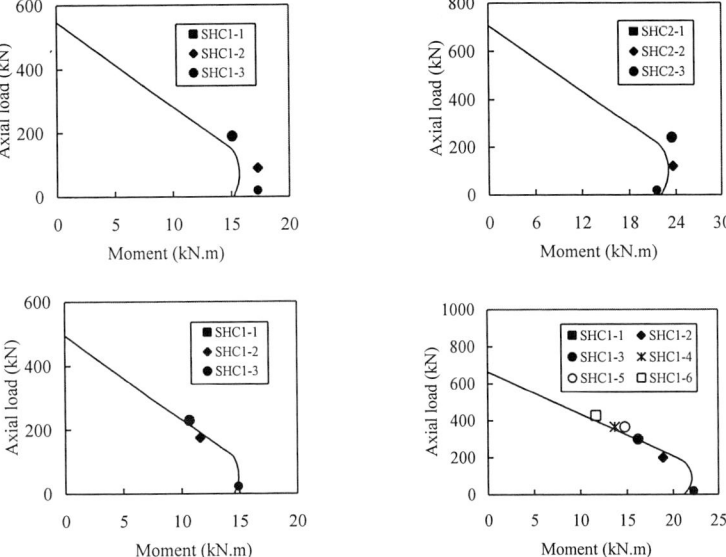

Figure 6: Effect of axial load ratio on the hysteretic curves

Fig. 8 shows the hysteretic stress-strain behavior that was assumed for the concrete fibers. In the figure, monotonic stress-strain curve was used to define the skeleton curve of core concrete, σ_p and ε_p are the cracking stress and strain, and σ_0, ε_0 are the peak stress and strain of concrete prism when subjected to compression. Additional details of these stress-strain curves are presented in Tao et al. (2001b).

The cyclic P-δ response from the fiber analyses are shown in Figs 2 and 3 (dash line). They indicate that the results from the fiber analyses of the cyclic beam-column specimens compared favorably with the experimental ones.

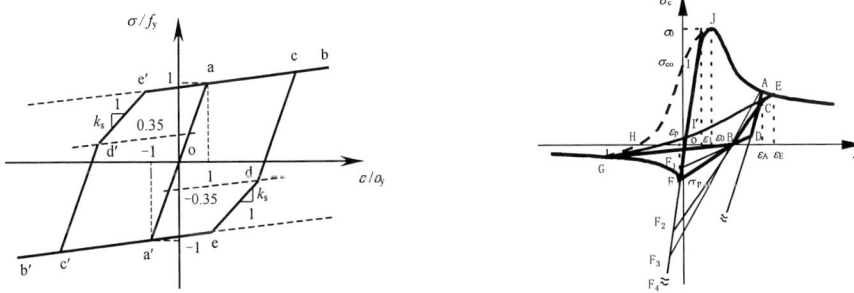

Figure 7: Stress-strain model for steel   Figure 8 Stress-strain model for concrete

CONCLUSION

The main conclusions obtained by this study can be summarized as follows.

1) Two series of experiments were conducted, the specimens in the first series were shorter while those in the second series were longer. There were no apparent difference between their collapse modes, the failure of all the specimens was of bending.

2) The main parameters varied in the tests are axial load ratio and tube width to thickness ratio. No apparent decrease in flexural capacity was observed with the increase of B/t ratio in the tests, further research work is needed. The axial load ratio had significant influence on longer specimens then on shorter ones. With the increasing of the axial load ratio, the specimens in series II exhibited less energy dissipation capacity as well as greater strength degradation.

3) Compared to conventional reinforced concrete columns, the SHS specimens exhibit richer energy-dissipation characteristics.

4) The moment capacities of SHS beam-columns can be predicted with reasonable accuracy from simplified interaction curves.

5) A mechanics model presented earlier can be used to predict the load-displacement hysteretic relations of the composite beam-columns, good correlation has been achieved between the analytical and the experimental results.

References

Hajjar J.F., Gourley B.C. (1997). A Cyclic Nonlinear Model for Concrete-Filled Tubes Cross-Section Strength. *Journal of Structural Engineering,* ASCE, **122:11,** 1327-1136.

Han, L. H. (2000), *Concrete Filled Steel Tubular Columns*, Peking, Science Press (in Chinese).

Han, L. H., Zhao, X. L. and Tao, Z. (2001). Tests and mechanics model for concrete-filled SHS stub columns, columns and beam-columns. *Steel & Composite Structures-An International Journal* **1:1,** 51-74.

Kang C.H. and Moon T S. (1998). Behavior of concrete-filled steel tubular beam-column under combined axial and lateral forces. *Proc. of Fifth Pacific Structural Steel Conf.,* Seoul, Korea, 961-966.

Sakino, K. and Tomii, M. (1981), "Hysteretic behavior of concrete filled square steel tubular Beam-Columns failed in flexure". *Trans. of the Japan Concrete Institute* **3,** 439-446.

Sakino K., Inai E. and Nakahara H. (1998). Tests and analysis on elasto-plastic behavior of CFT beam-columns - U.S.-Japan cooperative earthquake research program. *Proc. of Fifth Pacific Structural Steel Conf.,* Seoul, Korea, 961-966.

Shams M. and Saadeghvaziri M.A. (1997). State of the art of concrete-filled steel tubular columns. *ACI Structural Journal* **94:5,** 558-571.

Shanmugam N.E. and Lakshmi B. (2001). State of the art report on steel-concrete composite columns. *Journal of Construction Steel Research* **57,** 1041-1080.

Shiiba K. and Harada N. (1994). An experiment study on concrete –filled square steel tubular columns. *Proc. of the 4th Int. Conf. on Steel-Concrete Composite Structures,* Slovakia, 103-106.

Tao, Z. (2001). Several key issues for the behaviors of concrete filled steel tubular members with square sections. Ph.D. Thesis, Harbin Institute of Technology.

Tao, Z. and Han, L. H. (2001a). Tests on the hysteresis behaviors of concrete filled steel tubular beam-columns with square sections. *Earthquake Engineering and Engineering Vibration* **21:1,** 74-78 (in Chinese).

Tao, Z., Han, L. H. and Zhao, X. L. (2001b). Hysteretic Behaviors of Concrete Filled Steel Tubular Beam-Columns with Square Section. *Proceedings of The First International Conference on Steel & Composite Structures*, 14~16, June, Pusan, Korea, 1717-1724.

Tsuiki A., Kwaguchi J., Fukao H. and Morino S. (1998). Analysis of cyclic behavior of CFT beam-columns failing in local buckling. *Proc. of Fifth Pacific Structural Steel Confer.*, Seoul, Korea, 907-912.

Varma A.H. (2000). Seismic behavior of high strength square CFT beam-columns. *Proc. of 6th Inter. Confer. on Steel and Concrete Composite Structures,* USA, 547-556.

You J.T., Tao, Z. and Han, L. H. (2002). Cyclic tests of concrete-filled rectangular steel tubular columns. *Engineering Mechanics* **19:7**(to be published, in Chinese).

EXPERIMENTAL AND THEORETICAL STUDIES ON STEEL-CONCRETE HYBRID STRUCTURES

G.Q.LI[1], X.M. ZHOU[2] and X.DING[1]

[1] Department of Building Structure Engineering, Tongji University, Shanghai, China
[2] Department of Civil Engineering, the Hong Kong University of Science and Technology, Clear Water Bay, Kowloon, Hong Kong, China

ABSTRACT

The behaviours of steel frame and concrete tube hybrid structures for tall buildings are studied in this paper. Shaking table tests were conducted on a scale down model of a typical steel-concrete hybrid tall building. The dynamic properties, seismic responses and damage features of the model were observed and analysed. A simplified approach for elasto-plastic analysis of hybrid structure subjected to earthquakes has been developed. In the proposed model, the external steel frames are treated as a series of half frames while the concrete tube is modelled as a two-spring bending and shear system on each storey. The two parts are connected with horizontal rigid bars on floor levels and work cooperatively to resist horizontal loads. The proposed analytical model has been verified to be effective and accurate by comparing experimental and theoretical results of the tested model subjected to simulated earthquakes.

KEYWORDS

Steel-concrete hybrid structure, elasto-plastic analysis, scale down model, shaking table test, simulated earthquakes, simplified approach

INTRODUCYION

Hybrid structures composed of steel frames(SF) and concrete tubes(CT) recently have more and more applications in high-rise buildings in China due to the advantages such as high speed of construction, strong stiffness and low cost. Compared with pure steel structures, fire-resistant protections and welding works in construction sites can be reduced leading to less difficulty in construction. Compared with pure concrete structures, weights of hybrid structures can be reduced, leading to less cost to the foundations(G.Q.Li, 1997).

However, very little research has been conducted on aseismic behaviour of hybrid structures, and it is still questionable whether or not hybrid structures can be used in earthquake zones(ASCE, 1979). According to elastic analysis, SF only takes 2%~5% of the total shear force of the whole structure induced by earthquakes. It is therefore generally considered that seismic resistant behaviour of hybrid structures depends mainly on the concrete tubes, which generally show brittle failure modes when subjected to earthquakes. In this paper, a scale-down model of a typical hybrid tall building was made and tested on the shaking table at Tongji University.. The dynamic properties, seismic responses, damage features and displacement responses of the model were recorded and analysed. A simplified approach for elasto-plastic response analysis of steel-concrete hybrid tall buildings was developed and proposed. The effectiveness and reliability of the proposed analytical approach were verified by comparing the experimental and analytical results of the scale-down model subjected to simulated earthquakes.

EXPERIMENTAL STUDIES

The Tested Model

The simulation relationship is considered for manufacturing of the tested model (Sabnis, 1983). The length similitude coefficient (model vs prototype) S_l was equal to 1/20, while the density similitude coefficient S_ρ was taken as 4.0. Because CT contributed mostly to the lateral stiffness of hybrid structure, the whole similitude coefficients system was deduced from CT as Table 1.

TABLE 1
SIMILITUDE COEFFICIENTS OF THE HYBRID STRUCTURE MODEL

Variable	Symbol	Value	Variable	Symbol	Value
Length	S_l	1/20	Density	S_ρ	4
Time	S_t	1/8	Strain	S_ε	1
Frequency	S_f	8	Stress	S_σ	0.64
Velocity	S_v	0.4	Elastic modulus	S_E	0.64
Acceleration	S_a	3.2	Concentrated force	S_F	1/625
Displacement	S_u	1/20	Face-distributed Mass	S_w	1/5

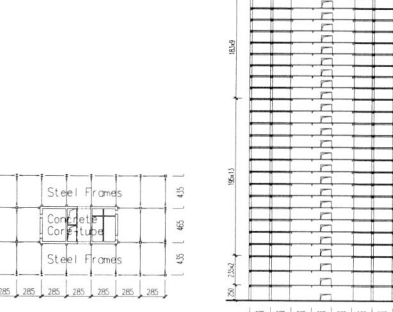 

Fig 1 Plan and Elevation of the model   Fig 2 The Model on Shaking Table

The 25-storey testing model finally constructed was 5.0m in height, with a plan dimension of 1.3m×2.0m. The rectangular concrete tube was located in the center with a plan dimension of 0.885m×0.465m, occupying about 15% of the floor area. The tube of the model was built using fine graded aggregate concrete reinforced with galvanized iron wire. The compressive strength of concrete used was 26.5MPa on average. The steel frame was made of Q235 steel with box-section columns and I-shape beams. The average tensile yield and ultimate strengths of steel are 236.1MPa and 350.5MPa respectively and the average percentage of elongation is 25.4%. Additional mass was put on the slabs of the model to satisfy the requirement of density similitude as well as simulate the vertical loads applied on the structure including dead load and 50% live load according to Chinese Buildings

Aseismic Design Code (GBJ-89). The additional mass added to each storey was 220kg and the total additional mass applied to the model was 5500kg. The plan and elevation of the model were shown in Fig 1 and Fig 2.

Shaking Table Test

The model was tested on the shaking table of the State Key Disaster-Prevention Laboratory in Tongji University. Two displacement transducers were put on the sixth floor of the model along X direction and another displacement transducer on the roof along X direction. Accelerometers were put at the base and floors 2, 6, 9, 13, 17, 21 and 25 floor respectively, as shown if Fig 3. Totally 8 strain gauges were put on the tube and steel frame columns at the bottom of the model.

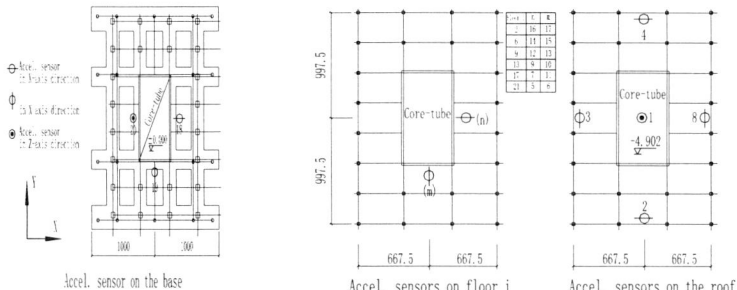

Fig 3 The arrangement of the accelerometers

The prototype structure was designed to resist intensity 7 (Chinese Code) earthquakes. Three seismic waves (acceleration records) were taken as input signals of the shaking table, which were artificial seismic wave of Shanghai (P), 1940 El-Centro seismic wave (E) and 1969 San Fernando (S) respectively. Since lateral stiffness of the X-axis was smaller than that of the Y-axis, the X-axis was selected as the primary exciting direction. When the model was excited at only one direction (X in this study), the horizontal component, having larger peak acceleration value of the real seismic wave chosen, was used as the X-direction input. When the model was excited in 2-D or 3-D, similar as the one-directional exciting case, the horizontal component, having larger peak acceleration value of the real seismic wave chosen, was used as the X-direction input and the other horizontal wave was input at Y direction. And the vertical component of the seismic wave was used as the Z-axis input signal of the shaking table. Moreover, the model was investigated through white noise (W) in different testing phases to measure the changes in dynamic properties. The designed peak acceleration values of the input seismic waves were 0.17g(intensity 6), 0.50g(intensity 7), 1.10g(intensity 8 more) and 2.00g(intensity 9) respectively for the scale-down model.

Behaviour of the model under different intensities of earthquakes

Under intensity 6 earthquakes, the model stayed in elastic state, and no obvious crack was found. When earthquakes increased to intensity 7, minor cracks appeared appreciably at the concrete walls at the bottom storey. As earthquakes further increased to intensity 8, cracks became more obvious and the longitudinal reinforcements of the corner columns began yielding and bulging outward. When earthquakes reached intensity 9, concrete at the bottom of the tube spalled, loosed and crushed in large area as shown in Fig 4. Apparent cracks were also found, as shown in Fig 5, around the embedded joint plates for the steel beam to the concrete tube connections, due to alternatively changed tension and compression forces generated in these joints during severe earthquakes. The displacement of the model roof increases up to 122mm, which was about 1/40 of the total height of the model under intensity 9

Fig 4 The concrete damages at the bottom of the core-tube

Fig 5 Concrete damages at the joints of SF & CT

earthquake. The steel frames sustained in good condition and played an important role of supporting the tube to survive, keeping the integrity of the whole structure, and preventing it from collapse under high intensity earthquakes.

White noise investigation indicated that the frequencies of the model decreased and damping ratios increased due to damage cumulation as the seismic intensity increases, as shown in Table2.

TABLE 2
THE DAMPING RATIOS AND THE NATURAL FREQUENCIES OF THE MODEL MEASURED

Seismic grade	X-axis direction		Y-axis direction	
	Damping ratios(%)	Frequencies (Hz)	Damping ratios(%)	Frequencies (Hz)
Non	3.84	4.31	4.61	5.66
6	4.58	3.91	5.19	5.47
7	6.50	3.52	5.18	4.69
8	6.56	3.32	5.78	4.30
9	8.85	1.95	7.66	2.93

THEORETICAL STUDIES

Disassembly and Simplification of Hybrid Structures

Two assumptions are adopted in this study. Firstly, slabs are assumed to be rigid in their own planes, and SF and CT have the same horizontal displacements on each floor level to ensure the two substructures resist earthquake loads cooperatively; secondly, the torsion is neglectable of the whole structure.

Then external steel frames are simplified as a series of plane frames parallelly connected. These plane frames resist earthquake loads in the direction of their own planes. The rectangular tube is first disassembled into shear walls parallel and perpendicular to the direction of applied earthquake loads respectively. If the shear walls have openings in their elevations, they are then transformed to equivalent solid ones with no openings under the rule of same lateral-resistance stiffness. These solid walls are modelled as bending and shearing multiple-spring systems. The multiple-spring wall systems are connected to steelframes(G.Q.Li 1995,1996), which are further simplified to a half frame model, with horizontal rigid bars on each floor level, as shown in Fig 6.

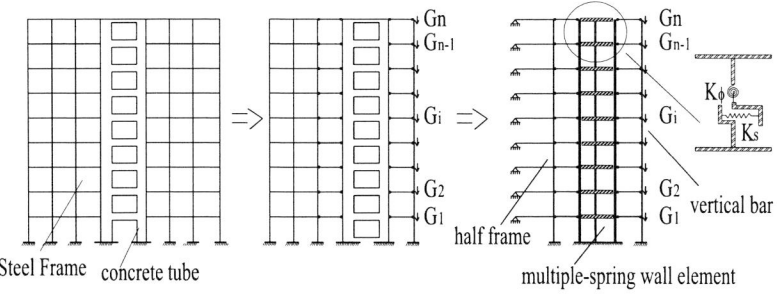

Fig 6 Analytical Model of Hybrid Structures

Bending-shear Two-spring Wall Element

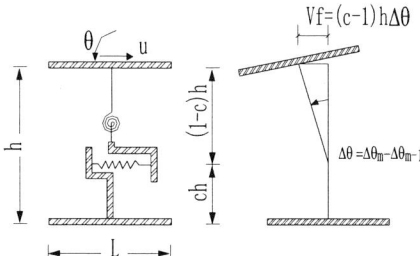

Fig 7 Two-Spring Wall Element Model

In the bending-shear two-spring wall element model(Fig 7), the shearing spring is considered to be elastic based on experimental results, while the rotation spring is employed a stiffness-degrading trilinear model to depict the hysteretic relationship between moment and cross-section curvature. The initial bending-resistant stiffness of the rotation spring is obtained as(P. Linde, et, al, 1994)

$$K_\varphi = 2(1-c)EI/h \qquad (1)$$

where E is the Young's modulus, I is the moment of inertia of the wall and c is a constant, generally being equal to 0.5. After cracking, the bending stiffness of the structural wall reduces to $\alpha_\varphi K\varphi$ and the reduction ratio α_φ is taken as:

$$\alpha_\varphi = (M_y - M_c)/(EI(\varphi_y - \varphi_c)) \qquad (2)$$

where M_y, M_c, φ_y and φ_c are the yield moment and the crack moment, and their corresponding curvatures respectively. After yielding, the bending-resistant stiffness is taken as 0.5% of the initial flexural stiffness in this study to show some strength hardening effects on structural walls.

The cracking bending moment and the yielding bending moment of the shear wall can be determined by (Vulcano, A, et, al, 1987)

$$M_c = Z_e(f_t + \sigma_0) \tag{3}$$

$$M_y = (A_s f_y + N/2) \cdot l \tag{4}$$

where Z_e is the section modulus of the wall including longitudinal reinforcements scattered in it; f_t is the tensive-resistant strength of concrete and σ_0 is the average compressive stress of wall cross-section; A_s and f_y are the total area and the yield strength of longitudinal reinforcements at one end of the wall; and N is the axial force in the wall.

The cracking curvature of the shear wall can be calculated as(P. Linde, et, al, 1994)

$$\varphi_c = M_c / EI \tag{5}$$

Yielding of a reinforced concrete component is defined as the first yielding in the tension reinforcement. Hence, the yielding curvature for a concrete structural wall was given as(Park, Y. J 1985)

$$\varphi_y = [1 + \frac{0.4}{0.84 + \rho_a} \cdot \frac{\sigma_0}{0.3 f_c'}] \cdot \frac{\varepsilon_s}{d - kd} \tag{6}$$

where $\rho_a = A_s f_y / A_w f_c'$, $f_c' = 0.8 f_{cu}$ in the above formula; A_w is the area of the cross-section of the wall, f_{cu} is the cubic compressive strength of concrete, d and kd are shown in Fig 8.

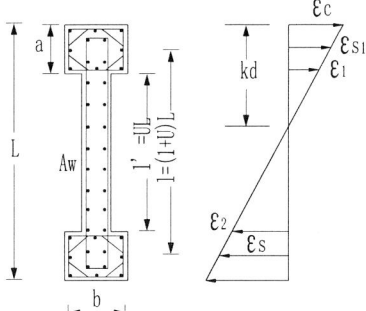

Fig8 The Typical Cross-section of Wall and the Strain Distributing

Lateral Stiffness Matrix of Hybrid Structure

In the proposed simplified analytical model in this study, the total lateral-resistant stiffness of steel-concrete hybrid structure $[K_m]$ is composed of three parts, which are the lateral-resistant stiffness of steel frame $[K_F]$, the lateral-resistant stiffness of concrete tube $[K_t]$, and the total geometric stiffness $[K_g]$ respectively. That is

$$[K_m] = [K_F] + [K_t] - [K_g] \tag{7}$$

$$[K_g] = \begin{bmatrix} N_n & -N_n & & & & \\ -N_n & N_n + N_{n-1} & -N_{n-1} & & & \\ & -N_{n-1} & & & & \\ & & & \ddots & & \\ & & & & & -N_2 \\ & & & & -N_2 & N_2 + N_1 \end{bmatrix} \quad (8)$$

where $N_i = \dfrac{\sum_{j=i}^{n} G_j}{h_i}$, G_i is assumption of vertical loads of ith floor and h_i is the height of ith floor.

Note that $[K_F]$ is the summation of lateral stiffness matrices of all planes of steel frames of a hybrid structure, i.e. $[K_F] = \sum_{i=1}^{n}[K_f]$ and n is the number of pins of steel frames comprising the hybrid structure.

COMPARISON BETWEEN EXPERIMENTS AND THEORETICAL PREDICTIONS

Natural Vibration Frequencies

Table 3 gives the first 3 orders of the natural frequencies in X direction and Y direction respectively measured before seismic excitation and those obtained through theoretical computation. Measured and computed natural frequencies of the model agree quite well. This indicates that the developed analytical model is reliable for elastic analysis.

TABLE 3
THE FIRST 3 ORDERS OF NATURAL FREQUENCIES OF THE MODEL (X AND Y DIRECTION)

No of Order	X Direction		Y Direction	
	Calculated	Measured	Calculated	Measured
1	4.297	4.314	5.664	5.592
2	20.117	19.819	25.391	26.939
3	41.602	45.484	45.117	51.457

Displacement Responses

The lateral displacements of the sixth floor and the top floor of the model in X direction were measured during testing. The displacements are also calculated from the time history records of acceleration responses of the model by integral method in frequency domain. Compared with the measured values, displacements by integral method had an acceptable accuracy. Table 4 lists the measured and calculated maximal displacements of the model under various intensities of earthquakes. It is shown that theoretical predictions agree well with experimental results, which indicates that the analytical approach proposed is also effective for elasto-plastic analysis of hybrid structures subjected to severe earthquakes.

TABLE 4
MAXIMAL ROOF DISPLACEMENTS OF THE MODEL IN X-DIRECTION

Earthquake Intensity	Measured (mm)		Computed (mm)
	Acceleration intergration	Direct measurement	
Intensity 5S	6.45	7.62	6.74
Intensity 5E	3.51	5.22	3.88
Intensity 7E	10.23	12.09	14.05
Intensity 7S	9.28	10.30	11.52
Intensity 8E	21.72	35.94	22.64
Intensity 8S	12.34	15.16	13.35
Intensity 9P	102.29	109.19	96.67
Intensity 9P	105.69	--	88.04

CONCLUSION

The following conclusions can be drawn based on the study conducted in this paper:

1. The maximum lateral displacement at the roof of the model reaches 1/45 of the total height of the model under severe earthquakes, which demonstrates the hybrid structure possess good ductility to resist earthquakes.

2. The internal forces in the joints between SF and CT are complicated; The concrete around the embedded plates and the corners of CT damaged seriously, partly due to large axial forces induced by earthquakes in girders of SF connecting to CT.

3. According to the proposed simplified approach, the hybrid structure is firstly simplified as a parallelly connected half frames and multiple-spring wall system. Compared with the complete FE model, this simplified model can reduce degrees of freedom of hybrid structures dramatically and hence reduces much computational efforts. And the proposed approach can predict the elasto-plastic responses of steel-concrete hybrid structures subjected to earthquakes with satisfactory precision.

References

ASCE, New York(1979), *Council on Tall Buildings, Structural Design of Tall Steel Buildings*
G.Q.Li(1993), A Simplified Model of Predicting Nonlinear Seismic Response of High-rise Buildings in Braced Steel Frame, *Proc. of the 4th East Asia-Pacific Conference on Structural Engineering and Construction*, Seoul, Korea, Sep.
G.Q.Li(1997), Developing Strategies of High-rise Building Steel Structures in China, *Journal of Building Structures (China)*, Vol.14, No.4
G.Q.Li, Z Y Shen(1995), *A Unified Matrix Approach for Nonlinear Analysis of Steel Frames Subjected to Wind or Earthquakes*, Computer & Structures, Vol.54, No.2
G.Q.Li, Z.Y.Shen(1996), 钢结构框架体系弹性及弹塑性分析与计算理论, 上海科学技术出版社
Park, Y. J. & Ang, A. H. –S. (1985), Mechanistic Seismic Damage Model for Reinforced Concrete, *J. Struct. Eng., ASCE*, Vol. 111, No. 4, Apr.
P.Linde and H.Bachmann(1994), Dynamic Modeling and Design of Earthquake-Resistant Walls, *EESD*, Vol.23, 1331~1350
Sabnis, G.M. (1983) *Structural Modelling and Experimental Techniques*, Prentice–hall, Inc., Englewood Cliffs, N.J.
Vulcano,A., Bertero,V.V., and Colotti,V. (1988), Analytical Model of R/C Structural Walls, *Proc. 9th, WCEE*, Vol. 6, p. 41-46, Tokyo
Vulcano,A. and Bertero,V.V. (1987), Analytical Model for Predicating the Lateral Response of RC Shear Wall: Evaluation of Their Reliability, EERC-87/19

SEISMIC DEMAND EVALUATION PROCEDURE FOR CONCRETE-FILLED STEEL COLUMNS

H. B. Ge[1], K. A. S. Susantha[2] and T. Usami[1]

[1] Department of Civil Engineering, Nagoya University, Chikusa-ku, Nagoya 464-8603, Japan
[2] Department of Civil Engineering, Aichi Institute of Technology, Yachigusa 1247, Yagusa-cho, Toyota 470-0392, Japan

ABSTRACT

The seismic demand predictions of concrete-filled steel columns mainly used in highway bridge piers are presented in this paper. To this end, two types of dynamic analysis procedures are proposed: (1) a SDOF system analysis method using force-displacement hysteretic models derived based on the static pushover analysis results; and (2) a fiber model analysis procedure that involves with individual cyclic stress-strain relations of steel and confined concrete. Two types of hysteretic models namely bilinear and trilinear models are used in the case of former method. The analyses are carried out using both procedures for specimens tested by Pseudo-dynamic testing procedure. The accuracy of each method is checked by comparing the predicted maximum and residual displacement demands with those of the test. It is found that the maximum displacement demand predicted by the fiber analysis and that involved with the trilinear hysteretic model exhibit good agreement with the test results. However, the predictions made using the bilinear hysteretic model do not agree well with the test results. In addition, the residual displacement demands predicted through both procedures are found to be unsatisfactory. Therefore, an empirical equation is proposed to predict the residual displacement demand in terms of the maximum displacement demand that can be computed by the either method proposed in this study.

KEYWOARDS

Concrete-filled steel column, dynamic analysis, fiber model, SDOF analysis, hysteretic model, maximum displacement demand, residual displacement demand.

INTRODUCTION

Concrete-filled steel columns have become very popular in bridge pier construction in seismically active regions due to their excellent earthquake resisting characteristics. As a result, capacity and ductility prediction procedures of concrete-filled steel columns are currently being subjected to a great attention in seismic design of steel bridge piers. The understanding and quantification of nonlinear performance of such structures through appropriate modeling and accurate analysis turn out to be an essential item in this regard. In bridge pier construction, most of the composite bridge piers are only partially filled with concrete so that the adverse effects coming from the additional dead weight can be minimized. The optimum height for the infilling can be 30 to 50 percent of the column height according to Usami & Ge (1994). Several experimental and analytical investigations on capacity and inelastic behavior of partially concrete-filled steel columns have been carried out in the past, for example, Ge & Usami (1996), Usami & Ge (1994), Usami et al. (1995) and Susantha et al. (2001; 2002). It is obvious that not only the capacity but also the demand prediction procedures should be focused on studies aimed at proposing any rational seismic design methodology. However, only a limited number of studies are available on seismic demand prediction methodologies of concrete-filled columns [e.g., Kobayashi et al. (1997); Yoshizaki et al. (1999)]. Therefore, further investigations on seismic demand predictions of partially concrete-filled steel columns have emerged as an important issue.

This paper presents two methods for seismic demand prediction of partially concrete-filled steel columns. First method is a single degree of freedom (SDOF) system analysis procedure based on two types of force-displacement hysteretic models. The hysteretic models are established using the results of pushover analysis and a certain failure criterion. The pushover analysis uses recently developed stress-strain models for concrete, as explained in Susantha et al. (2001a; 2001b), and, an elastoplastic stress-strain relation that includes a strain-hardening region for steel, as described in Usami et al. (1995). The analyses are conducted for 15 specimens tested at Nagoya University by a Pseudo-dynamic testing procedure or so-called hybrid test procedure. In the second method, non-linear time history analyses are carried out using direct finite element analysis procedure (FEM). Here, beam-column elements are employed for steel and concrete segments rather than shell and solid elements. A cyclic stress-strain relationship proposed by Susantha et al. (2002), which is an extended version of the model used in the pushover analysis, is employed for concrete. Steel behavior is represented by a kinematic hardening rule introduced to a bilinear stress-strain relation which is approximated from the same multilinear stress-strain model as adopted in the pushover analysis. Subsequently, the same test specimens as adopted in the first method are analyzed and results are compared with those of the test in terms of displacement-time histories, load-displacement hysteretic relations, and maximum and residual displacement demands.

ANALYTICAL METHODS

Analysis based on Force-displacement Hysteretic Models

The response of a bridge pier subjected to a ground acceleration can be obtained by solving the governing equations of motion of the SDOF system with the help of a load-displacement hysteretic model. In such a simplified model, as shown in Figure 1, the mass, M, which is approximated by the axial load plus 30 percent of the column mass, is concentrated at the column tip (JRA 96, 1996). The pier is assumed to be massless with a flexural stiffness of k. The damping of the structural system is expressed in the form of viscous damping, and the damping ratio, ξ, of 0.05 is specified in this study. Two types of hysteretic models namely "bilinear model" and "trilinear model" are proposed and employed in solving the equation of motion of the system.

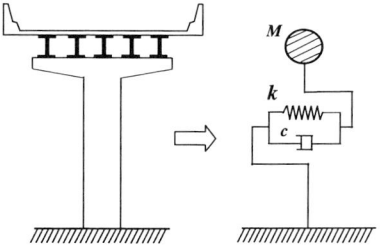

Figure 1: SDOF model for a bridge pier

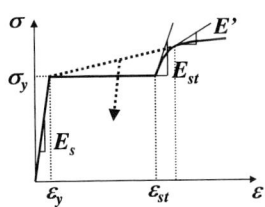

Figure 2: Stress-strain model for steel

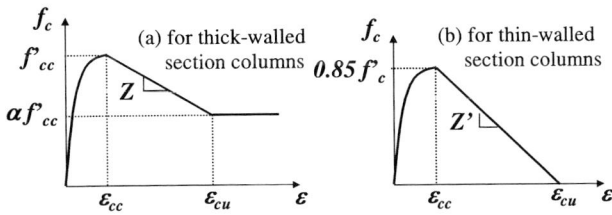

Figure 3: Stress-strain model for concrete (Susantha et al. (2001a; 2001b))

The bilinear model is the simplest version of the commonly used load-displacement hysteretic models. In this study, a bilinear lateral load-lateral displacement relationship is established from the pushover analysis results of the concrete-filled steel column. Pushover analyses are performed using a commercially available general purpose finite element program ABAQUS (1998). The detailed description of the static pushover analysis procedure, as adopted in this study, can be found in Susantha et al. (2001b) and Ge et al. (2001). The stress-strain relationships used for steel and concrete are shown in Figures 2 and 3, respectively. The steel material model includes a yield plateau followed by a strain hardening region, as explained in Zheng et al. (2000a). The details of concrete models can be found in Susantha et al. (2001a; 2001b). The load is applied in terms of lateral displacements at the column tip and the progression of lateral load-lateral displacement relationship is then recorded. A failure criterion, as explained in Susantha et al. (2001b), is used to determine the ultimate point, (δ_u, H_u). Subsequently, a bilinear approximation is adopted, as shown in Figure 4(a). Here, pre-yield elastic stiffness, k_1, is given by the initial slope of the pushover line. The post-yield inelastic stiffness, k_2, is determined by the intersection of the line having slope, k_1, and the one passing through the ultimate point (δ_u, H_u) satisfying the equal energy criterion. The unloading and reloading slope follows the hysteretic rule as shown in Figure 4(b), and no strength and stiffness deterioration is assumed with cyclic inelastic deformations.

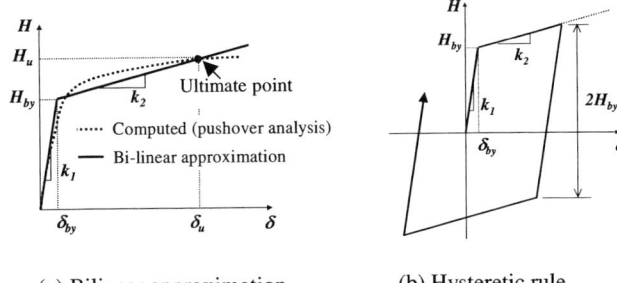

(a) Bilinear approximation (b) Hysteretic rule

Figure 4: Bilinear model for concrete-filled column

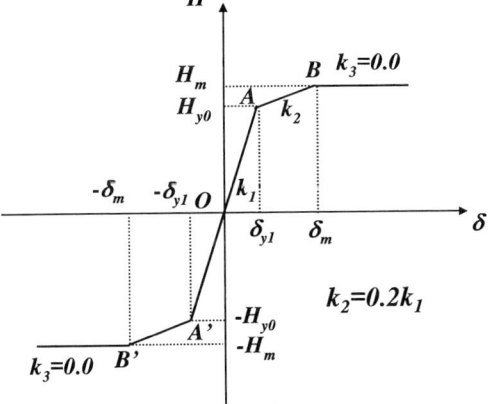

Figure 5: Skeletal curve of trilinear model

The skeletal curve of the trilinear model employed in this study is shown in Figure 5. This is a slightly modified version of the model proposed by Kobayashi et al. (1997). The model consists of three regions: (a) an elastic range with a slope k_1; (b) a plastic range having a hardening part given by slope k_2; and (c) perfectly plastic range beyond point B (δ_m, H_m), as illustrated in Figure 5. The elastic range OA is determined by the values of lateral yield load H_{y0} and the slope k_1, where H_{y0} is defined by

$$H_{yo} = \frac{M_{y0}}{h} \tag{1}$$

in which M_{y0} is the yield moment of the steel section without considering the effects of concrete and axial load, and h is the height of the column. Even though, the value of k_1 has been theoretically derived in the original model, here, it is directly taken to be the initial slope of the pushover curve. Then, the displacement δ_{y1} that corresponds to the displacement of the composite column at the yield load of hollow tube, H_{y0}, is directly available. The maximum lateral load, H_m, is determined by the lateral load-lateral displacement curve obtained from the pushover analysis instead of the test. In the absence of a clear peak point, H_m is defined as equal to the ultimate load, H_u. Then, with the slope k_2, which equals to $0.20k_1$, point B can be established. The hysteretic rule adopted here is as same as that proposed by Kobayashi et al. (1997) which accounts for both the stiffness degradation and change of peak point of the skeletal curve with cyclic loading.

Fiber Analysis using Stress-strain Models

Nonlinear dynamic analysis procedure using finite element method is presented in this section. The concrete and steel segments are modeled using beam-column elements and uniaxial cyclic stress-strain relations are employed for material behavior. The same analysis program ABAQUS (1998), as utilized in the pushover analysis, is used in computations. An example of the proposed analytical model is shown in Figure 6. The equivalent section concept described in Zheng et al. (2000b) is used in modeling of sections so that the stiffened box sections can be represented by the equivalent unstiffened sections. The mass, M, which is approximated by the axial load plus 30 percent of the column weight, is assumed to be lumped at the top of the pier. The element mesh size is decided according to the previously conducted mesh sensitivity checks using pushover analysis (Susantha et al., 2001b). The basic element configuration of the model adopted in this dynamic analysis and that of the static pushover analysis conducted in the first method are nearly the same. Two types of cyclic stress-strain models previously proposed by Susantha et al. (2002), as shown in Figures 7(a) and 7(b) are employed for concrete. A bilinear stress-strain relation approximated from the multi-linear elastoplastic model combined with a kinematic hardening rule, as shown in dotted lines in Figure 2, is employed for steel.

NUMERICAL RESULTS

A set of test specimens tested at Nagoya University are analyzed using the above explained procedures in order to check the validity of each method. Here are 15 specimens that have been tested using Pseudo-dynamic testing procedure under three input accelerograms having different characteristics of ground motions. The accelerograms used are: (1) Japan Meteorological Agency (JMA) (NS

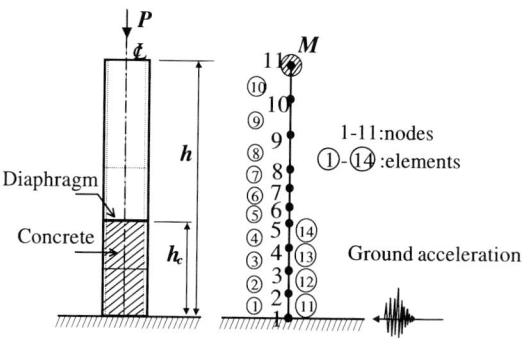

Figure 6: Analytical model

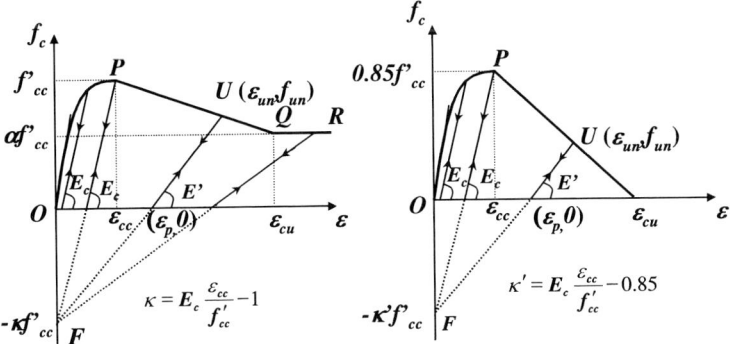

(a) For thick-walled section columns   (b) For thin-walled section columns

Figure 7: Cyclic stress-strain models (Susantha et al. (2001a; 2002))

component, Ground type I); (2) JR Takatori (JRT) (NS component, Ground type II); and (3) Higashi Kobe Bridge (HKB) (NS component, Ground type III). These three accelerograms are based on the ground motion records obtained in the 1995 Hyogoken-Nanbu Earthquake in Japan. The specimens were filled up to 20-30 percent of the total column height. Yield strength of steel is within 294.0 to 410.0MPa and cylinder strength of concrete is within 15.7-23.7MPa. Ranges of flange width-to-thickness ratio parameter, R_f (Susantha et al., 2001a), and column slenderness ratio parameter, $\bar{\lambda}$ (Zheng et al., 2000b) are 0.328-0.481 and 0.263-0.625, respectively. The axial load ratio, P/P_y, of 0.099-0.239 were used, where P_y is the squash load of the hollow steel column. The details of specimens are available in the literature published by Kobayashi et al. (1997), Yoshizaki et al. (1999), Saizuka et al. (1997) and Morishita (1996).

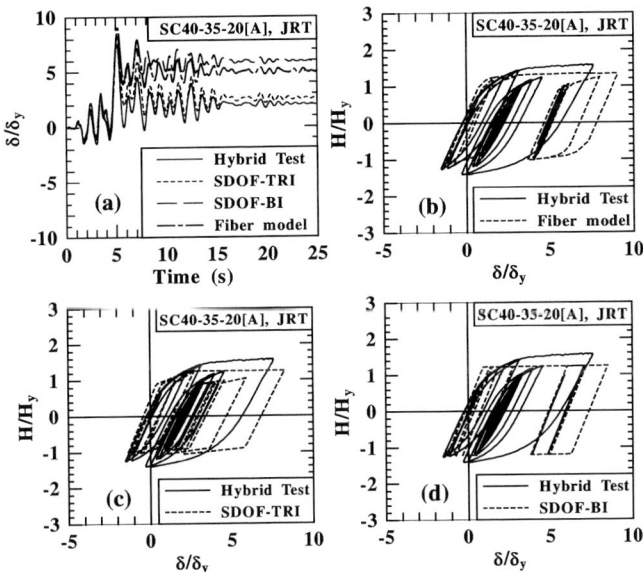

Figure 8: Examples of displacement-time and force-displacement responses

First, static pushover analyses are conducted for all the specimens and then the bilinear and trilinear hysteretic models are established. Subsequently, SDOF system analyses are performed for all the specimens using those hysteretic models and the fiber model under different accelerograms. The comparisons of computed and experimental displacement-time histories and force-displacement hysteretic relations of a specimen are illustrated in Figure 8. In this figure and hereafter, terms "SDOF-BI", "SDOF-TRI", and "Fiber Model" represent the results obtained using bilinear, trilinear and fiber models, respectively. The results show that the responses through all the models agree well with the test results until the maximum displacement is reached. Beyond that point only the result obtained from the SDOF-TRI model is fairly compliant with the Pseudo-dynamic test results. It is found from the comparisons made from all the specimens under the three accelerograms that the difference between the test and the prediction is somewhat large when SDOF-BI and Fiber models are used. This may be attributed to the simplicity of bilinear hysteretic model adopted in the case of SDOF-BI model and the simple bilinear approximation of steel material model employed in the analysis involving the Fiber model.

Figure 9 shows the comparisons made between the maximum and residual displacement demands (δ_{max} and δ_R) computed using the proposed methods and the test results. The results shown in Figures 9(a), 9(c) and 9(e) revealed that the values of δ_{max} predicted using the SDOF-TRI model are better than those of the other two models. Also, the results of Fiber model show good agreement with test results

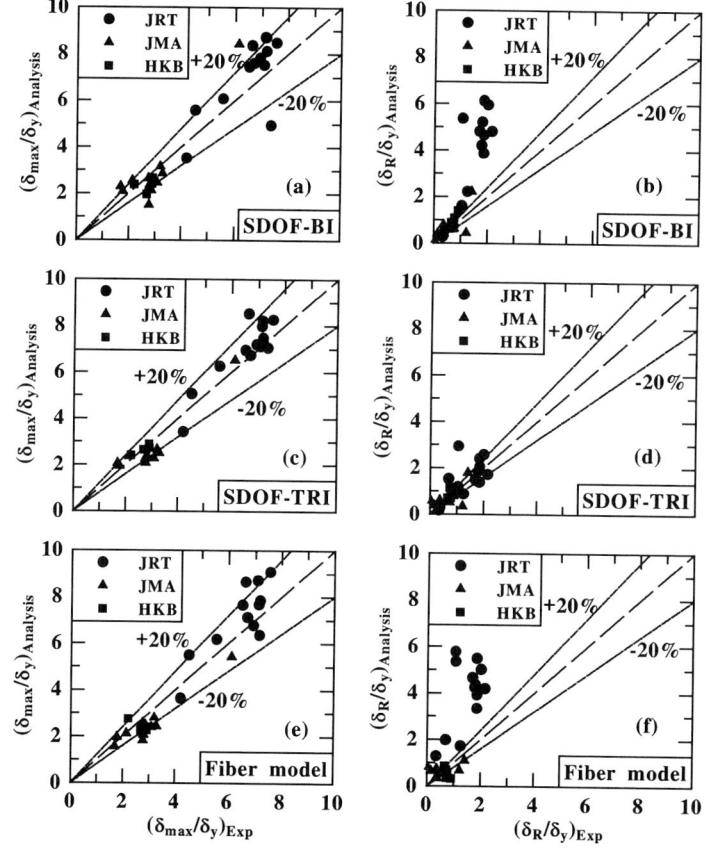

Figure 9: Comparison of computed and experimental maximum and residual displacements

than that of the SDOF-BI model. It is also revealed that under HKB accelerogram even SDOF-BI model yields fairly good results. In the case of the residual displacement demands, as shown in Figures 9(b), 9(d) and 9(f), it is clear that the results obtained using all the methods are not so accurate. Nevertheless, the SDOF-TRI model yields comparatively good agreement with test results. As a result, it can be concluded that the SDOF-TRI model is the most suitable for displacement demand predictions.

As previously mentioned, the simplicity of bilinear approximation has caused a big difference in the computed and observed δ_R values. It is understood that

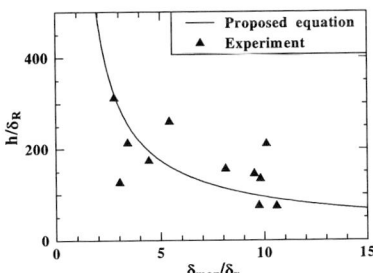

Figure 10: Proposed equation for residual displacement demand

a sophisticated hysteretic model is needed to obtain quite accurate predictions for δ_R. Such a model is not practicable in daily design use. Therefore, based on the experimental observation, as shown in Figure 10, a general equation is proposed to predict δ_R as shown below:

$$\frac{\delta_R}{h} = \frac{1}{400}(\frac{\delta_{max}}{\delta_y})^{0.7} - \frac{1}{500} \quad (S_R=0.00303) \quad (2)$$

Here, δ_{max} denotes the maximum displacement demand computed from previously proposed methods, and h stands for the height of the column (S_R is the standard deviation). It should be noted that the yield displacement δ_y should be computed by ignoring the in-filled concrete. The advantage of this equation is that the accuracy of predictions is considerably improved compared to the direct predictions from the analysis. This is clear from the results shown in Figure 11.

CONCLUSIONS

Two procedures for the dynamic demand prediction of concrete-filled steel columns are presented. One is based on the hysteretic models formed by means of the pushover analysis results whilst the other is a direct finite element analysis procedure that uses the individual cyclic stress-strain relations of concrete and steel. In the first method, two kind of hysteretic models, namely bilinear and trilinear models, are considered. A cyclic stress-strain model proposed in a previous study is employed in the second method. Thus, in this study, the effects of concrete and steel in demand predictions are treated in two different ways even though the same stress-strain models of concrete and steel are used in both methods. Moreover, an empirical equation is proposed to predict the residual displacement demand in terms

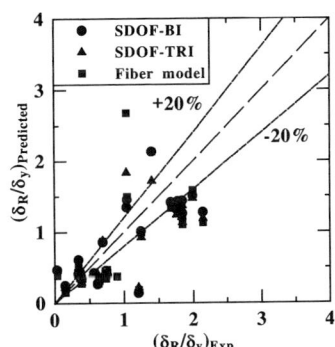

Figure 11: Comparison of experimental and predicted residual displacement demands

of the maximum displacement demand. The main findings of this study can be summarized as:
(1) The maximum displacement demands obtained using the dynamic analysis with the fiber model and using the SDOF system analysis with the trilinear hysteretic model exhibited better agreement with test results than those obtained using the SDOF system analysis with the bilinear hysteretic model.

(2) The residual displacement demands computed using the SDOF analysis with the trilinear hysteretic model are considerably better than those of the other two analyses. However, most of the predictions are still not so accurate when compared with the test results.
(3) The proposed empirical equation of residual displacement demand produces much better results than those directly yielded from the analysis. This equation can be used in design purposes when the maximum displacement demand is available from the proposed analytical procedures.

REFERENCES

ABAQUS/Standard User's Manual (1998). Ver. 5-8: Hibbitt, Karlsson and Sorensen, Inc.

Ge H.B. and Usami T. (1996). Cyclic tests of concrete-filled steel box columns. *J Struct Engrg ASCE* **122:10,** 1169-1177.

Ge H.B., Asada H., Susantha K.A.S. and Usami T. (2001). Unified earthquake resistance verification procedure for partially concrete-filled steel piers with thin- and thick-walled sections. *J Struct Engrg, JSCE* **47A,** 783-792 (in Japanese).

JRA 96 (1996). Specification for Highway Bridges, Part V. Tokyo: Japan Road Association.

Kobayashi M., Usami T. and Suzuki M. (1997). A hysteresis model for concrete-filled steel bridge piers and its application to elasto-plastic seismic response analysis. *J Struct Engrg, JSCE* **43A,** 859-868 (in Japanese).

Morishita K. (1996). Hybrid testing of steel bridge pier models using the Hyogoken-Nanbu earthquake accelerogram. Master Thesis, Nagoya University, Japan.

Saizuka K. and Usami T. (1997). Verification of proposed method for checking the ultimate earthquake resistance of concrete filled bridge piers based on the hybrid test results. *Struct Mech/Earthquacke Engrg, JSCE* **570/I-40,** 287-296 (in Japanese).

Susantha K.A.S., Ge H.B. and Usami T. (2001a). Uniaxial stress-strain relationship of concrete confined by various shaped steel tubes. *Engineering Structures* **23:10,** 1331-1347.

Susantha K.A.S., Ge H.B. and Usami T. (2001b). A capacity prediction procedure for concrete-filled steel columns. *J Earthquake Engrg* **5:4,** 483-520.

Susantha K.A.S., Ge H.B. and Usami T. (2002). Cyclic analysis and capacity prediction of concrete-filled steel columns. *Earthquake Engrg Struct Dyn* **31:2,** 195-216.

Usami T., Suzuki M., Mamaghani I.H.P. and Ge H.B. (1995). A proposal for check of ultimate earthquake resistance of partially concrete-filled steel bridge piers. *Struct Mech./Earthquake Engrg, JSCE* **525/I-33,** 69-82 (in Japanese).

Usami T. and Ge H.B. (1994). Ductility of concrete-filled steel box columns under cyclic loading. *J Struct Engrg, ASCE* **120:7,** 2021-2040.

Yoshizaki K., Usami T. and Honma D. (1999). Pseudodynamic test of steel bridge piers with purpose of reducing residual displacements. *J Struct Engrg, JSCE* **45A,** 1017-1026 (in Japanese).

Zheng Y., Usami T. and Ge H.B. (2000a). Ductility of thin-walled steel box stub-columns. *J Struct Engrg, ASCE* **126:11,** 1304-1311.

Zheng Y., Usami T. and Ge H.B. (2000b). Ductility evaluation procedure for thin-walled steel structures. *J Struct Engrg, ASCE* **126:11,** 1312-1319.

Zheng Y., Usami T. and Ge H.B. (2001). Seismic design method for thin-walled steel frame structures. *J Struct Engrg, ASCE* **127:2,** 137-144.

INDEX OF CONTRIBUTORS

Volumes I and II

Albermani, F. 429
Al-Mahaidi, R. 245
Anderson, J.C. 221
Ansourian, P. 713

Bakht, B. 773
Bambach, M.R. 617
Batista, E. 341
Bauer, D.B. 181
Beale, R.G. 303, 461
Beaulieu, D. 1209
Bechara, E. 205
Benaddi, A. 181
Berry, P.A. 487, 495
Bezkorovainy, P. 617
Bradford, M.A. 95, 625
Bridge, R.Q. 479, 511, 519, 667
Burns, T. 617
Bursi, O.S. 81
Byfield, M.P. 139, 261, 1201

Cameron, N.J.K. 1079
Camotin, D. 331, 341
Chan, S.L. 321, 543, 913, 1017, 1153, 1169, 1193
Chan, T.H.T. 791, 799
Chan, T.M. 1095
Chau, K.T. 1017
Chen, G.D. 641
Chen, H. 213, 285
Chen, J. 897
Chen, J.F. 755
Chen, S.F. 559
Chen, Z.Q. 849
Chiew, S.P. 1033
Chilton, J.C. 755
Chin, G.P.W. 351
Cho, S.H. 1193
Choy, S.C. 543
Chu, K. 763
Chung, K.F. 121, 351, 437, 445, 649
Chusilp, P. 69
Combescure, A. 659, 981
Coret, M. 981

Couchman, G.H. 261
Crawford, R.J. 1201
Cui, X.Q. 1161

da S. Vellasco, P.C.G. 253
Davies, J.M. 57, 401
Detandt, H. 783
Dezi, L. 535
Dhanalakshmi, M. 139, 261
Ding, X. 599
Dong, S.L. 15
Du, X.X. 453
Duan, J.X.J. 221
Duan, Y.F. 849
Dubina, D. 409, 989
Dundu, M. 383
Dymiotis, C. 321

Elchalakani, M. 567
Emi, T. 189
Espion, B. 783

Fang, Z.Z. 551
Feifel, E. 683
Fish, R.A.D. 357
Fujiwara, H. 1051
Fulop, L.A. 409

Gardner, L. 43
Ge, H.B. 607
Girão Coelho, A.M. 277
Gläsle, M. 713
Godley, M.H.R. 303, 461
Goto, Y. 171, 1051, 1145
Greiner, R. 667
Grundy, P. 237, 1043
Grzebieta, R.H. 567
Gu, J.X. 1153
Guan, D.Q. 1009, 1025
Guo, L. 799
Guo, W.H. 857
Guo, Y.J. 453
Guo, Y.L. 147, 155, 163, 551, 641, 873, 1161

Han, L.H. 583, 591, 1127, 1135
Han, Q.H. 229
Han, Y. 155, 551
Hancock, G. 3, 311, 421
Hao, W.Q. 155
Harada, H. 823
Harada, K. 815
Ho, H.C. 437, 445
Holst, J.M.F.G. 729
Hosain, M.U. 527
Howson, W.P. 1177
Huang, C.W. 921
Huang, X. 947
Huang, X.Q. 955
Huang, Z.F. 1111, 1119
Huang, Z.W. 1033
Hui, J.T.Y. 955
Huo, J.S. 1127

Iguchi, T. 197

Janjic, D. 831
Javidruzi, M. 755
Jullien, J.F. 675, 973

Kawanishi, N. 1145
Kemp, A.R. 383
Kennedy, J.B. 807
Kihara, S. 197
Kitipornchai, S. 429
Ko, C.H. 121
Ko, J.M. 791, 849
Kozlowski, A. 269
Kubo, Y. 815

Lam, D. 503
Law, S.S. 905, 913
Lee, G.C.M. 807
Lee, M. 357
Legay, A. 659
Leoni, G. 535
Leu, L.J. 921
Li, G.Q. 599
Li, J.J. 1169

Li, K. 453
Li, L. 1009
Li, Z.F. 213, 285
Li, Z.X. 791, 799
Liang, Q.Q. 625
Lie, S.T. 1033
Liew, J.Y.R. 1061
Lim. J.B.P. 391
Limam, A. 675
Lin, X. 737
Ling, T.W. 245
Liu, C.Y. 921
Liu, T. 155, 163
Liu, W. 583
Liu, X.L. 229
Liu, Y. 365
Lu, G. 931, 939, 947

Mäkeläinen, P. 1103
Mashiri, F.R. 237, 1043
Melcher, J.J. 1185
Mikami, I. 633
Milojkovic, B. 303
Miyazaki, Y. 633
Mufti, A.A. 773
Muramoto, Y. 823
Mursi, M. 575

Nagahama, K. 341
Nagai, M. 815
Namba, H. 189
Nethercot, D.A. 43, 391
Neves, L.C. 253
Ni, Y.Q. 849
Nip, T.F. 503
Niwa, K. 633

Obata, M. 171, 1051
Oehlers, D.J. 471
Outinen, J. 1103

Pan, Y. 147
Pashan, A. 527
Patrick, M. 511, 519
Petrovski, T. 205
Pi, Y.-L. 95
Pircher, H. 831
Pircher, M. 375, 667, 831
Prola, L.C. 331

Quispe, L. 421

Rafezy, B. 1177
Rasmussen, K.J.R. 357, 617
Ren, G.X. 873
Rotter, J.M. 27, 729
Rubal, M. 205

Saal, H. 683
Sakurai, T. 763
Samaan, M. 963
Sennah, K.M. 807, 963
Seracino, R. 471
Sha, W. 1095
Shen, Z.Y. 105
Shi, Y.J. 213, 285, 295
Shum, K.M. 865
Silvestre, N. 341
Simões da Silva, L. 253, 277
Song, C.Y. 693, 703
Song, Z.S. 105
Staquet, S. 783
Stratan, A. 989
Sun, D.K. 865
Susantha, K.A.S. 607

Tabuchi, M. 189, 197
Tadros, G. 773
Tan, H.B.A. 575
Tan, K.H. 1111, 1119
Tanaka, T. 189, 197
Tao, Z. 591
Teng, J.G. 693, 703, 721, 737, 745
Ting, S.K. 1111
Tong, G.S. 129
Tong, L.W. 237, 1043

Ukur, A. 567
Unterweger, H. 839
Usami, T. 69, 607
Usmani, A.S. 1079
Uy, B. 575, 625

Vafai, A. 755
Vaux, S. 311
Vincent, Y. 973

Vitali, A. 535
Voutay, P.A. 401

Wang, A.J. 649
Wang, B. 939
Wang, H. 873
Wang, Q. 1025
Wang, Y.Q. 213, 285, 295
Wang, Z. 295
Wheeler, A. 375, 479
Wilkinson, T. 205
Wong, C. 311
Wong, H.T. 745
Wong, M.B. 1071, 1089
Wong, Y.L. 543
Wright, H.D. 625
Wu, Z.M. 913

Xia, H. 889
Xiao, R.Y. 351
Xiao, Y. 221
Xiao, Z.G. 1001
Xu, L. 1135
Xu, Y.L. 857, 865, 881, 889, 897

Yamada, K. 1001
Yamaguchi, E. 815
Yamao, T. 823
Yan, Q.S. 889
Yang, D. 3
Yang, Y.F. 583, 1127, 1135
Yang, Z.C. 881
Yeh, S.H. 921
Yi, W.J. 1009, 1025
Young, B. 365
Yu, H.X. 1061
Yu, T.X. 947, 955

Zandonini, R. 81
Zhang, L. 129
Zhang, R.C. 897
Zhao, X.L. 237, 245, 567, 1043
Zhao, Y. 15, 721
Zhong, J.H. 559
Zhou, X.M. 599
Zhou, Z.H. 321, 1153

KEYWORD INDEX

Volumes I and II

Actively controlled platform, 881
Actuator dynamics, 881
AIJ-SRC, 543
Aluminum, 1209
Ambient vibration measurement, 921
Analysis, 1209
Anisotropy, 617
Antisymmetric, 95
Application, 15
Arches, 95
Arching, 773
Attachment, 1001
Automobile, 939
Axial compression, 737
Axial restraint, 1111
Axisymmetric, 659

Bare steel tubes, 479
Beam, 129, 939
Beam-column, 155, 163, 591
Beam-column connection, 213
Beam-column element, 913
Beam element, 1153
Beam-to-column joints, 989
Bearing capacity, 583
Bending load, 675
Bending moment, 823
Biaxial bending, 559
Biaxial compression, 625
Bifurcation analysis, 357
Bolted moment connections, 437
Bolted steel shells, 745
Bonded, 543
Box-girder, 807
Box section, 69, 163
Braced frame, 269
Bracing system, 807
Bridge deck, 773
Bridge deck modelling, 839
Bridges, 495, 535, 807
Bridge stay cable, 849
Buckle, 155
Buckling, 27, 57, 69, 95, 321, 401, 461, 575, 683, 693, 703, 721, 729, 737, 745, 755, 763
Buckling resistance, 703
Buffeting analysis, 865
Buffeting force, 889

Bumper, 939
Butt weld, 989

Cable-mass structure, 873
Cable-membrane, 1169
Cable-stayed bridge, 831, 857
Camber, 783
Cassettes, 57
Catenary action, 1111
Chimneys, 27
Circular tubes, 245, 947
Cladding, 429
Closed steel ribs, 511
Code and standards, 1201
Codes, 1209
Cold-formed, 429
Cold-formed section, 453
Cold-formed steel, 57, 205, 365, 391, 421, 461, 1103
Cold-formed steel beams, 331
Cold-formed steel cassette section, 401
Cold-formed steel C-section, 351
Cold-formed steel members, 341
Cold-formed steel purlins, 445
Collapse criteria, 229
Collapse properties, 229
Column buckling, 1193
Columns, 365, 575, 583, 1135
Combined loads, 675
Compatibility condition, 1161
Component method, 253, 277
Composite, 503
Composite beam, 487, 495, 511, 519, 551
Composite bridge(s), 783, 839
Composite columns, 559
Composite connections, 261
Composite construction, 261, 625, 745
Composite floor, 269
Composite floors in fire, 1079
Composite floor systems, 527
Composite steel-concrete girders, 535
Composite structures, 575
Compression, 3, 57, 401, 823
Computer simulation, 889
Concrete arch bridge, 831
Concrete-filled, 253
Concrete-filled HSS columns, 1127
Concrete-filled SHS or RHS, 1135

Concrete-filled steel column, 607
Concrete filled steel tube, 591
Concrete-filled steel RHS, 583
Concrete-filled tubes, 479, 567
Concrete shells, 745
Connection deformability, 535
Connections, 205
Constitutive behaviour, 973
Construction loading, 807
Contact elements, 351
Continuous composite beam, 487, 495
Contraflexure, 487
Corrosion, 1145
Corrugation, 713
Coupled instability, 57, 401
Cracks, 755, 1017
Cramping force, 189
Crash, 939
Crashworthiness, 963
Creep, 535, 583, 1111
Creep coefficient, 583
Creep effect, 1061
Critical location, 799
Critical moment, 129
Critical temperature, 1119
Cross beam, 815
Cross-section classification, 43
Cross-sectional distortion, 955
Cross stiffener, 641
Crosswind, 857
Cruciform joint, 1001
Crushing of tubes, 931
Cuplok, 311
Curved, 807
Cyclic, 409
Cyclic behaviour, 913
Cyclic loading, 763, 989
Cylinders, 683, 737
Cylindrical shells, 675, 713, 729, 737, 755
Cylindrical steel pier, 763

Damage, 221
Damage cumulation, 105
Damage theory, 1051
Damper installation configuration, 849
Deck slab, 773
Deck-type steel arch bridge, 823
Deformation capacity, 277
Deformation limit, 237
Design, 321, 421, 461, 745, 1089, 1209
Design criterion, 1185
Design development, 121, 445
Design evaluation, 807
Design method, 583

Design philosophy, 27
Design suggestions, 453
Destructive mechanism, 453
Development, 15
Dimple, 729
Direct strength, 421
Direct strength approach, 401
Distortional buckling, 341, 649
Distortional post-buckling, 331
Double-angle, 181
Double skin composite panels, 625
Double-skin composite sections, 567
Ductile crack, 197
Ductile fracture, 1051
Ductile tearing, 931, 947
Ductility, 69, 253, 261, 277, 591, 633, 763
Dynamic analysis, 409, 607, 1145
Dynamic interaction, 889
Dynamic magnify factor, 873
Dynamic programming, 905
Dynamic response, 799, 873
Dynamics, 755, 963

Early age shrinkage, 535
Earthquake, 1145
Eaves, 383
Eccentric discharge, 693, 703
Education, 1209
Effective flexural rigidity, 445
Effective stress concentration factor, 1009, 1025
Efficient algorithms, 659
Elastic analysis, 487, 495
Elastic buckling, 675
Elasticity, 95
Elastic-plastic analysis, 487,
Elasto-plastic analysis, 599
Elevated temperature, 1103
Elongation, 1095
Empirical formula, 921
Empirical mode decomposition, 897
Energy absorbing capacity, 955
Energy absorption, 931. 947, 963
Energy-dissipation, 591
ENV 1993-1-4, 43
Equilibrium state, 1161
Eurocode 3, 253
European, 27
Experimental, 253
Experimental investigation, 365
Experimental research, 213
Experimental study, 479
Experiments, 721, 737

Extended end plate, 253
External bending moment, 1119

Fabrication misfit, 729
Factors of safety, 303
Failure mode, 1043
Failure temperature, 1095
Falsework, 311
Fatigue, 1001
Fatigue damage, 791, 799
Fatigue life curve, 1025
Fatigue notch factor, 1009, 1025
Fatigue strength, 1009, 1025, 1043
FEA, 189, 197, 1001
FEM, 1001
Fiber model, 607
Field measurement data, 897
File protection, 1135
Fillet weld, 989, 1001
Finite element analysis, 285, 625, 939, 963
Finite element method, 105, 277, 551, 815
Finite element model, 799
Finite elements, 401, 617, 659, 675, 807, 973, 981
Finite shell element, 163
Finite strip analysis, 357
Fire, 1103, 1111, 1119
Fire code development, 1071
Fire curve, 1089
Fire duration time, 1127
Fire engineering, 1071, 1089
Fire modelling, 1061
Fire research, 1071
Fire resistance, 1061, 1095, 1135
Fire safety, 1061
First yield criteria, 641
Fixed/warping-free members, 341
Flexible supporting system, 295
Flexural behaviour, 383
Flexural buckling, 357
Flexural failure, 437
Flexural loading, 375
Flexural restraint, 1119
Flexural strength, 479
Flexural-torsional analysis, 1153
Flush end plate, 253
Footstep, 905
Force identification, 905
Formwork, 311
Fractionated casting, 535
Fracture, 205
Fracture failure, 213
Fracture mechanics, 1017
Frames, 321

Frictional joint, 913
Frontal collision, 963
Frozen-heated method, 1161
Full aeroelastic model test, 865
Full size tests, 527
Fundamental vibration period, 921

GBT, 401
GBT distortional buckling formulae, 341
General stress ratio, 1025
Generalised Beam Theory (GBT), 341
Generic lapped sections, 437
Geometric non-linear, 295
Geometrically nonlinear analysis, 1153
Girders, 69
Gliding cable, 1161
Global analysis, 839
Guardrail, 955
Gusset, 1001

Headed stud shear connectors, 527
Health monitoring, 791
Heat transfer, 1089
High performance steel, 633
High-strength bolt, 189
High strength concrete, 783
High strength steel, 3, 245, 575
High tech equipment, 881
High temperature, 1095, 1103
High-temperature,
Hilbert transform, 897
Hollow core slab, 503
Hollow section, 253
Honeycombs, 931
Hot-rolled girder, 783
Houses, 409
Hysteretic behavior, 591
Hysteretic model, 607

Impact, 955
Imperfections, 721, 729, 737
Imperfection sensitivity, 693
Imperfection shape, 375
Imperfection size, 375
Incremental method, 295
Incrementally launched bridge, 831
Initial imperfection(s), 375, 1193
In-plane bending, 237
In-situ vibration test, 849
Integrated analysis, 1169
Interaction buckling, 357
Interaction curves, 823

Interactive buckling, 147
Interface friction, 471
Internal pressure, 675
I sections with slender webs, 649

Joints, 253

Lapped connections, 445
Large deflection, 1169
Large deformation, 295, 955
Large displacement analysis, 1153
Lateral torsional buckling, 383, 391, 649
Life cycle, 1145
Light gauge steel, 409
Limit state, 1185
Limit state design, 27
Lipped channel beams, 331
L-load, 815
Load-deformation curves, 1043
Loading analysis, 1169
Local buckling, 3, 43, 197, 357, 375, 383, 453, 479, 625, 1051
Local forces, 683
Local moments, 683
Local plate buckling, 147, 391
Local stress, 1001
Long suspension bridge, 889
Longitudinal fillet welds, 245
Longitudinal shear, 511, 519
Long-span bridge, 791
Long-span structures, 15
Long-term sustained load, 583
Low cycles fatigue, 105

Magnetorheological (MR) damper, 849
Materials properties, 1095
Materials testing, 139
Mathematical modelling, 139
Maximum displacement demand, 607
Mechanical behavior, 1161
Mechanical properties, 453, 1103
Mechanics model, 1127
Metal deck, 773
Metal foams, 931
Microvibration, 881
Mill tests, 139
Misaligned welded joint, 1009
Modal parameter identification, 897
Mode superposition, 889
Modular construction, 745
Moment-connections, 391
Moment-curvature, 383

Moment resistance of connections, 437
Moment resistance ratios, 445
Moment-rotation. 383
Moment-rotation behaviour, 253

Narrow ribbed metal deck, 527
Natural frequency, 823
Non-composite, 471
Non-linear analysis, 303, 311, 321, 351, 429, 1061
Nonlinear buckling, 659
Non-linear homogenisation, 981
Nonlinearity, 873, 913
Nonlinear seismic response analysis, 823
Nozzle, 683
Numerical analysis, 1051
Numerical computation, 857
Numerical integration, 139
Numerical modelling, 783

Optimal design, 269
Overall buckling, 3
Overall comparison, 865

Pallet racks, 461
Parametric analysis, 583
Parametric instability, 755
Partial interaction, 471
Partial safety factor, 1185
Passively controlled platform, 881
Patch loading, 683
Patch loads, 351, 693, 703
Pedestrian, 905, 939
Perforated beams, 121
Perforated section approach, 121
Performance-based design, 1061, 1071
Phase transformation, 973
Piles and piling, 1201
Pinned/free-to-warp members, 341
Planar modelling, 1177
Plastic behaviour, 383
Plastic bending, 947
Plastic mechanism, 237
Plastic stretching, 947
Plasticity, 617
Plate rotation restraint, 331
Plates, 69, 617
Point supported glass curtain wall, 295
Pole design, 963
Post buckling, 401
Post-local buckling, 625
Precast, 503

Prediction model, 1009
Prestress, 913
Prestressed, 543, 551
Prestressing, 783
Pretensioned structures, 15
Probability, 1185
Proportional building structures, 1177
Prospects, 15
Pseudo-dynamic test, 171
Purlins, 421
Purlin-sheeting, 429
Push-out tests, 527

Rack-section columns/beams, 341
Random decrement technique, 897
Ratio of component plates, 147
Rectangular plate, 633
Reentrant open steel ribs, 511
Regression analysis, 527
Reinforcement, 683, 763
Reinforcing steel, 511, 519
Reliability, 659
Repair, 221, 1145
Residual displacement demand, 607
Residual strength, 1127
Residual strength index, 1127
Residual stress, 729, 1193
Residual stresses, 737
Resistance, 277, 1185
Restrained thermal expansion, 1079
Retaining walls, 1201
Retrofit, 221
Rib punch-through failure, 511, 519
Rib shearing failure, 511, 519
Ride comfort, 857
Rings, 721
Rivet hole, 1017
Road toughness, 857
Road vehicle, 857
Roofs, 745
Roof systems, 437
Rotation capacity, 261
Running train, 889

Saddle shade pavilion, 1169
Safety, 963
Scaffold, 303, 311
Scaffolding, 321
Scale down model, 599
Scaled-down testing, 955
SDOF analysis, 607
Second-order analysis, 321
Seismic behavior, 221

Seismic design, 763
Seismic design of bridge, 171
Self-excited force, 889
Semi-continuous composite beam, 487
Semi-rigid, 303
Semi-rigid connection, 913
Semi-rigid joint, 269
Shaking table test, 599
Shallow arch, 95
Shape finding, 1169
Shear, 57, 69, 409
Shear-carrying capacity, 641
Shear connector, 511, 519
Shear contribution, 543
Shear lag, 181
Shear-moment interaction curves, 121
Shear studs, 503
Shell elements, 351
Shells, 27, 659, 693, 703, 721, 745
Shrinkage, 583
SHS, 365
Silo loading codes, 693, 703
Silos, 27, 667, 693, 703, 713, 729, 737
Simplified approach, 599
Simplified design method, 559
Simulated earthquakes, 599
Slab, 773
Slab-on-girder beams, 471
Slenderness ratio, 163
Slip load, 189
Slotted-bolted-joint, 913
Smart damping technology, 849
Snap-through, 95
Soil characteristics, 963
Solid concrete slab, 527
Space frames, 1153
Spline finite strip method, 331
Stability, 3, 129, 311, 421, 461, 551, 667, 713, 721, 1209
Stability design, 69
Stainless steel, 43, 365, 617
Standard fire curve, 1127, 1135
Standards, 27, 1209
Static strength, 237, 1043
Statistical analysis, 783
Stay-in-place form, 773
Steady-state, 1103
Steel, 303, 503, 543, 1095, 1103
Steel beam-column connection, 285
Steel beams, 139, 261, 905, 1111
Steel bridge, 839, 1145
Steel buildings, 921
Steel column, 1119
Steel-concrete columns, 559
Steel frame, 913

Steel-free deck slab, 773
Steel hollow sections, 237, 567
Steel plate girder bridge, 815
Steel plate shear wall, 641
Steel plates, 625
Steel roof truss, 453
Steel silos, 721
Steel space structures, 15
Steel structures, 139, 181, 213, 261, 295, 321, 357, 575, 1061, 1071, 1089, 1201
Steel-concrete hybrid structure, 599
Stiffened box section, 823
Stiffener, 401, 713
Stiffener buckling, 391
Stiffening system, 815
Stiffness, 277
Stocky column effect, 1193
Strain, 905, 1095
Strain hardening, 139
Strain-hardening gradient, 633
Strain rate, 989, 1111
Strains distribution, 453
Strength, 285, 383, 591, 625, 1095, 1185, 1209
Stress, 285
Stress analysis, 139
Stress concentration factor, 1001
Stress field intensity method, 1009
Stress spectrum, 791, 799
Stress-strain model, 583
Structural analysis, 269
Structural design, 365, 503, 559, 1185
Structural engineering, 43
Structural Eurocodes, 1201
Structural hollow sections, 205
Structural instability, 649
Structural members, 3
Structural stability, 365
Structural steel, 1103
Structure optimisation, 831
Structures, 303, 1209
Stub columns, 567
Stud shear connectors, 625
Suboptimal control, 881
Substitute frames, 1177
Substructures, 1201
Support flexibility, 1169
Survival rate, 1025
Suspension bridge, 799, 865
Sway frame, 311
Symmetric, 95
System identification, 921

Tall buildings, 575
Tangent stiffness matrix, 1153

Tanks, 27, 667, 683, 729
Tapered, 155
Tapered I-columns, 147
Tapering ratio, 147
Tee section approach, 121
Temperatures, 981, 1089, 1135
Tendons, 543
Tensile testing, 1103
Tension, 205, 1017
Tensioned membrane, 1169
Tension web member, 181
Testing, 409
Tests, 357, 617
Textbook, 1209
Thermal bowing, 1079
Thermal-mechanical testing, 973
Thin-walled column, 1051
Thin-walled cylinder, 667
Thin-walled sections, 129, 237, 245, 1043
Thin walled steel tubes, 375
Three-dimensional analysis, 815
Three-dimensional degenerated curved shell element, 229
Time-varying structure, 873
Torsional analysis, 1177
Torsional buckling, 357
Torsional-flexural, 155
Towers, 27
Track irregularity, 889
Traffic-induced ground motion, 881
Traffic loading, 799
Transformation induced plasticity, 973
Transformation plasticity, 981
Transient-state, 1103
Transient tensile test, 1095
Transition junctions, 721
Transverse reinforcement, 503
Truss, 181
Tsing Ma bridge, 897
T-stub model, 277
Tubular members, 365
Two scale material model, 981
Typhoon, 791
Typhoon Victor, 897

Ultimate bearing capacity, 453
Ultimate behavior, 69, 1145
Ultimate capacity, 1079
Ultimate compression capacity, 229
Ultimate load capacity, 147
Ultimate strength, 237
Ultimate tension capacity, 229
Uniaxial compression, 633

Unit load method, 831
Unpropped construction, 261

Velocity, 905
Vertical stiffener, 815
Vibration, 755
Vibration control, 849
Virtual temperature load, 1161
Virtual work, 95

Wall panels, 409
Walls, 57
Warping effects, 331
Warping restraint, 331
WBFW connection, 189
Web openings of various shapes and sizes, 121
Weld connection, 197
Weld overlay, 221
Welded beams, 495

Welded connection, 181, 221, 989
Welded girder, 783
Welded hollow spherical joint, 229
Welded joint, 1025
Welding, 205, 973, 981
Wet concrete loading, 745
Wide ribbed metal deck, 527
Width-thickness ratio, 163
Width-to-thickness ratio, 197
Wind load, 667, 815

Yield line theory, 237
Yield plateau length, 633
Yielding, 713, 1017

Z-sections, 357

3-dimensional seismic behavior, 823